The Genesis of the Copernican World

Studies in Contemporary German Social Thought
Thomas McCarthy, General Editor

The Genesis of the Copernican World

Hans Blumenberg

translated by Robert M. Wallace

The MIT Press
Cambridge, Massachusetts
London, England

XC

This translation copyright © 1987 Massachusetts Institute of Technology. This work originally appeared in German as *Die Genesis der kopernikanischen Welt*, © 1975 Suhrkamp Verlag, Frankfurt am Main, Federal Republic of Germany.

The translation and the publication of this volume have been made possible in part by grants from the Division of Research Programs of the National Endowment for the Humanities, an independent federal agency.

This book was set in Baskerville by Asco Trade Typesetting Ltd., Hong Kong, and printed and bound by Halliday Lithograph in the United States of America.

Library of Congress Cataloging-in-Publication Data

Blumenberg, Hans.
  The genesis of the Copernican world.

  (Studies in contemporary German social thought)
  Translation of: Die Genesis der kopernikanischen Welt.
  Includes index.
  1. Astronomy—History.   2. Europe—History—1492–.   3. Copernicus, Nicolaus, 1473–1543.
  I. Title.   II. Series.
  QB28.B5813   1987   113'.09'031   86-27445
  IBSN 0-262-02267-2

Contents

Translator's Introduction

Interpretations of the Copernican revolution have been hampered for a long time by an unexamined assumption about the relation between the history of science, on the one hand, and the history of European 'consciousness' (in a broad sense that would include, for example, philosophy, religion, and metaphor, as well as science), on the other. Although Copernicus's work was predominantly scientific, its consequences, of course, extended far beyond science, inspiring philosophies, causing problems for religion, and producing powerful metaphors like that of man's "removal from the center" of the cosmos. But if we turn to what led up to this revolution, we find that although histories of science have mentioned some possible philosophical and religious "influences," they have not tried to show that any of these were necessary preconditions of the success of Copernicus's efforts, and consequently they have not really challenged the assumption, which is encouraged by the apparently self-contained, autonomous functioning of science in our own time, that the essential preconditions of Copernicus's work were all intrascientific. Thus what this case leaves us with is a picture of a one-way kind of causation, in which while developments in science can have major effects on culture, the reverse is apparently not true. Historical and anthropological accounts of pre-Copernican myth, religion, and philosophy have not effectively challenged this picture either, since (reflecting, in practice, the same kind of relations between 'science' and 'nonscience' that the history of science presupposes) they have tended to see Copernicus as a bolt from the blue, impinging on but not deriving from the history of their subject matters.

This kind of view of the Copernican event is especially troublesome if we believe, as is often said, that the Copernican revolution "created the modern world,"[a]* because then the modern world itself is seen as a product essentially of changes originating within the self-contained activity called "science." But whether its consequences are defined as broadly as this or not, this kind of one-way, 'exogenous' determination of important cultural phenomena by a process that is not affected by them in return has to provoke the kind of doubts to which all "reductionisms" are nowadays properly subjected. This is especially so when we remember that the notion of science as a self-contained and self-regulating long-term process itself originated only in the seventeenth century, at the same time that the Copernican revolution was taking hold.

In this book, Hans Blumenberg[b] breaks decisively with this traditional view of the Copernican reform as an event in the history of science that happened to have great repercussions outside science but had no essential extrascientific preconditions. Blumenberg undermines this view by showing how Copernicus's reform of astronomy was itself made possible, as a proposal that Copernicus and his readers could take seriously, by processes that involved the whole range of European thought, religious, philosophical, and metaphorical as well as literal and scientific. Bringing the history of science and the history of 'consciousness' together, in this way, also enables Blumenberg to clarify, for the first time, the relations between the Copernican revolution and the origin of the modern age as a whole—two phenomena that are not, in fact, identical, and of which neither was the cause of the other, though since both derived (at least in part) from the same antecedent process, their temporal coincidence was certainly not accidental.

As I mentioned, Blumenberg is not the first scholar who has considered possible connections between the Copernican turning and antecedent philosophical and theological ideas. A long series of writers has discussed the apparent influence on Copernicus of Renaissance Neoplatonism (which certainly would count as a philosophical influence); and others have thought that Copernicus may have benefited from the late medieval physical and cosmological

* Translator's notes, to the introduction as well as to the text, are cited by superscript lowercase letters, and are grouped at the end of the book. Author's notes, cited by superscript numbers, are also at the end of the book, after the translator's notes.

innovations of the Nominalists, which one might also classify, like Aristotle's physics, as more philosophical than (at any rate) experimental. Alfred North Whitehead went even further with his assertion—which Thomas S. Kuhn quotes with approval in his account of the Copernican revolution—that "faith in the possibility of science, generated antecedently to the development of modern scientific theory, is an unconscious derivative from medieval theology." [c]

It will become clear, however, that the connections that Blumenberg makes are far more systematic and thoroughgoing than these previous suggestions. What Kuhn and others offer are descriptions of the descent of particular possibly contributory ideas, and lists of possibly contributory circumstances (such as the religious ferment of the Reformation, the discrediting of Ptolemy as a geographer by the voyages of discovery, and the need for calendar reform), plus a few dramatic aperçus like Whitehead's. Among these proposals, Whitehead's is the only one that seems definitely to claim to have identified a *necessary* extrascientific precondition of something of which the Copernican revolution was a part; but it flashes by so quickly that it is difficult to evaluate it. Blumenberg, on the other hand, provides a detailed analysis of a systematic process in medieval Christian thought that cleared away certain key long-standing obstacles to a theory like Copernicus's and made it possible for certain crucial contributory ideas to come together in a synthesis that would have been impossible, whether or not the other circumstances had been present, before that process had been worked through. In this way, he shows how certain 'extrascientific' preconditions were in fact necessary to the Copernican revolution, rather than merely influencing (say) its timing or the form that it took; and he shows how these preconditions came about in the late Middle Ages. I will outline this process in sections 1–3.

In the remaining sections I shall go on to discuss what kind of "relativism" this account of Copernicus entails (section 4); Blumenberg's account of Copernicanism's consequences for the concept of time and for antiquity's idea of the cosmos as something whose "contemplation" could be man's highest fulfillment (sections 5 and 6); Blumenberg's typology of early attitudes to the new Copernican "truth" (section 7); the powerful (and much misunderstood) metaphors that we have extracted from Copernicanism (section 8); and finally Blumenberg's "revisionist" suggestion that a more consistent Copernicanism would require us to take seriously the possibility that we may in fact be unique in the universe—that reason may not be the logical culmination of nature's accomplishments, but an evolutionary anomaly (section 9).

1. Copernicus and Renaissance Humanism

In his preface to the *De revolutionibus*, Copernicus makes the remarkable statement that the reason why he set out to see whether the assumption of a moving Earth would produce better explanations of the celestial phenomena was that he was dissatisfied with the way "the philosophers could by no means agree on any one certain theory of the mechanism of the universe, which was constructed *on our behalf* by the best and most orderly Maker of everything"[d] [emphasis added].

If we are to believe what he himself wrote, then, the hero who "destroyed our anthropocentric illusions" was motivated precisely by a teleological, anthropocentric view of the universe: by the premise that we should be able to understand the universe because it was created *for us* by a supremely good and orderly Creator. Is this phrase just a bow to inherited pieties—or even a prevarication intended to disguise the revolutionary implications of the new cosmology? Probably the former is the assumption made by most modern readers, since Copernicus's reverence for the ancient authorities (including Ptolemy!) is too consistently maintained for us to be able to picture him as a cunning double agent whose real loyalty is to the future. It is easier to imagine—indeed, many writers do imagine—that his only real concern was with mathematical astronomy, and that while his philosophical and theological remarks, like his sketchy remarks about terrestrial physics, were no doubt sincere, they were mainly intended to placate potential critics. A dedicatory preface addressed to the pope is obviously a political gesture in any case, so perhaps its contents are not to be taken too seriously.

Blumenberg, however, takes not only this statement, and similar ones elsewhere in the *De revolutionibus*, but also Copernicus's whole literary manner—his references to Cicero, Plutarch, Hermes Trismegistus, and Sophocles—more seriously than previous commentators have done. As he says, "Copernicus himself places his work in

the Humanist tradition."[e] Renaissance Humanism turned to classical authors like these for assistance in combating the orthodox
Aristotelianism of the universities, which the Humanists saw as incompatible not only with "eloquence" (i.e., with a classical Latin
style) but with true piety and with wisdom. Copernicus, of course,
was certainly not an orthodox Aristotelian either! When Kuhn, describing Humanism as to a large extent antiscientific, writes that
"the work of Copernicus and his astronomical contemporaries belongs squarely in that university tradition which the Humanists most
ridiculed,"[f] he neglects the difference between the university tradition of astronomy, in which Copernicus certainly did stand, and the
university tradition of Aristotelian philosophy, in which he clearly did
not stand. While Humanists did often depreciate knowledge (*scientia*),
in favor of wisdom (*sapientia*) and piety, the knowledge they had in
mind was above all that claimed by the orthodox Aristotelians.
Astronomy, which was categorized as an art rather than a science and
which, since the time of Ptolemy, had claimed only to predict celestial
phenomena, not to describe the actual order of the universe behind
those phenomena, was hardly a threat to eloquence, piety, or wisdom,
and accordingly was not one of the Humanists' targets. As for astrology, to which some Humanists did indeed object on account of
its determinism, Copernicus's writings show that (unlike many Renaissance and even seventeenth-century thinkers) he was able clearly
to separate it from astronomy, in which the kind of knowledge that
mattered to him would be located.

But even granting that Copernicus may have been genuinely
sympathetic to Humanism, is it not impossibly paradoxical that his
application of Humanist anthropocentric teleology to astronomy
should have produced not a geocentric system (which seems to be the
usual, if not the universal, correlate of anthropocentrism) but a
heliocentric one? And, in the actual historical situation, how can an
anthropocentric premise have been a crucial element in Copernicus's
break with the medieval view of the world, when that view is known,
in retrospect, precisely for *its* anthropocentrism? It is probably questions like these that have deterred previous commentators from
taking Copernicus's anthropocentric premise seriously.

In part II, chapter 3 and 4, where Blumenberg deals with these
questions, he draws some very important distinctions. First, he tells us
that, contrary to a common assumption, the tradition of Greek

cosmos metaphysics did not produce a comprehensively anthropocentric teleology prior to Stoicism. Plato's and Aristotle's cosmoses
are indeed geocentric, but the only anthropocentric teleology they
contain is implied in Aristotle's emphasis on the usefulness of terrestrial things to man, an emphasis that is counterbalanced by the
teleological orientation of the cosmos as a whole toward the unmoved
mover. Augustine makes it clear that God created the world *not* "for
man," but simply *Quia voluit*, "Because he wanted to." Thomas
Aquinas adapts Aristotle's 'split' teleology, but in terms of an overriding goal: the *gloria dei*. Whatever content the Stoic concept of
providence, which the early Fathers had borrowed, still possessed,
was now restricted to arrangements on behalf of those who are chosen
for salvation—which does not affect how the world functions for
human beings in general. Thus, contrary to what the metaphorical
interpretation of our 'removal from the center' by Copernicus leads
us to imagine, there is no continuous or predominant tradition of
an anthropocentric interpretation of geocentrism. Only in the original, pagan Stoicism is the connection maintained. Medieval Christian thought, despite what the Incarnation, also, might lead one to
imagine, is not predominantly anthropocentric, but theocentric.

The second important point is that just as geocentrism does not
entail anthropocentrism, neither does anthropocentrism entail geocentrism. In the Renaissance, when Stoicism—through Cicero—
again becomes influential, it is its anthropocentrism, and *not* its
geocentrism, that catches the imagination. Petrarch, Ficino, and
Erasmus restore Providence—for man, "for us"—to its early Christian position of honor, but geocentrism has no special significance for
them. Man's position 'in the center' is "idealized," Blumenberg says,
and is no longer connected to a specific cosmological diagram. Pico
della Mirandola and Charles Bouillé (Bovillus), whose book was in
Copernicus's library, make this especially clear. Thus Copernicus's
nongeocentric Humanist anthropocentrism is not only not paradoxical, it was widely shared and was a natural source of inspiration
for a talented astronomer, who saw God's good intentions, in relation
to man's capacity for knowledge, being demonstrated by the new
"harmonies" he was discovering.

The impact of Copernicus's anthropocentrically ensured claim to
truth on the status and methodology of the astronomy that he inherited was more revolutionary than his heliocentric model, by itself,

could have been. For more than a thousand years astronomers had been satisfied to try to predict the celestial phenomena (to "save the appearances"), without making claims about the actual mechanism that produced them; and many continued to be satisfied with this goal in the decades after the *De revolutionibus* appeared. It was all that their tradition required of them, and it had proved to be more than sufficiently difficult to achieve by itself. This modesty was reflected in astronomy's status in the universities as a preparatory "art," to be learned prior to and apart from philosophy (which included "natural philosophy," the study of nature). When Copernicus announced that the Earth was a star, and had the same natural motion, and thus the same *nature*, as the other stars, that "university tradition" was irreparably violated. To the extent that his claim was taken seriously, astronomy became irrevocably intertwined with "philosophy," i.e., with physics. Rejecting the advice of his clever Lutheran admirer Osiander (who later inserted the notorious anonymous preface into the *De revolutionibus*, which in fact did disguise Copernicus's claim from some early readers), Copernicus refused to present his theory in the traditional manner as a hypothesis that would simplify calculations but was not meant to be a picture of reality. To present it in this way would have been to abandon knowledge that God intended us to have.

And it was precisely to this exemplary claim to truth, Blumenberg stresses, that "the epoch-making 'energy' that was set free by Copernicus" was due.[g] One can hardly imagine Kepler and Galileo persisting in their indispensable work without their conviction that Copernicanism was literally true, and not merely a convenient hypothesis. Nor are the conflicts with Catholic and Protestant authorities, and the potent metaphor of overcoming anthropocentric illusion, conceivable without that claim. It may well be that in the 'intrascientific' process by which a new theory supplants an older one, the claim to truth—rather than practical convenience—is methodologically irrelevant. But in our attitude to science, and to the human implications of scientific theories—what Blumenberg's analytical history of 'consciousness' is meant to grasp—it can make all the difference.

The consequences of Copernicus's claim reach an even more fundamental level in science itself than that of the motivation of its practitioners. There is no need to refuse to call the pre-Copernican astronomical tradition the tradition of a "science," since its objects and its techniques are to such a large extent continuous with Copernicus's, and even Newton's. But it lacked one characteristic that we now take for granted in all sciences, which is the ideal of unified science: that the sciences can (sooner or later) be systematically integrated with each other, as aspects of a single, homogeneous reality. An important historical point that Blumenberg makes is that the separation of astronomy from philosophy in the medieval university did not just go back to Hellenistic authorities like Ptolemy, but ultimately reflected an ontological and epistemological 'divide' that had separated astronomy from physics ever since Aristotle: the division between the Earth, with its four elements, and the heavens, which Aristotle said were composed of a fifth element, different from all the terrestrial ones. For Aristotle, man inhabits the terrestrial realm and can have knowledge of the higher realm of the heavens only through a special influence from that realm, called "active intellect" and identified with the intelligence that moves the lowest celestial sphere (that of the Moon). Consequently, "Reason does not, in the strict sense, 'belong to' man, but affects him, as a heterogeneous influx. The outcome of Aristotelianism is the subordination of anthropology to cosmology. Man is not really cut out for the contemplation of the heavens; instead, his theoretical curiosity confronts him with the appearance of an inaccessible and heterogeneous world region, for the understanding of which the world that he is familiar with supplies no assistance. Thus the systematic justification of astronomical theory's resigning itself to constructive hypotheses [as it did from, at least, Ptolemy onward] is sketched out."[h] So it was not just the "qualitative," nonmathematical character of Aristotle's physics that obstructed the integration of physics with astronomy. That character reflected an ontological division—deriving ultimately, Blumenberg says, from the transcendency of the Platonic Ideas, which in Aristotle becomes the transcendency of the celestial realm— that separated the mathematically intelligible realm of the heavens from the qualitatively comprehensible realm of the Earth. So it was not just a less philosophical bent on the part of the Hellenistic astronomers that led them to cease trying to interpret their models as physically real; they were operating consistently with the authoritative schema established by Aristotle.

It is this schema that Copernicus, the Humanist, cancels out with one stroke when he "makes the Earth a star." And in so doing he takes the crucial step toward the fundamental modern concept of a homogeneous reality, in which all special regions and disciplines must be capable of being systematically integrated. In this sense, he does make astronomy a science, in the full modern sense, for the first time; and in doing so he brings to the fore something that is a necessary condition for the other modern sciences, as well.

Perhaps this sketch also suggests why Platonism, as such, cannot have been the key motivating factor in Copernicus's efforts, as has been asserted by a long series of commentators.[i] Of course many Renaissance thinkers found "Plato" a useful authority to appeal to against the institutionalized authority of Aristotle; and the emphasis that Plato gave to mathematics is certainly shared by Copernicus, with his enthusiasm for harmony, circularity, etc. But the crucial Platonic doctrine of the duality of sensible appearance and intelligible reality, of "images" and "Ideas," is entirely absent from Copernicus's thinking. What Cassirer said of what he still called Galileo's "Platonism" applies equally to Copernicus's, that it would be "not a metaphysical but a physical Platonism," which was "a thing unheard of."[j] Even if it were the case (which it is not, as I shall show in the next two sections) that the only way to do justice to the philosophical aspect and motivation of Copernicus's reform, or of the scientific revolution as a whole, is to describe it as a reappearance and a transformed version of an ancient tradition of thought, it hardly seems appropriate to describe as "Platonism" something that contradicts what is universally recognized as Plato's central doctrine: the doctrine of the Ideas and their relationship to the world that we perceive. If the Platonism that is supposed to have inspired Copernicus (and Kepler, and Galileo) excludes this doctrine, it would seem more appropriate to call it, perhaps, "mathematicism." But without the direct connection to Plato, it would no longer appear as the recognized basic alternative to Aristotelianism, and the reasons for its influence in various circumstances would need to be analyzed much more carefully than the assumption of a continuous tradition originating in a great philosopher has required us to do.[k]

To return to Blumenberg's own account: Is he, then, proposing that we should interpret Copernicus's reform not as a result of the Renaissance's revival of Platonism but as a result of its revival of Stoicism? Such a proposal might be plausible to a certain extent, although the selectiveness of the revival (anthropocentrism—yes, geocentrism—no) might make one wonder how coherent the tradition would be. But if we are talking about a systematic process, rather than mere contributory ideas, we need to ask, What was the intellectual, or 'spiritual,' context in which this revival took place? Why was it appropriate to have recourse to Stoic anthropocentrism at just this point? Were there other aspects of the situation as a whole that were essential preconditions of Copernicus's project?

Blumenberg does in fact describe a process that produced the attraction and the plausibility of Stoic anthropocentrism for the Renaissance, and that also produced other essential preconditions of Copernicus's project; and in doing so he relates that project to the whole state of late medieval and early modern thought about God, man, and the world. The premise of his account is that a major cultural and intellectual transformation was under way that was independent of Copernicus's intervention, but helped to make it possible and to determine what its consequences would be. In order to understand Blumenberg's account of Copernicus it is necessary to have some acquaintance with his analysis of this broader transformation, which he presented in *The Legitimacy of the Modern Age*. I will therefore summarize the relevant aspects of that analysis, as briefly as I can, in the next section.

2. Modernity and Christianity

A central question to which *The Legitimacy of the Modern Age* addresses itself is the question of the origin and meaning of the modern theme of human self-reliance, or "human self-assertion," as Blumenberg calls it. Why, since the seventeenth century, have we been so preoccupied with security and survival, so conscious of scarcity, so intent on acquiring means of all kinds (capital, technology, mechanistic explanations of the world)—why have we been so 'worldly,' in general, and so unwilling to rely (as people in previous epochs seem to have relied) on a cosmic order, a providence, or a promised salvation in the next world?

Before proceeding to Blumenberg's answer to this question, I should register the obvious fact that if modernity is described in this way, Copernicus turns out—despite the central role he has been given in the modern age's self-image—not to be modern himself. His anthropocentric teleology—"the world that God made *for us*"—is unacceptable in principle to modern thought, beginning with Descartes and Bacon, which is why it has been consistently overlooked by people who want Copernicus himself to serve as the revolutionary who marks the beginning of the modern age. Copernicus is blissfully unaware that he will later be praised (and damned) for "removing man from the center," and thus abandoning him to his own devices.

The usual modern answer to the question of where modern 'self-reliance' comes from is that it is simply the natural, normal, and rational approach to human existence, which automatically asserts itself when hindrances like myth, dogmatism, prejudice—and anthropocentric illusions like cosmic and Christian providence, and promises of salvation—are cleared out of the way. As to why such 'hindrances' should have been predominant throughout the greater part of human history, we have no very clear idea. Finding 'anticipations' of modernity in the Renaissance, in the Middle Ages, and in antiquity—an effort to which much energy has been devoted—does not, of course, solve the problem. This innocent ignorance, on modernity's part, about itself, has given rise, since nineteenth-century Romanticism, to doubts as to whether modernity does not in fact represent an unfortunate rupture in some greater, more coherent mode of existence, which is variously identified with Christianity, myth, cosmic order, etc.

As the title, *The Legitimacy of the Modern Age*, suggests, this is not Blumenberg's approach. He takes considerable pains, in that book, to defend modern concepts against the suggestion that they are illegitimate because they are really "secularized" versions of concepts that were originally, and properly, medieval and Christian. But he does not accept the usual Cartesian/Enlightenment/positivist self-image of modernity, either, according to which it is simply 'normal' and unproblematic. Instead, he proposes a comprehensive analysis of 'where modernity came from' that, without making it a transformed version of something else, or something that was there all along (but obscured by other things), also makes it clear that it did not come "out of the blue," either. To put it in a nutshell: Blumenberg describes modernity as an attempt at solving a *problem* that was implicit in the whole state of late medieval thought about God, man, and the world. It was this problem that made modernity appropriate at a certain point in history, and the absence of which (the fact that it had not yet emerged) explains the absence of modernity at earlier points in time.

Blumenberg's rubric for this problem is "theological absolutism." What this refers to is the late medieval emphasis on God's omnipotence, with the associated idea of the "hidden God," whom we cannot expect to comprehend in any way. In theology, these ideas led to a stress on man's powerlessness to choose even to have faith, which is something that, like everything else, is freely and inscrutably given or withheld by God. The arbitrariness of this situation is often expressed by the idea of the eternal "predestination" of souls either for salvation or for damnation. In philosophy, this kind of thinking led to the ecclesiastical condemnation (in Paris, in 1277) of the Aristotelian proofs of the uniqueness, as well as the eternal existence, of the world, and later to Nominalism's denial of the reality of universals. These originally Greek doctrines were suspected of placing limits on the omnipotence that absolute divine sovereignty required. The consequence was, however, that just as man's faith and salvation were put in the hands of a completely inscrutable God, so was the world. It lost all the characteristics that, for classical and High Scholastic philosophy, had rendered it intelligible and reliable. "Providence" now had meaning only as God's provision for the elect, not for men or the world in general. This situation—a world that was made for God, not "for man," in which man could not be at home, but from which the only means of escape was also entirely outside his power— represented the extreme of human self-abnegation, and could not be endured indefinitely. And indeed, after a great struggle with these issues, extending through the Reformation and the wars of religion, many Europeans more or less consciously and more or less wholeheartedly replaced theological absolutism's human self-abnegation with modern "human self-assertion," in which people decide to see what they can make of the world "even if there is no God"[1]—and of which a key axiom must be that we cannot know the purposes, the "final causes," of phenomena (for to know them would be to know, and to rely on, God's will), but only the "efficient causes" by which we find that we can produce the same phenomena.

As to how things came to such a pass in the late Middle Ages, Blumenberg has, again, a fascinating story. Very briefly: The theology of divine omnipotence and predestination was first crystallized in the later works of St. Augustine, in the fifth century. "In many ways," Blumenberg writes, "the Scholasticism of the Middle Ages travels Augustine's path over again." How are we to understand that path? Once again, in terms of a problem: the problem of how "to hold the God of creation and the God of salvation together in *one* system." [m] The expectation of Christ's imminent second coming and the end of the world had put all the stress on the "God of salvation." As that expectation was increasingly disappointed, the God who created the world in which one was apparently going to remain became an issue again. Gnostic dualism said that the contrast between these gods was due to the fact that they were different gods, and opposed to one another. Blumenberg argues that this plausible interpretation of a difficult situation was the central problem that Patristic theology—culminating in Augustine's doctrines of free will and original sin (which made *man* responsible for the miserable state of the world), and then in the correlative ideas of divine free will, omnipotence, and predestination—had to try to deal with. But the means that were employed for this purpose wound up reproducing the problem, in that "the absolute principle's responsibility for cosmic corruption—the elimination of which had been the point of the whole exercise—was after all reintroduced indirectly through the idea of predestination. For *this* sin, with its universal consequences ... only the original ground of everything could be held responsible." This is the sense in which Blumenberg can describe the modern age as "the second overcoming of Gnosticism"; a thesis that presupposes that "the first overcoming of Gnosticism, at the beginning of the Middle Ages, was unsuccessful." [n]

3. Copernicus and Christianity: The "Opening Up of the Possibility of a Copernicus"

Clearly Copernicus as a philosopher, with his innocently anthropocentric Humanism, does not anticipate and could not survive modern "self-assertion" 's merciless critique of teleology. Neither does he anticipate its other characteristic themes: self-preservation, scarcity, the importance of technique (and experiment), method, and methodical doubt or critique. The ontological unification of reality that his position presupposes, which anticipates Galileo's and Descartes's homogeneous mathematization of the physical world, is never spelled out explicitly. Despite our retrospective heroization of him, and the fact that his work may have had a greater direct and identifiable effect on modern consciousness than all the philosophers of "self-assertion" put together, the philosophical propositions that Copernicus does articulate exclude his philosophical position, as a "transitional episode,"[o] from belonging to the modern age.

The tendency of much historical scholarship, since the nineteeth century, has been to push the origins, or the earliest indications, of modernity further and further back into what the eighteenth century had regarded as the "dark ages." While Blumenberg of course shares these scholars' desire to appreciate and understand late medieval and Renaissance thought, he maintains that its relation to the period beginning with Bruno, Galileo, Descartes, et al. exhibits such a fundamental discontinuity that it is not always useful to conceive of the Humanists as "forerunners" of the modern age. Instead, he suggests that we interpret their period as reflecting "the epochal crisis of the Middle Ages, their falling apart into Nominalism, on the one hand, and Humanism on the other."[p] Blumenberg's account of the thought of Nicholas of Cusa (1401–1464) (who has sometimes been described as the first Humanist, and consequently also as "ushering in modern thought"[q]), in Part IV of *The Legitimacy of the Modern Age*, is an example of the kind of subtle and powerful interpretation that Blumenberg's schema, in which "modern" thought appears much later, makes possible. Part II of the present book is another demonstration of this.

The reason why Humanism could only be an unstable, transitional formation is that it did not give its adherents means with which to resist the subversive doubts that Nominalism, and the Augustinian logic in general, promoted about the world and God's providence within it. Humanist anthropocentrism could administer a suitably "pious" rebuke to the orthodox (Averroist) Aristotelians in the universities, who were the Humanists' main targets. But in theology, as the success of preachers like Luther and Calvin suggests, theocentrism, divine omnipotence, and human helplessness were the compelling conclusions of medieval thinking. The Humanists, including Copernicus, relied, in effect, on God's goodness; but only human self-assertion—by ceasing to rely on Him for anything—could really deal with the way God's freedom, understood in accordance with Augustine's logic, problematized God's goodness, for man.

However, the fact that Copernicus avoided confronting this crucial late medieval problem does not mean that Nominalism had no influence on his endeavor. On the contrary, it played a very important (albeit indirect) role, not in determining the content of Copernicus's proposals, but in making them *possible* at all—that is, in enabling Copernicus to entertain them seriously, and to expect that his readers would do so as well. This was a crucial contribution for at least two reasons: first, because essentially the same theory had *not* been taken seriously when Aristarchus of Samos proposed it, in the third century B.C., and second because Copernicus's proposal "stands at the end of centuries that had born the imprint of the most closed dogmatic system of world-explanation, whose basic character, not accidentally, can be summarized as implying the impossibility of a Copernicus before his time."[r]

So the question is, what happened to that dogmatic system so that Copernicus finally became "possible"? Blumenberg's answer has two parts, which are presented in chapters 2 and 3, respectively, of part II. First of all, he tells us, the Augustinian logic that he described in *The Legitimacy of the Modern Age* undermined the Aristotelian geocentric cosmology and physics by insisting, as in the great Paris condemnation of 1277, that it was not necessary for God to arrange things in the way that Aristotle had thought they were arranged. This condemnation served as a license for thinkers to consider hypotheses that, in an 'Aristotelian world,' would have been impossible, and thus not worth considering. However, as the history of the debate over Copernicanism shows very clearly, this did not mean that Aristotle's doctrines ceased to carry authority. They continued to be the only comprehensively worked-out system of cosmology and physics that was available, which made it very difficult for people to imagine doing without them. But although they were the only comprehensively worked-out system, they were no longer the only available ideas. Thinkers who had availed themselves of the "license" of 1277 had in fact generated some partial alternatives, whose existence was an indispensable precondition of Copernicus's being taken seriously. There was the *appetitus partium*, the desire of parts to cohere as a whole, which, in lieu of the later concept of gravity, helped to explain why a rotating Earth that was no longer located "at the center" (so that Aristotle's idea of the elements' seeking their "natural places" could no longer be applied to it) did not simply fly apart. And there was the famous concept of *impetus*, which, in lieu of the Newtonian concept of inertia, helped to explain why objects such as falling stones and flying birds, which were not in contact with the Earth, did not have to be left behind by its rotation, as they would under Aristotle's theory of motion (which required a mover continuously acting, by direct contact, on the object that is moved).

There is a natural tendency to think of Jean Buridan, Nicole Oresme, and the other Nominalists who developed these ideas as engaged, essentially, in "science," or at any rate in "natural philosophy," which is really no different from what a pagan Greek or a secular modern scientist might engage in. Kuhn, for instance, describes the Scholastic criticism of Aristotle as "the first effective research in the modern world," and as exhibiting "an unbounded faith in the power of human reason to solve the problems of nature."[s] To counteract this natural projection of our own assumptions, and to help us to understand the very long period—from, say, 1400 to 1580—during which the Nominalists' theories were, indeed, taught in the Italian universities, but in which they led to no further concrete progress, it is important to realize how conservative the actual intentions of these thinkers were. Like all the Scholastics, they were working on a system whose internal coherence was their overriding concern. When they criticized Aristotle, it was not with the idea that entirely new ideas might eventually be needed (which might have directed their attention toward, for example, carrying out concrete measurements or experiments)—it was always with the intention of repairing the inherited system. Thus, as Blumenberg shows, "The clearest 'forerunner' of Copernicus in Nominalism, Nicole Oresme [who discussed the possibility of the Earth's diurnal rotation], deviated from Aristotelian *cosmology* precisely in order to be able to save Aristotelian *physics*"[t]—removing the ambiguity of Aristotle's account of the motion of comets by stipulating that even sublunar objects (and thus, necessarily, the Earth) *could* have a natural circular motion. And Jean Buridan, who also breached the Aristotelian division of the heavens from the Earth, but in the opposite direction, by applying the concept of *impetus* (which had previously been used in analyzing the motion of projectiles) to the motion of the celestial spheres, was not on the track of astronomical knowledge, as such; he was merely generalizing his critique of the Aristotelian idea of natural motion as goal-directed motion, a critique that was entailed by Nominalism's emphasis on the inscrutability of the divine will. As Blumenberg sums up the situation, "The exhaustive exploration of the speculative possibilities of the Scholastic system is aimed at displaying God's sovereignty and his *potentia absoluta* [absolute power], but not at expanding man's knowledge of the world; if it nevertheless proves useful for the latter, then this is a result of the efforts to maintain the system's stability in spite of that demand on it."[u] Thus while Nominalism created indispensable preconditions for Copernicus's great step forward, that step could not have been taken by a Nominalist, but only by someone who had reasons for thinking that knowledge about the heavens should in fact be accessible to man—i.e., by a Humanist.

The second part of Blumenberg's account of the way in which Nominalism helped to "make a Copernicus possible," which he presents in part II, chapter 3—"Transformations of Anthropocentrism"—has to do with the homogeneity of the world whose order Copernicus believes we can know, because it was made "for us." The problem here is that in antiquity, the anthropocentric providence on which Copernicus would later rely was intimately associated with an understanding of theory as an essentially *visual* relationship between man and the cosmos—the "contemplation of the heavens" (whose history Blumenberg discusses in part I of the book: see section 6 of this introduction). For such a relationship, it was appropriate for the "contemplator" to be located at the physical center of the cosmos, where the Stoics put him. The fact that Humanism was able to revive Stoicism's anthropocentric teleology without becoming entangled in Stoic geocentrism at the same time implies that in the meantime the theoretical attitude had become separated, in principle, from its traditional visual connotations—which is a result that cannot be adequately explained by the renewed influence, for example, of "Platonism," because, as I pointed out earlier, Platonism lacks precisely the conception of a *homogeneous* knowable reality.

Where, then, did Copernicus get the idea of a cognitive access to the entire, homogeneous world, which was not the visual access that the Stoics had assumed, but was not a Platonic access to a merely intellectual world either? This is the main question that Blumenberg addresses in the final pages of part II, chapter 3, where he brings out a surprising relationship between Nominalism's God and Copernicus's man. Copernicus's teleological premise, he points out,

cannot be read as a generalization of the partial teleology that, according to Aristotelianism, benefits man; it cannot be read this way because it excludes precisely the element of need. It is supposed to guarantee a possible access to truth by man that has nothing to do with the ways in which the art of astronomy is useful in life. A satisfaction is pictured that goes beyond man's needs—a satisfaction of a kind that, in the Middle Ages, only the *visio beatifica*, in the next world, could be. God is the final end of all natural beings, Buridan asserts, in a sense of sovereign independence [i.e., although they exist 'for His sake,' He does not *need* them], but not in the Aristotelian sense of His having no knowledge whatsoever of the world. Could one say that this relation too was now defined, *after* the Creation, as one of pure theory? And defined, in that theory, as a universal homogeneity of all objects in the world? In that case, this turning would already, by implication, have suspended Aristotelianism's split teleology and produced, objectively, the unitary character of the nature as the 'reference-person' of which man, again, makes his appearance in Copernicus's formula.[v]

Blumenberg's suggestion, then, is that the cosmos was first unified as an object of 'theory' by Nominalist *theology*, in keeping with Genesis's "And *God saw* all that He had made, and it was very good." Unlike any Greek god, and certainly unlike Aristotle's "unmoved mover," to whom Aquinas had tried to assimilate his Christian God, this (Augustinian/Nominalist) God took an interest in the world—simply because He had chosen to create it. And of course for Him distinctions of near and far, matter and form, Earth and heavens amounted to nothing. Everything without distinction existed (had been created, by Him) *for Him*.

Which is precisely the relationship that Copernicus asserts between man and the universe: Everything was created "for man" (though of course not *by* him), and is therefore equally accessible for him. "Without any blasphemous defiance, the Copernican world-formula grants to man a privilege, as an end, that only a theory that is beyond the realm of pressing needs could have . . . ." By claiming the right to such a theory, Copernicus crosses the boundaries of the human condition as the Middle Ages conceived it. He is able to do this because the change in the conception of God's relation to the world that Nominalism's Augustinian logic has forced upon Scholasticism—its dismantling of the idea of a God Who "remains external to the world, not in immediate relation to all of its elements" [w]—has set up a new framework, in which, given a "high degree of exertion against the late Scholastic Middle Ages," God's position (as far as knowledge of the world is concerned) can be reoccupied by man.[x]

Unlike its Stoic ancestor, the teleology "for man" that is produced by this reoccupation does not imply any particular cosmological diagram, because the relation to reality—the "idealized" position in the center, as Blumenberg calls it in the next chapter—that it entails is conceived on the completely nonspatial (and likewise nonvisual) analogy of God's position. This, then, is the ultimate model of a relation to reality—compatible with "Platonic" mathematicism, and compatible with Stoic anthropocentrism, but not dependent on either of them for its basic pattern—that underlies Copernicus's claim to knowledge of a homogeneous universe—the claim that motivated his whole effort. Thus in this essential way as well, Nominalism's dismantling of High Scholastic Aristotelianism is a necessary precondition of Copernicus's reform, although clearly there is no way in which one could undertake that reform and still remain a Nominalist. Although Copernicus does not "construct a position that can *confront* theological voluntarism,"[y] as modern "self-assertion" does, he does move irreversibly beyond such voluntarism by putting man, at least in respect to his type of knowledge, in the position of its God.

In the final chapter (chapter 6) of Part II, Blumenberg rounds out his account of how Copernicus's theory became (physically and metaphysically) "possible" with a hypothetical account of how Copernicus, as an astronomer, actually arrived at it. His idea, which he supports with a detailed reading of the *De revolutionibus* and the *Commentariolus*, is that Copernicus started from a partial heliocentric model, applied to Venus and Mercury only—which is a construction that would not have been original with him, having been proposed by Heraclides of Pontus, in the fourth century B.C., as a way of explaining the observed permanent proximity of these two planets to the Sun. Copernicus would then have generalized that model to include the other planets as well, and only inserted the Earth in the system—to fill the gap between Venus and Mars, and harmonize the whole picture—as the very last step in the process. The logic of this possible train of thought, in which the Earth's *annual* motion, as a planet, comes first, and its diurnal rotation is only, as it were, an afterthought, helps to explain why those "forerunners" of Copernicus (like Nicole Oresme) who considered a possible diurnal rotation did not get any further. They had not, in fact, taken even the "first step" on Copernicus's actual path.

To sum up, then: The tradition of astronomy contained ideas

that, when Copernicus combined them and worked them out in a thoroughgoing manner, yielded his new model of the planetary system. But, first, neither he nor his readers could have entertained that model seriously as a description of nature if it had not been for Nominalism's previous critique of Aristotelian physics and cosmology, and for the useful conceptual "spin-offs" (*appetitus partium, impetus*) from that critique. And second, he probably could not have believed in and pursued the kind of knowledge of a homogeneous nature that he did believe in and pursue if it had not been for the model of such a nature, and of a nonvisual 'theoretical' relation to it, that he could find in late medieval theology. (In this sense—which, however, is probably not the sense in which he meant it—Whitehead's remark about faith in the possibility of science being an unconscious derivative from medieval theology is quite correct.) Thus, Copernicus's effort and success were made possible by the internal logic of the Scholastic system and the modifications that that logic produced in the system. This is why the Copernican revolution was neither the result of an 'exogenous' input into European consciousness, nor the result merely of the return of one or more premodern philosophical traditions to prominence, nor the result of a combination of one or both of these factors with various social circumstances, but reflects, instead, a central systematic process in medieval Christian thought as a whole.

4. Does This Analysis Lead to "Relativism"?

Does this analysis of the process that "opened up the possibility of a Copernicus" imply, as it appears to, that Copernicus's theory was *impossible*—that it could not have been successfully advocated—prior to the late Middle Ages? Such a conclusion, based on Blumenberg's account of the key role of Christian theology in producing, through Nominalism, the conditions that made a Copernicus possible at that point in history, would be in sharp contrast to our usual conception of science as a self-contained, autonomous process, which generates outputs that affect our other activities but is not dependent on them, in return, for anything more than material inputs. It might also suggest a contrast with our usual assumptions about rationality. Taken together with Blumenberg's remarks about the novelty of the "self-propelling scientific attitude toward the world, which is projected as a guiding idea"[z] in the modern syndrome of self-assertion, his analysis of the Copernican reform of astronomy might encourage one to suppose, as Michael Heidelberger does, that for Blumenberg "the conditions which make internal relations between scientific theories at all possible did not arise prior to the emergence of modern science ... [so that] it is only natural ... that in general one cannot find a purely internal relation between Ptolemean and Copernican astronomy, which could be sufficient to justify the choice of the Copernican theory as rational."[aa] The 'choice' between Ptolemaic and Copernican astronomy, then, would not be one that could be argued rationally, but would depend upon one's epochal affiliation—on what historical experiences one had gone through. Such a view of science and of Copernicus would not only be in conflict with scientific rationalism's claims to universal relevance, intelligibility, and authority, but would make one wonder what kind of rationality, if any, is shared by humanity as a whole. If we read Blumenberg in this way, he would appear to be a rather extreme relativist, whose thinking might very well be open to the criticism that it puts in question at least the possibility of translation (between the languages of different epochs and cultures), and perhaps also (if its logic were pressed to the extreme) even the possibility of communication in general.[bb]

Now, in comparison with the Cartesian/Enlightenment/positivist tradition, Blumenberg certainly is a relativist. "Historical respect for the equal rank of the aids that man avails himself of in comprehending the world"[cc] is always his guiding idea, whether he is interpreting metaphor, myth, Gnosticism, Christianity, metaphysics, Romanticism, Idealism, *Lebensphilosophie*, phenomenology—or, for that matter, Enlightenment rationalism or positivism. None of these, he clearly thinks, has the right to judge the others and find them wanting. (Though each of them can certainly be assessed internally by how well it succeeds in dealing with the problems that it sets out to deal with. Such 'assessment,' as a process, is what history is made up of.) Such a view does, indeed, deny the special dignity of science presupposed by scientific rationalism.

Furthermore, Blumenberg is, I think, quite serious about the "possibility" of a Copernicus not presenting itself before the late Middle Ages. Of course the game of "What if?"—for example, What if Aristarchus of Samos had been Copernicus and Newton rolled into one?—is easy to play, and impossible to refute definitively. But there are good reasons why Aristarchus in fact was not Copernicus and Newton. For the preeminent ancient theorists of the cosmos, Plato and Aristotle, physics was not fully mathematizable because only whatever was perfect (the Ideas or the fifth element) could be perfectly mathematical. The main rival doctrine to theirs in physics—atomism—seems not to have been conceived as mathematizable either, perhaps for the same reason—as a description of a universe of chance, it was obviously 'imperfect.' The mathematical fictionalism that became the rule in Hellenistic astronomy took this dualism so seriously that it ceased to try to describe cosmic reality, as such, at all. It may be that this attitude even influenced Stoicism, if Blumenberg is right that Cleanthes, the leading Stoic, accused Aristarchus of impiety just because he presented his heliocentric model as more than a fiction, thus profaning the "mystery" of the cosmos. It appears, then, that a mathematical description of a homogeneous reality was not going to be possible until (again, through Nominalism) the idea of an omnipotent God Who creates "from nothing," and thus has the same 'immediate' relation to everything, had destroyed the dualism of matter and form that runs through these older doctrines. In response to the question why such a dualism should have been the first form taken by self-conscious reason in our tradition, Blumenberg suggests that it represents a pattern of "relapses" into "the double-layered relationship that exists, in mythical thought, between what one sees and what really happens—between the flat appearances in the foreground and a 'story' in the background."[dd] For whatever reasons, it does seem to be the case that the idea of a homogeneous and mathematizable reality is a unique characteristic of modern European thought; and it is certainly a necessary precondition of a theory like Copernicus's.

However, I think Michael Heidelberger is mistaken in concluding that, in Blumenberg's view, modern science is so different from everything premodern that premodern 'theories' are simply incommensurable with modern ones. It seems probable that Heidelberger arrived at this interpretation of Blumenberg by seeing Blumenberg's conception of the change from the 'prescientific' epochs to the 'scientific' one as a generalized version of Kuhn's conception of "paradigm changes" within science.[ee] But an epochal change, as Blumenberg conceives it, is not comparable to a Kuhnian paradigm switch, which, as he says in his second reference to Kuhn in this book, "neglects the way in which every possible discontinuity presupposes [an underlying] continuity." In the revised edition of *The Legitimacy of the Modern Age* Blumenberg compared the process of epochal change not to a paradigm switch but to the "logical situation that Aristotle put under the heading of *aporia* and that Kant discussed as the fundamental pattern of the 'transcendental dialectic.' In both cases the process of cognition itself forces the abandonment of its presuppositions and the introduction of new elementary assumptions, which, while they do create a way out of the problematic situation, do not require the shattering of the identity of the overall movement that gave rise to that situation."[ff] That is, Kuhn neglects the underlying continuity of the "process of cognition itself." "'Scientific revolutions,'" Blumenberg wrote, "if one were to choose to take their radicalness literally, simply cannot be the 'last word' of a rational conception of history; otherwise that conception would have denied to its object the very same rationality that it wanted to assert for itself."

Thus, rather than projecting the kind of incommensurability that Kuhn saw in science onto the plane of the history of Western consciousness as a whole, Blumenberg wants to find continuity beneath the great discontinuities of that history by broadening our understanding of "the process of cognition" as a whole to include what scientific rationalism, since Descartes, has condemned as irrational. He wants to find in his "object"—the history of Western consciousness—"the same rationality that he wants to assert for himself." Whether he succeeds in finding it, each reader will have to judge for himself. But it is clear that his type of 'relativism,' which expects to find rationality (in his own, broad sense) in more various modes of thought in our tradition than any rationalism (including Hegel's) has found it in, does not correspond to the usual picture of relativism as doubting the universality of reason.

Where it does agree with skeptical forms of relativism is in its estimate of the extent to which definitive rationality has been or can be *achieved*. In contrast with the classical spokemen of modern rationalism, Blumenberg underlines the negative side of the now generally acknowledged "endlessness" of scientific progress: the fact that, because definitive results can never be arrived at, practical conclusions also can only be temporary and provisional, and no decision (practical or theoretical) can ever be fully rationally justified. Aside from whether science, even in its ideal final state, could ever rationally determine our "ultimate ends," this actual insurmountable provisionality makes it necessary to supplement the scientific analysis that we are able to carry out at any given time with "rhetoric," which thus is "not an alternative that one can choose instead of an insight that one could *also* have, but an alternative to a definitive evidence that one *cannot* have, or cannot have yet, or at any rate cannot have here and now." [gg]

This, of course, is the entering wedge for myth, metaphor, religion, metaphysics, and everything else from which modern rationalism has wanted to free itself. Even if it has no scruples about condemning whole epochs for their predominant 'irrationality,' rationalism has to recognize that by itself it can never in practice be a sufficient guide to action. Once it has admitted this, Blumenberg suggests, maybe it can find some consolation in historians' and anthropologists' discoveries that the thought processes of other cultures and ages, which now turn out not to have been superseded as completely as rationalism had hoped, have not been entirely incoherent or arbitrary either.

Less surprising to rationalism than the enduring relevance of 'premodern' modes of thought in the "age of reason" is, no doubt, the presence of 'protomodern' scientific theories in earlier ages, especially in antiquity. Blumenberg's emphasis on the unusualness of modern science, and on the epochal break that initiates it, has led Heidelberger and perhaps others to suppose that Blumenberg sees the epochs as reflections of completely discontinuous paradigmatic relations to reality, so that a continuous scientific argument ("internal relations between theories") extending across an epochal break would be inconceivable for him. But the reason Blumenberg takes epochal breaks as seriously as he does is that modern, Cartesian rationalism, with its idea of "starting over from scratch," makes that kind of break—and the corresponding sharp division between science and 'irrational' modes of thought—into an essential part of its self-comprehension. Taking this claim at its word, and insisting on the 'abnormality' (in relation to mankind as a whole) of the result, is the first step in Blumenberg's effort to understand the experience that produced this version of rationality. But he does this for the sake of historical comprehension, not because he accepts the Cartesian claim that an absolute break did in fact occur. In a world as pluralistic as

Blumenberg's, in which Gnostic and Christian thought, for example, are in their own ways rational ("aids that man avails himself of in comprehending the world"), it is not surprising that antiquity had a version of rationality that could produce theories that make claims that are similar to modern scientific ones, and can be judged on similar criteria. (Though of course when one looks more closely at what those theories meant to their authors, one finds systematic differences between the ancient and the modern versions.) The "cognitive process" that we are all engaged in is, in an important sense, one process, despite the dramatic dissimilarities between its various traditions and epochs. That is why we are able to translate and interpret one epoch, or culture, to another.

It is also why Blumenberg's radical relativism is in its own way a radical rationalism. A question that has repeatedly been raised about his historical writings is why he never considers the possible influence of external factors—social change, modes of production, affective states, etc.—on the intellectual processes that he studies.[hh] The reason for this, it seems to me, is not that he is an idealist, for whom all significant change originates in the mind, but rather that he wants to find out to what extent human thought has in fact been rational in his sense: to what extent one can find something like a "cognitive process" in history. Once one had determined the answer to that question, one could go on to examine parallelisms and causal connections between thought and other aspects of human existence; but starting with the latter question would not facilitate, and might well hinder, progress on the former one. Blumenberg's procedure, in working on the first question, is to grant—hypothetically—to past thinkers "the very same rationality that he wants to assert for himself." (Not an omnipotent rationality, I repeat—merely a "sufficient" one.[ii]) It seems to me that the results, in this book and his others, indicate that this kind of hypothesis certainly deserves to be explored further.

5. The Consequences of Copernicanism for Time

The historical analysis that I summarized in sections 1 and 3 is laid out in part II of the book: "The Opening Up of the Possibility of a Copernicus." Part IV, on the position of Copernicanism in the history of the concept of time, provides more detail on Copernicus's preconditions and on his thought process. It focuses on time first as another aspect of the medieval undermining of Aristotelianism that contributed to the Renaissance "idealization of the center of the world," and thus helped to make Copernicus's project possible. And then it examines (I believe for the first time) the pains that Copernicus took, in spite of his own non-Aristotelian beliefs, to try to head off Aristotelian objections to his model by providing, like Aristotle, a "perfect," uniform cosmic motion—which would be the axial revolutions of a "perfect sphere" that in this case was not the outermost heaven, but the Earth—to exhibit and embody cosmic time. Thus another oddly 'medieval' aspect of Copernicus's work is made intelligible, as something more than a mere instance of the inertia of tradition, and at the same time our respect for Copernicus's skill in argumentation is enhanced.

Turning to consequences of the Copernican event, part IV then goes on to describe how the later development of Copernicanism totally obliterated Copernicus's effort in this connection, by postulating an "absolute" time or an ideal one (Newton, Leibniz), neither of which had a special connection to any particular heavenly body, and finally by demonstrating the truth of Copernicanism—that the Earth moves—precisely by proving the Earth's *im*perfect sphericity (Maupertuis).

6. Consequences for the "Contemplation" or "Intuition" of the Universe

The remaining four parts of the book—parts I, III, V, and VI—deal entirely with consequences of the Copernican event. So, with one exception, the material they cover is modern. The exception is the remarkable introductory chapters of part I, "The Ambiguous Meaning of the Heavens"—chapters that go back to Sophocles and Anaxagoras, and to the elementary physical preconditions of the practice of an astronomy based on vision, in order to clarify the significance of astronomy in its premodern and pre-Christian form, as the "contemplation of the heavens." For just as the perspective of the history of science has created the mistaken impression that the Copernican revolution, with all of its consequences for modern man's conception of himself and his relation to the universe, was an 'exogenous' input into culture, injected into it by a preexisting, self-contained, autonomous activity called "science"—so also in its assessment of "ancient science" that perspective tends (particularly in the case of astronomy, as a natural consequence of the very technical nature of much of its material) to isolate the processes of "observation" and "theory-formation" as though they were just as remote from the "life-world" of the Greeks as they seem to be from ours today.

On the 'anthropological' side, too, such a perspective does not encourage one to notice what Blumenberg calls the "remarkable improbability" of the circumstance "that we live on the Earth *and* are able to see stars—that the conditions necessary for life do not exclude those necessary for vision, or vice versa" (thanks to the Earth's unique atmosphere, which is transparent, but also protective against cosmic rays and particles); nor does it encourage one to notice the equal improbability of a creature on such a planet, bent on self-preservation, in fact "lifting its gaze out of the sphere of biological signals and drawing something inaccessible into its range of attention"—i.e., seeing stars. Thus the sheer *unlikeliness* of pure theory as a phenomenon in nature tends to escape the philosophy and the historiography of science.

A specific characteristic of such pure theory, as it was practiced and viewed until very recently—a characteristic that is implicit in the previous paragraph, and that will provide the theme of part I of this book—also tends to be overlooked by those who see modern science as 'normal.' Blumenberg underlines this characteristic memorably with his discussion of Poincaré's question as to whether, and if so, when, a Copernicus would have appeared if the Earth had been covered by a permanent blanket of clouds. The point is that theory, though it is (no doubt) conceivable as an activity entirely independent of vision, was originally and remained until not so long ago something intimately tied up with *seeing* the world (in this case, with seeing the stars). Vision was not just one source, among others, of (more or less reliable) "data"—accurate vision was theory's fulfillment as well as its source. And it was the confusion that was thrown into the way people saw the world, and eventually into the very idea of seeing it at all (i.e., of seeing anything more than a minuscule and superficial, unrepresentative aspect of it), that constituted the greater part of Copernicanism's historical impact.

It is only in the light of this 'anthropological' significance of vision (which evidently persists, as a need, even though science no longer thinks of itself, primarily, as serving it) that we can understand, or even take seriously, a statement like Anaxagoras's (from the fifth century B.C.), that a good reason to choose to be born, rather than not to be born, would be "for the sake of viewing the heavens and the whole order of the universe." The central role of "contemplation of the heavens," as a concept of human fulfillment, in Greek thought down to the Stoics (and in neo-Stoic thought down to the present) depends on the idea of true vision as a fulfilled state, a state of happiness (and thus, necessarily, on a conception of "theory" as an activity of individuals, rather than a cumulative, impersonal, institutionalized process). The real is what "shows itself," and is (indistinguishably) both known and *enjoyed* as such.[jj]

However, Blumenberg in no way romanticizes this relation to the "cosmos," as though it were a lost Eden or a golden age, from which the impersonal and unvisual mechanism of modern science has exiled us. Anthropologically, he makes it clear elsewhere that the improbable activity of pure theory does not distinguish mankind as a specially favored species, but is, if anything, an offshoot of a uniquely problematic situation faced by this species, to which it has responded with unique behavioral experiments, both "practical" and "impractical."[kk] On the historical side, his chapter "Cosmos and Tragedy" begins with the statement that the terms in its title constitute an unresolved antinomy, and goes on to take Anaxagoras's "reason for choosing to be born" with the utmost seriousness, as a risky answer to the (for the Greeks) very persuasive statement of the Chorus in the *Oedipus at Colonus* that the best thing is *not* to have been born. This "tragic" perception was expressed in the story of Prometheus's defense of mankind (which Blumenberg goes on to examine at length in *Work on Myth*), a fundamental premise of which was that (contrary to the image of the "cheerful" ancients, in their comfortable, geocentric cosmic housing) man does not *belong* in the cosmos that Anaxagoras wants to contemplate. Hence the antinomy. Any positive relation between man and the cosmos seems to be a matter of sheer luck, or (which amounts to the same thing) of unreliable divine favor, and thus a rather weak reed to lean on. Blumenberg shows how this perception, or this fear, is the focus of Epicurus's therapeutic efforts, how it is exacerbated by Stoicism's dogmatic optimism of "providence," and how it eventually feeds into Gnosticism's epoch-making demonization of the cosmos (which, according to his account in *The*

Legitimacy of the Modern Age, constitutes the problem that first Christianity and then modernity seek to overcome).

Thus we see that the "contemplation of the heavens" was not in fact the unproblematic source of fulfillment that its advocates presented it as, and to which neo-Stoicizing thinkers, beginning with the Renaissance, have wanted to return. But of course this has not prevented it from being a very strong orienting image in our whole tradition. It is to an examination of the effects of modernity and of the Copernican revolution on that image, and on the associated idea of "intuition" (*Anschauung*, as the Germans call it: knowledge as direct sensual/aesthetic contact with the world[11]), that the remainder of part I is devoted.

Blumenberg does not assign relative weights to the Copernican event and to the independent process of "human self-assertion" (which has its own far-reaching consequences in science) in undermining antiquity's linkages between theory and vision, and between both of these and human fulfillment (happiness). But the picture that emerges is one in which Copernicus's removal of man from the center, and the expansion of the universe (and the consequent invisibility of much of it), which resulted from the subsequent development of his theory, contributed to a process that was already well under way before their relevance to it became evident. So that Copernicanism perhaps had more to do with the way the process was experienced, imagined, and expressed than with its occurrence, as such. Blumenberg illustrates its early stages in Ockham, Pico, Francis Bacon (the *anti*-Copernican), Hobbes, and the "Cambridge Platonist" Henry More. (Incidentally, this range of examples makes it clear that it is not a renewed Platonism, either, that severed the links.) As far as the relation between vision and theory is concerned, Copernicus's own constructive/intellectualist epistemology is simply representative of the trend of the times. Also like most of his contemporaries, of course, he has no idea to what that trend will lead.

That realization begins to dawn in the eighteenth century, when Diderot himself begins to realize how the kind of knowledge that is gathered in his *Encyclopedia* can stand in the way of experience of the world, rather than enhancing it, as the Greek conception of *theoria* assumed it would. Science has become an endless, impersonal process. In his early *Universal Natural History and Theory of the Heavens*, Kant, responding to the boundlessness and disorder of the post-Copernican universe, tries to find a rational unity and totality in it by postulating laws of cosmogonic evolution, but he cannot pretend that there is any position in the universe from which this totality could be grasped in intuition. Blumenberg places Kant's later thoughts on "the starry heavens above me and the moral law within me," and his notion of the "sublime" as the symbolic presence of the ungraspable 'totality' in finite experience, in the context of this unsolved, and insoluble problem.

Kant's attempts, by means of ethics or aesthetics, to preserve some relation between the individual's 'intuition' of the universe and mankind's abstract knowledge of it are brushed aside by German Idealism and its heirs, for whom nature is arbitrary, and the only source of meaning is man. Paraphrasing Schelling, Blumenberg oberves that "man's localization in the universe is a matter of complete indifference, if he no longer stands within the whole, but rather stands over against it as subject." For the Enlightenment—as in Fontenelle's *Conversations on the Plurality of Worlds*—"the impact of the Copernican disappointment [had been] moderated by the guarantee that the cosmic presence of reason was not a matter of man alone." The other worlds were inhabited by other and probably more rational beings. Idealism, on the other hand, deals "with the Copernican disillusionment by revaluing it as a condition of the possibility of a new self-consciousness that pushes its eccentric position outward until it becomes an *exterior* one." [mm]

Of course, this dramatic solution did not carry lasting conviction. Neither has any of the post-Idealist proposals that address the problem. After discussions of Feuerbach's and Schopenhauer's attempts to rehabilitate 'intuition' as a fulfilling relation to the world (in the course of which Blumenberg expands his sympathetic account, in *The Legitimacy of the Modern Age*, of Feuerbach's theory of curiosity), he reaches the culminating episode of the whole story: Nietzsche. In his early "anthropomorphism," Nietzsche claimed to accept without reserve man's parochial and episodic status in the universe. "How can anyone dare to speak of the Earth's destiny? ... Mankind must be able to stand on its own without leaning on anything like that ..." (which is perfectly consistent with Idealism—but without the Idealists' convenient substitution of History as the thing to lean on). But in the end it turned out, Blumenberg says, that there was one thing Nietzsche could not endure: "... to be unimportant for the world, not

to be responsible for it . . . ."[nn] His solution to this problem—the myth
of eternal recurrence—raised the importance of man's actions to the
highest level by making him central, if not in space, then in something
even more important: in the world process. His actions were not mere
incidents, but sources of eternal law. But the result, ironically, was
that the superman's success depended, after all, on nature—in the
form of the overarching "law" of recurrence—and thus did not differ
in principle from the Stoics' reliance on cosmic Providence.

This demonstration of how the great advocate of human (or
superhuman) self-reliance—of the "will to power"—unconsciously
relapses into a pre-Copernican reliance on nature dramatically underlines the ambiguities of modern human self-reliance. Nietzsche is
probably not alone in being less self-sufficient than he would like to
think he is. Recognizing this persistent and perhaps constitutional
human 'neediness'[oo] helps to restore, once again, a continuity to the
human experience in our tradition, a continuity that is obscured by
the contradictory self-stylizations of admiration of cosmic providence, on the one hand, and hard-bitten, post-Christian human self-
reliance and rejection of teleology, on the other. (It is important to
notice, though, that this is a continuity not of inherited cultural
goods—truths, attitudes, *topoi*, or whatever—but of the reverse: of
constitutional *lacks*. It is the differing attempts to make up these
lacks—"differing answers to the same questions," in the "reoccupation" model in *The Legitimacy of the Modern Age*—that makes the
epochs so different from one another.)

If we admit that we have a 'need' for nature, after all, then the
question remains, how to satisfy it (if indeed we can). In the final
chapter of part I, Blumenberg describes the dilemma of an interest in
nature that, rather than, like Nietzsche, scorning both science and
(Romantic) personal 'experience' of the world (the one because of its
impersonality and tentativeness—its "asceticism"—and the other
because of its passivity), instead wants to embrace them both—and
seems inevitably to fall into anachronism in the attempt. Blumenberg
seems to suggest, though, that *conscious* anachronism need not be fatal
to thinking or to experience. After all, one of the lessons of modern
astronomy is that, because of the finite speed of light, no simultaneous
cross section even of the visible world is available to us, in any case.
And similarly, an ultra-Platonist "demand for ultimate authenticity
in all experiences" would overlook the fact that simulation, as in

the planetarium, can actually "augment what can be seen at all" (final sentence of part I). So whatever "intuition" we engage in is, potentially, a much more complex matter than Anaxagoras could have imagined. But that need not mean that its basic human significance—the need that it addresses—has changed.

7. Attitudes to the New "Truth"

Part III, "A Typology of Copernicus's Early Influence," might equally well have been entitled "A Typology of the Ethics of Claims to Truth." To read the title that Blumenberg actually gave it as implying a claim to present either a comprehensive survey or an exhaustive typology of Copernicus's influence, even in the period (bounded by Rheticus, on the one hand, and Galileo, on the other—i.e., from the 1540s to the 1630s) with which it primarily deals, would, I think, be a mistake. In particular, Copernicus's reception by astronomers (other than Galileo) is barely touched on. What this part does deal with thoroughly is the major ways in which thinking people in this period dealt with Copernicus's claim to have put forward a new truth about the world. There are those who are aghast, or are delighted, at Copernicus's "moving the Earth" out of its traditional repose and into orbit around the Sun. They—Donne, Ramus, Calcagnini, and many others—develop the (double-edged) metaphor of the theoretician as 'perpetrator,' which makes truth, in a sense, a function of power; which is a one-sided but a powerfully suggestive characterization of the new role of science in the modern age. Then there is Osiander, who, on the contrary, tries to mask Copernicus's claim to truth, by presenting it as mere "model building," in the manner of traditional astronomy (and anticipating modern positivism). There is Melanchthon, who, in the light of his own kind of neo-Stoic anthropocentrism, views the heavens so much as a divine *communication* that Copernicus's claim to understand their mechanism seems impious to him. There is Rheticus, who has such a powerful need for a completed truth, a definitive enlightenment, that even Copernicus's personal instruction does not equip him to tolerate the ambiguity of a 'truth' that is going to require a whole subsequent revolution in physics, as well as astronomy, to make it complete. There is Bruno, who is posthumously honored for going to the stake for this truth, when his central concern was not with it, at all, but with a parallel issue: the inconceivability of a finite divine incarnation within an infinite universe. And finally there is Galileo, for whom Copernicanism *is* the central concern, in a way that (together with his anachronistic faith in the direct visual apprehension of essential patterns, such as the 'Copernican' pattern of Jupiter and its moons) may even have inhibited his development as a theorist—but who is cast posthumously as a failed apostle, in contrast to Bruno, because he chose not to suffer for his truth.[pp]

Blumenberg's general point about all of these dramatic stories is that if a 'positivist' philosophy of science had prevailed during the period—a philosophy for which theories were merely more or less convenient 'models,' rather than candidates for Truth—then Copernicus's proposal would have had an honored place in the history of science, but would not have ignited the epochal conflict, and set free the "epochal 'energy'" of which these stories are early manifestations. "The important thing for Copernicus, in his changes in the world system, was to register reason's claim to truth; and the object of the anti-Copernican reaction, just as soon as it was fully formed, was to reject that claim" (introduction to part III)—which is why the Copernican reform could become the paradigm, for the Enlightenment, of the progress of reason. And it is also, at the same time, the reason why *both* fronts in that paradigmatic conflict could become so dogmatic and, to that extent, unenlightened. "The ethics of claims to truth consists in their having to demand of their adherents more power of conviction and testimony than can usually belong to a scientific theory (or, more generally, to a rational assertion). Thus those who are not convinced, or who hold back, quickly become traitors to the 'cause' ...," etc. (ibid.). So for an advocate of "historical respect for the equal rank of the aids that man avails himself of in comprehending the world"—that is, for the kind of "relativist" that I described Blumenberg as in section 3—it is important to understand the kinds of *dis*respect and intolerance that can be created by the "ethics of claims to truth" (and of claims to "enlightenment"). The trick is to understand these things in a way that does not add to the sequence of reciprocal denunciations a denunciation of the Enlightenment's claim to truth, by describing it (in its turn) as the primary source of obscurity, as a mere expression of a 'will to power,' or whatever. Blumenberg avoids that trap, with the result that part III (and part VI, which continues some of its themes) is an important contribution to an 'anthropology' of human attitudes to truth (as well as to the history of our age, in which the stories that he retells have been so potent and have become so paradigmatic).[qq]

8. Man's "Removal from the Center," and the Effort to Be "Still More Copernican"

The 'heroic age' of Copernicanism, which culminates in Newton's consolidation of its victory in astronomy and physics, is followed by the age of its metaphorical interpretation. The early metaphors of the theoretician as 'perpetrator' and of the 'stellarization' of the Earth (which Galileo celebrated) are followed, in this new period, by metaphors that bring out positive or negative implications of the astronomical reform for man in his relation to the universe. Fontenelle sees the Earth reduced to a salutary, 'democratic' equality with the rest of "the crowd of the planets" (a metaphor that accords with the Enlightenment's assumption of a cosmic community of reason, in which man participates). Goethe is more of two minds. For him, relinquishing "the enormous privilege of being the center of the universe" meant that "a world of innocence, poetry and piety, the testimony of the senses, the conviction of a poetic-religious faith ... went up in smoke"; but at the same time the new doctrine "justified those who accepted it in, and summoned them to, a previously unknown, indeed unimagined freedom of thought and largeness of views." Nietzsche, unambiguously on the negative side, asks, "Has the self-belittlement of man, his *will* to self-belittlement, not progressed irresistibly since Copernicus"—because his "existence appears more arbitrary, beggarly, and dispensable in the visible order of things?"[rr] Even those of us who (like Nietzsche in other contexts) are least inclined to see man's destiny as depending on his pregiven place in nature, as opposed to his own efforts—even such people find the diagram of man's "removal from the center" so persuasive—as, above all, the paradigmatic instance of our overcoming of anthropocentric illusion—that the supposedly historically interchangeable phenomena of geocentrism and anthropocentrism have served for three centuries, almost unchallenged, as the summary description of 'what modern science overcame.' So Blumenberg's critique (in part II, chapter 3) of the assumption that a geocentric cosmology always entails an anthropocentric worldview has pretty fundamental implications for both the negative and the (more or less self-congratulatory) positive interpretations of the Copernican revolution. We can no longer confuse the metaphor of man's "removal from the center" with a factual description of successive modes of thought in our history.

But, once again, Blumenberg would be the last person to dismiss our "removal from the center" as a *mere* metaphor. In his *Paradigmen zu einer Metaphorologie*, it was one of his leading exhibits in support of his thesis that metaphor is not something that can simply be eliminated in favor of literal, univocal terminology. And he takes it for granted that this metaphor relates to an important truth. As he says in this book, "To us, consistent Copernicanism seems to be the carrying out of the elementary insight that man's point of view and his optics, in relation to the universe, are arbitrarily eccentric, or, in the least favorable case, extremely unsuitable.'[ss] That is, that we cannot assume that the world is teleologically suited to our needs (whether epistemological needs or needs in general). The metaphor of our "removal from the center" expresses this very well. It is when we deduce literal, historical conclusions from it (e.g., that people who preceded Copernicus must necessarily have seen the universe as anthropocentrically teleological), and still more when we deduce conclusions about man's relative 'importance' from it (the standard deduction that we find not only in Nietzsche but in Bertrand Russell, Freud, Carnap—practically everywhere[tt]), that the misunderstood metaphor becomes pernicious.

Part V of this book, "The Copernican Comparative," which examines Johann Heinrich Lambert's and Immanuel Kant's cosmological extensions and philosophical applications of Copernicanism, needs to be seen against this background. Both Lambert, in his *Cosmological Letters*, and Kant, in his early *Universal Natural History and Theory of the Heavens*, saw themselves as extending into the universe as a whole the kind of thinking that Copernicus had begun, and Newton had perfected, for the solar system. Lambert's repeated reflection that "perhaps we are still not sufficiently Copernican" (whereupon he would postulate galaxies as rotating 'super solar systems,' or super-galaxies as 'super super solar systems'—etc.) sums up the endeavor that Blumenberg calls the "Copernican comparative": to transcend and subsume simpler systems in more comprehensive ones that are less constrained by one's initial, contingently given standpoint. Blumenberg points out that in spite of the radicalism of his materialistic cosmogony, the young Kant was less wholehearted than Lambert was in this kind of escalating 'Copernicanism,' because of his persistent attachment to a Stoicizing, teleological notion of man as contemplator of the universe.[uu] One might think, in view of the famous interpretation of his mature "critical" philosophy as carrying out a "Copernican revolution" in philosophy, that it represented his adoption of a more rigorous Copernicanism in general. It turns out, however, from Blumenberg's exhaustive analysis of Kant's references to Copernicus, in his introduction to the second edition of the *Critique of Pure Reason*, that although Kant did sharply curtail his earlier Stoicizing tendency, there is no evidence that it was a stronger influence of Copernicus's theory or example that led him to do so. (More likely, the major factor in the change was his effort to come to terms with Hume's skepticism.) Kant's intention in the main passage in the text was merely, as a result of his disappointment with the response to the first edition, to suggest to readers that they should 'experimentally' entertain the theses of the *Critique*, as Kant (probably following Osiander's anonymous preface) thought that Copernicus had done with his "hypotheses." Kant never compared his critical turning, itself, to Copernicus's reform of astronomy; on the contrary, when he talked about philosophy "entering on the secure path of a science" he probably had in mind people like "Galileo, Kepler, Huyghens and Newton," rather than Copernicus, as the comparable revolutionaries in the physical sciences. And, sadly, the Copernican analogies that generations of commentators have read into his text have only encouraged the same misinterpretations of the *Critique* as idealist that Kant tried to combat elsewhere in the second edition.

As Blumenberg says, "It is neither surprising nor reprehensible that a metaphor having the epochal importance that this one possesses sets the reader's constructive imagination in motion even before he has finished reading it. Kant is making use of a historical potential that seems, in the hindsight of historiography, to have been bound to become an independent force."[vv] This is the same unspecific 'potential,' of course, that impels all the thinkers I mentioned a moment ago—Nietzsche, Russell, Freud, Carnap—to found historical narratives on the metaphor of man's removal from the center. One would have to be pretty hard-nosed to resist it entirely. But it is particularly ironic that Kant should have got caught up in this process, when one realizes that his "critical" separation of theoretical reason from practical reason—which freed him from the Stoicizing tendency of his early thought—was intended precisely to prevent man from taking his cue, in "practical" matters, from his pregiven position in nature, so that while Kant the critical philosopher could certainly use Copernicus's theory as a source of *analogies*, to see it as defining man's relation to the universe (as many of us are chronically tempted to do) would be contrary to his most basic principles.

If, then, neither Lambert (with his nesting Copernican systems) nor Kant provides a model of how we can be "still more Copernican," i.e., how we can be more consistently modern, and if we are going to forswear self-congratulatory or self-depreciating historiographical applications of the metaphor of our "removal from the center," then the question of what consistent Copernicanism entails for us is, after all, a surprisingly open one.

9. Copernican Vision

The final part of the book—part VI, "Vision in the Copernican World"—addresses this open question first by continuing the examination of the 'anthropological' presuppositions and implications of the Copernican revolution, which was begun in the introduction to part I, and then by making some rather novel suggestions about the Copernican lessons of the contemporary "space age."

Unlike part I, which also had a great deal to do with vision, "Vision in the Copernican World" does not contrast ancient and modern conceptions of the relation between 'contemplation of the heavens' and happiness. The historical contrast now is simply with the premodern "postulate of visibility": that what is real must be in some way, at some time, visible. First Copernicus's expansion of the universe, and then the telescope and the microscope, explode this postulate. It takes a while for this fact to sink in. Galileo, entranced with the new things he can see with the telescope, does not realize what its indefinite improvability implies: that our faculties are in no way correlated with the scale of reality, so that the immeasurable range of what he *cannot* see is at least as significant as the measurable range of what the instrument has enabled him to see for the first time. Despite the revolutionary impact of his discoveries and his writings, Galileo is still tied to a premodern faith in vision as access to the eidetic patterns of reality—hence his inability to take seriously Kepler's purely quantitative investigation of the planetary orbits, and hence also the problems he gets into over his announcement of the moons of Jupiter. (Blumenberg examines in detail one of the critiques of that announcement, by Francesco Sizzi, and makes it very plausible.)

Pascal, with his two infinities—the great and the small—and Lichtenberg, with his conception of man as dealing merely with the "surface" of things, begin to conceptualize the real implications of the Copernican universe, as revealed by the telescope. Now, Blumenberg observes, "The visible world is not only a tiny section of physical reality, it is also, qualitatively, the mere foreground of this reality, its insignificant surface, on which the outcome of processes and forces is only symptomatically displayed. Visibility is itself an eccentric configuration, the accidental consequence of heterogeneous sequences of physical events." [ww] Blumenberg finds the first anticipation of the possibility of such a breakdown of the privileged position of human perception in Montaigne, who wrote without any knowledge of the telescope, but with a strong awareness—in the aftermath of late medieval "theological absolutism"—that no cosmic appropriateness can be taken for granted. Thus we are reminded that it was not the telescope, as a technological 'accident,' that gave rise to the modern doubts about the adequacy of man's faculties. Rather, the telescope entered a situation that was ready (in a way that antiquity, for example, was not ready) to receive it. [xx] Again we see how religion, philosophy, and culture in general are, or at least were, entwined with 'science and technology' in an interacting process, rather than a one-way process in which one of them 'exogenously' determines the other.

In his final chapter, Blumenberg turns to a recent example of such interaction: the exploration of space, and what we have ("reflexively") learned from it about ourselves and the Earth. He maintains that the experience of seeing the Earth—"seemingly alive"—in the heavens above the lifeless surface of the Moon (an experience that he takes to be 'real,' despite the elaborate technological linkages that alone enabled most of us to 'have' it) has enabled us, for the first time, to appreciate the apparent uniqueness of our "oasis in the cosmic desert"—has turned our "interest from the remote world to the proximate one, from a centrifugal direction to a centripetal one"— and, by showing us the unimportance of mere magnitude, has in fact "brought to an end the Copernican trauma of the Earth's having the status of a mere point, of the annihilation of its importance by the enormity of the universe."

Such a "centripetal" turning, if in fact it has occurred, runs the risk of being labeled a new "geocentrism," and being suspected of its twin sin of anthropocentrism. Blumenberg is not afraid of either label, but his series of skeptical questions about the search for extraterrestrial intelligence—the most striking "centrifugal" project that is currently being mooted—culminates in a suggestion that again (just like his analysis of the historiographical application of the metaphor of our "removal from the center") makes the usual applications of the "lesson of the Copernican revolution" look a little hasty. Blumenberg is afraid, he says, that "the burden of proving one's post-Copernican freedom from prejudice" lies not on those who doubt the existence of extraterrestrial intelligence, but on those who argue *for* it, because the assumption that rationality is the natural summit of organic evolution—so that if such evolution occurs elsewhere in the universe, it will eventually produce rational beings, with whom we might someday communicate—may itself be an anthropocentric illusion (a repetition of "the old favor that man rendered to himself of taking the crown of creation into his safe keeping"). If one follows "all the anthropological indications," Blumenberg writes, "reason would not be the summit of nature's accomplishments, nor even a logical continuation of them. Instead, it would be a risky way around a lack of adaptation; a substitute adaptation; a makeshift agency to deal with the failure of previously reassuring functional arrangements and long-term constant specializations for stable environments."[yy] In that case what the eighteenth century took to be the essence of Copernicanism—namely, the ubiquity of reason, in all the "worlds," as the telos of everything else—would be a fundamentally 'pre-Copernican' error; and an—*apparently* "anthropocentric"—acceptance of the possible uniqueness of the Earth and of human reason would actually be the long-overdue next step in consistent Copernicanism.

"Putting reason in its place" (both figuratively and literally), in this way, is of course also a further step in Blumenberg's ongoing "relativist" critique of the Cartesian/Enlightenment/positivist exaltation of scientific reason as (in contrast to other, merely human, modes of thought) an achieved transcendent perspective. A striking corollary of that critique is the way in which this final chapter takes as its point of departure the lived *experience* of the view of the Earth from space (more specifically, from the Moon). "Kepler had described it in advance, but in this case knowledge was not the important thing." Thus, in the centripetal, "geotropic" turning that Blumenberg thinks the "space age" is experiencing, visual 'intuition' reasserts itself as a key element in man's relation to the world. So that here too—once again, without denouncing or rejecting any phase in the history of our thinking on these subjects—Blumenberg reestablishes a continuity with premodern assumptions that modernity (often in the name of "consistent Copernicanism") has rejected as obsolete prejudices. As in both *The Legitimacy of the Modern Age* and *Work on Myth*, here again his endeavor is to distinguish essential modern accomplishments from their rhetorical accompaniments, and to show how those accomplishments grow out of, and do not invalidate, other phases of human experience.

I am grateful for grants from the Division of Research Programs of the National Endowment for the Humanities, an independent federal agency, and from Inter nationes, an agency of the Federal Republic of Germany, which made this translation possible.

Woyzeck, es schaudert mich, wenn ich denke,
dass sich die Welt in einem Tag herumdreht!
Was 'n Zeitverschwendung! Wo soll das hinaus?
Woyzeck, ich kann kein Mühlrad mehr sehn, oder ich werd
   melancholisch.

Woyzeck, it makes me shudder when I think that the Earth turns itself about in a single day! What a waste of time! Where will it all end? Woyzeck, I can't even look at a mill wheel any more without becoming melancholy.

Georg Büchner, *Woyzeck*, in *Complete Plays and Prose*, translated by C. R. Mueller (New York: Hill and Wang, 1963), p. 109

Part I

The Ambiguous Meaning of the Heavens

Introduction

The combined circumstance that we live on Earth *and* are able to see stars—that the conditions necessary for life do not exclude those necessary for vision, or vice versa—is a remarkably improbable one.

This is because the medium in which we live is, on the one hand, just thick enough to enable us to breath and to prevent us from being burned up by cosmic rays, while, on the other hand, it is not so opaque as to absorb entirely the light of the stars and block any view of the universe. What a fragile balance between the indispensable and the sublime.

One cannot help thinking of what we now possess in the way of continually increasing information about the other bodies in our solar system—about their defenseless exposure to the change of temperature between day and night, about the showers of particles and rays of every caliber that fall on them from space, about the leaden atmospheres lying on the surfaces of Venus and Jupiter, atmospheres that have definitively and impenetrably closed off any view of the heavens, and about the speeds of gales in the atmosphere of Mars, with the fantastic dust clouds of its plains. "Nature has been trying to wipe us off the face of the earth ever since ... the caveman ...,"[a] a pioneer of the technology of space travel said at the beginning of the decade of great extraterrestrial expectations, at the astronautical congress in Stockholm in 1960, so as to make the cosmic objectives and their dangers appear to be a natural continuation of our situation on earth. When, at the end of this decade, the inhabitants of Earth for the first time had their body made visible from a distance in space, it appeared to them to be comparatively livable and worth preserving. A soberer view of the cosmos has contributed to the will to preserve the Earth.

A thought experiment that the French mathematician Henri Poincaré suggested at the beginning of this century has acquired a
peculiarly new valence for us. He started from the question whether
there would ever have been a Copernicus if our Earth were continually surrounded by an impenetrable and always unbroken blanket of
clouds. Put another way, the problem ran: Would we know that the
Earth turned on its axis and went around the Sun if we had never
been able to practice astronomy? Poincaré did not yet know anything
of the technical possibilities of piercing even a blanket of clouds of this
sort with flying machines and rockets, nor did he know anything of a
nonoptical astronomy such as has arisen in the form of radio astronomy. How could a mankind located inside an atmospheric 'cave'[b]
ever have learned that the Earth belongs to a planetary system and
ultimately to a universe made up of worlds, and that it moves, in this
universe, in multiple ways? Without the view of the daily rotation of
the heaven of the fixed stars, would not any conjecture that was
directed against the overwhelming evidence that the ground on
which we stand and live is at rest have been impossible?

Nevertheless Poincaré comes to the conclusion, in his thought experiment, that even without astronomical optics men would have
achieved clarity about their cosmic situation and motion. Only, they
would have had to wait much longer for the advent of a Copernicus.
Poincaré wants to show that this delayed Copernicus would be
driven, as a pure physicist, to assert the motion of the Earth. Any
physics—so his reflection runs—that rests exclusively on observations that can be made in the immediate neighborhood of the Earth
gets entangled in ever greater difficulties. Such a physics of the realm
of experience that is close to the Earth would hardly be very different
from the Aristotelian and Scholastic tradition, with the assumptions and doctrines of which Copernicus himself, and still more
Copernicanism, in its discussions and oppositions, had to deal. Regarding the physicists of the optical 'cave' that is assumed in the experiment, and their 'Scholasticism,' Poincaré says, "They would get
themselves out of the difficulty doubtless, they would invent something which would be no more extraordinary than the glass spheres of
Ptolemy, and so it would go on, complications accumulating, until
the long-expected Copernicus sweeps them all away at a single stroke,
saying: It is much simpler to assume the earth turns round."[c]

The time for this delayed Copernicus would have arrived, perhaps, in the year 1737. In that year the French Academy of the Sciences caused to be empirically established, by angular measurements in Lapland and Peru, the flattening of the Earth that Huyghens had deduced as early as 1673. At the latest, the Copernican idea would have made its breakthrough with Foucault's famous pendulum experiment in 1850.

This solution of the imagined problem stirs doubts. Poincaré disregards the dependence, at their very beginning, of the physical observations and experiments that he cites, on the assumption—either already accepted or at least under consideration—of the motions of the Earth. It was, demonstrably, the specific conjecture that the shape of the Earth is deformed by its rotation, which people wanted to investigate when the plan of an angular measurement was not only raised but the necessary expense and the worldwide organization involved in effecting it could be considered as justified. It was only after one assumed that the Earth rotated that the task of measuring the curvature of the meridians, and thus finding an empirical confirmation of that assumption, presented itself.

But this objection is still a harmless one in comparison to the historiographical fact that modern physics received its essential impetuses and problems only as a result of the Copernican extrication of astronomy from confusion, insofar as this extrication could be definitively accomplished only with the aid of certain prerequisites in physics—such as the principle of inertia, and gravitation. To follow Poincaré's imagined situation one final step further, one can doubt, with reason, whether Copernicus the *physicist* could have existed before the turn of our century, if Copernicus the *astronomer* had not been possible.

Whenever a Copernicus would have become possible for a mankind located within the optical 'cave' of an impenetrable atmosphere —as a magnitude in the history of this mankind he would not have been able to achieve a cumulative impact comparable to that of the historical Copernicus and his influence. That is the decisive amplification that has to be given to Poincaré's experimental reflection. The change that was brought about by Copernicus in mankind's historical consciousness of itself is conceivable, in its radical quality, only against the background and as a consequence of the prior history—a history with no designatable beginning, and one that was never interrupted —of the human relation to the cosmic 'environment' [*'Umwelt'*].

This history disappears, for the kind of recollection that can be grasped through documents, into a mythic/magic primitive world. It will be difficult for us to comprehend the meaning of merely perceiving, and what was involved in merely perceiving, something that goes beyond the acute necessities of life and is incidental in relation to them: of lifting one's gaze out of the sphere of biological signals and drawing something inaccessible into one's range of attention. To see stars epitomizes the surplus that man was able, as a by-product of his upright carriage, to superadd to his pressing everyday concerns. Differently from a fleeting and occasional glance, but also differently from a view that relies on catalogues of positions and on photography, the early human observation of the starry heavens has to be imagined as the unhurried insistence on which even the slow displacements of the planets impress themselves as patterned trajectories, and the prospect of points of light articulates itself into a landscape of distinct configurations. In the leisure of Chaldaean herdsmen or in the zeal of Babylonian priests—both origins are named in the tradition, depending on which form of life one was more inclined to credit with the result—physiognomic observation may then have turned into effortful examination, establishment of periodicities, fixing of material to be handed down. Elementary accomplishments of rationality were first carried out, in opposition to the mythical readiness for stories, in relation to the phenomena of the world-prospect: accomplishments like the establishment of the identity of the Sun between its setting and rising, of the Moon across its dark phase, of the evening star and the morning star. While the heavens were already becoming an object of observation and calculation, they were still the tablet on which fate gave its intimations [of things to come], intimations that for their part soon appeared as interpretable and calculable and thus also gave a first presentiment of dependable order and regularity such as could not be experienced beneath the heavens. The opposed positions of (on the one hand) earthly realism and (on the other hand) the supposition (with an eye on the heavens) of a reliability in the background, of belief in subterranean divinities characterized by dull opacity and in star-gods characterized by higher intelligence and purer friendliness toward the world—these oppositions are involved in the formation of the cosmic background of the history of human consciousness.

Here, immediately, we must allude to the not unimportant circumstance that the great turning in this relationship between the universe

and consciousness did not take place under the ultradistinct constellations of the Oriental sky, but rather in the region of cloudy and
concealed skies, in that northern corner of the world in which the
astronomer is supposed never to have seen the planet Mercury, a
planet that had nevertheless become important for the genesis of his
rebellion against the astronomical tradition. It was hardly as an
observer that Copernicus, the canon of the cathedral in Frombork,
could become the reformer of the world system. To that extent his
position would come very close to Poincaré's imagined situation of the
optical 'cave,' if one could leave out of consideration the fact that
what he had before him was already the result of millenia of perceptual immediacy.

1

Cosmos and Tragedy

An antinomy in the early history of the European consciousness that may still be with us is contained in the fact that the Greeks discovered [or invented: *erfanden*]the cosmos *and* tragedy.

Even before Jacob Burckhardt's great thesis of the pessimism of the Hellenes, August Böckh expressed the dark background of the Greek admiration of the cosmos in more commonplace words: The Hellenes, he said, had been less happy than most people believed. The prodigious work of myth had transformed the terrors of the faceless superior powers into gods having human form, but had not mastered the caprice of these gods. "Mark the great cruelty of the gods. They beget children, they are hailed as fathers, and yet they can look upon such sufferings," complains Hyllus, in Sophocles's *Trachiniae*.[a] Tragedy is an expression of the fact that the gods are not responsible for the cosmos, have not devised or created it, and are, as it were, on the point of abandoning it in favor of transcendence or the interworlds,[b] while men, located only on the lower margin of the cosmos, in the zone in which its elements are most confused, do not so much belong to it as stand opposite to it. The answer to Burckhardt's riddle of how the Greeks' pessimism with regard to life correlates with their optimism with regard to the world can only be sought in this exterior situation of gods and men in relation to the cosmos. The Greeks could understand freedom only as something contrary to the cosmos, and consequently they were unable to associate either gods or men with what they admired.

Thus the two central statements of Sophocles and Anaxagoras, contemporaries in fifth-century Athens, belong inseparably together: the terrible saying from the choral song of the elders, in the *Oedipus at*

Colonus, that the best thing is not to have been born, and the other, reported in the *Eudemian Ethics* as a saying of Anaxagoras, who is supposed to have answered the question why one could decide that he would prefer to be born rather than not to be born by saying "for the sake of viewing the heavens and the whole order of the universe."[1] Precisely because he is not able to complete it as a totality, the cosmos can justify man's existence. The unattainable is the true reference point of contemplation;[c] the starry heavens—and only they—are the cosmos, because life cannot be it. The Greeks, at any rate, would not have accepted what Fontenelle writes in *L'Origine des fables* [*The Origin of the Myths*]: that men would have given to their gods, as their preeminent attribute, either wisdom or power, according to whether they admired the order of the heavens or feared the extraordinariness of their phenomena.[2]

Anaxagoras's idea makes the pagan substance of the antinomy of cosmos and tragedy visible. This is not because he answers the question as to what is the best life or the better lot in life—like, for example, Plato's myth of the choice of destinies in life by the souls in advance of their bodily existence—and still less because he answers the question whether man is destined for this or that form of existence, action, and attitude, for the happiness of theory or for practical virtue, but rather because he dissects out the naked question as to what could justify the fact of life, accepted (for once) as such, and without the mental construction of a possible choice preceding it. However matters may stand with regard to Anaxagoras's answer, the question that he poses for himself is part of the invisible underground of philosophy and comes to light, at the latest, when Kant deduces from the impossibility of obtaining the consent, in advance, of those who are to be born the conclusion that those who beget them owe them, by right, the compensation of reconciling them, after the fact, with the existence that they did not wish for, and thus enabling them to give their own consent to this fact.

One should take the time to contrast to Anaxagoras's question and answer what results from the logic of the biblical assumption of a Creation and a divine will that is realized in the world—a logic that no longer leaves open the possibility of an 'after the fact' justification of the fact of the world and of life, but instead implies the necessity of representing life and the world as a recognizably valid decree of divine power. Leibniz tried to answer this question of the reason for existence with what was, for him, the ultimate rationalization of the world that in fact exists: The actual world, he said, is necessary, because it is the best of the possible worlds. Whatever worth may be assigned to his argumentation in support of this statement, it was impossible for him to remove, by means of it, one final objection, namely, the uncertainty (which for him was not even formulable) as to whether the nonexistence of every possible world would not after all be even better than the existence of the best of the possible worlds, if the difficulties in rendering claims to happiness and the results of self-actualization compatible did after all continue to make inevitable the quantity of evil and suffering that is actually found in this world. That is why the more radical answer to the question of the reason for existence was given by Solomon Maimon in his commentary on Maimonides's *Moreh Nevukhim* [*Guide for the Perplexed*], where the contemporary and adherent of Kant writes, "If someone asks what is the overall purpose of the existence of the world, we answer him: The purpose of existence is existence, and nothing more." [3] In the closest proximity to Spinoza, this is the elementary form of a theodicy that has no need of the question of the quality of the world, if the world's sheer existence justifies whatever exists and occurs in it.

Entirely different is Anaxagoras's answer, which I quoted, to the question what purpose could actually justify man in deciding that he would prefer to be born rather than not to be born. This answer does not see man in the context of a cosmos that is justified as such, nor does it see him as carrying out a vocation with which he was endowed (by whom but himself?). It sees the cosmos as an offer of an opportunity to justify life, and to accept it for oneself, through the choice of theoretical action. The cosmos is a fortunate opportunity for man, although it does not exist for his sake.

Anthropocentric teleology is a late way of representing the connection between life and the cosmos. Anaxagoras produced the Greeks' first thought-out and unmythical cosmogony—in so doing competing, perhaps, with the cosmogony proposed by Democritus, whose enemy he is supposed to have become on account of Democritus's having refused him a personal conversation. This cosmogony can be called rational because it takes as its point of departure a homogeneous initial state characterized by the highest level of probability, a state in which all the constituent parts of what will later be the cosmos are mixed together. This original medley is separated and arranged into its constituent parts in the manner of a centrifuge, by violent rotation. The vast rotation of the cosmic revolutions also explains the solidity of the vault of the heavens, which would have to collapse if the centrifugal force weakened. Thus the fact that the cosmos endures proves the reality of the motion of the heavens. The heavens possess no special matter or nature, but are composed of the common fundamental material of the world, so that, as a sign of the possible downfall of the vault in case the revolutions should diminish, stones occasionally fall from the sky. The Sun, too, is nothing but a glowing mass of stone, and the Moon has homesteads, hills, and ravines like the Earth. So the cosmos's permanence is not a matter of course—there are signs of instability in it—and this cosmology's eschatological bent is perhaps most clearly expressed in the answer that Anaxagoras gave to the question whether the mountains at Lampsakos could someday disappear in the sea: "Yes, if time does not run out." The principle of motion is reason. It too is part of the original medley, from which it emerges only as a result of the centrifugal operation of the motion that it causes. Evidently the danger of a weakening in the heavens' centrifugal force arises from the fact that reason's kinetic energy is increasingly 'pressed out of' the rotating mass—that is, that reason is separated off into transcendence.

This sketch of Anaxagoras's cosmogony is only meant to show that the origin of the ideal of contemplation of the heavens is not tied to the attribution of a special metaphysical value to stellar reality, as it is in the later tradition. The value of the theoretical attitude does not lie in the dignity of its object, but in the observer's ability to make transparent for himself the mechanism of the world of which he himself is a product and by which he can be destroyed. The fact that the contemplator of the heavens is an episode, between their origin and their collapse, does not constitute an argument against his theoretical attitude as the sole justification of the life that is a result of, and never escapes, the total process. But contemplation does not yet convey any of the quality of the cosmos into life; still less is the polis a correlative of the cosmos. Anaxagoras was contemptuous enough of the state and its demands; he explicitly preferred the contemplation of nature to any involvement with public affairs.

In Burckhardt's "Overall Assessment of Greek Life" the Greeks' existential pessimism culminates in the mythical figure of Prometheus. If one asks oneself what constitutes the significance of this figure, one notices not only that the absence of a relationship in which the Olympian gods take care of men finds expression in it, but also, still more comprehensively, that it has to answer for the world's radical lack of consideration for man. One cannot fail to see that the Prometheus myth prefigures characteristics of Gnosticism, which comes so much later. Man evidently does not belong in the original scheme of nature, but is brought into existence by the demiurgic act of a Titan, a member of an 'illegal' generation of gods. Man is so far from belonging in the context of this cosmos, in which he appears as a delayed creature but also, no doubt, as an affront directed against Zeus's ascent, that only a supplementary and, once again, illegal provision of aid is able to procure for him the most necessary of his needs, fire, and to keep him alive. Burckhardt did not offer this explanation of the hiatus between man and the cosmos, but he describes the outcome, that the Prometheus figure had been able "to keep alive, deep in people's hearts, a mood of rebellious complaint against the gods and against fate," and that even during the gods' festivals "the image of the chained Titan on the rock must have surfaced here and there in the Greeks' thoughts," and they knew then "what position one was really in with respect to the gods." [4]

If we may take it as a premise that the deformation and misunderstanding of an idea or a figure that occur in the history of its reception can serve as evidence about the historical process that brings about such deformation and such a difference in understanding, then we can see from one characteristic of the late allegorical interpretation of Prometheus to what an extent the original antinomy of tragedy and the cosmos was concealed and made inaccessible by the formula, transmitted by the tradition, of man's having been destined for theoretical contemplation of the heavens. The chaining of Prometheus in the Caucasus, which first appears, for us, in Aeschylus, then no longer seems consistent with the allegorical image of the founder of culture and benefactor of mankind. Thus the first modern history of philosophy, which was read by both Goethe and Kant, advises us, regarding Prometheus, that his being chained to the Caucasus "is meant to imply that he devoted himself to astronomy, for a long time, on these mountains: although a celebrated man is of the opinion that all of this was derived from a misunderstanding of the story of Moses and the people of Israel." [5] The genuine Greek understanding of Prometheus has become so unintelligible in this late retrospect because, for it, Prometheus's devotion to man had gotten into conflict with the god's [i.e., Zeus's] ill will in such a way that, in connection with the one who was chained to the Caucasus, the observation of the heavens can only appear as a projection of a heterogeneous metaphysics.

Of course the modern age has also ceased to be able to put into effect the idea of the observation of the heavens as enjoyable leisure. The contemplation[c] that is done in repose[d] unexpectedly turns into consuming work. When Johann Christoph Gottsched gives his memorial address on Copernicus in 1743, he stylizes him as a restless observer of the heavens, whose astronomical passion does not leave him, day or night. For the exaggerated unrest of the new type of investigator, the overcast skies that are characteristic of his place of residence—which other biographers had held responsible for Copernicus's misfortune of never in his life having got a glimpse of Mercury—become a truly providential blessing: "The hours in which half the world lies on its couch in an insensible state resembling death were spent by Copernicus in the open air; and one would have had to say that Nature dedicated to sleep only those nights in which a dark fog and thick clouds veil the eyes of mortals and deprive them of the sparkling lights of the heavens."[6]

With this industriousness of theory, which Gottsched sees as the prerequisite for the removal of the scandal of the old world-system, Copernicus is set at the foot of a line of descent that Gottsched also sees as beginning with Anaxagoras, and in which the figures of Thales of Miletus, Alfonso of Castile, and the Emperor Rudolph II are the familiar standard figures. Gottsched ascribes the question of the reason for existence to Anaxagoras, though admittedly in a more harmless version, which is taken not from Aristotle's citation but from Diogenes Laertius's, which already conforms entirely to the normalized typification of the *contemplator caeli* [contemplator of the heavens]: "Being asked for what purpose he had been born, he replied, 'For the observation of the sun, the moon, and the heavens.'"[7] Just as Anaxagoras's challenge has paled into conventional metaphysical amiability, the characterizations of Thales of Miletus and Alfonso of Castile also lack any traits that would reflect the problematic of their audacity or their fall, so that Copernicus can be associated with them in a smooth, matter-of-course manner (and in words that recommend Gottsched's German—which was despised for so long—in its naive dryness):

Von eben dieser Art war nun auch unser Coppernicus. Nichts schien ihm der Betrachtung eines weisen Mannes, und eines unsterblichen Geistes würdiger zu sein, als das prächtige Gebäude des Himmels: welches zwar den Unwissenden als ein blaues Gewölbe vorkommt, das unsern Erdboden bedeckt, und mit viel tausend glänzenden Lampen geschmückt ist; von Sternsehern aber, als ein unermesslicher Raum für unzählige Weltkugeln, ja als eine prächtige Schaubühne angesehen wird, darauf sich die göttliche Allmacht und Weisheit, in aller ihrer Majestät und Grösse darstellen.

[Now our Copernicus, too, was just this kind of man. Nothing seemed to him to be more worthy of the contemplation of a wise man and of an immortal spirit than the magnificent edifice of the heavens, which does seem to the ignorant like a blue vault that covers our Earth and is ornamented with many thousands of shining lamps, but is regarded by astronomers as an immeasurable space for uncountable globes—indeed, as a magnificent stage on which the divine omnipotence and wisdom present themselves in all their majesty and grandeur.]

The Greeks' relation to the starry heavens was 'worldly.' Thus there were no 'functionaries' to administer it, like those Chaldaean priests of the third century before Christ, who developed a professional insistence on precision on the part of their "star-wardens," and consequently are admired even today, in the history of science (which seeks forerunners of our ideals and norms)—for example, for their exact determination of the length of the year. With the Greeks, the most important characteristic may be, in this as in other respects, that one did not allow oneself to be represented. But that also means, then, that astronomy lacks the asceticism of a professional code and has to deliver what might be called "values for life," as is always the case when theory takes one of its steps—which periodically become necessary—into broad propagation; when, as we would say today, it is (once again) forced out of its "ivory tower."

The Greeks insisted on extracting the unity from nature. To that end, they had to overcome the double-layered relationship that exists, in mythical thought, between what one sees and what really happens—between the flat appearances in the foreground and a 'story' in the background. We find it difficult to make some kind of a picture for ourselves, at this late date, of the point of departure of this insistence on the unity of nature, because it lacks all the categories that we now use without questioning, so that it is also only laboriously that we comprehend the path that has been traversed from there. Along the way, violent measures have never succeeded, not even the

Chapter 1

one that made intuition [*Anschauung*^e] absolute (with the program
according to which everything is as it appears)—in order, in this way,
to lay exclusive claim to the unity of theory by means of a skeptical
renunciation of every background story. It is the prescription that
we should avoid a relapse into myth—a prescription that Epicurus
will still consider worth recommending. Here we are dealing not
with the rationality that seeks maximally simple explanations, but
with the reduction of phenomena to the coincidence of their physical
nature with their two-dimensional mode of appearance. This can be
seen in the resistance to the assumption of physical corporeality in
cosmology—in the hesitation to associate material bodies with the
visible disks and points of light, when Anaximander regards the Sun
and the stars as holes, of a kind, in the dark ceiling of the sky, through
which a fire-world shines through. But then the whole starry heavens
is nothing but the permeability of the shield that separates us from a
'totally different' sphere—and with that a piece of mythical background is again admitted.

The homogeneity of the cosmos could not be gained by a *coup de
main*. The relapses into the double-layered pattern of story and appearance that I referred to only became more subtle, up to Plato's
"second astronomy" with its invisible heavenly bodies and its "true"
earth^f, and to Aristotle's division (which was dominant for a millennium) between the sublunar world of the elements and the supralunar upper world of the *quinta essentia* [fifth essence].

For an analysis of the transformation of consciousness that Copernicus produced, it is essential to gain clarity about a historical conjunction that arose between geocentric cosmology and the anthropology
that sees man as destined for pure theory in the form of observation of
the heavens. The unique role of the Stoics in establishing this conjunction is already made clear by the ancient report according to which
the only consistent advocate of a heliocentric system in ancient times,
Aristarchus of Samos, is supposed to have been accused of impiety by
one of the great leaders of the older Stoa, Cleanthes. The motivation
of what occurred then, around the middle of the third century B.C., as
a precedent for the post-Copernican events, has indeed remained
obscure, but in any case it was quite certainly not fought out with
passages from Scripture—which is one piece of evidence that after
Copernicus, too, it could have been only superficially and for the sake
of intensifying the issue that the conflict material of heliocentrism

was related to bibical objections. The only account of the ancient charge of impiety against Aristarchus that we possess is Plutarch's, which seems to imply, more than anything else, that here already the theoretician was understood as a doer who violates the cosmic order: "Cleanthes thought that the Greeks ought to lay an action for impiety against Aristarchus the Samian on the ground that he was disturbing the hearth of the universe because he sought to save the phenomena by assuming that the heaven is at rest while the earth is revolving along the ecliptic and at the same time is rotating about its own axis."

This may refer to Cleanthes's lost work *Against Aristarchus*.[8] It is puzzling especially because on the question of the central governing body of the cosmos, Cleanthes, going against his predecessors, had chosen the Sun. As the greatest of the heavenly bodies and thus the most powerful concentration of what the Stoics regarded as the formative power of fire, the Sun contributed the most to the conservation of the cosmos. Since Chrysippus took the heavens to be the chief organ of the cosmos, and others took the Earth to be it, it can be seen that, for the Stoics, functional importance has nothing to do with being located in the center.[9] But why, then, should Aristarchus's heliocentric system have been seen as sacrilegious by the very Stoic who regarded the sun as the chief organ of the cosmos? The reason cannot have lain in the specific arrangement of the bodies in the system, *as* an arrangement, but must be seen more generally in the attempt—which Cleanthes explicitly ascribes to Aristarchus—to "save the phenomena," and specifically by putting forward a construction that is asserted to be real.

The formula of saving the phenomena was a means by which to express theoretical resignation while claiming only phoronomic accuracy; but it could also have been interpreted as a formula of pious awe with respect to the mystery of the cosmos. That would make Cleanthes's indignation understandable because he, on his side, had characterized the cosmos, using a sacral term, as a "mystery,"[10] thus expressing his piety toward the world. One can go still further: The asserted equivalence of cosmos and mystery has a more definite meaning in view of the relation between appearance and 'background' reality. What was seen at the Greeks' mysteries had to be accepted as a 'sign' that was shown, and one could not interrogate it as to its actual objective status or the mechanism by which it was presented—still less could one see through it—without being accused of sacrilege. If one takes this as a metaphor for the saving of the phenomena by means of a realistic system—that is, by means of a more demanding claim to theory—then the will to cosmological knowledge that opposes the aesthetic immediacy of appearance becomes, as such, an incriminated act. At the same time the Stoa's geocentric cosmology is intensified into a more than theoretical assertion, namely, into the postulate of acceptance of the phenomena as what offers itself to man in the cosmic position to which he is assigned, and in relation to which it is true that 'looking' ['*Schauen*'] would have had to be required, but not 'seeing through' ['*Durch-schauen*']. One can regard initiation into the mystery by mere looking, and the instruction to save the phenomena without the demanding-ness of realism, as two different forms of expression of the Greek concept of theory, which is defined both by the ideal of the observer in repose and also by the concept of a reality that is adequately compre-hensible by means of ideas. To that extent, the starry heavens—the self-presentation, free of all intermediate shadings of gray, of an object as a mere configuration—are congenial to the ideal embodied in this concept of theory.

The idea of the world-observer is not as such immediately anthro-pocentric. The *contemplator caeli* [observer of the heavens] can also be enlisted in order to provide the cosmos with its admirer, because the beautiful evidently cannot be conceived without someone whom it pleases. This idea was first articulated in Pindar's (almost entirely lost) "Hymn to Zeus"; it motivates the begetting of the Muses by the incapacity of the other Olympian gods to let the grandeur of the world-order pass by in any way but in silence[11]—this too being an expression of the way matters stand, for the Greeks, with the relation between the gods and the cosmos. The fact that it is not primarily man's striving for happiness that requires its fulfilling theoretical object but rather, in reverse, the magnificence of the cosmos that requires its receiver and integrates him into itself through his function was only expressed, in Stoicism, at the end, when the analogies to the self-presentation of political majesty already obtruded themselves for nature as well; but it was already inherent in Cleanthes's indignation against Aristarchus.

Seneca writes that when we say that the highest thing is to live in conformity with nature, it must be stated that nature produced man for theory as well as for practice. "Nature has bestowed upon us an inquisitive disposition, and being well aware of her own skill and beauty, has begotten us to be spectators of her mighty array, since she would lose the fruit of her labor if her works, so vast, so glorious, so artfully contrived, so bright and so beautiful in more ways than one, were displayed to a lonely solitude." [12] Seneca was the first to articulate the idea that Kant was both to appropriate and to correct when he says that even the most vulgar understanding could not free itself of the judgment that "without man the whole creation would be a mere desert, vain and without purpose," but that the mere theory of the world—without the assumption of a final purpose that observation of the world ultimately serves—could not give existence a value.[13]

From nature's desire to excite admiration and enjoyment, Seneca deduced that man—and, for his benefit, the Earth too—was positioned in the center of the cosmos: "That you may understand how she wished us, not merely to behold her, but to gaze upon her, see the position in which she has placed us. She has set us in the center of her creation, and has granted us a view that sweeps the universe; and she has not only created man erect, but in order to fit him for contemplation of herself, she has given him a head to top the body, and set it upon a pliant neck, in order that he might follow the stars as they glide from their rising to their setting and turn his face about with the whole revolving heaven." What one could call the anthropological semantics of cosmology is an application of the Stoics' fundamental idea of the world-reason and its providence. One should not forget that, in Hellenism, this was also a piece of philosophical 'enlightenment,' since it countered the advance of a fatalism of 'signs' of all kinds by attempting to make nature itself into the single real sign. For Seneca, the cosmological alternatives of the Earth's being either at rest or in motion—alternatives that seem to him to constitute a question deserving consideration (*digna res contemplatione*)—are immediately bound up with man's reflection on his position in nature, on God's way of dealing with men: whether he drives the universe around us or drives us around: "circa nos deus omnia an nos agat." [14]

Still more important than the ways of reading geostatics or geokinetics, in the Stoic doctrine, is the fact that the heavens are essentially an object of *pure* theory, because they are at an absolute distance from man. This exclusion means that the heavens are the only thing that cannot be a product of human skill or even an object of human intervention. Since that time, a guarantee of inaccessibility lies on the heavenly bodies which becomes the criterion of the utmost ambiguity of the heavens as soon as the insistent occupation with the sphere of the unmakable becomes the scandal of man's fixation on the unalterability of his situation in the world. This is the ancient position, against which—at the latest—Francis Bacon's negation, with its identification of all theory as power, will be directed. In the older Stoa, Chrysippus had formulated this delimitation of what man is not capable of and what, by virtue of his origin, he has to recognize as superior to him: "Si enim est aliquid in rerum natura, quod hominis mens, quod ratio, quod vis, quod potestas humana efficere non possit: est certe id, quod illud efficit, homine melius" [If there be something in the world that man's mind and human reason, strength and power are incapable of producing, that which produces it must necessarily be superior to man].[15]

The view of the heavens, which Gnosticism was finally to demonize, is drawn into the realm of the questionable by Epicurus. His liberating deed, in the words of Lucretius, is to raise his eyes audaciously to the heavens, not in order to let his gaze rest there, but in order to calm the affect of fear or of admiration that was originally bound up with this view. Here, too, then, theory and rest [or: calm[g]] are connected, but the theoretical attitude does not arise from a self-possessed contemplation of the object, but rather from thinking through the possible explanations of it and establishing the equivalence of all the explanations, for human consciousness.[h] Self-possession is not a precondition of theory, but only emerges as its result—as the triumph in which theory suspends itself; it opposes the admiration of an object that was formed by the accidental encounters of atoms, and thus reduces its own motive to the *pure* disinterest with which Epicurus's gods let the world take place around them. Distance from the object is its lack of interest, not the impossibility of making it, which contains no sting for human consciousness if only because the gods, too, are only happy because they have not made the world.

Looking up to the heavens always involves the danger of forgetting the gods' inactive mode of existence and relapsing into the *antiquas religiones* [ancient religions], because these heavens in fact operate on man like evidence of those old powers. A dispassionate gaze at the starry heavens stands only at the end of the passage through philosophy, not at its beginning. For initially the view of the heaven's motion—comparable in this respect only to the images in dreams—is the main source of belief in the gods. Whereas with Epicurus himself we must see in the background the relation of divination and astrology to the heavens, with its impact in the form of fear and hope, with Lucretius the Stoic program of the *contemplatio caeli* [observation of the heavens], with its metaphysical/theological implications, is the polemical target. The 'naive' wise man of the Stoic type is imperiled, in his self-possession, by the view of the heavens. What can disconcert him is the seemingly irresistible conjecture that behind the great spectacle there stands an overwhelming power (*immensa potestas*) in the face of which man would have to feel like nothing and whose favor he would have to court in fear and hope.

Behind the universe [*der All*] there stands—if one has not immunized oneself against it with philosophy—omnipotence [*die Allmacht*]. As soon as the Greek gods step behind the cosmos, to which they do not originally belong, they deform the physiognomy of the heavens into a hideous grimace. That is the upshot of Lucretius's review of what (in his view) had been an unsuccessful argument of philosophy with religion, "quae caput a caeli regionibus ostendebat/horribili super aspectu mortalibus instans ... " [which displayed her head in the regions of heaven, threatening mortals from on high with horrible aspect]. Appearances speak against the gods being indifferent, and these appearances can only be overcome by a 'theory' in a new sense, at the end of which man is a match for any object, and consequently also for this one: "nos exaequat victoria caelo" [the victory makes us equal to heaven].[16]

The gods must be mere unmoved onlookers at the world, if man is to be able to be this. The incidentalness that that requires presupposes a [literally] eccentric location. The gods, in their 'interworlds,' have such a location; man has to procure it for himself. For atomism, the central point of view on the world is already suspended presumptively, even before this could be achieved as a theoretical result. Seen in this way, it becomes quite clear how, in the sectarian opposition between the Stoa and Epicurus's garden, the positions of the early modern age by which the supposed semantics of the Copernican reform will be defined are found, preformed.

The doctrine of the plurality of worlds, which Epicurus revives, from presocratic atomism, to combat Plato and Aristotle, also serves to disenchant the heavens. It makes them into a mere instance of a type, like other objects are too—a "parvula pars summae totius" [tiny part of the whole sum], and thus reduces the disproportion between man and the universe that is accessible to him, which has become nothing more than an arbitrary corner of the world.[17] The whole is no longer given to man to be contemplated through the senses, but now only as a product of rational extrapolation: a totality of worlds in comparison to which what he can see is nothing.[18] Here an idea is introduced into the world that will also occupy people in the most active possible way in connection with the consequences of Copernicus: consideration of the relation between man and universe as a proportion that can be grasped quantitatively. As the post-Copernican universe becomes distended, man again and again has to rescue himself from the situation—which Epicurus had in mind in a different way—of amounting to nothing, in the infinite. Nietzsche saw, in this consequence of Copernicanism, the nihilistic impetus of an immense devaluation and eccentricity of man. Nietzsche's "will to power" can be understood as a conclusion drawn from the failure of the Epicurean attempt at indifference, from the breakdown even of the toughest position of an uninvolved onlooker. The complaint about the indifference of the heavens to human fate was even further from falling silent in the modern age, because the hope that the world would participate in man's happiness, a hope derived from the Stoic tradition and its adaptation by Christianity, could not be overcome by any renewal of atomism. The indissoluble kinship between hope and fear (the latter as the price we pay for the former), which Epicurus asserted, never convinced the enthusiasts of hope, who believed themselves to be allied with the universal law either of nature or of history.

2

The Heavens as a Cave

The Christian tradition oscillates between the extreme initial values represented by Stoicism and Gnosticism. In Gnosticism, for the first time, the starry heavens have become the scene of a deceitful spectacle: The dazzling splendor of supposedly exalted arrays suggests to the viewer that he is in the place where he was meant to be, while the truth of his position—that he is banished and exiled into the benighted state of forgetfulness of his origin—remains withheld from him. The spheres of the heavens become, for him, impenetrable walls of his captivity in the world, walls that are watched over by demonic powers. The true God is not made evident, but is hidden, by the view of the heavens. Oswald Spengler already, and after him Hans Jonas, described how Gnosticism differed from the ancient world by its consciousness of space.[19] For Gnosticism's experience of space, reality is not centered in the bodies around which space extends as the mere possibility of their motion and their ultimate delimitation; instead, space itself has concentrated itself into a reality, it is "demonized," "a cavity that is charged with forces, that is active in the sense of its function of enclosing, and against the limits of which—seen from within—the gaze that is confined in it and directed outward runs up." The abhorrence of the heavenly bodies that mark the limits of the world-cave is in keeping with this consciousness of space. What we can infer, for the end of the Middle Ages, only from the triumph with which, in his self-glorification, Giordano Bruno breaks through the closed vault [of the heavens], becomes tangible in Gnosticism's mythological constructions as the lengths to which the powers governing the world go to prevent man's salvation. The "iron wall" of the firmament "let fear, rebellion, longing, entreaties, and scorn rebound from it without feeling." Nothing changed in the ancient cosmology, no theoretical reform took place, but "despite identity in its form, the vision of the world became completely different." Not despite, but on account of its insurpassable distinctions, the cosmos had turned into the epitome of a glittering tyranny.

The fact that a powerful stream of ancient, especially Stoic, elements of admiration of the cosmos flowed into the early Christian literature first becomes intelligible when one recognizes their function of repelling Gnosticism. Maintaining the Creation as the work of a true and good God against the suspicion that it derived from a demiurge and had a demonic function represses the original biblical fear of the heavenly bodies' being turned into idols and also the idea, which was no less authentic to theology, of man's foreignness in this world. That is the basis of the fact that the theology of the Creation begins seriously to compete with that of the Redemption—a process that left its imprint on the entire Middle Ages. There would not have been any opportunity, in the Christian resistance to Gnosticism, for a new insensitivity to the impression made by the cosmos—an insensitivity such as Epicurus had recommended—because it would have had to sever the connection between the main dogmatic themes of redemption and creation (which could only be united with difficulty in any case).

Only the African, Arnobius, who walked on the edge of the abyss of heresy, presented man's natural situation in the world—in his apologetic work, composed at the beginning of the fourth century—as one of not being surprised by the world. For that purpose, Arnobius went back, once again, to the outline of the (initially Platonic, then Aristotelian) allegory of the cave, and opposed to it an antithetical variant.[20] His thought experiment imagines, in the interior of the Earth, a habitable room, which is shielded from the change of the seasons and from the sounds of nature, illuminated only by a constant twilight, and isolated by an entrance that is obstructed by a labyrinth. In this unstocked experimental space of thought a human child grows up who is wordlessly fed, tended, and watched over by a naked nurse, and whose entire rearing is aimed at not allowing needs and agitations to arise, and thus at preserving the unneedy poverty of the human prehistory depicted by Lucretius in his 'history of culture.' The question toward which everything is directed and in which the confrontation with both the Platonic and the Aristotelian cave allegories culminates is the question of the effect of emerging from the cave. At an arbitrarily chosen age—the later it is, the more instructive [the experiment]—this person is brought out of his isolation, confronted with the outside world, and brought together with other people. How will he behave, what will he say, if he is expected to testify about himself and about where he comes from? He will, so Arnobius confirms, meet every demand for comprehension stupidly and with indifference, and will let the world be, without fear and without amazement.

This demonstration subject is the exact opposite of Plato's slave in the *Meno*, who was capable, without education, of proceeding from his innate endowment of anamnesis to mathematical comprehension. What is important to Arnobius is to dispute man's natural disposition toward theory and toward the truth; the less the ancient concept of man is validated, the clearer the part played in man's actualization by the Christian blessings of salvation becomes. There is no mediation between this unrefined human nature and that superaddition that is promised him, and least of all is it the starry heavens that could disclose a transcendental direction to man. The result of the thought experiment is above all directed against the idea of immortality that is implied in the doctrine of anamnesis. For Arnobius, immortality becomes the sum of what can be gained, in Christianity, as an addition to man's mortal nature. Arnobius works with an anthropological minimum. He is a kind of Christian Epicurean. His cave man corresponds to Lucretius's original man, before he strayed into culture: provided by nature with meager nourishment and constantly in flight from nature's dangers, and consequently without an upward glance at the heavens, and without the corresponding idle affects of amazement and fear.

Arnobius betrays no horror at the outcome of this thought experiment. On the contrary, he has a sort of pre-Rousseauvian sympathy for his figure's insensitivity toward the world's seductive magnificence, which could only seduce him into superfluity. For man is a creature of superfluity, a superfluous creature in a world-constitution that is finished and complete without him. This *animal supervacuum* [superfluous creature] is a kind of practical joke played on God's work by unknown heavenly courtiers, so that man has no part in the legitimation of the creation, does not bring with him and cannot find any relation to it, but instead finally has to be extracted from it by a pure act of grace.

Chapter 2

From the same basic idea of man's 'surplus' status as the last member of the Creation, Pico della Mirandola was to draw, more than a millennium later, the conclusion of the *dignitas hominis* [dignity of man]: freedom for self-definition and for the continual change of his point of view in contemplating the world. For the Renaissance philosopher man will no longer need validation by the world's quality of order; for Arnobius, man's worldlessness is still his absolute neediness. If the great spectacle of the heavens does not lead him to the Stoics' amazement, it throws him back onto the single question that he can ask himself (even though he cannot answer it unaided): What else, then, could he be meant to do, now that he is allotted no meaning in the world? Here lies the starting point for concentration—undiverted by worldly tasks—on himself and on a possibility (foreign to the world) of his own salvation that might offer itself. The turning back to the self is presupposed with naive obviousness, in the interpretation of the 'third allegory of the cave,' as the alternative to the cosmic theory that failed.

Here a historically lasting mechanism is discovered, in which disinterest in nature is supposed to realize man's self-interest automatically. From this perspective, Gnosticism's demonization of the starry heavens appears as the mythicization of an elementary state of affairs: man's loss of his chance of salvation as a result of aesthetic/theoretical captivation by nature. In the process, man's central position in the cosmos becomes the hypostatization of a cunning trick that is able to prompt, in man, a false vocation and a false familiarity, the dangerous forgetfulness of his not belonging to the cosmos. Consequently the Gnostic deliverance can only be presented as a disturbance, if not a destruction, of the cosmos. Thus in Gnostic texts the deliverer, in his ascension, breaks the power of the guardians of the spheres over the earthly universe, by altering the paths of the heavenly bodies and the times of their revolutions, so that along with the astronomy that binds the world together it also makes any astrology impossible. Valentinian Gnosticism already connects the breaching of the world-order, which brings deliverance, with Jesus's birth, because this was announced by a new star and thus by a change in the constellations in the heavens.[21] A piece of eschatology is anticipated, because man's deliverance can only be carried out in opposition to the stability of the cosmos.

Gnosticism is radical anthropocentrism combined with a negative characterization of man's position in the cosmos. In this regard Christianity, for its pagan opponents, differed little from the Gnostic milieu. Origen quotes the imputation that Celsus, in his anti-Christian treatise, put in the mouths of the Christians regarding their God: "He has even deserted the whole world and the motion of the heavens, and disregarded the vast earth to give attention to us alone; and He sends mesengers to us alone and never stops sending them .... "[22] To be able to gain human salvation only against the stability and order of the world, and to want to gain it, too, without regard to this implication, is the offensiveness to their surrounding world that Christianity and Gnosticism share.

A dualism of the Gnostic kind is ambivalent for the satisfaction of the demands of rationality. On the one hand, it allows one to tell a story of the struggle of the two principles, to categorize the present state as an episode of the predominance of the negative over the positive principle and thus to produce, admittedly, no theodicy, but certainly a utopia. On the other hand, every dualism presses toward monism as the end stage that, it is true, says nothing, but by that very token exhausts the potential of questioning and silences the dynamic of dissatisfaction. The historical attempts to unite the efficiency of dualism and that of monism in *one* system have their obligatory structure in the emergence, from an original monism, of a secondary dualism and pluralism, which, during an intermediate phase, unfolds its whole explanatory power, in order, in the end, to return, in a promised reunification, to the initial stage of the one. The most important differentiation between such systems is whether this return brings a gain with it. But even if, on the circumstantial, roundabout route to its self-comprehension (which has to be denied to it in the beginning), the world-spirit takes the history of this proceeding back into itself and preserves it, nevertheless such system-building magic always suffers from the inexplicability of the first step, of the self-loss and self-surrender of the unity of the One, without any other necessity than the one that is laboriously defined into it, namely, that it is a spirit without proper marks of spirithood, a consciousness that is not really conscious, a thought without an object.

The problematic characteristic of Gnostic dualism is not that it permits one to tell a story, but that in the process the potential for stories rapidly explodes and degenerates into arbitrary production, unless the rigorous discipline of an ecclesiastical system of dogma

should get this process under control. Otherwise Gnosticism is potentially mythology again. What it is not able to accomplish is a justification of the world as creation; in order to be able to tell its story it has
to separate absolutely its world from its God, who may not be made
responsible for the world and who justifies the hope that is placed in
him only if he finally destroys it. Eschatology is the surrogate for
theodicy. In contrast to the Greek antithesis of cosmos and tragedy,
Gnosticism achieves its 'intention' [*Intention*: envisaged objective]
when, through the apocalyptic tragedy, it attains the finality of a
worldless God. The logic of the separation of creation and deliverance
consists in making intelligible something that, as a contradiction,
pervades the biblical world, namely, the possibility of being seduced
into opposition to the God of the world by the world that He created:
"And beware lest you lift up your eyes to heaven, and when you see the
sun and the moon and the stars, all the host of heaven, you be drawn
away and worship them and serve them, things which the Lord your
God has allotted to all the people under the whole heaven." [23]

The opposite pole to the idolatry [*Vergötterung*] of the world is its
deification [*Vergottung*];[a] modern pantheism, as the abolition of the
Christian separation of the generation of the Son of God from the
creation of the world,[b] can only be understood as the attempt to finish
finally with the residue of Gnosticism in Christianity and to throw off
the two burdens of theodicy and dualism at once.

The Christian Patristic literature opposed to the Gnostic demonization of the view of the heavens a cautious elucidation of their quality
as expression. A God who reveals Himself legibly to man [i.e., in
Scriptures] cannot have remained dumb in His role as Creator. The
concept of a God Who speaks and makes things speak, Who becomes
Word and to Whom words matter, implies the metaphor of the book
of nature. Of course, from the point of view of the priority that the
ancient world gave to contemplation, word becomes appearance and
obedience becomes knowledge. The starry heavens take on a sort of
physiognomic quality, the perception of which leaves no room for the
indirectness of mere deductions like the proofs of God's existence:
"Man is induced to observe the world by his upright posture and his
upward-facing countenance; he exchanges looks with God, and reason recognizes reason." [24] This attitude of man to the world is also
formed by the biblical offer of dominion; he is no longer only an
observer, who comes to rest in contemplation—instead, he knows in

order to exercise dominion.[25] This relation of dominion does not yet involve any demiurgic/technical ambition; it is the form of dominion that is practiced in a military review. That too has a function that is directed against Gnosticism; it contains a denial of the tyranny of the spheres and of man's impotence against the world's fatefulness. The use of fire, which according to the ancient myth of Prometheus was a theft from heaven, becomes—in the framework of a Stoicizing physics, for which fire is the all-controlling, both constructive and destructive, element—a pledge of man's universal dominion. Whereas the other living creatures only know how to use water, man, as the *animal caeleste* [celestial creature], possesses a means of proof of his own immortality (*argumentum immortalitatis*), in the form of something that comes from heaven, as the realm of what is imperishable.[26]

In spite of the defense against Gnosticism, the Patristic formulas for the admiration of the heavens in which the *contemplator caeli* [contemplator of the heavens] engages remain more cautious than the Stoic ones. The desire for happiness may not be definitively satisfied by the world and be at home in it. A cautious formulation is, for example, that of Irenaeus, who in spite of his anti-Gnostic identification of his God with the creator of the world speaks only of a quite unspecific "assignment" of the world to man,[27] or restricts its anthropocentric teleology to those who are already saved, and considers the heathens to be totally incapable of raising their eyes to the heavens.[28] In any case, the Gnostic experience prevents the Fathers from constructing an all too sharp mutual exclusiveness between the contemplation of the heavens and the summons to transcendence.

The Jewish Alexandrian, Philo, had provided the model of how to build the viewing of the heavens into man's path to salvation, in his allegorical interpretation of the travels of Abraham. Abraham's path leads him from Chaldea to Egypt, and that means: out of the land of astronomical curiosity and avidity for knowledge, by way of the stations of self-knowledge and the discovery of his own ignorance, to the recognition of God, and a view of the world that is bound to this recognition and anticipated by it.[29] The turning point of the allegorical journey is marked by the requirement of self-knowledge. Opposing it to the contemplation of the heavens becomes a permanent *topos* [theme] of the Patristic literature. But such self-knowledge is not automatically inner experience; rather, it has the character of a consideration of the external world with a view to ascertaining man's

cosmic 'position' on a scale between above and below. Self-knowledge
is setting oneself right in the cosmos and is thus above all a correction
of relations of subjection and domination, a correction of the false
instrumentalization of what is above man and of the equally wrong
instrumentalization of man for that which should be at his service.
Even self-knowledge, as the grasping of one's 'place in the order,' is a
specific form of looking at the cosmos.

How little one could be concerned, in this context, with the issue of
whether one's position was central or eccentric—that is, with the
post-Copernican problematic—is shown by Lactantius's verdict
against the assertion that the Earth is spherical; this, he says, would
have the consequence that for every location heaven would not only
be above, but also below—and to the Church Father that seems
intolerable: "Nullo modo fieri posse, ut coelum terra sit inferius ... "
[In no way can it come to pass that heaven should be lower than the
Earth].[30] For the Christian epoch, the distinction between above and
below overlies that of center and periphery; to a large extent it makes
possible the success of Aristotle in the Middle Ages, and at the same
time explains the fact that, for the subsequent interpretation of the
Copernican turning, the more natural 'language game' will be the
one that had been developed by the Pythagoreans and Stoics and that
interpreted all cosmic relations as relations to the center. With the
change in orientation, even the devil must change his location:
Gnosticism's guard of the spheres, who watches over the cosmic
housing so as to prevent the penetration of the good messenger, as well
as the escape of the emprisoned Pneuma, becomes the ruler of Hades
and its gatekeeper, who finally—in Dante—has his position precisely
in the middle of the Earth and thus of the universe.[31]

The linguistic difficulties with the two different cosmic ground
plans are reflected, even at the height of Scholasticism, in Thomas
Aquinas's commentary on Aristotle's cosmology: People call heaven
both that which is outermost in the world and also that which is
above. Here "above" could not be taken in the strict sense in which
it is used in natural science, because this orientation designates the
direction and the place of the lightest element, which is fire, with
which, however, one is still far from reaching heaven. The situation of
the latter is "above" in a different sense, namely, as what is most
remote from the center. But it is only by a further leap, out of this
orientation as well, that "above" is the place of divine things—in

which connection one must say explicitly, against surviving pagan ideas, that one cannot think of the heavenly bodies, but only of immaterial and incorporeal substances, and in the strict sense only of God Himself.[32]

The competition between concern for salvation in the next world and pleasure in seeing the stars becomes obligatory for the Middle Ages largely in the formulas that Augustine had coined in his discussion of *curiositas* [curiosity]. Reserve or even mistrust in relation to theory's turning toward the starry heavens is also a factor promoting the centuries-long stagnation of astronomy. In the case of the Middle Ages one should not overlook the fact that the immutability and imperishability of the starry heavens, which were affirmed by the ancients and were almost manifest to the eye, were incompatible with theology's teachings regarding the Creation, miracles, and the end of the world. The aspect of the heavens could not be the demonic grimace that Gnosticism saw there, nor could it strengthen confidence in the world's durability and reliability too much. Contingency and perishability, dependence on the *creatio continua* [continuous creation] as the abyss of the world's readiness, at any time, for annihilation, were not manifest—as characteristics of the world—in the starry heavens. The conflict between the ideal, which was transmitted by literature, of the ancient onlooker, and the consciousness (drawing its vitality from other sources) of men's fleeting pilgrimage on Earth, for which the contemplation of its mysteries was always only impending, remained virulent or became virulent to the extent that the mere patience that is implied in a state of waiting was subjected to the intensified tests of submission to the absurd, to the expectation of sheer obedience, to ascetic renunciation.

The insertion of the elementary curriculum of the "liberal arts"[c] into the medieval Christian educational framework, an insertion that Augustine had spelled out in his programmatic treatise *On Christian Teaching*, implied, for astronomy, the unusual new situation that the highest of the ancient objects was now functionalized on the level of the "lower grades." For Augustine it had not been sufficient that a science did not confuse man's will to salvation; he required of it the opposite, that it must help that will to clear out of its way the insignificant things—in the midst of what is essential—that confuse him (like the puzzles in Scripture), "ut non sit necesse christiano in multis propter pauca laborare ... " [so that it is not necessary for the

Christian to toil through many things for the sake of a little].[33] But
even where the usefulness of a discipline for the secure attainment of
eternal salvation can be assumed, the general rule holds that one
should avoid excess, especially where it is a matter of objects of
experience that carry out motions in time and that are situated in
space.[34] Astronomy is imperilled by the transition from contemplating physical constellations to interpreting them as signs: by the *sig-
nificatio superstitiosa* [superstitious/prophetic significance]. Heavenly
bodies were mentioned only rarely in Scripture; consequently, dealings with the science of those bodies were not very useful or not useful
at all, in dealing with the divine Scriptures, and were more likely
to interfere, by uselessly burdening one's attention. Consequently
it was more benefincial and honorable to desist from astronomy
altogether.[35]

With increasing distance in time from the acute danger of Gnosticism that defines Augustine's spiritual biography, the need to extract
from astronomy a more salutary importance for man and to preserve,
at least, its position in the canon of the "liberal arts" must have
grown. When Boethius and Cassiodorus, in the first half of the sixth
century, sought to gather together the valid contents of the ancient
world and to convey them into the new epoch, they only took from the
wrongful possessors—following Augustine's juristic allegorical interpretation of the theft of the Egyptians' vessels of gold by the departing
people of Israel—what those possessors had gathered from the treasures of Providence and had removed from its intended function in
the service of the truth: "... debet ab eis auferre christianus ad usum
iustum praedicandi Evangelii" [... the Christian ought to take it
away for the righteous use of preaching the gospel].[36]

Here we have before us one of those abominable theories of historical self-legitimation according to which those who are in full possession of a truth may place in their service everything that those
people are and possess who know nothing of this truth and therefore
are declared to be its wrongful beneficiaries. It is the morality of the
chosen, who claim to make everyone their debtor and everything
their material. One may say that in history we are not concerned with
the theory of legal titles; but if one thinks, precisely, of the *artes liberales*
in the Middle Ages, one encounters the marks of an unearned possession, a piece of booty that now can only be displayed, and that,

from generation to generation, people have less and less idea what to make of.

Boethius writes about the threefold utility of geometry, for human skills, for health, and finally for the soul: If we investigate this art strictly, carefully and with a moderate spirit, it illuminates our senses with great clarity and allows us to approach the object above us— that is, the heavens—with our spirit, to explore the whole heavenly mechanism with an inquiring reason, and, with penetrating acuteness of understanding, to grasp it, to a certain extent, and to recognize the maker of the world, who veiled from us mysteries of this kind and this magnitude.[37] One notices how in such a text apprehension of a sudden turning toward Gnosticism is entirely absent, and trespassing into the veiled mysteries of the Creation seems not to involve any blasphemy, although the impact is reduced when the whole inquiry is made to serve the knowledge of God. Following Boethius, around 544, Cassiodorus gave the *septem artes* [seven arts] the firm organization of an elementary stock that has been abridged at the expense of all methodological understanding and all cognizance of problems. With regard to astronomy, he quotes almost word for word what Boethius had written in his *Geometry*, but with the noteworthy omission of Boethius's reaching out beyond astronomical knowledge to the knowledge of God, and also—connected with this—of the agent who was responsible for masking the world from man.[38] God neither reveals himself to the astronomer through his object, nor did He wisely and providentially conceal the mysteries of this object from him (two things that are in any case hardly compatible).

The need for a justification of astronomy will remain a problem for the Middle Ages. But it is nevertheless already evident here how few excisions had to be made in a text in order to relieve this "liberal art" of its cramped functional role. The motive for doing so is discernible in Cassiodorus: The disciplines of the liberal arts have become settled in such an elementary role as the preliminary stage prior to higher insights, that they do not need to be legitimated by anticipations of what is yet to come. But no doubt this role is bound up with a kind of artificial reduction: What is abandoned to the category of 'school' and its transmission of knowledge loses contact with the problematic quality that is found on the margins and thus loses the motivation for any theoretical progress.

The image of amiable stagnation is almost touchingly visible in an

Chapter 2

account of the school of Odo of Tournai, which tells how the master, in addition to the pedagogical methods of the Peripatetic school (in walking around) and the Stoic (in sitting and responding to questions), also practiced the nocturnal observation of the heavens, following the course of the stars with his scholars until late in the night and pointing it out to them with his finger—although this is combined with the assurance, which (as it were) domesticates these manifestations of curiosity, that this sort of thing took place before the doors of the church.[39]

Until the revaluation of the Faculty of Arts as the guardian of the true heritage of Aristotle, astronomy was firmly linked, by its insertion into the elementary curriculum of the liberal arts, to the canon of a completed body of material to be learned, material that one did not, in fact, encounter 'afterward.' The absence of this linkage was an advantage for the Arabian tradition of astronomy and made it possible for it to develop further. If one compares the motives that made theoretical dealings with the heavenly bodies suspect for the Latin Middle Ages, then the greatest weight does not belong to the proneness of *phenomena* to turn suddenly into *signs*. Instead, it belongs to the evident indestructibility of the starry heavens, together with the orthodox Aristotelians' theorem of the eternal existence of the world, a theorem that was cherished—precisely—in the Faculties of Arts, but that also caused the less orthodox Scholastics grief by its rational irrefutability.

The view of the heavens stands for a world that runs according to unbreachable laws, that is unalterable, indestructible, and betrays no readiness for eschatological collapse. The appearance of new stars, which Tycho Brahe with amazement first establishes in 1572, or the disappearance of familiar ones is just as unattested in this period as are the difficult reflections that will result from the combination, in the seventeenth century, of increasing expansion of the universe and a speed of light that is small in relation to that extent. For William of Ockham, the place of such prior developments is taken by thought experiments about the relation between omnipotence and perception: God could bring it about that man sees a star in the heavens, although this star has already ceased to exist.[40] Naturally, this reflection could also have been undertaken in regard to any other randomly chosen object, in order to exemplify the *intuitus non-existentium* [observation of nonexistent things]; but it is only the choice of an object that, for the Aristotelian tradition, possessed a unique stability, that gives the consideration the character of an intellectual demand that verges on the absurd, a demand that is likewise contained in the assertion that even for the heavenly bodies a continuous divine *conservatio* [conservation], over and above their *creatio* [creation], is necessary.[41]

3

At the End of the Observer in Repose[a]

The revival of the formulas describing man as the *contemplator caeli* [contemplator of the heavens] in the Renaissance has an anachronistic quality, which does not disappear even when one has noted the traces of the deforming forces operating on the traditional material. As an opposition to Scholasticism, the Renaissance is an episode in which the expectations or apprehensions that had been harbored since the time of the Fathers with regard to the relation to the starry heavens are fairly accurately fulfilled. The new interest in astrology is only in part a substitute for faith, and much more frequently a stylized pagan pose, like the dealings with the decorative old names of the gods. But the motives immediately become a matter of indifference for the stabilization of a new interest that makes use of the sanctioned formulas in order to sanction itself. There was never a renewal of the ancient world; it was an invention of those who had rhetorically professionalized themselves in the role of its renewers. Daring strokes of transformation occur precisely where a conservative anxiety is predominant, as in Nicholas of Cusa, or where, on the biblical ground plan, ancient materials are put to work, as in Pico della Mirandola's *Oration on the Dignity of Man*.

In the contrast between the School of Padua and the School of Florence, the Platonizing Academy has for a long time drawn the historians' attention almost exclusively to itself, because it seemed to present the features of a new type of individuality, which one associated preeminently with the concept of the Renaissance. But the School of Padua, which gave precedence to ancient medicine instead of ancient literature, led to the development of a concept of scientific method that is able to produce a clearer continuity with Galileo than can be constructed from the anthropology of the Florentines. Although the latter idealize the old concept of man's central position in the world, and thus make it superfluous for physics, they certainly continue to accept the confirmation offered by cosmology, and thus deprive the need for an astronomical reform of any impetus. The relation of the two schools to the early history of modern science is instructive in regard to the deceptions that threaten the history of science in this area. There is no evidence that Galileo might have appealed, in questions relating to his methodology, to his predecessors in the School of Padua. His appeals to philosophy lead in an entirely different direction from that of an Averroist Aristotelianism, within which even such an important opponent of his as Cesare Cremonini arose. Copernicus, on the other hand, furnishes all the evidence one could wish for in order to assume, as a given, a connection to the Platonism of the Renaissance, and its supposed heliocentrism. But here we have by no means to deal with a methodological or material influence, but rather with the referring back that typically occurs in the literary preparation of a work whose methodology and philosophical possibility are founded in other contexts than those that provide its apparatus of quotations. Nevertheless, the results of such reaching back cannot be described as purely ornamental. They take into account the capacity and readiness for reception possessed by the audience to which the work is directed.

One will be able to see the Renaissance's share in the formation of a new concept of reality in the fact that it also forced into the open and elucidated man's autonomy in the world because it was able to regard as superfluous an all too drastic anthropocentrism's providential teleology. In fact, the rational anthropocentrism that Copernicus was to uphold as his world-formula only becomes isolable precisely as a result of the circumstance that reason cannot be made into the mere organ for reading off what is pregiven in nature, or merely drawing on the reservoir of the Ideas and concepts. The Renaissance's concept of nature is, to a large extent, a release of the world from a preimprinted subservience to man, just because man is seen as being capable, by virtue of his autonomous reason, of proving that he is a match for a nature that is occupied with itself—all the way to the idea of overpowering and commanding it, which can no longer be the fulfillment of a teleological structure. Such a character of being their own master and their own purpose was asserted in the Renaissance,

explicitly and in a way that was meant to be exemplary, of the
heavenly bodies. Moses is right to say—writes Pico della Mirandola—that the stars were created in order to shine in the sky and
to illuminate the Earth. This, he writes, expresses the fact that serviceability for us is not the primary goal of their existence—nor,
admittedly, is the *gloria dei* [glory of God], any longer—but rather,
their purpose is first to illumine themselves, and finally us as well.[42] It
is to be observed, in such formulations, how difficult it had become, at
the end of the Middle Ages, to legitimate the object of theory in any
other way except as having a definite function for man, which was to
be actualized in it by knowledge.

To translate astronomy's concept of its object, that stars are points
of light in the heavens, which move in accordance with laws, into the
language of the theology of Creation in such a way that one answers
the question of what use and what task God intended the heavenly
bodies for by saying that their functions are to move and to shine,
means precisely to set astronomy's object free both from an immediate teleology and from the suggestion that there must still be a
concealed communication to be extracted from this great display for
man.[43] The opportunity for the autonomy of reason consists precisely
in nature's not having the meaning of a text that is directed at man or
an instrument that lies ready for him. Pico constructs his critique of
astrology on the Platonic idea that men's souls were taken from the
same material as the stars and that consequently neither the instrumental nor the astrological relationship of subjection could exist
between them.[44] Jacob Burckhardt interprets the saying that Ludo-
vico il Moro caused to be mounted on a cross in the cathedral of
Chur—"Vir sapiens dominabitur astris" [A wise man will rule the
stars]—as an "onset of determination" to break away from the power
and disfavor of the stars.[45] But this susceptibility to appearances as
signs is much too tightly bound up with the outgoing Middle Ages'
consciousness of reality for "determination" to have been enough to
overcome it. After all, the inscription in Chur agrees almost word for
word with what Luther expressed in the words: "Et tamen nos sumus
domini stellarum" [And yet we are rulers of the stars].

In spite of the criticism of astrology, Copernicus's contemporaries
are still largely subordinated, in their astronomical interests, to astrology. The pope to whom Copernicus dedicates his principle work,
Pope Paul III, never scheduled a consistory without seeking astrolog-

ical assurance regarding the date, and on the other side Melanchthon did nothing in ecclesiastical politics without astrological advice. Copernicus himself appears in astrological contexts, in documents of the time, as though it were a matter of course, when Johann Apel, of Nuremberg, points out in a letter accompanying the horoscope that Joachim Camerarius had cast for Duke Albrecht of Prussia that if there were need for expert interpretation one could make do: "Thus there is also an old canon in Frauenburg, if it cannot be found elsewhere."[46] That is the dusky background of the age, from which the figure of Copernicus stands out, just good enough to serve as a stopgap in astrological difficulties.

In the dedicatory foreword to his work, Copernicus defined the location of his life with the words that it was the "remotest corner of the earth": "in hoc remotissimo angulo terrae, in quo ego ago ...." Thus he explains his dedication [of the *De Revolutionibus*] to the pope as addressing it to a central embodiment of the public sphere, by means of which he could reach both learned and unlearned and submit to their judgment: "Ut vero pariter docti atque indocti viderent ... " [So that learned and unlearned alike may see ...]. But at the same time, with this differentiation between his own position and that of a center for the public sphere, he reflects the cosmological differentiation between the parochial perspective of his terrestrial 'corner' and the central point of construction from which the universe cannot, indeed, be viewed, but can be thought. The [literal] eccentricity of one's location does not imply any discouragement of reason's claim, which, according to the world-formula contained in this same dedicatory foreword, can only appeal to the fact that the world was made for man. For precisely where Copernicus speaks of the vexation of lacking a reliable and rational presentation of the motions of the world-structure—this lapidary "Coepit me taedere, quod nulla certior ratio motuum machinae mundi" [It began to offend me that philosophers could agree on no more certain theory of the mechanism of the universe]—he adds immediately, and in antithetical association with the finding that is the source of the vexation, that this is after all precisely the world-mechanism that was created on man's behalf by the best and most accurate master builder.

The parochial 'corner' perspective is what provokes reason to make sure of a principle of the constructive presentation of the cosmic system that is adequate to and accessible to itself. The surrender of the

physical central position in favor of the rational one may owe its
possibility to the idealization of the world-center by the Platonic
Renaissance, but over against the cosmic pathos of the latter's anthropocentrism (which, if it is not an anthropocentrism of space, is still one
of signs and powers) it is a disappointment for the onlooker role for
man that had been attempted once again. In his *Dialogues of the Dead*,
Fontenelle will make the realist, Moliére, say to the Renaissance philosopher, Paracelsus, that the truth is, after all, disappointing, and he
is persuaded that "if most people saw the order of the universe as it
really is, so that they noticed neither characteristic virtues in numbers, nor influences of planets, nor fates that are attached to certain
times or certain revolutions, they would not be able to prevent themselves from saying, about this admirable order, 'What! Is it nothing
more than that?'"[47]

At this point the unity of theoretical and aesthetic vision, since
the ancient world, is shattered, and in a way that, in spite of the
Renaissance's (admittedly halfhearted) attempt at renewing it, is
final. Since, at the latest, Copernicus, comprehending and enjoying
can no longer be carried out in *one* act. When, in the preface to the
first book of the *De revolutionibus*, Copernicus cites, among the motives
of the astronomer, that of the beauty of the heavens, as well, one
imagines for a moment that everything is still, at least verbally,
together after all. What could be more beautiful than the heavens, the
rhetorical question asks. But the reason given for the hyperbole is,
unexpectedly, formulated from the viewpoint of thought rather than
of intuition [*Anschauung*, 'viewing']:[b] The heavens, he says, are in fact
the most beautiful thing because they contain everything beautiful
—that is, as the outermost world-container, they enclose it.[48] The
appeal to the ancient formula of the "visible god" does not alter this
'constructive' characterization of beauty. The idea is introduced from
the point of view of the Scholastic concept of formal causality: If there
are beautiful things in the world, then they must be contained, in
their form, in the first cause of the world's processes. Of course, to
articulate this as a train of thought could not be allowed from the
beginning of a work whose content finished off the theory of the
heavens' being the first cause of all processes in the world, even if
Copernicus himself did not yet draw the conclusion (which his theory
suggested so naturally) of dissolving the first heaven as a physical
unit.

Copernicus can still say that to recognize the *ratio ordinis* [systematic order] it is enough to look at the facts themselves with both eyes open; but that is already a metaphor taken from optics, and relates to the inferences that must be drawn from the disposition of the planets in relation to the Earth and from the changes in their apparent magnitude.[49] Galileo, who would be overwhelmed by the surprise constituted by the telescope and the Copernican analogies that present themselves in it, and would be inclined to overrate the mode of intuition, was nevertheless able to recognize in the Copernican system the turning point in the dissociation of intuition and construction: "For let no one believe that the reading of the lofty concepts that are written in those pages [of the heavens] would be completed by merely seeing the splendor of the Sun and the stars and their rising and setting (which is as far as the eyes of animals and of the vulgar penetrate). But inside those pages there are such profound mysteries and such sublime concepts that the vigils, the labors, and the studies of hundreds and hundreds of the most acute minds, in investigations continuing over thousands of years, still have not been able to penetrate them entirely . . . . So what the mere sense of vision presents to us is like nothing in comparison to the marvels ... that the ingenuity of those with understanding discovers in the heavens."[50] The appearance of the telescope on the scene, and then later of photography, and finally the replacement of optical objectification by different methods, like that of spectroscopy, and different information sources, like that of radio astronomy—all of this follows logically from the first step of putting the eyes, in their natural capacity for viewing the heavens, in the wrong, and leaving them only the residual function of aesthetics, as though as a compensation for the agreement between anticipation and confirmation, between the receptive capacity and what is given [to it], that is forfeited in the scientific attitude.

In Francis Bacon, the optical presentation of phenomena, as the disappointment of a definite anticipation of order, is already the symptom of the fact that constructive reality, too, will not correspond to this anticipation of order and that rational preconceptions must always collide with the empirical sifting of the facts. Bacon appeals to the Epicurean, Velleius, in Cicero, who criticizes the Stoic idea of god with the comparison that here it is a question, with the universe, of something like the Roman aedile, whose office included decorating the city on festive occasions: What occasion could there have been for this god to emerge from his eternal darkness and create an illuminated world? In saying this, the Epicurean had not even called in question the premise that the lights in the heavens were worthy of a festive occasion. But Bacon denies that the heavenly bodies have been brought into a somewhat beautiful and tasteful arrangement; and he does this not because he doubts the existence of a rational creator of the world, or his ability to present the heavens in an aesthetic manner, but because he suspects the presence of an elementary disharmony between the reason according to which the world is constructed and the reason from which man takes his standards for measuring the world.[51]

Bacon's thesis goes further than that of the Epicurean whom Cicero introduces. Velleius sees on the one side the accidents described by atomism, and on the other man's futile claim to order, which is projected on the origin of the world and ascribed to a god as a motive for distinguishing the heavens with stars: "Quid autem erat, quod concupisceret deus mundum signis et luminibus tanquam aedilis ornare?" [Also, why should god take a fancy to decorate the firmament with figures and illuminations, like an aedile?][52] But what, conceivably, could the aesthetics of a god be, who refrained through endless ages from applying it: "Quae ista potest esse oblectatio deo? Quae si esset, non ea tam diu carere potuisset." [How can a god take pleasure in things of this sort? And if he did, he could not have dispensed with it so long.] As is not the case for Cicero's spokesman from the Epicurean school, for Bacon there stands behind the thesis of the discrepancy between [the divine and human] concepts of harmony the idea that a divine origin of the world, in particular, cannot guarantee that it will be comprehensible, in its rational principle, to the human intellect. Bacon's objection to the aesthetic quality of the starry heavens only furnishes a symptom of this incongruence, which reaches deeper. Bacon's whole distrust of Copernicus can be described as being directed at a world-formula that postulates agreement between man's theoretical rationality and the world's constructive principle, and elevates it to the criterion.

The only reason why the discrepancy between [the divine and human] concepts of harmony, which Bacon takes as his point of departure, does not signify—to him—a catastrophe for theory is that the same argument that had elevated the heavens to the object of theory as such for the ancient world—namely, their inaccessibility for human action—means, following Bacon's equation of *verum* [the true] and *utile* [the useful], that the theoretical importance of this object for man is excluded. Astronomy's membership in the *artes liberales* has become senseless as soon as Bacon's antithesis, *non caelum, sed artes* [not the heavens, but the arts], is accepted. He relentlessly plays off the influence that the mechanical arts and their inventions have on human relations against the impotence that is embodied in a star—almost ironically, Bacon uses astrology's language of the *influxus* [influence] to contrast this new power to the old power.[53] For Bacon, the contemplation of the heavens was characterized by paltry unprofitableness for man because this object tangibly marked the limit of his equation of knowledge and ability. For all his polemics against the sterility of Scholastic final causality, Bacon in his turn is nevertheless only a rigorous anthropocentrist because the derivation of the *utile* from the *verum*, as the condition of dominating nature, becomes perfectly reversible for him, since for man what cannot be made serviceable for him can no longer have any importance for him in regard to truth either. That makes Bacon an anti-Copernican as a result of the same arguments that made him an anthropocentrist. Here the question remains entirely open whether the constituents of the regained paradise might not include what does not contribute anything to its regaining: If nature, through science, should again become the garden of pure subservience to man, the starry heavens may belong in the wings of this state of affairs; but there is no evident reason why there should then still (or again) be a science of that which overspreads the state of enjoyment. Here, too, the dissociation of theory and happiness—contrary to all ancient premises—is palpable.

Part of this is the fact that an immediate relation to the object no longer represents superiority in a theory. The modern age will in any case have to come to terms with the fact that precisely in the case of stellar objects, presence and simultaneity no longer mean anything, in view of the finite speed of light. Bacon saw the foolishness in the behavior of Thales of Miletus precisely in the fact that in observing the stars he looked directly toward them, with the result that he fell into the spring,[c] when he could after all have been satisfied with the image of the stars in the spring's reflecting surface, in which case he would not have been exposed to the danger of falling. It is a fundamental experience of the modern age that the object first becomes accessible when one turns away from the immediacy of perception to indirect experience and construction. The acts of intuition [or 'contemplation': *Anschauung*]^d are emptied of their self-fulfilled quality as ends, and instrumentalized as elements in a process of methodical observation. Nicholas of Cusa had first articulated this idea in his simile of the cosmographer: He furnishes a clear representation of the entire world that can be apprehended by the senses, by remaining at home, having data and information brought to him from outside, and organizing and reducing them to a common scale, but then, having closed his doors, turning his gaze inward, to the creative principle of a world that is present in himself (*ad conditorem mundi internum*), which alone makes it possible for him to process the empirical material that is brought to him from outside.[54]

Hobbes will intensify the nominalistic tendency of this thought experiment dramatically by making the standpoint of natural philosophy independent of the actual existence of a world. He bases its possibility on something that to Descartes could only have appeared as a source of insecurity for reason, namely, on mere recollection or representation in imagination. True, Hobbes has not yet conceived Laplace's universal intellect as the ideal subject of knowledge of nature; but when he experimentally suspends the actual world, an imaginary human subject survives, who is now freed from the burden of intuition and for whom what remains in his memory is perfectly adequate for a theory of nature.[55]

Every epoch invents its imaginary standpoints, from which it thinks that it can bring its characteristic type of knowledge to its most advantageous execution. The change that these imaginary standpoints undergo is also instructive in regard to the differentiation of concepts of reality. Hobbes's thought experiment is directed against the rationalistic assumption that the world could already be deduced *before* its existence, and independently of it, from the principle of the *ens perfectissimum* [most perfect being]. It would then be in principle superfluous for a world actually to come into existence, and even when it has come into existence it remains somewhat superfluous to inspect it more thoroughly. For empiricism, of course, such an inspection shows that what presents itself can hardly be the result of a deduction. That holds above all for the starry heavens, the image of which presents itself to empiricism, since Bacon, as the stigma of contingency and resists every rational principle. Indeed, it is precisely in that respect that the ideas possessed by the postworldly subject that Hobbes conceives *a ficta sublatione* [by pretended annihilation (of the world)] will bear the mark of the independence of nature from the spirit that pursues the theory of it (*a virtute animi minime dependentia*). If the prospect of the starry heavens satisfied our demands for an intelligible arrangement, then the certainty of reality's independence of our own faculty of imagination could not be conveyed with such stringency. Even the imaginary postworldly considerer remembers what makes one accept a world as 'real,' in the elements of rationality that it lacks.

That anticipates a concept of reality that will manifest itself in the aesthetic realisms, and that already deveops a theory of the stigmata of the real in the aesthetics of the subsequent century—for example, in Diderot—as when in an idealized portrait tiny traces of ugliness, such as a wart or a pockmark, appear and break the dawning expectation of the ideal.

But Hobbes's concern was to show that the physical presence of the world made no difference to the fact that in cognition only ideas are processed, ideas whose relation to reality cannot in its turn become an object [of cognition]. The starry heavens exemplify this in a special way, because this object is one that we simply cannot manipulate when we undertake the processes of objectification—classifications and measurements—on it: "Non enim si coeli aut terrae magnitudines motusque computamus, in coelum ascendimus, ut ipsum in partes dividamus, aut motus ejus mensuremus, sed quieti in musaeo vel in tenebris id facimus." [For when we calculate the magnitude and motions of heaven or Earth, we do not ascend into heaven that we may divide it into parts, or measure the motions thereof, but we do it sitting still in our closets or in the dark.]

Thus the theorist of the starry heavens finds himself in the same situation as Hobbes's imagined considerer of the world *a ficta sublatione*; to him it makes no difference that his ideas are not derived from memory, but from current perception—he gains his knowledge precisely and only when he has withdrawn into his cabinet, which is screened off from the view of the heavens, and developed his 'theory,' which now hardly derives its name any longer from the onlooker's viewing,[e] but instead signifies the stock of methodically confirmed propositions. The Copernican reform was not the result of more intensive and more accurate observational activity, as Kepler's surrender of the postulates of circular orbital motion (which represents the result of the observations he inherited after Tycho Brahe's death in 1601) was to be—which was quite contrary to Kepler's taste in theory, and consequently was perceived by him as a forcible imposition. For Copernicus, the Platonizing formula is by no means false that describes him as having solved the problems of the heavens by not looking at the heavens. To interpret the Copernican reform as a product of the difference between phenomenon and Idea, sense-datum and reason, is too clear an invitation to the Platonists for them to have been able to overlook it. In the English renaissance of Platonism it was above all Henry More, who was to influence Newton's concept of absolute space as an existing idea, who made the most of the Copernican model for his dualism.

In 1642, More published a Platonic poem on the immortality of the soul, entitled *Psychathanasia platonica*, which depends, in its formulation of questions and in its content, on Marsilio Ficino's *Theologia platonica*. This early work is the complex result of More's turning away from his exclusively natural-scientific interests and of his attempt, now, to put his knowledge of nature in the service of a question that concerns man more closely. The problem of personal immortality seems to constitute this closer concern for More. If he is able, in connection with this topic, to go beyond his dependence on Marsilio Ficino and Florentine Platonism, then the stimulus to this came from his acquaintance with the Latin translation of Galileo's *Dialogue on the Two Chief World Systems*. This translation, by Matthias Bernegger, had been published in Strasbourg in 1635 and appeared as early as 1641 in a new edition revised by Galileo himself, an edition with which Henry More became acquainted in that same year. It is characteristic of the history of Galileo's influence that it is based on this Latin version—which circulated without hindrance—and not on the Italian original. As More read Galileo, he was in search not, indeed, of new proofs, but certainly of new corroborations of the immortality of the soul—that is, of what Marsilio Ficino had called its *signa* [signs, indications]. More finds such a *signum* in the Copernican theory, as it is presented to him by Galileo. He sees, in the replacement of the geocentric by the heliocentric system, the overcoming of the power of sense-appearance by rational insight. In it, the human spirit liberates itself from its dependence on the immediacy of the physical organs. Not much later, admittedly, More was to encounter a no less plausible support for his view of immortality in the Cartesian dualism of *res cogitans* [thinking thing] and *res extensa* [extended thing], and the dignity of such an orientation to "signs" is not much strengthened by the fact that the sequence of references could place Descartes after Galileo.

In the first two cantos of the third book of the *Psychathanasia platonica*, which largely correspond to the ninth book of Ficino's *Theologia platonica*, More treats the soul's independence from the body by summoning up all the traditional arguments, after which, in the third canto, he introduces the Copernican theory. The "signs" strengthen one in the conviction of what has already been proved, or is supposed to have been proved. Where Ficino had still exploited the arsenal of the ancient authors, More restricts himself to the—for him, overwhelming—"sign" set up by Copernicus. To oppose the sense-appearance of the motion of the heavens and the motions in the heavens, and finally to prevail over this appearance, demonstrates the primacy of a rational soul and the finality and independence of its faculty of judgment. The nobility of the stellar objects, which had come to be accepted in the philosophical tradition, now no longer consists in the greatness of their appearance and in the capacity for being admired that goes with that, but rather precisely in the cancellation of intuition by construction, in the neutralization of the power of the immediate impression by the objection lodged by rational judgment, which spurns the inferior authority of perception. Thus the new astronomy becomes the paradigm of a required catharsis of the soul, with which the theoretical interest justifies itself by transcending the state of being affected by its object. How unspecific this argumentation is can easily be gathered from fact that in the seventeenth century the theory of atoms achieves a new acceptance because it is presented as an assertion of constructive reason that cannot be confirmed by anything given in sense perception.

Ralph Cudworth describes atomism's accomplishment in terms of the same Platonizing 'topos' [theme] of the triumph of reason over the senses. That is also the main characteristic of the *Democritus platonissans*, which Henry More published in 1646. This work is followed just a year later by the second edition of the *Psychathanasia platonica*, to which More has added numerous new notes.[56] These are meant to make the poem's argumentation on behalf of the Platonic philosophy more easily comprehensible. But they refer almost exclusively to astronomical theory and to the Copernican system as it is represented by Galileo. Here one is struck by the fact that More appears to have no awareness of the importance that astronomical discussions had had all along, inside Platonism. Thus a perspectival occultation comes about, which reveals More's mistaken impression that the pre-Copernican theories—represented by Ptolemy's *Almagest*—had been of an entirely different type from Copernicus's, that is, less rational and constructive.

The *renunciation* of intuition is a precondition of modern science; the *loss* of intuition is a necessary consequence of any theory that systematizes itself, that is, that consolidates and arranges its results in such a way that, by virtue of their heterogeneous order, they place themselves in the way of access to the original phenomena[f] and finally take the place of these. Results of science have, to an ever-increasing degree, the characteristic that they contain knowledge as a terminal state that can no longer be related to any sort of previously familiar object. It is only by way of intermediate stages that a scientific statement, if it is (for once) interpreted as an answer to a question, can be related to a very general question formulation that could also be localized outside the theoretical discipline, in an experiential space.

Already in the first instance in which the provision of science to serve the needs of a broader public was recognized and accepted as a task in its full contemporary importance—namely, in the French *Encyclopedia*—the presentation of knowledge in alphabetical order, and made accessible by a system of cross-references, threatens to lose its function of making access to the objects of nature and culture possible in such a way that the intelligibility of the phenomena themselves retains its primacy. The auxiliary function that knowledge could have in relation to intuition [in fact] takes the latter's place. The history of science leads not only into a situation in which its results are isolated from the comprehensibility of their methodical derivation, but still more into a situation in which theories are separated from their original motivation, which was to disclose man's life-world, to make it more visible and more perspicuous—that is, to enhance our capacity for experience. The remonstrance that the great stocktaking of results and procedures in the French *Encyclopedia* could obstruct access to the objects themselves does not come first from Herder or Goethe, but already stands before the eyes of the *Encyclopedia*'s editor, Diderot, as the serious dilemma of the "monstrous work" (as Goethe called it).

The *Encyclopedia*'s systematization and criteria for selection seem to suffer from the defect that man possesses no normative standpoint over against reality, that the universe of things as well as that of the spirit permit an endless number of perspectival accesses, which would have to have an equal number of systems corresponding to them. A standpoint that was distinguished from all the others could only be had at the cost of a concession of reason to theology, for only the system founded on God's will (*dans la volonté*) would exclude any arbitrariness (*d'où l'arbitraire seroit exclu*). Insight into the highest principle, and only it, could bring the system of knowledge of the world, as a deduction from that principle, to the certainty that would have to exclude all competing perspectives.

At this point in his reflection, which (as it were) compares the ordering of an ideal encyclopedia with the real one that, with this article in the fifth volume, lies before the reader, Diderot introduces a comparison with the Copernican system, which on the one hand exiled the Earth and man to an eccentric and arbitrary standpoint, but on the other hand, precisely by doing so, found the distinctive point of departure for the construction of the system, in the position of the Sun. The ideal system—unattainable for the *Encyclopedia*—"in which one would descend from that first eternal being to all the beings that have issued from its womb in the course of time, would resemble the astronomical hypothesis in which the philosopher transports himself, in thought, to the center of the Sun, so as, from there, to calculate the phenomena of the heavenly bodies around him." [57] Now Diderot not only has the objection that the price for this insurpassable encyclopedia of the true system would be too high, with its connection to theology, because the work thus begun would have to fall short of its claim to universality for all men and all ages; more than that, he radicalizes the issue by questioning even whether the ideal system that is brought forward in the thought experiment would have an arguable advantage.

This is the decisive point in his reflection: Would the universal system, removed from every possibility of perspectival arbitrariness, and which we shall never possess, really be a gain for man? The question is unusual because it at least admits the possibility that the privileged standpoint for theory might not be the distinctively quali-

fied locus for intuition. That is, Diderot's idea is that perfected theory
would represent a sort of doubling of reality; to study an encyclopedia
that contained the presentation of this theory would be equivalent, in
extent and in effect, to the exploration of the universe itself, but by the
same token would also impede that exploration. Here, in Diderot, the
old metaphor of the "book of nature," which has been recast again
and again since Augustine, has found its strangest consequence. The
nature that is finally made legible[g] in the form of the great book of
a definitive encyclopedia would be the overwhelming surrogate for
the intuition of nature itself.[58] Diderot assumes that a finite being's interest in the totality, however favorable the conditions of access to it
might be, would not be uniformly distributed, that is, that the attention devoted to the encyclopedia would not follow the presumed
'systematics of nature' any more than it does in its allocation among
the objects themselves.

The great book, too, is something that we could only partially read
and understand. Then impatience and curiosity, which do rule us and
which already so easily interrupt the course of our investigation of
nature, would also bring disorder into the reading of that great
encyclopedia. Thus the state of our knowledge would be just as
contingent and fragmentary, in spite of our secure possession of the
'natural system,' as it already is without this. The presentation of the
universal system could not be more perfect than this system itself, but
the use that we could make of it would differ only in the fact that it
would have to produce knowledge without intuition.

At first sight one might think that this rejection of even the possibility of a distinctively qualified systematic standpoint for human
theory was based on the suspicion that any attempt of that sort would
open the door to theology or to metaphysical speculation. When one
takes a second look, however, it becomes clear that the theoretical
optimum has no allure for the encyclopedist above all because that
volume immense [immense tome] would have to stifle any aesthetic
quality in the human relation to the world, a quality that for Diderot
continues to be bound to the perspectival situation of the human
subject. A god's view of the world is no temptation for him, because
only the existence of man in the world gives it a dignity that cannot be
reproduced objectively: "C'est la présence de l'homme qui rend
l'existence des êtres intéressante . . . . " [It is the presence of man that
makes the existence of things interesting . . . .]

This man, as Diderot sees him, who cannot make his theoretical patience prevail against his theoretical curiosity, so as to follow the chain of inductions 'systematically,' and who thus always himself violates the program that he gave himself in virtue of his curiosity,[h] does not endure the consequences of his premises, that is, does not endure the world from which the intensity of subjectivity (in which it presents itself, even if it is dimmed by objectivity) is excluded. "If one banishes man, or the thinking and contemplating being, from the surface of the Earth, then this stirring and sublime spectacle of nature is nothing but a sad and mute scene. The universe falls silent; silence and night take possession of it." Everything is transformed into *une vaste solitude* in which phenomena run their course obscurely and dully. Here we are already close to Kant's saying in the *Critique of Judgment* that "without man, the whole creation would be a mere desert, in vain and without purpose." For the editor of the *Encyclopedia* this insight means initially only that man must not only be a topic for the work as a creature in nature, but that he cannot be omitted from the *Encyclopedia* in regard to his subjectivity and perspectivity either. "Why should we not introduce man into our work, as he was installed in the universe?" But just what man cannot be in the world after Copernicus, and what in Diderot's opinion he should not even want to be—namely, the real center of the system of reality, which is the immovable fixedness of his standpoint—he can be in the *Encyclopedia*: "Pourquoi n'en ferons-nous pas un centre commun?" [Why not make him a common center?]

Now the *Encyclopedia*'s anthropocentrism has a very remarkable effect, one that is almost paradoxical in relation to its presuppositions. The retraction of the claim to universality, in favor of man as the measure (*à ce qui convient à notre condition d'homme* [of what accords with our human condition]), aestheticizes not primarily the intuition of nature, but rather the relation to the *Encyclopedia* itself: It becomes the landscape on which the light of the Enlightenment falls. "A universal dictionary of the sciences and the arts must be regarded as an immense countryside, covered with mountains, plains, rocks, waters, forests, animals, and everything that makes up a great landscape." Now it is part of this metaphorics of landscape that the light of the heavens lies on everything, in other words, that the Enlightenment has lost its 'spotlight' character, the loneliness of light in the darkness, the quality of shining out from an obscure background, so that the impression of atmospheric brightness arises: "La lumière du ciel les éclaire tous; mais ils en sont tous frappés diversement" [The light of the heavens brightens them all; but they are all struck differently by it.][59]

To keep in mind the temporal comparison between [the various] Enlightenments, which is always interesting, and at the same time the state of our topic in relation to that, it is worth noting that Diderot wrote this article in the year 1755. In precisely this same year, Kant's *Universal Natural History and Theory of the Heavens* appears.

The Nonsimultaneity of the Simultaneous

Kant's early book on cosmogony struggles to conceal the difficulties that result from the new dimension of time for the intuition of the starry heavens as the (from then on) accidental simultaneity of its elements, once this simultaneity had already begun, as a result of the discovery of the finite velocity of light, to become a foreground appearance. The ideal of theoretical immediacy is bound up with the preeminence of the perfected and (in its perfection) entirely present form [*Gestalt*]; it furnishes one with the whole in the present moment. Since antiquity the starry heavens were the appearance of the whole, the cosmos itself—excluding, above all, any possibility that what is invisible could be important, and making the actual moment in time a matter of indifference for the comprehension of this totality. The modern age's cosmogony has to abandon definitively these advantages of the *contemplator caeli* [contemplator of the heavens]; to that extent it too is a logical consequence of Copernicanism in regard to the observer's actual position in time. For the earlier atomistic-mechanical cosmogony of Lucretius, and even for that of Descartes, the observer's advantage that accords with the ancient theoretical ideal was, of course, preserved by the fact that these cosmogonies had, quite as a matter of course, regarded the moment of the human presence in the world as the conclusion of cosmogony (either the cosmogony of this one cosmos among the infinitely many ones, or even that of the universe as a whole); so that in spite of the idea of development, it was always the *eidos* [form] of the finished world that presented itself to the observer. Kant could not completely maintain this privilege of theory, but he located man, in the infinity of the universe, in a zone of relative completion, on the narrow edge between coming into being and decay.[a] The result, admittedly, was that the foreground of a 'finished world' became the appearance of an evolution that flows toward and terminates in the world in which man exists. The [physical] possibility of man's life, in a zone of consolidated nature, coincides with his need for intuition of a reality that presents itself, in stable forms, as reliable. Perhaps man as a cosmogonic episode, here, was already bearable for Kant only thanks to the idea that the immortality of the soul could be connected with its moving through space to other heavenly bodies.[b]

To begin with it seems to me necessary to glance forward across four decades to the late Kant and to become aware of the remarkable and instructive circumstance that that daring thought experiment of Poincaré—which is formulable as the general question, what could we know of the heavens without ever having seen them—was already outlined by Kant, after the Nominalist, Nicole Oresme, had for the first time stepped back to consider what we would know of the world if we could not see the heavens ("Suspose que nul ne veist le ciel ... ").

In his polemic against a contemporary neoplatonism, "Von einem neuerdings erhobenen vornehmen Ton in der Philosophie" [Regarding a Tone of Superiority That Has Recently Arisen in Philosophy], in 1796, Kant rejected the myth of the immediacy of intuition. To want, by means of the instantaneous grasp achieved by an intuitive act, to avoid the troublesome process of connecting intuition to concepts, or even to *promise* such a possibility, appears to him as the impudence of an enthusiast. The consequence of such presumption would have to be contempt for empirical intuition, as opposed to a higher kind. One finds oneself reminded of Plato's demand for a second, and real, astronomy.[c] Now, the false superiority of the tone that is struck in such claims consists in the fact that the possession of the capacity required for intuition on this level can only be asserted, but cannot be imparted or made testable. It is a question of a relation to reality that lacks the constitutive element of intersubjectivity— that is, it is a question, for example, of a starry sky that would no longer be the most public of all objects. Kant had already written to Johann Georg Hamann on 6 April 1774, with regard to Herder's *Älteste Urkunde des Menschengeschlechts* [*The Oldest Document of the Human Race*], requesting some information to aid him in comprehending the cosmic symbolism of the work, "but, if possible, in human language. For I, poor mortal, am not at all organized for the divine language of intuitive [*anschauend*] reason."

The divine language of intuitive reason and the tone of superiority that has recently arisen are both derived from an absolutism of intuition which characterizes, for Kant, above all a "philosophy in which one does not have to work." This strong statement does not go wide of the facts if only because it points back to the ancient ideal of the combination of theory and leisure, of intuition and a position of repose in the center of the universe, as the origin of the absolutism of intuitive reason. If Kant had ever made use of the Thales anecdote, then he would doubtless have attributed the first Greek philosopher's tumble into the spring to an offense against "reason's law that one must acquire a possession through work." Investigating the starry sky while taking a walk would undoubtedly have appeared to him as a prototypical expression of the temptation to pursue theory under the easiest conditions of mere intuition; the 'first astronomy' of empirical intuition would then have been no better than the 'second astronomy' of pure intuition.

What Kant reproaches the "would-be philosophers" with is their claim to a cognitive faculty with which "they elevate themselves above others of their guild and violate the inalienable right of the latter to freedom and equality, in matters of reason." Plato, of course—whom Kant separates, as the Plato of the Academy, from the letter writer—is not to blame for this philosophy, "for he used his intellectual intuitions only in a backward direction, to explain the possibility of a synthetic cognition a priori, not forward, to expand it by means of those ideas that can be read from the divine mind." To be sure, the author of the "Seventh Letter" became, through it, "the father of all enthusiasm in philosophy," while Aristotle receives the contrasting praise that his philosophy was, "on the contrary, work." Now, precisely in connection with astronomy, this separation between the Plato of the doctrine of anamnesis and the author of the "Seventh Letter" cannot be carried out as cleanly as the opponents of the letter's authenticity would have to want it to be. For the demand for a second astronomy based on pure intuition is in fact precisely the transgression of the boundary at which the "backward" use of intellectual intuition had been abandoned.

It is the limit that Idealism only needs to mark with aesthetic disapproval of the starry heavens as soon as the interweaving of fact and necessity is sought less in nature than in history. For the offensiveness of the starry heavens for the expectation of intuitive necessity lies precisely in the fact that while they do represent the extreme object of intuition as it proceeds toward totality, at the same time they make one miss the appearance of reason, in the acute contingency of the distribution of their elements.

Kant had become acquainted with Plato's "Seventh Letter" through the treatise by J. G. Schlosser, *Platos Briefe über die syrakusanische Staatsrevolution* [*Plato's Letters on the Revolution in Syracuse*]. From this book by the author who, though Kant never mentions him by name, is the target of his polemic, Kant quotes a peculiar version of the simile (which has been varied again and again, since Plato) of the Sun: "All human philosophy can only sketch the blush of dawn; the Sun must be an object of presentiment [*muss geahnet werden*]." If the simile is meant to make clear the relation of conceptual philosophy to intellectual intuition, as a special faculty of presentiment, then—to put it in a simile—it gives this higher intuition a black eye: When one sees the blush of dawn, one *infers* the Sun, because prior experience has stabilized the expectation that the reddening of the sky will be associated with the sun's rising. There is nothing to have a "presentiment" of here if one has seen nothing—that is, there is nothing that one has not seen. Kant uses the failed simile in developing his early form of Poincaré's thought experiment. On the Earth, the regular succession of day and night could very well appear, and man's life could be regulated by the change of times of day and seasons of the year, "without one's ever getting a chance to see the Sun, on account of the sky being continually overcast." Then too, and precisely then, there would be nothing of which to have a presentiment, nothing for that intellectual intuition to do; one would only have to make use of the concept of causality. Here "a true philosopher would not, indeed, have a presentiment of a Sun (for that is not his business), but would perhaps be able to guess its existence, after all, so as—by assuming the hypothesis of such a heavenly body—to explain that phenomenon; and he might in fact be able to hit upon it in that way, with luck."

The passage reminds one of the Socrates in Plato's *Phaedo*, who, dazzled, turns away from the intuition of nature and toward the *logoi* [concepts, Ideas], if only because Kant lets his derision of the Sun simile in the mouth of the "anticritical philosopher" pass into a Sun simile of his own, which comes very close to that Socratic turning from intuition to concepts: "It is true that it is not possible to look into the Sun (the supersensible) without being blinded; but to see it in the reflection (of the reason that morally illuminates the soul), and even to an extent that is sufficient for practical purposes—as the older Plato did—is quite feasible; whereas the neoplatonists 'certainly give us only a theater-Sun' . . . . ''

The reason why this text of the late Kant is so instructive is that it throws a sobering light on the much-invoked linkage of the starry heavens with the moral consciousness.[d] In relation to this formula, the thought experiment with the obstructed heavens implies, after all, a relativization: The steller phenomenon becomes contingent because it can be replaced by a causal hypothesis, while the immediacy of the imperative consciousness remains absolutely irreplaceable. Intuition continues to be threatened by the imputation that it is possible for it to change into reason itself; that defines the change that had taken place, for Kant, in regard to the intuition of the starry heavens, since his early cosmology and, indeed, as a result of it. What degree of consistency with the mechanistic principle—one must ask, from the perspective of his late anti-Platonism—did the early idea have, in the "Conclusion" of the work of 1755, that the nocturnal view of the starry heavens only contains a promise of increasing pleasures, still in the future? Did this view really have, for him, to some degree, the kind of promise that is contained in a merely preliminary acquaintance with something that would require closer proximity in order to satisfy the desire for knowledge better, "in a new relation to the whole of nature"?[e] Or are those only the obligatory formulas with which the Enlightener clothes an unbeliever's longing for the possibility of immortality?

To think of the faculty of intuition that belongs to immortal beings as being transferred to other heavenly bodies means, in Kant's cosmogony, to soften the (fundamentally post-Copernican) contradiction that while this productive universe does bring forth man in the center of its formative powers—and thus also in the center of the part of it that is formed and deserving of intuition—still, nevertheless (and for that very reason), it cannot keep ready any optical standpoint that is especially appropriate for the whole and for the riches of its phases of development. The *Critique of Practical Reason*'s doctrine of postulates does not yet exist,[f] and this makes it doubtful whether the cosmogonic hypotheses really harmonize with the moral requirements in the way that Kant says, namely, that "the view of a starry sky on a calm night furnishes a kind of pleasure that only noble souls are sensible of." One perceives how much resignation will be bound up with the claim to turn reason critically against itself, so as to be completely in earnest in one's Copernicanism.

The opposite position to the harmonization of the view of the heavens and the quality of the soul, in Kant's early work, is the more difficult thought that it was the increase in astronomical knowledge that intensified the intuition of the starry heavens to the degree of delight and dread, of rapture and annihilation, that defines a hitherto unknown ambivalence of experience, in its ability to turn suddenly, at any moment, from admiration to contempt.

If one leafs backward from the "Conclusion" in the *Natural History and Theory of the Heavens*, then the permeation of intuition by the knowledge of the chaotic background, and of the simultaneous existence of natural formations that are still below the higher level of orderliness, comes into threatening relief. Cosmogony has created a linear systematization in time, in relation to which every configuration of spatiotemporal simultaneity carries the mark of arbitrariness in the choice of a cross section. What Copernicus had examined and reformed—the solar system—turned out to be the result of a process that one can only understand when one includes the background of its story.

At the beginning of the first part of the *Natural History and Theory of the Heavens*, Kant does not mention Copernicus by name. But it is manifest that the elimination of the appearance of the daily and annual motions of the heavens has made cosmic intuition uncertain insofar as the connection between the fixed stars, in their periodic motions, could no longer bring about an experience of order, as soon as it had been shown to be a mere appearance. While, for the world of the planets, a rational system seemed to have been rescued by Copernicus, the world of the fixed stars was robbed of its unity as a result of this rescue, and was convicted of having an accidental order, so that it could seem "as if the regulated relation, which is found in the smaller solar system, did not rule among the members of the universe as a whole." If Copernicus had succeeded in correcting intuition, the latter was disturbed, henceforth, not only by its capacity for illusion with regard to the reality of motions, but also by the loss of the rationality that had for so long been perceived in the simultaneity of the motions of the fixed stars: "The fixed stars exhibited no law by which their positions were bounded in relation to each other; and they were looked upon as filling all the heavens and the heaven of heavens without order and without intention."

With this important observation Kant reveals how the task of a cosmogony stands, for him, in the context of working out the implications of Copernicanism: Only by adding the dimension of time could the difference in rationality between the planetary system and the world of the fixed stars be reconciled. One must picture to oneself how post-Copernican astronomy had displaced the ancient experiential qualities of the sublime and of the cosmos's worthiness of admiration; the exponentially increasing amounts of quantitative information had to compensate for the impoverishment of intuition as far as accessible order was concerned. It is clear that Kant intended to oppose the increasingly pressing suspicion of contingency with the aid of a different kind of theory, namely, a theory of the cosmic process's form of temporal unfolding.

To that extent, the idyllic picture of the calm, morally fortified observer of the heavens, at the close of this early work, presupposes the success of the theoretical process of the new cosmogony, as having deprived its original situation of its bite. In the seventh chapter of the second part, the suspicion of the absence of order, as the background that has in the meantime been mastered and cleared up by means of the new-found lawfulness of the origin of the universe from homogeneous prior conditions, is still present when Kant says, "All the fixed stars which the eye discovers in the hollow depths of the heavens, and which seem to display a sort of prodigality, are suns and centers of similar systems."

The transfer of cosmic rationality into the dimension of time has created a model that no longer permits any satisfying standpoint for intuition, but has to turn everything over to constructive theory. That prevents the idea of immortality, and its connection with cosmic transpositions, from continuing to have any consoling quality; there no longer exists any hope that any station on the path of the immortals could furnish a higher satisfaction, thanks to the optics of a standpoint that they gain. The example that Kant chooses—that we would perhaps someday be able, on Jupiter, to take pleasure in the fourfold motion of its moons—makes this entirely clear. For the fourfold multiplication of the familiar satellite of the Earth cannot alter one bit the contingency of what is intuitively given to someone on Jupiter.

Here, as elsewhere, Kant helped himself by distributing to different faculties what he could not unite. He put down as the accomplishment of the understanding the success that the power of imagination could no longer provide except in the quantitative realm: "The universe, by its immeasurable greatness and the infinite variety and beauty that shine from it on all sides, fills us with silent wonder. If the presentation of all this perfection moves the imagination, the understanding is seized by another kind of rapture when, from another point of view, it considers how such magnificence and such greatness can flow from a single law, with an eternal and perfect order." The moral quality of the observer is not required in order to accomplish something that can, after all, never emerge from the greatness and magnificence that Kant praises in the simultaneous view of the starry heavens. Instead, what is required in order to satisfy a claim that is no more fulfillable on Jupiter, or anywhere else, than it is here on Earth, is construction based on laws operating in time.

Very much later, after the completion of the *Critique of Judgment*, Kant will formulate the result of this last *Critique* as follows: He had shown "that without moral feeling, there would be, for us, nothing beautiful and sublime: that this, precisely, is the basis of the (as it were) lawful claim to approval that belongs to everything that ought to bear this name . . . ." [60] When, at the end of the *Critique of Judgment*, Kant investigates the question why the proof of God's existence that proceeds from nature's serving a purpose is, in spite of its lack of validity, the only one that could gain "the powerful influence which this proof exerts upon the mind, and exerts especially on a calm and perfectly voluntary assent arising from the cool judgment of reason," then this is due, for him, to the connection between the aesthetic intuitability of nature and the moral quality of its observer.

For reason, physical purposiveness [*Zweckmässigkeit*, teleology] as such cannot be fully satisfying as long as the question of a purpose that does not, in its turn, serve a purpose has to remain open—and precisely such a purpose is never found in looking at nature, nor is it found in the nature of man, who would have to permit himself to be interrogated as to the purpose of his existence. If each element in the whole has its purpose, what is then the purpose of the whole? This need to continue inquiring can only be put at ease by something that is an end in itself. "To suggest that it was made for enjoyment, or to be gazed at, surveyed and admired—which if the matter ends there, amounts to no more than enjoyment of a particular kind—as though enjoyment was the ultimate and final end of the presence here of the world and of man himself, cannot satisfy reason. For a personal worth, which man can only give to himself, is presupposed by reason, as the sole condition upon which he and his existence can be a final end."[61]

So the meaning of the naming of the starry heavens and the moral consciousness in one breath follows from the fact that the senselessness of every question about the purpose of nature can only be removed by pointing to the existence of the moral consciousness. The disappearance of the aesthetic qualities of the *contemplatio caeli* [contemplation of the heavens], as a result of Copernican and Newtonian science, opens the horizon within which a connection can be created between cosmogonic construction and human self-consciousness. The loss of the totality of the world-concept, which had taken place, for Kant—in the face of the supposed defenses in his early work—as a result of his transcendental critique of reason, could only be gotten over if the person, as an end in himself, stands over against the empirical fragments of this totality as the answer to the question of the meaning of the unattainable whole. This observer is no longer dependent on deciphering his topographic position in the world, and nature is freed from the compulsion to give man signals for his self-comprehension.

Kant saw the curtailment of man's expectations of finding meanings for himself in nature as it appears to him as analogous to the curtailments of the pretensions of reason by transcendental criticism. At the beginning of the thoughtful "Conclusion" that he gave to the *Critique of Practical Reason* there is not only the famous formula "the starry heavens above me and the moral law within me" but also a less noted remark that for the first time makes clear what this formula covers in the history of reason. Kant speaks of the immanent threat to the contemplation of the heavens if it does not receive its critical curtailment, too, through the concept of the moral person as an end in himself. Here he says, "The observation of the world began from the noblest spectacle that was ever placed before the human senses and that our understanding can bear to follow in its vast expanse, and it ended in—astrology."

Astrology is compared, in its relation to the intuition of the heavens, with reason in its pure use, when it goes beyond experience. Grasping at the phenomena hastily, astrology relates the universe to man, makes it the sum of signs for him, and thus makes him the reference point of all physical processes. As the first *Critique* points out the limits to the precipitance of reason, so the second does to the precipitance of intuition, since moral consciousness makes man its end only in his autonomous self-constitution. The result is that every external definition of an end, by which man wants to provide himself with an intuitive confirmation of the anthropocentric meaning of the world, against the "view of a countless multitude of worlds" (of which it is said, in the well-known statement, that it "annihilates, as it were, my importance")—every such external definition is rejected. This direct [*unmittelbar*] route from the intuition of the starry heavens to [man's] self-consciousness no longer exists—although Kant seems, in his language, to retain the feature of immediacy [*Unmittelbarkeit*] for the correspondence that he describes, when he can in fact say of that which is above me and that which is within me (which fill "the mind with ever new and increasing admiration and awe") that "I see them before me and I associate them directly [*unmittelbar*] with the consciousness of my own existence." True, the observation of the starry heavens "begins at the place I occupy in the external world of sense, and broadens the connection in which I stand into an unbounded magnitude of worlds beyond worlds and systems of systems," but only the moral law "exhibits me in a world which has true infinity but which is comprehensible only to the understanding—a world with which I recognize myself as existing in a universal and necessary (and not only, as in the first case, contingent) connection, and thereby also in connection with all those visible worlds." [62]

"The sublime is the infinitely great, to which we draw ever nearer in our thoughts, but which we can never fully attain." [63] The aesthetic quality of the starry heavens, by means of which the finite power of the imagination can enter into a relation with the rational accomplishment of the idea of a whole, whereas concepts fail when confronted with the demand of totality, is directed at saving the human meaning of nature after the end of anthropocentrism. At the same time, however, it reveals the dilemma that is the point of departure of Idealism's negative evaluation of nature. The universe's reality is experienced as resistance to concepts. It is the impossibility of complying with the demand that is given in the compulsion of the Idea,[g] in the ambiguity of "negative pleasure" [64] as independence from the conditions that are set for the conceptual grasping [*Begreifen*] of objects in nature and at the same time as the incapacity of concepts [*des Begriffs*] to comply with this demand. If one briefly defines the sublime as the symbolic presence of the infinite in the finite, then it still continues—right into the latest phase of Kant's thought—to contain the problem from which the early cosmogonical speculation started: The insufficiency of the intuitive presence of the universe, at any given time, to the concept became the occasion for the construction of the history of the universe as the dimension in which totality is conceivable.

This actualization of history in the form of the cosmogony of 1755 is still—is again—an answer to a question that had found its most precise expression in the anecdote about Alfonso the Wise, of Castile: The reproach that he would have been able to provide more reasonable advice for the construction of the universe, if he had only been present at the Creation, expresses in an image the more abstract thesis that the only reason why the totality of nature cannot be conceptualized is that it did not originate in concepts. Confidence in one's capacity to generate, originally, a concept of the universe from which a permanently intelligible reality could have arisen is, in view of the poverty of human imagination, an amazing assumption. The cosmogony confirms the suspicion that nature and concept came together only late and only partially. The upshot of this confirmation inevitably favors the exclusive and exhaustive dominion of the concept in history. One can understand the avoidance of a decision between these two alternatives as a pervasive and homogeneous motive of Kant's thought.

Into the modern age, the observer of the heavens had believed that he could put himself into a direct relation to the totality of the universe, the totality that had been unquestioned in the ancient and medieval relation to nature. The Copernican break introduced a world optics that not only implied the contingent facticity of the human perspective but also, in its logic, terminated in the absolute and irreducible partiality [i.e., nontotality] of experience. True, the predicate of infinity emerges, at the beginning of the modern age, as one of the great worldly emphases, but it loses its fascination as a result of the inescapable consequence that all experience now carries with and around it a background of inaccessibility. Even if science's idea of method had put in the place of the individual who had contemplated the starry heavens as the appearance of the totality of nature a new supersubject of experience extending across space and time, still the hiatus between the objective accumulation of knowledge and the subjective need for intuition and totality—in short, between concept and Idea—becomes continually more oppressive.

Kant made one of the last great attempts not, indeed, to *overcome* this falling apart into the collective possession of theory, on the one hand, and the individual claim to intuition, on the other, but to *justify* it as a positive tension. His transformation of the idea of the sublime must be seen in this context. Unattainability becomes the mode of consciousness of the Idea, which can only be actualized in the contradiction between the finitude of experience and the infinity of the experienceable. In the post-Copernican universe, man does not find his interests confirmed in nature; precisely this turns in Kant into the aesthetic quality, insofar as we are to accept as "sublime" only "what pleases immediately by reason of its opposition to the interest of the senses." [65]

For its observer, the universe satisfies the absurd need to have it violate his interest in being the central point and the point of reference, and nevertheless force his assent. The aesthetic preserve to which the contemplator of the heavens is assigned requires a peculiar historical 'extraterritoriality,' a sort of reduction of the stock of theory that goes into and permeates intuition: "So, if we call the sight of the starry heaven *sublime*, we must not found our estimate of it upon any concepts of worlds inhabited by rational beings, with the bright spots, which we see filling the space above us, as their suns moving in orbits prescribed for them with the wisest regard for ends. But we must take it, just as it strikes the eye, as a broad and all-embracing canopy: and it is merely under such a representation that we may posit the sublimity which the pure aesthetic judgment ascribes to this object."

It is part of the logic of Copernicanism that the *contemplator caeli*, if he wants to remain this, must have forgotten Copernicus just as he must have forgotten the new distances measured in light-years. Is this reduction of cosmology to the state of innocence of a pretheoretical life-world, is this aesthetic Rousseauism of Kant's the reason for his disinclination ever to permit a reprinting of his *Universal Natural History and Theory of the Heavens*? Of course, readers of the *Critique of Judgment* will hardly have read that early work, which perished in its publisher's bankruptcy before it was distributed. So one cannot expect that beyond the third *Critique*'s problem of mediating between theoretical and moral reason, they had before them Kant's own opening of the dimension of history for nature when they read that "a feeling for the sublime in nature is hardly thinkable unless in association with an attitude of mind resembling the moral." Just so, indeed,

Kant recommends that the observer of the ocean should not regard it as a factor influencing the weather or as a medium of world trade, but rather should admit nothing that is "not contained in the immediate intuition." It is the poets, Kant thinks, who view such an object "merely according to what the impression upon the eye reveals."

5

The View of the Heavens and Self-Consciousness

For Idealism's negative assessment of nature as that which is finished and is nevertheless contingent 'fact,'[a] Kant's early attempt at portraying the incomplete universe was in vain: "The creation is never finished or complete. It has indeed once begun, but it will never cease. It is always busy producing new scenes of nature, new objects, and new worlds."[b] This new *creatio continua* [continuous creation] is always performed by nature, never by man.

It is almost palpable how the cosmogonic total conception had to become a metaphor in Idealism. From the early Friedrich Schlegel, we have a fragment from the Jena *Transcendental Philosophy* on the thesis "that the world is still incomplete": "This proposition, that the world is still incomplete, is extraordinarily important in every respect. If we think of the world as complete, than all our doings are nothing. But if we know that the world is incomplete, then no doubt our vocation is to cooperate in completing it. Experience is thus given an infinite latitude for variation. If the world were complete, then there would only be knowledge of it, but no action."[66]

The "incomplete world" legitimizes man's demiurgic inclination; as if this unfinished quality were also a decrease in determinism, it appears to Schlegel, in the *Philosophy of Life*, as a correlate of human freedom: "Man is free, but nature, or the world of the senses, the material Creation, is quite unfinished, thoroughly incomplete."[67] And in the *Philosophical Lectures from the Years 1804–1806* we read, "Only if the world is thought of as in the process of becoming, as approaching its completion through an ascending development, is freedom possible."[68] The idea of the unfinished state of the world is combined, in Schlegel, with the Romantic predilection for organic metaphors. That is a remarkable incompatability, since the organic system does not have the character of a substrate; that is, it does not tolerate artificial intervention: "If the world is thought of as a series of necessary laws, then predestination is inevitable. It is entirely different in our theory, where the world is an organism, a nature. We do want our action to succeed, we want something to emerge from it, we do not want everything to be already foreclosed; but the mechanistic system prevents this. Our point of view also supports the importance of the moment and of the present in general." [69] Schlegel tried to reduce, again, the acute lack of determination that belongs to the idea of the incomplete world, by means of an overarching teleology of man's role in the world: "Even though the end, like the beginning, of human history is supernatural and mystical, nevertheless it remains philosophically certain that man's strength and active cooperation are counted on as necessary to the true completion of the world." [70]

If one wants to make intelligible the changes that are characteristic of the Cologne lectures of the years 1804–1806, not much is explained by saying that after the Jena lecture on transcendental philosophy Friedrich Schlegel freed himself from Fichte's premises and thus no longer needed to abide by the correspondence between active Ego and incomplete world. Rather, it is necessary to observe the resistances and contradictions that inevitably had to arise in carrying out the original idea of the "incomplete world"; they determine the Romantic modification of Idealism. It is, on the one hand, the resistance that the concept of organism opposes to all demiurgic constructiveness with which Romanticism defines a good part of its difference from the outcome of the Enlightenment. But the distance of the part of nature that is absolutely inaccessible to demiurgic treatment also stands in the way of the idea that the cosmogonic substratum of the worlds that are still in the process of becoming has something to do with man's claim to creativity. In the outcome of this intellectual process, the starry heavens can present their ambiguity on a new level: As that which has asserted its unavailability, in the universal materialization,[c] they are at the same time that which convicts man anew of his finitude and that which, in his inevitable resignation, either mocks him or presents itself to him as the mere foreground of a metaphysical transcendence.

Romanticism, as it presents itself in Friedrich Schlegel's Cologne lectures, is, in its relation to nature, a reduced form of Idealism: The foreignness of the heavenly bodies points to a background of hopes that are of a different kind from those that could be realized by human action. "The dispassionate mind will have to perceive, in the starry heavens, the reign of peace that it fails to find anywhere in the stormy, conflict-filled life of the earthly element. Therefore the contemplation of the pure, serene brightness of the silent peaceful clarity of those heavenly lights also fills the heart of the observer with that higher longing that elevates him far beyond the noisy confusion of the world, to the dwellings of eternal peace." [71] Now, the observer of the heavens can apparently ignore, after all, what he has known for at least half a century—the pitiless mechanism of cosmogony, this "great, sublime story" whose "struggles" seem to have come to an end, for the earthly onlooker, in view of the "peaceful state" that he has before his eyes. What the cosmogonists of the previous century had become conscious of as the mere appearance induced by an all too brief present period of observation, and which had to be seen through, is anthropocentrically revalued again with the aid of the premise that it is not an arbitrary moment in the infinite process. Instead, it is its successful result, in "this fullness of peace and magnificence," which justifies the mythical cosmogonies in speaking of "many millennia, in which the world was gradually formed . . . since only an immense length of time could bring about such an incalculable sequence of developments and revolutions."

It looks as though in Cologne, Schlegel, who was approaching the date of his conversion,[d] had again discovered the old Aristotelian/ Scholastic hiatus between the stellar and the terrestrial world, and thus restricted the characteristic of incompleteness to the "imperfection of the earthly element," which is now no longer metaphysically final since "it must in the end be possible to overcome it, if the world is to be completed." [72] This incompleteness—remaining, as a result of the difference of elements—of the world insofar as it is already, and only, 'our world,' no longer requires the active Ego that carries the imprint of Fichte, because it has been redefined as an organic condition: The organic is, preformatively or epigenetically, already what it will become, and becomes pathological as soon as it is only supposed to be a substrate to be acted upon. To say that "organization" is the "foundamental characteristic of the Earth" and that here the usually asserted contrast between the organic and the mechanical and chemical no longer exists, means—anticipating Johann Wilhelm

Ritter's *Fragmente aus dem Nachlass eines jungen Physikers* [*Fragments from the Literary Remains of a Young Physicist*], of 1810—to designate the center of Romanticism in the philosophy of nature (and not only of this but of every subsequent Romanticism) as the portentous "productive power of the Earth."[73] This interests us here only insofar as it makes man into one who waits, who wants to assure himself of powers that he cannot possess, and makes him look up to the starry heavens as the assurance of a completion that awaits him. Besides self-detachment and self-preservation, "organization" is also self-presentation; one might think that this principle is most perfectly accomplished in the appearance of the starry heavens, if Schlegel did not insist—in the interest of the already mentioned transcendence—that it is not permissible "to want to introduce organic laws into the whole of nature."[74]

It is a new tone, which will continue through the whole century, when the poet Heine indicts the indifference and muteness of the stars toward man, their coldness and refusal of any answer. Here we get only a foreground explanation of the change in the aspect of the heavens when it is described as depending on youth and age, as Heine describes it in the *Memoirs of Herr von Schnabelewopski*. It is the night sky over Hamburg that the "I" of the memoirs takes into his youthful enthusiasm for the world: "... till the heaven grew dark and the golden stars came forth yearning, hope-giving, wondrously and beautifully tender and transformed. The stars! Are they golden flowers on the bridal bosom of heaven? Are they the eyes of enamored angels, who with yearning mirror themselves in the blue streams of earth below and vie with the swans?"[75] That was, he says, long ago; the world has changed, and with it the meaning of the heavens. Twelve years later, and it again becomes dark and the stars come out brightly, "the same stars who once so warm with love wooed the swans on fair summer nights, but who now looked down with frosty brilliancy, and almost scornfully on them. Ah! I now perceive that the stars are no living, sympathetic beings, but only gleaming phantasms of night, eternal delusions in a dreamed heaven—mere golden lies in a dark blue Nothingness."[76] Could twelve years' advance in age transform the charm of the view of the heavens into its opposite in such a way?

If the plan for the *Memoirs* goes back to Heine's years in Berlin, from 1822 to 1824, then one need not regard as far-fetched a connection between it and a key experience in Heine's evolution during this period, an experience that is connected with the name of Hegel. In his *Confessions*, of 1854, Heine tells of a nocturnal, almost unreal encounter with the Berlin philosopher, whose influence he views, from this late perspective, with the same distance he exhibits toward his entire spiritual biography. "Frankly I seldom understood him, and only by mulling over his words did I manage to grasp them. I believe he didn't *want* anyone to follow him, whence his labyrinthine lectures . . . . The baroque quality of his expressions often struck me, and many of them have remained in my memory."[e] The late Heine describes Hegel's effect almost scornfully as one of inciting a frivolous apotheosis of his own self: "I was young and proud, and it gratified my self-esteem to learn from Hegel that, contrary to what my grandmother thought, it wasn't the Lord in heaven, but I myself here on earth who was God."

The account of the nocturnal scene of the encounter with Hegel reads like the final word on Kant's attempt yet once again to maintain an underlying connection between the starry heavens and the moral law. One evening they had stood side by side at a window, Heine remembers, and he, the twenty-two-year-old, had spoken enthusiastically of the stars, calling them the abode of the blessed. "The Master, however, grumbled to himself, 'The stars, harrumph, the stars are only a gleaming leprosy on the sky.'" The younger person indignantly disagreed and asked whether, in that case, virtue would not be rewarded up there after death. "But he, staring vacantly at me with his pale eyes, caustically replied, 'So, you want a bonus for nursing your sick mother and for not having poisoned your brother?'" With these words Hegel had anxiously looked around him and had only composed himself when he saw that another confidant had joined them in order to invite him to join a party at whist.

That was written after the 'conversion' in February 1848, when Heine, according to his own testimony, had come to his senses again during the days of general insanity. By a curious coincidence, we have another, earlier description of this encounter, from the pen of Ferdinand Lassalle, a description that had already been made public at the end of 1845, in the seventh *Proceedings* of the Philosophical Society, and was published by Michelet in his journal, *Der Gedanke*, in 1861. Lassalle claims to have personally heard Heine's account in the year 1840. Here, too, the anecdote is connected to Heine's confession that he understood nothing in Hegel's philosophy, but had nevertheless seen in it "the true spiritual culminating point of the age." This very insight had come to him, we are told, in the nocturnal hour when, during his period of study in Berlin, he paid Hegel a visit. Hegel was occupied, just then, with work; so Heine went to the open window and was seized, in view of the starry heavens, by a romantic disposition toward this expression of divine love and omnipotence. "Suddenly he, who had entirely forgotten where he was, felt a hand laid upon his shoulder, and at the same time heard the words: 'It's not the stars, but what man puts into them—that's the real thing!' He turned around and Hegel stood before him. From this moment on he knew— Heine concluded—that in this man, however impenetrable his doctrine was for Heine, the pulse of the century beat. He never lost the impression of this scene, and whenever he thought of Hegel, he always remembered it." [77]

It is certainly useless, in the face of the difference between the two accounts, to ask what is the historical truth. One can regard them both as different aspects of one and the same set of circumstances, since Heine's own recollection, in the *Confessions* of 1854, is no less indirect and refracted than what the writer in 1845 is able to tell us about what he heard half a decade earlier. Heine's own text emphasizes the aspect of the meaninglessness of the view of the heavens, the false sublimity of the stars and of nature, in which what man confronts is only spirit that is alienated from itself, in its lack of consciousness and historylessness. Lassalle's report puts the accent on man's participation in the meaning of the physical phenomenon, on the projective character of its quality as an experience—what Nietzsche will call anthropomorphism in the relation to nature. Only because nature is the unconscious substrate can it be made serviceable for man's understanding of himself—can its original meaninglessness nevertheless clarify such a thing as the human 'position in the world.'

One cannot fail to notice that Idealism's devaluation of the starry heavens is not only directed against Kant's attempt to bracket them together with the moral consciousness, but also includes the whole connection between astronomy and rationality, as a characteristic feature of the Enlightenment. For this we have, once again, Heine's testimony, which presupposes astronomy's function as an element in Enlightenment pedagogy and in doing so reminds us of many similar statements that document the role of, for example, Fontenelle's dialogue *On the Plurality of Worlds* in people's lives and educations. Heine writes in the third part of *On the History of Religion and Philosophy in Germany*, in connection with his presentation of the philosophy of Kant, about his original religiousness and its early perplexities: "I was quite puzzled over the astronomical lore with which in the 'Enlightenment period' even the youngest children were tormented, and there was no end to my amazement on learning that all those thousand millions of stars were spheres as large and as beautiful as our own earth, and that over all this glittering throng of worlds a single God ruled." [78] Part of the turning away from the Enlightenment's emphasis on astronomy, and particularly from its expectation that reason would be ubiquitously distributed in the universe, is the resigned abandonment of the idea of the plurality of inhabited worlds, the sober acceptance of man's contingent givenness [*Faktizität*][f] in the universe, an acceptance that now has the triumphant undertone of an assertion of his singularity. Among the bizarre nocturnal visions of an overstimulated imagination there is also the waking dream of a flight to the stars, in the second book of [Heine's] *Ludwig Börne*: " . . . and so I flew from one star to another. But they're not inhabited worlds, as other people dream, but only shining stone spheres, desolate and sterile. They don't fall down because they don't know what they could fall on. They float up and down up there, in the utmost perplexity." [79] Nature's disregard for man is only the correlate of its monstrous difficulty as to how to acquire meaning outside man and his history.

When the Greeks invented or discovered the cosmos *and* tragedy, the relation of the gods to the admired world-order remained unclarified and doubtful for them. It was to emerge that Christianity, too, could not overcome this ambivalence—indeed, that it intensified it even further because it had to claim the identical God for the creation *and* for its redemption. How the prefection of the first act could allow the second act's desperate intervention to become necessary was so far from being satisfactorily explicable that the Gnostic separation of the responsibilities for the world and for salvation had to remain the most tempting solution to this radical dilemma.

With the sharp separation of history and nature, which is nothing but an explication of the Cartesian difference between thinking and extended substance, the thought pattern of dualism returns with a new content. At the end of this explication stand man and his history as what was necessary in order for God to attain His self-consciousness and thus the highest level of possible being, in contrast to the unconsciousness and thus nullity of nature. They are what the Necessary One found necessary because it could not "do without the finite," and in that way they even take on its dignity, which opposes nature, "the accidental and the falling, by being alien, indeed by alienating itself, historically." What had incited and provoked postmedieval pantheism, up to the point of a fatal contradiction, as in the case of Giordano Bruno, was the intolerability of the thought that the creation that was produced by an infinite power should be, precisely, what Hegel called the "aggregate of finitude," that which is capable of falling and of accident, and that it could be in need of man and of God's becoming man.

If, following a remark of Karl Löwith, "the arrogant depth of Hegel's contempt for nature" is evidenced by his being able to say that the stars are only a sort of "rash of light, no more admirable than a rash on a person or than the multitude of flies," then this is still the consequence of the concentration of thought on the event of the Incarnation of the biblical God—an event that did not integrate the divinity into the cosmos, but rather broke man out of nature, so as to propel him into [first] an extraworldly, and then an unworldly position, in which, in any case, he could find no satisfaction or contentment (nor could he find his final state) in looking at nature and the starry heavens.[80] Paul Lafargue writes in his *Erinnerungen* [Memoirs], about his father-in-law Karl Marx: "I have often heard him repeat Hegel's dictum: 'Even the criminal thought of a scoundrel is nobler and more sublime than the wonders of the heavens.'"

The affinity between the metaphysics that is concentrated on man and contempt for nature is no longer based on the questionable status of the creation in relation to redemption, but rather on the need for man to be an end in himself, not limited by any natural order, and for matter to be at his economic, technical and aesthetic disposition. The gesture of Prometheus, who forms creatures for himself out of ownerless clay, in the manner of a potter, and cares for them and shields them from the wrath of the nature-God, is at the same time a gesture of scornful turning away from the presence of an alien reality:[81]

Was haben diese Sterne droben
Für ein Recht an mich
Dass sie mich begaffen.

[What right do those stars up there have over me, that they stare at me?]

The demiurge, in the midst of his creatures, at the foot of Olympus, demands a reversal of the old direction of gaze, in which one looked up from below:

Sieh nieder Zeus
Auf meine Welt sie lebt.

[Look down on my world, Zeus. It lives.]

This Prometheus already is what Fichte insisted on being described as, in his defense against the accusation of atheism, in 1799: an "acosmist."[82]

Goethe was alarmed when, from the forgotten youthful fragment, first the ode surfaced again, being drawn into the "Spinoza dispute" by Friedrich Jacobi in 1785, and finally also the two acts of the dramatic fragment, from Lenz's literary remains, in 1819.[g] "Don't let the manuscript become too public," Goethe writes to Zelter on 11 May 1820, "so that it won't appear in print. It would come quite opportunely as a gospel for our revolutionary youth, and the High Commissions at Berlin and Mainz might take a punitive view of my young man's whims." Ten years later Goethe himself still takes the ode, and the fragment into which it was incorporated, into the *Ausgabe letzter Hand* [the final edition prepared under his supervision] of his works. This is half the way to two fragments from Nietzsche's "Dionysus Dithyrambs," which seem to conclude the context of the revolt in favor of acosmicness. The first fragment:[83]

Diese heitere Tiefe!
Was Stern sonst hiess,
zum Flecken wurde es.

[This serene depth! What used to be called a star became a spot.]

And the second:

Trümmer von Sternen:
Aus diesen Trümmern baute ich eine Welt.

[Wreckage of stars: Out of this wreckage, I constructed a world.]

I have reached back and forward so as to make clear what is included within the arc that has to be set up here, an arc that includes the process by which the bracketing together of the view of the heavens and self-consciousness that Kant, once again, attempted, was dissolved. Schelling spoke of the "false imputation of sublimity" that had persisted, from the early *Theory of the Heavens*, in the contemporary influence of the Kantian juxtaposition of the starry heavens and morality. The questionableness of this resonance emerges from the connection that Schelling sets up with the Copernicus analogy in the preface to the second edition of the *Critique of Pure Reason*: "Kant compares idealism to Copernicus's conception." [84] But idealism, in Kant, must always be seen as the inescapable result of the Transcendental Dialectic—as the avoidance of the contradiction in the attempt to grant infinitude (or finitude) to the universe as a real characteristic. The supposed overwhelming greatness [*Übergrösse*] of nature cannot, any longer, have entered into Kant's late dictum about the starry heavens and the moral consciousness, even if it did enter, as Schelling suspects, into the dictum's potential influence. The importance of idealism as eliminating "falsely imputed sublimity" consists, Schelling says, in the fact that it is "the means by which to get rid of the many infinites that are still found, up to the present, in the natural sciences." But a meaning of idealism that Kant would have strictly denied would be the extended one according to which man stands opposite the whole of nature. "He is the universal being, a characteristic that is as little invalidated by his present localization or restriction to one heavenly body as his being distributed over all of them (which is still so widely assumed) would make him the universal being, if he were not that being already, by nature. Man's true home is in the heavens, that is, in the world of ideas, which he shall also attain again and where he shall find his permanent abode." [85]

All the anthropological conclusions at which the Enlightenment thought that it had arrived by interpreting Copernican astronomy are now invalidated precisely by the consequences of Kant's Copernican analogy. Man's localization in the universe is a matter of complete indifference, if he no longer stands within the whole, but rather stands over against it as subject. In the perspective of an Idealist anthropology, the speculation on the plurality of inhabited worlds and of rational world-inhabitants does not alter man's quality as a universal being, however much the manifest reality of his physical presence may contrast with that. The pure feeling of nature's overwhelming greatness is only the first step after the Copernican turning, from which, in the second step, we must derive the admonition not to seek human self-consciousness in the wrong context: "... but the first sensation of the man who feels himself so alien and so distant from that world, but at the same time, however obscurely, is still conscious of his original vocation, might nevertheless be that of the loss of his position in the center—which is only followed by the elevating feeling that this present situation is only a [temporary] condition, and that a new reversal is imminent, an order of things in which justice, that is, the right and true relationship, will be stable and lasting . . . . " The Enlightenment had been in tune with the idea of the plurality of the inhabited worlds because man's contingent givenness [*Faktizität*] was supposed, after all, only to be the price of the necessity of reason in the universe—of the necessity of the reason whose agent (if only partial and provincial) on the heavenly body called "Earth" man, even in his accidental quality, was and remained. The impact of the Copernican disappointment was moderated by the guarantee that the cosmic presence of reason was not a matter of man alone. One can regard Idealism as carrying out this beginning, dealing with the Copernican disillusionment by revaluing it as a condition of the possibility of a new self-consciousness that pushes its eccentric position outward until it becomes an exterior one.

With Schelling, the physical cosmology of the early Kant has become a metaphysical one. His theory of matter, which seeks the mean between Plato and Aristotle, allows him to attribute a special substance to the [nonterrestrial] heavenly bodies again. Inertia, he says, is the original characteristic of matter and the only characteristic of that metaphysical matter, as the *quinta natura* [fifth nature]; inertia is mere resistance to movement as the execution of an idea, self-activity, striving for a goal. Extension, which since Descartes had been the essence of matter and of nature, first arises, here, from the opposition between inertia and self-activity, as a secondary characteristic of matter when it has already emerged from its pure materiality. The stellar phenomena represent something like the transition from that primary materiality to the extension and impenetrability of physical world-bodies this side of the heavenly bodies. Pure matter is not supposed to be something extraworldly, something fully removed from us by transcendence of the kind represented by Ideality[h]; rather, it appears in nature as its demiurgic 'background,' insofar as the universe is and remains incomplete. There are, in this universe, places (and Schelling explicitly considers it possible to prove that they must exist) "where nature must, for the first time, prepare the material for new creations—places where, instead of real extension, there remains, not indeed so-called atoms, but mere potencies for extension."[86]

This speculation about matter is one of Schelling's attempts to deal with the pointlike reality and contingency of the heavenly bodies as they appear. His other attempt is in the context of his critique of Kant's Transcendental Aesthetic. The latter, he says, made space and time into characteristics exclusively of appearances, whereas the Ideal world too must have an intelligible space, and each being must have a place that belongs to it in that space. The postulation of a spatiotemporal structure of Ideality as well makes it possible for Schelling to regard the starry heavens—in spite of their contingent givenness [*Faktizität*], which has been unsatisfying for all idealists since Plato—as an approximation to an Ideal spatiotemporality. "In fact, among all visible things the stars are, in their form of existence, still most like the Ideas, even if they seem, viewed partially, to be composed of corporeal things: That which drives them, the real heavenly body in them presents itself as a purely intelligible thing." [87] Speculative cosmogony makes it possible to picture transitions; intelligibility is defined, from the perspective of impenetrable and extended nature, as a reduced level of contingency, and this distinguishes the stellar nature as a sort of foreground Ideality.

Schelling returns to the eighteenth lecture's concept of matter when, in the twenty-first lecture, he remarks about the stars that they had "preserved their intelligible relation most of all," and that it could even be considered possible "that some of these beings fully preserved their intelligible location, that they have remained in the state of mere metaphysical materiality, and represent, in contrast to the world that has lapsed into accidental and transitory materiality, a sort of immaterial world...." [88] If one asks oneself what such a speculation means in the context of reaching back to Kant's Copernican turning and his double formula of the starry heavens and the moral law, as well as to the exclusion of false imputations of sublimity, then one finds that the direction of the argumentation is against the treatment of infinity as a source of antinomies, in the Transcendental Dialectic of the *Critique of Pure Reason*.

Just as little as Schelling could be pleased by the false sublimity of the cosmogonic speculation, in Kant's early work, according to which nature could be produced by mechanistic processes, no more did he like to accept the outcome of the cosmological antinomy—a degradation of the forms of the universe's appearance to mere determinations of sensibility—if this was only a means of 'getting around' the uneliminability of the antinomy. The dialectic of reason is dissolved into the metaphysical process. "Would this perhaps be the simplest means by which to settle the dispute regarding the unboundedness or boundedness of the universe, which Kant sought to present as a conflict of reason with itself—just as the same contradiction in regard to time can only be settled in an analogous manner?" The contradiction takes place, for Schelling, not in reason but in nature itself, in the metaphysical cosmogony that makes appearances like the shining points in the firmament emerge into the foreground of intuition from the background of a pure matter.

Now all of this would still have to be understood as what it presents itself as, namely, being conceived in conformity with ancient models, if it went no further than the mediation between Platonism and Aristotelianism by means of the attempted new concept of matter and even the reintroduction of the *quinta natura*. But the main thing for Schelling is something else. As Kant had made man appear, among the wave sequences of the formation and disintegration of worlds, at the zenith of matter's progressive acquisitions of form, so that in this schema the background of still unformed or again disintegrating matter always remained, inexhaustible, behind the foreground of form—so for Schelling the Earth, as man's residence and perspective, is unexpectedly distinguished by the fact that it is not, itself, a heavenly body—that it is not that transition between primal stuff and nature— but rather a contraposition that is withdrawn from that process.

One result of Copernicanism is the uninterrupted modern effort of resistance against its relentless execution. Schelling's both *post*-Copernican and (in desperation) *anti*-Copernican formula reads: "So from the fact that those beings have scarcely, or not yet, withdrawn from the intelligible world, we can understand the fact (which is enough to give offense to some) that it is not those proud lights of the heavens, which can in a certain sense be thought of as exalted above the human realm, that are man's homes, but rather the lowly Earth; for here too the rule is, He gives grace to the humble. God esteemed man so highly that man on the Earth only was sufficient for Him . . . . To be sure, man is the goal, and in this sense everything exists for his sake. Something ultimate is to be arrived at, but this does not prevent it from leaving room for other things; on the contrary, the broader the basis above which it elevates itself, the more its uniqueness stands out. The paths of the creation do not lead from the narrow to the wide, but from the wide to the narrow."

Schelling considers it a defect in Kant's early cosmogony that it pictured the organization of the whole system of the universe in accordance with the analogy of the solar system, that is, as a projection of the Copernican schema onto the totality of the systems of bodies. To Schelling it seems fair to see the force (based on familarity) of such analogies, "the mind-killing and unproductive uniformity of the world-system," as being breached by the discovery of double stars, which are to be regarded as double solar systems, or at any rate as systems of bodies having equal rights. But the interpretation of all world-systems as constellations around a central body is not only an extension of familiarity over the unknown, but also a logical expression of the mechanistic character of matter that is turned over to its mere qualities, and is unable to produce anything but the same thing over and over again. Intuitive evidence [*Anschaulichkeit*] has been absorbed into the model that is applied again and again and that seems to furnish satisfaction, although the production of the model was a process of the most complicated alteration of what was intuitively immediate. The model is like a matrix for our power of imagination. For it, the model takes on the same evidence that, before Copernicus, geocentric perception possessed, so that every attempt to breach the repetition compulsion can appeal to the Copernican paradigm precisely when it does not accept its world-model.

What has taken place is the dissociation of intuitiveness and immediacy: Making intuition of the starry heavens accessible is not a process of producing immediacy. For Schelling it is important that, according to the accounts of missionaries from, for example, the interior of Africa, peoples living under the clearest skies, to whom the heavenly bodies appear incomparably more brightly than they do to European man, had developed no idea of God as a result of this brilliant intuition. He contrasts this barbaric insensitivity not to pure theory's Stoic world-admiration, as its opposite pole, but rather to the erection of the Tower of Babel, which according to the biblical myth was supposed to reach to heaven; in connection with which, the important thing is the explicit explanation that those who took part in it had wanted to make a name for themselves. Not the contemplation but the storming of the heavens appears as the original human attitude. In one breath with the building of the Tower of Babel, the Caucasian race of Prometheus is named, for "only he, it seems, could also be the One Man whose deed broke through the world of Ideas, separated man from God, and opened the world to him, in which he was free from God and for himself."[89] Prometheus's demiurgic construction of man seems just as adequate as an answer to an incomplete universe as does the building of the Tower of Babel as the myth (thus understood at a late date) of a figure of man that is no longer reliably placed in the center, but instead breaks out, centrifugally.

If I were asked whether Schelling's metaphysical cosmogony can still be taken seriously, one possible answer would certainly be that a speculation that works up the process of the history of philosophy in regard to a particular problem can only be read as metaphor. In the case of Schelling's concept of matter this becomes evident insofar as the constituent elements of the concept, which are gained *through history*, are *genetically* realized and integrated, and as phases of this process always have correlates in matter's highest form of organization—in the organic realm. Thus the original characteristic of inertia finds its correlate in disinclination, extension finds its correlate in physiological turgor, and impenetrability (as the maintenance of a piece of space) corresponds to the instinctive self-preservation of the living system. True, these features of organic nature do appear as continuations of those fundamental characteristics of matter, that is, as, in their turn, further phases of the process, but they can be read as metaphors for the abstract predicates of matter.

Now one can object, What then is the value of a metaphorical cosmogony or theory of matter? The answer is, None at all, unless it contains (or reminds us of) something that has been forgotten in methodically disciplined theory, or that cannot be considered in it at all. The metaphor conserves or restores the motivating context of the life-world, the context in which an interest in theory originates and from which it derives a claim whose (of course only partial) fulfillment was purchased by the renunciation, defined by method, that is implicit in the history of science. Science flourishes at the expense of the questions to answer which it was set in motion; every uncontrolled extravagance interferes with its success, but at the same time brings its motive up for discussion again. In Schelling's cosmogonic myth, the stars, as marks of the process of an incomplete world, are important only in relation to man's self-understanding—be it only in relation to his supposed needs and orientations. The demand embodied in the Copernican renunciation, whether as the prototypical expression of the discipline of scientific reason or as the final transformation of Platonic/Christian asceticism, has never been accepted in its full rigor and without new kinds of anthropocentric compensations. If one considers Idealism's key position in relation to the nineteenth century, then at least what it did not want to permit, but also what it was bound directly to provoke, by its refusal of renunciation, is clear as day.

If one takes a metaphysically intended cosmogony like Schelling's as metaphor, then the crucial weakness of every form of post-Copernican metaphorics becomes discernible in the fact that it seems not to be able to get by without the construction of a topographic-schematic center and concentric relations to it. That is certainly laid out in the Copernican language, when the center of the system remains the privileged standpoint from which the problems of construction resolve themselves, even if this point of view can never be realized for an observer. Schelling sees in this Copernican state of affairs an analogy for a construction of the concept of matter, such as (he says) Leibniz had already undertaken and such as seemed to him to come closest to the function of clearing up the history of the concept.

In the *Monadology*, a monad that was placed in the center would lose the distorted and confused perspective on the universe of monads from which the concept of matter, as the concept of the disordered state of ideas of monadic elements, is acquired. The idea of physical bodies as such conglomerates of monads relates to a position starting from which the rationality of the overall system cannot be adequately and clearly penetrated. So "matter," as the summing up of such an immature state of ideas, is a function of one's standpoint, but that does not make it pure appearance, just as, in Copernicanism, the movements of the stars and planets are not pure appearance, because they can be traced back to their preconditions in reality. Just as, in the Copernican reform, the disorder of these movements is dissolved by means of systematic construction, so in Leibniz the appearance of matter is dissolved by the monism of the monads. Although Schelling expressly refers to this attempted construction of matter,[90] nevertheless he does not accept, for his metaphysics, the prescription that is contained in it, that after Copernicus one should forgo, in speculation too, the realization of the privileged standpoint. To him, as to others, it is intolerable not to see man at the center of a reality that is concentrically directed at him, even if it should not be composed of the orbits of heavenly bodies.

6
Pure Intuition[a] as an
Anthropological Utopia

When one considers how ambiguous the effect of Copernicanism on
human self-consciousness was, since it could just as well be the humiliation of losing the central position as the triumph of the reason that
penetrated foreground appearances, one will not be able to avoid the
question whether a decrease in the privileges of intuition had to be
such an unambiguous post-Copernican tendency as it has so far
appeared. Every (in the broader sense) idealist epistemology had
almost unavoidably to appeal to an interpretation of the Copernican
paradigm as the radical separation of construction from intuition. If
one looks for authorities on the other side of this question, one will
have to pay attention to idealism's zones of fracture and points of crisis
as indicators of the intuition that is pressing for its rights. What
happens to the enthusiasm of Ludwig Feuerbach's earlier invoca-
tion of Copernicus when, in 1839, he renounces Hegel's "rational
mysticism"?

He had still written in 1838, in a review of F. Dorguth's *Kritik des
Idealismus* [*Critique of Idealism*], "The Copernican system is the most
glorious victory that idealism has won over empiricism, and reason
over the senses. The Copernican system is no truth of the senses, but a
truth of reason. It is a system that contradicts the senses, that is
absolutely transcendent, extravagant, and incomprehensible for
them. Only the thinking spirit, not sense and not imagination, is a
match for the sublime objects of this system." [91] Across the history of
mankind, thought destroys images, dreams, imagination, fancy; the
intuition of the heavens had already diminished, in its effectiveness in
the life-world, when geocentric immediacy was still unbroken. "As
long as the Greeks did not think, that was how long the heavenly

bodies were animated, divine beings for them. Only philosophy transformed these images into things." Philosophy sets up an intention that pushes forward the break between thought and the power of imagination, an intention whose beginning is marked by, perhaps, Parmenides, and which reaches a culminating point in Copernicus. Copernicus fulfils a standard that was already set up by philosophy, and he in his turn sets such a standard for modern science. Feuerbach, still thinking with Hegel, sees actual science as failing in relation to the idealistic rigor of this pretension.

If Feuerbach's conversion must be seen as an execution of the principle of dialectics, in that the new absolutism of the senses is dependent on the prior absolutism of ideality, then the model of the one becomes a condition for the other. Feuerbach loves the passage in Galileo's *Dialogue on the World Systems* in which he has his Sagredo, in an expressive exclamation, cry to Copernicus, across the century that has passed between them, what a confirmation of his system the telescopic intuition of hitherto undiscovered phenomena in the heavens would have given him: "Oh Niccolò Copernico, qual gusto sarebbe stato il tuo nel veder con si chiare esperienze confermata questa parte del tuo sistema!" [Oh Nicholas Copernicus, what a pleasure it would have been for you to see this part of your system confirmed by so clear an experiment!][92] But Galileo makes Salviati, in the dialogue, dampen and almost reject this very same rejoicing, by pointing to the incomparable sublimity of intellect (*sublimità del suo ingegno*) that was required in order to arrive at the new system on purely rational grounds, against the overwhelming verdict of sense-experience.

But Feuerbach's admiration of reason's running ahead of intuition is not without the reservation that only the latter brings the former to its goal. When he appends to the Galileo quotation that thus "speaks the man from the real 'next world,' the man of the future, to the man of the past," this is preparing the way for the idea of reason's serving the purpose of intuition, in a temporal schema of past and future. True, the foreground of interest is still occupied by the basic idea of reason's destruction of an unbroken orientation to the senses, but that does not prevent reason from functioning as an anticipation of sense-experience if, in particular, the preeminence of the anticipatory faculty relates to the position in the temporal context from which the expansion of intuition is prepared—if, that is, reason's systematic construction contains directions as to what could become intuition and under what conditions it would become realizable. One must reach ahead to the Feuerbach who follows the proclaimed break with Idealism, in order to perceive what is potentially involved in his use of the Galileo quotation: "Thinking is only a sensation that is expanded, extended to remote and absent things, a sensation of what is not really, actually sensed; the seeing of what is not seen." [93] Before his turning away from Hegel, the idealist and the presensualist functions of reason can be paired—symmetrically, with respect to the present—with, in the one case, the past, as the sphere of the dominion of unconscious sense-experience, which has not yet been overcome, and, in the other case, the future as the sphere of expanding sense-experience.

In the Dorguth review in 1838, what Feuerbach finds in the call to Copernicus that I quoted is above all the occasion for an apostrophic criticism of the age, which relates more to the loss of the idealist standard than to the guidance to be got from a concept of the future. Galileo, Feuerbach says, was able to speak in this way to his age, but now one would have to formulate the call differently: "Oh Nicholas Copernicus, be glad that you did not end up in our age, the era of the spirit's enslavement, of historicism, of empiricism, of positivism; if you were to come forward today with such a daring, such a heroic apriorism, you would have to atone for your idealistic chimera, as a shocking anachronism, in—at least—a madhouse!" The question is not whether his own age would have made a Copernicus possible, but whether it would have tolerated him. Feuerbach immediately turns to his contemporaries in this historicist and postitivist era and unfolds the concept of an apriorism that is not interested primarily in the conditions of the possibility of experience, but rather in anticipating its content. Such a material "anticipation of experience," carrying the title of an "act of genius," and in connection with which the accent does not yet fall on the intuitive verification that follows after, is represented by the figure of the astronomical reformer, whose greatness is measured by the resistance that he overcomes, not by the expansion [of sense experience] that he achieves: "So take this Copernicus as your example and learn from him the truth that every discoverer, every inventor, even if only of a machine, was an idealist, because he did not let himself be deterred, or be disturbed in his faith in the reality of his idea, by the hindrances of matter or by conflicts with what had been experienced up to his time—until he finally succeeded in transforming his thoughts into things." [94]

Just because the a priori anticipation's confirmation by intuition adds nothing essential to it, but in fact for the first time makes its daring stand out, in contrast to its subsequent trivialization—just for that reason, the field of sensual immediacy (since it is no longer the primary taken-for-granted component of the life-world, but is relieved of the admiration that gives rise to metaphysics, an admiration that now only falls to the "act of genius") can, in reverse, be set free for enjoyment. Here, what lies on the far side of the break with Idealism has, at least, its anticipatory license. When Feuerbach, a great deal later, in *Die Naturwissenschaft und die Revolution* [*Natural Science and Revolution*], pronounces Copernicus the "first revolutionary of modern times," who "deprived mankind of its heaven" and "thereby gave full license to all the Earth's subsequent and differently located revolutions," then the violence no longer consists above all in constructive thought's anticipation of possible later experience, but in its leading the way "to the intuition of the real world, to the simplicity of nature."[95] What Feuerbach here latterly perceives in Copernicus is exactly what Marx had already reproached him for in 1844, in the fifth of the "Theses on Feuerbach": "Feuerbach, not satisfied with abstract thinking, wants intuition; but he does not comprehend sensuousness as practical, human-sensuous activity." Not many years later, in the great "Feuerbach" polemic in *The German Ideology*, intuition as a relation to nature is a flight from the exigencies of the "world of man," and a flight that takes refuge precisely "in nature which has not yet been subdued by man." The negativity of the starry heavens is unsurpassable in this relationship. It is the limit of what cannot be subdued by men, and, as such, the abiding presence of the nature "which first appears to men as a completely alien, all-powerful and unassailable force, with which men's relations are purely animal and by which they are overawed like beasts...."[96]

What for Idealism is the highest offensiveness—the rational unfathomability of what is contingently 'given' as fact—is, for sensualism, something like the stigma of inexhaustibility, owing to which the sensual datum can become an end in itself, the epitome of all intentionality.[b] Thanks to this relationship to reality—a relationship that is free of instrumentality—man as "the living superlative of sensualism" is enabled, by means of reason's anticipatory accomplishment, to realize a gain in pleasure from the given. He is the one who "imbibes heavenly bliss from the purposeless viewing of the stars."[97]

This sensualism is not an epistemological but an anthropological position. It is the absolute sensualist who raises himself above the objectifying attitude of experience and experiences totality no longer as an idea, but as a suspension of the finitude of enjoyment, and to whom therefore "the world, the infinite—and in fact purely for its own sake, that is, for the sake of aesthetic enjoyment—is the object of his senses and of his feelings."

However contradictory it may sound, this is the ultimate logical consequence of Romanticism in the nineteenth century: the *coup de main* of immediacy, issuing from the sharpest reproach against Hegel that Feuerbach could formulate, that while the word *immediate* had been used by Hegel countless times, it was precisely the immediate that was entirely missing from his philosophy. To produce immediacy nevertheless, by force, is one of the great anachronisms of this century, in which what had, after Idealism, become contingent 'fact,' or had been left to itself as such, is now revalued as what is significant and fulfilling. The principle of insufficient reason^c is only thinkable after the principle of sufficient reason; here it presents itself as the principle of absolute sensuality.

One sees this consequence issue directly from Hegel's arrogance vis-à-vis the starry heavens. They are already no longer offensively contingent fact as soon as, through the isolation of sensuality, they offer themselves to the pure capacity for enjoyment. In the *Todesgedanken* [*Thoughts on Death*] of 1830 the starry heavens had already become, for Feuerbach, the pure expression of the principle of insufficient reason. "In your intuition of the stars you only collide with sensually independent existence, with the magnitude and quantity of these heavenly bodies." The thought with which the Enlightenment had provoked itself—that it could preserve the universe's teleological character, after abandoning anthropocentrism, only by letting the whole universe be inhabited by rational creatures—now appears as a leveling off of the pure contingent givenness of the starry heavens, of their surrender to the intuition that does not inquire after any purpose. To populate the heavenly bodies was precisely to subject the purest sensual data to rationality's longing for justification. "Laboring under this error, you accordingly overshadow and populate the distant heavenly bodies, without exception and without distinction, with the insubstantial shadows of your abundant imagination. What astounds you and leads you astray is only sensual existence—merely the unlimited space of these heavenly bodies." The consequence of measuring the universe against the standard of rationality would have had to be that "all space, all existence, all nature appears to be for nothing," that everything becomes empty and futile in view of the unanswerable query regarding its sufficient reason—a principle that is devastating for intuitive fullness and repleteness. If one obeyed the radicalism of the demand for the justification of reality as something that was, after all, rational, then the whole universe would have to be reduced to a single point, because "what goes beyond the domain of an atom is wasted, superfluous, purposeless existence."

What justifies the violation of rationality—what provides the true cosmodicy [justification of the cosmos]—is intuition as the epitome of all intentions. Superabundance [*Überfluss*] is an element of theory only, again, in the ancient sense; in the modern sense it is exclusively an aesthetic element. We can see in the paradigm of the starry heavens what the new sensuality, the post-Copernican orientation to the senses, can be at all, or can be again, and what distinguishes it from the initial situation of the 'life-world,' but also from the "original data" constructed by the various epistemological sensualisms. Precisely what claims to be pure naturalism postulates a historical quality in its discovery of meaning: namely, its position, which it could not exchange with any other, at the end of the effectiveness of the principle of sufficient reason and after the failure (and in the consciousness of the failure) of efforts extending over centuries to justify God and His work.

It is an endangered sensuality, not a 'matter of course' one, a sensuality that supports itself above the abyss of the unanswered and unanswerable question of the reason for being, on the edge of the nihilism that is avoided by only a hair—and, after Idealism, nothing is separated from nihilism by more than this hair. Feuerbach expresses precisely this fact, as an epilogue to the Enlightenment expectation of a universe that was inhabited, everywhere domesticated, and subjected to culture by rational beings: "Truly, if one extended your teleological ideas—which, strangely, you apply only to the stars—to everything, then one would finally come to the conclusion that all sensible, material existence is really mere waste, is purposeless expenditure, and that consequently the best thing would be if nothing at all existed; for only in nothingness would all lack of purpose be avoided and abolished." [98]

Chapter 6

Post-Copernican sensualism consists in one step of resignation—
the forgoing of any justification of reality—and a next step of human
self-intensification, since what could not be rationally justified nevertheless is given, so that all that would remain would be to intuit it. As a
program it is in undiluted contrast to the entire century, though the
only way in which it leaves a trace is (once again) that it was thought.

If one may say that Feuerbach takes the Copernican schema as
his guide, one must reckon with the fact—which already holds for
Copernicus himself, and is found again and again—that the surrender of cosmic anthropocentrism presupposes or compels people to
adopt conceptions that are centered on man in a different respect. If,
for once, one disregards the ritualized metaphors according to which
the center is the privileged place, at which one can best be seen from
every direction and around which everything else revolves, then the
superiority of the cosmic center was only that it constituted an
advantage for theoretical intuition, but not without further ado that
it was also advantageous for an aesthetic intuition. If one examines it
more closely, the preference given to the center is a compensation for
a deficiency of finite optics, which can be a match for the universe
only because it reaches equally in all directions. But for just that
reason this would not be the location for a god. And precisely this is
the salient point for Feuerbach's anthropological absolutism: If he
wants to reveal anthropology as the secret of the old theologies, he
also has to provide man with a god's mode of intuition, and that is
much easier said than done.

It means above all that the contingent givenness of the world's
aspect not only drives one into sensualism as an epistemological
escape route, and a form of resignation, from the perspective of which
one *cannot* inquire further, but it also makes possible a localization of
consciousness from the perspective of which one does not *need* to
inquire further—so one does not need to inquire, either, whether this
that is given to the senses is, after all, everything, and whether
everything is the way it is given to them. Seriously to accomplish this
freedom from need, which all gods possess, for man insofar as he is a
creature of the senses gives rise to the greatest difficulties. It is easier
to resist the functionalization of man in the service of the absolute,
as Feuerbach does in his critique of Hegel, than to make man himself into the absolute—really to carry out the 'reoccupation,' without
again making use of physical legitimations (since the highest being

would only be the one in relation to which everything else has to legitimate itself).

That is why the introduction of the "knowledge drive" into the great Hegel critique of 1839 is a questionable argument with which to elevate the human form to the highest one, which it has always appeared to be in art. It is the highest because it cannot be transcended, because every attempt to go beyond nature and man is as futile as "the art that wants to give us something higher than human form, but gives us only caricatures." The proof that man himself has succeeded to God's metaphysical position is manifest in the incapacity of the imagination to surpass man's form. It is a demonstrative application of the formula of the thing "id quo maius cogitari nequit" [than which nothing greater can be conceived].[d]

Thus Feuerbach makes mighty preparations in his thesis, only to beat a pitiful retreat in his argumentation: "The human form cannot be regarded as limited and finite, because even if it were so the artistic-creative spirit could easily remove the limits and conjure up a higher form from it. The human form is rather the genus of the manifold animal species; it no longer exists as species in man, but as genus. The being of man is no longer a particular and subjective, but a universal being, for man has the whole universe as the object of his drive for knowledge. And only a cosmopolitan being can have the cosmos as its object."[99] One notes how in this passage Feuerbach anthropologizes the classical attributes of the theory of God, not only omniscience, which is concealed in the knowledge drive as something stretched out over time, but also omnipresence, whose true correlate is sensualism.

One necessarily expects the stars to be the test case in this *translatio imperii* [transfer of power]. Nor does Feuerbach refuse to meet this expectation; but his argument is simply the rational axiom of a homogeneity of nature, which allows us to infer from what is immediately accessible to us to what is not, or not yet. "It is true that the stars are not the objects of an immediate sensuous intuition, but we know the main thing—that they obey the same laws as we do." This feeble escape into what we "know"—but which is after all only one of the assumptions we make that are necessary preconditions of the possibility of knowledge, which we cannot, ourselves know—is one of the many disillusionments of the creature that has become the highest being for itself. Enjoyment, for the observer of the starry heavens, is no go if he must first deduce that they are subject to the same laws as he is himself; here Feuerbach's appeal to the old maxim *Simile simili gaudet* [Like takes pleasure in like] does not accomplish much.

What is important to me is to point out the continual reversion, in the post-Copernican relation to the world, to the pre-Copernican procedure of legitimizing human self-consciousness with the aid of a physical schema.

It is not so easy to *be* the God Whom one had previously only *conceived*—and that is precisely what is at stake in Feuerbach's transformation of theology into anthropology—when the time has really come (finally and definitely) to decide the self-comparison with theism's God, which has been pushed to extremes for so long, in one's own favor. The difficulty lies in assigning [*bestimmen*] the world to man's enjoyment without allowing it to have been intended [*bestimmt*] for that. For man, who has been reclaimed from theology, cannot be taken care of as he was before (what after all would a God be Who was the object of providence?), any more than he can fall back into [an *attitude* of] care, which disturbs and destroys all enjoyment of existence through intuition by instrumentalizing intuition's objects. Feuerbach did not want to recognize that the dissolution of the ancient connection between theory and teleology could lead into the absolutism of care. Copernicus had still not believed that the truth about the world, which he sought, could exist without the world's having been arranged for man's benefit by a providential reason; as a result of this distinction accorded to man, a distinction that established only itself, geocentrism had become a matter of indifference, and unnecessary, to him, and together with it intuition—as the self-confirmation of a new anthropocentrism—had become problematic. The world is not edible, not enjoyable in this most intensive sense; if it was, Feuerbach would have relinquished intuition. But could he have conceived of a God Who eats, after the tradition of European theism had done so much to get nectar and ambrosia off of the gods' table and, instead of it, to get God onto men's table?

The historical importance of theism, for Feuerbach, does not lie simply in the fact that it is man's self-consciousness 'by other means'; if it had only been this, then the counterquestions would remain: Why should the process be so laborious? Why should we not have the result right away? Self-consciousness's detour by way of the concept of God would have been useless and superfluous if the quality of the outcome had to be the same as that of the substance that was subjected to the process in the beginning. But this self-consciousness underwent an organization *in vitro*, in theology, that it would not have been able to achieve in the sealed-off solipsism of the species. To that extent, Feuerbach agrees with Hegel in seeing consciousness as in need of history, but (in his case) without hypostasizing its subject—even if it is only the species that can fulfill the ideal that its individuals have created for themselves.

To make this clear, Feuerbach sets up a comparison between the interchange of subjectivity and objectivity that is presupposed by the anthropological result that is extracted from theology, on the one hand, and Copernicus's astronomy, on the other. The latter precisely did not exclude or devalue the experience of subjective appearance; instead, by rationally constructing the process by which that experience is produced it freed it, for the first time, from the supposedly higher anxieties of inquiry into the nature of reality (so we see that the aesthetic aspect is in fact still subject to the encroachment of cares related to theory). "Just as astronomy distinguishes the subjective, apparent world from the objective real one, so 'atheistics'—in reality, theonomy, which is distinguished from theology in the same way that astronomy is distinguished from astrology—has the task of distinguishing between the divinity, which theology regards as an objective reality, and the subjective reality. Theonomy is psychological astronomy." [100]

Feuerbach sketched the thought experiment of imagining a thinking being on a planet or even on a comet who happened to run across, in a Christian dogmatic text, the paragraphs on God's being. What would be the result of this examination, from an eccentric perspective, of the core component of terrestrial theology? The inhabitant of alien worlds would infer the existence of beings on Earth who had defined themselves in the divine being. And quite rightly so, as Feuerbach says, because "in the case of this object the distinction between what the object is in itself and what it is for man comes to nothing." Only objects of equal rank, Feuerbach says, can be objects for one another, especially as they are in themselves. The highest being can only be man's object, because he himself is the highest being. In this context we find a statement that could have provided the most concise post-Copernican expression of the anthropocentrism that the Stoics put into circulation: "God, however, is an object only for man. Animals and stars praise God only in a human sense." [101]

The Heavens as Charming Landscape; Photography and Anthropomorphism

The most important admission, in the middle of the nineteenth century, that immediate intuition and theory in the form of science were no longer alternatives was Alexander von Humboldt's *Kosmos*. Once again at least the impossibility of renouncing the claim to totality is put before us, but not, as in Humboldt's great travel narratives, through the presentation of immediacy, but rather through a compensatory realization in the form of literature, as that form was determined by the intended and interested audience. The world is grasped as a unity when it is represented as a physiognomy. But while for a face, one look is enough, this huge delayed task[a] that is still attempted at the last moment requires five volumes. Humboldt himself defined for this undertaking—which he regarded as his "life's work"—a pretension that exhibited a Faustian disregard for norms, and it is surely not an accidental index of the needs of the time that Goethe had hardly completed *Faust II* and sealed it for posterity when Humboldt set to work on his *Kosmos*: "The whole is not what has hitherto been commonly called 'physical description of the earth,' as it comprises all created things—heaven and earth."[102] The title, on which Humboldt had labored repeatedly, betrays, in its combination of the lapidary with the provisional, the specific situation from which it comes and for which it is meant to take responsibility: "I began it in French fifteen years ago, and called it *Essai sur la Physique du Monde*. In Germany I at first intended to call it *The Book of Nature*, the sort of thing one finds in the Middle Ages, by Albertus Magnus. All these, however, are too vague. Now my title is *Kosmos. Entwurf einer physischen Weltbeschreibung* [*Cosmos. Outline of a Physical Cosmography*] .... I know that 'cosmos' is very grand and not without a certain tinge of affectation; but as a title it conveys, with one striking word, both heaven and earth .... " Thus he explicitly asks his trusted correspondent, Varnhagen von Ense, before the first volume goes to press, for support in his choice of the great title.[103] At the same time, the expression for totality is meant to proclaim the intended "total impression," within which the starry heavens can acquire "the charm of a landscape." It is language that brings about this unity.[104]

But it is not only and not primarily the range of the subject matter in which the pretension is manifested; rather, it is the will to make present, to create a surrogate for what the world traveler had continually sought, throughout his life, on the spot: the presence of nature *in propria persona* ['in person']. One of the oldest European metaphors, that nature is like a book, has produced the amazing inversion that this book must be like a nature for its readers: "A book about nature must produce the same impression as nature itself." [b]

Thus while it is true that the author of *Kosmos* is a proxy for his readers, he is one who would like to share this privilege of a life of travel and investigation with his readers, ex post facto. So the language of the work can no longer be instructions for intuition that could be carried out at any time and by anyone; instead, it takes the place of this intuition. One will be able to find that surprising only if one leaves out of account how little had really been accomplished up to that point in the way of description—how scanty, for example, the descriptive accomplishments of the French *Encyclopedia* had turned out to be, in spite of, or perhaps because of, the magnificent plates. Only in this kind of retrospect can one comprehend how extraordinary his plan could seem to Humboldt himself: "The mad fancy has seized me of representing in a single work the whole material world, —all that is known to us of the phenomena of the heavens and of terrestrial life, from the nebulae of stars to the geographical distribution of mosses on granite rocks, and this in a work in which a lively style shall at once interest and charm." As Humboldt conceives it, this presentation does not draw its life from a timeless relation to nature, from the pure presence of its objects, but from the history of experience of it and comprehension of it; the world does not consist only of facts but also of ideas, and indeed not only of the ideas that enable us to understand the facts, but also of the motives for interest in and attentiveness to those very facts—that is, of the "inducements to the study of nature to be found in the spirit of the age." "Each great and important idea, wherever it lurks, is to be mentioned in connection with facts. It must represent an epoch in the mental development of man as regards his knowledge of nature."

The resistance of the investigator's material (as it was accessible to him) against this original intention makes itself significantly noticeable in two points. The exact facts could no longer be fully integrated into the book's mode of exposition, if it was to maintain the precision of the model of Laplace's *Exposition du système du monde* [*Exposition of the System of the World*]: "As these details are not capable of being treated from a literary point of view, in the same way as the general combinations of natural science, the merely factual material will be stated in short sentences arranged almost tabularly ...." As the work proceeds, it sounds—in spite of its public success—almost melancholy when Humboldt once again sums up his expectations for it: "I would like the work itself to be characterized by universality of design and breadth of view, written in an animated and even, where possible, a graceful style, and with the technical expressions transformed into others which should be happily chosen, descriptive and pictorial .... The book should thus become the reflex of myself—my life—my person, now grown so very old. With such a free manner of treatment I can proceed aphoristically. I want rather to suggest than to lay down. Much will be perfectly intelligible only to those who are well versed in some one branch of natural history. My mode of treatment, however, is, I think, calculated not to disturb those who know less. My aim is to soar above the things which we know in 1841." But the normative idea is still not given up, that one should proceed like nature itself, keeping the contemplative and the exact sides of theory together with the power of rhetoric—the idea of what he calls "the simple and scientifically descriptive": "It is so in nature herself. The glittering stars delight and inspire us, and yet everything beneath the vault of the heavens moves in a path of mathematical precision." [105]

The second respect in which the formlessness of the material persistently resisted the elegance of his plan was the size of the work, for which, to be sure, he had radically misjudged the length of the life that he still had before him. He realizes early on, "I have not succeeded in compressing the whole into one volume; and yet it would have left the grandest impression with that kind of brevity. I hope two volumes will include the whole." [106] Humboldt had anticipated having to leave the work behind him as an unfinished fragment; he wanted to write it, even as he completed the individual parts, in such a way that "those who see me buried will have something complete in each fragment."[107] But he had underestimated his time and his powers. Many years later the really old man had to write to his decade-and-a-half younger friend Varnhagen the shocking words, "Beware of such long patience in living."[108] A month later Varnhagen was dead.

In regard to the impression produced by his work, Humboldt spoke of the "*Kosmos* mania" that spread from the Prussian court, through the society of the capital, all the way to the countryside and the parsonages. The echo that reached him in the form of piles of manuscripts about speculative worldviews, constellations, and planetary disturbances caused him to remark sarcastically: "Everyone wanted to be pregnant with a *Kosmos* at once . . . . " It is in accordance with this audience's interests that it should not have to pursue the direct experience, but should be able to have the additions to man's knowledge of the world delivered to it at home. This need corresponds to the mastery of 'vicarious' intuition that makes *Kosmos* a work of world literature: " . . . in the salons of fashionable ladies, *Kosmos* lay uppermost on the coffee table, and opened, so as to show visitors that one had actually *read* in the learned work of the great philosopher!"[109] But Humboldt did not see the success of his work only as questionable; even in the late continuation of the work he held fast to his conviction of the influence of science on men's thinking, of the possibility of mastering the great masses of material, and of language's efficacy for this task: "How is it that *Kosmos* has achieved such unexpected success? Partly, I suppose, from the train of thought which it awakens in the reader's mind, and partly from the flexibility of our German language, which makes it so easy to make something vivid, to paint it in words."[110] This remarkable fact—Humboldt's total lack of mistrust of language—is the last result, in relation to the language of the natural sciences as well, of the great accomplishments of Classical[c] and Romantic literature.

At the same time, Humboldt shares his century's inclination to interpret the Copernican postulate as implying that the illusion of being at the center is the fundamental pattern of man's situation in reality, even if it appears in subtler and more insidious forms than that of geocentrism. One of the cases in which he forwent the projection of familiar models was when he insisted that outside of our solar system, no phenomenon analogous to this system had yet been detected. On the contrary, the discovery of double stars had created an inducement to regard other types of systems, besides those having central bodies, as common. The expectation, which dominated the eighteenth century, that the Copernican diagram would be encountered everywhere in space as the basic form of all cosmic systems and their supersystems, yields, for what had once been the strong demands placed on the imagination by Copernicus, an unexpected degree of familiarity. For the endeavor to rediscover, in the unknown, the basic pattern of the solar system—for instance, when the changes in nebulas are thought to reveal cosmogonies that will produce additional systems built around central bodies—already has "something comfortable" about it, as Humboldt puts it in expressing his reserve toward expectations of familiarity. Totality must again and again be wrung from dissolution. That is why, as late as 1843, Humboldt uses Encke's "giant refractor" so as to be impressed "by the immediate viewing of the heavens."

For Humboldt, the question of the finitude or infinitude of the universe remains undecided. The historical form of science as progress makes it seem more appropriate "to assume an infinity as a limit of this progress." The interminability of experience of the world, an interminability that is experienced in time and that Kant elevated to the status of an "Idea," [d] is projected onto an advantage in view of the antinomy of space. But this very openness of the experiential context is bound to a state of affairs that becomes ever more concentrated: the dependence of every expansion that is still possible in our access to the world on the increased effectiveness of tools. The contemporary importance of this observation, as a result, for example, of the resolution of certain supposed nebulas into multitudes of stars, had pushed even the speculative assumptions about their cosmogonic changes in time into the background. "The sphere of inquiry is expanded only by the perfection of instruments." [111]

One could call this dictum of Humboldt's the axiom of the instrumentalization of pure theory. It was uttered only two decades after the death of Goethe, who had once again resisted optical instrumentation so as to preserve the purity of intuition, and thus also to give preference to the *intensity* of experience over its *expansion*. If Humboldt's dictum about the dependence of scientific experience on instruments was also resignation in relation to his own will to see things for himself, Goethe's "Observation Regarding Subjective Aids in the Sciences," which he had entrusted to his diary on 7 April 1817, in Jena, was already pure anachronism, because it objected to the combination of physics and mathematics as impairing the "phenomenon." It is true, Goethe says, that physics benefits from mathematics, but it is not protected from material errors, because "its hypotheses and analogies are concealed anthropomorphisms, parables and the like. They think that they express the phenomenon by that means, instead of which they should concern themselves about the conditions under which it appears, since then what is true would soon be as clear as daylight." One must add a diary note, "Thoughts on Fiction and Science," from 10 June of the same year, in order to be able to make the rootedness of Humboldt's effort at synthesis stand out: "The mischief that they cause is merely a consequence of the needs of the reflective power of judgment, which provides itself with some image or other to use, but subsequently establishes this image as true and objective, as a result of which then what for a while was helpful becomes detrimental and a hindrance further on."

In defining the position of *Kosmos* in historical time it is also instructive to note that although photography was invented during the period in which the book was being written, it only gained its great importance for astronomy toward the end of the century, when the Paris Congress of Astronomers in 1887 organized the observatories all over the Earth to scrutinize the heavens photographically. One will almost have to say that the modern city dweller, for whom the floods of illumination of his streets make the heavens pale, now knows the starry heavens (from countless widely distributed reproductions) only as an object of photography—that is, as an aggregate of objects that he could not even locate, that not even the heavily equipped eye will ever get to see, because they can only be 'constituted' at all by exposing very sensitive plates for long periods. One may still describe the photographing of a star cluster or a spiral nebula as an indirect process of intuition, but the stellar spectrum that has become well-defined and measurable by means of the chemistry of the photographic plate, and that makes the Doppler effect applicable, is no longer in the same category. Humboldt's *Kosmos* is also localized by the fact that the work of Bunsen and Kirchhoff on spectral analysis, which extracted the first non-'intuitive' information from starlight, was still taking place as it was being written. Kirchhoff's

Chapter 7

Untersuchungen über das Sonnen–spektrum [*Investigations Regarding the Spectrum of the Sun*] were published in 1861, two years after Humboldt's death.

Astronomical photography raises to a higher power the simultaneity of the nonsimultaneous; it now completes the Copernican differentiation of appearance and reality by pursuing the logic of the finite speed of light, also, to its conclusion: the technical analysis and display of the heavens, as a section through time, which no longer has anything to do with the equation of intuition and presence. The product of the chemical darkening of a plate by a source of even the faintest light is, in a certain respect, no longer an auxiliary means, but has become the object itself, of which there is no other evidence but just this. Only for the limiting case of our natural optics does it hold that photographs are images of something else [*Abbildungen*], images that one can compare to what is also still there apart from them and was there before them. In other cases they are no more like that than a microtome section made under a powerful microscope is something that has, in addition, an organic reality that one could find somewhere and compare with the dissection. A series of additional actions and data, such as specifications regarding optics, length of exposure, and position, is required in order to 'objectify' such dissection, which means (here) precisely to make the outcome of a procedure repeatable. The observer has become the evaluator. To that extent, classical astronomical photography does not differ from the radiation diagrams that are provided by modern radiotelescopes and for which there is no longer an unambiguous association with optical 'objects' if only because the radio waves come precisely from the optical gaps, so that they transform the heavens into a continuum of radiation.

Schopenhauer was the first to reflect on the relation between astronomy and photography.[112] The old valuation of theory by the magnificence of its object, he says, would cause perplexity in the case of astronomy if only because astronomers "for the most part have mere calculating minds and are in other respects of second-rate ability." Schopenhauer suspects that the special veneration accorded to an astronomer like Newton is not least of all a result of the fact that "people take as the measure of his merit the magnitude of the masses whose motion he traced to their laws and these to the natural force that operates in it." Why should not Lavoisier, for example, be esteemed just as highly as Newton? The chemist's objects were, after all, of an incomparably higher degree of complexity than the simple tasks of astronomical mechanics, whose most remarkable accomplishment, in the eyes of the world, had been the prediction of never observed planets like Neptune.

Now, contempt for astronomy has been connected, since the laughter of the Thracian maid,[e] with an emphasis on its uselessness for practical life, and the traditional feeble response has been to point to the necessity of exact ephemerides for seafaring. Schopenhauer could not have constructed a significant contest if he had restricted himself to playing off the everyday utilitarian value of chemistry against the discovery of Neptune, which was useless even for seafaring. His confrontation with Daguerre's invention is instructive because to people at the time the utilitarian value of this invention can hardly already have been clear. The contest is full of deeper meaning not so much in view of the uselessness of the discovery, in the one case, and the usefulness of the invention, in the other, as, rather, on account of the difficult character of the relations to reality in both cases.

Schopenhauer makes a value decision in favor of the creation of images [*Abbildung*] and in favor of Daguerre's invention. This is a decision against the Platonism of our entire tradition, the Platonism that had sought reality, in the strict sense, beyond the appearances, and that is always still involved when the background becomes the background constituted by an invisible mechanism. This determines Schopenhauer's disapproval of astronomy: "From the point of view of philosophy, we might compare astronomers to those who attend the performance of a great opera. Without allowing themselves to be diverted by the music or the contents of the piece, they merely pay attention to the machinery of the decorations and are so pleased when they find out all about its working and the sequence of its operations." Since science is trained to seek out the background of appearances, and believes that it has achieved the triumph of its arts in tracking down the planet that was hitherto unseen and that remains invisible for the viewer of the heavens, it ceases to be capable of allowing validity to the foreground (as the product of that mechanism), and of disclosing appearance as the reality of perception that is pertinent to us.

For Schopenhauer intuition is once again the only possibility of happiness, but theory has come to be suspected of being an aggregate of what prevents intuition. One must extrapolate the idea to apply it to photography; in photography, science turns back to what it

originated in—to the foreground of appearances, the outermost surface of reality, which it now helps us to skim off and to conserve. The
loss of intuition of the foreground of nature—the loss that Schopenhauer confirms, in the case of astronomy, with the theater metaphor—
appears as one of the reasons for the rise in the importance of photography and of the techniques that result from it: The possibility of
procuring pictures acts as a substitute for the old ubiquity of intuition
for the privileged *contemplator caeli*.

Here too one must look back in order to be able to define the
point that has been arrived at. Schopenhauer himself, more than
three decades earlier, in *The World as Will and Representation*, had still
spoken of what he now regards as the false glamour of astronomy—
of the "magnitude and importance of the objects" that it brings,
with its plain cognitive instruments (spatial intuition, inertia and
gravitation), to the certainty and simplicity of its "very interesting
results." The constructive indirectness of astronomical knowledge is
only a surrogate for the immediacy that would be possible for it
in principle—a consequence of the contingent fact of our eccentric
position and immobility, which in their turn result from the metaphysical given of the body, as the "appearance of the unfortunate
will." The salvational importance of intuition, which must accept the
body as its accidental condition and as an arbitrary restriction of its
potential ubiquity, is comprehended in exemplary fashion in astronomy as the insignificance of the place where it is fixed in space, since
"this truth could be established even directly through a single empirical perception, if we could freedly pass through universal space, and
had telescopic eyes." [113] From the general proposition that "the whole
world of reflection rests on, and is rooted in, the world of perception,"
Schopenhauer infers that "it must be possible in some way to know
directly, even without proofs and deductions, every truth that is
found through deductions and communicated by proofs." Concepts
have only "borrowed light, like the moon"; the theorist must again
and again turn back " to silent looking [*Anschauen*, "intuiting"]." In
astronomy, in spite of the extreme conditions—the inaccessibility of
its objects—under which it operates, all nonintuitive operations are
just as much "so to speak, merely bridges from one intuitive apprehension to another." As soon as the body is seen as a mere hindrance,
by which the theoretician's factual localization in space is enforced,
our parochial perspective [*Winkelperspektive*] on the cosmos loses its

aesthetic function and has to allow itself to be measured against the ideal of the ubiquity of intuition, and that means, once again, against the ideal of a god. Self-consciousness, and its continual exposure to the possibility of care, is here simply antithetical to self-oblivion, which is or would be the condition of pure intuition.

The cognitive ideal of an intuition that freely traverses the universe is conceived as the opposite type to the agency of theory that was projected by Laplace and the imprint of which is evident in the nineteenth century's concept of science. The productivity of this agency culminates in determining, on the basis of analysis of the total cosmic system in *one* given moment, the state of the system at any other arbitrarily chosen moment, by means of *one* formula. Laplace went on from this idea to say that in the degree of perfection that it had reached through astronomy, the human intellect offered a feeble sketch of that universal intelligence, and all its efforts were directed at unceasingly approaching it.[114] However, Laplace himself, in his exposition of the system and the origin of the world, almost two decades earlier, had doubted whether the human spirit could be satisfied by means of analysis. This was how he had tried to explain the peculiar methodological dualism of Newton, who had indeed *found* his results by means of the analytical method, but had *exhibited* them by means of synthesis. This, he said, had been done only out of respect for the geometry of the ancients, but on account of the advantage of geometrical synthesis, "that it never makes one lose sight of its object, and that it illuminates the entire path that leads from the first axioms to their final consequences, instead of the way in which analysis quickly makes us forget the principal object, in order to occupy us with abstract combinations, and only returns us to it again at the end." [115] In this way, Newton, so Laplace says, combined the intuitive or graphic quality of synthesis and the fruitfulness of analysis, and "rendered the intellectual operations of the latter sensible by means of the images of the former." The result has an almost optical quality, in that it can "reveal to our eyes" the past and future states of the universe; and thus "the sight of this sublime spectacle makes us feel the noblest of the pleasures that are reserved for human nature." The perfection of what Emil Du Bois-Reymond later called "Laplace's spirit" consists, in contrast to this, precisely in its remaining in an arbitrary momentary position and, from it, disposing effortlessly over past and future by means of formulas. So for Schopenhauer, too, this perfection was unsatisfying and paltry in comparison to his ideal of an intuition that is ubiquitous across space and time, and which does not need to make use of mathematical means.

In his valuation of astronomy Schopenhauer remains an idealist, however he may set up the relation between concepts and intuition. Copernicanism, since it fixes man to the parochial perspective of his physical position, belongs among the implications of bodiliness as the incarnation of the will. The body localizes man in space and time, and reason is the totality of the devices by which he tries to extricate himself again from the fortuitousness of this fixed position. In relation to the astronomical object the Copernican perspectival configuration [*Aspekt*] is inessential, capable of being surpassed from the point of view of the metaphysical principle, and functionally already a refusal of the merely contingent and factual.

But just this becomes, for Nietzsche, the "fundamental question": "Whether perspective is essential." Or whether it is only a "form of observation," a—precisely—contingent and factual relation between the observer and the observed.[116] Before Nietzsche made this note on the question of the essentiality of perspective, the mechanism of reflection in man's theoretical and aesthetic attitude to the starry heavens had been his continuous theme under the rubric of an anthropomorphism that had outgrown any primitivism. With the physical necessity with which the spider spins, man throws the net of his ideas and claims over the world. When he finds it admirable, what he executes is the reflexive admiration of himself: "All that conformity to law, which impresses us so much in the movement of the stars and in chemical processes, coincides at bottom with these properties which we bring to the things. Thus it is we who impress ourselves in this way."[117]

So in this respect the pre-Copernican attitude to the world does not differ from the post-Copernican; the later attitude only refines the procedure of the earlier one. "The process of all religion, philosophy, and science vis-à-vis the world: It begins with the grossest anthropomorphisms and unceasingly becomes more refined. The individual person considers even the celestial system to be serving him or in connection with him."[118] Nietzsche sees the earliest form of consciousness in which this state of affairs is exhibited in the fundamental mythical idea of metamorphosis. "It was as if [the Greeks] regarded nature merely as a masquerade and a disguise for anthropomorphic gods." Consequently astrology, too, he tells us, is not a degenerate form of astronomy. It is in fact the model for the degree of involvement between man and the world to which philosophy, also, has to orient itself, if it undertakes "to regard the highest evolution of man as the highest evolution of the world."[119] Man's reflection, by way of nature, on himself, the reflection that Nietzsche is discussing, is the result of a deeper 'carrying over,' of the preconscious execution of a set of metaphors—an execution that even precedes their linguistic (not to speak of their aesthetic) articulation. And it is certainly more original than any theoretical/conceptual system, which represents only "an imitation of temporal, spatial, and numerical relationships on a basis composed of metaphors."

It is certainly not arbitrary when Nietzsche makes the unique attempt to characterize the fundamental tenor of his philosophy by a relation to the phenomenon of the heavens that differs from the traditional ones. It is no longer the starry sky at night whose ambiguity, as between admiration and irritation, even now still had to be decided, as a firmly prescribed pair of alternatives, by any metaphysics. Instead, this obligatory preestablished choice is to be leaped over in a philosophizing whose aesthetic equivalent would have to be provided by the extinguished stars of the daytime sky. It is the "clear sky" that people hate in him (and, of course, in the music of the addressee of his letter, Peter Gast). Nietzsche writes to Franz Overbeck on 14 September 1884, "An Italian said recently, 'In comparison to what we call sky, *cielo*, the German *Himmel* is *una carricatura*.' Bravo! That contains my entire philosophy!" The thesis suddenly makes us realize that the metaphysical tradition had never troubled itself about the daytime sky. Only that makes possible the provocative metaphor for a late philosophy that wants to see itself exempted from the obligatory tasks of the tradition and that finds the signature of this status in the daytime sky.

In doing so Nietzsche took up once again, in a surprising way, the negativity of the starry heavens, which he expressed (not accidentally) almost in the words of Heine's anecdote about Hegel, in the fragment from the "Dionysus Dithyrambs" that I already quoted[f] — now, to be sure, not as an essential status, but as a transformation that turns what used to be called a star into a spot. This starlight is in any event the stigma of a great imprecision, a lack of appropriateness in the relation between the process of nature and human self-consciousness. Nietzsche is one of the few who have deduced from the fact of the finite speed of light, and the nonsimultaneity of appearing objects with the observer's present, which follows from that, the consequence of the indifference of the present. Presence cannot enable us to apprehend the necessity of what is given in it, because it is only an accidental section through reality. The irregularity of appearances in space turns out to be a projection of the fateful delays onto the plane of what is just now visible; it is a paradigm of the distortion of reality by time, not only, and not most painfully, in nature but also in history.

'Contemporaries' [in the literal sense] live right past the greatest events and thoughts; they do not experience them. This is where writing history in terms of its 'imperishable' phenomena finds its hidden justification. Reality lies in a different dimension from that of the contingent factuality of appearances. Here the pitfall of Platonism yawns, for one who wanted to avoid the alternatives of the tradition. "What happens is a little like what happens in the realm of stars. The light of the remotest stars comes last to men; and until it has arrived man denies that there are—stars there. 'How many centuries does a spirit require to be comprehended?'—that is a standard, too; with that, too, one creates an order of rank and etiquette that is needed—for spirit and star." [120] It is only a matter of centuries, in both cases—a comparatively comfortable analogy; here one can do something to assist in the synchronization of what is significant, at least in the case of history. The hopelessness of the explosion of light-years still seems to be remote.[121]

The path that had been traversed since the ancient conception of the manifestation of nature as something that occurs of its own accord and something that shows itself is made clear not only by the quantitative predominance of the invisible in the world but also by a displacement of the accent: The visible, as such, has also diminished in its function; now it is only an indicator of the invisible, which is what really matters and for which what is perceivable is a sort of screen. The invisible, in the narrower, optical sense, becomes a metaphor for what is withheld, what is kept secret. The anthropological model assimilates itself to the cosmological one, or the latter turns out to be only a projection of the former. "Countless dark bodies are to be inferred beside the sun—and we shall never see them. Among ourselves, this is a parable; and a psychologist of morals reads the whole writing of the stars only as a parable- and sign-language which can be used to bury much in silence." [122] Whereas Kant had set up a correspondence between the starry heavens and the moral law, now the surface character of the visible prospect of the heavens, on the one hand, and the outwardness of human self-presentation (even to one's own self), on the other hand, are seen as related. One senses that this formulation is directed against Kant's, even though Kant had been the first to draw the conclusion that inner experience was composed of mere appearances in just the same way that external experience was.

This also holds for Kant's other formula, that the view of a countless multitude of worlds annihilates, as it were, man's importance, as an animal creature, which must give back to the planet (a mere speck in the universe) the matter from which it came.[123] For Nietzsche, the view of the worlds already has to be mastered in the universal anthropomorphism. It continues to present a challenge to test self-consciousness against the stock of knowledge, to let it measure itself against appearance. Only in this way could the astronomer also be the sage: "As long as you still experience the stars as something 'above you' you lack the eye of knowledge." [124] That is still entirely in line with the reproach of man's self-belittlement that Nietzsche raises against Copernicus, who, he says, got man onto an inclined plane on which he rolls faster and faster away from the center into nothingness, or who (more precisely) gave man the consciousness of this nihilistic process and displaced him "into a penetrating sense of his nothingness." [g] This self-belittlement is part of the typification of the opposite of that which, in the idea of the superman, would have to prove itself a match for the greatness of nature.

At first glance this censure of Copernicus seems to contrast with Nietzsche's aversion to all teleology, to any hint of an interpretation of his type of anthropomorphism as a new anthropocentrism. Looked at more closely, this superficial contradiction has a deeper basis. For Nietzsche, every pregiven coordination and serviceableness, every preexisting subjection of nature to man's needs means precisely a hindrance to what he demands and expects of man: his self-empowerment over a refractory reality, his violent assault on an untamed nature. The supposed anthropocentrism had to discourage man from pressing his knowledge beyond his stature, from reaching for the "tyrannical rule of the spirit," as long as "one believed that, because everything in the world seemed to be accommodated to man, the knowability of things was also accommodated to a human time-span .... From this it follows that by and large the sciences have hitherto been kept back by the moral narrowness of their disciples and that henceforth they must be carried on with a higher and more magnanimous basic feeling." [125]

Nietzsche did not think in the category of secularization. Otherwise he would still have interpreted the Copernican self-belittlement as a cunning transformation of the Christian ideal of man's turning against himself. Nietzsche could not think in the category of secularization because for him the simply prototypical case of the ascetic turning already lay in ancient Platonism, of which Christianity appeared to him as the vulgarized form; and then nothing remained to be 'made worldly' [i.e., in the literal sense, secularized]. Corresponding to the universal anthropomorphism in Nietzsche's epistemology is the fact that man can learn absolutely nothing about himself from the world that he has always already worked up, except that the world reflects the destiny that he defines for himself. "How can anyone dare to speak of the earth's destination? ... Mankind must be able to stand on its own without leaning on anything like that ...." [126]

Such a philosophy, which has made anthropomorphism cosmological, must, with pre-Copernican linguistic means, not only present a post-Copernican state of consciousness but make it into its own principle. Nietzsche assimilated Copernicus, and functionalized him for his historical metaphysics of self-intensification, in a way that is perfectly analogous to what he had done with Darwin (or what he had even read off from Darwin's schema).

In the end, Nietzsche did not manage to stick to his radical renunciation of cosmological guidance of self-consciousness, to his combination of acceptance of man's parochial perspective with universal anthropomorphism. His late myth of the eternal recurrence of the same is an escape into a central position of man in the world process, a position that while it is no longer demonstrable in relation to space is ensured in relation to time. To the same extent that this myth was meant to conceive things in ancient terms again, it had to be pre-Copernican. It is true that stars, planets, and satellites no longer rotate around a central position belonging to man; but the new emphasis (beyond every morality) that was meant to be put on human action could only be established by making every action consciously and intentionally irrevocable, so that what is contingent and factual is already an eternal law when it is posited. Man cannot be an episode and cannot be a peripheral bystander in the cosmos. The reciprocal confirmation between his parochial existence in space and his episodic status in time is cut off by a compensation, in which the irrevocability of every act at its point in time causes man's spatial eccentricity to be forgotten.

What is forgotten here, and what had to be forgotten, can be brought out by a glance back at an early myth of Nietzsche's in which he introduces his variety of Laplace's spirit as an external onlooker at the great human efforts in the category of "world history," "truth," and "fame"—a "heartless spirit" whose job is to pronounce, in the language of Copernican shock, the net result of man's appearance in the world: "Once upon a time, in some out of the way corner of that universe which is dispersed into numberless twinkling solar systems there was a star upon which clever beasts invented knowing. It was the most arrogant and mendacious minute in the history of the universe, but nevertheless only a minute. After nature had drawn a few breaths, the star cooled and solidified, and the clever beasts had to die. The time had come, too, for although they boasted of how much they had understood, in the end they discovered to their great annoyance that they had understood everything falsely. They died, and in dying cursed truth. Such was the nature of these desperate beasts who invented knowing." [127]

Nietzsche understands this myth as an admonition to renunciation, to renunciation of truth—resigning oneself to self-deception and to being denied information from nature about man: "Does nature not conceal most things from him, even the nearest things—his own body, for example, of which he has only a deceptive 'consciousness'?" But to testify so recklessly about the key that nature threw away is already to renounce renunciation, to break with the illusion. Nietzsche describes not only what others cannot bear but also what he himself will not be able to bear to the end: to let the contingent fact be a contingent fact and to accept "how miserable, how shadowy and transient, how aimless and arbitrary the human intellect looks within nature"—the fact that there were eternities in which it did not exist and that when it is all over with it nothing will have happened. And because he can bear it less than anyone else, "the proudest of men, the philosopher, supposes that he sees on all sides the eyes of the universe telescopically focused upon his action and thought." [128] For the superman is certainly not a figure of resignation in view of nature's deceptions; he counts on the very great assistance of this nature, however unfeelingly it may look on at his revolt.

The acceptance of the fact that perspective is essential is not the acquiescence of the theorist, who is the one who will always have to suffer from this, but the result of an already accomplished scorning of theory, a scorning that at the same time makes what is essential unimportant. Optics and self-consciousness are forcibly disconnected, so that one need no longer be affected by the fact, if it should be a fact, that "the apparent stupidity of the course of the world, the character of wastefulness, of useless sacrifices, is perhaps only a parochial view, a perspectival view for little creatures such as we are." [129]

Just as Descartes had earlier risked his doubt, as the consistent consequence of all skepticisms, only because he believed that he already possessed a novel means by which to establish the process of knowledge securely, so Nietzsche permits the complete Copernican defeat of every anthropocentrism by just this realization—that "The further our knowledge extends, the more man perceives his parochial location" [130] —because, and as soon as, he thinks he is sure of the great antidote to it.

Thus all talk of what man must be able to endure without leaning on anything turns out, as things progress, to be in its turn a lesson about what he is not able to endure: namely, to be unimportant for the world, not to be responsible for it—a possibility that is definitively eliminated with the aid of the myth of eternal recurrence. Nietzsche demonstrates his inability to comply with a prototype of the wise man that is unique in the history of philosophy and that was also admired by him above all else: that of Epicurus's unconcerned gods in the spaces between the worlds. Nietzsche, too, leaves to them the pure aesthetic pleasure with which they may see stars and worlds in their sky, to which they relate in the same disinterested way as the onlooker in the second book of Lucretius relates to the shipwreck that he observes from his secure place on the shore.[131] Once again, here too and on this side of all Idealism, it is very difficult to be a god.

The failure of the aesthetic attitude is the essential motive for the genesis of the myth of eternal recurrence. True, man's new distinction in the universe is only established by his becoming of immense, eternal, and ineffaceable importance for the world without the world having to exist on his behalf, in the sense of the old anthropocentrism.

But this importance is again by the grace of nature, of its unalterable fate, which vouchsafes human history its repetitions. And so the one who is distinguished in this way stands there no differently from the ancient Stoic with his expectation that the cosmic revolutions turn around his location so as to make him their observer and, as such, to favor him. In this way, the imperious disregard for petty facts, operating on the scale of a revolution of values, falls short of the full pretension of self-exaltation. Man acts in his dimension, and the transposition of his action follows by virtue of an alien lawfulness. The idea of producing by human means something that is not based on human interests, is not regulated by them and goes beyond man's parochial existence—this idea of acting for a new morality—is after all only constructed, once again, in accordance with the schema by which we understand art as something that surpasses man's handi-crafts in the same way. When the greater part of the art and book collections of the Louvre had been destroyed by the burning of the Tuileries in May 1871, Nietzsche had written to von Gersdorff, about just metaphysical value of art, that after all it "cannot exist for the sake of poor human beings." [132]

Anachronism as a Need Founded in the Life-World: Realities and Simulation[a]

If Idealism's fundamental thought, in relation to the observer of the starry heavens, was the incomparability of the natural phenomenon to human thoughts and human deeds, then the expressive formula for Nietzsche and after Nietzsche is more nearly that of its unbearableness, but also that of its irrelevance in view of man's own difficulties.

The grandeur, indeed, the sublimity, of the view of nature is retained, or revived, so as to make tangible its whole irresponsible unfeelingness toward mankind.

Hegel's arrogance lies far back. Even among the guardians of his posthumous fame, in the greatest philosophical school of the nineteenth century, that attitude was not widely shared, nor was it much understood in its systematic context. Varnhagen von Ense, always an alert observer of the philosophical scene, notes in his diary on 21 April 1854, about Hegel: "There are splendid things in his *Encyclopedia*, which no one reads any more. One should put together an anthology of them!" Then there follows a single reservation, and it is significant that it is directed precisely at Hegel's relation to the starry heavens: "Hegel is no admirer of the starry heavens, which give him only abstract points of light, the light there being (he says) in its first unassimilated crudity. 'The story has gone around,' he says, 'that I compared the stars to an eruption, a rash, on organic bodies, or to an anthill, which also contains intelligence and necessity. Indeed I do think more of something concrete than of something abstract, more of an animal nature, even if it produces only jelly, than of the host of stars . . . . One can honor the stars on account of their peace, but in dignity they cannot be equated to the concrete individual. The filling of space unfolds in endless substances; but that is only the first efflorescence, which can delight one's gaze. This rash of light no more deserves to be admired than a rash on a human being does, or a multitude of flies.' "[133] Three decades later Hegel's posthumous fame comes to an end, for the century, with the local Berlin event that the last Hegelian, Michelet—who in 1870 was still able (but was also compelled) to write *Hegel der unwiderlegte Weltphilosoph* [Hegel, the unrefuted world philosopher]—loses the chairmanship of the Philosophical Society. In 1873 Friedrich Albert Lange had gone from Zurich to Marburg, in a prelude to the rise of the Marburg school of Neokantianism.

That school had its own arrogance in regard to the view of the starry heavens. Hermann Cohen formulated it when he said that the stars were not given in the heavens but in the science of astronomy.[134] It is Plato's procedure of going behind appearance, but in the opposite direction: The pure heavens that are worthy of theory do not lie behind the visible firmament, but in front of it, in man's possession, in the construction of hypotheses. Neokantianism is a re-Platonizing of Kant. By shifting the accent of Kant's epistemology more and more from the constitutive fundamental concepts, the categories, to the regulative concepts, the "Ideas," [b] it reabsorbed cosmological infinity into the infinity of the task of science. So the "Ideas" no longer imply the toil of extracting concepts from the appearances, because the latter are now in any case only material to be consulted occasionally, and are in any case only visible far outside the great cities with their floods of light and their firmaments that have grown pale. The world of intuition is narrow—it is, once again, the old closed cosmos—and the alternative to it is the realm of pure theory. The superabundance of knowing does not implement intuition—it opposes it.

The scientific material no longer in any way performs as its task, or at least as its secondary task, that of setting intuition on the right path, clarifying it, helping it to achieve a greater depth—not to speak of the fact that there is no intuition that could deny that it is backed up by science.

This is all contained, linguistically, in the narrow confines of the first subordinate clause of a text that brings out the intensification of the conflict between intuition and knowledge so clearly just because it comes from a writer who was still capable of being deeply affected by nature, namely, Adalbert Stifter. In his sixth *Winterbrief von Kirchschlag* [*Winter Letter from Kirchschlag*], one of his last statements on the theory of art, we read, "When one regards the starry heavens and when one knows that the light, which travels 40,000 leagues [*Meilen*; 184,000 miles] in a second, requires thousands of years to reach us from some stars; when one knows that there would be room for a million terrestrial globes in the Sun, or that the Earth together with the orbit of the moon would fit inside the Sun; when one knows that our whole solar system, with its most distant planets, could without any trouble occupy the hollowed-out interior of the star Capella; when one knows that the ring of the Milky Way consists of nothing but solar systems, at immense distances from us, and that the nebulas that we see in the telescope are again galaxies, still more distant and perhaps even greater than our Milky Way; when one discovers with the telescope more new nebulas in the deepest distances of the heavens . . . . " The first condition that was named—that one should regard the starry heavens—is immediately buried under an abundance of further conditions consisting of sheer knowledge, of the exhaustion of a body of data that rapidly becomes obsolete, and that scarcely still clarifies, extends, or deepens the immediate view.

But initially the text—so far as we have quoted it up to this point—only deals with the projection of the progress of the telescopic investigation of the heavens onto a simultaneous representation of space. But Stifter now leaps over this process of the amazing raising of data to higher powers (a process that could in principle be pursued without limit), and poses the dialectical problem of the cosmological antinomy,[c] of the conceivability of a limit. Here, for the first time, the ground finally vanishes from under the will to intuition: " . . . and when one now asks whether all of this has a limit, and if one wants to imagine that beyond the limit empty space continues, and if one cannot imagine that, and if one wants to imagine that the universe goes on into the infinite, and if one cannot imagine that either . . . . " Yes, what exactly happens then? Does one then travel, once again, the path that led Kant through the Transcendental Dialectic? The surprising turning in the text consists in the fact that it breaks off from the abstract self-suspension of reason's immanent compulsion to totality and gives itself an aesthetic point. After the greatest possible detour, by way of the totality of the thinkable, a new basis for relating to an object emerges, because, immediately after dialectically establishing what is no longer imaginable, Stifter's text is again dealing with what stands before us: " . . . and if one cannot imagine that either—then a beauty rises up before us that delights us and makes us shiver, that blesses us and annihilates us." [135]

The artful naiveté with which this paradox is prepared—the extraction of the feeling of the beautiful, in its ultimate universally human simplicity, from these studied delays—makes the contribution of the mass of knowledge, ending in the failure of the questions that it excited, appear unforeseeable: "All human beings have this feeling, from the dullest, who gets pleasure from a red rag, to the sage who contemplates the starry heavens." It is supposed to be possible to develop the feeling for beauty through knowledge, to experience a higher beauty on a correspondingly higher plane; but just this is refused by the text, on account of the dialectical break, which it does not get past. "There human thought and human imagining go no further." After that it sounds like a stubborn assertion when we read, introduced with an "And still," that "on the peak of a mountain under the immense bell of the heavens, when on a clear winter night the many millionfold world burns above our heads and we walk along beneath it in contemplation, a feeling can come into our souls that majestically covers over all our little sufferings and troubles and makes them fall silent, and gives us a greatness and peace before which one bows down." For one who does not arrive at this point, because that knowledge may not have produced this higher feeling, there stands ready, unexpectedly, the little secondary idyll: "And he who cannot feel this, because he lacks the development, is nevertheless gently moved when, on a still night, the nearly full moon shines down, and over everything, both clouds and meadows, the silver mist of light spreads, continuing the twilight, silently."

Since he accomplishes effortlessly the transition from the sublime to the idyll, from the overwhelming to the small form, the author has no trouble making the beauty that men produce arise from the same feeling out of which they experience the beauty of nature: "Because the feeling for beauty is so implanted in them, men have also attempted to produce beauty themselves. They have invented the arts." This reciprocal substitution is so unproblematic that we can no longer be surprised at anything, including the assertion that art, insofar as it imitates nature, is meant to function as a sort of didactic guidance from this artificial beauty to that natural one.

Here we are approaching the point where reality and simulation complete with one another and become interchangeable, the point at which technique only has to be perfected in order not only to call the original experience to mind but to intensify it. And from the point of view of the object, has it not become a matter of indifference whether what we experience really exists, when after all the objects themselves, whose light signal we sense, can be long since gone at the moment of this experience? The principle that art is the imitation of nature was valid, in its original meaning, for every kind of 'technique'; but its use can be seen to reach a point at which its turn into anachronism becomes palpable: "The arts imitate nature, both human and extra-human, and because in the arts the beauty of nature seems to be more limited, smaller, and produced only by men, it is apprehended by most people much more easily than it is in nature; indeed, it is very common for a person to be led only from first perceiving beauty in art to perceiving the infinitely greater beauty in nature."

The domestication of the post-Copernican world into the idyll of Kirchschlag makes anachronism and banality the almost unavoidable fate of such a text. What is the relation between our catch phrases and the experiences that are possible for us? Even the unequal potency of the anachronisms that emerge for the post-Copernican situation is instructive. Their order of magnitude fluctuates between overexertion and harmlessness, the former in the recourse back to myth as the recurrence of the same, the latter in the retreat into the obligingness of a firmament that is now only separated from its integration into the 'landscape' by our knowledge of how its dimensions are measured in light-years. The overextension of the scope of what can still be grasped and equipped with experiential qualities prepares the way for the collapse into the mere banality with which a "love of the starry heavens," as an attribute, can be conferred on subjects of obituaries or biographies—and can hardly still be denied. The Assyrian king Ninus was the prototype of the warrior-hero who had boasted of never having looked at the stars or felt any desire to do so.[136] "The sword even of Napoleon the world-conqueror," Hebbel writes in his diary in 1862, "would have fallen from his hand if he had ever regarded the Milky Way and formed a clear idea of what the Earth represents over against the universe."[137] The propaganda of the tyrants of the more recent past supplies them with love of children, animals, and stars as the cheapest cliché, from which the historian still thinks he can extract a significant character trait when he informs us that Hitler was deeply affected by the starry heavens.[138]

The modern city is not favorable to the contemplator of the heavens. It wraps itself in its own light. That is a point that was not foreseen in the tradition of the metaphors of light,[d] and one that makes things still easier for all sorts of frivolous 'cultural criticism': not to be able to see the lights in the sky on account of too much light. It was probably Kierkegaard who first found an image for this form of self-darkening by means of one's own light. In 1845 he notes the following comparison in his diary: "When a rich man goes driving in the dark night with lights on his carriage, he sees a small area better than the poor man who drives in the dark—but he does not see the stars; the lights prevent that. It is the same with all worldly understanding. It sees well close at hand but takes away the infinite outlook."[139] The "rich man" here is not a social but a religious category. It is the person who relies on his means, on his realism of what is nearest at hand, and his light is unquestionably that of rationality: It does indeed enable him to see better, but it hides the stars from him.

Kierkegaard is one of the anachronistic figures in his century. The fact that he could be, and had to be, 'rediscovered' reveals that the conditions of the reality that was contemporary with him were not those under which it became impossible to ignore him. Kierkegaard could not yet anticipate that the city, by illuminating whole landscapes and turning the night into day, would extend its screening off of nature to the starry sky as well and would create, if not a 'blinding system' ['*Verblendungszusammenhang*'],[e] at least one that screens things off [an '*Abblendungszusammenhang*']. This system deprives the old direction of one's gaze upward (a direction that is by no means a matter of course) of its meaning, of its objective correlate in something inaccessible, unconvertible, nonnegotiable—that is, of that limiting value of every culture at which the 'relation to practice' fades out. The city is a competing reality; it is so noisy and so bright in order to make everything other than itself, and nature above all, forgotten, even when it encloses and conserves it, in nature preserves.

The rich man of whom Kierkegaard had spoken in his simile can get out of his carriage or extinguish his lights so as to look at the sky. This is, in fact, the admonishing, preaching meaning that must always be understood as part of this thinker's intention. For genuine or chosen poverty, Kierkegaard thought, there continues to be the primeval and timeless possibility of "driving in the early twilight and seeing one star, until the darkness deepens and one sees more and

Chapter 8

more . . . . " But as an alternative of human existence, the one Kierkegaard describes is no longer part of modern reality. Entirely independently of the attitudes and willingness of individuals, the city
constitutes a secession from one of the most human possibilities: that
of disinterested curiosity and pleasure in looking, for which the starry
heavens had offered an insurpassable remoteness that was an everyday phenomenon.

One of the absolutely cynical thoughts in the letters from Satan
that Mark Twain prefixed to his *Letters from the Earth* is that not a
single star stood in the sky over the biblical Paradise; only three and a
half years after the completion of the Creation did the light of the first
star reach the Earth. "At the end of the first hundred years there were
not yet twenty-five stars twinkling in the wide wastes of those gloomy
skies. At the end of a thousand years not enough stars were yet visible to make a show. At the end of a million years only half of the
present array had sent their light over the telescopic frontiers, and it
took another million for the rest to follow suit, as the vulgar phrase
goes. There being at that time no telescope, their advent was not
observed." [140] The satanic point of this account, given by the Devil, of
the results of the divine experiment is that the view of the heavens
could not possibly have been a privilege granted to man if even in his
paradisaic state it had not been present for him at all; this part of
nature, too, would have had nothing to do with man's happiness.

More important, though, is the indication that the bitter satire
contains of the problem of reality: By putting the cosmic foreground
in the perspective of something whose simultaneity as a given for
perception is accidental, it confirms positivism's reduction of reality
to naked sense-data in a way that makes technical simulation cease
to take second place. It has become a matter of indifference which
reality corresponds—and whether any reality at all corresponds—to
the points of light that move in an exactly describable manner in the
heavens, and the measurement and classification of whose light makes
up the sole content of a possible theory. "What then is real in, for
example, a star that we see shining in the night sky? Is it the glowing
matter of which the star is composed, or is it the sensation of light that
our eye receives from it? The realist says the first, the positivist says the
second. Each of these propositions has its attractions and can be
defended with convincing arguments. But neither of them can claim
to be the only legitimate one. But if one considers both propositions

admissible, then the word 'real' no longer has any definite sense." [141] But then the dispute about the truth of the Copernican world-system would also be deprived of its impact. "Accordingly, Copernicus is to be valued not as a pathbreaking discoverer but as a highly gifted inventer. Positivism takes no more notice of the great spiritual revolution that his doctrine brought about, and of the bitter battles that were fought over it, then it does of the feeling of silent awe that the sight of the starry heavens inspires in an attentive viewer . . . ." [142]

The absence of the view of the heavens under the conditions of the modern city and the attenuation of the theme of reality under the conditions of modern science produce the medium in which the technical simulation of the starry heavens could become actual in the museum. This scientifically prepared experience is unique in comparison to the protection of other parts of nature in the enclaves of the exotic—zoological gardens and botanical parks—in which nature presents itself in solitary and weakened form, in isolated examples, whereas the starry heavens, in their abstract character, seem almost 'strengthened' by the technical simulation. The representation of nature overcame the advantage enjoyed by the objects in technical exhibitions—their superevident and exponentially intensified presence—when finally, with the Zeiss planetarium at the world's fair in Montreal, even the impression of motion in space (for example, that of drawing near to the planet Saturn) could be produced. But the effect of space travel in the planetarium entirely removed from consciousness the idea that this one object of theory is not at man's disposal. It has become possible to choose one's standpoint in the universe: In the planetariums of the Sixties the viewer could already be given the impression that one receives while circling the Earth, on any particular orbit, in a satellite, or the view of the starry heavens from the Moon or from a planet. Just as one can carry out experiments about the nighttime orientation of migratory birds in planetariums, astronauts received their astronomical training in the Morehead Planetarium in Chapel Hill. Simulation prepares the way for access to and traversing of its objects.

The planetarium is the mausoleum of the starry heavens as the ideal of pure intuition. As a technical phenomenon it is deeply rooted in the nineteenth century's longings for a popular knowledge of the starry heavens, longings that expressed themselves in "people's astronomers" and "people's observatories." They retrieved the reserved property of science as a relic for a 'natural' mass religion of the solved 'riddles of the universe' and of ersatz emotions. To that extent the planetarium too is an end, an end of what Ernst Haeckel wrote in his *Welträtsel* [*Riddles of the Universe*] (a book that was distributed in many hundreds of thousands of copies): "The astonishment with which we gaze upon the starry heavens and the microscopic life in a drop of water, the awe with which we trace the marvellous working of energy in the motion of matter, the reverence with which we grasp the universal dominance of the law of substance throughout the universe—all these are part of our emotional life, falling under the heading of 'natural religion.' " Modern man, Haeckel went on, does not need the narrow enclosed space of a special church in order to live in this religion; he finds his church "through the length and breadth of free nature, wherever he turns his gaze, to the whole universe or to any single part of it . . . . " [143] It is harsh, but indispensable in order to display the arc of this theme's development, to quote immediately after the enthusiasms of this certainly important zoologist and theoretician of "family trees," from 1899, what Hitler said on the subject in conversation during the noon meal in his headquarters on 5 June 1942: He had "directed that every town of any importance shall have an observatory, for astronomy has been shown by experience to be one of the best means at man's disposal for expanding his view of the world and thus saving him from any tendency towards mental aberration." [144]

Under the artificial skies of the planetariums, the upright carriage of the observer of the heavens can be practiced sitting down, with the gentlest constraint to adopting the attitude of the onlooker in repose. Here, if anywhere, one should inevitably expect the demonization of the technical surrogate for the most sublime object—of the projected heavens as the false heavens. If one disregards the context of the [particular] concept of reality,[f] into which this simulation fits as one of its logically most consistent elements, it is easy to make sarcastic fun of the false starlight and the false salvations that are sought under it. Nevertheless, this marvel has seldom been so little marveled at as in the work of Joseph Roth, who had his "first encounter with Antichrist" under this technical backdrop. [145]

Roth writes a book of unmaskings. He follows the old pattern of the Platonic discovery that the realities with which we deal are only shadows and imitations; but he goes a step further beyond this schema when he establishes that everything that is even capable of being imitated is thereby lowered in its rank in reality. It is an attempt to oppose even the concept of reality that allows imitations to be real [*wirklich*] because they are efficacious [*wirksam*], without prejudice to what they may be derived from. Not only the shadows of the Platonic cave are convicted of their existential weakness, but the Ideas themselves are too, because it is still possible for those shadows to be their final derivative and the extreme indicator of their origin. What we have before us is a mirror-image reversal of Platonism: If in it the null grade of reality, in the shadows, was only possible because as images they were subordinate to the essentially imageable Ideas, now the unreality of the projections is only possible because their 'originals' already suffer from unreality, so that "the reality that they imitate so deceivingly was not at all difficult to imitate, because it is not real." This description of the cinema could in its turn be an imitation of the classic of this sort of culture criticism, Max Picard's *Das Menschengesicht* [*The Human Face*] of 1929: "Indeed, the real human beings, the living ones, had already become so shadowlike that the shadows on the screen had to seem real." The unreality of reality is responsible for the artificial reality of unreality.

What Joseph Roth calls "the Antichrist" is the sum of the false realities. The boy encountered them for the first time at the beginning of his *paideia* [education], in his Platonic cave: Not only the shadows but the cave itself was, so as to make the shadows possible, an artifact. "In those days a great wagon came along, drawn by invisible powers, and remained standing on an open space before the city. To begin with it sent a great machine forward, which was covered with a little tent made of linen, and on this a great tent, also made of linen, was spread out and set up like a dome, and if one went inside, the inside of the dome was a blue sky with many gold and silver stars .... The dome was blue, and the stars were just as inaccessible and just as close as real stars are. For since a human being is not even tall enough to reach the roof of a circus tent erected by others of his kind, it did not matter to the person who sat beneath the dome whether it was the genuine sky or a copy of it. He could grasp neither the one nor the other with his hands. Consequently he was glad to believe that the one was the other, or vice versa. And since it became quite dark beneath and inside this dome made of tent linen, he was convinced that he sat in the midst of a clear, starry summer night ...."

Of course, under false heavens one can encounter false salvations. But they come from false expectations of an 'authentic' and ultimate reality, of the genuine substance of nature that, because it is genuine, is at the same time not ready to hand. The demand for an authentic reality presupposes that one could tell by looking at the real that it is not the unreal—as long as one does not have to deal exclusively with the latter. But the production of this exclusiveness is what the Platonic cave and its technical successors imply.

The modern age added to this premise a further one. In Descartes's consideration of doubt, the possibility is accepted that all the characteristics of the real could be imitated without the production of these characteristics having to generate, at the same time, the objective equivalent of reality. Leibniz was the first to urge, against Descartes, that the complete simulation of reality would in the end no longer be deception, because a deception requires both the implication of an assertion of what does not exist and that the person affected could suffer from being disillusioned, neither of which is the case here. The Baroque idea that life could be a dream has no terrors for Leibniz because expectation is determined by a new concept of reality in which the internal consistency of everything that is given is identical with all the 'reliability' of reality that is still possible.[g]

There is something questionable and productive of misgivings in the demand for ultimate authenticity in all experiences, for an unmediated relation to the original, in a world that is characterized by overcrowding and can no longer keep open all paths to everything. This is no longer and not only a matter of the sincerity of one's desire, not least of all because simulation surpasses artificially unaided [natur-wüchsig] intuition. The starry heavens of intuition in the life-world are motionless for their viewer; if one also assumes that the everyday opportunity to view the heavens occurs at about the same time of day, there remain only the gradual seasonal displacement of the constellations, the Moon's changes of phase, and the (even more difficult to perceive) motion of the planets. It is just not true that the natural heavens rotate soundlessly around the viewer; only the herdsmen of Chaldea were credited with having this experience without having any professional interest in having it. In contrast to this, the planetarium is a short of temporal telescope, which puts the static heavens in motion and by means of technical projection makes visible things that were never seen, that were really only disclosed by comparison of observations. Here it is a question not of duplicating experience that, with some effort, would also be possible 'in the original' for anyone at any time, but rather of augmenting what can be seen at all.

Part II

The Opening Up of the Possibility of a Copernicus

1

The History of What Led Up
to the Event as Conditioning
the History of Its Effects

In our working up of the temporal dimension of our orientation in reality, a change of interest is beginning to be evident that combines satiety with the long-dominant pattern of 'histories of what led up to' something ['*Vorgeschichten*,' literally "prehistories"] with a turning toward the pattern of 'histories of [the thing's] effects' ['*Wirkungsgeschichten*'].[a] Histories of effects seem to meet halfway the need for current practical relevance, if only because in them the process that is examined at least approaches the present in which we examine it, whereas the direction of questioning in those 'histories of what led up to...' loses itself, with every step in the successful identification of new preliminary stages and forerunners and additional factors and ingredients, in the temporal distance of the past, of which one gladly believes that with the distance of the historical self-experience, its informative value must decline. Thus the 'events' of history, defined in the broadest possible sense, have data assigned to them in a symmetrical way within their temporal horizon, which embraces on the one hand the totality of what prepared the way for them and made them possible and on the other hand the aggregate of the changes that they cause or that unfold in response to them.

Any analysis of this change of interest and of its methodological implications, however, leads one to doubt whether there can be an exclusive set of alternatives corresponding to the terms that we are being urged to embrace. The meagerness of many 'histories of effects' is evidently a result of the superstition that in order to present the magnitude that we call something's 'effects' (as the bundle of data lying exclusively on the temporal side of it that is closer to the present) adequately and, in the ideal case of completeness, definitively, the documentable occurrence of a name or of a reference to a fact is sufficient. This procedure is comparable to the equally superficial procedure of a history of a concept in which the occurrence of particular words, as names for the concept, is taken as the exclusive guide.

In the case of the genesis of the Copernican world we certainly are dealing with a case of powerful effects, but also one in which the history of these effects is complicated. One cannot study it without confronting the problematic of the supposed alternatives of the history of what led up to it and the history of its effects, and in fact doing so not only in anticipation and abstractly but also in continuous contact with the concrete demands of the analysis of this case. Then the appearance of an exclusive set of alternatives, in regard to which we could decide in favor of the more 'modern' type and thus be sure of the applause of our contemporaries, quickly vanishes. This does not prevent us from being able to admit that even in the formula that the history that leads up to something conditions decisively the potential for the history of its effects, the interest in the 'history of [its] effects' is determining, and thus the required renunciation is bestowed on the antiquarian historiography of antecedents that disappear in the gray mists of time.

However, I do not want to pass over the fact that I regard the absolutism of whatever happens to be the present as precisely the kind of superstition—making itself into the center—the destruction of which will always be the business of the theory of history.[b] That is why I cannot accept an 'interest' in history as a preeminent criterion in historical cognition. It is entirely possible to object that in the present case, with this turning away from a specific form of anthropocentrism *in time* a particular element from the history of the effects of Copernicanism is made into the principle of the description of that history and of the solution of its methodological problems. To this I can only reply that the anthropocentrism that affirms and posits itself must be transparent to itself and should not be based on the illusions of the organization of our equipment for experiencing the world, if that anthropocentrism is meant to be constituted as a rational option for consciousness.

The history of what led up to the Copernican event—an event that is datable with all desirable precision by the appearance of the *De revolutionibus* in 1543—cannot be reduced to the diagram of an angle composed of lines converging on this point. For this 'prehistory' contains not only conditions determining the reforming work itself, but also those determining its reception, and indeed even the preconditions, which the author could only surmise, of its acceptance. The strict separation between 'prehistory' and 'history of effects' is promoted in this case by the impression that Copernicus's work was such a 'foreign body' in its spiritual or intellectual environment that the violent reactions with which it was rejected by the powers that were, culminating in the fates of Giordano Bruno and Galileo, practically define its typical effect. But precisely this is a mistake. The susceptibility to irritation that characterized the reaction to Copernicanism at the beginning of the seventeenth century has novel preconditions that were not present at the time of the event or during the preparation for it that we can comprehend in terms of Copernicus's biography.

So as to document this very palpably I cite the earliest historical testimony to an encounter of the highest spiritual power of the age with Copernicanism's basic idea: As early as 1533, the German jurist and orientalist Johann Albert Widmanstadt explains Copernicus's doctrine of the motion of the Earth to Pope Clement VII during a walk in the Vatican gardens ("... in hortis Vaticanis copernicam de motu terrae sententiam explicavi") and receives from the pope, in acknowledgment of this communication, an illuminated Greek parchment codex, in which he makes a note not only of the occasion for the gift but also (as a jurist) of the witnesses of the event as well.[1] That there was no supreme displeasure, in the cheerful environment of the gardens, at the offense to the Earth, is not mere Humanistic or devout glossing over. When, ten years later in Nuremberg, Andreas Osiander added his much-criticized preface[c] to be printed with Copernicus's manuscript, he must, from his standpoint, have seen the situation for the work's reception differently from the German scholar in Rome, who could quite confidently believe that he was putting an interesting and inoffensive novelty before the pope. The expectations that an author has in regard to his audience's reactions are quite certainly part of the 'prehistory' of his work, but at the same time they are also part of the complex of factors conditioning the history of its effects.

Judging by Widmanstadt's report, Copernicus cannot have appeared to Clement VII as one engaged in violent transformation. Of course everything would depend on knowing what Widmanstadt's account "of the motion of the Earth" really contained, because this singular [form of the noun: motion] is deceptive and definitely admits the possibility that it was restricted to the daily rotation, which had already been considered and proposed many times and is easy to make plausible, but which can leave the geocentric systematics untouched. Contrary to the external sequence of the presentation in Copernicus's work itself, the Earth's annual motion around the Sun is the heart of the reform and the cross roads of its reception. For here in principle, if not for the accuracies of observation that were possible at the time, it is all over with the pure equivalence of modes of describing identical phenomena. Lichtenberg correctly said, in reference to Bradley's proof of the annual motion by means of the aberration of the light of the fixed stars, that by it every star in the heavens proves that the Earth revolves around the Sun and thus proclaims the glory of Copernicus.[2] If one interprets that metaphorically, as it is intended, Copernicus changed our vision of the world so radically that this change confronts us everywhere and in everything. This makes its effect an epoch-making one.

Goethe, who supplied the quotation most often used regarding this epoch-making effect,[d] made the still more general statement about the history of the sciences that in it "the ideal part has a different relation to the real than it has in the rest of world-history."[3] It is in fact our central problem that, in the history of science, none of the things that a theory of history has been able to assure itself of in other areas holds true—the case in point, here, being the derided (or amiably accepted) late arrival of the ideal in comparison to the real: the twilight nature of thought, which is hardly able to catch up with the things that it after all wanted to have and was supposed to have produced. Whatever the nature of these relations may be and under whatever conditions they may hold, in the history of science, at any rate, the reproach that it has to catch up in this way—which has given philosophy's claims for itself so much trouble—does not hold. Here theory does not merely read off what practice already possesses.

People have thought that they could carry over the model of technical invention as a reading off of mechanizations that have already been realized in the process of production—the model that Marx provided in chapter 13 of *Das Kapital*[e]—to the history of science. The fact that Galileo's visit to the arsenal of Venice was just as prototypical as Descartes's concealed, surreptitious dependence on the achievements of vulgar mechanics, in the form in which they were familiar to the engineers of his time, seems to speak in favor of this analysis. But Galileo described his experiences in the arsenal of Venice by means of the Platonic doctrine of anamnesis, and tried to hold to this schema. It is based on the unmistakable fact that to learn in the arsenal is only possible for one who is already on the track of the regularities that are realized in the multiplicity of the equipment there. Anyone else would be blind, condemned to the mere imitation of external data. This state of affairs becomes clearest of all at the moment when Galileo receives the news of the telescope and is immediately in a position to 'invent' it himself as well. The priority of thought over reality; perhaps also only the easiest summoning of thought by a specific experience; the immediate leaping in of generalization—these are securely established in the case of the history of science.

The ease with which we confuse 'preconditions' and 'effects' in the history of science is connected with this same phenomenon. The priority of thought is not necessarily that of the specific occurrence of the central idea of the new conception. Copernicus himself valued highly, in relation to his potential influence, the fact that he could point to forerunners and predecessors, especially ancient (real and supposed) predecessors. But an analysis of the relation between his work and its effect shows, rather, that with his references to the Pythagoreans and to Aristarchus of Samos he was able, at most, to head off the superficial reproach of having a mania for innovation, while the mass of difficulties did not have to do with the possibility of entertaining the new thesis, as such, but rather with the background conditions of its compatibility with physical and metaphysical assumptions of the Aristotelian and also of the Platonist tradition. An astronomical reform that was this radical just could not merely be an affair of the discipline of astronomy. In the narrowest area of his specialty, Copernicus would not have had reason for dissatisfaction when even Erasmus Reinhold, the son-in-law of his resolute opponent Melanchthon, based his *Prutenic Tables* on the Copernican system. Since Copernicus demanded more than the agreement of the authors of astronomical tables—namely, the acknowledgment of a new cosmological truth—he would himself have put in question the success of his work if he had not been sure that he had something else to offer, as argumentation, besides a model on which to base calculations.

This wholly inescapable logical consequence in regard to Copernicus's own assessment of his possible effect gives rise to the task of describing the preconditions and effects of what he did in such a way that the matter does not end with the 'prehistory' leading from the central idea of reconstructing the planetary system to the signs of his having caused an Earth that had hitherto been at rest to tremble. The task is to identify the background conditions of the assurance that enabled Copernicus to have any expectation at all that his readers would consent to the work's full claim to truth. Even if, biographically, an encounter with the heliocentric proposal of an ancient predecessor—were it only an encounter with the Pythagoreans' misleading pyrocentrism—had given the first impulse or the decisive impetus toward his new scheme of the world system (a hypothesis that, to be sure, no evidence supports), he would still have had to have the central concern, the inevitable problem: How could he, for his part, avoid his forerunners' and predecessors' manifest failure of encountering an audience that is not only uncomprehending but committed to incompatible assumptions?

I shall try to clarify this by means of the most important example that exists in this connection. One can describe the completion of classical mechanics as satisfying the needs to which Copernican astronomy had given rise, insofar as it wanted to hold to its emphatic claim to truth and to achieve the alliance with physics that was necessary for that purpose. Then it is correct that the principle of inertia, in its final formulation by Newton, was a necessary consequence of Copernicanism, and also presents itself that way in Newton's intention of providing the final confirmation of Copernicanism. "Looking backwards, we can see how the acceptance of the Copernican concept of a moving earth necessarily implied a non-Aristotelian physics." [4] But it can be shown to be no less correct that while the Copernican reform could perhaps have been thought up without an equivalent of the law of inertia, it could hardly have been accepted in such circumstances, and still less could it have been made the principle that forced the emergence of a new physics. Here lies the problem of the intermeshing of the history of what comes before and the history of effects: Without a certain amount of the respect that is shown to reality, Copernicus would have become the father of many works of astronomical calculation, but not the instigator of new considerations in physics.

The principle of inertia is Copernicanism's logical outcome exhibited in pure form. It comprises the removal of all the theoretical difficulties that stood in the way of acceptance of the new astronomy. Nevertheless, one can doubt whether the principle of inertia would not, historically, have been developed just as quickly, as an empirically unprovable, rational axiom of physical autarky, even without the Copernican reform of astronomy.[5] Without being able to assume, in the background of consciousness at the time, the equivalent, however vague it might be, of a principle of continuance [of *Beharrung*, the corollary of inertia], Copernicus would have been able to count at most on founding a new sect of lovers of paradoxes. Here the important thing is that the objections that Ptolemy had already brought forward against giving up the immobility of the Earth could no longer be effective for Copernicus's contemporaries in the same way that they were for the Aristotelians of medieval High Scholasticism.

In exploiting the increased scope for variation in dealing with this question, Copernicus everywhere chose the path of the greatest economy in argumentation, the least possible commitment to the opinions of specific schools. I would like to contribute something to increasing our admiration for this miserly style of argumentation and for the governing consideration, in it, for what he could expect of his contemporaries. Often the means with which Copernicus seeks to widen the latitude for what he has in mind do not exceed the degree of intensity of an allusion—for example, when, by emphasizing the spherical shape of the Earth, he establishes its equivalence (which at the time was unheard of) with the spherical shape of the outermost heaven [the sphere of the fixed stars], so as to make the predicate of natural motion transposable, even before he has provided good reasons for actually carrying out the transposition. The fact that Copernicus does not go into these presuppositions in the area of natural philosophy more deeply—which is something that he was immediately to be reproached for—not only gives him the advantage of being able to keep out of the arguments between the philosophical schools but also is the only thing that makes it possible for him to imitate Ptolemy's *Almagest* very closely in the structure of his work and in its technique of demonstration, and thus to avoid the odium of having a mania for innovation. At the same time it is clear that with an audience with different expectations this mode of presentation, with its strict parallelism, could have had exactly the opposite effect, that of underlining the difference.

Of course there are also 'effects' that do not need to be under suspicion of having already been preconditions beforehand. There are cases in which the carrying out of an implication did not have to overcome any rigid blockades, or where the use of undefined concepts conceals the demands [on the audience] that had to be contained in them. Something of this sort holds for the enlargement of space that emerged as a consequence of Copernicanism, because the new doctrine had to deal with the difficulty of explaining why the annual motion of the Earth around the Sun does not produce any optical parallax in the fixed stars. Here Copernicus could argue that one had to assume that the heaven of the fixed stars was at an immense distance, in comparison to which not only the size of the Earth but even the diameter of the Earth's orbit remains so insignificant that no ascertainable optical deviations are produced at points that are diametrically opposite to one another on the orbit. This has nothing to do with the much-discussed Renaissance need for infinity; no doubt, though, the need for this inexpensive expedient in argumentation is made less of a difficulty and is finally picked up as inspiration by the new enthusiastic dealings with the concept of a universe that is hemmed in by no transcendence and that itself approaches the attributes of divinity. In order to make quite clear to oneself what an encouragement this operation with the scale of the universe represented, one only needs to imagine how much more difficult it would have been to take the same way out in the High Scholastic thirteenth century. It was, after all, only the first step in a sequence—uninterrupted up to the present day—of multiplications that become necessary (and that mostly take the form of leaps) in the magnitude of the empirically accessible universe that is assumed in each case and that was already, shortly before, regarded as immense.

The difference between the two examples I have adduced is obvious. In the case of the assumption of inertia, or its equivalent, it is a question of an indispensable condition of the possibility of raising and answering realistic questions in regard to the physical constitution of an Earth that is now thought of as being in motion, for here there is no sort of 'transition' between the state of rest and motion, and the order of magnitude of the annual motion is fixed as soon as it is considered at all. So this problem is not quantitative, as in the case of the necessary enlargement of the distance of the fixed stars, where after all it was only a matter of introducing a new degree of indefiniteness into people's idea of the magnitudes. One only had to think of what was great as somewhat greater yet—to declare the unmeasured to be the unmeasureable—in order to arrive at the desired protective assertion.

The Copernican reform cannot be presented as the product of various series of factors that had their point of convergence in the epoch-making work, the *De revolutionibus*. The latter's historical inevitability is more that of the traditional system's incapacity to function, its contradictoriness and unwieldiness, which is less palpable in its quality as a theory than in its applications, perhaps most of all in the application that in the meantime has become the most contemptible: in astrology. But the pressure of the inevitability of an innovation does not in itself create the latitude in which it can be discovered and successfully introduced. The proposition that necessity is the mother of invention has scarcely any convincing illustrations in the history of science. What is the mother of invention is liberation from obstructive fixed principles all around the critical area, principles that previously left no room for facility in carrying out thought experiments and playing through possibilities. It may also be possible to show how the innovative conception positively came about; but since for ideas of lasting importance there is often no history of what led up to them to be found, or only a very artificial and forced one, we are more intensively occupied with the question—which is indispensable for all knowledge of history—as to how what was impossible, intolerable, and systematically excluded at some earlier time was even able to get within the radius of the attainable, of what people could consider.

This 'prehistory' of increased potential, of the opening up of freedom for theory, is, in its inseparability from the history of the event's effects, the strictest definition of a historiographical 'interest,' if such an interest should survive the constriction of the temporal dimension under the pressure of the spatial diffusion of interests at all. Our subject is the Copernican nootope (if this construction is permitted, by way of exception, in imitation of "biotope"): the conditions of the possibility of the fact that there is any such thing as a history of Copernicus's effects—which is by no means a matter of course, since there had not been such a history in the case of Aristarchus of Samos.

To speak, in the history of science, of a "latitude" for possible changes means to determine the breadth of variation within which certain theoretical actions are possible and others are excluded. The narrowness or broadness of this enclave of the possible inside the occupied territory of supposed necessities, with the restraint or the freedom that it imposes on or grants to intellectual motions, is determined by the stability or instability of the system of world-explanation in which it inheres. Such systems are not primarily aimed at leaving lacunae for possibilities, but rather precisely at shutting down the motives for laying claim to heterogeneous, extrasystematic potentials. A system is meant to be, quite simply, satisfying, to achieve the lasting calm of homeostasis. People commonly imagine that it is precisely the state of satisfaction, when it is achieved, that causes the emancipators—first the amateurs, then the professionals—to appear; however, historical experience seems rather to tend to allow systems to decay of their own accord, as a result of their internal contradictoriness and their lack of elasticity in the face of unmastered reality, as a necessary precondition of skeptical or dogmatic negation. In connection with this question, too, the pre-Copernican glacis provides manifold indications of subcutaneous processes.

But to discuss the latitude for the Copernican reform as a question of 'background' determination does not mean to speak of Copernicus's (supposed, and perhaps real) 'forerunners,' except to the extent that they can be indicators of the expansion, in each case, of the horizon of possible variations. That is so, if for no other reason, because most of the forerunners of Copernicus who have ever been named never became known to him, either directly or indirectly. Thus, in regard to the anticipation of the Copernican diagram, it would only be a matter of historically insular phenomena, to the extent that one could not make progress on the question of their rhizome.

At this point the question inevitably arises as to whether one can draw conclusions, and if so what conclusions, from the history of science in the early modern age that apply to the later, or even already the 'late,' phase of the process of science. I am not very optimistic in this regard. Science has not only reached an overwhelming tumescence, quantitatively; it has also changed its overall state qualitatively. If one were permitted to put it this way, I would describe it as having "less and less history," or being less and less historical. Since the nineteenth century the tendency of the development of problems at least in the natural sciences is more and more subject to an *immanent* lawfulness of the individual disciplines and of their influence, dependence, and overlapping on one another in a compound that in its turn is also immanently regulated. The more rigorously this inner logic of the unfolding of a theoretical totality is developed, and the more dependent the acute questions at any time are upon the results already arrived at, the less the course of the theoretical process can be influenced by heterogeneous conditioning factors and the less important that background of more general prior assumptions and that free space of the possible become. To put it differently: The realization of knowledge is driven from behind, by the inherent pressure of its workings, and is not (as it were) challenged from in front by the destruction of outdated assumptions.

The present-day study of the history of science is stamped by what we encounter in the reality of science as the modern schema of problem-construction and arriving at results. It is not difficult to foresee that the history of science in the twentieth century, if it ever can be and ever is written, will have to be of a very different type from that of, for example, the seventeenth century: It will have to dissect out the institutionalized dynamic of the theoretical events and es-tablish the impotence of all external motives, and of everything in the 'background,' in contrast to that iron logic. Here it is indeed touching to see how much futile effort has been expended in trying to demon-strate that it is great and powerful interests that keep the workings of science in motion. In contrast to that, one might rather speak of the helplessness of all those who consider it possible to influence the course of the scientific dynamic, or who would like to do so (from motives that may range from the humane to the base), or who even think that they already have such influence. This institution has an autonomy that is just as impressive as it is alarming, even in cases where it has been possible to accelerate the derivation of by-products by forcible means.

To verify that for oneself one need only think of the surprising simultaneity in the appearance of particular [scientific] milestones in political systems that are painstakingly shielded from one another, a simultaneity that, in the face of radically different backgrounds of world-interpretation, epistemology, and social self-understanding, and with dissimilar scientific organization, again and again produces unexpected (but in retrospect seemingly matter-of-course) conver-gences and analogies. If science ever had the function of a "superstructure" of socioeconomic relations—which is methodologically very difficult to prove or to disprove, but for that very reason makes an impressive assertion—then this vertical structural connection must certainly have increasingly weakened. That explains the doubts about this theory precisely where it should, in view of its own 'system-immanence,' have been preserved. The example of Lysenkoism shows how in spite of the pure conformity of a far-reaching interdisciplinary premise to a reigning system, and despite rigorous shielding and insistence, it is able to endure against the immanent logic of the theoretical process only for a relatively short period.

Whether or not one wants to call this the late phase of the formation called science, it no longer has much in common with the configurations of elements and factors that have to be analyzed in the case of the relation between the 'prehistory' and the history of the effects of Copernicanism. The preponderance of internal logic over determinability by external influences is not the structure that we have to assume in the case of the beginning of modern science.

2

Loosening of the Systematic Structure through Exhaustion of What the System Can Accomplish

Since the investigations of Pierre Duhem and Gerhard Ritter on late Scholasticism, the Nominalists of the fourteenth-century Parisian school have occupied a prominent position among the conjectured or supposed forerunners of the Copernican reform. It is true that in Parisian Ockhamism the geocentric system was no longer accepted as a matter of course as it was in the Aristotelian tradition. But the tendency of the discussion was not toward considering, or even undertaking, a reoccupation of the central position of the universe, but was rather to test, by thought experiments, the systematic agreement between the accepted astronomy and the premises of Aristotelian physics.

In Aristotle's system the placement of the Earth in the center is connected to the doctrine of the "natural places" of the elements. The four elements are located in concentric regions, one above the other, between the lowest sphere of the heavens—that of the Moon— and the center of the world. Earth has the lowest place and the most inferior rank, but at the same time it has the central position in the universe. Discussion was occasioned by the difficulty that, in contrast to the regions of the other elements, this lowest and central place of the element earth could, strictly speaking, only be a point. So it was simply not possible to conceive of a complete congruence between the extended mass of the terrestrial body and its natural place. To put it differently: Every bit of the element earth continued to strive to reach its natural place—continued in potential motion, that is, continued to be 'heavy' in the direction of the center. This difference between the geometric center (*centrum quantitatis*) and the physical center (*centrum gravitatis*) intensively occupied the Paris Ockhamists. The uneliminable imprecision of the *locus naturalis* [natural place] was thought to have, as its consequence, a constitutional unrest in the terrestrial body, since the unappeased striving of all of its parts could only be active at the cost of the instability of the total mass.

Therefore the Ockhamists ascribed to the Earth a state of tectonic instability, which did not lack a providential appropriateness either. This was seen in the fact that in this way the element water was prevented from being distributed in a uniform concentric manner over the surface of the Earth. A clear trace of this discussion of the relation between the elements earth and water is found in Copernicus's work. For the Ockhamists, the pure form of a sphere—which was supposed to be proper to heavenly bodies and which Copernicus therefore also had to assert as belonging to the Earth, so that he could make it into a star—had to be excluded, in the case of the terrestrial body, precisely because the corresponding uniform covering of the sphere of the Earth with water would have prevented human life, which became possible on the dry land masses.

Even though one must allow that the use of the term "motion" in Scholastic terminology is very broad, the idea of a motion of the Earth's sphere about its axis or in an annual course around the Sun is neither spoken nor dreamed of in this whole debate. This discussion was a vibration in the rigid structure of the system, but not the beginning of the idea of reconstructing it, nor even one of the underlying conditions for this.

We do find in Jean Buridan, the most daring of the Paris Nominalists, around the middle of the fourteenth century, the explicit question whether the daily rotation of the heaven of the fixed stars could not be explained by a corresponding axial rotation of the Earth.[6] Buridan distinguishes between the natural philosopher's and the astronomer's ways of answering this question: Philosophically, no positive decision is possible, while for astronomy the answer remains open and is left to a mathematical solution. The reason for this is characteristic. The astronomers, with their purely mathematical models, did not need to take the question of truth into consideration; in saying which Buridan implied that the truth was in any case inaccessible to cognition in astronomy.

With the assumptions that Buridan expresses in this passage, the possibility of a Copernicus was not moved one step closer. One must in fact go so far as to say that astronomically, too, with the admission of the hypothetical daily rotation of the Earth the 'prehistory' of the Copernican reform got into a blind alley. It is easy to show from other examples that the assertion of the Earth's axial motion in its place in the center of the universe made the approach to the abandonment of geocentrism and to the postulation of the Earth's annual motion around the Sun more difficult rather than easier.

This circumstance helps to clarify the question of what *had to* happen, and what may in fact *have* happened, in order to make something like the Copernican solution of the problem of the arrangement of the Earth, the Sun, and the planets in relation to one another permissible and adoptable for someone who did not want to restrict himself to making astronomical calculations as simple and reliable as possible. So the question is not where the constructive idea may have come from but how it could become acceptable and systematically tolerable. The result of an investigation of this question cannot, then, by any means claim to have 'explained' Copernicus's reform historically.

The Copernican reform had to demonstrate its possibility in relation to the background of the accepted natural philosophy, or at least not to lay itself open to the charge of being in insuperable contradiction to it. Precisely in this regard it is simply not correct when in the (in other respects lively and fruitful) American Copernicus research it was occasionally asserted that Copernicus could also have appeared at any other arbitrary point in time after Ptolemy. Instead, the latitude for a radical critique of the geocentric world scheme was only opened up by a far-reaching change in the foundations of the explanation of nature. Here it was not a matter of details and secondary elements of the Scholastic system, but of the fundamental principles of the Aristotelian natural philosophy itself, which Scholasticism took over: its ideas of space and time, of motion, of causality.

In this connection the historiographical statement that the Scholastic understanding of the world was as a matter of fact expounded by means provided by Aristotelian natural philosophy is not sufficient; rather, this conception of nature was built into and embedded in a system of premises that may indeed, in their ancient origins, have been one sectarian option that could be exchanged for another, but that, in their medieval reception, had to meet certain needs that led to their being almost exclusively preferred and incorporated. Perhaps it may be helpful, for one thing, to point out that someone who needed, and wanted to develop systematically, the proof of God's existence that Aristotle's metaphysics offered could hardly avoid the physics on which it was based.

So I have to demonstrate that the natural philosophy that was sanctioned by the Scholastic total system would not have permitted one to solve the fundamental astronomical problem in the way suggested by the Copernican reform. Beyond that I have to show how the internal logic of the Scholastic system and the modifications that that logic produced in the system led to suspension of this blockage against a correction of the basic cosmological picture, or at least made it possible to outflank it.

In the *summae*, collections of "questions," and commentaries of the twelfth and thirteenth centuries—that is, in the Scholasticism that had resulted from the appropriation of the works of Aristotle—physics appears to be united with metaphysics in such a way that both of them unequivocally lay down the *direction* in which, in the whole universe, all processes are causally connected. This direction runs inwards from outside; it is centripetal. For Aristotle just as for Thomas Aquinas, all events in nature have their ultimate cause in a normal cosmic movement, which is ascribed to the outermost sphere of the heavens. This motion in turn is induced by an extracosmic unmoved mover, the being that is absolutely at rest, the thought that thinks itself. Scholasticism had to accept this unmoved mover as the philosophical equivalent of Christian theology's concept of God, if it wanted to be able to make use of Aristotle's proof of God's existence and of the deduction of his God's pure metaphysical attributes.

The unmoved mover God performed His causal function for the world only through His existence, not through His participation or His will, and indeed He exercised it only on the outermost and highest sphere of the heavens, which in its rotation lovingly imitated the reflexive circle, closed in on itself, of pure thought. The primary sphere moves all the other heavenly bodies and indirectly, through them, keeps all the processes of the world in motion. By way of the planets, the Sun, and the Moon, it reaches the region of the four elements in a diminished efficacy that is far removed from the purity and regularity of what it comes from, being, as it were, multiply refracted. The cosmos of Scholasticism, moved and guided from without and from above, had its weakest place—because it was most remote from the divine origin of all causality—precisely in its center, which was occupied by the inert mass of the Earth, exposed to every kind of accident. In Scholasticism the orientation from outside to inside, from above to below, is at the same time the scale of the value and dignity of physical reality. This orientation is governed by the Aristotelian principle that the causal direction of motion determines where "above" is: "Sursum unde est motus" [Above is where motion comes from].

It is easy to see that this principle would already have made the Copernican reform impossible as an intra-Scholastic event. After Copernicus the majority of the cosmic motions—in particular, the appearances of the daily rotation of the heaven of the fixed stars and of the Sun's annual cycle—were to be brought about not from outside to inside, from above to below, but the reverse, beginning in the system's interior, that is (in Aristotelian language), 'from below to above.' The motion—the revolution of the heaven of the heaven of the fixed stars—that for reasons in reality was supposed to possess the most sublime regularity, because it represented the closest and most suitable effect of the mover God, was to turn out, in Copernicanism, to be a mere appearance that was caused by the rotation of the lowliest of all the heavenly bodies. The unbearableness of such a reversal for Scholasticism and for its cosmological metaphysics is obvious.

This conclusion enables us to see how and why the modern age could find in Copernicus's work the most conspicuous turning point between itself and the Middle Ages and the clearest sign of its break with Scholasticism. Copernicus had concentrated the real motions in the cosmos in its innermost region. It was possible to see that as a paradigm for the gaining of immanence, for the shielding of nature against transcendent uncertainties and interventions. That the moving heavens were a mere reflection of a reality that was bound to the Earth—that was an unmistakable guide to inferences.

The paradigmatic function of the Copernican reform becomes still more evident when one considers another basic characteristic of the medieval conception of the world on the assumptions of Aristotelian physics: The universe was necessarily a sort of energy subsidy system. Aristotelian physics knows no conservation laws. Only absolute rest is conserved; motion must be explained, in every case and for every moment, by a causal factor. The importance of physics for metaphysics results from the key principle of the Aristotelian theory of motion according to which everything that moves is moved by something else: "Omne quod movetur, ab alio movetur." The Scholastic axiom imitates, in its linguistic form, the original Greek construction, in which there is no discernible difference between what for us is distinguishable: the reflexive 'moving oneself' and the passive 'being moved'; the [Greek] "middle" expression is like the passive. That permits and even encourages one not to conceive of motion as a state but rather to perceive in it a continuous dependence on a reality or an action that ultimately can no longer be conceived as motion.

That harmonized not only with the medieval proof of God's existence, but also with an understanding of the world that was based on the salvation story, an understanding for which it was only possible for the Earth to become the place of man's temptation, fall, exposure, and trials because in its nature and placement in the cosmos it did not embody the pure and direct expression of divine power and of the accomplishment of the divine will. In the diagram of the cosmos it was as remote from God as was in any way possible, so that in it the world-order could encounter contradiction. When in Dante Lucifer resides in the center of the world, this—contrary to all assumptions about the center being a privileged position—is logical (precisely) in the Scholastic schema of the orientation from above to below, so as to mark the adversary of God's nature unambiguously even by topography. To put it differently: Nature's lawfulness is not 'distributed' homogeneously in the universe, but instead is clearly and unambiguously executed only in the motions of the heavens, while sublunar matter combines resistance and disorder with the purity of form. As a location, the Earth is suited to bringing man to consciousness of his need for salvation, and to preventing him from expecting the fulfillment of his notions of happiness.[7]

The medieval consciousness of the world is characterized by one of the few concepts of which it was the original producer, even though it used a term that had originally belonged to Aristotelian logic: "Contingency" is the ontic condition of a world that was created from nothing, is destined for destruction, and is retained in its continuing existence only by the divine will—a world that is measured against the idea of a necessary entity and that proves, when that is done, to have no implication one way or the other with respect to its own existence.[a] If one chooses a different language, one can also say that the world's continued existence has taken on the quality of [an act of] 'grace.' In the conception of contingency, the principle of the dependence of motion [on simultaneous causation by something else] turns out to be simply universalized: The world—in the origin of its reality, in its entire contents, and in every one of its states at every moment—is dependent on an act of transcendent concurrence. To that extent, the Aristotelian principle of motion had revealed itself ex post facto as a symptom of a metaphysical state of affairs that the ancient world had not been able to arrive at, if only because it regarded the world as the exhaustion of everything that was possible at all. Possibility was the justification of reality.

But since Augustine had answered the question why God created the world with a naked "Because he wanted to," not only the question of the justification of its origin but also that of the justification of its continued existence was barred. It followed unavoidably that one could not adequately explain by means of the biblical concept of the Creation the fact that the world that arose through an act of power did not sink back, in the next moment, into the nothingness from which it came. Here too one had to fall back on the "Quia voluit" [Because he willed it]. The doctrine of the one-time Creation, the *creatio ex nihilo* [creation from nothing], turned out to require a crucial supplement in the form of the assumption of a *creatio continua* [continuous creation], a creation that is renewed at each moment. This is the most extreme consequence of what I described when I characterized the world as an "energy subsidy system."

A Copernicus of the thirteenth century would not have been able to break through the internal consistency and stability of this connected system, which, however, he would have had to break through credibly in order to be able to cherish even the slightest hope of encountering comprehension and agreement. The provision of the world's continual 'energy requirements' from the outside inward, by way of the mediating agency of the heavenly spheres and all the way down to the terrestrial processes of coming to be and ceasing to be, would have been incompatible with the Copernican implication that the primary motion of the heavens and the path of the Sun were illusory. With its 'centrifugal' direction of the derivation of the phenomena, Copernicanism was to establish the precise opposite of the medieval conception. Copernicus himself was aware of this opposition only as it affected the means of argumentation that he needed in order to introduce and to protect his reform. Nor will one be able to say that Copernicus's opponents quickly achieved clarity about the epoch-making importance of this reversal. Instead one has the impression, as in so many great altercations in history, that we are dealing, in the historical documents, with only a surface layer of expression, in the argumentations of which a deeper disturbance is concealed.

Kepler was to be the first to comprehend the fact that the logic of Copernicanism entailed not only reversing the relation of appearance and reality in the relation of the heavens to the Earth but also adapting the energy relationship to this reversal. He recognized this when he made the dynamics of the system emanate from the center of the universe, that is, when he made it proceed from the inside outward. Kepler makes something that had still been unknown to Copernicus—the rotation of the Sun, as the new central body, around its axis—into the cause of the planets' motions. The bridging—which the pious Kepler still needed in addition to this—of the gap between transcendence and immanence lay in the idea of a symbolic correspondence, which was his substitute for Aristotelian causality. In a letter to Michael Mästlin, on 3 October 1595, he writes, "So by standing in the midst of the wandering stars, itself at rest but nevertheless a source of motion, the Sun provides an image of God the Father, the Creator. For what creation is to God, motion is to the Sun." Ten years later, on 5 March 1605—still before the discovery of the laws of planetary motion that are named after him—Kepler writes to the same addressee, "The Sun's body is magnetic, in circles all around it, and rotates in its place, carrying the sphere of its power around with it. This power does not attract the planets, but moves them along .... What I mean by magnetism is an analogy, not exactly the thing itself." [b]

Looking back at Copernicus, one notices that he certainly described the solar system's construction in a new way, but not its internal connections; he had been satisfied to deduce the mere possibility of the Earth's motion from its spherical form, but for the rest to regard it as an isolated circumstance. Kepler's claim to truth is based on a higher level of explanatory accomplishment and reveals the task that Newton will set for himself in order to confirm the Copernican system by means of the causal relationships within it.

In connection with our exclusion of the possibility of a Copernicus in the Middle Ages, but also in connection with our comprehension of the situation of the reception of Copernicus's work, it is important to understand that neither specific questions of faith nor the well-known critical passages in the Old Testament played a role in the preference for geocentrism. Faith had indeed always accepted and made use of philosophy's aid, but at the same time it had included in this relationship the proviso that the act of belief must contain an element of unsupported and unprotected submission, which could not make use of any basis in knowledge about the world. Philosophy aimed at a systematic consistency and an external intelligibility whose necessity, for the contents of theology, could not be admitted; this prevented the religious sphere, in principle, from being endangered by the scientific one. In his short work addressed to John of Vercelli, Thomas Aquinas expressed theology's capacity for tolerance with respect to attempted changes in cosmological systematics in one sentence, according to which, the reverse of the accepted doctrine could also be accepted without danger to faith: "Videtur tamen mihi contrarium posse tolerari absque fidei periculo."

The problem of the possibility of the Copernican reform in the Middle Ages does not lie in its theological acceptability but rather in the impossibility of successfully pressing its claim to truth in the face of the consistency of the overall system, however much that system might consist of the putting into effect of very general theological premises. The important thing for me is to show that the resistance of both the Reformation and the Counterreformation to Copernicanism, with their appeals to individual biblical texts, is not simply the continuation of the medieval exclusion of changes in the cosmological system, that is, that it does not merely express something like the inertia of the Scholastic system itself. Instead, and this is still more important, it was the liquidation of this system from within—and that must mean precisely from the center of its theological motivations—that turned the impossibility of a medieval Copernicus into his possibility. This process of the opening up of a latitude as the result of loosening of the systematic structure—as a result of the accenting of specific theological propositions—must be described.

Seen from the point of view of the medieval system, the change of places that Copernicus undertook between the Earth and the Sun, as well as the exchange of the predicates of rest and motion that was bound up with that, could not be the decisive and unmanageable element of his reform. Instead, that element was the real standstill of the heaven of the fixed stars, whose motion was revealed—in spite of its proximity and immediate relation to the *ens realissimum* [most real thing, i.e., God]—as mere appearance. Geocentrism or heliocentrism—this choice of alternatives, with which the radical character of the Copernican reform was to be described, and toward which the opposition to Copernicus was also directed, was not, in any case, the great difficulty that the Middle Ages would have had with Copernicus. The real difficulty, which initially appears as a difficulty of relations of spatial distance, is very closely connected to the concept of causality, which medieval Scholasticism, following Aristotle, had developed and tried to maintain.

In a system, the possible and even the inevitable questions are not posed at once. It is often the postponement of problems that in retrospect appear to have been long since due that makes possible the system's continued existence, and even the classical culminating form of its elaboration. One can describe High Scholasticism, with its much-admired structures, as the absence of complete recklessness with regard to the heterogeneity of its presuppositions, a recklessness that was only able to develop fully as a result of the division of school activity into tendencies corresponding to [different monastic] orders. The classical character of an epoch as a high form can be a period of protection for its still concealed inconsistency. It takes time before the faith that with such reliable foundations the system's possibilities of questioning can be fully exhausted in all directions brings to light its contradictions and causes its capacity for self-regulation to break down. The autolysis of systems begins when they have to submit their productivity completely to proof.

The Aristotelian-Scholastic concept of causality does not recognize a process as a state; it requires, for every phase of a motion or change, a directly engaged agent. Here I designate this basic idea as that of "accompanying causality." For cosmology it implies not only the already mentioned necessity of the new production, at every moment, of even a constant process, but also the impossibility of *actio per distans*, action at a distance. Taken together, these implications exclude the possibility that the *metaphysical* cause of all motions could have been located outside of and beyond the cosmos while the primary *physical* cause of the phenomena of motion was nevertheless located in or near the center. The system of agencies transmitting motion and change from the outside inward is firmly anchored in the physics of 'accompanying causality.' Since the Christian God had taken possession and made use of the ancient world's cosmos, He restricted the "orderly"ᶜ use of His omnipotence to the conditions contained in this physics. In fact it was better suited than has often been asserted to a theology in which the *actio per distans* of divine deeds of salvation on behalf of human beings seemed in any case to be impossible, and redemption could only be transmitted in the form of accompanying causality, so that God Himself had to betake Himself not only to the place of His action but also into the form of the objects of this action, by incarnating Himself as a human being. Aristotelianism was able to answer one of Christianity's long-unposed questions, which Anselm of Canterbury had raised laconically with his *Cur deus homo?* [Why did God become man?]

The idea of accompanying causality makes it understandable why the Middle Ages were so hopelessly distant from both the principle of inertia and the law of gravity. These two elements made the cosmological construction definitively independent of accompanying causality and of the schema of transmitting agencies. But we know that both of them were discovered only as a consequence of the Copernican challenge to physics. How, then, was it possible, nevertheless, for Copernicus to assert his reform as cosmological truth? The answer must run thus: Because in the late Middle Ages there were already equivalents for both elements—for gravity, the idea of the striving of similar parts to come together (*appetentia partium*), and for inertia, the *impetus* concept, which I introduce here as the more general concept of 'communicated [*übertragenen*] causality.'

Initially it will be enough to illustrate the replacement of the concept of accompanying causality by that of communicated causality with a case study of the earliest steps in the process. The sample that we choose in so doing is extremely characteristic of the way in which, within Scholasticism and using the means that were specific to it, developments were set in motion that seemed to be entirely in the interest of the consolidation of the system but that then severely fatigued it and made excessive demands on its elasticity.

The problem in relation to which this development began was, not accidentally, the theological question of the mode of operation of God's redeeming grace, more specifically in the form of the sacraments. In order to understand this one must try to put oneself into the paradoxical situation of Christian theology, which, because of the impossibility of action at a distance, makes its God become Man and enter the world but immediately makes Him disappear from it again, without the fulfillment of the promised destruction and transformation of what exists. In this originally unexpected situation, everything did depend after all on the credibility of the action of the absent God, that is, on the institution and the instruments by which His earthly presence is prolonged. In that sense the question of the mode of operation of the Church's sacraments was not a marginal problem, but the core of all the difficulties that resulted from Christianity's specific historical character. In the obligatory commentaries on the sentences of Peter the Lombard, it had its fixed place at the beginning of the fourth book.

With the Lombard, the sacraments in general were defined as instruments of the operation of divine grace. But this definition specifically left open the question of the immediacy of the operation, that is, whether at the moment the sacrament was administered the power operating in the instrument derived directly from God or whether it was to be regarded as a capacity for operation that was deposited in it (*virtus inhaerens*). Now in 1320 this question was discussed for the first time, in a *Commentary on the Sentences* by Franciscus de Marchia, in such a way that in the central part of the article a sort of model is offered, which represents a borrowing from natural philosophy: the theorem of projectile motion. This model almost comes to stand by itself, in an excursus that makes a long preparatory detour.[8]

It is easy to see how the theological inquiry could gain orientation from the mechanical problematic of the motion of a projectile. Throwing, as a goal-directed action, is the use of an instrument across spatial distance. The thrower remains at the place where the act of throwing was performed, while the body that is thrown leaves there and approaches its goal, so that at the moment when the effect takes place it does not seem to emanate directly from its cause, from which, at any given moment in which it is observed, it is separated by a definite stretch of its path. This is plainly not a case of accompanying causality, and very roundabout constructions had already been needed in order to maintain the basic idea of contact causality even for the explanation of projectile motion.

The solution that Franciscus de Marchia tries to provide for the mechanical problem should not be regarded as the result of independent theoretical investigation. Instead we must reckon on the theological interest having already been directed at rendering the divine operations independent, when it endeavors to adapt the physical model to the real separation of the moving force from the thrower, from whom the motion originates, and from the medium through which the projectile passes. Theologically, this process of rendering operations independent implies the basic idea, which is set out in the doctrine of the sacraments, of a deposited store of grace that is at the disposal of those who administer it. This typical medieval tendency is expressed precisely in the exclusion of a direct action of God on the recipient of the sacrament.

For the history of its effects, the functional context in which the physical model was developed (very much 'to serve a purpose') is unimportant. In the tradition of the required writing of commentaries on the *Sentences*, which was in any case an unsystematic work, fixed locations are developed at which, in completely heterogenous contexts, certain questions are discussed. The consequence of this is that a transplant from physics, like the excursus on projectile motion, does not grow together with its new environment but is used there only as a key. Thus it continually remains possible to carry it over to a specific context—say, in a commentary on Aristotle's physics or cosmology—while preserving the adaptation that had been undertaken elsewhere. It is this constructive permeability of the Scholastic system that made possible the influence of theological premises on very remote inquiries, as the exploitation of a heteronomous pressure of variation.

The thrown stone, whose motion is, according to the Aristotelian classification, "violent," appears in Franciscus de Marchia as the simplest way of presenting the difficulties that arise in the explanation of the mode of operation of the instruments of 'technology' [in the broadest sense: *Technik*]. The motion of any heavy body upward, that is, in the opposite of the direction of its "natural place," is violent; if a pick or a hammer is swung up, experience shows that on account of the tool's own motion, it has to be restrained before one can bring about, with it, in the opposite direction, the operation that is proper to the tool. This sensation, which is immediately accessible to anyone, raises the question "whether the instrument that is moved by an artisan takes a force into itself, and whether the stone that is moved upward, or any other heavy body, takes into itself a force that continues the motion (*virtus continuans motum*)." In order to protect the principle of contact causality, Aristotle and his interpreters had had to invent a mechanism of grotesque complexity, which actually discredited what it was meant to save.

Aristotle imagines the matter ("imaginatur enim Philosophus . . .") in such a way that air and water have a more rapid motion than a stone or another similar heavy body that moves in air or water as its medium. This specific difference between the elements causes a motion that is bestowed on the medium by the hand always to hasten ahead of the stone, propagating itself in the form of vortices that one after another take up the stone and hinder the execution of its tendency to fall, until, as they weaken, they are no longer able to accomplish this. "Whence Aristotle, and Averroes too, imagine that just as when a stone is thrown into water and there are thereby formed and generated in the water certain circles, so similarly a stone thrown in air makes in the air certain invisible circles, the first of which— because it is moved more quickly than the stone would by itself descend—carries the stone along to the second circle . . . . And because these circles follow one after the other and are not continuous, Aristotle concludes that violent motion of this kind is not continuous but 'one after the other' (*consequenter se habens*, i.e., in jerks), and that everything that moves in this way is not moved by itself (*per se*), but rather by something else (*per accidens*), as a sailor relates to the motion of the ship that carries him." [d] Franciscus de Marchia opposes to this awkward theory his own cautiously formulated thesis that the motive force, in a case of violent motion, should be assumed to be in the moved body itself rather than in the medium through which its motion is accomplished.[9] This result is not arrived at empirically, but by a process of exclusion of all other possible explanations, the outcome of which is that the motion of the thrown body must result directly, in every phase of the process of the throw, from the act of the thrower. So the force that is to be supplied to the thrown body must, as it were, have been left behind in it by the initial impulse (*virtus derelicta ab ipso primo motore*).

The Scholastic author wants the force that puts the stone in motion to be distinguished from the one that keeps it in motion: "One force begins the motion or determines the heavy body to some motion, and this force is the force of the hand; the other force completes the motion that has begun and continues it, and this force was either (directly) produced by the first one or (indirectly) left behind by the motion that it brought about." [10] The whole train of thought is based on Aristotle's assumption, the validity of which Franciscus de Marchia does not curtail, that the state of motion constantly requires a force that moves it. The deviation from Aristotle consists in elimination of the mechanism of contact between the first cause and the state of motion that is brought about. The causality is transferred to the moved body itself, and in fact with the thoroughly Aristotelian argument—whose rigorous application is, however, characteristic for the first time of fourteenth-century Nominalism—of economy: "quia frustra fit per plura quod potest fieri per pauciora" [because it would be in vain for something to be done by many (causes) that can be done by a few]. By applying the principle of economy, he uses Aristotle to argue against Aristotle—a pattern of thought that is quite typical of the destruction of Scholasticism.

One must also keep in mind that the fundamental idea of 'accompanying causality,' which had suited so well the thirteenth century's needs for proof and the fundamental theological schema of 'transmitting agencies,' no longer satisfied the needs of the fourteenth century in the same way. Scholasticism's own sharp criticism of the quality of the cosmological proof of God's existence had freed the theory of causality from all too binding interests. The idea of 'communicated causality' was able to appeal to the fourteenth century because, while it did seem, in terms of its systematic location, to perform better service in the treatises and articles on the means, promises, and administration of grace, it also de facto offered a model of transcendence transferred into immanence, and thus preformed, with the idea of the deposited treasure of salvation and the delegated disposition over it, the possibility of a world that is substantially lasting in and of itself.

This is already discernible in Franciscus de Marchia's formulas. He sees the superiority of his theory of the communicated motive force in the fact that "by assuming that the force is in the moved stone as a form we can save the appearance that the stone is in some way moved by itself (*movetur aliquo modo per se*), and does not depend (*per accidens*) on the motion of the air as the sailor does on the motion of the ship." [e] The saving of the appearance refers to the continuity of projectile motion, whereas according to the Aristotelian hypothesis this motion would inevitably have to take place in thrusts (*consequenter se habens*) from vortex to vortex. While the actual continuity of projectile motion does not affect the Aristotelian assumption that projectile motion is a "violent" motion, it does require an interpretation that allows one to understand the relation between the force that moves the projectile and the initial force of the thrower as a "natural" one, insofar as in the process itself the initially external action on the projectile has become an internal action.[11] The communicated causality makes the originally "violent" motion, as a continuing one, into a "natural" one—or as a natural one into a continuing one. That will become important, as a possibility (even) in the Scholastic system, for the Copernican argument that the motion of the Earth must be acquitted, as a "natural" motion, of having destructive consequences.

This is the connection, which simply cannot be missed, between the physics of projectiles and cosmology. Franciscus de Marchia already establishes the connection between his explanation of projectile motion and the cosmological problem of the rotation of the primary sphere of the heavens. This connection would have made it possible to think of a motive force that was communicated to the spheres at the Creation, if it had not been entirely a matter of course for the commentator on the *Sentences* to regard the communicated force, like any other motive force, as—by analogy to organic forces—capable of growing tired, and finally as exhausting itself. With projectile motion the resistance of the air relieves the speculative discussion of any thorough consideration of the relation between the supposed process of growing tired and the resistance of the medium. But even where the body that is moved would encounter no resistance—as had to be taken for granted in the case of the heavenly spheres, if only because of their special substance—the communicated force was a form that dwindled and came to an end in time (*virtus inhaërens ad tempus*). Thus Franciscus de Marchia also holds fast to the intelligences, which Aristotle had introduced, that move the spheres, although he is able to make their activity easier by his assumption of a communicable motive force: "The heaven that is moved by its spiritual principle receives from it a force or form that has no contrary but that is nevertheless only accidental, that is not identical with local motion, and that is inherent in the heaven in the manner of a forming principle."[12] If the intelligence were to cease moving the celestial sphere, the sphere would continue for some time in its circular motion, as a potter's wheel continues to turn for a while after the power that drives it has been interrupted. In both cases the same thing seems to be happening, namely, a force is being left behind (*virtus derelicta*) in the bearer of the motion from the stimulus that first moved it. This communicated moving force is designated here, as in all subsequent Scholastic physics, as *impetus*.

There has been much fine writing on the prehistory of the impetus concept. In evaluating this material one must, however, keep in mind the fact that the term does not need to be used in a terminological sense, but is close to the level of simple description of the 'liveliness' ['*Schwung*'] that occurs when something is thrown; and one must also keep in mind, in reverse, that avoidance of this imaginative element must be attributed to the coercive effect of the Aristotelian system. The difficulties that had arisen, for Aristotle and his successors, in the explanation of projectile motion must be seen as the price of the advantage that had readily resulted, for comprehension (in the absence of knowledge of gravitation) of the phenomena of falling and the acceleration of falling, from the idea of a thing's 'tendency' toward its natural place, and from the goal directedness of its process of arriving at that place. As an element in the system, the explanation of falling is primary, and that of projectile motion is secondary and highly artificial. In comparison with this, the introduction of the impetus, whatever its motivation may have been, is a late step of correction and of reduction to what description suggests—to a factor producing the continuity of projectile motion. Consequently, little is gained when evidence is produced of the appearance of the term *impetus* in the time before the complete reception of Aristotle's physics, as has been done for Thierry of Chartres in the first half of the twelfth century,[13] and even for someone as early as Augustine.[14] There, one will have to say, there is still (or already) too little Aristotle present for it to be necessary to deal with the disadvantages of his advantages by reaching for a simplification.

Franciscus de Marchia, in any case, with his transfer of the impetus theory to celestial mechanics, immediately lands in a dilemma, which he does not yet dare to resolve decisively. It consists in the fact that the moving force, in the realm of the matter of which the heavens are made, is regarded as capable of becoming exhausted, and thus as mutable, although the premises of Aristotelian cosmology do not permit mutability and corruptability in the stellar objects. The same argumentation that requires one to postulate the eternal existence of the cosmos must require pure permanence in a communicated causality that is understood as form. One immediately perceives the threatening character of this for a system of contingency.

And Franciscus de Marchia does in fact choose the (on Christian assumptions) lesser evil: He admits the mutability that is involved in the disappearance of the communicated motive force, so as not to risk the enormity of allowing that finite beings (and the movers of the spheres, as created things, must be such) could produce infinite effects.[f] The argumentation exhibits the loveliest medieval naiveté: If in this way mutability is brought into the region of the heavens, this need not shock the believer ("ista ratio non debet movere fidelem"), because after all for him it is well established that the song of praise of the blessed in heaven will be not only spiritual or intellectual but also audible, and the saints will even converse among themselves; but this would only be possible by means of sounds that are produced and die away. Thus Aristotle's heaven becomes a stage for the Christian image of the beyond, which makes its metaphysical attributes even more severely problematic than the concept of creation does.

The simplification that the concept of the impetus had provided in the secondary theory of projectile motion now again becomes, in the (tertiary) theory of the motions of the heavens, a disadvantage, because the capacity to become exhausted must be asserted as a constitutive element here too. The auxiliary construction that is mobilized for this purpose is only of interest because it indicates a difficulty that can be seen as analogous to the one that was just removed on the level of projectile motion: "To the objection that that force perishes and is destroyed, and to the question how that can happen, I reply that it is not by the perishing of its bearer (*per corruptionem subiecti*), nor by the supervention of a contrary action (because it has none), but by the cessation of the agent (*per cessationem agentis*) or of the moving force .... This can happen in two ways. On the one hand, something perishes as a result of the cessation of the mover's force in such a way that it perishes and is destroyed simultaneously with that cessation, as with motion, which perishes and becomes extinct at the same time and even in the same instant as the mover's force ceases. The other possibility is that something perishes and ceases to exist as a result of the cessation of something else not by perishing simultaneously with the other thing's cessation or perishing but by not being able to continue for a long time after the other thing's

cessation, because of the imperfection of its own being (*proper imper-
fectionem suae entitatis*) . . . ." [g]

A special characteristic of Aristotelian metaphysics—and not the
inner logic of the impetus theory—creates the inducement to a further step that removes these complications in one stroke, namely,
daring to assert the persistence of communicated causality in, at least,
the motions of the heavenly bodies. It is this step that merely by its
pattern presses beyond the Middle Ages. It removes the production of
the infinite as motion from the exclusive preserve of theology and
makes this possible even in a world that is still characterized by
finitude. It is the first hint of the reoccupation of medieval systematic
positions that will culminate in the line through Nicholas of Cusa,
Giordano Bruno, and Spinoza.

I would like to question whether Franciscus de Marchia could already have taken this step, although a sentence that accords with it
can be found in the manuscript edited by A. Maier. At the conclusion
of the discussion of the doctrine of motion we read that one could perhaps also say that such a force, produced in the region of the heavens, would be imperishable: "posset etiam forte dici quod huiusmodi
virtus causata in caelo esset incorruptibilis." But this sentence is so
incompatible with the amply discussed and resolved difficulty of introducing a perishable force into the imperishable heavens, and is
so isolated in its laconicism, that it is natural to assume that the idea,
which appears quite unexpectedly and in a passage that is not arranged very much to suit it, could be a marginal comment of a user
of the text, into which it was inserted later; in which case the user
should be dated after Jean Buridan, at the earliest.

It was in fact this most important and most daring thinker of the
Parisian Nominalist school around the middle of the fourteenth century who first systematically established and developed the connection between the impetus theory and cosmology. However, it would
be a mistake to see in this a process above all of the system's drawing closer to empirical data. What has to be brought into mutual
harmony is, instead, the elements of the system itself, so that the
whole process presents itself as one aimed at regulating the self-
preservation of the system. In Buridan this becomes evident in the
way in which he initially applies the doctrine of the impetus to the
phenomenon of the acceleration of fall, that is, in which he gives up
the isolated consideration of projectile motion in order to generalize

the results that emerged from it. This would be a comparatively harmless and cautious extension if the Aristotelian theory of motion had not drawn the line between "natural" and "violent" motion at just this point.

Projectile motion, as a violent and thus (in relation to the nature of the things) accidental event, could have no paradigmatic significance for the explanation of physical processes because the teleological element, which can only lie in the thrower's intention of a target, remains external to it. But if a form of "natural" motion like that of falling could once be described by the theoretical means that had first been gained in reference to "violent" motion, then this boundary would be crossed and the distinction between naturalness and violence would itself have become problematic.

What was most in accordance with the tendencies of the Nominalist system here was the abandonment or disregarding of the element of teleology: Acceleration of fall no longer needed to be explained, as it predominantly was in Scholasticism, by the object's *approaching closer* to its "natural place," but could instead be attributed to its *increasing distance* from the point at which the fall began and to the summation of the impetus that was conserved in each phase of its motion. This did not by any means bring the Nominalist close to a discovery of the law of falling bodies, but it certainly did provide an anticipation of the form that that law could have. Note that it is not free fall as such that is deteleologized (for it continues to be the return of a part to the whole that is the region of its element), but only the acceleration that we observe in the free fall, acceleration that, as a change in velocity, is most nearly analogous to "violent" motions.[15] So the convergence between this way of viewing the acceleration of falling objects and the interests of the Nominalists consists in the fact that the teleological factor is now preserved only in the direction of the motion of falling.

Teleological elements in natural philosophy contradicted the fundamental theological position of voluntarism, which did not permit one, in explaining nature, to have recourse to a knowledge of the purposes God meant nature to serve. Knowledge of acts of God's will must be reserved for revelation. If the Creator's intentions cannot be known, neither can the goals of natural processes. Even knowledge itself cannot fall back on the premise that it was intended to serve a purpose, and deduce from that its reliability and capacity for being evident: "Cum nihil scias de voluntate dei tu non potes esse certus de aliquo." [Since you do not know God's will, you cannot be certain of anything.][16]

Once the basic conception of communicated causality had penetrated into the understanding of "natural" motion as well, the Aristotelian cosmology was already, in principle, suspended. So Buridan was also the first to articulate the consequence that the doctrine of the impetus made the doctrine of the intelligences, as movers of the spheres, superfluous. Here he could consider the idea that the motion of the heavenly bodies was brought about, as a continuous state, by a solitary act of the communication of motive force to them at the moment of the Creation: "For it could be said that when God created the celestial spheres, He began to move each of them as He wished, and then they were moved by the impetus which He gave to them because, there being no resistance, the impetus is neither corrupted nor diminished." [17] The elimination of the impetus's tendency to become exhausted, together with the absence (which had been assumed all along) of any resistance, in the case of the heavenly bodies, produced, speculatively, precisely the same conditions that were to be introduced almost three centuries later, in Galileo's mechanics, in the form of the equally speculative normal condition of the vacuum. The region of the heavenly bodies could become, in this way, a field for experiments conducted in imagination, even though the assumption of empty space continued to be strictly forbidden. In the case of the spheres, the rule was the constancy of a motion, once it was induced, because the circular form of their revolutions continued to be accepted as the simplest form of "natural" motion. Thus the presentiment of the principle of inertia was valid for celestial mechanics exclusively, and the possibility that it could be a sublunar state of affairs as well could at most dawn upon one in connection with the rotation of millstones or potters' wheels.[18]

Buridan had pushed the impetus theory forward across the hiatus between natural and violent motion; but he had allowed the problematic of this distinction, which was not to become fully transparent even for Galileo, to remain latent. For Buridan, too, the tendency of the motive force, in every case of "violent" motion, to grow tired had remained beyond question, because he already regarded the concept of violence as including a sort of inner resistance of the body against change. This assumption was important in blocking theoretical progress because it did not permit one to base any judgment of natural (and still more stellar) processes on insight into technical objects and phenomena—that is, it excluded precisely what Galileo was to do.

For natural motion, what Buridan assumes—in contrast to that inner resistance—is actually an inner inclination (*inclinatio*) toward the state of motion, an inclination that ensures the permanence of the impetus's effect. Consequently he could say of the heavenly bodies that they are, as it were, self-moved bodies (*quasi per se mobilia*). At this point one really touches the possibility of the Copernican argument that makes all objections to the motion of the Earth irrelevant by showing that the Earth has the form of a heavenly body, and thus almost attributing self-motion to it.

One must envisage clearly the destructive consequences that this development of thought contained for the overall medieval system. It is true that the inference from cosmic motion to the prime, unmoved mover continued to be valid. But this inference was no longer adapted to carrying out a somewhat useful proof of God's existence. The moving force of the prime mover now no longer had to exhibit an attribute that was indispensable for the production of the concept of God in the Middle Ages, namely, the attribute of infinity. Infinity, as an outcome of the proof of God's existence, had resulted from the fact that the force of the prime mover had to be at least potentially sufficient for a whole eternity of carrying on the propulsion of the world, because the concept of time that was connected to the motion of the first sphere made a beginning or an end rationally impermissible. Buridan's cosmological impetus theory had allowed the most important divine attribute, for the Middle Ages, to drop out of the stringency of the causal proof of God's existence. I do not want to go into the fact, here, that the tradition in which the Nominalists stood was brilliantly prepared for this failure of the Aristotelian instrumentation by Duns Scotus's critique, which also disposed it to abandon that instrumentation.

So Buridan was the first to state that a *finite* force was sufficient to produce the one-time impetus for the motions of the heavens. In the final version of his commentary on Aristotle's *Physics* we find the formulation—which seems one that could not have been expected of the Middle Ages—that even a very small force, for example, that of a fly, would have been enough to set the rotation of the first heaven going, even if the velocity were faster by any amount than it actually is.[19] This conclusion, however shocking it might seem, could count on protective arguments from theology's inexhaustible arsenal, arguments that become unexpectedly plausible. A piece of the biblical story of the Creation that had previously not come into play very much could be given a new accent and 'taken literally,' namely, God's day of rest after the six days' work. With all the predilection for emphasizing contingency in its most rigorous form as *creatio continua* [continuous creation], it was easy to overlook the necessity that was implied in this detail. In a system that was governed by the idea of communicated causality, the divine day of rest now became conceivable even as a physical reality. Of course the inescapable consequence was that the rigor of the concept of contingency, which necessitates an uninterrupted demand for divine concurrence in the world's continued existence, could not be kept acute. It seems that Buridan found this [biblical] support and for the first time suggested to the theologians, with cautious words, that God could be allowed to rest, after completing the work of the Creation, by 'abdicating' his causality (*committendo aliis actiones et passiones ad invicem* [by entrusting acting and being acted upon to others instead]).[20]

In the last form in which it was received from the ancient world, as natural philosophy, philosophy unexpectedly reformulated its medieval role as *ancilla theologiae* [servant of theology]. In the process, it gained a sort of right of proposal, vis-à-vis theology, a right that contains a high degree of imaginative freedom precisely as a result of its employment of elements that have not been theologically dogmatized. Reason's self-assertion against theological absolutism is not carried out primarily by throwing off obligations but by a subversive twisting of the functional orientation of the theological contents themselves. Thus nature's permanent need for God could be at least partially abrogated, and weakened in its effect on consciousness, through a detour by way of an entirely appropriate theological critique of the doctrine of the movers of the spheres—that is, by means of a neutralization of the apparatus of the administration of the world. The beginning of the modern age could be marked by the insistence, which had become acute, on the almost anecdotal biblical feature of God's rest after the Creation, which had to be made possible by the assumption of an initially 'commissarial' [i.e., deputized] autonomy on the part of the world's powers. From this starting point one can anticipate the tendency toward the principle of inertia and toward physics' assumptions of conservation, the end point of which was to be the turn from the maximalization of theology to its minimalization.[21]

Parisian Nominalism has, in its whole intention, a conservative attitude toward the Aristotelian-Scholastic system. The changes that were proposed or attempted in the physics and cosmology of the system can be understood as repairs. Of course this original aim does not prevent the expedients that were introduced in the process from coming to stand on their own. If I said that Buridan was the most daring representative of Parisian Ockhamism, it was above all because of the demolition of the classical differentiation between sublunar physics and supralunar cosmology, which he accomplished by introducing the impetus into the mechanics of the spheres. It may seem paradoxical when I have to say that this same daring prevented Buridan from seriously coming to grips with the astronomical thesis, which was well known to him, of the motion of the Earth. His attention was fixed on the problem of the explanation of the motions of the heavens; allowing these to be considered as mere appearances, produced by the Earth's own motions, would have forced him to apply the impetus theory to the terrestrial body, which would inevitably have had to bring into play the tendency—which is evident in projectile motion—of the matter of the inferior elements to grow tired. Because in the Aristotelian system circular motion was not regarded as "natural" to the sublunar elements, one could not lay claim for the Earth to the argumentative advantage that was present in the case of the heavenly bodies in their lack of resistance to the force that moves them.

Nevertheless, for another Parisian Nominalist the carrying over of "impetus" to the heavenly spheres seemed so intolerable that he believed he should accept the Earth's own motion, and harmonize it with the system, instead. Thus the clearest 'forerunner' of Copernicus in Nominalism, Nicole Oresme, deviated from Aristotelian *cosmology* precisely in order to be able to save Aristotelian *physics*. While Buridan projected the impetus theory, which had been gained from the theory of projectile motion and of fall, onto celestial mechanics, Nicole Oresme proceeded in the opposite direction. He attempted to 'recover' for sublunar physics a fundamental characteristic of heavenly bodies, namely, a natural circular motion. This reversal of the theoretical direction is the reason why Oresme ignored the impetus, in spite of its importance for the whole Parisian Ockhamist school.

In the context of Aristotelian natural philosophy there was a spectacular problem that could not easily be solved by the usual means: the question of comets. In order to preserve immutability and eternal existence in the region of the heavenly bodies, the comets could only be sublunar objects, falling, consequently, into meteorology's realm of responsibility. And that was in fact the heading under which Aristotle had discussed them. For him, the uppermost part of the stratum of air, in which a hot and dry exhalation collects, borders immediately on the concave underside of the rotating sphere of the Moon, creating a situation in which, as a result of friction, sparks and lumps of fire are created, which are carried along in the sublunar medium, on account of the rotation of the sphere, as comets. While this rotation of the sublunar stratum that was concentrically layered around the globe did explain the comet's apparent path in the sky, it created difficulties because it could not be conceived as a "natural" motion. This would have been in conflict with the fundamental Aristotelian thesis that each element can have only *one* specific proper motion, leading from its initial situation to its "natural place," which it strives to reach by the shortest possible route. That is why air bubbles in water rise upward. In his *Meteorology* Aristotle had been forced to grant an additional inherent motion to at least air and fire, on behalf of comets.

Thus almost incidentally, and against his own principles, the Philosopher had introduced into sublunary nature a circular motion that was permanent and that depended on the motion of the heavens, so that it could hardly be described as "violent." That at least made it easier for Nicole Oresme to extrapolate the principle of circular motion downward, from above, and to 'put it into effect' all the way down to the terrestrial globe. With that, the most important requisite for the equal status of the Earth as a heavenly body was already implicitly fulfilled, namely, that it should be possible to lay claim to the attribute of "naturalness" for its proper motion. Oresme did not go quite so far. He contented himself with adding to the classical dichotomy of motions a third class of motion *hors nature* [outside of nature]. But Copernicus will argue brilliantly and economically with this "naturalness": "Verum si quispiam volvi terram opinetur, dicet utique motum esse naturalem, non violentum." [Yet if anyone should hold the opinion that the earth revolves, he will surely assert that its motion is natural, not violent.][22] At the same time, one must say that the permissibility of circular motion in the case of the Earth had been gained only in passing and as a consequence of the integration of a foreign body into the overall Aristotelian system.

More important was the argumentation with which Nicole Oresme was able to overcome the contradiction of a 'doubled' natural motion in each of the sublunar elements. His new insight is that the expression "natural motion" does not characterize an event unambiguously. It is clear that for Aristotle circular motion has, in this regard, an unconditional preeminence and is the purest form of what he meant "natural motion" to describe: Specifically, it has no opposite motion, as do the two other natural motions, the motions of the sublunar elements "downward" and "upward," that is, toward the center of the world and away from it. But these two elementary motions are natural only if and only because they presuppose a separation of a body (a part) from its "natural place" and thus from the total mass of the region from which it derives. So the natural motion that we observe as free fall is, in spite of being "natural," an extraordinary event, a process of repatriation, which always genetically presupposes a "violent" motion. For in view of the eternal existence of the world, for Aristotle, all parts of elements would otherwise have had to be in their natural places long since. To put it differently, natural motion, in its straight-line form, exists only by the grace of violent motion. One could, then, conceive of a condition of nature in which all natural motions except those of the heavenly bodies would have come to an end. Admittedly, that is not easily compatible with the concept of "naturalness" within the Aristotelian system.

So it was quite logical to inquire into the possibility of a kind of natural motion of the sublunar elements, as well, that would belong to them by virtue of their quality as elements and independently of a prior violent separation from their original region. Since the regions of the elements are concentrically arranged around the middle of the world—so that if they were seen from a perspective not tied to the Earth they would have exactly the same shape as the sphere of the heavens—it is only consistent to grant the honorific quality of "natural motion" primarily to these total structures, and to see in the return motions of separated parts only a secondary phenomenon, integrated into the total process.

From this higher-level perspective, straight-line natural motion becomes the exception that it should have been recognized as all along. The fact that the immanent correction of the Aristotelian system was not a lasting gain in relation to the future development of physics and cosmology was not to emerge in connection with Copernicus, but certainly did so with Galileo. If Nominalism had already been able to arrive at and to persuade people of the special rational status of straight-line motion as a state of objects in motion, people's receptiveness to the Copernican circular motions of the Earth would have been seriously impaired. At least for us, the possibility of Copernicus's mode of argument, in relation to his audience, would hardly have remained imaginable.

Nicole Oresme did something that most of the real or supposed destroyers of the medieval formation called Scholasticism intended to do: With a deep incision, he repaired the system, and tried to prove Aristotle right by preserving the axiom that each element can have only *one* natural motion. Admittedly, if he had proceeded consistently in his terminology as well he would have had to describe the straight-line return motions of bodies (parts) of elements as *hors nature*. In this harmonization of the Aristotelian system, the Earth's proper motion is only (as I said) a by-product, if only because it is not induced by astronomical or cosmological considerations at all, but is only part of a rearrangement of the physics that is derived from the doctrine of the elements.

Not even the famous thought experiment with a tube that penetrates all the layers of elements, the experiment from which Oresme gained his insight into the relativity of the predication of "natural places," has cosmological dimensions—this tube is only permitted to reach to the underside of the Moon's sphere. But the thought experiment belongs quite generally to the kind of considerations by which the relativization of the 'order of essences' that was characteristic of the medieval world system was brought about.

"Je pose par ymaginacion" [I posit in imagination]—thus Oresme begins his consideration—that in that tube, which reaches from the center of the Earth up to the Moon's sphere, if it were filled only with the two lighter elements, fire and water, the air would be pushed down to the lower end—that is, in Aristotle's absolute scale, it would arrive at earth's "natural place"—while the fire, in the layer above it, would be able to occupy the places of all the other elements. So the arrangement of the elements in the tube depends only on their difference from one another, not on their situation with respect to the center of the world or the lowest sphere. Although this remains unstated, the thought experiment undermines geocentrism as well, although the imagined tube would as a matter of course contain a certain amount of earth in the lowest and thus the central position. But each of the four elements can be imagined as being missing, and then being replaced by the one that is next to it in weight. It is not the elements' relations to specific places in the world but rather their relations to one another that determine their actual arrangement.[23]

One may not infer from this discussion that along with the diurnal motion of the Earth, Oresme even thought of the 'next step' on the way to Copernicanism, that of removing the Earth from its position in the center of the universe. What is pertinent to that step is only the fact that the relativization of the doctrine of "natural places" opened up, for it, a latitude that had previously been denied to it. For the rest, one should not let oneself be misled by the expression "the next step," and suppose that when the annual motion would only have had to be added, as a second motion, to the already admitted (if not yet definitively asserted) diurnal motion, one already has before one the logic of the genesis of the Copernican system, in the manner of Haeckel's fundamental biogenetic law.[h] On the contrary, the way Nicole Oresme advanced to the diurnal motion and not one step further is one of the examples that show that this construction represents a blind alley. If one keeps this in mind, the difference between talking about "forerunners" and about what is here called "scope or latitude" ["*Spielraum*"] is clarified further.

If I said of Oresme's discussion of the motion of the Earth that, seen from the point of view of access to Copernicanism, it was a blind alley, this is also due to a specific deficiency in the motivation of Nominalism. The exhaustive exploration of the speculative possibilities of the Scholastic system is aimed at displaying God's sovereignty and His *potentia absoluta* [absolute power], but not at expanding man's knowledge of the world; if it nevertheless proves useful for the latter, then this is a result of the efforts to maintain the system's stability in spite of that demand on it. Oresme's thought experiment with the world-tube is an example of this. It is one of the long series of attempts to test, in regard to their binding force over omnipotence, the essential necessities of nature that Aristotle asserted. In the process it became evident that with the things that he laid down with the concept of 'nature,' Aristotle had committed Scholasticism to restrictions of divine freedom that were much too far-reaching. One merely had to exoticize sufficiently radically the conditions under which Aristotle had elevated everyday experience to natural necessity in order to make the thought experiment into the equivalent of a cognitive will whose potential is evaluated negatively. Scholasticism's autocatalysis is not a phenomenon of burrowing, progressive self-liberation from the Middle Ages—not a model instance for future emancipators. Though the principle is certainly correct in this case too that man makes history, this instance provides no evidence for the assumption that the outcome of making history has some resemblance to the ideas that people have while they are making it.

So it is not simply that Nicole Oresme would not have dared to adhere to the thesis, which he posited experimentally, of the Earth's diurnal motion; rather, to press forward to a position of that kind was not at all in accordance with the intentions of the work that was to be done on Aristotle in a commentary. The freedom that the author thought he was winning for his God, against the authority of Aristotle, might be rediscovered in the function of a latitude for human cognition. The removal of hindrances to mental operations presents itself as the divine will's not being constrained by the cosmos's official 'channels,' which, beginning with the Aristotelian causality that proceeds along the radial scale of the finite cosmos, had transformed themselves into the medieval *ordo creaturarum* [order of created things]. What people have been fond of calling the Reformation principle or the early modern principle of immediacy is precisely that suspension of a supposedly naturally necessary code of rules governing all proceedings between Creator and creation.

The founding document both for the principle of immediacy and for Parisian Nominalism was the decree of Bishop Tempier, in 1277, with its condemnation of 219 propositions representing, in most cases, orthodox Aristotelianism—that is, representing cosmos metaphysics as something that is able to regulate assertions about God's power. The practice of securing the cosmic order by asserting that infringement of it would be contrary to nature had to be abandoned: Nature was a contingent given, not the unique and exhaustive possibility of divine action. The new questions that arose here could be summed up in the single formula of Paul of Perugia in his *Commentary on the Sentences*

(1344), which asked whether God could make something that He did not make: "Utrum deus potest facere quod non fecit." From this point of view there was no reason to prefer the motion of the primary heaven, as opposed to the daily motion of the Earth, in respect to the distance across which the moving power had to operate, and the sequence of intervening agencies that it had to adhere to in doing so. Even the respect accorded to astrology in the Middle Ages still depended on the axiom that God had to make use of the cosmic order in order to act upon earthly and human events. The Paris decree of 1277, on the other hand, is an instance—surprising to later posterity—of a comprehensive license that is granted exclusively in the form of a prohibition.

The Middle Ages' symbiosis with Aristotelian philosophy had depended on an inequality in the attention given to God's attributes: on neglecting omnipotence, and still more omnipresence, in favor of spiritual self-sufficiency and of the world's continual and constitutive dependence on the divinity's inner life. One will not go wrong if one says that the attribute of omnipresence interested the Middle Ages the least and the latest, and indeed that it no doubt did so only circuitously, by being connected to omnipotence. It is clear that the speculation about omnipotence undermined, in principle, a schema that had not only restricted causality to the cosmic order, but had restricted even the Christian God's acts of grace to the official channels of the ecclesiastical administration of the treasure of grace, and of the hierarchy of offices. In Aristotle one could understand why the unmoved mover was operatively connected only to the outermost sphere. The Aristotelian God was so exclusively occupied with Himself and sufficient for Himself that His causality vis-à-vis the world could not even be described as "efficient."[i] But the Middle Ages could ask and had to ask why the *potentia absoluta* [absolute power] should adhere to supposedly official channels in its operation in the world. Symmetrically corresponding to this question was the other one, why man should not be able to address himself directly to his God.

The correspondence between the two questions makes it manifest that there is a connection between the Reformation's principle of immediacy, as a consequence of Nominalism, and the Copernican reform. The latter presupposed that a kind of direct 'encroachment' by God into the low levels of His world had occurred at least at the beginning, when the Earth was set in motion. If this influence was to be restricted to the beginning, it presupposed either the impetus or "natural motion." However, the concept of the impetus had originated in the effort to secure more effectively the principle of indirectness [*Mittelbarkeit*, the opposite of "immediacy," *Unmittelbarkeit*]. That may also have been the reason why the real 'revolutionary' at the beginning of the fourteenth century, William of Ockham, rejected the impetus doctrine. This partial immanence may have seemed to him more like a diminution of the principle of omnipotence. Still more remarkable is the fact that Nicole Oresme, too, makes no use of the impetus theorem. Exactly corresponding to this omission is Oresme's detailed treatment of the divine attribute of ubiquity, insofar as it no longer permits one to make the remoteness of bodies from the outermost sphere (as the boundary of transcendence) into the criterion of the proximity or distance of the divine influence, pointing out that God is not present to a higher degree in the heavens than He is in any other region of the world, and that His presence involves the same kind of efficacy everywhere: "Dieu n'est pas ou ciel plus que en un autre lieu." [24] Equivalent language to this emphasis on ubiquity is found in the assertion that it is "nature" alone, and in the same manner, that moves all the parts of the world. Then for all "natural motions" this element of naturalness is in fact enough to make recourse to an unmoved mover just as superfluous as it had already been for, for example, the phenomenon of free fall.[25]

The principle of immediacy is also stronger than the original motivation behind the concept of the impetus. This is evident in Jean Buridan, who both inclines toward the new theory of motion and also brings out (in a way that clearly implies an attack on the idea of the *actio subordinata* [subordinate action]) the possibility of every effect being related directly to the primary world-cause. To make room for the exclusive coordination of all effects with the primary, divine cause, he denies that there is any intraworldly necessity for a hierarchical ordering of causalities.[26] This assumption at least justified the thought experiment in which Buridan, by assuming that the first heaven (and, with it, the other revolving spheres) stands still, 'tested' the independence of sublunar natural processes with respect to the motions of the heavens. For this it was already sufficient to assume other types of influence by the heavenly bodies on the processes of terrestrial nature, types of influence that would not be dependent on their motion, such as, for example, the warming effect of the Sun, the continuation of which, even if the Sun were motionless, would not allow terrestrial life to become paralyzed. This reflection stops short of the conclusion that the apparent motions of the heavens could also continue, in spite of the assumed standstill of the sphere of the fixed stars, if only the Earth were imagined as being set in motion, to compensate for it. This has become a more plausible consideration, in terms of its systematic compatibility, even if for Buridan the distance is still too great for him to have been able to leap it. But even the thoughts that no one thinks are no longer impossible thoughts.

What we have described as a preparation, in natural philosophy, of the scope or the latitude for a Copernicus—as a change in people's readiness to receive such a proposal—becomes, in the post-Copernican history of his effects, a factor in people's affinity for him. At the same time it becomes evident that the new structure of ubiquity and immediacy contains, potentially, both Giordano Bruno's radical implementation of pantheism and the delayed understanding with Reformation theology, the most important and widely circulated evidence of which was the *Discourse Concerning a New Planet* of the Bishop of Chester, John Wilkins, which first appeared in London in 1640 and then in German translation, after another great delay, under the title *Vertheidigter Copernicus* [*Copernicus Defended*] in 1713. Here the state of the argument regarding the Copernican problem is still, or again, very much that of early Nominalism, though of course without the central motivation of preserving the boundary conditions of Aristotelianism.

Wilkins, who is more inclined toward Cambridge Neoplatonism than toward a strict Calvinism, conducts a polemic against the intelligences that are supposed to move the heavenly spheres, a polemic that is entirely on the level that had been reached three centuries earlier: They are "both superfluous and very improbable. Because a natural Power, intrinsical to those Bodies, will serve the turn as well. And as for other Operations, which are to be constant and regular, Nature does commonly make use of some inward Principle." [27] It would have been a needless and consequently an unworthy arrangement on the part of Providence to choose intelligences to perform something "which might have been done as well by the only Will of God." Now no violence is required in order to urge this conclusion as a way of refuting the objections brought against Copernicus: "Now, as it would be with the Heavens: so likewise it is with the Earth, which may be turned about in its Diurnal Revolution, without the help of Intelligences, by some motive Power of its own, that may be intrinsical unto it." Copernicus is defended because he had provided a new cosmic confirmation of the immediate relation of terrestrial and human realities to their origin.

My objective was to show that in the history of philosophical thought and of its role in the foundation of modern science it cannot only be a matter of presenting the derivation and development of particular ideas and hypotheses, and of bringing to light what stimulated them, and their early forms. Instead, we need to begin one level lower, with the origin of the scope or latitude in which those new conceptions first became possible at all, and within which both the affinities that gave them an effect and the means by which to formulate them arose. That does not answer the question as to how the Copernican system arose. It only removes the isolation of that issue from the preconditions of the fact that Copernicus did not become the Aristarchus of the sixteenth century, a thinker without any effect. What I had to present, in outline form, is the history of Copernicus's freedom, but also of the ways in which he takes into account his contemporaries' powers of comprehension—a masterly performance that is, quite simply, one of the things that still remain to be discovered in his work.

From this angle, the 'prehistory' of the Copernican reform does not present itself as the gradual consolidation and convergence of sets of motives into what is finally an irresistible historical necessity. To me, the exciting historical problem of this epochal turning is precisely not the explanation of the *fact* of Copernicus's accomplishment, or even the affirmation of its *necessity*, but rather finding the basis of its mere *possibility*. It stands at the end of the centuries that had borne the imprint of the most closed dogmatic system of world-explanation, whose basic character, not accidentally, can be summarized as implying the impossibility of a Copernicus before his time.

But that also means that Copernicus was not the original liberator that the self-consciousness of the modern age was to understand him and heroize him as, over against the Middle Ages. However much gratitude may be owed them, liberators have, for the onlooking historiographer, something distressing about them: However well one may understand their motivations, they are events that break into the structures of unfreedom from outside, even if they are located within those structures in terms of space and time. The modern age has favored the category of the creative power of genius for its outstanding individual achievements and also for its understanding of the origin of its own epoch. But the 'beginning' of reason is a Münchhausen tale. Copernicus's greatness is not diminished by the fact that he stepped into an already existing freedom, which the late Middle Ages had in fact themselves achieved, in the logic of their internal discussions and the working out of their systematic tensions, and had put into effect despite concern about their own stability (and, not least of all, on account of that concern). The turning came about neither as the result of an act of the mythical spontaneous generation of reason nor as the result of a forcible intervention, 'from outside,' of genius. Examination of this historical process cannot from the outset subject itself to the modern age's historical self-understanding, which, in regard to the mysteries of its own origin, only repeats the metaphor of a transcendent light streaming into the world—a metaphor that it had found already present in the schema of the Christian story of salvation and its zero point.

3
Transformations of
Anthropocentrism

Science misleads man into a peculiar arrogance toward himself. Our pronounced sensitiveness in regard to all arrogance on the part of our contemporaries in what has become the confined space of our world, and in regard to claims to exclusive rights on the part of those who inherit a head start in progress, as against those who need aid in order to catch up—these attitudes melt away, in the dimension of time, into a willingness to grant as a matter of course the existence of every kind of exotic backwardness. Such injustice no longer directly affects those who came before us, but it is no less corrupting in its effect on those who practice it.

In the history of science, failure to achieve equity in the dimension of time not only calls for the customary 'discussion' but for continual work on the destruction[a] of standards. Here we must be able to deal even with the supposed absurdities found in history. "Mais quelles erreurs!" [But what mistakes!] Voltaire notes in regard to what was practiced as science before and at the same time as Newton, despite all the "beautiful discoveries." Then Voltaire is noticeably taken aback when it occurs to him that after all Kepler, *dans un livre ridicule* [in a ridiculous book], had laid the foundation for Newton's admired work. That decides him, after so much throwing around of *sottises* and *folies* ["nonsense" and "follies"], to still grant a generous amnesty for the disreputable forerunners: "Il faut pardonner à ses erreurs, et jouir de leurs découvertes." [We must forgive their errors and rejoice in their discoveries.][28] That is a passage from the eighteenth century, I will be told; it is a caricature if one relates it to the history of science in our own century. But is it really so remote from what is done today, in the attitude and the manner of evaluation that it typifies? Of course today it is done more subtly, with more 'historicism.' [b]

Speaking of man's arrogance toward himself—of the arrogance of later men toward the earlier ones—inevitably leads to the almost mythical elementary theme of the elimination of the teleological worldview, of exposing it even in its more refined late forms. Here there seems to be a contradiction, since arrogance after all receives a direct hit when man does not remain the center and the end of nature and of nature's arrangements. It was after all this that first made scientific inquiry possible and that found, in its results, confirmations that really made man feel modest.

But again it is only the others, those in the past, who are convicted of presumptuousness and found to have been mistaken in it. The final and most subtle forms of anthropocentrism always remain hidden from their critics.

Let us not delude ourselves; just because this is a subject in regard to which we think we have got beyond ourselves and our illusions in a way hardly equaled in any other case, we fall into the trap of historical arrogance. It consists in what one could call temporal 'nostrocentrism,' [c] which establishes the status of each present moment as a goal, in relation to time, precisely by believing that it has definitively invalidated anthropocentrism in all of its open and concealed varieties.

The seriousness of this statement becomes evident when one understands the granting of a privileged historiographical status to the present, which Voltaire introduced and Hegel made into a system, as the principle of a relation to people. True, they are bygone people, but surely they are only presumed to be the entirely different people who would have deserved such treatment. The forms of contemptibleness and foreordained condemnation by the tribunal of reason may change. The principle is of the same kind that for centuries, without being contradicted, determined people's attitudes toward a hoped-for final justice of which it could be assumed as beyond question that 'for higher reasons' it would condemn the bulk of all those who had ever lived to an Orcus of eternal torment. I think that in our relation to history we should also be concerned with the idea[d] of a universal equity.

That concerning ourselves with history must always make us tremble for man is a conclusion reached by the early Enlightenment that

is represented by Fontenelle, a conclusion that we lost as a result
of our expanding retrospective view of the process of science and its
definitive accomplishments. If we followed Fontenelle, we would
never see man get beyond himself or arrive, in history, at the "point
of no return." ᵉ It is the problem of every enlightenment; it becomes
insecure as soon as it becomes aware that all historical moments after
each present one are equivalent with regard to man's radical potentialities. It is true that the rhetoric of enlightenment, a rhetoric that
cannot be avoided, refuses to admit that; this is the root of the hostility
between rhetoric and historiography. Fontenelle had taken ridicule
of man away from man and turned it over to the higher standpoint of
cosmic beings; purer reason may be permitted to be arrogant.

When the earnest conviction that everything is centered on man
and related to him, if not in fact created on his behalf and intended
for him by Providence, is accused of a presumptuous willingness to
submit to illusion, it will certainly not be possible to exonerate it, but
it will hardly be any easier to exonerate of the same charge the
consciousness of critique and rationality that, in the opposite direction, works itself up to engage in ruthless self-defoliation—because
it has paid no attention to the need for help that is also involved,
and perhaps is the main thing involved, in all anthropocentrism.
The expression "need for help" [*Hilfsbedürftigkeit*] in this statement
is meant to cover the whole spectrum from needing consolation to
requiring temporary constructions of any kind. We are also marked
by what we smile contemptuously over. Who still dares to say that
man is always a creature in need of consolation, when after all one is
supposed to believe that he is just now or will shortly be in the act of
dealing with the causes of that need for consolation, an intention in
which 'anthropological' statements of that kind could only hinder
him?

Augustine wrote that the stars were created for the consolation
of people who are obliged to be active at night, and in Linnaeus's
opinion the song of the birds at the first light of morning was instituted
as consolation for the insomnia of the old. To be sure, these ideas
cannot be renewed, and one need not even mourn them on account of
their lack of fear of sentimentality; but we must distinguish between,
on the one hand, the thought prohibition of teleological elements in
modern theories and, on the other hand, historical respect for the
equal rank of the aids that man avails himself of in comprehending

the world. History is precisely the theoretical attitude that allows us to do this. History of science is the specific form of history that makes it especially difficult for us, but which for that very reason can become a test of what is possible in our relation to history. I only mention what an immense job had to be done in order to remove magic and ritual from the sphere of the purely contemptible.

A contemporary of ours who has an interest in the history of science and who opens Copernicus's chief work will perhaps be most disturbed in his expectations by the dedicatory preface to Pope Paul III, and within that, again, above all by the explicit appeal to an anthropocentric teleology. Copernicus has recourse to this formula— that the world was created by the best-disposed and most systematic maker of all, and that it was created for man (*propter nos*)—in order to present astronomy's failure, in spite of its painstaking investigations, up to his time, as something by no means God given and unavoidable, but rather as a scandal that is to be vividly felt as such. This phrase has left the commentators, who elsewhere have pursued even the smallest indication of an element of tradition, totally unmoved. In the mouth of the destroyer of cosmology's anthropocentric illusion, this clause must have seemed to them to be pious talk for the benefit of the Roman addressee. But precisely in such a case the courting of favor would not have been squeezed into a subordinate clause, nor would it remind us of other and undoubtedly older sentences in the work.[29]

In order to understand the context of this preface one must keep oneself free of the idea that Copernicus wrote to the pope in a situation that had anything to do with the fear of pressures such as were to determine Galileo's situation more than half a century later. It is true that he expresses the expectation that some of his readers could disapprove of the thesis of the motion of the Earth,[30] but his choice of terms does not reveal any connection to specific objections or reactions of ecclesiastical authority. Instead, the teleological world-formula is clearly directed against an epistemological reservation in regard to the possibility of true knowledge in astronomy. While it is true, he says, that the philosopher's reflections are not subject to the judgment of the general public, still even in seeking the truth about the objects that he investigates he may only go so far as has been permitted by God to human reason. Only within this scope, then, would there be a question of keeping oneself free of erroneous opinions.[f] Here we arrive at the critical point: Could not astronomy's dismal state be an indication that the human spirit is simply not granted an adequate acquaintance with this realm of objects?

It is clear that if the teleological world-formula is assumed, there cannot be such a reservation, insofar as the formula's anthropocentrism is to be understood as not only regional and terrestrial but universal and cosmic. But this is exactly what Copernicus proclaims when he applies it to the totality and unity of the *machina mundi* [mechanism of the universe]. His appeal to the world-formula is not so much in support of his particular thesis of the Earth's motion as it is in support of the unrestricted claim that he makes for astronomical cognition.

The commentators' lack of interest in the dedicatory preface's world-formula could very well be a simple result of the fact that what they see being used here not only is talk that is intended for purposes of courting favor but is also an insignificant commonplace. It would then be a case not of argumentation but of rhetoric, even if its being such does not mean that it deserves to be disdained. But this supposition that the world-formula lacks specificity is, in fact, superficial. That the world was carefully made in the interest of man is not, in the overall Christian tradition, such a settled premise, especially not if it is understood as implying the possibility of a theoretical relation to the universe as a whole. In the two premises of the dedicatory preface—on the one hand, that reason's permitted competence might be restricted, and, on the other hand, that the world is homogenously intended for man, so that he is capable of theory—theology and teleology stand in a relation of tension that is characteristic of the early history of science. This relation is not simply medieval. To determine the derivation of the factors at work in it is indispensable for an adequate interpretation of Copernicus's presuppositions.

The Latin Patristic authors between Minucius Felix and Augustine already acquired the teleological world-formula from the same source from which Copernicus claims to have received the first impulse toward his reform, namely, from Cicero, this most influential transmitter of ancient philosophy to the early Christian literature in Latin. The device that conceals the appropriation of pagan teleology and assimilates it to the biblical idea of the Creation can be seen in a passage from the *De officiis ministrorum* [On the Duty of Ministers] of Ambrose of Milan, a Christian paraphrase of Cicero's treatise on duties, written toward the end of the fourth century.

Ambrose polemicizes against the philosophical conception of justice as respect for property, both public and private. Property, he says, is not at all *secundum naturae* [by nature], because nature produced everything for everyone and by God's will the Earth was meant to be a sort of common possession of all. Private rights first arose through appropriation (*usurpatio*). The train of thought is verified by a literal quotation from Cicero, according to which the Stoics had taught that on Earth everything was produced for men's use. Now the authenticity of the origin of this premise is disputed. Where else should the Stoics have got such a doctrine except from the sources of the Christian religion: "Unde hoc, nisi de nostris scripturis, dicendum adsumpserunt?" [31] For evidence there follow the Bible references Genesis 1 : 26 and Psalms 8 : 8. The Bishop of Milan does not conceal the fact that these Bible passages do not say the same thing as the Stoic doctrine about nature. The governing idea in the Bible is not creation *for* man but the subsequent subjugation of nature *to* man. It is a momentous difference, if only because it provides the only reason why the world is not Paradise itself, but rather the reality with which man has to deal as a result of his expulsion from Paradise. Ambrose of course blames this difference on the Stoics' lack of comprehension of their secret source, a lack that made the subjugation of things to man appear to them as their being originally intended for his benefit.[32]

How lastingly successful the harmonization, by Ambrose and others, of the biblical postulate of dominion and pagan teleology was can be seen when the Ciceronian tradition is picked up again. To Petrarch's ear, the result is again a reversal of the Patristic inversion, which makes him observe with amazement (in his *De sui ipsius et multorum ignorantia* [On His Own Ignorance and That of Many Others]) that in Cicero one does not hear the pagan philosopher but the anticipation of an apostle. As in much else, here also the preparation of the modern age depends on the decay of the harmonization that had been accomplished in the first centuries of Christianity, in the zeal of wooing the late ancient world. First of all, the important thing is that Copernicus, with his formula of the anthropocentric teleology of the world, did not simply make use of a customary Christian phrase in his address to Paul III. The interpretation of astronomy as a technique of calculation could have been sufficient for the biblical command of subjugation, though this undervaluation of the human claim to the world could not be adequate to Stoic anthropocentrism as soon as cognition, too, was seen as a natural need.

For the formula of a universal teleology for the benefit of man, the Christian reception meant a misunderstanding of its original sense. It can be described as a reversal of the relation of foundation between the concept of God and the concept of the world. For Stoicism and for Cicero, the gods and men were beneficiaries of cosmic teleology; theology is ultimately still based on cosmology. From the point of view of the biblical idea of creation, such a conception was intolerable. The world could not be the condition of the bliss of the divinity, just as, eschatologically, it could not even be this for man, inasmuch as he was supposed to be blessed only as a result of its destruction.

In Cicero we can observe still another connection that leads forward to the givens that Copernicus inherited. The teleological world-formula had sanctioned the geocentric construction of the cosmos. This would have had to exclude its being used by Copernicus, if it had remained what it had originally been. In Cicero, the idea that the world was created for man and was comprehensively apportioned to meeting his needs was no eclectic flotsam but rather the central doctrine around which and by means of which the heterogeneous elements of his rich source extractions arrange themselves meaningfully. The Skeptic's abstinence from theory presupposed (while concealing this presupposition from itself) that man is able to be so sure of the conditions of his existence that he has no need whatever of theoretical prerequisites to the provision and ensuring of those conditions. This primary teleology permits the Skeptic to do without the luxury of a physics, such as had still been a pressing subject for the Stoics. For the Roman it is a practical, not a theoretical, certainty that the world, as a structure of order that is oriented toward man—that is, as the old cosmos—has remained intact.

It is also in connection with Copernicus the reader of Cicero—who encounters the Roman Skeptic through Niccolò Leonico Tomeo in Padua shortly after 1500 and thus becomes acquainted with the note on Hicetas in the second book of the *Academica priora*—that it is important to consider the complex of theoretical Skepticism and Stoic practice that Cicero offers for him to receive.

The divine manifests itself to man through the cosmos; the traditional attitudes of *pietas*, *sanctitas*, and *religio* are responses to the definite sense that is manifest in the world—to the favor, through the world, that is granted to the human race by the gods.[33] In this way religious behavior becomes a piece of justice, a *iustitia adversum deos* [justice toward the gods], as the fulfillment of an intracosmic reciprocity, on which alone the moral transformation of the natural order is based.[34] Justice to the gods!—that is like a superlogical articulation of the basic Old Stoic idea that gods and men are citizens of the one polis constituted by this cosmos and that everything else was produced for, and, accordingly, to be used by both the former and the latter. For Cicero the state cult becomes the product of an officially administered function of distributive *iustitia* toward the gods, who for their part have always already done their share toward men. The prerequisites of happiness, in this world community, are shared. Man's disadvantage consists only in his mortality.

Part of the inner logic of this foundational arrangement is that cosmology can change directly into theology, since the whole of the world, as the most perfect being (*ens perfectissimum*), is able to satisfy the concept of the divine itself.[35] The world is divine because gods and men live in it and because it meets their needs. There is not supposed to be any contradiction in the fact that this cosmos simultaneously regulates and satisfies life. The world, on the one hand, is the perfect and insurpassable thing that binds man, as a part of the whole, to this whole, and, on the other hand, it is offered to man for the 'intuition' [*Anschauung*: viewing] that makes him happy.[36] The ethical and the aesthetic cannot be isolated from one another. Virtue, happiness in life, and aesthetic enjoyment are only different aspects of man's localization at the point of reference of the cosmic teleology. Because the cosmos is there for man, man is obliged to make his behavior "similar to the well-ordered cosmos."[37]

This anthropocentric systematics makes it understandable that the geocentric and the anthropocentric world-diagrams were combined and based on one another for the first time in Stoicism. That is why the Stoics defended this cosmology, as an expression of their piety toward the world, to the point of intolerance. When Aristarchus of Samos tried, for the first time, to remove the difficulties of astronomy by means of a heliocentric system, the head of the Stoic school, Cleanthes, demanded that a public charge be brought against him of impiety—this reproach that is always also politically plausible, and with which the subtlety of a theoretical disagreement could be intensified into the clarity of a scandal. This case of *asebia* [impiety] dramatizes the fact that what Ernst Goldbeck called "geocentric religiosity," which cannot by any means be ascribed, already, to Aristotle, really first comes into existence with the Stoics.

Anthropocentric piety toward the world is also the basis of the Stoics' position in the history of astronomy. They needed geocentrism as the truth, not as an element in a calculation. This is the goal of the Stoics' attempts at integrating [into their system] the theoretical development that had taken place since Plato and Aristotle, attempts that already contained equivalent—that is, empirically undecidable—corrections of Aristotelian homocentrism.[g] In the middle period of Stoicism, above all in the Syrian, Posidonius, around the beginning of the last century before Christ, one can see once again an effort to compensate for the disadvantage that the great authorities of classical philosophy, Plato and Aristotle, had had at their disposal only what was still an underdeveloped astronomy, which became part of the canonized world-system and to contradict which was to become, for long centuries, something that one could not risk. One has to see it as a shock to philosophy that theoretical constructions had appeared in astronomy that were equally respectable, that adhered equally to the supposed Platonic requirement,[h] but that worked with eccentrically superimposed circles or with epicycles and that achieved their effectiveness precisely by disregarding classical physics.

When the older formula that theory must "explain" the phenomena was now, no doubt for the first time, transformed into the more important one (in its consequences) according to which it must "save" the phenomena, the higher degree of complication of astronomy's relation to physics—up to its complete isolation—can be gathered from the language alone. The astronomers' resignation to wanting to be nothing more than technicians of calculation was the consequence—intolerable to the Stoics—of the equivalent constructions that had been found. But at the same time it was the only license for theoretical progress that was still possible, however modest the scope might be that it left open.

What is important here is to show that the appearance of the anthropocentric world-formula of the Stoic type in Copernicus's dedicatory preface is not a rhetorical accident, but is connected with his work's claim to theoretical truth in such a way that its localization in the tradition can be unambiguously defined. Copernicus wants to assure himself, for his work, of exactly the same argumentation that had previously been exclusively (and apparently inseparably) at the disposition of geocentrism. But if that is so, we must also clarify how the anthropocentric dogma was coordinated with the Christian tradition. Only thus can we achieve clarity about the sources of the opposition to Copernicanism.

The Christian reception of the Stoics' world-formula did not serve primarily to manifest a distinctive status of man within the world. Instead, it was meant to oppose tendencies, in late antiquity, toward the divinization of the starry heavens. If the heavenly bodies were created to serve man—Lactantius, for example, argues—they could not be regarded as gods.[38] Here the teleological principle, which at the beginning of the modern age was to become the focus of almost all the tendencies of criticism of traditional metaphysics, itself had the function of a critical principle against the pagan inclination to fill the world with gods. The fact that geocentrism can be maintained only at the cost of awkward and seemingly fantastic constructions did not make its way from the astronomy of late antiquity into the consciousness of the early Christian authors. Consequently, it could not make the idea of anthropocentrism suspect either. Only the return of Hellenistic astronomy, by way of the Arab tradition, into the Latin spiritual world of the Middle Ages was able to set up connections with the Christian doubts as to whether the manifestation of absolute omnipotence must be characterized by any congruence at all with man's existential objectives in the world.

The incompatibility of the elements that were appropriated in the Christian reception of antiquity emerged earlier. It had become unmistakable in the many layers of Augustine's spiritual biography. The young African graduate in rhetoric is saturated with Cicero. He unhesitatingly takes over into his early treatise *On Free Will* the Stoic axiom of the perfection of the world. In view of this, he does not doubt that human reason must have the desirable insight into the intentions of the Creator's reason. The inner necessity of human reason is also the outer necessity of the world. Man is familiar with the rationality of the Creation in advance of any experience. So what his theoretical reason outlines cannot represent an arbitrary system, but must contain the reliable ground plan of reality itself as an intelligible summation of the optimum of what is possible.[39] For reason, which, in line with the Platonic tradition, is essentially defined as *memoria* [recollection], the contingently given position in the visible universe to which body and eye remain bound cannot have any importance. The ancient unity of the theoretical and the aesthetic disintegrates, as does that of the theoretical and the ethical. Stoicism's relation between cosmology and theology is reversed; the former becomes a function of the latter. Human participation in the reason of the Creation does not, however, mean, without further ado, the friendliness of the Creation to man. The tradition of the world's teleological functioning on behalf of man could not be continued unchanged.

The turning toward a rationalization of the teleological principle was at the cost of an impoverishment of the human relation to the world. Man was no longer considered and provided for, by means of the world, in the full range of his needs, of his striving for happiness and his desire for beauty; instead, the claim of his reason was minimalized into what, more than a millennium later, was still to be decisive for Descartes and his foundation of the theory of knowledge: not to be grossly in error about the world. Such rational asceticism in relation to the world has a certain appropriateness in the context of Christianity. The cosmos could no longer signify the comprehensive fulfillment of human existence, if one's sojourn in the world was only a prelude to access to truth and happiness. Only God was supposed to be able to satisfy man's essential needs. He alone was supposed to be the object of the enjoyment of the blessed, of a *fruitio* [enjoyment, delight] as an absolute end, to which everything else had to be subordinated as means. And "everything else" now also meant the cosmos itself and as a whole.

This subordination of the teleological relation under the theological one was formulated, in Augustine, with the aid of the separation of use and enjoyment, of *uti* and *frui*. The cosmos was no longer the sum of human existential ends, but rather of the means to them, and also the obstacles to them—in a combination that was not clarified because the person who arrived at his end was still, in a certain respect, a means to the end of the Creation.

The instrumentalization of the world that is introduced with the Augustinian separation of use and enjoyment is not comparable to ancient teleology and its detection of anthropocentric functions because, by its transcendent radicalism, it inserts into the relation to the world an element of proceeding 'at all costs' [of '*Rücksichtslosigkeit*'], an element that cannot deny its equivalence to Gnostic prototypes.

In Cicero, but also still in the Patristic literature and in Scholasticism, *utilitas* [usefulness] means a quality of nature's 'self-presenting' readiness for service, which is taken up and 'put into effect'—that is, conducted to its intended end—when man makes use of it. The character of this relation unmistakably acquires a harsher tone as a result of the biblical metaphors of subjugation and dominion. The source documents are interested in concealing this distinction, if only because the implication of heterogeneity permitted and encouraged linkages with Gnosticism. Just as the world cannot have a disposition to be, or to be able to become, Paradise, no more can it take on the hostile gleam of the product of a Gnostic demiurge. When Augustine detaches the qualities of enjoyment from the world and projects them into transcendence, he reduces the world's teleology to an aggregate of 'means of living'—to the dull utility of materials. Only on this assumption could he adapt Cicero's principle that everything that was made, was made for man's use: "Omnia quae facta sunt in usum hominis facta sunt." [40]

Right alongside this Cicero quotation opens the abyss of the question whether man can still inquire for a reason and purpose of the world at all, when the idea of the Creation is presupposed. [41] The answer is concise and stinging: The concept of an absolute will excludes the question of the reasons for its acts. What philosophy irresistibly has to ask about a personal world principle—namely, why it produced a reality beyond itself—is warded off by theology: The question "Quare fecit deus caelum et terram?" [Why did God make heaven and Earth?] is answered by the laconic and (for the human need for truth) sterile "Quia voluit" [Because he wanted to].[42] In this short sequence of question and answer two spiritual worlds collide. The attempt to connect the ancient cosmos to the new concept of God—if not, certainly, as its precondition, then at least as its consequence—is frustrated by the early emphasis on the element of will in this concept of God. Therewith, the teleological world-formula is relegated back to paganism. Its renewal will be connected to claims that run counter to the intentionality[i] of medieval thought. This must be kept in mind in connection with Copernicus's preface.

Augustine offers an anticipatory summary of the subsequent history of philosophical teleology and theological transcendence, a history whose peripety coincides with the zenith of medieval Scholasticism.[j] The comprehensive reception of Aristotle in the thirteenth

century—especially that of his physics and cosmology—concludes
the attempt to bring biblical theology and ancient metaphysics to a
systematic synthesis. In connection with Copernicus, this process is
important not only because it bases the acceptance of the geocentric
system on the foundation of Aristotelian metaphysics, and because it
still preforms the position from which the new astronomy is opposed
for almost two centuries after Copernicus. No less important is the fact
that, precisely in its cosmology, Aristotelianism seemed to make
possible a surface connection between theology and teleology, which
could serve to smooth over, systematically, the difficulties that had
arisen.

The suitability of the Aristotelian theory of the heavens as a schema
for such a synthesis is a result of the fact that it had 'split' the cosmic
teleology, and had split it into world-spheres that were (respectively)
theocentrically and anthropocentrically organized. Man remains the
center of a regional goal-relation of the terrestrial and especially the
organic world, from which he derives his nourishment and other
services. Only in this space that is sublunar and at the same time
central to the universe did nature, which does nothing without a
purpose and in vain, have man as its object.[43] But man is not the
highest-ranking being in the cosmos; he is far surpassed by others,
and specifically, we are told, by the heavenly bodies. If Socrates, in
Xenophon, had still been able to say that the motions of the Sun, as
the causes of the seasons, took place on man's behalf, in Aristotle that
is metaphysically excluded; man is now only a regional goal in nature,
and no longer the highest *telos* [goal] of the whole cosmos.[44]

If one takes it more strictly, one cannot even speak of a partial
teleology on behalf of man in Aristotle's universe. This is because the
regional anthropocentrism is only an aspect of the overall economy of
a nature of which it can be said that it does nothing in vain. For
terrestrial nature, this principle of teleology does permit one to assert
that everything that man needs is present in nature; but that is not yet
identical with the statement that everything in terrestrial nature is
there for the sake of man. The difference between these two assertions
is decisive precisely when, going beyond the conditions of the naked
preservation of life, claims in the area of theory and aesthetics are
supposed to be based on the principle of teleology. So Copernicus
could not even have introduced a treatise on physics with his appeal
to the anthropocentric world-formula, not to speak of a treatise on

astronomy, if he had wanted to keep strictly within the horizon of Aristotelian metaphysics. The Stoics had gone substantially beyond Aristotle when they saw nature's teleology as aimed not only at the preservation of man's existence but also at its fulfillment.

In Aristotle the world of the celestial spheres is turned away from man, in its meaningful relation. It has its *telos* in the pure perfection of the first, unmoved mover, which it imitates through its circular motions. Between this supralunar world and that sublunar one there is an absolute division of directions of teleological reference. This division prevents the universal centering of a homogeneous teleology of the cosmos, entirely apart from the question whether man could be this center or be in it.

The coincidence of the center of the universe and the middle of the Earth in the Aristotelian cosmology is a contingent fact having physical, not metaphysical importance. It has no value as an indicator of how man should assess himself and classify himself within nature. Disagreeing with the Pythagoreans, Aristotle declares the container to have precedence over that which is contained, and thus the spheres over their center. That the center of the world is not a specially distinguished place already follows from the Aristotelian doctrine of the elements, which assigns to the element earth both the center of the world and the lowest rank in terms of value. The Earth is not made of the material of the stars, so it cannot be a star among stars, as the Pythagoreans maintained and as Copernicus will determine that it is.[45]

The way things stand with Aristotle's world-system has epistemological consequences. The highest form of knowledge—the philosophical—cannot refer preeminently to man, to his moral insight or his political action. The objective rank of objects in the world and the subjective quality of knowledge of them are correlated with one another. Consequently, philosophy has to have to do with the objects that are of the highest rank—with the heavenly bodies and the metaphysical causality of their motion. As human knowledge, however, its accomplishment would have to conform to man's low rank in the world, if Aristotle had not traced reason itself back to an influence of the world of the heavenly bodies, in the form of the "active intellect," which, as the organ of principles and concepts, is identical with the intelligence that moves the sphere of the Moon. Thus reason's capacity for generality is attributed to a factor that is heterogeneous to man's terrestrial lowliness and his organic nature, and withdrawn from any individuation. The result is that reason is primarily a cosmic potency, external and almost accidental in its relation to man. It is still the Platonic anamnesis of the transcendent Ideas, which here produces an equivalent to itself that avoids the implication of individual immortality and thus prepares itself to serve as the scandal of the orthodox medieval reception of Aristotle.

In regard to the teleological world-formula, the doctrine of the cosmic "active intellect" is an experiment with the idea that there is only a contingent connection between man and the highest objects in the world. Reason does not, in the strict sense, 'belong to' man, but affects him, as a heterogeneous influx. The outcome of Aristotelianism is the subordination of anthropology to cosmology. Man is not really cut out for the contemplation of the heavens; instead, his theoretical curiosity confronts him with the appearance of an inaccessible and heterogeneous world-region, for the understanding of which the world that he is familiar with supplies no assistance. Thus the systematic justification of astronomical theory's resigning itself to constructive hypotheses is sketched out.

To medieval High Scholasticism, the Aristotelian world-schema, with its split teleology, seemed suitable for giving systematic expression to its understanding of reality. This attempt ultimately failed because the Christian Scholastics, like the Islamic ones before them, could not accept one of "the Philosopher's" inescapable premises, that of the eternal existence of the world. It had been inescapable for Aristotle because he could conceive of the highest and pure being only as thought that is turned in on and referred to itself, which is indeed able to move the world, but not to produce it. Pure thought was "pure" precisely as a result of its exclusion of any intention or volition that is related to the world. But in that case the teleology could only be thought of as running in the opposite direction, from the world to God, as its unmoved mover. In the context of the Scholastic systematics, the world's eternal existence is the price of the possibility of the proof of God's existence that was provided by Aristotelian metaphysics. Now what is at stake here is not only a choice, for dogma, between the proof of God and the idea of creation; what is at stake is the possibility of obstructing the meaning-questions regarding man's importance and position in the world, questions that oppressed him and that could not be posed on the Aristotelian assumption of the world's eternal existence any more than they could on the atomistic assumption of infinite space.

Here, already, we can see something that Leibniz was to be the first to urge, against Newton's absolutes of space and time: that in the infinite the principle of sufficient reason, as the epitome of all meaning-questions, can no longer be applied.[k] The eternal world that is postulated by those who adhere to a strict Aristotelianism derives its 'life,' if one may put it this way, from the equivalence of all essences that have qualified themselves for their permanent existence by the fact that they were always already there, and from the absence of significant differences between the individuals that these essences 'run through' merely in order to exhibit their substantial identity. Averroism, after the fleeting thirteenth-century synthesis, isolated and displayed this pagan logic in its pure form: There can only be one unique immortal intellect for all men (the intellect that belongs to the sphere of the Moon), if the eternal duration of the process of individuation is not to produce continually new individual immortal souls—a consequence that appears absurd as long as Humboldt's statement has not become possible at least by approximation: that the formation of individuality is the ultimate purpose of the universe.

About the time of the turn from the fourteenth century to the fifteenth century, Master Blasius of Parma advocated, at the University of Pavia, the Averroistic thesis of the mortality of the individual intellectual soul. Otherwise, as a result of the eternal existence of the world there would have to be an actual infinity of individuals, which would not only violate Aristotle but also the still higher authority of the principle that nothing in nature happens in vain: "Quaero statim, ad quid deserviret haec multitudo animarum infinita?" [I ask immediately, what purpose would that infinite multitude of souls serve?]. And now there follows, at least an an idle reflection, what it might mean to speak of such a far-reaching individualization of the intellectual soul: if, for example, it were intended to move the heavens or any one of their parts, or even this lower world or one of its parts. That it could, for its part, be the goal of a teleology remains out of the question, it is so manifestly unthinkable. The conclusion can only be " ... ergo ista multitudo videtur esse frustra" [therefore this multitude is seen to be in vain].[46]

Now one would think that the rejection, within Aristotelianism, of an overall cosmic teleology benefiting man was not such a great

misfortune, since a sublunary one would imply enough, if not everything, for his chances in life, and would in any case still assert so much
more than experience seems to document. The difference only becomes important if the teleology is supposed to be interpreted and
spiritualized in regard to man's position in the world as one who
engages in theory, and if it is to be set up against the formula of
resignation according to which what is not useful to man does not
concern him either. Just here is also the difference from the biblical
idea of the subjugation of the world to man, which inevitably has
to make accessibility the criterion of the claimed dominion. So the
splitting of the cosmic teleology contains virtually, and precisely in
connection with the biblical idea of subjugation, a demarcation
between the reality that man can know and the reality that is withheld from his *curiositas* [curiosity]. As applied to the question of the
derivation of Copernicus's world-formula this means that the teleology that was restricted to the sublunary realm could not be transformed into a justification of the universal claim to truth. Instead, it
made this claim appear as an overstepping of the threshold of man's
competence, a threshold that could only have been marked off within
the treatment of teleology.

Our task now is to investigate this connection a little further. An
instructive illustration of the medieval recasting of the Aristotelian
formulas on this subject is provided by a disputation conducted
around the middle of the thirteenth century in the presence of King
Manfred, of which we possess the final *determinatio* [conclusion], which
was given by the presiding master, Peter of Ireland, the teacher of
the young Thomas Aquinas.[47] The king himself had set the topic for
the academic exercise, with the question of the relation, in living
creatures, between the organs and their functions: "Utrum membra
essent facta propter operationes, vel operationes essent factae propter
membra." [Whether the members were made on account of their
functions, or the functions on account of the members.] The trick of
the question lies in the ambiguity of the preposition *propter*, which can
be read either causally, as "as a result of," or teleologically, "for the
sake of." The master's decision is that this is more a metaphysical
question than one belonging to the philosophy of nature, and it is
a question in which what is at stake is "the concern of the first cause"
for the world (*quaestio de sollicitudine causae primae circa res quae sunt in
universo*). It is a mark of the incomplete reception of Aristotle, prior to

Thomas Aquinas, when the almost affectionate characteristic of solicitous worry about the things in the world is ascribed to the "first
cause"—a quality that, as an "unmoved" mover, it can after all on
no account be permitted to have.

What, then, is the aim of the first cause's concern for the world? A
wise and all-powerful Author of the world cannot allow anything bad
to persist in it, we are told, but must arrange everything in the best
possible way so that the continuation of the whole—and more than
that, its eternal endurance—is guaranteed: *ad permanentiam aeternam
universitatis.* Here the fundamental theological idea of the world's
origin in a Creation and of the continuous influence exercised over the
way it runs its course stands immediately alongside the Aristotelian
assumption of its eternal duration in the future. Such an everlasting
permanence, as the highest accomplishment of the divine care for the
world, really makes the objection based on the evils in the world a
troublesome one, whereas when the duration of the world is an
episode between its Creation and its Judgment, the problem of its
defects is blunted.

In the disputation before King Manfred, the objection was brought
to a point as the antithesis between the intended permanence of the
whole and the continual reciprocal destruction of living things, as its
parts. In the Aristotelian view of the constitution of the physical world
this antithesis is unavoidable because motion and materiality are
coordinated with each another. Motion, as the succession of forms, is
the fundamental metaphysical characteristic of reality; the succession
of forms is accomplished on the basis of the enduring substratum of
the primary matter. Since there is no formless matter in the world,
every motion must depend on the destruction, in favor of a new form,
of a form that has been realized in the substratum. Thus the occurrence of evils[1] in the world is an aspect of the possibility of motion in
it. The evils can be justified only through the justification of motion.
To return to the initial question, it is true that a living thing that is
eaten loses its essential form, but the matter that is freed in the process
enters the form of the living thing that does the eating. This metabolism can become a justification of the destructions that serve it only
if there is a teleological connection in the relationship between living
things—that is, if the matter changes from lower forms to higher ones.

The Scholastic master, who knows of no evolution in the forms
themselves, shows no hesitation about elevating the process of a

metabolism directed toward higher forms, in a spatially finite and temporally eternal world—as an evolution of matter, rather than of forms—to the status of a principle. But the justification of the divine care for the world on the basis of the teleology of the sequence in which creatures eat one another is not only contradictory but also falls short of the goal, which is obligatory for it as part of the reception of Aristotle, of understanding, to some extent, the motions of the heavenly bodies as well in relation to the *pugna animalium* [combat of living things] in the sublunary world. Here too the *malum* [evil] of lower levels of being would have to be able to become the *bonum* [good] of higher levels, if this is supposed to be the metaphysical meaning of all motion, because the motions of the heavens exhibit the sense of the divine care for the world—that of guaranteeing its stability and endurance—in the purest possible way.

In his final judgment in the disputation before King Manfred, the Scholastic master has evident difficulty in systematically harmonizing the placement of man at the summit of the teleology of the eating sequence—that is, the basic idea of the instrumentalization of all living things for the benefit of one of them—with the idea, which is realized in the heavenly bodies, that eternal stability can be achieved without paying the price of the destruction of other creatures. For this distinctive character of the heavenly bodies at the same time brings out the fact that the maintenance of the sublunary creatures, and particularly of man, in existence is not an outstanding and not even physically an unambiguously necessary solution of this problem. So the consistency of the text of the *Determinatio* also contains a discernible leap when Master Peter comes to the point where he has to address the question of the position of man in his conception of the world.

Peter of Ireland's thesis on this question is that the whole of nature has provided aid and sustenance for man to the highest degree.[48] For this assertion he claims the authority of "the Philosopher," who said (we are told) that we too are the purpose of everything—if not, indeed, the purpose for the sake of which everything exists, at least the purpose that the user of an object attributes to it when he finds it useful to him. In using objects we find in them a preexisting appropriateness for our purposes that we do not have to interpret as an intention in their origin. In the final judgment of the disputation this is formulated as follows: "Sumus et nos finis omnium, non finis propter

quem omnia sunt, sed ut illud cuius dicunt esse omnia propter aliquam utilitatem." [We too are the purpose of everything; not the
purpose on account of which everything exists, but as people say of
something that everything exists on account of some usefulness.]
Anthropocentrism is not a physical fact but more nearly a kind of
juridical state of affairs that not only allows man to preserve his
existence with the aid of nature but also certifies the undisputedness
of this usufruct. One might say that this fits the biblical formulas of
subjugation and dominion, of the *dominium hominis*, better than it fits
the Stoics' doctrine of the teleology of nature. Above all, the idea of
the unchallenged utilization of a preexisting usefulness excludes an
attitude that was evidently, if not detected, then suspected as an
implication in the pagan formulas of anthropocentrism: the conversion of the right to self-preservation into a claim of right to the world,
which would be perhaps the strongest antithesis to eschatologically
flavored expectations: "If the world has indeed been created for us,
why do we not possess our world as an inheritance? How long will this
be so?" The Apocryphal apocalypse of the Fourth Book of Esdras[m]
had, after all, already reached this provocative logic, which is hardly
to be found in such an explicit form anywhere else.

Master Peter of Ireland does not intend anything of this kind. He
quotes the passage from a translation of Aristotle's *Physics* in which
the focus is not on the use of things as simply drawing a benefit from
a given stock, but rather on the relation of human art to the matter
that it presupposes and works up. We find in nature everything that
we need in order to practice our arts, and thus "we use everything as
if it were there for our sake." From this Aristotle infers that "we also
are in a sense an end [a *telos*]," even if we do not occupy the highest
rank among the things in the cosmos.[49] The misinterpretations of this
passage gave rise to an entire pseudotradition of teleology in medieval
Aristotelianism.[50] It is easy to see Peter of Ireland's need to 'enrich'
the ancient document metaphysically. The Arabic-Latin version
of the *Physics* that he had in front of him had rendered the text, in
its indefinite modality, quite faithfully: "Nos enim ipsi etiam sumus
finis quodammodo . . . ." [For we ourselves are in a certain way an end
. . . .] The Scholastic master replaced the attenuating "in a certain
way" with a decisively intensifying "of everything," which he found
nearby in the text, and thus produced his universal formula, "Sumus
et nos finis omnium." [We too are an end of everything.] In this way

the reader is led to forget Aristotle's split teleology. One can see that
the unifying trick then serves to pick up the original logic of the text
again and to proceed from the paraphrase of Aristotle's *Physics* to
the establishment of a real teleology in nature as a whole, in which
everything finds its highest goal-definition in its relation to the one
mover of the universe. The text takes up the *omnium* once again: "Sed
omnia sunt propter unum motorem omnium, primum scilicet." [But
everything exists on account of one mover of everything, namely, the
prime mover.] Here something that in Aristotle explains only the
motion of the heavenly bodies is drawn, by means of the detour by
way of man, into a homogeneous picture of the world.

As long as the reception of Aristotle is still diffuse and incomplete,
it brings the motives of the receiving author into prominence, in
comparison to those of the author who is being received: The material
that is handed down is deformed, even if only (in each case) in
accordance with the principle of the least possible modification. With
the growth of knowledge of the text, of the author's status as an
authority, and of needs relating to systematic proof, this relation is
reversed. In the course of the process of reception, Aristotle becomes
so indispensable that on no account can he be judged to be mistaken.
Now, the commentaries on the passage of Aristotle's *Physics* that Peter
of Ireland interpreted have to give themselves still more trouble. An
example is Thomas Aquinas's commentary on the *Physics* (written
after 1261) in which he is able to base his discussion of the crucial
passage on an adequate Latin version: "Et utimur tanquam propter
nos omnibus quae sunt (sumus enim quodammodo et nos finis ...)."
[And we use everything that exists as though it existed on account of
us (for we too are in a certain way an end ...).][51] He relates the
quasianthropocentrism of this passage exclusively to the realm of
artifacts: Human art, he says, prepares what it finds in the world
for use, and the success of such treatment allows us to infer that
an inherent usefulness [*eine Zweckmässigkeit*, a 'teleology'] has been
utilized and converted. Just as Averroes's commentary (which had
already been available to Peter of Ireland) had interpreted the passage in a juristic manner, Thomas employs the potency/act schema in
order to infer from nature's utilizability by art that it contains, at least
as a predisposition, a teleology that is favorable to man.

Admittedly, this would have to lead as a consequence to a conception of matter that on Aristotle's assumptions would be nonsensical.

The bare *hyle* [matter] was supposed to be the most pliant raw material of every art precisely by virtue of its lack of any definite characteristic, without needing a particular disposition to that end, not to speak of an anthropocentric teleology. It is clear that following the tendency of dogma, a matter that is created cannot satisfy the strict rule that it should lack all definite characteristics; but neither can it be permitted to have the mere incompleteness of raw material that must first pass through the hands of man in order to produce a correspondence between his needs and the facts. It turns out to be appropriate to distinguish between, on the one hand, nature's fitness to serve our purposes and, on the other hand, what it was intended for: A house is built with a view to its being habitable, but it is only in a certain sense that the actual inhabitant can infer from that that it was erected for him.[52] We no longer hear of nature being intended for man; strictly speaking, what is conceived and made for man is only everything that is produced by man's arts and techniques.

Thomas holds to the Aristotelian hierarchical order that gives the heavenly bodies priority over everything beneath the sphere of the Moon, and thus also over man. In doing so he encounters the biblical conception of the stars as signs and lights in the service of man (according to Deuteronomy 4:19). Lactantius had already harmonized this passage with the assumptions of astronomy by suggesting that it was precisely the strict regularity of the paths of the heavenly bodies that disqualified them from being divine, since the behavior of a god was, characteristically, impenetrably arbitrary. Thomas uses an ingenious construction to harmonize the Bible passage, which is directed against the worship of the stars, with the Aristotelian dogma of the lunar dividing line.[53] Can it be, he asks himself, that the heavenly bodies, which are higher in dignity, were created so as to illuminate and to give signs to the lowly Earth and its inhabitants? The way out is not to identify man with his position in the cosmos. At least in his nature as having a soul he enjoys a precedence over the luminous heavenly bodies: "secundam animam praefertur corporibus luminarium." In any case, this teleological question is only of secondary importance. It is surpassed by the basic idea that every part of the universe serves, above all, the durability of the whole, and that in that connection the functional order could very well breach the order of dignity.[54] Thus the discord between the Bible and Aristotle is resolved by the principle that by virtue of the world's comprehensive teleology, reversals of lower-level goal-relations are possible.

However, this does not overcome the difficulty that Christian dogmatics produces for the hierarchical ordering of things. The continuance of a physical universe that is destined to perish cannot be superior, in the teleological order, to the existence of man, who is destined to survive the world. Besides, it remains unclear what Aristotelianizing Scholasticism can still make of the concept of providence (which originated in Stoicism), when the preeminence of the heavenly spheres consumes the divine efficacy, so that the divine administration of the world can no longer continue to be *providentia omnium* [providence for all]. For there is no guarantee that this influence penetrates the cosmos all the way down to its center and thus to the Earth.[55] The lack of a divine world reason that remained powerful all the way to the innermost region of the cosmos had already been raised in antiquity as an objection to Peripatetic metaphysics: The Philosopher's 'mover' God was weaker, it was said, than Homer's Zeus. Viewing the same state of affairs in the opposite direction, Dante makes *la contingenza* [contingency] reach only from the Earth to the sphere of the Moon, which runs counter to the Christian idea of a world that, as a whole, is nonnecessary on account of its origin from nothingness.[56]

The importance of the lunar boundary line is reversed as soon as it is not a matter of ensuring metaphysical dignities, but instead the topic becomes the relevance of the partition of the world to epistemology. Then the question does not relate to the power of divine Providence to penetrate from the outside to the inside of the world, but rather to how far human cognition can reach out of its sublunary low position up into the region of the movements of the heavenly bodies. In Thomas Aquinas this question surfaces in connection with drawing the boundary between philosophy and theology. Astronomical cognition, with its restricted claim to truth, is treated as analogous to reason's attempt to penetrate deeper, by its own means, into the mysteries of the Divinity. To the question, which had been raised again and again since Augustine's *analogia trinitatis* [analogy of the Trinity], as to whether the doctrine of the Trinity of the divine Persons could also be apprehended without a special revelation, Thomas gives a firmly negative answer, and he denies the usefulness of Augustine's anthropological triads. One can only speak of true knowledge where, beyond analogies, a sufficient proof (*ratio sufficiens*) is presented. In order to clarify this requirement, Thomas cites examples from natural philosophy.

Thus, he considers the proofs of the uniform velocity of the motions of the heavens to be sufficient—no doubt because he regards Aristotle's deduction of this uniformity from the concept of time, and time's cosmic basis, as compelling. On the other hand, the constructions composed of eccentric and epicyclic orbital elements that are used in astronomy do not seem to him to be sufficiently established merely because the motions of the planets in the heavens can be described and made calculable by means of them. These constructions do offer one possible explanation, but it is not sufficient because it is not able to exclude other possible constructions.[57] The proof of God's existence corresponds to the proof of cosmic uniformity, and the Trinitarian speculation corresponds to the hypothetical method in astronomy.

If one pursues this reflection only one step further, one arrives at what appears in Osiander's foreword to Copernicus's work as God's reservation of a true astronomical knowledge that we could only have if it happened to be one of the mysteries that He revealed. Aristotle had blamed the inadequacy of astronomical cognition above all on the great remoteness of its objects, and the consequent restrictions on our perception of them.[58] Thomas takes that both literally and metaphorically. He makes the spatial distance represent, more as a surface expression, the specific difference of the [celestial and sublunary] natures: The distance between their natures is greater, he says, than the spatial distance; the attributes of the heavenly bodies cannot be expressed in concepts derived from terrestrial ones, and they stand in no representable proportion to them at all.[59] In other words, our natural optics are not only unclear and unfavorable; they are not, in their nature, a match for the object. Since it is not a question of quantitative disproportion, the idea of an optical magnifier would have to be excluded on these assumptions.

The astronomical resignation of the Middle Ages stands within this horizon. It was not only and not above all the torpidity of a naive self-reassurance derived from the consciousness of geocentric privilege and security within the world that Copernicus had to break through but also the reinforcement of sense-appearance by the metaphysical assumption that attempts to overcome it would be in vain. Only in what is nearest at hand does man find what is compatible with his cognitive power, most unambiguously where he himself becomes the author of the sphere in which he lives. As an instrument of orientation for him, astronomy is one of the useful things [*Zweckmässigkeiten*, 'teleologies'] that he himself sets up; that is why it does not belong in the realm of theory (*scientia*) but in that of technique (*ars*). Its accomplishments are adapted to the economy of human needs for time measurement, navigation, calendar reckoning, and the regulation of holidays, but they have no objectively securable value as knowledge. The purposes they serve in life justify us in forgoing the satisfaction of theory that is provided by truth. This epistemological finding affects the teleological character of the world only secondarily. For while the Scholastic author can admit that within the world there exist certain relations of serviceability between one object and another, he cannot relativize the totality of the world, in the same way, to a distinctive creature within it, as its purpose. This totality has its purpose outside itself, and Thomas defines it as the *gloria dei*, the praise of God.

Whatever import this formula may have behind it from its biblical and Roman origins, in the medieval systematic context it is indispensable for the recasting of the relation that Aristotle had sketched between the world and its "unmoved mover"—that relation whose reality was absolutely one-sided. There could not be such a one-way relation between the medieval created world and its Creator any more than a world that was supposed to be destroyed in the end (and, according to the fears or hopes of many at the time, all too soon) could be an end in itself. Still, for the Aristotelian that world remained the whole that cannot be in vain. Thomas emphatically points to the fact that Aristotle argues for the impossibility of the heavenly bodies' revolving in opposite directions from the premise that "God and nature do nothing in vain." Thomas is extremely gratified by the double subject in this sentence, in which he thinks he can see Aristotle going beyond the pure purposelessness, vis-à-vis the world, of that unmoved mover (though to go beyond this would have had to be inimical to Aristotle's system, because indifference toward the world constitutes, in philosophical abstraction, the essence of the autarky of the ancient gods).[60]

The issue here is not the incorrect interpretation of Aristotle but the way in which it was possible to fit the need for God to have an 'intention' for the world into the context despite its inimicalness to

Aristotle's system. To that end it was necessary above all not to violate the purity of the highest being's relation to itself and freedom from need. The idea that in producing the world and its processes the Unmoved could only have considered and intended Itself, and that It did this in the form of praise, accomplishes the adaptation of the basic Aristotelian condition to the heterogeneous assumptions of Christian dogma. To elucidate this once more in terms of the example of the theory of stellar revolutions: The fact that, in an eternal cosmos, certain regularities prevail, such as that of the uniformity of circular motions, is a result of the immanent conditions of the cosmos, such as the connection between the concept of time and uniform motion or the fact of its not being destroyed after an infinite duration; but the fact that in a world of finite duration certain contingent resolutions of possibilities are found, such as a uniform preference for one direction of travel on a circular path, can only be due to a goal-determining intention, so that it forces one to assume an active relation between the Unmoved and the world that depends upon Him. Only, this intention cannot 'end' in and with the world; otherwise it would contradict the assumption of the Divinity's exclusive relatedness to Himself, for which every goal of His action can only be, once again, a means of referring back to Himself.

The world is the objectivized glory of God. Since it is now a question of an intention belonging to the highest being, and not of the intention of the most sublime bodies in the universe, the world can be this glory only as a whole and without remainder. Aristotelianism's split teleology can, indeed, divide up the 'directions' of the processes in the world, but cannot divide the destinies of creatures in the world according to their natures. Thus only when one views the circumstance in isolation can the Aristotelian axiom "nihil fit propter vilius se" [nothing is made for the sake of something lower than itself] be infringed when, according to the testimony of the Bible, the heavenly bodies are supposed to have been created to serve all the nations (*in ministerium cunctis gentibus*). This 'mistaken connection' is put right again by being integrated into a double overall purpose: ... *vel propter totum universum, vel propter gloriam dei* [either for the whole universe or for the glory of God].[61] In other words, every form of anthropocentrism can only be partial, and subordinate to the overall cosmic purpose. Even the Old Testament citation confirms this when it makes the subservience of the stars to man into an argument against idolatrizing them.[62] So the anthropocentric formula is, as a rhetorical means to the end of not endangering the *gloria dei*, only a part of the truth, and certainly not to be used to found claims in the area of theory.

Always more difficult than understanding what is there, in a text, is perceiving and establishing what was left out and passed over. In the context of the use of biblical texts by Scholastic authors, one who is familiar with the post-Copernican scene is struck by how seldom Joshua's command that the Sun stand still is cited—the citation that was to become one of the weapons of the anti-Copernicans.[n] The difficult thing for the Aristotelian Scholastic to cope with in this biblical passage had nothing at all to do with the difference between geocentric and heliocentric theories. It concerned the reckless dealing with heavenly bodies in the interest of human concerns of which this text speaks. As far as I can see, the great Robert Boyle was the first to extract the rigorously anthropocentric point of the Joshua text, without bothering about its cosmological implications.[63] Even if the abandonment of geocentrism appeared to the delayed Scholastics to violate precisely this biblical authority, the passage was hardly especially dear to the Scholastic Middle Ages, because it commanded them to count an essentially impossible irregularity among the things that are possible.

Thomas Aquinas attempted to present Aristotle's split teleology as a nested one, whose comprehensive objective is the *gloria dei*. If one examines this more closely, what emerges from the refined language of High Scholasticism is only, once again, Augustine's crude idea that God created man in order to fill up the heavenly choirs that had been decimated by the fall of Lucifer's followers, and indeed to fill them up only with the picked voices of those who were predestined for salvation.[64] This mythical explanation of the existence of human beings already contains the basic theme that presents the final purpose of the whole production of the world as the restoration of the song of praise that was interrupted by the angels' presumption, so that it pictures the glory of the Divinity as the sole goal of His actions. Only this goal-definition makes possible Augustine's monstrous idea that the tiniest part of mankind will perform that function of praise, which at the same time is its only chance of avoiding calamity and damnation. That would not need to be discussed here except that it clarifies the obstacles that prevent us from seeing in Copernicus's anthropocentric world-formula only a repetition of medieval platitudes. We cannot avoid making clear what an opposition to fundamental medieval ideas the almost incidental insistence of Copernicus's dedicatory preface on anthropocentrism entered into.

The medieval restriction of the favor granted to man to those who are chosen for salvation does not alter the mediation by which this teleology, too, continues to be tied to the divine self-reference. But it decosmicizes Stoicism's providence into an understanding, which can no longer be exhibited in the world of experience, but instead is actually disguised by it, between God and those to whom He grants His grace. Therefore it can have neither theoretical nor practical consequences when, at the zenith of Scholasticism, the Franciscan Duns Scotus makes the statement that the visible world was created to serve the humans who were chosen for salvation, and thus appends another interpretation to the passage from the second book of Aristotle's *Physics* that we already discussed.[65] The hidden God's salvational will is made into the principle of a teleology that no longer leads to empirical particulars and that can no longer have consequences for man's self-consciousness. Just as little can we discern, any longer, how the 'remainder' of the world that is now split through mankind will make out with the world's 'teleology' (which in any case could only be a side effect of that side effect).

The logic of the Scholastic system leads to the 'carrying out' of the partial theocentrism of Aristotle's metaphysics as a universal theocentrism: "Deus est finis omnium naturalium" [God is the end of everything in nature], Jean Buridan writes in his commentary on the *Physics* in 1328.[66] Buridan too insists that in no action of God in relation to the world could this lower being become a final cause and thus make the Unconditioned into something conditioned. The goal of an action, as the final state that is meant to be brought about by it, determines an action in a stricter sense than does the one who benefits from it, whom one might at most call the indirect end of the action. It becomes evident that benefiting by an action always presupposes a degree of need and necessitousness; that is the only reason why there can be a universal teleology benefiting man. Man builds a house and he is himself the end for the sake of which he builds it, and he is the final purpose of his house in the sense that he built it for his own sake and for the sake of his welfare. Usefulness depends on need, and in this sense God cannot be the final purpose of the universe.

This distinction of Buridan's leads to an important observation in regard to Copernicus's world-formula. It cannot be read as a generalization of the partial teleology that, according to Aristotelianism, benefits man; it cannot be read this way because it excludes precisely the element of need. It is supposed to guarantee a possible access to truth by man that has nothing to do with the ways in which the art of astronomy is useful in life. A satisfaction is pictured that goes beyond man's needs—a satisfaction of a kind that, in the Middle Ages, only the *visia beatifica* [beatific vision], in the next world, could be. God is the final end of all natural beings, Buridan asserts, in a sense of sovereign independence, but not in the Aristotelian sense of his having no knowledge whatsoever of the world. Could one say that this relation too was now defined, *after* the action of the Creation, as one of pure theory? And defined, in that theory, as a universal homogeneity of all objects in the world? In that case, this turning would already, by implication, have suspended Aristotelianism's split teleology and produced, objectively, the unitary character of the nature as the 'reference-person' of which man, again, makes his appearance in Copernicus's formula.

Seen from the perspective of the Middle Ages, talk about the Copernican turning receives a more radical sense than that of the interchanging of the attributes of motion between the heavens and the Earth. Without blasphemous defiance, the Copernican world-formula grants to man a privilege, as an end, that only a theory that is beyond the realm of pressing needs could have, a theory that therefore could not be enjoyed from the center of the world—that is, from the side of its greatest rational indistinctness and natural confusion. For the Aristotelianizing Middle Ages did not, as one must continually remind oneself, share the Stoic premise that the center of the world was at the same time the privileged position for theoretical consideration of it and rational access to it.

The medieval God remains external to the world, not in an immediate relation to all of its elements, and every possible conception of the "ontological truth"° of the world relates to this external 'standpoint.' One had not been permitted to think of the Aristotelian God, the Unmoved, as permeated by care for the cosmos, but neither could one even think of Him as a mere observer and onlooker at its cosmic revolutions, contemplating them from a rationally privileged angle. He would not have had to contemplate what He had neither wanted nor made, and what was not even moved 'by' (though it was moved 'owing to') Him. It was unimaginable that He could have had any obligation toward the world—the Old Testament Covenant relationship was unable to affect Scholasticism, and the latter's still antique God does not even know the fidelity that had been ascribed, in the Greek myth, to Prometheus in relation to his ceramic creatures, and for which he had to undergo persecution and pain inflicted by his Olympian cousin.

The Prometheus myth shows clearly the difference between what the Greeks had understood by a god, who was under no obligation, and what, for them, a demiurge had to undergo who discovered a sort of obligation as a result of having put creatures into the world. To that extent the negative 'characteristic' of not having been a creator was an essential part of the autarky of a god, in the ancient world; this is the only way he can possess that unconcern (which is often described as immoral) toward everything for even the existence of which he was not responsible. As Epicurus's idle gods show, the philosopher did not credit a god with the ability to remain carefree even if he only had to view this world, even though he had not made it nor would have been able to make it. That is why even the gods of Greek philosophy did not practice theory, unless the content of those endless conversations of the gods in Epicurus's interworlds had to do with physics. The whole difficulty of the medieval system is due to the fact that it had to reduce the biblical substance to a consistent order of propositions within these prescriptions, from antiquity, as to what a god was like—so little could the Middle Ages choose for themselves how to judge what a god is and what he would have to do.

One simply cannot, as with this little excursus, take enough pains to grasp the conditions under which the Middle Ages sought to make and keep their realities accessible to them. Only if one could have the illusion that the modern age was a result of spontaneous generation, like Athena springing from the head of Zeus, would one not need to care from what intellectual and moral difficulties the epochal initiative pushed to escape. The prescription that a god must have the purest freedom from need and fullness of being requires that he should have created everything for his own sake; but at the same time he was no longer permitted to take no notice of that which, coming from his hand, was supposed continually to direct itself and move in relation to him, while he simultaneously circled and reposed in himself. The blissful carefreeness of a god could now be 'theory,' insofar as it was the ultimate foundation of that "ontological truth" without which no secondary "logical truth" could be acquired. That the goal-directedness of the whole of reality could be exhausted in its being contemplated and approved by God was already authoritatively dictated by the Creator's 'retrospective' endorsement of His work (which was so difficult for Aristotelianizing Scholasticism to integrate): "Viditque deus cuncta quae fecerat: et erant valde bona" [And God saw every thing that He had made: and they were very good], we read in the Creation story in the Vulgate.

The immanent tendency in all these toilsome efforts could only be to imagine the world as the correlate of a theory [i.e., of a theoretical attitude, a 'viewing'], which not only forced an 'alteration' in the God of pure thought that thinks itself, but also forced the world to assume a homogeneity suitable to theory. This is a process that could benefit man only indirectly and after a historical delay, and that did not have to do so as long as the teleological formula was not resolutely applied to man. This is the turning, underlying the Copernican one, that still remained to be taken. I do not think that one can say that it was in the air; rather, it involves a high degree of exertion both against the late Scholastic Middle Ages and against the astronomical tradition, which was epistemologically congenial to them.

4
Humanism's Idealization of
the Center of the World

Even though the readers in those days could not know when they received Copernicus's *De revolutionibus* from the Nuremberg press of Johannes Petreius in 1543 that the anonymous preface "To the reader" was an unauthorized insertion, still they had to be struck by a certain contradiction between the two prefaces, the one directed to the reader and the other directed to the pope—unless the difference in the fundamental attitudes of the two prefaces could already, at that time, have been imputed to a distressingly necessary cunning with an eye to ecclesiastical authority. Leaving aside, for once, the difference that manifests itself in regard to the work's intention, it remained unmistakably clear, for the thoughtful reader, that the prefaces differ fundamentally in what they allow to man and believe him capable of. The two prefaces, the authentic and the illegitimate, make antithetical decisions on this subject.

This contradictory situation reflects the epochal crisis of the Middle Ages, their falling apart into Nominalism, on the one hand, and Humanism, on the other. Andreas Osiander, the author of the preface *Ad lectorem* [*To the reader*], translates Copernicus's accomplishment into the language of Nominalist epistemology by making use of the traditional concept of astronomical hypothesis, so as to satisfy both the book's professional audience and the Scholastics, in the wings, who could only take offense at the basic plan of the new thesis. Copernicus himself places his work in the Humanist tradition—and with reason, as one can judge if one sees as the essence of his intention the saving of the cosmos, the reestablishment of the metaphysical faith in an orderly shape of reality that is accessible for man.

Copernicus did not only use Humanism's formulas or appeal to

Humanism's authorities; with the self-comprehension of his astronomical reform he defined the genuine meaning of the Humanist movement of the declining Middle Ages more accurately and realized it more substantially than did many of those who are associated with this current as its classic representatives. The origins of Humanism are inseparable from opposition to the Aristotelianism of High Scholasticism, with its split cosmology and metaphysics—and to the doubling of truth that that Aristotelianism produced in its orthodox, final form—and to the corresponding phenomenon of Nominalist fideism, which Andreas Osiander's preface, too, will pass on. Humanism organizes itself against Latin Averroism as the orthodox final stage of the reception of Aristotle. It takes aim at the dogmatic center of this philosophical orthodoxy, which consisted in the theses of the eternal existence of the world and the unity of the "active intellect." The consciousness of the counterposition that was sought and found— a consciousness that, while appropriate to the facts, was mostly not explicit—was that these dogmas had to eradicate both the metaphysical and the anthropological questions of meaning.

Humanistic wisdom becomes a program directed against the Aristotelian concept of science, which is seen as a form of knowledge that levels down the preeminence of man and of his existential interests. The field of meaning of *sapientia* [wisdom] contains almost all the motives and the complaints that come forward again and again, up to the present day, in our doubts about and our satiety with the idea of knowledge as *scientia* ['science']. In the preface to his translation of Plotinus, Marsilio Ficino reproaches the Aristotelians with destroying the foundations of religion, which they have done by taking man out of the focus of God's purposes for the world: "quia divinam circa hominem providentiam negare videntur" [because they seem to deny God's providence in regard to man]. This rebuke repeats the relationship of foundation that Cicero had set up between the adaptation of the world to man, on the one hand, and man's duty toward the gods, on the other.

Humanism, for example, in Erasmus of Rotterdam's partiality for Lactantius, brings into currency again those early Christian writers who had borne the imprint of the reception of Cicero and who for that reason had emphasized the idea of anthropocentric teleology. Petrarch, who through immersion in the early Augustine became, as a result of Augustine's dependence on Cicero, the founder of a renewed understanding of Cicero, was the first to see again the center and the unity of Cicero's philosophical aims in the universal centering of the world's meaning on man. On this subject, Petrarch writes in *De sui ipsius et omnium ignorantia*, "And all this he does merely to lead us to this conclusion: whatever we behold with our eyes and perceive with our intellect is made by God for the well-being of man and governed by divine providence and counsel." [a]

Humanism's contradiction, if not of the Middle Ages in general, then at least of the consequences of their intensification in the course of their disintegration, always has to do—to put it pointedly—with the idea of Paradise. Francis Bacon will be the first to formulate this by saying that the vital direction of the sciences is toward the recovery of Paradise, or, to put it differently, toward compensating for the disadvantages and detriments of the human situation in the world that had been attributed, in the language of theology, to man's expulsion from Paradise. However much he also writes about the empirical procedure of the sciences, Bacon understands this recovery largely according to the schema of a magical act: The paradisaic restoration is characterized by domination by means of the word, by means of the knowledge of the correct name. What emerges here characterizes the result of the Humanistic movement in that the expulsion from Paradise is not accepted as a final situation in this world, nor as the harsh seriousness of man's self-assessment.

The diagnosis that could be found in High Scholasticism is reversed: If, there, the suppression of anthropocentric teleology by means of the Aristotelian schema was the consequence of the fact that the loss of Paradise as the *locus congruens homini* [place suited to man],[67] the uncoerced agreement of nature with man, had to become and remain lastingly palpable, Humanism, on the contrary, assures man again of his privileged position in the world—or sees it precisely in its not being a position but an option, as Pico della Mirandola does in his *Oration on the Dignity of Man*. That is, Pico shows that the central position of man in reality could not return with the innocent naturalism that the Stoics had given to this thesis. What happens can be described, instead, as the idealization of the position in the center of the world.

Where man is domiciled in the universe, cosmologically and topographically, becomes a matter of indifference. It can no longer be related to his comprehension of himself—just as it had become senseless, in general, to draw conclusions about the metaphysical dignity of an entity from its location within the totality of the world. It was only a rearguard action when Nicholas of Autrecourt, the Nominalist, had to recant the proposition that in the world as a whole no ranking of creatures could be demonstrated.[68] It is true that Humanism contradicted this, but not by the traditional means of connecting anthropocentrism to geocentrism as its proof.

In his dedicatory preface to Paul III, Copernicus not only participated in this idealization, by Humanism, of anthropocentrism, but brought it to its logical conclusion: For him, the fact that the world was created for man does not guarantee primarily the security of his life, but rather the performance of his reason in relation to the whole—a guarantee that becomes critical in the classical borderline case of astronomy. The senses have lost their Paradise, not reason.

One should not forget that Copernicus addresses his preface to the pope who, by convoking the Council of Trent, initiated the theological turning against the demonization of fallen reason and of its autonomy, and that Copernicus's work appeared in the same year, 1543, for which the council had been summoned. In this connection we shall have to free ourselves, for once, from the continual extrapolation of the so much later case of Galileo to Copernicus's situation, so as not to impute mere opportunistic considerations to Copernicus in the dedication of the product of his life's work. Seen in historical perspective, Alessandro Farnese (the addressee of the dedicatory preface), who rose out of the Borgia epoch and extricated himself from its corruption only with difficulty, produced a surprising result and one that is in keeping with Copernicus—he brought about the turning back of pagan Humanism to the Roman Church, and thus the earliest beginning of the Baroque spirit, which will agree with Copernicanism in its reevaluation of spatial infinity. Here there are significant convergences, not only superficial conformities, between the dedicatory preface and the history of Copernicus's effects.

With his piece of theoretical innovation Copernicus also shows, anticipating the Tridentine profession of faith, that man's situation in the world did not accord with Nominalism's epistemological resignation and with the late medieval and Reformation theologians' absolutism of transcendent salvation—that there was more 'natural' truth for man than was to be admitted.

Paul III is said to have been "partial to Platonism." Copernicus will have made the right move when he referred to Cicero as the source of the initial suggestion for his reform. For the Platonism of the Humanist movement is traceable, to a large extent, to the declaration of Platonism that Cicero had made on his own behalf and on that of his Academic Skepticism. It had caused Petrarch, after having gone back from Augustine to Cicero, to go back, beyond that, from Cicero to Plato. Perhaps that would not have had such important consequences if the opposition to orthodox Aristotelianism had not needed the authority, apparently of equal rank, of Plato. All of this is important on account of the anthropocentrism that cannot derive from Plato but was found in Cicero's "Platonism," which was deformed by Stoic influences. Of course the philosophical motives of the Humanists' reception of Cicero were overlaid by a strong interest in rhetoric.

The theoretical foundation of rhetoric contains an authentic connection between a stylistic principle and a view of the world. Man needs the art of persuasion only to the extent that he lacks an access to reality that can convey truth; where he is credited with such an access, truth requires only a transmission that is appropriate to it. Of course, as the art of persuasion, rhetoric is also distinguished by the fact that it distances itself from the mere power of command. In the cosmology of his *Timaeus* dialogue Plato had portrayed a decisive act of the demiurge as the "persuasion" of *Ananke* [Necessity], the figuration of original matter—not, that is, as a command, in the manner of the biblical act of creation, but as a rhetorical appeal to the sense of order. In his argument, as a Roman, with the Greek ideal of idleness as the prerequisite for theory, Cicero does not emphasize the pure antithesis of activity in public service; instead, he derives political action from the intuition of the universe, which bases the obligatoriness of active citizenship precisely on the fact that it is the totality of what is useful to man.[69] Political action is free from violence in that it only takes care of and puts into effect what was already prescribed and provided for by nature: the *hominum utilitas* [utility for man].

This 'cosmic political science' is the basis of the kind of 'gentle' transformation of involvement with philosophy into public activity that takes place in the form of rhetoric, the art of appropriate and well-formed speech, and that differs in that respect from the Sophistic type of political technology (or better, from that type with which Plato charged the Sophists). So for the Roman philosopher a 'private' type of thinking—the solitary theoretical contemplation of nature, renunciation of the Forum, or, to sum up, a type of thinking that lacks eloquence—is not acceptable as exemplary.[70] But the prototype of such a type of thinking—turned in on itself, free of needs and goals—in Greek philosophy was Aristotle's God, the Unmoved Mover of the *Metaphysics*. It was natural for the polemic against strict Aristotelianism to make use of Cicero's aversion to merely internal thinking. In Epicurus's idle gods it was converted into a less abstract and metaphysical variant.

What Humanism sought to take over from Cicero and to introduce into the argument of the epoch that was coming to an end was an 'application' of an understanding of the world according to which everything relates to man, so that the world's serviceableness to him does not need to be initially 'set up,' but only needs to be 'disclosed' in the appropriate statement. Of course we should not overlook the fact that while this approach on the part of Humanism might be a match for Aristotelianism, it did not make it possible to construct a position to confront the radicalism of theological voluntarism.[b] Here, more than a theory of rhetoric with a cosmological background was required.

During his stay, as a student, in Italy, Copernicus had been in the midst of the conflict between Aristotelianism and Humanism. Having studied, first in Cracow and then in Bologna, the *artes liberales* and both types of law,[c] he had pursued the study of medicine in Padua. This was the stronghold of Latin Averroism, the center of a naturalistic medicine that Petrarch had been the first to recognize as opposed to Humanism, and had attacked in his *Invective contra medicum* [*Invective against a Doctor*]. The university of the Venetian Republic permitted itself a philosophy not only of a distinctive type but also of demonstrative scrupulousness (not always with perfect consistency, as the exclusion of Pietro Pomponazzi shows). In Padua Copernicus learned Greek from the important Aristotelian John Leonicenus, whose inclination toward Pythagorean ideas cannot be entirely disregarded as an influence on the young Copernicus. At any rate, Copernicus refers to him very much later with his remark that he had waited not only nine years but four times nine years to publish his work. This teacher of ancient literature also refers him to the mention of the Pythagorean, Hicetas, in Cicero, and to his doctrine, handed down by Theophrastus, that no object in the universe moves except the Earth. Copernicus wrote this in the margin of his edition of Pliny's *Natural*

History (here, as he does later in the *De revolutionibus*, he spells the name "Nicetas"). So he found both things in Cicero: the teleological world-formula, which did not allow one to forgo truth in astronomical cognition or to ignore the confusions in it, and the most radical form of the thesis opposing the assumption that only the Earth, in the universe, is at rest—a thesis that might at least stimulate doubts about the acceptance of Ptolemaic geocentrism. If this was so, then the decision against the philosophy of Padua took place in Padua.

The world-formula in the dedicatory preface to Paul III becomes comprehensible to us as Copernicus's expression of his position in regard to the spiritual alternatives of his age. It enables us to understand why he had to take such strong offense at the state of astronomy at the time, and where the intellectual energy of his insistence on the "true astronomy" came from. The little that we know about influences on Copernicus from the Humanism of the time confirms this assumption. Among the scanty remains of Copernicus's library that are in Uppsala is the volume of Carolus Bovillus (Charles Bouillé) that was published in Paris in 1510, which contains, among other philosophical and mathematical treatises, the *Liber de sapiente* [Book of the Wise Man]. Through his teacher, Faber Stapulensis, the editor of the *Opera* [*Works*] of Nicholas of Cusa, Bovillus was conversant with the Cusan's speculative anthropology. He cautiously detached it from its author's theocentrism, and produced the most pronounced form of Humanistic anthropocentrism. Just because he holds to geocentrism only as the most external confirmation of man's central and universal importance in the world, he makes it superfluous just as soon as it is opposed to the universality of man's knowledge. To that extent Bovillus, who Copernicus, by the evidence of his marginal notes, did actually read, speaks to the latter's intellectual situation: If man is supposed to remain the center of the world by virtue of his rational accomplishments, he must forgo the physical manifestation of this state of affairs in geocentrism.

In Bovillus Copernicus could read the thought-image, which carries the imprint of the Cusan's paradoxes, according to which man, precisely when he is in the center of the world, is located outside everything, as though the rest of the world constituted a firmament for him: "... in medio mundi extra omnia factum esse hominem ...."[71] The universality of the human relation to the world is due to the fact that man is not subject to the old regionalization into heavens and Earth, but instead is no longer a member of this totality. The idealization of the concept of the center of the world makes the anthropological schema secure against the cosmological revolution, but at the same time makes it unconditionally binding for the claim to adequacy of every assertion in the science of nature. Man is no more subject to enforced localization by the world than the mirror is characterized by the objects that it reflects. The relation between man and the world is one of opposition.[72] The open or secret basic idea of every idealism, that without man the world would be as desolate and empty as if the rest of the Creation had never taken place, is not only conceived here but is also brought to an end in its claim that this *publica creatura* ['universal' creature] has a right to everything, that nothing may be withheld from it. The proof of the idealization of the center of the world is that in cosmology Bovillus can articulate, without any change of attitude, a heliocentrism in connection with which it becomes quite clear that what is in question is the central position between the planetary spheres, that is, between the heavens of the fixed stars and the Earth. This heliocentrism reflects the way in which anthropocentrism is conceived here: not as the creature's being distinguished by its topographical location, but, in reverse, as the defining of the neutral 'position' by the function of the substance that occupies it. The Sun 'makes' its position into the center of the world.[73]

Finally, we also find in Bovillus a close imitation of Pico della Mirandola's *Oratio de hominis dignitate* [*Oration on the Dignity of Man*], when Bovillus concludes his work on the wise man with an exhortatory discourse whose fundamental idea—just as alarming, for its addressee, as it is elevating—is that man is not man by nature, and thus 'guaranteed' as such, but becomes man only by his own effort: *igitur, o Homo, qui natura Homo non Homo es . . .* [therefore, o man, who, though a man by nature, are not (yet) man . . .].[74] He who is an end in himself is the world's end, and consequently knowledge of the world is only the means to self-knowledge: "Finis ipse es mundi ac tuiipsius notitia mundanae cognitionis est finis." [You yourself are the world's purpose, and to get to know you is the purpose of knowledge of the world.] Of course, the instrumentalization of knowledge of the world for self-knowledge certainly does not diminish its rank as theory, but rather gives the ideal of 'intuition' the new urgency—beyond any suspicion of irrelevance to man's existential and salvational needs—

of being necessary to his existence. As a medium of self-knowledge, the world was made for man: "Propter te factus est mundus . . . ." Any forgoing of knowledge of the world leaves man in a state of remoteness from himself, of incomplete realization of himself. Here, too, Copernicus could find his formula word for word—especially its connection with his claim to truth, of which he may perhaps have become confident for the first time with Bovillus in his hand, even if the conception of the cosmic reform was already accomplished or even written down, in the form of the *Commentariolus*.[d]

To put it differently: The dedicatory preface to Paul III reflects the considerations with which Copernicus sought to come to an understanding with himself as to why he could not be satisfied not only with the technical state of astronomy but also, and above all, with its character as knowledge, its theoretical self-understanding.

Among his biographers, Lichtenberg grasped with the greatest affinity, and in lapidary fashion, the moment when Copernicus comprehended more than the intradisciplinary intolerability of the dismal state of astronomy: "It could not be that way. The order of nature and the ordering intellect do not announce themselves to each other in that way when they meet one another in the open."[75]

5

The Intolerability of Forgoing Truth in Favor of Technique

Copernicus's Humanistic horizon also determines his relation to the unavoidable astronomical authority of Ptolemy, both in his obligation to it and in his departure from it.

In its origin, Humanism is an opposition to Scholasticism. Only secondarily is it a 'renaissance' of new authorities, of a rediscovered relation to the authentic texts, especially Greek ones, which became accessible as a result of the ecumenical Council of Ferrara and Florence and especially as a result of the exodus from Constantinople. A century earlier, Petrarch had possessed a Greek codex of the Platonic dialogues as a relic, without being able to read it. If the opposition to the Averroism of Padua understands itself as Platonism, it only gradually finds out what this meant. Above all, it meant that it was possible to hold to the postulate of *nihil novum dicere* [to say nothing new]. Not only did one need an equivalent authority in order to resist the Scholastic authority of "the Philosopher,"[a] but one also found confirmation in the new texts that no novelties whatever were required in order to say what could be opposed, on the one hand, to the absolutism of the eternal world and, on the other hand, to the absolutism of transcendent revelation.

If Copernicus's astronomical masterpiece is also a work of art in its internal organization, then this is above all because it maintains the postulate of *nihil novum dicere* in the most ingenious manner and makes it the principle of its simplicity in argumentation.

Copernicus's only student and first biographer, Joachim Rheticus, in his *Narratio prima* [*First Account*], described his teacher in an entirely medieval manner when he applied to him the principle that it is foreign to the nature of the philosopher to depart from the doctrines that are handed down, and that Copernicus did so only under the compulsion of astronomical necessity (*urgente astronomica ananke*) and in obedience to the things themselves (*rebus ipsis efflagitantibus*).[76] Copernicus lacks the stirring gesture of the break with tradition, as it had been expressed a century before him by the great Regiomontanus. Regiomontanus had written, regarding the question of determining the magnitude of the precession of the heaven of the fixed stars, that the accuracy of the new observations would be of no use if the data of the ancients were inaccurate. In that case it could be necessary "to liberate posterity from the tradition."

For Copernicus that would be an impossible idea. It is precisely the accuracy of the tradition of Greek astronomical observation, as digested in Ptolemy, that he jealously maintains in his controversy with the Nuremberg astronomer John Werner.[77] For his part, he holds largely to the body of data used by Ptolemy. Above all, though, he takes the structure of Ptolemy's *Almagest* as his model in constructing his work. This procedure makes it especially easy to check the minimum of innovations with which he gets by; and probably the effect is characteristic of a historical sensitivity that was foreign to his contemporaries when this appears to us precisely as making his own contribution stand out more. The way in which he nevertheless withdraws from the tradition and gives it his turning can perhaps be characterized by saying that he opposes the tradition with the aid of the tradition.

He destroys the Aristotelian hiatus between the superlunary and the sublunary worlds, but he does this by applying Scholastic premises more consistently—for example, when he bases the "naturalness" of the Earth's motion on the perfection of its spherical form. When he takes his professional audience's forms of thought into account in this way it is always with the least possible specificity in relation to the teachings of the different schools—with just as much Plato or Aristotle as had become their common property.

If one wanted to name names that represent the antitheses of the tradition more precisely for Copernicus than the distant horizon of the great philosophical teachings, then one would have to put Ptolemy and Cicero foremost. It is not accidental that Copernicus remembers having found in Cicero the first hint of the interchangeability of motion and rest in the universe. Likewise in Cicero the teleological formula often presented itself—the position from which

Copernicus confronts not only Scholastic Aristotelianism but also the authority of Ptolemy. I see the peak of this confrontation not in his constructive innovations but in his epistemological opposition to astronomy's forgoing (and to the metaphysical justification of its forgoing) of truth.

The author of the handbook of astronomy that was authoritative for almost a millennium and a half—the *Syntaxis mathematica* [*Mathematical Collection*], which the Arabs called the *Almagest*—had adduced, as a justification for the high degree of complexity of his construction of the heavens and for the deficiencies in accuracy of the orbital calculations based on it, the essential split in the Peripatetic cosmos. It was a philosophical prevention of what Copernicus was to cite, in order to justify his reform, as the vexatious state of the accepted astronomy.

Ptolemy devotes a fundamental excursus to the question of the metaphysical possibility of astronomical knowledge. The criterion for the degree of complexity of astronomy's description of the heavens cannot, he says, be derived from the limited constructive capacity of the human faculty of imagination.[78] Earthly and heavenly things, which relate to one another as human to divine, are in principle incomparable. So there is no philosophical guarantee that our cognitive means can be a match for astronomy's object. They are necessarily and insurmountably drawn from the terrestrial realm of experience, so their application to the world of the stars can only be inadequate. We might describe the process of this 'application' with the name of another inexactitude, as 'carrying over.'[b] The hiatus between the substances, Ptolemy says, separates from one another beings that, on the one side, are capable of behavior that is eternally the same, and of undisturbed regularity in their mutual relations, whereas on the other side, they continually disturb one another and never behave in a way that remains uniform. When Ptolemy finally advises us that in the choice of a hypothesis we should prefer the simpler one, if it is able to accomplish the same thing, this can only have a methodological and immanent value and must not be interpreted as an objective law. There cannot be any guarantee that rationality was permitted to dominate the choice of the pattern of real nature—on the contrary, metaphysics and method can be aimed in opposite directions: 'Technique' need not contain any 'theory.'

The epistemological excursus culminates in a sort of general metaphysical stipulation that is very similar to what was to be imposed on Galileo as a theological proviso: to admit, in connection with every explanation, that for God nothing is impossible. In Ptolemy, the omnipotence of the heavenly bodies guarantees that he will forgo any type of celestial mechanics.[79] One should not have in mind too ambitious explanations as examples of such 'mechanics'; for Ptolemy, even the simple demand that the orbits, in the individual planetary spheres, should be so constructed that the spheres fit within one another and make it possible for motion to be transmitted by simple contact has to be excluded. It is important, in connection with the difference that it is our purpose to consider here, that Copernicus wants at least to keep open the possibility of a simple mechanics of the celestial spheres, by making the fitting within one another of the spherical shells into a test of his construction; thus the point nearest to the Sun in the orbit of one planet is not allowed to be lower than the altitude of the farthest point from the Sun in the orbit of its neighbor.

Ptolemy protects himself from such considerations not only epistemologically, by emphasizing the heterogeneity of empirical matter vis-à-vis stellar matter and denying to the human intellect an adequate concept of objective simplicity (not even the daily rotation of the outermost heaven, he says, can be subsumed under the human concept of simplicity), but also by granting the heavenly bodies divine or miraculous qualities—not only freedom from compulsion and resistance but also the capacity, which is unavoidable when the spheres overlap one another without constraint, to pass through one another: "Thus, quite simply, all the elements can easily pass through and be seen through all other elements, and this ease of transit applies not only to the individual circles, but to the spheres themselves and the axes of revolution." Ptolemy takes aim above all at the dependence of the human faculty of imagination on mechanical models (*epitechnēmata*); one cannot check, by means of them, whether the bases of the mathematical description are too artificial. But the decisive thing is not that mechanical models cannot attain the degree of complexity that the arrangement of the heavens possesses, because our materials and skills are not capable of that, but that they cannot do this because earthly materials simply do not have the characteristics—which are described as miraculous—that such a model would require. Where everything that can be described by mathematical means is possible, a theoretical explanation is not possible.

Without his explicitly saying so, Ptolemy also seems to hold to the old idea that the motions of the heavenly bodies not only are brought about by their own intelligences but are also guided by the will of each of these intelligences. This compensates in a certain way for the high degree of complexity of the whole machinery. The heavenly bodies acquire the ability to make way for each other's motions. Here two different basic ideas seem to converge: the idea of the spheres passing through one another and the idea of the planets making way for each other by means of intelligence and volition. The fundamental idea in this text is evidently that the heavenly bodies execute their motions by nature, but that as a consequence of the overlapping of their orbits we would encounter—purely in terms of kinematics—the possibility of their impeding one another, if they did not have in addition to their established natural orbital motions the possibility of yielding to (*eikein*) or making way for (*parachorein*) the equally natural motions of other heavenly bodies, "even if the motions are opposed to one another." From this statement we can get an idea of the licenses for the construction of astronomical models that could only be interpreted mathematically that had been granted in Ptolemy's *Almagest* and thus in the pre-Copernican tradition, and with which Copernicus was to break.

The history of the acceptance of the *Almagest* showed that the difference between a mathematically and a mechanically interpretable model was not an issue. Twenty years after this principal work, which he wrote in his youth, Ptolemy had actually also set up a physically interpretable world model, in the *Hypotheses planetarum* [*Hypotheses of the Planets*], the second part of the first book of which was rediscovered, in an Arabic version, only in 1967. In this work he was evidently less concerned with a theory of the heavens than with the construction of an instrument, comparable to an astrolabe, with which the motions of the heavens could be represented in reality. The great prototype of this sort of 'still mechanically possible' cosmology is found in the myth of the demiurge in Plato's *Timaeus* dialogue, which seems, following Cornford's plausible thesis, to have had as its pattern and source of orientation the production of a metal model of the heavens. But the description of such a model is still far from the claim made by a realistic theory, although unrestricted use can no longer be made of the freedoms of exclusively mathematical representation. The mechanical model is distinguished by the mutually unobstructed mobility of the orbital circles, without having to provide a representation of the transmission of motion from one sphere to its neighbor—true to the principle that the manner by which motion is transmitted between terrestrial bodies cannot be applied to the heavenly bodies.[80] The mechanical model, too, still stands in the context of the tradition's predominant interest in astrology, by which —through the Middle Ages and beyond—astronomy's theoretical accomplishment was, above all, gauged, and to the needs of which its theoretical tolerance was adapted: The reading of signs in the heavens does not require a sufficient explanation of how they are produced.

In Ptolemy himself, it is not astrology's lack of interest in pure theory that induces his epistemological excursus as the argument for the general license, but rather the historical experience of the undecidability of theoretical alternatives within astronomy. To put it another way: Ptolemy had experienced vexations with the previous history of astronomy that were comparable to those that were to become the motivation for Copernicus's efforts at reform, but he had drawn from them the opposite conclusion, that the human intellect was inadequate, in principle, for cosmological questions.

We can determine where the point of resignation lay for Ptolemy. It may not have been so much a matter of inaccuracies, in the sense of deviations in data over long periods of time, as rather of the impossibility of deciding between equally possible constructive principles, to which the criterion of simplicity evidently could not be readily applied.

Put briefly, the undecidability here has to do with the choice between representing the apparent motions of the Sun, the Moon, and the planets by means of multiply superimposed circles—in other words, the system of epicycles—or by means of an eccentric arrangement of individual spheres with respect to the center of the system. It is easy to see that the effectiveness of the two modes of construction has to differ depending on whether one is dealing with the inner planets, which in each of the historic systems have to remain coordinated, in the epicyclic manner, with the Sun, or whether it is a matter of the planets that are external in relation to the Earth's heliocentric orbit, the planets whose apparent retrogressions only reflect the Earth's orbital motion. Geocentric astronomy could not make this distinction

between inner and outer planets, and consequently it had to generalize one of the two methods of representation for the entire system.

Given this assumption of the homogenization of means of construction, the criterion of simplicity has to fail. Ptolemy seeks the
reason for this fact, as the text of the excursus makes clear, in a
deficiency of concept formation. He is obviously not a Platonist:
otherwise he would not trace the concept of simplicity back to the
initial material of terrestrial experience, so that it has to fail in its
application to the stellar phenomena. His theory of concept formation is, without any doubt, that of the Aristotelian *aphairesis* ['taking
away': abstraction], whose application to objects that are not, or are
only indirectly, accessible to our sense organs (as is the case with the
spherical bodies of the heavens, which we can perceive only through
pointlike phenomena) has to be problematic. To make the criterion
of simplicity applicable again for astronomical objects is a central
task for Copernicus's truth-claim founded on the teleological world-
formula.

Ptolemy discovered the importance of treating the criterion of
simplicity as problematic when he was unable to maintain homogeneously a choice of one of the two constructive procedures. It is true
that he proves the kinematic equivalence of the two possible constructions for all the bodies of the system—for the Sun, for the Moon,
and for the planets (as a group)—but for the Sun he decides, in
contrast to his epicyclic construction of the orbits of the planets, in
favor of the eccentric orbit (that is, one proceeding around a center
different from the Earth), which was introduced by Hipparchus. His
argument is that while here, too, either of the previously discussed
hypotheses could be employed, "it would seem more reasonable to
associate it with the eccentric hypothesis, since that is simpler and is
performed by means of one motion instead of two." [81]

In connection with the dogmatic stability of Ptolemy's astronomy
across the Middle Ages, the question arises whether the epistemologically problematic quality that he had found and had pointed out
in his solution of the dilemma of alternative constructions also had to
penetrate to the central element of geocentrism and at least make it [a
variable] at the astronomer's disposition. The eccentric construction
for the Sun, at least, would have had to reflect on the question of the
central body, once the Sun no longer moved around the Earth as the
center of its orbit at all, but now only enclosed the Earth's position.

Here one will have to say, first of all, that Aristotle's geocentric cosmology had the rank of a scientific theory, whereas Ptolemy's astronomy only had that of an art, and in the end, above all, the function of an auxiliary agency for astrology. But there is still something else as well. Ptolemy's epistemological reservation does not hold for the Earth. The Earth is an object of physics and thus of adequate experience and its concepts. It is true that Ptolemy also designates as "hypotheses" what is imparted in the first eight chapters of his work, but here the terminological usage is not the technical one from astronomy but rather the Platonic and philosophical one signifying "assumption" or "principle." This accords with the designation of this part of the work as a "preparatory" one: The Earth as a heavenly body is not an object of astronomy. In spite of its inferior cosmological position, it has the epistemological advantage that it can be the object of an unequivocally true theory. Quite consistently, then, the criterion of simplicity breaks down in the case of the Earth, but it does so for an entirely different reason than the one that is decisive in the case of the constructions for the heavens: The empirical truth that is possible [in the case of the Earth] does not need a criterion, so it is not affected by Ptolemy's observation that the simpler constructive solution would be to assume that the Earth moves. Referring to "some philosophers," and without naming Aristarchus of Samos, Ptolemy disposes of any truth competition between physics and astronomy: "They do not realize that, although there is perhaps nothing which would count against that hypothesis [sc. that of the Earth's rotation], at least from simpler considerations, nevertheless from what would occur here on Earth and in the air, one can see that such a notion is quite ridiculous." [82]

Copernicus will give up the Earth's empirical priority: He considers it from outside, as a globe in the universe, recognizable as such by the form of the shadow that it throws on the Moon in eclipses. The epistemological hiatus has to disappear if only because the Earth can no longer have a special character as a physical field of experience at the moment when it itself becomes an example of a heavenly body, under man's feet. This transformation involves a conjunction of two distinctive characteristics: the advantage of stellar phenomena, that they are susceptible to purely mathematical treatment, and the advantage of terrestrial physics, that it serves as a basis for empirical concept formation. A contradiction is resolved that Ptolemy could not

escape, namely, that in spite of its being part of mathematics, astronomy could not share in the latter's advantage of certainty, although
in his preface to the first book Ptolemy had held out precisely this
prospect.

Ptolemy had followed Aristotle's division of theoretical philosophy
into physics, mathematics, and theology, and had assigned three
regions of cosmic reality to the three disciplines as their objects: To
theology he had assigned the first cause of the motion of the world as
a whole, the unmoved mover, as that which is both invisible and exclusively thinkable; to physics, the moved and transitory constituents of
the world below the sphere of the Moon, as that which is both visible and
accessible by thought; and to mathematics, finally, as the object belonging to it, that which lies in the middle between the other two realities, and which can be apprehended both by sense perception and by
thought, and is also in that respect something intermediate, but at the
same time belongs to everything that exists—the transitory as well as
the permanent—as a characteristic. Ptolemy follows this Aristotelian
arrangement with his own observation that the two disciplines of the
world-regions that are polar opposites can offer only uncertain knowledge. So there is no prospect of uniformity in the views advocated in
them: not in theology on account of the invisibility and intangibility
of its object, and not in physics for precisely the opposite reason—on
account of the sensual unclarity and instability of its material object.
Mathematics, on the other hand—here, too, something intermediate
—offers reliable and incontestable knowledge, through proofs in
arithmetic and geometry, but also in "the theory concerning divine
and heavenly things. For that alone is devoted to the investigation of
the eternally unchanging. For that reason it too can be eternal and
unchanging (which is a proper attribute of knowledge) in its own
domain, which is neither unclear nor disorderly. Furthermore it can
work in the domains of the other two divisions of theoretical philosophy no less than they do." [83] In spite of its caution with regard to
the problem of certainty, this declaration stands in a peculiar contrast
to the epistemological excursus in the thirteenth book, especially
because there the sublimity of the object directly deprives the mathematical apparatus of its objective reference and relegates it to the
immanence of man's technical makeshifts.

The Aristotelian diagram of the sciences and their competences,
which Ptolemy describes, suggests what the 'methodological' con-

sequences of a suspension of geocentrism would have had to be. It reveals the following alternatives: Either the astronomical object is released from its relationship to mathematics and drawn into the competence of physics, or else the objective region belonging to physics is assimilated to that of astronomy, in that the Earth, following the formula of the Pythagoreans, is made a star among stars, and thus in principle susceptible to treatment by mathematical means. Post-Copernican science will more and more follow the path of physicalizing astronomy. In his critique of the epistemological excursus in the *Almagest*, Kepler explicitly denies that there is nothing in terrestrial, near-at-hand experience that is comparable to the motions in the heavens.[84] He may have been thinking especially of the possibility of employing the phenomenon of magnetism, as described in William Gilbert's *De Magnete* (1600), for a causal explanation of the processes of motion in the solar system. Whereas in Copernicus the Earth had received an orbit around the Sun for which, in a consummate act of integrating the terrestrial body into astronomy, Hipparchus's eccentric construction (transmitted by Ptolemy) could be adopted, in Kepler this orbit is produced by the magnetic *orbis virtutis* [orb of force] that rotates with the Sun—that is, by a force that was known from terrestrial experience and that seemed to be investigable only in it.[85] In the context of his opposition to Ptolemy's 'general cancellation' of epistemological obligations, Kepler introduced the first empirical evidence of the Earth's cosmic homogeneity, where Copernicus had only been able to appeal to the perfect roundness of the Earth's shadow in eclipses of the Moon, and Galileo was to offer, initially, only the evidence of the Earth's 'shining,' as demonstrated by the Moon's secondary light.

So when Kepler inserts the text of Ptolemy's epistemological excursus, in his own translation, in his *Epitome of Copernican Astronomy*, he has a more definite support than that of a metaphysical world-formula in contradicting the stipulation of human reason's incapacity for astronomy. His most important argument is that the logic of Ptolemy's argumentation would not so much have excused astronomy as destroyed it: "et videtur Ptolemaeus nimis longe extendere hanc excusationem, adeo ut universam rationem astronomicam confodiat ... " [and Ptolemy seems to draw out this excuse to such lengths that he undermines the whole possibility of astronomy]. That epistemological 'exculpation' can be imputed, he says, neither to astronomers nor to philosophers nor to the Christian point of view. It cannot be imputed to astronomers because with its empirical-abstractive foundation of the concept of simplicity it abolishes every criterion for the evaluation and comparison of theories in their discipline, just as it does for the objective reference of any theory at all with its thesis of the inapplicability of terrestrial concepts to stellar objects. All theories would be equally subject to the suspicion of falsehood if reason could not even apply the criterion of simplicity to its geometrical constructions.[86]

The demands made of philosophers appear comparatively harmless when Kepler does not in fact ascribe the contradiction of Ptolemy's general license that was contained in Copernicus's world-formula to them, but rather to the theologians: It has become intolerable, now, not only that the ancient astronomer had attributed divine characteristics to the heavenly bodies but also that he asserts a difference between the nature of the stars and the human capacity for comprehension such as does not even exist between God and man in respect to their rational capacity for evidence. For God, Kepler says, wanted to give man, as [the creature made in] His image, a common intellect with His own—but in that case the orbits of the heavenly bodies could not have that asserted simplicity, according to the 'Platonic' precept, if our intellect is able to comprehend only their complex nature.[87] In other words: The transcendence of the stellar objects with respect to man's rational faculty of comprehension must not be greater than the transcendence of the divine reason with respect to the human one. And for the latter, Kepler excludes any qualitative deficiency on Christian premises, as Galileo will do.

One can see why Kepler does not use Copernicus's anthropocentric formulation in his argument, but instead considers the biblical characterization of man as made in the image of God as the stronger position for the purpose of bringing the human intellect onto the side of the divine one with its capacity for insight into the mechanism of the world's workings. Kepler sees the Copernican reform's claim to truth as killing the alliance between Nominalism and the Ptolemaic tradition's reservation about astronomical knowledge, even though he himself is more inclined to use voluntaristic ways of speaking ("intellectum communem secum esse voluit")[c] than Copernicus's teleological-anthropocentric language. He may still not have realized that his first argument—the argument from the objective of making astronomy possible as a theory at all—represented the type of self-assertion that was more in keeping with the epoch that was beginning.

Kepler opposes not only the epistemological deduction, from the hiatus between stellar and terrestrial reality, which allows the special matter of the heavenly bodies to justify the idea that 'everything is possible,' but still more the Nominalist thesis of a more refined, systematic and intentional opaqueness of processes in the heavens, the thesis according to which God used incommensurable proportions in the heavens, proportions that only He could reduce, precisely in order to repulse the human desire for knowledge. Kepler's third law of planetary motions[d] has to be seen against the background of Nicole Oresme's suggestion that there are no rational proportions between the motions of the planets. The late medieval Nominalist's affinity for an inaccessibility of celestial processes that was substantiated rationally as well as materially was due to its potential positive significance in theology, as a theoretical good that was reserved to the divinity, and as a boundary of human pretensions, entailing humility. So it was not unimportant that Kepler opposed trains of thought of this type not only in the language of Humanism but also in that of Christian dogmatics. To be sure, Copernicus the reader of Cicero could already have chanced upon an appropriate passage in the *Tusculan Disputations*, in which the human spirit's capacity for astronomy is traced back to its kinship to God, because the constructive invention of theory appears equivalent to the act of the production of its objects themselves, as when Archimedes, as the constructor of the astrolabe, has become comparable to the world's demiurge.[88]

Of course Kepler's intention is only to go further in the direction that Copernicus had entered upon with his revival of the idea of a cosmic symmetry. This expression goes beyond uniformity in the choice of means of construction (that is, either eccentrics or epicycles), which Ptolemy held to, and also goes beyond obedience to the 'Platonic' requirement,[e] insofar as "symmetry" motivates one to overcome the isolated representation of individual planetary orbits and to portray the relations between them in accordance with principles that embrace them all.

This train of thought, too, can still be understood as a way of adhering to Aristotelian premises, because for Aristotle the motion of the planets in their orbits was supposed to be related to their distance from the moving principle, so that the relations between them could be understood as the successive diminution, from the outside inward, of the moving power, until it is exhausted when it reaches the Earth, in the center—as long as, in all of the subordinate motions, one had allowed for the daily rotation of the heaven of the fixed stars by means of counterrotating spheres. My hypothetical genetic account of the process by which Copernicus constructed his theory will show that while, on the one hand, his integration of the Earth into astronomy begins with a solution of the problem presented by the peculiarities of the "inner" planets, it is based, on the other hand, on carrying out the systematic principle of a commensurable *ordo orbium* [order of spheres]. Human reason, if it is supposed to have a genuine claim to understanding of the universe as the product of an intelligence that is kindred to and means well by it, must be able to comprehend it under the idea of its unity in accordance with a rule that is valid throughout.

This claim made it evident that Ptolemy had not only divided the universe at the lunar boundary but, beyond that, had also divided the stellar region into cells by his isolated kinematic constructions of the paths of the planets. The Nominalist, with his thesis of the incommensurability of the planetary orbits, had merely given this splintered situation a more recondite significance instead of simply letting it stand as a theoretical difficulty. He had given it the function of a rationalization of the imputed irrationality.

So when Copernicus, in the *Commentariolus*, already objects that the representation of the events in the heavens that Ptolemy had given was only provisional and not sufficient to satisfy the demands of rationality,[89] then, all things considered, that cannot refer primarily to the inaccuracies that had become obvious in traditional astronomy, and the confusion that these produced. Instead, it concerns the systematic weakness of the classical theory even when viewed merely from inside, that is, independently of the evidence of what it is able to produce. The metaphor, in the dedicatory preface of his chief work, of the patched-together pictorial representation of a body, which produces more of a monster than a man, has to be related to astronomy's systematic structure—not only to adherence, within the construction for an individual planet, to the 'Platonic' postulate of uniform revolutions, but also to the relations of all the partial constructions to one another and to the whole.

This is also what the preface's reproachful use of the term "symmetry," in relation to the mathematical astronomers, refers to: "Rem quoque praecipuam, hoc est mundi formam ac partium eius certam symmetriam, non potuerunt invenire..." [Nor have they been able to ascertain the principal thing—namely, the form of the universe and the unchangeable symmetry of its parts]—as does the (in its turn) 'symmetrical' satisfaction at the conclusion of his own reorganization of the system of the planets: "Invenimus igitur sub hac ordinatione admirandam mundi symmetriam ac certum harmoniae nexum motus et magnitudinis orbium...." [So following this arrangement we find an admirable symmetry in the universe, and a firm bond of harmony in the motion and magnitude of the spheres.][90] The point toward which Copernicus's efforts converge is not kinematic success but the confirmation of the unprovable hypothesis expressed in the phrase "in optima sunt ordinatione constituta" [they are established in the best arrangement].[91]

He had already stated with satisfaction at the end of the *Commentariolus*, referring to the thirty-four circles remaining in his scheme, that with them all the workings of the universe and the starry host had now been put in order: "...quibus tota mundi fabrica totaque siderum chorea explicata sit" [by which the entire structure of the universe and the entire dance of the planets is explained]. The language of the essay is filled with such expressions of expectations that have been satisfied, of enhanced rationality, and of the achievement of confidence that the universe is not opaque.[92] Even in the case of the strange orbit (*mirabilissimus cursus*) of the planet Mercury he believes that a complete clarification would be possible if only people would apply themselves to it with greater acumen.[93] The anthropocentric intention behind the world has to be converted into a universal understanding of the world, an understanding that no longer permits any justifications of the failures and inadequacies of cognition.

When the Wittenberg mathematician Joachim Rheticus had been in Frauenburg with Copernicus in order to become acquainted with the new doctrine and to urge Copernicus to publish his book, Rheticus's teacher Melanchthon complained in a letter to Joachim Camerarius of 25 July 1542, that it was only on account of Rheticus's youth that he excused Rheticus's allowing himself to be carried away, by something like enthusiasm, into the kind of philosophy that he was now pursuing; he had told Rheticus more than once that he would like to see more of the Socratic type of philosophy in him— perhaps that would change if he married.[94] Melanchthon already seems to see that there is more involved here, and more at stake, than the foolish reversal of the whole "Kunst Astronomiae" [art of astronomy] that Luther had contemptuously mentioned, over a meal, three years earlier. Rheticus's *Narratio prima* [*First Account*]—which, with the greater courage of the disciple but under the eyes of the master, had embedded the astronomical reform in its philosophical context—must have made this unmistakably clear to Melanchthon. Copernicus's chief work, appearing later, was to indicate these things only sparingly and hesitantly. Rheticus's pamphlet had been published in Danzig in 1540 and in Basel in 1541, and had led people to anticipate a "second report," the existence of which Pierre Gassendi still assumed in his biography of Copernicus, written in the next century—but the master's late decision (which no doubt Rheticus himself no longer expected) to let his work be printed made it unnecessary to continue with mere reports.

It may be that, considering his own caution in raising his new claim, Copernicus was glad to see his programmatic utterances anticipated by the young Rheticus. In this way the anthropocentric world-formula in the dedicatory preface to Paul III could be played down, to the level of a mere gesture toward a context that had already been made visible. For probably only this preface was still written after Rheticus's arrival and after his *Narratio prima* was published.

Rheticus again and again allows the Copernican consciousness of the "restoration of the astronomical truth" to show through, despite the pains he takes, in deference to the addressee of his report, the Nuremberg astronomer Johann Schöner, to preserve the authority of the astronomical classics. "Reborn astronomy" had merely been brought by Copernicus to the theoretical certainty that was owed to the object.[95] What Rheticus had taken upon himself with such proclamations (for which, after all, he could not also supply the argumentation) is occasionally perceptible in the uncertainty of the not yet thirty-year-old author as to how far he should go, vis-à-vis his first teacher of astronomy, Schöner, in emphasizing the validity of the new theory, and to what extent he should yield to his fascination by the heterodox astronomy. Thus he seems to leave open, or at least to leave it to the reader, whether "most of the appearances in the heavens are produced by the mobility of the Earth or can certainly most conveniently be saved" in this way.[96] Rheticus uses the old formula of "saving the phenomena" as an alternative even though everywhere in this prospectus of the new doctrine more than such saving is explicitly claimed.

It is true that the Earth is no longer the center of the cosmos, but it is important to Rheticus to emphasize that this involves no loss of distinction, but instead actually emphasizes the Earth's cosmic function. It is in accordance with the wisdom of the Creator to let the complex multiplicity of the celestial motions derive, for the most part, from *one* source; in this regard one could not attribute less skill to Him than common clockmakers possess. So it is in keeping with the idea of the divine art that it satisfies the criteria of human skill. In that case it must be possible "to explain an almost infinite number of appearances by this one motion of the Earth." [97] When, immediately afterward, the motion of the Earth is pronounced the "efficient cause" of the apparent path of the Sun, the heterogeneous expressions *hypotheses* [hypotheses] and *causa efficiens* [efficient cause] stand side by side in the astronomical context like a challenge to the well-versed reader to recognize a new relation to reality in the promised theory.[98]

What Rheticus must have understood in Frauenburg is, above all, the new importance that Copernicus had given to anthropocentrism—his comprehension of it exclusively as an uninterrupted rational coordination [of the world with man]. That Rheticus cannot have inspired the world-formula in Copernicus's dedicatory preface can be gathered from the derivative and clumsy 'application' that he gives it: God, he says, embellished the heaven of the fixed stars for our sake with so many shining little globes so that we can establish the positions and motions of the other enclosed spheres and planets by comparison to them (since they are undoubtedly fixed in place).[99] In this almost intentionally paradoxical version of the relation between reason and the universe, even the outermost and remotest reality is arranged and equipped with a view to man's theoretical activity, so that the standstill newly decreed by Copernicus for the final sphere appears as a sort of guarantee of the reliability of astronomical determinations. The heavenly bodies are not coordinated with man for his admiring enjoyment, nor for the practical purposes of seafaring and the calendar, nor for the illumination of the lowly world, but for the certainty of his theoretical orientation in the world. It is only for the satisfaction of this need that he is once again called and intended to be the *contemplator caeli*, whose allotted purpose converges with nature's. "We shall therefore gratefully admire and regard as sacrosanct all the rest of nature, which is enclosed by God within the starry heaven, and for the study and knowledge of which He has endowed us with many methods of investigation and with innumerable instruments and gifts, and made us capable; and we shall advance to the point to which He desired us to advance, and shall not attempt to transgress the limits imposed by Him."[100] Here Rheticus has gone one step further yet in the legitimation of the astronomical object: Not only does the arrangement of the heavens favor the theoretical attitude, but the resources and methods employed by that attitude were also contributed by Providence. Many years later this idea will lead the aging Rheticus to the cult of the obelisk as a revelatory astronomical instrument, an *acheiropoieton* [something not made by hands].

Now, before the appearance of Copernicus's principle work, the crucial thing is that the acknowledgment of limits that are imposed on the human spirit no longer confirms the boundary between the sublunary and the supralunary realities, but instead reason's limit is pushed outward until it becomes identical with the limit of the universe itself. Rheticus does just this: Only what lies "beyond the concave surface of the starry sphere" cannot be investigated and must be left to revelation.[101] Here lies the difference between Rheticus's position and the reservation against astronomical curiosity that Osiander will lodge in his preface to Copernicus's book. For Rheticus, the Copernican theory is in fact a confirmation of reason's universal cosmic competence—of the freeing of astronomy's claim to adequacy—when he can write in regard to the new theory of the motions of the planets, "And if it is possible anywhere else to see how God has left the universe to us to examine, it surely is eminently clear in this case."[102]

Of course, on this point too an authority like Ptolemy did not err out of weakness or frivolity; even in regard to his (and other people's) epistemological resignation, it turns out that a higher dispensation constrained their eyes, when Rheticus writes that "God permits Ptolemy and other illustrious heroes to dissent on this point." Rheticus still shares, to a large extent, the respect that Copernicus keeps for Ptolemy; that too was to change with the discovery of the Egyptian astronomy.

But Rheticus already sees that in Copernicus's epistemological point of departure, he does not continue in the "imitation of Ptolemy" for which he is to be praised in other areas, and that the difference is philosophical, that is, concerns the possibility of and the pretension to truth. Vis-à-vis his teacher, Schöner, to be sure, Rheticus finds himself compelled to maintain, in the interest of harmony, that Aristotle and Ptolemy would also have had to concur with the "construction of a sound science of celestial phenomena," while he says more cautiously of Averroes that he would not have received the Copernican hypotheses "harshly."

These efforts do not have the laconic economy with which Copernicus will make use of a few Aristotelianisms so as not to give offense to the Scholastic audience. Rheticus appeals to a passage in Aristotle's *Metaphysics* about the hierarchy of truths, according to which the measure of the importance of a truth consists in the quantity of truths that can be derived from it. But if the contemporary reader of the *Narratio prima* had looked up the context in which this thesis belongs in Aristotle he would have encountered a definition of the scientific concept of truth that Copernicus still could not satisfy and that Rheticus therefore had to exclude from his quotation. For the sentence preceding the quoted one says that we do not know the true at all as long as we do not know its reason and cause.

In regard to the fulfillment of this condition, Rheticus overestimates Copernicus's accomplishment. Or he suggests such an overestimate to his readers, by the verbal device of presenting the demonstration that the real motions of the Earth "produce" the greater part of the apparent motions in the heavens, as though this demonstration already satisfied theory's fundamental question as to the causality of the celestial processes. But it is only in a metaphorical sense that there can be a *causa efficiens* for something that is not real. Then it resembles the hyperbolical formula according to which the astronomical reformer put the Earth in motion and made the heavens stand still. This formula only puts in the form of positive greatness the reproach against the mathematical astronomers that they endeavored "to pull all the stars together around in the ether in accordance with their own will, as though in leading strings...."[103]

The history of Copernicus's reception is accompanied by the development of the consciousness that language is not able to keep step with reason's accomplishments—that rather than being able to follow the progress of understanding, it remains persistently bound to intuition. Copernicus himself did not need to make any comment on this difficulty because he made use of the technical language of astronomy and, in the Greek motto on the title page, refused access to the work to those not versed in mathematics. There is a difference between the *Narratio prima* and the *De revolutionibus* in the anticipation and assessment of their reception: Copernicus nowhere indicates that his scientific action could ever induce excitement outside the circle of his colleagues, whereas Rheticus included the interest of a much wider horizon of contemporaries to whom the Copernican transformation of the world could not remain a matter of indifference. This is the only way to explain his remark about the differing degrees of acceleration, as between rational progress and linguistic mastery, such that "the remarkable symmetry and interconnection of the motions and spheres—which, as maintained by the assumption of the forgoing [that is, Copernican] hypotheses, is not unworthy of God's workmanship and not unsuited to these divine bodies—can be conceived by the mind (on account of its affinity with the heavens) more quickly than they can be explained by any human utterance." [104] Language belongs to the sphere of the senses; it does not participate in the coordination between reason and the heavenly bodies that is guaranteed by their kinship. It is not unamusing to observe that in this passage Rheticus cannot appeal to the anthropocentric world-formula; if he did, he could not make language an exception to man's providential adaptation to the object of knowledge.

Nowhere in his work did Copernicus give way to the temptation to describe the origin of his innovation by means of the "language game" of inspiration or of illumination. At first glance, a pious set phrase might serve to soothe a disposition that might be irritated by the work; on second thought, however, it could only emphasize the consciousness (on Copernicus's part) of innovation and overturning if it laid claim to a special divine influence on behalf of the reform. Also it may not have accorded with his own experience of the formation of the heliocentric theory to employ for it the model of an idea that abruptly occurs to one (which is the model that goes with an interpretation in the language of inspiration or illumination). But above all, Copernicus may have hesitated to endanger what today we would call "intersubjectivity": One who appeals to inspiration and illumination must not be surprised if the audience to which he has laid claim excuses its failure to understand him by presuming that it has not yet experienced comparable influences from above.

There is no basis for assuming that Rheticus is a reliable historian when he says, in a later passage and no longer consistently with his own premises, "For in the examination from afar of those divine objects so remote from us, of what avail is the strength of the human mind? Of what avail dimsighted eyes? Accordingly, if God in His kindness had not imbued the astronomer with the heroic ideas of the divine motions and led him by the hand, as it were, along a road otherwise inaccessible to the human intellect, the astronomer would not be, I think, in any respect better circumstanced and more fortunate than the blind man...."[105] This remark already tends more in the direction that Rheticus was to follow after the master's death and that was finally to lead him to the Egyptian obelisk and to an astronomy of the 'occult science' type. Here, as it is in most cases, this path is connected to a initial disappointment; what its nature was we can only infer from Rheticus's reaction, much later, to Petrus Ramus's magical phrase about an "astronomy without hypotheses." As will have to be shown later, Rheticus must have been disappointed with Copernicus too. The disposition for such a result is already evident in the language of the *Narratio prima* in the ingenuous use of the concept of efficient causality. Before there was any hope of the feasibility of a celestial mechanics, the pretension to truth that went beyond astronomy's self-interpretation as a technical skill still had to run the risk of disappointing, with 'too little truth,' the great expectations of youthful intellects, which are always ready to leap.

To be sure, when Joachim Rheticus returns to Wittenberg from Frauenburg, none of this future process is as yet perceptible. In his prospectus to the Wittenberg students for the winter semester of 1541/1542 he meets the curiosity for news of the new doctrine on the part of those who had remained behind with all the ceremony of one who has just been initiated. He announces an exposition of by far the most beautiful of all human works. It is—Ptolemy's *Megale syntaxis* [*Great Collection—Almagest*]. It is not because he had seen in Copernicus only an interpreter of Ptolemy,[106] but because commenting on a canonical author was the unavoidable form of an academic communication even of the most unusual innovation, that Rheticus does not mention the name of Copernicus—even though everyone knew that he would be the issue in this *enarratio* [exposition]. How delightful the knowledge that was to be communicated (*suavem et dulcem esse cognitionem*) must be was supposed to be evident from Rheticus's explanation: "The human mind, which originated in the heavens, is as delighted by the knowledge of the things in the heavens as by the view of its native country." [107] If the stars were set in the heavens as marks by which to measure the years and the seasons, God certainly also intended that they should be regarded and taken heed of, that the confusion of life should be avoided by means of the signs that divide up years, epochs, and chronologies. It would be absurd to ascribe this admirable and most dependable variety of motions to the divine architect as something constructed in vain.[108] "So I begin expounding Ptolemy next Thursday," Master Rheticus concludes.

6

A Hypothetical Account of the Way Copernicus Arrived at His Theory[a]

The historiography of science has submitted, almost as a matter of course, to the criterion that it found already established by its object: It has made the present, and thus the final state of science at each present moment, into the legitimizing point on which all the efforts that preceded it converge. As a result it has become largely a process of describing what are in part interesting, in part at least charming (even if by now scarcely comprehensible) errors. And it is more difficult than it seems at first glance to extricate it again from the spell of this optics of progress.

Thus Ptolemy's system could already appear to the Enlightenment writers of the seventeenth and eighteenth centuries as an almost perverse aberration representing a kind of autism of the human race. They failed to realize that this reproach itself merely displayed a new form—this time a historical rather than a cosmological form—of egocentrism. One saw oneself as having arrived at the pinnacle of possible development and of self-liberation precisely in that one was able to bear reason's stern decree that one could no longer see oneself as being in the center of the universe. In return for that, one could charge the dark Middle Ages with geocentrism as the expression, in the form of a system, of a superficial self-deception—as though the Middle Ages had invented it expressly to meet their own needs.

Over against such obscurity, it seemed more than enlightening [*einleuchtend*] and deserving of complete agreement that with the beginning of the modern age the Sun had been set in the center of our world system. In its function it was the central organ that animated and illuminated the whole system of the planets, just as, according to Kepler's speculation, it was finally to move them as well. Copernicus had already tried to use this evidentness of heliocentrism to make it easier to agree to his innovation when he summarized his result in the lapidary sentence, "In medio vero omnium residet Sol." [But at rest in the middle of everything is the Sun.][109]

Impressed by this enlightening choreography, it is all too easy to forget that the astronomical tradition before Copernicus—from Ptolemy right through into the Renaissance—had been no less able to leave the center of the universe to the Sun, despite its assumption of geocentrism. Of course this had been a different localization of the "center of the universe," one that had started from the sequential order of the spheres, between the heaven of the fixed stars and the Earth. It made the Sun stand symmetrically "in the middle" between the orbits of Venus, Mercury, and the Moon, on the one hand, and Mars, Jupiter, and Saturn, on the other. This was the most favorable position so as to be able to reach all the bodies of the system equally with its brightening and animating influence. This symmetry had nothing at all to do with the Earth. This was how un-'geocentrically' (in the sense that was later imputed to it) one could talk in the language of that traditional system. Of course this heliocentrism on the world radius only held for the astronomical tradition that, contrary to Plato and Aristotle, held fast to the arrangement—which the Babylonians and Egyptians already used, and then the Pythagoreans—that put the "inner planets" in the "inferior conjunction," that is, placed Venus and Mercury between the Moon and the Sun. If one adopted the solution of the "superior conjunction," this sort of central position of the Sun did not result.[b] Among the Stoics, only Cleanthes seems to have distinguished between geocentrism as defining the optically privileged position given to man, the world-onlooker, and heliocentrism as the placement of the central organ of the world in the position that is functionally appropriate to it.

One must consider this kind of interplanetary heliocentrism from another point of view as well. For the tradition, Aristotle had laid it down, metaphysically, that the Earth could not be a stellar body, so that for astronomy it could only be a propaideutic object. But in that case the kind of 'heliocentrism' in which the Sun followed a path that divided the other spheres symmetrically was, as a 'purely' astronomical solution, the more impressive of the two. The Earth, which was 'only' an object of physics, was not included here at all; its central position, as the observer's location, had only a methodological significance. Ptolemy had definitely employed the Sun's central position on the radius between the Earth and the outermost sphere as a factor in his argumentation. He had used it as the basis of as important a component decision in his system as the one in favor of the "inferior conjunction" of Venus and Mercury. This disputed question, Ptolemy had argued, could not be decided by astronomical means, because the two planets exhibited no observable parallaxes (and because, for optical reasons, there were evidently no observations of transits in front of the Sun's disk). In order nevertheless to come to a decision, another criterion had to be brought in: In Ptolemy's opinion, the "inferior conjunction" solution conformed more naturally to the Sun's central position because with it the remaining planets were separated from the Sun and coordinated with it on both sides in accordance with a specific characteristic, the criterion being whether they can stand in opposition to the Sun or whether they have to remain within a narrow angle of elongation with respect to it.[110] Thus the Sun, with its path, separates two 'types' of planet that are clearly differentiable astronomically—if, with Ptolemy, we suppress the 'minor blemish' that the Moon is after all one of the members of the system that can stand in opposition to the Sun. If it is correct that, in antiquity's way of viewing the spatial relations in the world system, the radial orientation of up and down between the heaven of the fixed stars and the Earth predominates over everything else, then this type of heliocentrism, on the radius, was also the one that was in keeping with the system. Paying attention to the Earth's central position presupposes a different cosmological point of view, a diagram of the positions in the universe that looks at it 'from outside.' When one thinks about it this way, it becomes questionable whether Ernst Goldbeck was right when he wrote that "from Aristotle onward, in fact, one can speak of a geocentric religiosity of the Greeks. This state of affairs did not change until Copernicus's times."[111]

Even if the Platonism of the Renaissance, with its metaphysics of light, gives the Sun a privileged position in its cosmological doxography, this could not contribute much to any preparation of the way for the Copernican reform because the traditional astronomy in fact already contained a heliocentrism that was perfectly in keeping with its system. Copernicus even had to expunge its traces when, in arguing for the "inferior conjunction," he avoids the prototypical argument presented by Ptolemy and restricts himself to indicating that the "superior conjunction" would leave too great an empty space between the sphere of the Moon and that of the Sun. Thus, far from offering support for the Copernican solution, the heliocentrism that derives from Platonism leads into a blind alley.

Plato himself had decided in favor of the "superior conjunction" in the *Timaeus*, by making the Sun travel on the second orbital path (counting outward from the Earth), admittedly with the argument that in that way it could best illuminate the whole heavens.[112] This was probably not Plato's final word. According to the account that Plutarch took from Theophrastus, Plato is supposed, in his old age, to have regretted having given the place in the center of the world to the Earth.[113] This is supported by Proclus's statement that in his late lecture "On the Good," Plato explained the systematic connection between the Sun, Venus, and Mercury by means of a metaphysical kinship "in the forecourt of the Good," a combination in which the Sun represents truth, Venus beauty, and Mercury symmetry.[114] If we assume that at the time of this late lecture there existed the close association between Plato and Heraclides of Pontus that Diogenes Laertius asserts,[115] then Plato could have been on the verge of deciding to describe the apparently indissoluble and close linkage between the Sun, Venus, and Mercury with the aid of a partial heliocentric construction, as Heraclides had done. Then, to be sure, the Earth was isolated over against this huge subsystem, and the Aristotelian solution—to remove it entirely, by means of a metaphysical sanction of its nonstellar nature, from the questions about astronomical systems, and thus also to devalue its placement in the center of the world, as being equivalent to the 'lowest' position of all—was a natural one.

For a long time research on Copernicus has favored the approach according to which, within the Copernican reform, the new placement of the Sun in the center was the outstanding change, because the intensification of pagan solar enthusiasm that was already immanent in the Renaissance had prepared the way for it. But even if that had been the case, we cannot overlook the short duration of this condition affecting Copernicus's reception: All these fine ideas were deprived of their importance by, at the latest, Newton. What celestial mechanics left in place was the disappointingly accidental—because purely quantitative—circumstance that the Sun far surpasses all the other bodies of the system in mass. From this time onward, that is what causes its central position, which it would occupy even if it were only a cold and lusterless body. Or even one step further: For celestial mechanics, a central body is only a limiting case of a many-body system, and its position could just as well be unoccupied. In view of this disillusionment by physics, all rhetorical or metaphysical embellishments of heliocentrism were to a certain extent misleading.

Of course such considerations cannot diminish the fact that Copernicus himself claims credit for having given the Sun the central position, with his reform. How was that possible, when after all it had already occupied the central position in the traditional system, and furthermore it had done so in a sense that was more plausible astronomically? The question may seem naive and of secondary importance, but I think it leads a step further. One must only remember that Ptolemy regarded the problem of whether the Sun's sphere should be placed above the spheres of the Moon, Mercury, and Venus, or beneath all the planetary spheres and immediately above the Moon's sphere, as one that could not be decided by astronomical means. After him, too, this problem remained unclarified. The disposition of Mercury and Venus in relation to the Sun constituted the real scope of variation *within* pre-Copernican astronomy. And it was one of the few problems that, if one accepted the premise of geocentrism, could be discussed—and that also inevitably had to arise.

Now my thesis is that this lack of a deciding difference between the superior conjunction and the inferior conjunction was Copernicus's starting point. He sought and found the reason why, in spite of Ptolemy's authority, the two propositions had remained open and undecidable: It was possible for Mercury and Venus to obey both the hypothesis of the superior conjunction and that of the inferior conjunction because they possessed no independent orbit around the Earth at all, and thus no geocentric sphere. Bound, in appearance, to the neighborhood of the Sun, they also revolved around it in reality, as if they were on epicycles coordinated with it.[c] One could say that it is the history of astronomy, with its indecision between the solutions for the system of the Sun, Mercury, and Venus, that Copernicus considers in the light of his principle of man's teleological competence for the universe and declares that there must be an answer that both preserves Ptolemy's authority and also justifies the irrefutable doubts about this authority. Here too it is astronomical reason that Copernicus defends, by withdrawing the close connection of the two inner planets to the Sun from the suspicion of being a voluntaristic, contingently given 'arrangement.'

The fact that this first step toward the universal reform presented, in its content, the "Egyptian" construction (and that of Heraclides of Pontus) became important in the horizon of its reception in the sixteenth century, but cannot be appraised as a 'stimulus' operating on Copernicus. It was not the constructive step as such, which after all was too obvious to be capable of being characterized as an 'idea' that 'occurred' to him, that still required daring, but rather the weighing of the argumentational prerequisites for deciding in favor of it, insofar as that decision was not supposed to avail itself of Ptolemy's general license. Consequently where Copernicus found stimuli or suggestions, or where he could have found them, can only be of secondary importance when we are concerned with the central strand of the development of his theory. The fact that such an author has nevertheless offered a rich yield to those seeking literary sources only demonstrates the Humanists' procedure of continually referring to exemplary authors for support and for protection against the charge of having a mania for innovation—it does not demonstrate that the content of his reform depended on them. Let us not forget that Copernicus could hardly have an advantage over his learned contemporaries in knowledge of those canonical authors.

So we must look for the logical sequence that integrated those 'stimuli' or 'suggestions' into itself as testimonials. If Mercury and Venus did not have independent geocentric orbits, but rather constituted a system harnessed to the Sun, then in the system as a whole the Sun could no longer occupy the center between two sets of three spheres each, but only now a partially central position at the center of the adjacent spheres of Mercury and Venus. This partial system now represented, in its turn, the diagram to which the geocentric cosmos, viewed from outside, had hitherto conformed: circular orbits arranged concentrically around a central body. Once one had accepted this step, the result was that the classical type of 'heliocentrism' was destroyed, and doubtless the motivation that had promoted it was set free to reestablish this systematic advantage, that is, to realize it in a *new* manner. When, at the end of his train of thought, Copernicus exults in the fact that the Sun now stands in the center, this is gratification with a result that, by way of a daring detour, satisfies Ptolemy's requirement again after all. Of course this is a supplementary confirmation, because this inducement should not be overestimated in comparison to the immanent consistency that had become accessible as a result of the partial solution for the system of the inner planets.

If we are going to talk about a logical sequence in the constructive reshaping of the traditional system, we must allude to an aspect that is indispensable for understanding the architecture of Ptolemy's system. A striking fact about the arrangement of the two planets that are always near the Sun is that in spite of Ptolemy's decision in favor of the inferior conjunction, with its 'heliocentric' symmetry, they are arranged in such a way that starting from the Earth, the sphere of the Moon is followed first by that of Mercury and then by that of Venus (before that of the Sun), even though Mercury's possible distances from the Sun always lie within the breadth of variation of Venus's positions. The reason for this arrangement of Mercury and Venus lay in the fact that Ptolemy granted a higher value to time as the ordering criterion than he did to space. In the interests of a strict sequence of the spheres according to the time they required to make one revolution, he puts up with inconveniences in the form of spatial complications. In this way Ptolemy arrives at an unambiguous coordination between distances from the Earth (i.e., the diameter of the spheres) and orbital periods: For its revolution around the Earth, the Moon requires $27\frac{1}{3}$ days; Mercury, 88 days; Venus, 225 days; the Sun, 1 year; Mars, 2 years; Jupiter, 12 years; and Saturn, 30 years. It is only by reference to this remarkable interchanging of Mercury and Venus, which above all neglects their apparent brightnesses, that we can grasp how Ptolemy adheres to a homogeneous systematic principle. It is just from this point that Copernicus, once again conservative vis-à-vis the great authority, extracts the fundamental principle of his own systematic structure.

That is, he retains time as the ordering criterion; it is just that he is no longer as willing to accept spatial complications as the price of this criterion. A condition of the success of the principle was that the Moon could be taken right out of the series. Together with the Earth, the Moon will move, on the time scale of revolutions, into the position of the Sun's one-year orbit, while Venus and Mercury change places. After this operation, strict adherence to the sequence of orbital periods (seen from the point of view of the central Sun) was assured. It was the reestablishment of the Ptolemaic principle of construction at the end of the whole transformation. It was necessary to anticipate this result here because only the traditionalness of the procedure enables us to understand what saved Copernicus from going no further than Heraclides of Pontus's partial solution for the system of the Sun, Mercury, and Venus, or from becoming a Tychonist before Tycho Brahe.

When I speak of the logical sequence of Copernicus's construction, I would also like to derive from this consideration some guidance in relation to a secondary but instructive problem. For someone who reflects on the research relating to Copernicus, the question arises, in view of the material about forerunners and anticipatory forms, why, with so many discussions of a possible 'motion' of the Earth, the step that Copernicus was to take had not become possible earlier. Certainly there was the blockage of the Aristotelian-Scholastic system here, its cosmology and its metaphysical concerns. But our genetic reflection on the relations of logical consequence that become visible in the case of Copernicus also reveals why the speculations about possible motions of the Earth, however daring they were, had inevitably led into blind alleys. The starting point of all those who had tried assuming the diurnal rotation of the Earth had been the very evident circumstance that the celestial phenomena participate, without exception, in the diurnal motion, as an addition to their own proper motions; and this additive component could most easily be dissected out and presented as the *one* motion of the Earth. Now one would think that with that the most difficult blockage in the way of astronomy's progress to the new system—namely, the Aristotelian assumption of the Earth's inactivity and natural motionlessness—would have been so effectively abrogated that the assumption of additional motions would now have seemed like a smaller step. But just this further step—the supposed 'next' step—always remained untaken. The most impressive example of this is Copernicus's contemporary, Celio Calcagnini.

We cannot accept as a mere contingent datum the fact that in none of the geokinetic speculations does even a sign of the attainability of a further stage of the constructive process—in other words, of the Earth's annual motion around the Sun—become noticeable. It may be that precisely the solution to the apparently central problem that had resulted from the enormity of the assumed motion of the so much greater body of the outermost sphere, instead of the little Earth—that is, success with the palpable foreground matter—concealed or diminished the continuing problems of the traditional theory as a whole. In this way the inducement to further interventions may have been deprived of its effectiveness, as can in fact be observed in the case of Calcagnini.

If this observation should be correct, it would have to find its strongest evidence in Copernicus himself, and would have to permit at least a negative hypothesis as to how Copernicus's progress can *not* have come about. In that case, though, the universal assertion of his early historians and late interpreters, that the "first motion" that he attributed to the Earth must also have been the 'first idea' in the genesis of his reform, would have to be wrong. Instead, if our statement of the case should be correct, the thesis of the Earth's diurnal motion would have had to have emerged only as a logical consequence of the solution of the primary question regarding the arrangement of the bodies in the planetary system, with the assumption of the annual motion that became necessary there. It is also quite legitimate, seen from the point of view of the astronomical tradition, that astronomical solutions proceed by way of authentically astronomical problems and that in doing so the Earth is left out of the question as long as possible and its motion is only admitted as something enforced 'from outside.'

This accords with a remark that Copernicus already makes in the *Commentariolus*: He protests that he has not assumed the motion of the Earth rashly (*temere*), with the Pythagoreans (*cum Pythagoricis*). This protest is not so much against being taken for or being designated as a Pythagorean as it is against the insufficient justification for assuming a motion of the Earth as a result of describing the daily rotation of the heaven of the fixed stars as merely phenomenal—the argument that was ascribed to the Pythagoreans. In other words, the threshold for meeting the burden of proof involved in successfully asserting the motion of the Earth against the traditional theory is placed too high by Copernicus for it to be surmountable on the basis of the phenomena that can be inferred from the Earth's diurnal motion. The predicate of motion should continue to be interchangeable between the heaven of the fixed stars and the Earth—and thus an undecideable question should persist—as long as one had not arrived at more far-reaching astronomical or even physical arguments (of which the latter had to appear to Copernicus to be unattainable).

He could not overlook this equivalence in connection with his own argumentation for the natural capacity for motion (the *mobilitas*) of the Earth, for if its perfectly spherical form was supposed to prove this as an essential characteristic, then the immobility of the outermost heaven, which he was to assert henceforth, became a new offense against the "natural" relation between spherical form and motion. No doubt this equivalence in connection with the diurnal motion led Copernicus to the idea of making the 'form' of the universe problematic by seeing an evidently spherical form as given only in the case of the *inner* surface of the outermost heaven. The perfect hollow sphere could very well be inserted in a body the boundaries of which are entirely different, and unknown. A mere thought experiment is enough to shift the preponderance of probability onto the inference from the Earth's spherical shape. If the heavens did not have the form of a sphere from the outside, as they do from the inside, then the premise of the harmlessness of every "natural" motion would no longer be applicable to them, and the application to the heavens of the argument that Ptolemy directed against the Earth's rotation—that it would have to destroy the Earth by the centrifugal forces that it would generate—could no longer readily be prevented.

Copernicus pushes this idea to the paradoxical point that, given the *immensitas* [immensity] that he grants it, the periphery of the universe could expand to the *infinitum* [infinite]—an idea that he may have derived from the Cusan. But in that case a motion of the outermost heaven would be impossible, because no point can move infinitely fast, as this would entail. Is this only a sophisticated playing with ideas? No, because it immediately turns out that it is possible to assume a heaven that is infinite on the other side of its concave inner surface, as soon as it does not need to be in motion, since its finitude was assumed only so as to make its motion noncontradictory—that is, its finitude was deduced from its motion. The whole difficulty of the question that arises in Aristotle's cosmology as to what space the outermost heaven, for its part, moves in—the difficulty that had driven Aristotle to the concept of the container space—disappears when the outer spherical boundary is put in question. To put it differently: The heavens can be infinite, or indefinitely limited, to the extent that they can be unmoved.[116] Copernicus does not entirely abandon the thesis of his first chapter here, that the world is spherical in shape, but he restricts it to the inner surface (which alone is empirically accessible), so as to bring to bear an argumentative advantage of the motion of the Earth. Assuming that the former *primum mobile* [first moved thing—the outermost sphere] is immobile unburdens us of the unanswerable limiting question as to the *figura mundi* [shape of the world]. At the same time, with elegant logic, one of Ptolemy's objections against the motion of the Earth is transformed into an argument for doubting the motion of the heavens.

But the whole train of thought shows that it was only with difficulty that one could get past the equivalence of the assumptions of the motion of the heavens or the motion of the Earth, and that in spite of its superficial plausibility, this approach was not capable of initiating the overall cosmological reform. Copernicus's historical strength is based on the fact that he did not approach the matter from this side. In fact, in the *Commentariolus* he classifies the annual motion around the Sun as the first of the Earth's motions. That is also logical inasmuch as Copernicus uses the arrangement of the spheres (*compositio circulorum*) for his argument, basing it on their sequence according to the relation between the sizes of the orbits and the time required for each.[117] If that *compositio* agrees with the data and observations (*numeris et observationibus*), then the Earth 'fits' perfectly between Venus and Mars. Although Copernicus already has at his disposition here the argument for the *mobilitas telluris* [mobility of the Earth] that the terrestrial body's spherical form made the diurnal motion appear most in accordance with its nature (*sibi maxime proprius*), nevertheless this is not, for him, an astronomical proof of comparable importance to the one that he can present for the annual motion.

Ptolemy's system had rigorously accepted and bracketed out the complications of the spatial arrangement of the spheres, in order to 'save' the regularity of their sequential arrangement in accordance with their periods of rotation. The monstrous constructions that arose in the process of doing this appeared to Copernicus as at least indications that something was not in order here, as he emphasizes in his dedicatory preface, asserting—over against the purely mathematical solutions for the orbital relationships—that the "main thing" (*res praecipua*) is to preserve the form of the world and the symmetry of its parts. Naturally the monstrosity that he censured stood out in connection with both of what were later called the "inner planets," especially Venus, whose epicycle had to be constructed as large enough to enable it to represent both its superior and its inferior conjunctions with the Sun. What must have struck Copernicus first is the connection between the unwieldy hugeness of Venus's epicycle (*ingens ille Veneris epicyclus*) and the problem of the arrangement of the spheres of the planets (*ordo orbium*). If this could be established, then the thesis of the Earth's diurnal motion would only be the final result of this initial move toward reordering. By it, a final gap in the rationality of the *ordo orbium* would be closed by eliminating the disproportion between the extreme rapidity of the supposed rotation of the outermost sphere, in one day, and the extreme slowness of the revolution of Saturn—the planet next to it in sequence—in thirty years. This disproportion had made it extremely difficult to assume a moving influence of the *primum mobile* [outermost sphere] on the planets in their sequence beneath it, even if one assessed the magnitude of the moving force not by the orbital period but by the magnitude of the spherical body to be moved (a procedure that was inadmissible, as being "physical," on account of the special subtle matter of the heavenly bodies).

If the outermost sphere could be assumed to be at a complete standstill, the *ordo orbium* could be reversed and one could follow it from the inside outward: the result, going outward from the Sun to Saturn, was a diminishing series of angular velocities. This reversed way of looking at the *ordo orbium* was to enable Kepler to see the whole of the system as connected mechanically, after all. If the Sun was not only the topographical center of the system but also the center of its moving force, then the given 'fact' of the angular velocity that declined as one went outward became a logical result.

Even an introduction, by itself, of a diurnal rotation of the Earth in its old central location would have been satisfactory from the point of view of an energy-related ordering of the spheres and would have secured the unity of the system. But the result would have been to stabilize that same monstrosity in the case of the "inner planets," since the sequence Moon, Mercury, Venus, Sun, which Ptolemy had introduced, would have had to be retained. But the confidence with which Copernicus passed over the various stages of a partial heliocentrism is based on the fact that the diurnal motion of the Earth had not been his first, but his last idea. After all, the Tychonic solution only seems anachronistic to us because it appeared half a century after Copernicus's death.

Now, of course, the reversed way of viewing the *ordo orbium*, of which we were speaking, is not only and not in the first place oriented toward the 'energy' aspect, which is only preponderant in connection with the history of its influence. What is more important is the element of [the direction of] the 'slope' of rationality. The Aristotelian-Scholastic universe was not only driven from the outside inward, it was also 'thought' from the outside inward: The unmoved mover, beyond the outermost sphere (and mediated only through the latter, as the *primum mobile*) was not only the first cause of all the processes in the world but also the principle of their rationality (even if it was a rationality that was hidden from man). This was confirmed by the way the heaven of the fixed stars displayed its diurnal motion and, with that, the homogeneous basis of time, and by the way all the other heavenly bodies joined in this motion in addition to their proper motions. With the increasing mixture of portions of different causalities, as one proceeded inward from outside, a falling off of rationality took place whose limiting instance was displayed in the ultimate irregularity of the complex of cosmic factors that is found in terrestrial nature. In the perspective of this metaphysical assumption, one could still make clear why the greatest difficulties for astronomy's mastery of the phenomena had to arise precisely in the interior of the universe, that is, the closer one got to the Earth. For the medieval system's theocentric point of view, it was just as plausible as it was consistent that at a distance from the boundary where divine power impinged, the pure presentation of this origin underwent 'worldly' losses.

For Copernicus, such a falling off of rationality as one proceeded inward from outside contradicted his anthropocentric world-formula in the preface to Paul III, according to which the world had been established on man's behalf by its supremely perfect and supremely orderly Author. Following this structural principle, the connection between causal descent and a falling off in the efficacy of order would have to be inadmissible, because an increasing opaqueness in the construction of the world as one approached man's abode was precisely the opposite of what the teleological hypothesis made one postulate. The Copernican world is thought outward from inside, and the only reason this is possible is that in contrast to the medieval system of effects that are always mediated, it is based on the immediacy of God's action on every point in the world.

This reversed way of viewing the matter could give the clarification of the local systematic question about the relation between the Sun,

Mercury, and Venus precisely the priority that it has in the genesis of Copernicus's theory. The challenge implied by the phenomenon of the connected paths of these three heavenly bodies—the way the two planets near the Sun travel along with the main body—could be ignored by an astronomy that did not claim to 'explain,' if it had not been given an accent of offensiveness to rationality by the abnormality of Venus's epicycle, whose diameter had to be fixed six times larger, by Copernicus's estimate, than the distance between the center of the Earth and the closest point to the Earth on Venus's path as determined by this epicycle.

The rationality of a 'true astronomy' did not necessarily mean that all of its phenomena could be explained, but it certainly did entail the exclusion of the idea that 'anything is possible.' What will henceforth be called "possible" is also determined by the fact that Copernicus holds strictly to the classical idea of a spherical body supporting each planet. However subtle, spiritual, and 'ethereal' he may, in accordance with traditional precepts, have considered the matter of these spheres to be, it was in any case contrary to his strict interpretation of the 'possibility' of astronomical constructions that they should be able to overlap one another and pass through one another at will—in other words, that they should have nothing to do with the minimal concept of matter. But in that case even a purely constructive astronomy, if it did not want to give out everything as possible, could not ignore the requirement that it should make the 'thickness' of each sphere agree with the extreme points on the paths determined by the epicycles that were housed in it.

This is why the figures that Copernicus gives in connection with Venus's unwieldy epicycle are scandalous. The thickness of the sphere in which Venus is placed is determined by, on the one hand, the shortest distance between the center of the Earth and the closest point of Venus's orbit to the Earth and, on the other hand, the thickness of Venus's orbit, as the sextuple of the first distance.[d] Here the assumption is that space is filled quite compactly by the nesting shells of the spheres, an assumption that contains or permits the quasi-mechanical implication that the communication of motion takes place on the contacting surfaces of the spheres. To that end, they must fit so perfectly inside one another that the point in the path of each planet that is, in each case, the most distant from the Earth lies on the convex surface whose concave counterpart at the same time contains the nearest point to the Earth on the path of the next planet.

Admittedly, with Copernicus's decision in favor of the "inferior conjunction" this constructive principle could not be exactly carried out with the distance figures that had been handed down.[118] Consequently Copernicus—still arguing on the basis of the traditional system, and without violating the classical procedures of the discipline—had to deal somewhat generously with his figures if he wanted to avoid letting Venus, at its apogee, be farther from the Earth than the Sun is at its perigee, which would have meant that the two spheres could not fit into one another. Initially, all that this procedure proves is that in details, too, Copernicus has abandoned Ptolemy's epistemological standpoint (the stipulation that the question of material possibility is not to be taken into account).

In implementing this intention, it was natural to widen the inclusion of consideration of quasi-mechanical modes of functioning. Somewhere in this process one takes the step of not only taking into consideration what is conceivable as a mechanism but also not ignoring the evidence that a mechanical nexus exists. That is what happens in Copernicus's critique of Ptolemy's representation of the apparent connection between the Sun, Venus, and Mercury. This manifest relation of dependence must have persuaded Copernicus that he was justified here, at least, in posing the question of how the system functioned, and in clarifying it at first without regard to the principle of arranging the spheres according to the orbital periods of the bodies associated with them.

What cause (*Quam vero causam . . .*), Copernicus asks, can be alleged for the fact that the two planets do not have the same independence in their positions with respect to the Sun that the other planets have—so far as their order has not also been falsified as a result of the criterion of orbital periods?[119] If a 'true astronomy' is required, the possibility of setting up a causal connection, however inexplicable it may be, must take precedence over any construction that accepts the appearance of a combination of phenomena as a mere factual 'given.' At the same time, though, Copernicus suggested the conjecture that the implementation, within the Ptolemaic system, of the principle of basing the order of the planets on their orbital periods—and, consequently, the peculiar reversal of the positions of Venus and Mercury above the sphere of the Moon—could have impeded insight into the connection of the two planets with the Sun. To put it differently: Once one has set up the requirement that one should comply with the mode of functioning of this partial system, the result can either be that one abandons the classical principle that bases the order of the planets on their orbital periods or that one satisfies both requirements by means of a new overall system.

If Copernicus wanted to 'explain' the connected motions of the Sun, Venus, and Mercury by making them into their own indepen-dent, concentric system, he had to shift the sphere of Mercury, on account of its smaller angle of apparent elongation from the Sun, *inside* Venus's sphere around the Sun. Then of course he forfeited Ptolemy's arrangement in accordance with orbital periods in either case, that is, both in the inferior and in the superior conjunctions—both of which now became equally possible, depending on where the planets were on their orbits, so that the history of this great difference was justified. Thus the rearrangement of the relations between these three heavenly bodies by isolating a partial system could not be ac-complished without violating a principle that Copernicus not only calls the "first," but of which he explicitly says that it cannot be replaced by any more suitable one.[120]

This dilemma of losing the basis of a principle of rationality as a result of a tangible theoretical advance must have induced Copernicus not to stop at Heraclides of Pontus's solution and also not to be satisfied with the next possible step, which consists of making all the planets, rather than just the two innermost ones, circle around the Sun, and then making the Sun circle the Earth. This later system of Tycho Brahe, like a fossil from before the Flood, allows us to recognize still the daring of these first steps of Copernicus's in which, in order to represent the special characteristics of Venus and Mercury, he in fact envisages the possibility of an 'alien' center of the paths of heavenly bodies—whereupon of course with Tycho this daring is brought back into the realm of the acceptable by the assertion that that center in its turn circles around the Earth, which is thus the ultimate systematic reference point.

Now one cannot say that the way was already prepared for the partially heliocentric solution by Ptolemy's system, in that the latter's epicycles in any case represent circles mounted on circles. For the point on the primary circle around which those circles turned was not occupied by a central body, so it was not a physical reality but only an imaginary point. It is not only that Copernicus, in that first step, had to consider a partial system in which the two innermost planets each circled on an epicycle around a deferent^e that was occupied by the Sun; he was also confronted by the novel consequence that in this case for the first time two epicycles were nested inside one another, so that the path of one planet concentrically enclosed that of the other. However, not much is gained by understanding how this intermediate step had come about if it does not become clear at the same time why it could not be sufficient—why what Copernicus had entered here was not another blind alley.

We can only infer how Copernicus arrived at his theory. Not, however, without the aid of some clues in the text of his chief work, above all in the tenth chapter of the first book. The formula *Hinc sumpta occasione* . . . [Seizing this opportunity . . .], with which Copernicus leads into his new paragraph, can be translated, in its connection with the reflections about Venus and Mercury, as follows: Once one has gone this far, it is natural to bring Saturn, Jupiter, and Mars, as well, into relation to the same center (as Mercury and Venus), nor will one err in doing so (*non errabit*) as long as one makes their orbits big enough that they can contain and encircle, together with them (Venus, Mercury, and the Sun), the Earth as well as lying within them (". . . cum illis etiam immanentem contineat ambiatque terram").[121]

The next clause states the constructive condition for taking this step: Only in this way can the well-established order of these motions be made intelligible. What does he mean by this? The *canonica ratio* [well-established order] is the rule that with the upper (or outer) planets their greatest proximity to the Earth (perigee) and their greatest distance from the Sun (aphelion), on the one hand, and their greatest distance from the Earth (apogee) and greatest proximity to the Sun (perihelion), on the other hand, coincide. The circumstance that the outer planets can sometimes be located 'behind' the Earth, so that a line planet-Earth-Sun is formed, confirms that the Earth is placed between the orbits of the outer planets and the Sun-Mercury-Venus system. But no more than that is said. The localization of the Earth between the spheres of Venus and Mars *is not* yet a localization on a path on which it moves, even if it seems almost as clear as day that it *could* be that.

Copernicus's laconic manner of argumentation gives us a hint that the placement of the three outer planets, Saturn, Jupiter, and Mars, on concentric orbits around the Sun originally represents an analogous formation to the prior solution for the complex of the Sun, Mercury, and Venus. This would still be a typically Scholastic operation, relying on the identity of the stellar species "planet," to which the same substantial predicates have to be ascribed. Therewith, at the same time, the old heliocentrism is reestablished in a different way: "Quae satis indicant centrum illorum ad Solem magis pertinere et idem esse, ad quod etiam Venus et Mercurius suas obvolutiones conferunt." [Which sufficiently shows that their center pertains rather to the Sun (than to the Earth), and is the same as that around which Venus and Mercury also perform their revolutions.]

But the Earth itself is still not a planet, and no Scholastic analogical procedure brings one closer to the decision to ascribe to it, too, the status of a planet and to make it revolve on a path around the Sun. The argumentation has not yet overcome what one might call the "Tychonic" barrier. But now two other principles come fully into force. The first is the view of the spheres as real, combined with the *horror vacui* [dread of a vacuum]. Once all the planets have been related to the Sun as the center, it turns out that between the convex outer surface of the sphere of Venus and the concave inner surface of the sphere of Mars there remains an unoccupied space, which the body of the Earth, at rest, cannot fill up, but an additional spherical shell certainly could. The question of the whereabouts of the Moon commands one to accept such a shell. The Moon alone cannot be released from geocentrism; it becomes a heavenly body *sui generis*— and the figures for the housing of its orbit, as an epicycle, in a terrestrial sphere between Venus and Mars are correct. Here Copernicus had to overcome a weightly obstacle against a single heavenly body's having a path that, contrary to the homogeneous constructive principle of the traditional world-system, does not relate *immediately* to the center but only *indirectly*, by way of the Earth, by sharing the center of its orbit. The resistance that resulted from this *singularity*, for an epoch that thought in terms of conceptual *specificity* [i.e., the characteristics of *species*], was only fully refuted by Galileo's discovery of the moons of Jupiter.

It is impossible to overestimate the weight, at this stage in the logical sequence of the development of Copernicus's theory, of the assumption of a compact fit between the spheres. But it would certainly still not have been great enough without the other, most important element in the argumentation: the *ordo orbium* [order of the spheres] that was once again fully attainable by means of the new localization of the Earth and that did not tolerate an Earth that was at rest in this position, but required an Earth that was in motion, so that it could reestablish the splendidly unfolding proportionality between the succession of the planets' spheres and their orbital periods by 'taking over' the Sun's one-year orbit. One sees how important the removal of the Moon from the classical sequence of concentric spheres became in allowing Copernicus to apply the supreme systematic principle.

The fact that the last step, with which the Earth is not only placed between Venus and Mars but is put in circular motion, was at the same time the most difficult and the riskiest is still betrayed by the rhetorical signal of daring with which Copernicus finally introduces the *orbis ille magnus* [that great sphere] of the Earth and its Moon: "Proinde non pudet nos fateri . . . . " [Therefore we are not ashamed to assert . . . .] He softens that only with the one soothing observation that at least in regard to the Moon he has stayed with geocentrism.[122]

One of the results of our analysis of the genetic context is that the question of the center of the world remains merely incidental, and is only decided, as though (at that point) it were a matter of course, in the final step. It is only in retrospect that the alternatives of geocentrism and heliocentrism became the dominant way of characterizing the world systems. In the *Commentariolus*, in spite of the extreme simplicity (which he still asserted there) of his construction, Copernicus had already anticipatorily attenuated the importance of the change in the center, by means of the thesis that there was no common center of all the orbits and spheres. The expression "center of the universe" was now defined only in relation to the heaven of the fixed stars. Even the Sun did not become the central body in the strict sense: " . . . circa Solem esse centrum mundi" [the center of the universe is in the vicinity of the Sun].[123] In this respect as well, that against all philosophical pressures he left the center of the universe empty, Copernicus had followed Ptolemy, who had located the center of the universe in the unoccupied center of the sphere of the Sun.

The Earth could receive astronomical dignity only by its center's ceasing to be identical with the center of the universe (which Aristotle's cosmology had required it to be for the sake of the ordering function of weight). But then this *physical* datum—the center of the heavenly body, the Earth, as the reference point of the weight of the element, earth—could become an element in *astronomy*, as the center of the Moon's orbital path.[124] In the Aristotelian cosmology, the Earth had had the attribute of being the center of stellar revolutions only on account of its identity with the center of the universe, as the absolute reference point of all "natural" motions. For the first time a heavenly body circles around the center of gravity of another heavenly body that is clearly remote from the center of the universe, without its being possible to see this as involving a physical relationship.

The dissolving of the identity of the center of the universe with the center of gravity[f] was a necessary step where terrestrial physics could no longer make relations to the center of the universe empirically perceptible. At bottom, this correction was already due with the appearance of the 'imprecision' of Ptolemaic astronomy in contrast to Aristotelian cosmology. But Nicholas of Cusa was the first to draw the consequence that the little word "almost," as a limitation of the identity of the center of the Earth with the center of the universe, already implies the equivalence, in principle, of the Earth and all other bodies in the universe. The fact that the Earth, unlike the stars, seems to be without any light of its own is due, according to the Cusan, to the merely perspectival difference that we see the stars from outside, so that what we see is their envelope of elemental fire, whereas on the Earth the layered arrangement of the four elements is seen from within and from below. "If someone were on the Sun, its brightness would not appear to him as it does to us. For if one regards the body of the Sun, one sees that toward the middle it contains something like [the element] earth and that a brightness like that of fire lies around the outside, and between both a sort of water vapor and transparent air; that is, it has just the same arrangement of elements as the Earth has. If one were outside the fire region of the Earth, then the Earth would appear to him in the whole extent of the fire region as a brilliant star .... "[125]

This homogeneous construction, from the elements, of all the bodies in the universe including the Earth, requires a solution of the problem of the maintenance of their spherical shapes that does not have recourse to the Aristotelian assertion of the 'gravitational' operation of the center of the universe. The spherical shape represents indifference toward any change of state, so that it becomes an expression of self-preservation. 'Gravity' is the relation of the parts to the self-preservation of this whole. While, in the *Commentariolus*, *gravitas*, as a 'characteristic' of the center of the universe, had already passed over to the element earth, in his chief work Copernicus makes it into the mode of appearance of the *appetentia naturalis partium* [natural appetite of parts]. Gravity becomes an expression of the immanentized providence of the Author of the world, "ut in unitatem integritatemque suam sese conferant in formam globi coeuntes" [in that they gather themselves into their unity and wholeness, coming together in the form of a globe].[126] It is to be assumed, he says, that the same inclination (*affectio*) also maintains the spherical forms of the other heavenly bodies, although these carry out their revolutions in the most diverse ways.

Here, without any great expense, Copernicus has abrogated the Aristotelian physics of the elements and their natural motions, has assumed the existence of homogeneous conditions governing the self-preservation of all world bodies in spite of motion, and has started the argument for the admissibility—which is now also the physical admissibility—of motions of the Earth. Of course one should not overlook the fact that with this procedure of homogenizing the bodies in the universe physically and rendering them independent, Copernicus neutralized the center of the universe; and the result is that, until Kepler, the possibility of drawing conclusions for a 'celestial dynamics' from the transfer of the Sun to the center of the universe was disposed of. One can also say that precisely by its proving itself to be unoccupyable, the center of the universe loses every importance except its metaphorical one.

In the fragment (unpublished in his lifetime) of an *Essay on the History of Astronomy*, Adam Smith—whom Markus Herz could refer to in a letter to Kant as "your favorite"—set up, as one of the generalizations about scientific investigations that could be derived from the historical process that astronomy had gone through, the thesis that the conviction that a system needs correction already prepares the way for its downfall: "When you have convinced the world, that an established system ought to be corrected, it is not very difficult to persuade them that it should be destroyed." The fine structure of history does not corroborate this statement. On the contrary, the proposals of the most penetrating correction, the introduction of the diurnal rotation, were not capable even of merely shaking the traditional system. Copernicus's reform, I dare say, was successful only because he had the correct maxim, of starting from the most obscure point of the prevailing theory and not relying on the most natural and most prominent possible correction. In proportion to the problems that had to result from ignorance of the principle of inertia, the introduction of the diurnal rotation was too modest an astronomical success to be able to outweigh the new burdens. For the same reason, there is not much to be said for the conjecture that Copernicus could have been stimulated to seek a nongeocentric system by a passage that he cites from Ptolemy. For this purposefully introduced passage also speaks not at all of a reconstruction of the planetary system, but only of the possibility that the Earth might rotate in its old location.[127]

The genetic hypothesis that I have presented here is based above all on the tenth chapter in the first book of the *De revolutionibus*. Does the historical elucidatory value of this text stand a comparison with the so much more frequently analyzed preface to Paul III? Only the preface speaks explicitly of the genesis of the theory. But, at the same time, it does so with the most minimal freedom for personal testimony, in view of the addressee, of whom it could be expected that he would most readily understand and approve the work's origin if it was closely tied to the suggestions of authors from antiquity. But for that very reason a parallelism in the formulation is striking. In connection with the mention of the Pythagorean, Nicetas, in Cicero, and of Philolaus, in Plutarch, the preface says, "Inde igitur occasionem nactus coepi et ego de terrae mobilitate cogitare." [Therefore, having obtained this opportunity I, too, began to consider the mobility of the Earth.] The important transition in the tenth chapter, *Hinc sumpta occasione* ... [Seizing this opportunity ...], which I discussed above, is very similar, though without the modesty of the *et ego* [I, too].[128]

But the tenth chapter is the text of an author who had not yet made up his mind to publish, a text that belongs among the oldest parts of the work, which we can be sure was composed before 1522, so that it was probably at least two decades older than Copernicus's self-presentation to the addressee of the dedication. The tenth chapter is also older than the completed version of book I, which will have been produced as late as 1525. The transition from the ninth to the tenth chapter, at least, reveals this: The two last sentences of book I, chapter 9, anticipate, in a compressed formulation, an evidently already finished text—its procedures and its result. One will convince oneself, he tells us, that the Sun occupies the center of the universe; and we are taught all of this by the *ratio ordinis* [principle of order] according to which the heavenly bodies are arranged with respect to one another, and by the principle of the harmony of the universe.[129] This transition prepares the reader for the fact that in what follows the argument will operate differently, and under a different pretension, from the way it has operated up to this point. While the last sentence of the ninth chapter anticipates the demonstration that is to follow, its first sentence summarizes the result of the process of argumentation in the first eight chapters, in its preliminary character: "Cum igitur nihil prohibeat mobilitatem terrae ..." [Since then nothing prevents the Earth from having the power of motion ...]. If up to this point the inquiry had dealt with the Earth's *mobilitas* [mobility], now it was to deal with its actual *motus* [motion].

The distinction (which is not always maintained, terminologically) between *mobilitas* and *motus* marks this change of modality, which had been made necessary for Copernicus by his imitation of the construction of the *Almagest*, with its opening propaedeutic treatment of the Earth. For the other heavenly bodies the essential characteristic of mobility was taken for granted, but for the Earth it had to be explicitly laid claim to, in a discussion of its possibility. This becomes most evident, terminologically, in the exchange of roles between the Sun and the Earth, when the motion of the Sun, as apparent motion, is explained by the motion of the Earth, as real: "Quicquid de motu Solis apparet, hoc potius in mobilitate terrae verificari." [Whatever apparent motion the Sun has can be better explained by the mobility of the Earth.] Here the difference between *motus* and *mobilitas* is joined by that between *apparere* [to appear] and *verificari* [to make true].

In Ptolemy and Copernicus, the function of the first eight chapters of their chief works is, in spite of the difference between their systems, entirely comparable. Ptolemy secures the astronomical theory that will follow against any possibility that the physics of the Earth could influence the astronomical discussion and make the Earth leave the mere indifference of the observer's zero point. Copernicus shields his astronomical theory, which will follow the propaedeutic, from the danger that the physics of the Earth could make it unsuited for the integration into astronomy that is inevitably impending for it. In both structures of proof the point is to prevent physics from being a disturbing factor for astronomy, in the one case by resolutely restricting it to the neutralization of the Earth, in the other case by just as resolutely, even if extremely economically, extending it to all the heavenly bodies. Only the modality of this 'propaedeutic' procedure, in the two cases, is fundamentally different: Ptolemy can (without reflection) proceed categorically in physics, whereas in astronomy he claims the epistemological 'general proviso'ᵍ for all his hypotheses; Copernicus, on the other hand, gives his initial physical discussion the modality of a mere consideration of possibility, which is available for the use of the subsequent astronomical demonstration's claim to truth. It is correct that "with the annual revolution and the heliocentric ordering of the heavens Copernicus leaves the ground of natural philosophy and continues his argument as a purely astronomical one,"[130] but this change in the level of the argument involves no diminution of its theoretical dignity, no restriction of its claim to truth. On the contrary, from the point of view of the situation in regard to the burden of proof, the tenth chapter's passage through astronomy now carries the accent of the decisive stage of the proceedings. The *prima pars quaestionis* [first part of the question], as the last sentence of the eighth chapter summed up what went before it, did not even reach a decision regarding the diurnal rotation of the Earth, because this could only be of importance, for Copernicus, as the final consequence of the astronomical reconstruction.

So on no account does the composition of the first book of the *De revolutionibus* reflect the genesis of the Copernican theory. It is a product of assimilation to the *Almagest*, and this assimilation, in its turn, is an extremely economical adaptation to what his contemporaries were ready to receive. In the *Commentariolus* Copernicus had shown the course of his reflections more openly, by not beginning with the spherical form of the Earth as the equivalent of the spherical form of the heavens, with their mobility, but instead introducing the diurnal motion only as the "other" motion, in spite of its 'naturalness.'[131] There the argumentation initially aimed at securing the principle of the regularity of all celestial motions, as a premise for deductions, and at doing so by deriving '*regularitas*' [regularity] from the '*rotunditas*' [roundness] of all bodies, not specifically of the Earth. In that 'Platonic' postulate the requirement of regularity proved to be a mere consequence following from the primary requirement that the bodies and orbits should be circular. So astronomy's task of saving the apparent motions in accordance with regularity (*sub regularitate salvare*) is no longer deduced from the special nature of the heavenly bodies or from their nearness to mathematical ideality, but from what the Scholastics had called the *forma substantialis* [substantial form]: "Valde enim absurdum videbatur coeleste corpus in absolutissima rotunditate non semper aeque moverit." [For they thought it absurd that a heavenly body which is absolutely round should not always move uniformly.][132]

Seen in this light, the argumentation for the mobility of the body of the Earth, insofar as it too possesses a perfectly spherical shape, appears as an argument a fortiori: If the spherical shape of the heavenly bodies allows one to deduce that their motions on their orbits proceed utterly uniformly, then the minimal attribute of mobility cannot be simply excluded in the case of the Earth. When Copernicus already says in the *Commentariolus* that with his principles as laid down in the seven *petitiones* [assumptions] the uniformity of the motions can be saved '*ordinate*,' it is not enough to translate this as saying that he wants to show "how well" they can be saved. This is not only an intensifying epithet but also a reference to the principle of the *ordo orbium* [order of the spheres], as the reference to the lengths of the radii of the spheres (*quantitates semidiametrorum orbium*) and their relation to the *circulorum compositio* [arrangement of the circles] shows. But argumentation based on the order of the spheres has not yet prevailed over that based on adherence to the 'Platonic' requirement.

The real motion of the Earth is *one* concession that had to be made in order to generate the rational order of the planetary system. An additional concession that followed from this one and that carried no less weight was that of a new and immeasurable expansion of the universe. In relation to the distance between the Earth and the heaven of the fixed stars, the diameter of the Earth's orbit did not manifest itself—was not reflected—in any parallax. It is not a matter of course that this concession could be made, in order to overcome— indeed in the end only to reduce—the constructive complications of the traditional system, to which Copernicus unhesitatingly attributes infinitude (*infinitum*), whereas for the new order of magnitude of the universe he concedes only a cautious intensification to immeasurability (*immensum*)—which here means only that this magnitude had to be greater than the value that would still produce an observable parallax. The almost infinitude of the multiplication of spheres is intolerable because it overstrains the human intellect's power of comprehension and thus runs counter to the world-formula that asserts the commensurateness of reason and the universe.[133] The enormousness of the new world seems, in comparison to the defects of organization in the old world, not to challenge the principle of rationality.

This circumstance announces one of the fundamental tendencies of the modern age. The expenditure of space and then, in cosmogonies, of time in pursuit of gains for theory is of no importance rationally. This means of explanation is the most readily available one. But it is not yet a matter of course that in the first leap of the expansion of space, Copernicus does not anticipate the traditional objection that nature avoids the superfluous, the useless, and the empty. Here, already, space and time prove to be rationally indifferent magnitudes that are outside the purview of the principle of sufficient reason and consequently are at our unlimited disposal.

Part III
A Typology of Copernicus's Early Influence

Introduction

The crisis of geocentric astronomy is supposed to have reached its blasphemous eruption, a millennium after the system's codification by Ptolemy, in the dictum of the king of Castile, Alfonso, who was called "the Wise," that the world would have turned out better if he could have been present at the Creation. This legendary censure of the creation became a *topos* [standard theme] that was varied in many ways. The most startling fact that one can establish in this connection is that the wording of the arrogant claim initially had to do with an objection only to the nature of man.

In the article "Castille" in his *Dictionnaire historique et critique*, Pierre Bayle devoted one of his famous footnotes to the king's verdict. Following Roderic Sanctius's history of Spain, Bayle reproduces the text of the remark in the third person, as indirect discourse: "Si a principia creationis humanae Dei altissimi consilio interfuisset, nonnulla melius ordinatiusque condita fuisse." [If he had participated, through advice to God almighty, from the beginning of the creation of man, man would have been better and more properly fashioned in many respects.][1] Franz Hipler took that, only transposed into the first person, into his description of supposed forerunners of Copernicus, and added to it—following the apocryphal tradition—that the remark became the king's doom.[2]

It is natural that the restriction of the censure of the Creation to the constitution of man could not last, when one considers that the king's posthumous fame as a stimulator and promoter of astronomy, in the *Alfonsine Tables*, survived the centuries and at the same time made the inadequacy of the theory underlying those tables increasingly manifest. How inevitable it was that the anthropological criticism would be translated into the cosmological one becomes tangible in Bayle's article: His procedure here is entirely contrary to his 'critical' method of demonstrating the contradictions in what is said in the tradition and by the historians and thus dissecting out the hard core or the remainder of the historical reports that does not come under suspicion, but not adding anything suppositious that is unsaid [in the reports]. Here it is true, Bayle says, that the silence of the Spanish historian carries some weight, but it cannot dissuade him from assuming that Alfonso had included in his daring critique the celestial spheres as described in Ptolemy's system. The state of astronomy at the time could not possibly have allowed its notorious amateur to restrict his censure of the Creation to man.[3] To be sure, Bayle does not go so far as to offer a conjecture of his own. He is satisfied to quote an author who can hardly be described as an "original source" on this question, namely, Fontenelle, who only makes Alfonso say, quite generally, that he could have given God good advice if God had consulted him when he was creating the world.[4] In the passage that Bayle quotes, Fontenelle does call the idea *trop libertine* [too wayward], and its author *fort peu devot* [very undevout] (which in the printings from 1687 onward was softened to *peut devot* [not very devout]), but he also and especially calls the Ptolemaic system *une occasion de pèché* [a cause of sin].

The blameworthiness of the Castilian king is reduced not only by Bayle's doubts about the legends according to which Alfonso was punished by lightning and thunder from heaven but also by his attempted demonstration that Alfonso's political downfall was not due to the disfavor of the stars but rather to his having believed in that disfavor. Bayle tells us that this distorted Alfonso's behavior so much, preventively, that he created for himself the enemies that he thought he had to be prepared for: " ... une Prédiction, qui n'est en soi qúune chimere, devienne un mal très réel par la conduite qu'elle fait tenir" [a prediction that in itself is nothing but a chimera becomes very real as a result of the behavior that it gets one into]. One sees the strategy of the early spokesman for the Enlightenment: to make the bold venture of censuring the Creation appear as great and as comprehensive as possible, while on the contrary making its supposedly transcendent consequences appear only as the immanent logic of its unenlightened medievalism. The very great sin becomes a premature

application of reason to the historical fact that Copernicus still lay in
the distant future.

Leibniz integrated the astronomical reform into his theodicy: The
king is excused by the fact that he did not know Copernicus's system.[5]
Leibniz could say this because he no longer assumes the ubiquity of
reason in time. For him, whether man can overcome the deficiencies
in his knowledge does not only depend on the world's objective
transparency—in other words, on the character of its origin and its
author; instead, this is, above all, a process that is conditioned by
development in time. King Alfonso's mistake is not due to the fact
that he could have seen something that he did not see. God, too, is not
responsible for the fact that with men everything takes time and does
not come before its time.

Goethe turned theodicy around in such a way as to unburden man
by making the censure of the Creation depend on the uncertain
answer to the question whether understanding of the world was man's
lot: "If God's concern had been that men should live and act in the
truth, then He would have had to arrange matters differently." [6] Here
everything depends on seeing that precisely this was something that
Copernicus could not have said. For him, the world was arranged so
that it conformed—without recourse to force, though not without
exertion—to man's interest in the truth. Goethe, on the contrary, in
the "Incidental Observation" (which no longer needs to be quoted)
in his *Materials Illustrating the History of the Doctrine of Color*, saw
Copernicus's effect—what he "demanded of mankind"—more in
the renunciations and forfeitures that he required of man's sense
experience than in the validation that he thereby brought to reason.[a]

The vehemence of the indignation in Alfonso's remark is also due to
the high rank that astronomy's object had possessed since ancient
metaphysics. Precisely because what was in question was the most
sublime objects, God should not be forgiven for having failed in
connection with their perfection and its manifestation. Only, the
indignant Castilian would have overlooked the fact that ancient
astronomy, in whose tradition he stood, was able to view the problem
in an exclusively epistemological light, because it did not need to
worry about the intention and the intelligence of an Author, and to
what extent He took man into consideration. It had regarded the
disproportion between the object's special character and the
inadequacy of human understanding as being involved in the fact

that the world was a 'cosmos.' For what could not be otherwise, no one needed to be blamed or exculpated.

It is true that Alfonso is anything but a forerunner of Copernicus; but he is nevertheless a significant pre-Copernican figure. He grasps for the first time the whole offensiveness of the fact that the God of the Middle Ages was supposed to have created the heavens in such a way that man is only able to perceive confusion in them. For this is the dilemma that Copernicus will aim to remove by relieving the Creator of the reproach that it is the object that confuses cognition, rather than cognition that fails when faced with its object. Seen in this way, the modern age begins with an act of theodicy. In it we are called upon not to rest satisfied with a state of theory that renders questionable God's intentions with the world and with man.

The dispute about the world-systems is not of the same rank as the dispute about man's access to the truth. Thus also the antithesis that came into the world through Copernicus and was to become disturbing for the century after him is not originally an antithesis of cosmological schemes but rather one of anthropological preconditions for the possibility of truth. The important thing for Copernicus, in his changes in the world-system, was to register reason's claim to truth; and the object of the anti-Copernican reaction, just as soon as it was fully formed, was to reject that claim. The condition imposed on Galileo in 1616, that he should remain in the hypothetical mode of speech, will not be the culminating point yet—rather, that point will be reached in the concluding formula that was dictated to him for the *Dialogue on the Two Chief World Systems* of 1632.[b]

Copernicus's break with the tradition of the astronomical system could have been almost unnoticed if he had kept up the tradition of astronomy's distance from the truth. He was not the kind of critic of the existing system who would have had to defend himself against the dogmatists of the old; on the contrary, the legacy that he left behind him was not to let the generous procedures of traditionalists pass— not to tolerate nonchalance in dealings with the truth.

It must be admitted that Copernicus's opponents immediately concentrated on what had in fact been the important thing for him, and that they would have gladly let his partisans retain what was to appear to historians of science, so much later, as his revolutionary turning. Without his claim to truth, the anthropological and metaphysical conclusions drawn from the abandonment of geocentrism (conclusions that Copernicus himself had not drawn) would hardly have been articulated; still less could they have been accepted with the embitterment of a humiliation or the triumph of a self-liberation. The claim to truth did not help Copernicus's system to become accepted any more than it was able to add even a jot to the honor accorded to Copernicus in the history of science. But the epochal 'energy' that was set free by Copernicus was, nevertheless, due only to it.

Claims to truth have their own inevitabilities and their own ethics. Precisely to the extent that the past lives on, or is even predominant, among the contemporaries of these claims, they cannot avoid criticizing the past more sharply, and accusing it of a more thoughtless inertia, than it deserves and than is conducive to the noninflammatory conveyance of the new into the public's comprehension and common possession. The ethics of claims to truth consists in their having to demand of their adherents more power of conviction and testimony than can usually belong to a scientific theory (or, more generally, to a rational assertion). Thus those who are not convinced, or who hold back, quickly become traitors to the 'cause,' and those who are not prepared to make the ultimate or penultimate sacrifice for it are put before the bar of history as having failed. On the other hand, one who exposes himself to danger for that truth (albeit only very indirectly for it), one who does not recant when faced with the torture apparatus, or who even goes to the stake, shares in all the glory and the triumph of that same 'cause.' Only such behavior seems adequate to the importance of the theoretical event, even if previously every lack of enlightenment, every persistent—or even preponderant—obscurantism has to be excused.

It will be evident that I am speaking here of Osiander, Rheticus, Giordano Bruno, and Galileo, but also, and first of all, of the image of Copernicus himself that is constructed—of the language in which his action in the realm of theory is soon represented as the 'deed' of a 'perpetrator.' When an effect intervenes so penetratingly in consciousness, what was able to bring it about cannot have been mere thought; it can no longer have been illumination or inspiration, but must have been theoretical action as a violent deed.

1

The Theoretician as 'Perpetrator'[a]

On the base of the Copernicus monument in Torun stands this inscription: *Terrae Motor Solis Caelique Stator* [Mover of the Earth and Stayer of the Sun and the Heavens]. The kings of Prussia had owed the monument to Copernicus for a long time. On 12 August 1773— that is, in the year of the astronomer's 300th birthday—Frederick the Great had made this promise in a letter to Voltaire. The only small blemish in this for the king was that the year before, in the first partition of Poland, of all towns the town of Copernicus's birth had escaped him.[7] It was still more than half a century after the third partition of Poland before the monument could be erected in Torun. One of the bitter witnesses of political developments after 1848, Karl August Varnhagen von Ense, saw Friedrich Tieck's sculpture in Berlin, before it was shipped, in 1852, and noted in his diary, "Saw the colossal bronze statue of Copernicus, which is to go to Torun, in the Münzstrasse. Broad, ponderous, devoid of expression . . . ."[8]

The inscription on Tieck's monument has, one might think, precisely the "fine, grand ring to it" that Droysen, in the same century, considered characteristic of the linguistic formations of his time. But the formula does not belong to the lofty style of the nineteenth century. The metaphors of the 'perpetrator' can be followed back to close to Copernicus's time, and, with a characteristic interchange of subjects, right to Copernicus's own writing.

At first the image of Copernicus the 'perpetrator' who stops the heavens and puts the Earth in motion seems to be a rhetorical element in the polemics against Copernicanism, in connection with which the important thing is not so much the kind of changes made in the world construction as rather the act of overthrow and the forcibleness of its intervention. The Jesuit Melchior Inchofer, who participated as a consultant in the proceedings against Galileo, composed an accusation against the Copernicans—designated as "neo-Pythagoreans"—whom he characterizes as *terrae motores et solis statores* [movers of the Earth and stayers of the Sun].[9] The designation of Copernicus's adherents as Pythagoreans is meant to evoke the odium of a heretical sect and may indeed have succeeded in conveying that impression, whereas a century earlier, when Copernicus himself appealed to the Pythagoreans, as a Renaissance gesture, he could count instead on goodwill.

The formula of abjuration that was imposed on Galileo in 1633 contains not only the antithesis of rest and motion but also an emphasis on the central position: "... ut omnino desererem falsam opinionem, quae tenet Solem esse centrum mundi et immobilem, et terram non esse centrum ac moveri" [that I must altogether abandon the false opinion that the Sun is the center of the world and immovable, and that the Earth is not the center of the world, and moves]. In contrast to this double definition of Copernicanism's "false opinion," the demonizing description concentrates on making the exchange of the predicates of rest and motion appear not only as a transaction within theory but also as a criminal deed. The consequence of this prominent linguistic and metaphorical appearance is that the event is moralized.

In his satire *Ignatius His Conclave* of 1611, John Donne sketched a vision of hell in the course of which Copernicus, together with other audacious revolutionaries, appears before Lucifer and obstreperously beats on the doors of Hell. In an ironical lament he complains over the fact that these gates should remain shut against one to whom the mysteries of the heavens were open and who had been able to give the Earth motion, as though he were a soul to it. Should these gates—Copernicus complains—be open to small innovators but resist him, who had reversed the frame of the world and had thereby almost become its new Creator?[10]

Even if it is just as suitable for disparaging as for heroizing, the formula of the theoretician as the perpetrator who makes the Sun stand still and the Earth move could still be mere rhetoric. But even what one describes so contemptuously precisely as "mere rhetoric" speculates on the thoroughly real pregiven potential of its audience, which either is or can be filled with fears about the dependability of what-it stands on, or with admiration for the daring of the performer. Perhaps the triumph of Copernicanism, as it was to be expressed on the nineteenth-century monument [in Torun], was only secure and unambiguous when, with the domestication of comets in the astronomical system, the connection between motion and uncertainty was dissolved. Until then one could also appeal to people's fears in order to make Copernicus's violation of the sanction of geocentrism appear not only dogmatic but also inhuman.

One is given grounds for wondering about the assumption that the formula represents mere rhetorical or poetical exaggeration by the fact that Galileo, when he sets out in the Second Day of his *Dialogue on the Two Chief World Systems* to deal with the emotional resistance to Copernicanism, sees its source in the fact that the more or less naive onlooker at the change in astronomy confuses the inversion that takes place in the assertions of theory with the real unrest of the ground under his feet, which he now imagines that he feels. Galileo is referring to the objection (already brought forward by Ptolemy) against the rotation of the Earth that rocks and animals would have to be flung toward the stars by the centrifugal force that would arise, and that no mortar could keep a building connected to its foundation while it was under this influence. Engaging in a certain amount of hairsplitting, in his dialogue Galileo makes Salviati point out that Ptolemy's argumentation already presupposes the existence of animals, people, and buildings on the Earth, so as to discuss the effects that the Earth's motion would have on them, whereas after all the preexisting motion of the Earth would not permit the existence of animals, people, and buildings in the first place if Ptolemy's assumption were correct. So the opposing argumentation would have to assume that the Earth at first stood still, so as to allow those things to be formed in reality and to colonize the Earth, and that a cosmological reformer (the talk is always of Pythagoras) then set it in motion.

More important in this context is what Salviati says about an observation that he has been able to make a thousand times, and not without amusement. Almost everyone, when they heard for the first time of the thesis that the Earth moves, had the impression, which they received with a shock, that the Earth's millenniums-long rest had been converted into motion for the first time by this assertion "of Pythagoras." As though reality itself obeyed the decree of the theoretician: "... quasi che, dopo averla egli tenuta immobile, scioccamente pensi, allora, e non prima, essersi ella messa in moto, quando Pitagora o chi altro si fusse il primo a dir che ella si muoveva" [as if such a person, after having held it to be motionless, foolishly imagined it to have been set in motion when Pythagoras (or whoever it was) first said that it moved, and not before].[11] And (so Salviati continues his report of his observations) not only uneducated and superficial people (*uomini vulgari e di senso leggiero*) had this silly idea that the Earth, from its creation up to the time of Pythagoras, had stood still and had only been put in motion by him—even Aristotle and Ptolemy had not escaped such childishness (*incorsi in questa puerizia*).

Even against the doubts of his partner in the dialogue, Sagredo, Salviati insists on his suggestion that the arguments against the motion of the Earth assumed as resulting phenomena things that could only occur if the Earth, having previously been at rest, were suddenly forced by a violent action to rotate. It should not be forgotten that Galileo held to the idea of circular motion as a "natural motion," and thus at bottom still availed himself of the argument, which was discovered by Copernicus himself, that what is natural cannot have destructive effects, and the motion of the Earth is, precisely, natural. It could be a "violent" motion only if at some time it had had a beginning. So Copernicus cannot be described as a perpetrator, precisely because that would have implied what the Copernicans had been charged with, as an objection to the motion of the Earth.

Apocalyptic fears—even those that are connected with the actions of the intraworldly agent of scientific innovation—are counteracted by means of the concept of nature and its unconditional reliability. As the appearance of what is essential, it is immutability. In relation to its object, knowledge of nature is subject to the condition that innovations put it in the wrong. Copernicus himself knew that the suspicion of a mania for innovation would burden his work more heavily than its difference in content from the traditional cosmology and its geocentric pattern, and also more than the difficulties with individual Bible texts. An appeal to the immutability of nature was one of the strongest means of argument in the Renaissance. It is even the assumption underlying the literary reaching back to the texts of antiquity. It ensured the possibility not only of reading and understanding the newly accessible writings of the ancients but also of assuring oneself of their validity, as though they had been written in the present.

Machiavelli had given expression to this ahistoricism during Copernicus's lifetime in the *Discorsi sopra la prima decade di Tito Livio* [*Discourses on the First Decade of Livy*] (1513), when he writes that imitation of the ancients in law as well as in medicine and in politics depends upon the immutability of nature. This holds, he says, in opposition to those who regarded such imitation as anachronistic—as though the heavens, the Sun, the elements, and men were different in their motion, their order, and their power from what they had been in antiquity.[12]

This language contains an elementary realism of metaphors—a constitutive confusion of reality with the way in which it is apprehended. There is no awareness of the fact that the immutability of nature does not of itself legitimize the immutability of our ideas of it. The screening off of this objection is one of the preconditions—which are difficult to get at—of the not merely rhetorical identification of making a change in theory with making a change in reality. Depriving the revolution that theory inflicts on the object, nature, of its plausibility and probability from the very beginning had the fundamental character of self-preservation. Theory is subjected to the criterion of the characteristics of its object, so that even a reformer with Copernicus's genuinely conservative discretion lays himself open to the metaphor of violent action. The reproach that Copernicus overturned the whole order of nature in order to preserve his hypothesis[13] is serious enough, independently of the actual changes contained in his system. As long as people held to the dependence of all human ideas of order on nature, they could not admit that man could act and could succeed, theoretically and technically, 'against nature.'

Carried over to the dimension of time, this is the Stoics' idea of the *consensus omnium* [consensus of everyone] as the teleologically founded mark of truth—a mark that only the Enlightenment thinkers will come to regard as a mere appearance. Copernicus knew what he was saying when he wanted to make his daring undertaking plausible to Paul III: He had risked it, he says, both against the received opinion of the experts and also *propemodum contra communem sensum* [almost against common sense], after he had realized that the astronomers had no consistent tradition (*mathematicas sibi ipsis non constare*). The consensus of common sense could only be invalidated by the 'dissensus' of the professionals.[14] Undoubtedly Copernicus's strongest, even though highly indirect, effect was that he broke the natural guarantee of the *consensus*; but in order to be able to break it he could not escape the odium of one who forcibly violated what it vouched for.

In Galileo's dialogue—to return to it once more—he tries to find the root of the defensive reaction against Copernicus in a realism of metaphor. In the antique costume of the shock supposedly produced by Pythagoras, the contemporaries' emotional resistances are presented. More important than this costuming, however, is the argumentative advantage arising from the fact that the ancient philosopher's analogous role as 'perpetrator' makes it possible to present Aristotle and Ptolemy as theorists who already represent this emotional resistance. They are not systematizers of an authentic geocentrism founded on the common sense of the life-world; instead they are the prototypes of reactive shock. This trick makes hostile Scholasticism appear as a secondary formation of all-too-human fears in the face of the painfulness of an old truth. The tradition that Copernicus's opponents defend as though it were an original revelation is only, for its part, the result of the intolerability of Pythagoras for the authoritative figures of the system of cosmological dogma. On this assumption, the conflict appears not to be tied to biblical and Christian premises. It is reduced to a more general need for cosmic security. It is of course human to oppose Copernicus's demand, but for that very reason it is not specifically Christian.

We have before us here an excellent example of the basic pattern of modern 'paratheories' in which, for a given system of assertions, one is immediately provided also with the explanation of why this system is bound to attract the opposition of its contemporaries. Only from the perspective of Galileo does one perceive how dexterously and appropriately Copernicus had argued when he took the 'natural' character of the Earth's motion as his point of departure. Galileo's parrying of the assertions about the ruinous consequences of the Earth's motion is directed precisely at showing that this objection would only hold in relation to a *motus violentus* [violent motion]—that is, only if (expressed in Christian terms) the Earth's motion had not been set going immediately at the time of the Creation. The introduction of the concept of creation simplifies the argument once again, since everything that is posited with this divine act is *eo ipso* 'natural,' however 'violent' this act of God's may happen to be represented, linguistically, as being.

Without its being explicitly mentioned, Galileo's argument is also directed against the anti-Copernican objections that appeal, with Luther and others, to the standstill of the Sun in the book of Joshua. If Copernicus had only asserted something that was supposed to have proceeded naturally and regularly since the beginning of the world, this seems like the lesser demand on the human power of comprehension, in comparison to the violent deed of cosmic dimensions that had taken place when the Sun stood still over the battlefield of Gibeon.

This biblical passage, which was to achieve a certain melancholy fame in the course of the Inquisition's proceedings against Copernicanism, had never had any prominent importance before Luther referred to it against the fool, Copernicus.[b] One will have to understand the slight pleasure taken by the exegetes in this great miracle as a result not least of all of the fact that this sort of divine display did not fit the alliance of biblical theology with antiquity's cosmos metaphysics very well.

After the disappointment of the earliest eschatological expectations, which were able to repress all secular conformism, the proclamation of the Christian gospel turned out to have an elementary interest in ridding itself of any suspicion of hostility toward the cosmos and its security. For a long time Christianity remained sensitive to the reproach that by its original eschatology it epitomized unreliability and irresponsibility in relation to the continued existence of the world, and that by its belief in miracles it was permeated by an inclination toward the catastrophic disordering of natural courses of events. And it is in fact in such a context—however surprising it may sound—that the formula of the standstill of the heavens and the motion of the Earth is used for the first time.

Without presenting a cosmological assertion, as such, Tertullian (around the turn from the second to the third century A.D.) grimly formulates the animosity of the persecutors of Christianity as their supposed right of self-defense against the Christian promoters of chaos who are intent on the end or at least the confusion of the world: "... si caelum stetit, si terra movit, si fames, si lues, statim: Christianos ad leonem acclamatur!" [... if the heavens have stood still, if the Earth has moved, if there is famine or plague, the cry is immediately: To the lion with the Christians!][15] The alliance with a metaphysics that seemed to ensure the reliability of the world to such an extent that it made the Bible's catalogue of signs of the coming of the apocalypse into an enumeration of perceptual illusions more than of real possibilities was finally able to reverse the direction of confrontation in such a way that Copernicus and his followers are pushed, or are meant to be pushed, into the role of the early Christians who were suspected of wanting the world's downfall. The metaphors of the 'perpetrator' continue to imply the utmost irresponsibility with regard to the world's stability, for Copernicus's mark is none other than the one coined by Tertullian: that the heavens stand still and the Earth moves.

Seen in the perspective of the turning away from the early eschatological expectations and of the rationalization of their disappointment, the Joshua text becomes instructive even in the scanty instances where it is given exegetical attention prior to the Copernican turning. Once the command to the Sun to stand still is drawn into the arguments about Copernicanism, its interpreters differ above all in their decisions as to whether they will regard the Sun's standstill as real or as only apparent. When the test is economy in the attainment of the goal to be achieved in the biblical scene, mere appearance is regarded as sufficient. The pre-Copernican interest in the Old Testament text sees it from a different angle: It weighs the display of the biblical God's power over the immutability of antiquity's celestial spheres against the distinction conferred on Joshua, as the sole speaker in the battle scene, by having the divine ability delegated to him to exercise power by the word alone. Of course this choice between alternatives requires a resolute exegetical realism.

Neither the demonstration of divine power nor the demonstration of its having been conferred on man had to assume this reality and thus require literal faith if the production of a deception of the senses that was sufficient for the purpose (and thus for reason) could reduce intervention that was contrary to order to a superfluous exercise of force. With this consideration, which opposes voluntarism, Kepler tried to remove the text's awkwardness for the Copernicans. Joshua himself, he suggests, would have admitted that "the important thing to him was only that the day should be lengthened for him, irrespective of how this was accomplished .... Of course it seemed to him as though the Sun stood still, since the object of Joshua's prayer was that it should seem so to him, regardless of what actually happened."[16] How, though, would Kepler avoid docetism in relation to Christianity's central phenomenon, if in the peripheral zones of miraculous events he was ready to sacrifice realism, and with it literalism, in the interest of weakening the opposition to Copernicus?

Christianity's burden of proving that it did not, at least, *promote* the end of the world disappeared to the extent that the cosmos metaphysics that it received from antiquity itself appeared as a constituent part of the Christian system. Then theology's interest in the demonstrations of divine power against the background of the consolidated cosmos could return without embarrassment, without emphasizing, again, the world's constitutive decline toward its speedy end. What the pre-Copernican exegesis of the Joshua episode had made manifest was not the preference given to the geocentric cosmology but the way nature was at the disposition of the divine purpose of salvation, which suspended the apparent necessity of what indeed *could* have an end, but above all *had to* have a beginning. So interpretation of the passage stressed the disturbance of the cosmic lawfulness that was possible at any time: "Quod si ita est, diaphoniam passa est caelestis illa harmoniae suavitas .... " [But if it is so, then that sweetness of the celestial harmony suffered discord.][17] It is not the ordinary course of nature but its extraordinary interruptions that become, for the believer, the auguries of his salvation, just as for Joshua there had been no contradiction between his interest in his army's victory in battle and the intervention in the motion of the heavenly bodies for which he asked God.

Nature's serviceability to man relates to the salvation story rather than to physical teleology. It was not adapted to serving as a principle from which an expectation of theoretical understanding and reliability could be derived. If a thinker like Nicholas of Cusa could find a positive clue in this text, it was a clue to a Christian form of rhetoric. In a sermon, *Debitores sumus*, extracted from the collection made by Aldobrandinus de Tuscanella, the Cusan links to Joshua an idea about the power of the word, an idea that opposes philosophy's long tradition of denouncing rhetoric. In view of a God who exercises His power of creation through the word, and His will to salvation again through the word, the 'art' of mastery through words cannot be illegitimate. True, the celestial bodies are still moved by the old powers of the spheres' 'intelligences' (which have become angels), but speech proves to be more powerful. It is able to stand up to those secondary causes and bring the motions that they produce to a standstill: "Est enim oratio omnibus creaturis potentior: nam Angeli seu Intelligentiae movent orbem, Solem, stellas, sed oratio potentior, quia impedit motum; sicut oratio Josue fecit sistere Solem." [For speech is more powerful than any created thing; for instance, angels or intelligences move the spheres, the Sun, the stars, but speech, which checks their motion, is more powerful, as when Joshua's speech makes the Sun stand still.][18] The Cusan's metaphysics cannot refuse to man what it ascribes to God as His self-expression, if it is to stick to its intention of linking theology, cosmology, and anthropology. The world is not a contingent 'given' if it is supposed to have proceeded from the word, and nothing arbitrary or violent can be done to it through man's word. In this sense the action of the word, which is superior to all creatures, is still not a 'deed.' If the Cusan nevertheless speaks here in a language that presses toward man's being a 'perpetrator,' through the word, over against nature, then this is an autonomous tendency, one that evades his intention of preserving the medieval. It approaches the fundamental conception of Francis Bacon, who will present human domination over nature through the word as the essence of the Paradise that was lost and is to be regained. But Bacon will draw from the spirit of magic what the Cusan thought he could arrive at from the spirit of theology.

The fact that the Joshua episode contains a cosmological dogmatics could only emerge when the interchangeability of the real predicates of motion and rest between the heaven of the fixed stars and the Earth was at least considered, speculatively. Strict Scholastic Aristotelianism, which regarded the motion of the first sphere as the effect of the divine *causa finalis* [final cause] and thus as the principle of the world's preservation and its processes, could not allow even the thought experiment of the heavens' standing still to be accepted as meaningful. Nominalism, for the first time, is cautious about asserting 'necessities' in the relation between God and the world. If Nicole Oresme wants to play through his cosmological variations, he has to shield himself against the Aristotelian assumption that the motion of the heavens follows necessarily from God's existence. Here he makes use of the figure of thought, which was central for the Middle Ages, of the *gloria dei* [glory of God]: True, God is, as in Aristotle, the cause of the world's motion, as its *finis* [end, goal], but that makes Him the *causa finalis* [final cause] primarily, and necessarily, only for Himself. Only secondarily and only by His will does He become this also for the world: "... it does not follow that, if God is, the heavens are; consequently, it does not follow that the heavens move. For, in truth, all these things depend freely upon the will of God without any necessity...." [19] The ordinary course of the heavenly bodies is determined by the same volition as is its infringement; miracles are, by their nature, only 'phenomenal.' Consequently the Joshua episode can become, for Oresme, the license for precisely what was to be prohibited, on the strength of that episode, to Copernicanism. And with some reason, it must be said. Copernicus himself, by his rationalization of anthropocentric teleology, refused precisely the freedoms that medieval voluntarism had appropriated to itself after 1277.

Toward a typology of post-Copernican uses of the Joshua text, I would like to offer three characteristic examples. First of all Hermann Samuel Reimarus, with his *Apologie oder Schutzschrift für die vernünftigen Verehrer Gottes* [*Apology for the Rational Worshippers of God*], which after more than two centuries has finally been published in full. The scandal that was bound to be created, and was created, for the Enlightener by the fact that Joshua "in the pursuit of his victory told the Sun and the Moon to stand still," and still more that "God actually granted his wish," becomes evident in its full severity. [20] Miracles of this kind, which violate the essential tenor of nature as the completed and self-maintaining work of the Creation, are (he says) no longer even relished by the theologians. "Then gradually the theologians themselves begin to be ashamed of such monstrosities, which throw the whole of nature into disorder. No doubt they comprehend that the interruption of the motion of the great heavenly bodies has to mean more than if someone wanted to make the pendulum of his clock stand still for a day, and that if this *perpetuum mobile* [perpetual motion] stood still for even a moment it would cost the life of everything that breathes." The reproach that makes the perpetrator of the Sun's standstill into a monster has its rationality in the fact that this offense is no longer one of violating the sublimity of a sanctioned world-order so much as it is one that would have to endanger the sensitive mechanism of self-preservation that life represents.

The legend of the Sun's standstill gives the philologist Reimarus the occasion to distinguish sharply between its source, in an oriental work of poetic fiction, and its misuse. For the Old Testament author's having "taken the matter seriously and literally," there is no excuse. Such an excuse would always be available for the poet who described the following "natural experience": "If the Israelites at the battle of

Gibeon had, from the rays of the declining Sun and the rising Moon, an evening and a night that were as clear and luminous as though the day continued beyond its time—then it was poetic to enliven the event as though Joshua had commanded the Sun and Moon to stand still and to remain above the horizon for another whole day." The historiographer, "out of a love for the miraculous," took literally what was not too audacious for the poet, and thus "made the poem into a history that we, as rational people, must liberate again from its theatrical dress."

Herder, too, distinguishes the poetic source from the momentous piece of supposed historiography. But with greater partiality for the poetic, even in its displacement, he sees the inexcusable abuse only in those who made the text from a remote time into an instrument of the Inquisition. In his essay "Über die verschiedene Schätzung der Wissenschaften nach Zeiten und Nationen" ["On the Different Valuations of the Sciences in Different Periods and Nations"], Herder describes the effect of the legend of the Sun's standing still at Gibeon as one of the dangers to the progress of astronomy that, by good fortune, have been surmounted: "The Sun's standstill in the Book of Joshua would almost have imposed a standstill on genuine astronomy if Galileo and Kepler had not remained faithful, in spite of all persecutions, to Copernicus . . . ." [21] Herder concludes his plea against any plan to assign a goal to "science's spirit of inquiry" by means of a dogma with the rhetorical question, "That you apprehend (and interpret) unpoetically an enthusiastic exclamation of Joshua's, of which a heroic song sang—is the world system supposed to accommodate itself to this obtuseness?"

Voltaire gave a witty culmination to the history of the effects of the Sun's standstill at Gibeon. In one of his invented anecdotes under the title "Jusqu'à quel point on doit tromper le peuple" ["Up to What Point Is One Permitted to Deceive the People?"] he has a Dominican and an English philosopher meet in the street in Rome. The monk abuses the philosopher for teaching that the Earth moves and forgetting, in doing so, that after all in Joshua the Sun was brought to a standstill. The philosopher answers that it was just that that he merely endorsed. For precisely since that moment the Sun has in fact stood still. Reconciled, they fall into each other's arms. Since that time even in Italy people dare to believe that the Earth moves. [22]

It had been Luther who first invoked Joshua as an authority

against Copernicus. In his *Table Talk* we find on 4 June 1539—no doubt occasioned by the same reports that had decided Joachim Rheticus, a month earlier, to set out for Frauenburg—this laconic objection: *nam Josua iussit solem stare, non terram* [because Joshua commanded the Sun to stand still, not the Earth].[23] The position of one who carries out cosmic changes not only was stamped in advance by Joshua but was also occupied by him. The only way in which such a thing could be legitimated was firmly established. Laying claim to the language of the command to the Sun to stand still, as being descriptive of a theorist's action, seemed to be something reserved for the rhetoric of condemnation. Nevertheless, however surprising it may sound, Copernicus himself prepared the way for these metaphors. In doing so, he departed from his cautious efforts to soften the forcible demands that he was making on his contemporaries and colleagues and to wrap those demands in a protective raiment of humanistic citations and allusions.

Only in a late passage of his chief work, where he makes the reader look back at the most important part of his presentation, does Copernicus choose the language of metaphorical action. The actor is, admittedly, not himself, but the Earth. Although it had in fact only become the complex that conditions *appearances*, it is addressed as a factor that generates *realities*, as a subject in a network of forces and a relation of power—in other words, precisely in the language in which Copernicus was to become the demonized actor who changes the world. At the beginning of the sixth book of the *De revolutionibus* he speaks of the Earth as, by its motion *(revolutio terrae)*, not only exercising force and efficacy *(vim effectumque)* on the apparent longitudinal motion of the planets and forcing their regularity upon them ("in quem ea omnia cogat ordinem") but imposing its command and its laws on the planets' paths ("exercet imperia legesque praescripsit illis") also in respect to their motion in latitude.

One must not overlook the fact that Copernicus could not use such a language in any but a metaphorical way. Ptolemy's epistemological license retains its linguistic equivalent at least where the talk is supposed to be of cause and effect in the connection between astronomical objects. The first conjectures about an active connection in reality between heavenly bodies were advanced by Kepler. So when Copernicus speaks of the Earth's force, efficacy, commands, and legislation in regard to the phenomena in the heavens, only the choice of metaphors is instructive, not the metaphorical manner of speaking itself. One could call this a classic case of "absolute metaphor." c The speaker is simply not free to mean what he says realistically. The choice of metaphors of action violates above all Aristotelianism's dogma that the Earth consists of the most sluggish element and consequently, as a body in the universe, can only be motionless and passive, that is, without any action upon other bodies in the universe. When Copernicus metaphorically makes the Earth into the agent in an action, this is only in accordance with the theoretical change by which the Earth has become a star. The function of these metaphors has established itself in an intermediate realm; it makes it possible to experience something that is not sufficiently substantiated in the domain of objective experience: that the Earth has become a cosmic agent.

This language neutralizes the change in the Earth's position from the center of the universe to an eccentric location, by replacing the criterion of topography by the criterion of efficacy, of action in a story. If Copernicus had been intent on justifying his reform in the manner familiar to the Renaissance, by providing the Sun with the specially distinguished position in the center, then in view of his ignorance of the importance of the greatest mass in the system this proves to be a misjudgment of the criteria by which 'importance' was to be assigned in the future. Once again Kepler enables us to see how little the position in the center, as such, still meant, if it could not be equipped with the central function of the power that causes motion. As long as one did not have the concept of action at a distance, the criterion was satisfied by the complexity of the proper motions of a heavenly body, because the multiplicity of the astronomical phenomena, and thus 'importance' in the theory of those phenomena, went with it. The Earth had not only become an object for astronomy but had immediately become its most important object.

Petrus Ramus expressed Copernicus's accomplishment in a single sentence: "Astrologiam non ex Astrorum sed ex Terrae motu demonstravit" [He demonstrated astrology (i.e., astronomy) not from the stars but from the motion of the Earth.] At the same time, in the place where he promises his professorial chair to the person who satisfies the requirement of an astronomy without hypotheses, he says that this would have had to be an easy matter for Copernicus, after he had accomplished the gigantic labor of putting the Earth in motion.[24]

Whatever it may have been that the Paris professor of rhetoric understood by an "astronomy without hypotheses," it is evident from the formulation of his spectacular offer that he meant the metaphor of the theorist's gigantic labor, in regard to Copernicus, to be understood above all as describing the breaching of the epistemological barrier. Evidently for him Osiander's anonymous preface, with its proviso of the hypothetical nature of Copernicus's results, stood in the way of the perfection of astronomy. He believed, on the strength of it, that Copernicus—a perpetrator who lacked the utmost resolution—had not dared to take the step of laying claim to truth. The metaphor of the perpetrator is in the service of the break with astronomy's epistemological timidity—of promoting the astronomical realism of which Copernicus had supposedly fallen short and that still had to be demanded. The realistic use of the metaphor according to which Copernicus put the Earth in motion proves to be only a source of rhetorical support for the possibility of proceeding to unambiguous knowledge.

In the field of Copernicus's early influence, Petrus Ramus is one of the few who do not share the eschatological basic mood of the century. When he has Copernicus the 'perpetrator' still miss his goal by an amount reflecting the absence of a thoroughly consistent effort, he nevertheless takes Copernicus as the point of departure for his anticipating and soliciting the execution of what remains to be done. Almost at the same time, it is possible to perceive in the figure of the astronomical perpetrator what is already the lost grandeur of the past. This is the characteristic mood of the poem by Nikodemus Frischlin that takes as its subject the astronomical clock at the Strasbourg cathedral and the portrait of Copernicus, situated on the housing of the clock weights, which was painted—on the basis of a self-portrait that has disappeared—between 1571 and 1574 by Tobias Stimmer.[25]

The Tübingen Humanist's *Carmen* [*Song*] has a melancholy and culture-critical disposition. The astronomical clock appears to him as the 'Gesamt-Kunstwerk' ['total work of art' (R. Wagner)] of a late age, which once again pulled together, and displays, the powers and the potential in the world's old age.[26] The portrait of Copernicus shows him in his youth, with a lily of the valley in his hand as the symbol of the medical art. Not accidentally, the astronomer's portrait is mounted just where the clock's mechanism moves the circles of the heavens and the signs of the time. It may have been the preexisting symbolism of this spatial proximity between the portrait and the mechanism that decided the poet to find in the figure of Copernicus also the element of the power of causing motion, of which the restrained portrait itself wants to show so little. In the portrait that was copied here, the canon of Frauenburg wanted to draw himself as a medical helper of men; the Humanist poet, on the other hand, wants him to be imagined as the researcher and teacher, as he appeared at the moment when he commanded the stars and the heavens to stand still, represented the Earth as rotating, and placed the Sun in the center of the universe.[27] The poet, wearied by the spirit of the age, claims to know and wants us to know that the present no longer has the stature for such things.

A half a century, at the latest, after Nikodemus Frischlin's *Carmen*, the repertory of Copernicus the 'perpetrator' is gathered together in full mythical superrelief in Simon Starowolski's biography of him, which was published in 1627.[28] Jupiter sees how man, through Copernicus's spirit, creates the world for himself, halts the heavens, and puts the Earth in motion; and in watching this, Jupiter uneasily remembers the battle of the giants, and fears that such a god could have survived on Earth. More important than the employment of the mythical scenery is the formula that Copernicus did this *contra naturae iura* [against the law of nature]. This formula, in the wording *naturae iura mutare* [to change the law of nature] or *contra naturam facere* [to act against the law of nature], reaches back as far as early Scholasticism. In Peter Damian it is related to God himself, the Creator, for Whom the "philosophical" concept of nature, deriving from antiquity, is supposed to be proclaimed as "no longer" binding.[29] When Copernicus is elevated to the role of 'perpetrator' *contra naturae iura*, the possibility of reoccupying the position for whose occupant the concept of nature is not binding is announced.

The restraint that Copernicus himself had practiced in regard to the stirring gesture of the perpetrator—so as to accent still more sharply the enhanced status of the Earth—can be made clear from still another angle if one brings in the parallel that is provided by Celio Calcagnini's treatise on the motion of the Earth. This work was dated by Franz Hipler—on the admittedly untenable assumption that it was influenced by reports of Copernicus's early essay—to the period between 1518 and 1524.[30] If we disregard anxiety about protecting Copernicus's priority, the work could also have been written before 1518, or later, between 1524 and 1541, the year of Calcagnini's death. The public was first apprised of it in 1544, when it was printed in Basel. This posthumous printing could indeed have been a result of the interest in similar manuscripts that had been awakened by Copernicus.

The life paths of the later canons of Frauenburg and Ferrara had twice come very close to one another. After his studies in Bologna and Padua, Copernicus had gained the degree of a doctor of canon law in 1503 in Ferrara. He could very well have known Calcagnini, who had been born there in 1479 (so that he was only a few years younger), and could have acquainted him with the Hicetas quotation that he noted in his Pliny. If that could be verified, it would at least inform us regarding the point that Calcagnini was not to go beyond. But nothing in the contents of the little treatise on the motion of the Earth allows us to infer that Calcagnini could have turned to advantage the much later accident that in 1518 he came to Cracow, as an attendant of his duke, in order to take part in the festivities at the wedding of King Sigismund of Poland with a princess of the house of Sforza. If Calcagnini had received information, during this visit, of the way in which Copernicus's ideas had matured in the meantime, he would not in any case have benefited from it in any discernible way.

The sole motion of the Earth of which Calcagnini speaks is the diurnal motion, as an alternative explanation of the rotation of the heaven of the fixed stars. The system of the planets, on the other hand, is not discussed at all; indeed it is not even mentioned. Nor will one be able to say that Calcagnini went just halfway—in which case the imagined encounter of the young men in Ferrara could just as well have inspired Copernicus as Calcagnini. The solution of the problems of the system of the planets is the key to a cosmological reform of which the first partial accomplishment must, by its genetic logic, be the Earth's annual motion. If one has not found an access to this solution, the assumption of the Earth's diurnal motion is a dead end. Calcagnini would not only have made poor use of any information he might have gained, during his stay in Cracow, from the *Commentariolus*—he would simply not have understood it if he had appropriated only the element of the diurnal motion.

Celio Calcagnini's treatise on the motion of the Earth has to be seen in the context of a process of disintegration that made theoretical assertions frequently take the form of mere paradoxes. That is still the case with Giordano Bruno at the end of the century. When one mentions this name, one must be struck by the ease with which Calcagnini expounds his paradox, an ease that is an index of the *leggierezza* [lightness] with which cosmological speculations could still be expounded at the beginning of the century. The demonization of Copernicus—and a fate like that of Giordano Bruno—already presuppose the altered conditions of institutionalized irritability that are associated with the advancing Counterreformation.

Calcagnini employs the formula of the theorist as perpetrator with the most charming ingenuousness. It only proves how little he has to shun the stirring gesture of the cosmological overthrow. In his dedicatory preface Calcagnini compares the effect of his thesis of the Earth's rotation to the accomplishments of the great artificers of antiquity, among these especially Archimedes, who were able to move very great weights with small mechanisms: "...iure admirati sunt, quod parvis machinis ingentes moles agitarent" [they are justly admired for moving huge masses by little devices].[31] The highest admiration (*summa admiratio*), which, according to the author's explicit declaration, the little work deserves, is supposed to refer to the act of force with which it brings the irresistible course of the Sun and the six planets, and especially the incredible impetus of the rotation of the sphere of the fixed stars, to a halt ("incredibilem impetum sistit"), while on the other hand it abolishes the inertia of the Earth and incites it to the most rapid motion: "...ita impellit atque urgit, ut citatissimo cursu deferatur."

More amazing than *what* the little book (which, as the agent of the action, deputizes for the author) changes in the world-order is *how* this change is pictured. This action is carried out not with levers and screws, as the mention of Archimedes, who had promised to change the world if he were given one fixed point, seemed to announce, but only by the aid of the magical efficacy of the word: "Solo magicae orationis beneficio tantum nobis miraculum constat." [By the aid of magical speech alone, we achieve such a great marvel.] The addressee of the dedication could, its author writes, describe that, without fear of contradiction, as a paradox.

The execution of Calcagnini's deed of theoretical violence does not measure up to this preparatory rhetoric of the cosmic perpetrator. He introduces it with the skeptical observation that we cannot trust our eyes, which means initially only that the appearance of the heavens' rotation does not speak against the motion of the Earth. By assuming a hiatus between the senses and reason, between appearances and valid reality, Calcagnini at first presents himself as a Platonist. However, the heart of our ignorance is not, he says, that we do not know that which truly exists, but rather that we do not perceive our own human situation in the prodigious rotation of the heavenly body on which we live: "Tanta in mole constituti rapimur atque agimur ... nostrae humanae conditionis ignari." The placement of the accent shows where the author's interest lies and what cannot be expected from him. Man's ignorance about himself is more surprising to him than what could take place with the ground under his feet.

Platonism, for Calcagnini, means above all to make a metaphysically grounded decision as to how motion and rest are to be apportioned, in the hierarchical system of cosmic objects, as real predicates. What is supposed to be eternal, immortal, and immutable cannot be endowed with the most rapid of all motions, while the lowest sphere, the sphere of instability and mutability, is conceived as being at a standstill. If it should not be possible to adduce proofs (which was the situation sanctioned by the epistemological tradition of astronomy), then, in case of doubt, the motion must be awarded to the Earth: "Hoc vero multo iustius ac rationabilius terrae quam coelo irrogaveris." [You will in truth have imposed this much more properly and rationally on the Earth than on the heavens.]

The form of the terrestrial body at least offers no argument against assuming that it moves. Here Calcagnini is momentarily in accord with Copernicus's argumentation for the *mobilitas terrae* [mobility of the Earth]. Calcagnini, too, speaks of the Earth's *absoluta rotunditas* [absolute roundness] and extracts from it a principle guaranteeing the persistence of its revolving—precisely not, however, for the stationary rotation that he asserts: The perfect ball that rolls on a flat, smooth surface pauses in its course only when it encounters an obstacle or an unevenness.[32]

Not only are Calcagnini's and Copernicus's arguments from the Earth's globular shape and its equivalence to the spherical shape of the heaven of the fixed stars not sufficiently improbable, in their specific forms, to allow us to assume a dependence of one on the other —they are also employed too differently in their argumentative functions. Copernicus only shields his further considerations on the correction of the planetary system by introducing the idea, from natural philosophy, of the connection between form and motion, whereas when Calcagnini presents it he has already reached the center of his demonstration. Everything beyond this point is a search for a corroborating investment with meaning. The fact that the Earth turns around its axis is an expression of its heliotropism, which it shares with the whole of organic nature. Its turning toward the sunlight, by rotating, is only a special case of the most universal conservation drive of all things, "ut se suaque tueantur" [that they should preserve themselves and theirs]. So Archimedes's fixed point and lever are not really needed in order to make us (in Calcagnini's fine image) become our own antipodes every day ("nos ipsi nobis quotidie efficiamur antipodes").

The organic explanation, by means of heliotropism, is only one aspect of the Earth's motion; the other one is mechanical. The Earth does not change its position, which not only represents the center of the universe but also, according to the Aristotelian schema, the lowest position and the one subject to the most pressure. Calcagnini uses both views in one single formula: "Tellus in medio mundi constituta, loco scilicet omnium infimo ac pressissimo...." [The Earth is fixed in the center of the universe, that is, in the lowest and most pressed upon place of all.] He derives the Earth's stationariness *and* its heliotropic mobility from the premise that, on the one hand, the center of the world is the point at which all material heaviness is directed, while, on the other hand, the Earth's matter cannot reach this point by means of its heaviness, but can only revolve around it. Just as in the relation between the first sphere and the unmoved mover in Aristotle's *Metaphysics*, earlier, here too the circular motion arises precisely from the fact that the absolute goal of accomplished rest cannot be arrived at. For the Earth's matter, the center of the world is just as transcendent as the immobility of the unmoved mover had been for the first sphere. The basic pattern of the argumentation corresponds to the *tonos* [tension, force] of Stoic cosmology, although the latter had not derived from the centripetal cycle of fire any circular motion of the terrestrial body, which represented only the stationary framework for the cycles of the other elements.

Just as Calcagnini underpins the Earth's axial rotation with a Stoicizing argumentation, he uses an Aristotelianizing reflection to strengthen still further the argument for the heavens' being at rest. It is precisely the fifth substance, of which Aristotle had said the heavenly bodies were composed and by virtue of which he was able to ascribe to them freedom from fatigue and resistance, that requires absolute immobility, Calcagnini says, on account of its simplicity and purity and its lack of any weight. This train of thought shows that cosmic motion arises, for Calcagnini, only from weight operating toward the center of the world, and that he lacks the idea of the metaphysical superiority of circular motion as the, in the strict sense, "natural" motion. As is so often the case, here too the mention of Plato on behalf of a theory that diverges from the Scholastic tradition is a self-deception.[33]

Calcagnini's universe remains geocentric if only because he sees no other possibility of saving it as anthropocentric—and it is so by his explicit statement. The Earth, he says, is placed in the center of the world so that by its motion it can stimulate the other elements to fertility and thus preserve the living creatures that the immortal God created to contemplate the divine objects.[34] If one thinks of the teleological world-formula in Copernicus's preface, it lacks precisely the connection that is set up here between man's being intended for contemplation and the Earth's being placed at the center. It is a connection, once again, that the Stoics had established. Calcagnini was not able to carry out the program, which he regarded as Platonic, of relating cosmic anthropocentrism only to reason and not to sensual 'intuition.' It is true that he intensifies his assessment of the certainty of his argumentation up to the formulation—which he admits he would only venture when he had been provoked and his spleen aroused—that hardly anything more absurd had been invented by the philosophers than the proposition that the heavens move.[35] But this strong language is only justified by the long-familiar reflection about the dimensions of the eighth sphere and the resulting incredible speed of its rotation. The circumference of the sphere is "immense," and to determine it exceeds mortals' powers of comprehension.

Toward the end of his treatise Calcagnini takes account of the objection that the authorities and the great names are on the side opposed to his. His opinion too has its gods, he says, varying Heraclitus with a play on words: *numina* [gods], here, in place of *nomina* [names]. Here he takes up once again the figure of Archimedes, from the preface, and gives it an original twist. Could the Syracusan have made his famous remark that he would remove the Earth from its place with his mechanical arts (*architectonica arte*) if only he were given a fixed point on which he could station himself—could he have said this if he had not regarded the Earth as a movable body ("nisi terram mobilem existimasset")?[36]

It is remarkable that Calcagnini lays claim to Archimedes's remark on behalf of an idea of the motion of the Earth by which a change in its location in the cosmos was supposed to be *excluded*, and the fixity of its position in the center of the world was supposed to be ensured, in spite of the diurnal rotation. One would think that the figure of Archimedes would have to be more readily related to Copernicus's "deed."

That this had indeed occurred is something that, at any rate, we find already presupposed by John Wilkins, who deals with the appeal to Archimedes's remark in his *The Discovery of A World in the Moon*, which appeared in a German translation, under the title *Vertheidigter Copernicus [Copernicus Defended]*, in 1713. This, no doubt, most widely read Copernican of the turn of the century writes, "To this purpose likewise is that Inference of Lansbergius, who from Archimedes his saying, that he could move the Earth, if he knew where to stand and fasten his Instrument; concludes, that the Earth is easily movable: whereas it was the intent of Archimedes, in that Speech, to shew the infinite power of Engines; there being no Weight so great but that an Instrument might be invented to move it."[37]

What could not be used as an argument remained as a rhetorical figure to use in remembering the one who, in two centuries, had changed more than the planetary system. By the anniversary of Copernicus's death, in 1743, the metaphor of theorist as perpetrator was so familiar that in his memorial address, Johann Christoph Gottsched was bound to make the most of it. "The news went out that the keen-eyed Copernicus had found, as it were, on the tower of his cathedral, where he was accustomed to observing the heavens, the fixed point outside the Earth that Archimedes had wanted—so as, from that position, to move the whole terrestrial globe from its place by means of his lever mechanism. With his daring hand he had shattered the crystal spheres of the heavens, so as to assign to all the planets a free path through the thin celestial atmosphere. He had released the Sun from the paths that it had traveled for so many thousands of years, and had as it were anchored it and brought it to a standstill. In short, he had transformed the terrestrial globe, also, into a runaway top, which was supposed to whirl once annually around the Sun, in the midst of the planets' courses. The entire learned

Occident heard with horror of a cathedral canon who had made men's hitherto secure and firmly established residence insecure and unsteady."[38]

Finally Lichtenberg, the most thoroughgoing of the eighteenth-century Copernicans, gave the rhetorical figure of the Archimedean deed a very individual, unsurpassed turning. Raising the question—which, in view of the impossibility of seeing to the end of scientific progress, has obtruded itself from time to time, and will obtrude itself in the future—as to whether we do not already have enough knowledge of the universe, Lichtenberg communicates some of the latest information about the Sun. He contradicts the remonstrances of theoretical resignation, arguing that we cannot determine what is enough for the human spirit, because, for all the finite determinateness of its capacities, the extent of the objects that it can comprehend remains undeterminable. The disparity between man and the world is the fundamental post-Copernican experience whose continual intensification constitutes one of the main lines of the history of consciousness in the modern age. Lichtenberg points out that the oppressive disproportion is the result of a one-sided way of regarding the question. One must insist, he says, that man has been able continually to expand his sensorium by providing himself with, in the place of his body, a cosmic reference system of his experiences. "Man's body is a point, in comparison to the Earth, just as the Earth is a point in comparison to the planetary system and the planetary system is a point in comparison to the whole structure of the universe. Only that first point, the seemingly circumscribed apparatus of sense organs, is connected to the whole by relationships that, having been spied out and ordered by our spirit, give it a range that has no limits but nature. Thus the spirit, by studying nature, constructs a body for itself, and the Earth, with all the powers of the materials that make it up, becomes its organ; and the powers that previously seemed unimportant become of great importance, only on account of this machinery."[39] What is only a point in relation to the planet on which it exists, and still more in relation to the cosmic system in which this planet, as a particle, also disappears, nevertheless provides its 'punctuality' with the importance of the Archimedean self-consciousness.

This self-consciousness of course immediately involves Lichtenberg in the ambivalence of his view of man. Motion and annihilation turn out to have an unsuspected affinity. The magnification that man was

able to give to his powers and his faculties has its apocalyptic dimension: "He who previously still had to doubt whether he could carry
twice the weight of his body can now say, Give me a place to stand
and I will move the Earth for you or, if you would rather, burst it in
pieces for you." In a footnote Lichtenberg analyzes the boast, as he
formulated it, as a combination of two elements, of which, for him, the
first one stands at the beginning of a historical sequence and the other
one at the end: "The first was said by an ancient Greek as early as two
hundred years before the Christian time reckoning began, and the
second by a Frenchman three years after this same time reckoning
had ceased in his native country. The first speaker probably thought
of mechanics, of levers and gears, while the other perhaps thought of
chemistry and fulminating silver." The figure of the theoretician as
perpetrator need not find its point of application outside the world
that it removes from its state of rest, but can just as well have it inside a
reality to which the theory that creates a disposition to active deeds
refers. The remark that is ascribed to Archimedes had excluded just
this possibility. The only one who could gain power over the Earth
as a whole would be someone who was in a position to approach it
from outside: "Archimedes, the Sicilian, demanded a site outside the
Earth, in order to be able to turn the whole Earth upside down,
because as long as he was in it, he had no power over it." [40]

One can see in the case of Lichtenberg that alongside the pride
of the Copernican 'mover' there had also been growing, since the
middle of the eighteenth century (especially after the earthquake at
Lisbon), an uneasiness. The modern age's first 'perpetrator' in the
realm of theory is joined by a second: the mover by the preserver.
It was a reassurance in the grand manner when Laplace, with
his celestial mechanics, which solved the "many-body" problem,
had provided a theory ensuring the endurance of the solar system.
Beginning in 1799, the five volumes of the *Mécanique céleste* [*Celestial
Mechanics*] appeared. The first volume, with its general theory of the
motions and shapes of the heavenly bodies—in particular, the periodicity of their perturbations of earch other's orbits—already made
Laplace, in a new manner, into a 'perpetrator.' The work that he had
already carried out in the years from 1773 to 1784 on the stability of
the solar system was meant to show that the planets' mutual disturbances had no influence on their mean motions and distances and
that the great orbital axes returned, in thousand-year-long rotations,

to their old positions again. Thus the solar system was integrated into the great rational principle of self-preservation.

Herder, of whom Goethe said that his "existence was a perpetual blowing of bubbles," and whom Kant called a "great artist of illusions," described in his way the intermeshing, in the realm of theory, of the deed of violence and the need for security: "Just as Copernicus overthrew the heavens of the ancients, and Kepler did the same to Copernicus's epicycles, so Newton's simple law drove Descartes's vortices from the empty aether . . . . The whole century adorned itself, in astronomy, with Newton's name, and pursued its calculations according to his law, and at the end of it there arose a second Newton—de la Place—who perfected, by the most profound analysis, what Newton and his successors had left unfinished. He not only harmonized the heavenly bodies' disturbances of one another—their long-term inequalities, and so forth; he also calculated the effects of the universal law of gravitation on all the bodies, both fluid and solid, of our solar system, and thus secured our universe through eons. By showing that the mean motions and the great axes of the planets' orbits are constant, he as it were commanded the universe to endure." [41]

The astronomical reformer's role as 'perpetrator' can be measured by the opposition that he is said to have had to overcome, and also to have aroused. Feuerbach writes that Copernicus was the "first revolutionary of modern times," and that is said on the premise (which is taken for granted) that he "overturned the most universal, the oldest, the most sacred belief of mankind, the belief in the Earth's immobility, and with this blow shook the old world's whole system of faith." The metaphorical cosmic deed of violence in its turn becomes a metaphor for interference with a system of tradition shared by all mankind, as a model for all subsequent interferences of this kind. "As a genuine 'subversive,' he turned the lowest into the highest and the highest into the lowest . . . gave the Earth the initiative of motion and thus opened the door to all the Earth's subsequent and different revolutions . . . . With his audacious hand he burst open the celestial firmament, which before him was closed and impenetrable to even the great intellects of antiquity, with the exception of a few heretical thinkers, and served only as the breastwork of human narrowness, thoughtlessness, and credulity. . . . It is Copernicus who deprived mankind of its heaven." [42]

Chapter 1

The history of metaphors of the 'perpetrator' in the reception of Copernicus is more than anything else the history of a great modern need to see concepts as guarantors of reality—to realize the idea, which was invested in the medieval conception of the Divinity, that thought as such, and without delay, could determine (if not, in fact, produce) the world. Consequently the history of this Copernican metaphor is a history of the most welcome confusions between theory and practice: that theory, if indeed it cannot be practical, might at least be an assurance of the possible effects of thought on action too. Finally, the history of the metaphor is also the history of the disappointing discovery that the ease with which man 'makes' the history of his knowledge of the universe implies little or nothing in connection with the making of history in any other sense—to that extent, the history of science, precisely as the history that is most nearly 'made' by men, is the least specific history of human actions. The Scholastic axiom that "Praxis primo dicitur de actu voluntatis" [It is the act of the will, first of all, that is called action] hardly holds for one who knew and intended as little of his effects (indeed, even of the possibility of his effects) as Copernicus did.

2

Consequences of an Instance of Well-Meaning Misguidance: Osiander

For a witness to the truth, it is intolerable to be praised for the great utility of what he advocates. Disregarding entirely the question of what advantages the Copernican theory could offer in relation to the accuracy of astronomical calculations, the acknowledgment of their superiority in calculation was not only not enough, for Copernicus and the Copernicans, it was precisely the wrong seal of approval, an obstinate avoidance not merely of accepting but even of taking notice of their claim to truth. An instance of this attitude was, ironically, already in Copernicus's hands on his deathbed, in the first copy of the Nuremberg printing of the *De revolutionibus*, of 1543.

Prefixed to the first edition was the preface, which was not contained in the manuscript, "To the Reader on the Hypotheses in This Work." The author remained anonymous; in the copy that Kepler owned, the name is added in handwriting: Andreas Osiander. Rheticus had handed over a copy of the manuscript to Osiander, the principle pastor of the Church of St. Lawrence in Nuremberg and a protagonist of the Reformation in this city, so that Osiander could supervise the printing on the spot. The history of what led up to this mandate does not permit us to assume that Rheticus had no idea of Osiander's attitude to Copernicus's reform. The breach of trust that posterity has blamed Osiander for did not occur so unexpectedly, and was, rather, in its motivation, an attempt to force the happiness of success upon the author and his book by limiting their claims.

Osiander had already written to Copernicus in the summer of 1540, when Rheticus's *Narratio prima* [*First Account*] had just reached the public, to induce him to formulate his innovations in the modality of traditional astronomy. This first letter, and Copernicus's reply to it, have been lost. A second letter from Osiander, dated 20 April 1541, has been preserved, in which he repeats his representations to the effect that Copernicus should exercise caution with regard to the claim to truth that Rheticus had already proclaimed. Hypotheses, he says, are only bases for calculation, not articles of faith, and even if they are false, so long as they exhibit the celestial phenomena exactly it does not matter.[43] Osiander wishes (admittedly in a very weak form) that Copernicus might express himself on this subject in a preface to his work; in that way he would make the Peripatetics and the theologians—whose opposition he did, after all, fear—better disposed toward it.[44] This formulation does not permit the interpretation that Osiander is suggesting that Copernicus should fear opposition to his reform, or that he merely conjectures that he does; instead, an independent remark on Copernicus's part must have preceded this—which would be quite consistent with the way he delayed publication of the *De revolutionibus*. Only one should not read into the term "fear" the fear of the Inquisition and the stake; the failure of the outsider faced with the resistance of the world of the specialists and schools would already be enough, even then. Osiander is full of good will—a fact of which we must not lose sight.

On the same day the Nuremberg chief pastor had also written to the author of the *Narratio prima*, to the same effect. The Peripatetics and theologians would be easy to placate if one left open the possibility of various hypotheses for one and the same apparent motion; that is, if one put oneself under the protection of Nominalistic pluralism: "Fieri posse, ut alius quis alias hypotheses excogitet." [It is possible that someone else may devise different hypotheses.] The Copernican theory would then have to recommend itself by its advantages for computational procedures. If the author would forgo the rigor of his claim to persuade (*severitas vindicandi*), soon everyone would follow him of their own accord, after they had sought in vain for better methods.[45] Osiander is not indifferent to the Copernican cause. But Copernicus and Rheticus, whose answers have not been preserved, were certainly not disposed in favor of his intercession. It is out of the question that their lost replies to Osiander could have offered him any pretext for thinking that his offer of assistance to the reader was authorized. Evidence against such an assumption is provided by the fact that Copernicus ultimately wrote a different preface to his work than the one that Osiander had called for, and that in this very preface he revealed more of what he was claiming, and of the reasons for it, than in everything that he had written previously. The speech of the preface to Paul III comes alive for the first time when Copernicus refers (inexplicitly) to what had at first only been Osiander's suggestion and his offer of assistance, but was to become (unexpectedly and without authorization) his addition to the text. When Copernicus charges his predecessors in astronomy with having been unable to provide any certainty for the system that they advocated, he bases his complaint on his own conviction that he is able to get to the bottom of the most important problem, namely, the question of the true form of the world (*forma mundi*).

It is amazing that Osiander nevertheless dared to put his own text before this authentic preface, when fate gave him the opportunity as a result of Rheticus's departure and mandate. In spite of its papal addressee, he did not dare to meddle with the dedicatory preface; there the interests of the publisher, Petreius, were involved. But the fact that he suppressed the *prooemium* [introduction] to book one, which is so instructive in regard to Copernicus's intentions, with the result that it was also omitted from the editions published in Basel in 1566 and in Amsterdam in 1617, shows that distorting infringements of the text were not outside his moral horizon. This prooemium was printed for the first time in the Warsaw edition of 1854, which was based on the autograph. So it has to be kept apart from the history of Copernicus's influence. Neither a consideration for Copernicus's apprehensions nor his own apprehensions in regard to future opposition can explain Osiander's behavior. It is even less probable that with his emphasis on the hypothetical modality Osiander wanted to commit Copericus to practicing astronomy in a way consistent with its status as an "art." Looked at in the other direction, though, the traditional concept of astronomical hypothesis is paradigmatic, for him, of the distance of worldly forms of cognition from the truth, in contrast to the certainties that are accessible through faith. The antithesis between "bases for calculation" and "articles of faith" is the key to the obstinacy, if not fanaticism, of his behavior. Osiander's Nominalism is not an independent magnitude; it is only a correlate of his theological realism. How important this correspondence was to him was to become evident in the shock that he was to cause for Protestantism, in the form of the "Osiander controversy," shortly after he was called to a professorship in Königsberg in 1547. Here the issue was in fact once again the opposition between calculation and reality, that is, it was Osiander's assertion—against Luther's "imputatio"—of the substantial nature of Pauline 'justification.'ᵃ So Osiander was one who not only thought in the category constituted by this opposition but was prepared to risk a great deal on behalf of it. This connection of ideas makes the astronomical concept of hypothesis appear solely in the function of making the realism of truth an exclusive attribute of the realm of theology.

Now, it is easy to agree about the pedantic element in past dogmatic controversies. Even the title that Copernicus meant to give his work seemed 'heterodox' to Osiander. We know from a letter of Johann Praetorius to Herwart von Hohenburg, written in 1609, that Copernicus wanted the title to read *On the Revolutions of the Bodies in the Universe* (*De revolutionibus orbium mundi*). Osiander intentionally altered this and put *On the Revolutions of the Heavenly Bodies* (*De revolutionibus orbium coelestium*).[46] The philologically overstimulated mind of the modern hermeneut might think itself hypercritical in registering this difference if it were not for the fact that there is clear evidence that people at the time already suspected or knew that the title had been changed, and registered their objection to this, as they did to the anonymous preface, by means of handwritten corrections.[47] Thus in the copy of the first edition that is in the University Library at Uppsala, the *orbium coelestium* part of the title is crossed out in a red ink of the same period. This copy was a present given by Joachim Rheticus to the canon Georg Donner, and we can assume that the giver and the receiver were sufficiently familiar with the circumstances to make this cancellation significant.[48] It is also found in the copy of the Nuremberg first edition that Kepler possessed and that conveyed to him, by the notation of the previous owner, Hieronymus Schreiber, the information that the person concealed behind the anonymity of the preface was Andreas Osiander.[49] The autograph of the *De revolutionibus*, which is now in Cracow, tells us nothing about the question of the title because it lacks all the introductory material —the letter from Cardinal Schönberg and the dedication to Paul III—and begins immediately with the prooemium to the first book. Only in the explicit of the fifth book does it contain a hint of what may have been Copernicus's older, authentic version of the title: "Quintus revolutionum liber finit" ["End of the Fifth Book of the Revolutions"], while none of the other books has a title or an explicit.

The autograph cannot in any case be called upon as unambiguous evidence of what was turned over to Osiander to be printed, because it represents an earlier state of the text.[50]

Is the question about the title important? It may be that it only became important as a result of all of Osiander's infringements, taken collectively. The editors of the critical edition published in 1949 consider the difference between the two versions of the title to be slight (*minimi momenti*), especially (they remark) since Copernicus himself uses, in the text, the form that Osiander inserted, and indeed this was how Osiander hit upon it. Everything indicates that, even if Osiander had made sure that his alteration would be unobtrusive, he acted consciously and skillfully. The significance of the fact that, in Johann Praetorius's judgment, Osiander altered the title *citra mentem autoris* [without regard to the author's intention] depends on our determining that he could know that Copernicus would not regard his alteration as trifling or even as insignificant, and would object to it even if he could recognize its intention as well-meaning and useful in promoting the dissemination and approval of the work. But Osiander acted in an even higher interest than that of the success of the work that was entrusted to him.

The concept of astronomical hypothesis had constituted a realm of reality for which there could be no unambiguous theoretical certainty. The insertion of the expression "heavenly bodies" into the work's title holds to this association of partial access and reduced certainty, whereas Copernicus already included in his title his thesis that the heavenly bodies have no specific character that is radically distinguished from that of the Earth, and that their common feature is their *revolutiones*. If the Earth was taken up into the class of objects studied by astronomy, then this class could no longer be characterized by the uncertainty of what is inaccessible and most remote. It was no longer separated, in its degree of knowledge, from the stringency of physics. The homogeneity of the "bodies in the universe" no longer permitted a boundary on the other side of which lay a region characterized by a constitutional deficiency of truth. But this was just what Osiander wanted to get around by means of the three measures that he took: altering the title, inserting the preface, and deleting the prooemium. Before his eyes, Copernicus destroyed the paradigm of astronomy's traditional distance from the truth. Perhaps Osiander even knew that Copernicus, by the late choice of the expression "bodies in the universe," wanted to obviate Osiander's intervention and make more precise the technical term "heavenly body" that he had used without reflection in the text of the book. Copernicus laid it down that a unified theory of the world, with thoroughgoing rationality, was possible; Osiander tried to inhibit this proclamation in its very title and to banish the work to the province of astronomy, though of course as a fruitful work in that area.

He lived long enough to see that the assimilation and evaluation of Copernicus's work seemed to proceed entirely in accordance with his anonymous preface. In the *Prutenic Tables* of Erasmus Reinhold, Melanchthon's son-in-law, the advantage of the Copernican system was exhibited, for the first time, as a calculational one. On 8 January 1544, from Wittenberg, Reinhold gives a prospectus of his "new tables of the motions of the heavens" to Duke Albert of Prussia. He plans "to follow the observations of Copernicus, as compared to the observations of other older and recent men of learning."[51] On the recommendation of Melanchthon, the duke had given Reinhold a subsidy toward the completion of the tables. Reinhold delivered the printed work to the duke on 14 October 1551, with a letter in which he suggests that the tables should be submitted, for examination, to Andreas Osiander, "who does also have a great repuation, in our age, on account of his erudition in mathematics." The theological controversy in which Osiander, who was now teaching at the newly founded university in Königsberg, had become entangled may have permitted him to take only a cursory interest in the work produced by the son-in-law of his theological opponent, the author of the *Calumnia Osiandri* [*Osiander's Calumny*]; in any case, Duke Albert acknowledged the receipt of the *Tables* on 21 March 1552 with (among other things) the words: "Nor have we omitted to show your work to Andreas Osiander, to whom, after he had looked it over in haste, it was very agreeable ...."[52] If one considers that Reinhold had to work under Melanchthon's supervision, then the first recognized evaluation of Copernicus can appear almost as a successful stratagem of Osiander's.

Thus do experiences of success come about. Osiander will never have seen the letter that Copernicus's friend Tiedemann Giese sent to Joachim Rheticus on 26 July 1543 to acknowledge receipt of the *De Revolutionibus*.[53] This letter contains the beautiful, 'pregnance'-producing legend[b] of the arrival of the printed opus on the day of

Copernicus's death: " . . . in extremo spiritu vidit, eo quo decessit die" [with his last breath he saw (it), on the day he left us]. But it also contains the bitter statement that he could have overcome the pain of the loss of his friend by reading the book that seemed to bring him back to life for him, if the astonishing misrepresentation had not leaped out at him when he first looked into it: " . . . in primo limine sensi malam fidem . . . " [at the very beginning I sensed bad faith]. In the letter accompanying the copies he sent to Giese, Rheticus must have ascribed the infamous act (*impietas*) of alteration to the Nuremberg printer, Petreius, for Giese quotes this as the correct characterization. The agitation of one who was so close to Copernicus is a good indicator of the incompatibility of Osiander's preface with the intention of the author, who was no longer able to defend himself, but whose restrained animation by the truth now seemed to burst out in response to the attempt to deny it.

Who could remain unmoved, Giese writes, in view of such a shameful action (*flagitium*) committed under the cover of fidelity? But he is not sure whether behind the printer there is not someone who, out of envy or of fear for his professional reputation, wanted, by taking advantage of the printer's ingenuousness, to prevent the book's having an effect. Evidently Giese does not know who would be a candidate for this role. Otherwise he would not place his complaint before the senate of Nuremberg against Petreius, who (he says) lent himself to the unknown person's fraud. We do not have the letter in which Giese asks for the necessary measures to be taken to restore the author's credibility (*ad integrandam auctori fidem*). Rheticus is supposed to pursue the matter before the Nuremberg City Council, for the rectification would be in his interest as well as that of the author. But above all he has to compose a new preface (*praefatiuncula*), so that even the copies that have already been distributed can be purged of the falsification (*a calumniae vitio*). Also, a description of Copernicus's life, completed up to the day of his death, and an *opusculum* [short work] by Rheticus about the compatibility of the Earth's motion with the Bible should be added now. We know nothing else about this composition, but its existence, which is presupposed in this letter, does betray the direction of an apprehension to which Copernicus himself had paid too little attention. Finally Giese mentions a possibly important, puzzling fact, not without criticizing the negligence of his deceased friend: Copernicus had made no mention of Rheticus in his preface, and the introductory material that Giese is calling for would make up for that.

This letter is the first testimony to Copernicanism after Copernicus. Nothing is known about the proceedings against the printer, before the Nuremberg City Council, that Giese sought. But it is reasonably certain that Rheticus did not accede to the wish that he should make himself the leading advocate of an emendation of the *editio princeps* [first edition]. We may cautiously infer a certain sobering process in Rheticus, the Copernican, who in the meantime had been appointed a professor in Leipzig; it probably was due chiefly to the fact that in the meantime the anonymous preface's contribution to the reception of the book had become evident. At the same time, a process of change had begun in Rheticus's attitude, by which the finality of the masterwork's claim to truth receded in the face of the astronomical difficulties that it had left unresolved.

With Rheticus, something that is discernible only in traces in Copernicus's work will be brought to an issue: dissatisfaction with regard to Copernicus's original premises. Copernicus had not given his work the definitive summing up that was promised. He could not, at least, have appealed to the simplicity that had been his program in the *Commentariolus*. In the end he had had to give the Earth no less than eight motions. The center of the Earth's orbit was not a fixed point, but traveled in its turn on an eccentric in the neighborhood of the Sun. The Sun, for its part, did not remain in the center of the universe; it is not the center of any of the planets' orbits; indeed the center of Saturn's orbit lies outside even the circle described by Venus's orbit, while the center of Jupiter's orbit lay near Mercury's.[54] Rheticus's hesitation to grow passionate, in keeping with Giese's invitation, on behalf of a correction of the *impietas* that had been perpetrated in Nuremberg, must have to do with the unresolved terrestrial remainder of Copernicus's theory; otherwise Copernicus's only student's further path into "Egyptian astronomy" would hardly be intelligible.

The note that the Bishop of Kulm, Tiedemann Giese, had struck two months after Copernicus's death, in speaking of the unauthorized action of the anonymous person (or, in the foreground, of the printer), does not, indeed, determine the character of Copernicus's immediate impact, but it does characterize the formation of Copernicanism as a view of the world that is inseparably bound up with the question, "What is truth?"

In his dialogue *La Cena de le Ceneri* [*The Ash Wednesday Supper*], Giordano Bruno, who was to give Copernicanism's inspirational side its literary eminence and its enthusiasm for infinity, has Teofilo say that that preface came from an ignorant and presumptuous ass, who made himself, at the entrance to the *De revolutionibus*, into a door-keeper of doubtful fidelity. Without any regard for its professional 'correctness' as a reflection of the "art" of astronomy, or for its success in facilitating the book's reception, Johannes Kepler, who himself possessed one of the first physical interpretations of Copernicus's planetary system and was thus allied with its truth claim in a way that was pregnant with the future, calls Osiander's preface, in 1609, simply a *fabula absurdissima* [most absurd story]. Seven years later Galileo learns that he may—but also must—continue in this story-telling kind of Copernicanism. Almost at the same time, on 13 August 1617, Kepler declares to the Estates of Upper Austria, regarding "Copernicus's so outstanding hypothesis," that he considers it his "duty and task to defend outwardly as well, to readers, with all my intellectual powers, that which I have recognized in my mind as true, and the beauty of which fills me with incredible pleasure when I contemplate it." He speaks in one breath of a "hypothesis" and of the "truth," although in a letter to Mästlin in September 1597 he had already laid claim to Petrus Ramus's prize for an "astronomy without hypotheses," and from this time he repeatedly used this formula for himself.[55]

The important thing for Kepler, as it had been for Copernicus, is the *forma mundi* [form of the world]; but even it is only a beginning from which to "press forward to the true causes of the motions"—the investigation that makes an astronomy without hypotheses possible. For that purpose, for Kepler, heliocentrism as a hypothesis is the lever by which to eliminate hypotheses from astronomy. For his colleagues, that is just as much a violation of the code of the profession as Copernicus's claim had been for Osiander. On 1 October 1616, Mästlin warns Kepler that he should "leave the physical causes entirely alone, and explain astronomical things by astronomical methods only, with the aid of astronomical causes and hypotheses, not physical ones." On the same day Kepler defends himself in a let-ter to David Fabricius: "You blame me for the delays resulting from my investigation of those causes, and you demand tables from which you can make predictions, whatever the situation may be with philosophy ...." It was one of the ironies in which the history of science is not lacking that in order to arrive at his astronomical truth Kepler had to abandon precisely the criteria the final satisfaction of which Copernicus announced and thought, no doubt right to the end, that he had produced: the requirements, combined in the "Platonic" postulate, of circular orbits and uniform motion, which were abandoned one after the other in Kepler's first and second laws.

However far he may still have been here from Newton's insights, Kepler saw that this abandonment did not represent a passage from one fiction to another because it opened the way to a physical interpretation of the 'deviations' from the classical postulate. Looking back on Mästlin's warnings, Kepler writes on 28 February 1624 (to Crüger), "Maestlin used to laugh at my efforts, when I would trace everything back to physical causes, even in the case of the Moon. But in truth it is my delight, and the main consolation and boast of my work, that I succeeded in this ...." If Osiander had written that the philosopher could not be satisfied with astronomy's cognitive resignation, that he would demand more "verisimilitude" (*veri similitudinem*), but would be no more able than the astronomer was to obtain or to convey anything certain if it was not revealed to him by God, then Kepler overstepped just this limit that Osiander had drawn when he made the choice of an astronomical system depend not on its advantages for calculation but on its physical interpretability.

If, in his first and second laws of planetary motion, Kepler proclaimed the realism of his claim to truth by deviating from the classical requirements (which were more than ever valid for Copernicus), he made a kind of amends to Copernicus for this in his third law. It is a law about the *ratio orbium* [system of the spheres] in a tighter and more precise sense than Copernicus could have reestablished. At the same time it is directed against Nominalism and the thesis of the incommensurability of the planets' motions. In formulating this law Kepler forgoes the deformations and nonuniformities of the first two laws. The third law compares the orbital periods of the planets as wholes, that is, without taking into account the differences in their velocities in different sections of their orbits, and sets the squares of their orbital periods in the same proportion as the cubes of the mean distances of their orbits from the Sun. It is true that the expression "the mean distances of their orbits from the Sun" presupposes the statements made in the first and second laws, but it disregards, precisely, their specific content, by permitting us to idealize the distances of the orbits as the radiuses of circles whose [common] center is occupied by the Sun. When it is idealized in this way, the third law achieves what Copernicus had still aspired to in the *Commentariolus* but had failed to do in his chief work. Kepler's (initially) speculative retrogression behind even the late Copernicus, to a strict heliocentrism, is connected to his physical conception of the Sun as the agency that moves the planets—a conception that could not provide Kepler with an explanation of the elliptical shape of their orbits, even if, assuming that shape, it could explain the different velocities in relation to the planets' distances from the Sun. The elliptical shape of the orbit of Mars bore the seal of realism, of the truth that is unaccommodating and is mastered contrary to all expectations of what is "natural."

In Newton's dictum, "Hypotheses non fingo" [I do not devise hypotheses], the idea of an astronomy without hypotheses finds its final expression, but at the same time undergoes a reversal of its content. Whatever is not supposed to be a hypothesis must be deducible from the phenomena: "Quicquid enim ex phaenomenis non deducitur, hypothesis vocanda est." [For whatever is not derived from phenomena ought to be called a hypothesis.] Here the phenomena are already those of the Copernican system—the central body and the planets that move around it in accordance with Kepler's laws—so the hypothesis is no longer a model of the relations of the bodies to one another, but is rather an explanation of those relations, going beyond mathematical description, by means of a cosmic circumstance such as the power of gravitation. Thus the very assumption that for Kepler seemed to justify Copernican realism definitively, namely, the Sun's power, across empty space, to produce the revolutions of the planets, becomes for Newton a mere limit concept, which, on the one hand, is forcibly required by the description of what happens, but, on the other hand, has to do without any predicative characterization.

Copernicus had wanted to demonstrate the truth content of his theory precisely by exhibiting his observance of the classical rule of construction, that one should only employ circles and uniform motions, which were deduced, as requirements, from the "essence" of the heavenly bodies and from the "natural" character, in accordance with that essence, of their motions. With this he returned to the original foundations of that supposedly "Platonic" requirement,^c which had been leveled off since the introduction of the epistemological difference into more of a conventional rule, although Aristotle had still considered it possible to make assertions about the "natural" character of the motions of the heavenly bodies around the center of the world in spite of the transcendent character of the celestial region. There is something correct in the designation of the requirement as "Platonic," insofar as Plato had regarded the circle and the sphere as the patterns of motion and the corporeal forms that were most in conformity with reason (as in the form of the head, as the seat of reason), and at the same time had ascribed them to the heavenly bodies, as the characteristics that are appropriate to their essence. Copernicus quite clearly reaches back to this original connection between truth and essence. Anything that is supposed to be true astronomically has to conform to this essence of the heavenly bodies, because any "violence" (*motus violentus*) is excluded in connection with them. What is crucial is that he includes the Earth in this deduction of characteristics from essences, and that he does so by virtue of its demonstrably spherical form. The difference between his use of the term "hypothesis" and Osiander's is that Osiander denied human reason any knowledge about the essence of heavenly bodies and thus left it nothing but "hypotheses," which were an exclusively "artificial" means of description.

When Newton says that he devises no hypotheses, he does not place himself entirely on either Copernicus's or Osiander's side. At first glance that must seem questionable, because Newton was a Copernican not only in that he took the Copernican system as his basis for calculations but also in his clearly recognizable intention of providing this system with its definitive proof and thus with the confirmation of its truth claim. But his rejection of everything that cannot be deduced from the phenomena unquestionably conflicts with Copernicus's procedure, in his argumentation, of basing the validity of the "Platonic" postulate on the essence of the heavenly bodies and the natural quality of their motions. For Newton, the approximately circular motion of the heavenly bodies has lost its relation to 'reason' and thus its essentially natural character. The only thing that has remained, in the strict sense, "natural" is a body's inertial state, which as a form of motion means its unaccelerated and uniform straight-line motion in absolute space. Any deviation from this is caused by the action of one body on another—that is, in the traditional language, it is "violent"—so that a circular orbit is the combined result of a natural quality and the operation of an external force.

So to devise no hypotheses means, here, to make no inferences from the unknown essence of bodies and from the equally unknown essence of force. What remains as a permissible inference from the phenomena is the establishment of the causal relationship between the masses of the solar system, that is, between their magnitudes and distances from one another. In other words, astronomy no longer has an independent rule of construction. Osiander is wrong because in a physically interpretable many-body system not every kinematically possible construction can be permitted. But Copernicus is not vindicated, vis-à-vis Osiander, because one cannot deduce any more from the gravitational force that has been inferred than was already there as the initial basis from which one inferred it. Thus only the phenomenon of the orbits of comets induces Newton to grant that the gravitational effect of the Sun's mass extends beyond the outermost planet, Saturn. The only meaning, for Newton, of talk of how far gravitation reaches is to make it possible to describe motions that are known to take place. That is not only methodical caution, or even 'positivism,' but also, on the contrary, an almost embarrassed concession to Osiander's standpoint: Newton did not want to and could not exclude the possibility that the law-governed motions in the solar system were not the result of a homogeneous force but instead could be executed by God's guiding power operating directly and in accordance with the phenomenal relationships.

Knowledge of actual reality in the space of the heavens is not denied to human reason, as Osiander thought, but an unequivocal explanation of it certainly is. The Copernican system is the only true one, but it does not contain a complete account of the reasons for its truth. Respect continues to be accorded to a reservation for transcendent majesty. Newton needs this reservation, if for no other reason than that he can only trace the part played by linear inertial motion in the visible circular motions back to a divine intervention at the beginning of the world by which the planets and comets would have been given their initial acceleration.

Nevertheless, Newton did not put astronomy back into its pre-Copernican status as a mere art. The laws that he discerned are indeed de facto, contingent, but they are not fictive. The arrangement of the masses in the solar system that they determine is not a "hypothesis" in Osiander's sense. It is not the assumed force but rather the relationships that result from its spreading through space in inverse proportion to the square of the distance (that is, in the geometry of spherical surfaces) that are the realistic content of Newton's theory—that are, to put it in Copernicus's language, its *ratio orbium* [system of the spheres]. Kant was the first to be able to interpret 'realistically' the force of attraction that Newton had introduced, because he had shielded the concept of the "phenomenon" from any possibility of a transcendent intervention. The result is that force itself becomes, for him, a phenomenon.

In the controversy between Osiander and Copernicus, which only presented itself posthumously, the term "hypothesis" had lost its innocence. In Newton, the pairing of the two terms *hypothesis* and *fingere* already points to Osiander's preface. In order to characterize the astronomical knowledge that he regards as unattainable, Osiander uses the compound *hypothesis vera* [true hypothesis], while what is in fact possible for humans is characterized and given the go-ahead with the verbs *excogitare* [to invent] and *confingere* [to contrive]. Copernicus was supposed to find his place, and to be excused for the novelty of his hypotheses (*de novitate hypotheseon huius operis*), within this space where we can do what we please. It is not the essence of things but the sovereign freedom of divine action that decides what the true relationships and motions of the heavenly bodies will be; only through revelation could we learn anything about this. Fundamentally this is also Newton's conviction, but with him it becomes the background of his refraining from the use of hypotheses.

Osiander had gone so far as to say that in astronomy constructions and spatial relationships were permissible that would have to be called *absurda* if one wanted to assume their existence in reality and as the real foundation of the appearances. As an example of such absurdity there is the explicitly mentioned improbability of the epicycle of Venus. But Newton, too, uses the same term when he labels the assumption of a force that is essentially proper to bodies and that operates across empty space as "absurd." From this we may infer that the talk of a force could only have been intended metaphorically and that he inclined toward the assumption that nature's lawgiver will also be the executor of his laws by means of his continual active operation upon nature. This too—though he could no longer express it this way—would be the kind of "hypothesis" that he refuses to devise. Explicitly, Newton leaves such a consideration to his reader— not, I dare say, without a confident expectation as to which "agent" the reader will decide in favor of.[56]

It was not always Newton's tendency to leave access to the theological background open. One can see this in the fact that in the same *Scholium generale* he allows, without special comment, what he had just beforehand rejected in the case of gravitation, when he explains the cohesion of bodies by means of a "universal spiritual substance," which is also supposed to be responsible for electrical attraction and repulsion, for the propagation of light, and for transmitting organic stimuli. Newton sets up no connection between that *vis gravitatis* [force of gravity] and this *spiritus subtilissimus* [most subtle spirit], although he employs, for the operation of the latter, the terminology of mutual attraction across what is, admittedly, a minimal distance. So he does not do what Kepler, stimulated by Gilbert's treatise on magnetism, had done. Could it have been only the immense magnitude of the empty space across which one had to imagine the force operating in the planetary system—in contrast to the minimal distances of which he talks in connection with the cohesion of bodies—that prevented him from setting up a connection?

Looking at the epistemological side of it, one should not overlook, either, the phrase "not yet," which Newton uses twice and which implies that his not engaging in hypotheses is a temporary state of affairs that is subject to change with the progress of knowledge.[57] So for Newton a hypothesis is precisely not an anticipatory outline of a problem solution, requiring if not confirmation at least refutability. Instead, he follows Osiander in regarding hypotheses as fictions including or at least permitting the inclusion of absurd elements (as, in this case, action at a distance, across empty space), the only difference being that, unlike Osiander, he thinks that hypotheses, and thus the imputation of absurdity that they involve, can be avoided.

One need not assume that Newton's refusal to devise hypotheses was only intended to improve the impression that would be produced by the *Principia*. It was published too late for that, because the *Scholium generale* was not contained in the first edition of 1687 and constitutes the conclusion of the work only after 1713, with the second edition. In this second edition Newton also no longer called the propositions at the beginning of the third book *Hypotheses*, but rather *Regulae Philoso-phandi* [Rules of Philosophizing] and *Phaenomena*. This caution can no doubt be explained by, more than anything else, the abundance of speculative questions that were inevitably raised about an influence on the heavenly bodies that operates across empty space—questions that occupied his contemporaries more than everything else. Newton more and more avoids responding to this interest.

A characteristic of modern reason presents itself here whose importance cannot be overestimated. It is the execution of the insight, which may not even be very clear, that gains in truth and accuracy may be attainable only by means of renunciations—that one cannot urge every question if one wants to get sound answers to at least some of them. The first great batch of renunciations was that of questions about final causes, the favorite instance of the new rationality in its attitude of distancing itself from the Scholastic past. In connection with Newton it becomes clearer what the renunciations would be like. They consist in a divergence between the formation of concepts and the posing of problems.

"Force" is a term that Newton uses to designate the causes of changes in states, especially of changes in the inertial states of a body, interpreted as acceleration. Here one always thinks of something acting on something through contact with it, as in throwing. But now Newton has to accept a type of change in state that is not brought about by perceptible contact—namely, that brought about by the gravity of one body with respect to another. Since it changes the other body's inertial state, it is legitimate to designate this cause also as a "force" and to subsume it under the general definition of the concept. That is a matter of the formation of a concept, not of the posing of a problem—even if it is undoubtedly a concept formation that 'tempts' us to set up a problem and to ask for an explanation.

Newton's half-sentence, "et hypotheses non fingo," comes in at just this point: It expresses a renunciation for the sake of a gain, a renunciation in relation to the quality of part of our knowledge of gravity that remains below the level of the concept formation, which is the part dealing with its spatial diffusion and its relations to the masses. Renunciation of the question of whether there is a power that operates at a distance across space, with or without a medium, and what (if there is such a thing) its essence would be, not only avoids a premature and provisional answer, but also defines a region of inevitable 'loose talk' [*Leichtfertigkeit*]. In retrospect, Copernicus's claim to deduce truth about the heavenly bodies from the supposed natural quality of their essence—to deduce the simplicity of the elements of their orbits from the simplicity of their substance—moves into the realm of such loose talk. The renunciation of questions about the nature and the essence of things had proved to be the only means by which to get answers to a more limited desire for knowledge. The medieval prohibition of curiosity lived on in the form of a postulate of rational moderation—which admittedly was to turn out to be the instrument of a boundless curiosity as soon as the initial renunciations of questions had been forgotten.

The disagreement about the concept of truth, which had broken out posthumously between Copernicus and Osiander, still feeds into the great argument between Newton and Leibniz at the beginning of the eighteenth century.[d] It is easy to foresee that the application of the principle of sufficient reason had to make it impossible for Leibniz to concede that the way in which the edifice of the world is really arranged is kept out of reach of all human comprehension, in God's wisdom, while human thought remains restricted to the phenomena and their consequences. As soon as Leibniz realizes that his principle of universal rationality would not be applicable to Newton's absolute space and absolute time, because the equivalence of all points, in both of them, makes rational decisions impossible even for a divine faculty of reason, he drops the reality of space and time and makes them into ordering forms used by the human spirit in relation to nature. This too, notwithstanding all the claims on behalf of the principle of sufficient reason, is a solution that is above all meant to make it possible to avoid questions. Among these is the question of the truth of the great world-systems. Leibniz perceives more acutely than Newton does that in the landscape of denominationalism not much is to be gained by proofs in favor of Copernicus, even a century after Galileo's conflict.

Long before Leibniz gave this solution (as radical as it was elegant) of the truth question by extirpating it, he had intimate contact with the post-Copernican problematic. Before he published his *Tentamen de motuum coelestium causis* [*Essay on the Causes of the Celestial Motions*] in the *Acta eruditorum* in 1689 and thus for the first time aroused Newton's suspicion that he could not have written this without being acquainted with the *Principia*,[58] Leibniz attempted, by means of a document that he passed to the Inquisition during his stay in Rome, if not to suspend, then at least to 'refine,' the continuing application of the measures directed against Copernicanism.[59]

In this *Promemoria* [Memorandum] Leibniz develops the idea that the bitter conflict about the cosmological systems was futile if the relativity of statements about motion to their systems means that "strictly speaking, every system can be defended ...."[60] Leibniz wants to test whether someone who writes this sort of thing, who while retracting the Copernicans' claim to truth also relativizes that of their opponents, would still fall into the grip of the censor's prohibition of the unrestricted defense of the Copernican system (*contra absolutam Copernicani Systematis defensionem*). It is a cunning piece of writing.

In nature, Leibniz argues, what is realized in at least one place cannot be called absurd and philosophically impossible; thus if, in Jupiter's system of satellites, heavenly bodies move around a different center from that of the universe, this has to be permitted by the philosophy of nature. This had also been accepted, in the meantime, by Jesuit authorities. As far as the language used in the biblical Scriptures is concerned, it cannot be interpreted more strictly than the language used by Copernican astronomers, who when speaking technically defend the motion of the Earth and the Sun's being at rest, but in their everyday manner of speaking will continue to say, and must say, that the Sun goes down or comes up. Even if one acknowledges that all the treasures of wisdom are contained, in a hidden manner, in the holy Scriptures, so that only correct things are said even about the objects of astronomy, this does not affect the new system; the inspired authors could not have expressed themselves at all differently about what was given to the senses and avoided nonsense, even if they had already wanted a thousand times to represent the new system as true. They not only wanted, as was often said, to remain comprehensible to their contemporaries; they wanted to say the truth. And it is the truth, because the relativization of motion restores their original truth to the senses.

Did this put Leibniz on the side of Osiander, who also considered many hypotheses equally capable of being defended? The difference is that for Osiander hypotheses are a characteristic of man's backwardness in comparison to an absolute knowledge and what it could reveal, whereas for Leibniz the perspectival character of every individual position cannot be dissolved and surpassed by the contradiction presented by an 'absolute standpoint,' so that a revelation could not be expected to provide anything but another perspective. Even for a divine intelligence, there cannot be an absolute distinction between motion and rest, but only a coordination between them and particular locations. The concepts of motion and rest depend on one's system, and there is no higher-level system, of the kind represented by Newton's absolute space, from which a final assignment of predicates could be undertaken. To that extent, the *Promemoria* of 1689 already contains the *Monadology* of 1714.

One only has to define clearly, Leibniz explains, what it means to declare the Copernican hypothesis to be "true," which is nothing but this, that it is the best, that is, the one that is best adapted to explaining the phenomena (*ad explicanda phaenomena aptissima*), and that no other "truth" can replace it in the face of a state of affairs that by its nature irrevocably includes the choice of a point of reference (*neque aliam in re quae sua natura respectum involvit veritatem locum habere*). He who understands this correctly can, without prejudice to the censorship, be a supporter of the new system and be a Copernican without qualification: "Atque his recte perceptis quilibet salva censura novum systema sequi et summo gradu Copernicanus esse potest." At first sight the text leaves us in the lurch in regard to the question of how one could be a Copernican "in the highest degree" [*summo gradu*]. This formula undoubtedly refers to the warning issued by the Congregation of the Index in 1620, which explains the mildness of the censorship of Copernicus (*suspensio donec corrigatur*) by the fact that the book "contains much that is very useful for the common welfare," but at the same time designates the precise passages in which definite assertions had to be retracted into the modality of hypothesis.[61] So what could be called the highest degree of Copernicanism could be recognized by its modality.

More important, however, is the problem that is posed by the expression "adapted to explaining the phenomena." One has to remember that for Osiander all hypotheses that could be invented for representing the celestial phenomena were equivalent in their truth value and were distinguishable only by their usefulness. The art of astronomy, constitutionally condemned to untruth, was not supposed to be able to distinguish its hypotheses by their epistemic quality. In a universe that is determined throughout by the principle of sufficient reason, indistinguishability cannot exist in any form other than that of absolute identity. It must be possible to distinguish hypotheses by means of their relation to the truth. Of course, in the meantime the term "explanation" had strayed a long way from the astronomical model that Osiander had taken as his basis. The examples that Leibniz adduces for the success of the Copernican hypothesis even among its original opponents reveal this.

Jupiter's system of satellites is an exemplary "phenomenon." As such, it requires at least the admission that the "natural" revolutions in the heavens do not have to be related exclusively to the center of the world. In this negative function, the analogy only removes objections to the new theory. But after that the view of Jupiter's system does also suggest the *explicatio* [explanation, explication] that the revolutions of heavenly bodies do not relate at all to an abstract "point" in the world—irrespective of whether that point is occupied by the Earth or the Sun, or whether it remains altogether empty—but that their paths have something to do with the corporeality of the reference point around which they revolve. But that, since Kepler, was the core of the cosmological realism of the planetary laws as the expression of an active connection between the Sun and its "hangers-on." It was also the basis of Leibniz's *Tentamen*, which he claims to have put on paper before he had any acquaintance with Newton's *Principia*.[62] Leibniz deliberately chose, for the treatment of the phenomena, the vague term *explicatio*, which he is able, significantly, to apply at the same time to the "interpretation" of that "earlier"—which no doubt is supposed to mean obsolete—formula of the Roman censorship (*censurae pristinae explicatione*).

Leibniz works cautiously with the warning that by insisting on the old interpretation, the Congregation of the Index would be resisting in vain the irresistible power of the spirit of the age (*torrenti proficientis seculi*). It would deprive believers, and the Italian nation, of the enjoyment of the illumination of the age (*lux seculi*), but at the same time it would obscure the *gloria dei* [glory of God] through ignorance of the true magnitude of the creation. Light metaphors, in the style of the early Enlightenment, are abundantly employed here, in particular for the (explicitly included) physical causes of the planetary motions and their universal laws, which are described as a *nova quaedam lux exorta* [a newly appeared light]. The phrase about the hypothesis that is best adapted to explaining the phenomena, which was at first introduced so timidly and vaguely, acquires its precision through the—in the traditional terminology—unequivocally "physical" description of the astronomical phenomena. The gravity of the planets with respect to the Sun as the central body—a gravity that is comparable to the effect of magnetism—can in fact no longer be comprehended under the concept of hypothesis (*per modum hypotheseos*), Leibniz says, but has the quality of a proof by deduction, by the laws of geometry, from the phenomena (*per regressum geometricum ex phaenomenis demonstrantur*). The best adapted hypothesis is, as such, no longer a hypothesis.

The "best system" is marked by more than and by other things than its technical productiveness; it is, rather, a system that hardly leaves a doubt any longer: "... de optimo systemate vix amplius dubitandi locus aliquis relictus esse videatur." This text of an intervention with the censorship authorities of which the outcome is unknown to us is a masterpiece of rhetoric above all in the way that, in its last move, it makes the kinematic relativization of the systems, which was introduced at the beginning, disappear again, quietly, in favor of their unequivocal physical distinguishability.[63] Nevertheless, Copernicus's question about the "form of the world" can no longer (and cannot again) be posed: "Nulla est idea universi, quemadmodum figura quoque eius nulla est." [There is no idea of the universe, just as there is no figure of it either.][64]

Perhaps it would be possible to develop and to display the modern age's problematic of truth using the disagreement between Copernicus and Osiander as the point of departure. When one seeks to establish a distant point toward which Osiander's "infamous act" seems to be aimed, one still encounters the ambiguity of the predicate of truth for the cosmic systems. Max Born formulated the position of equivalence from the point of view of the theory of relativity: "This gives us freedom to return to Ptolemy's point of view of a 'motionless Earth.' This would mean that we use a system of reference rigidly fixed to the Earth .... From Einstein's higher vantage point, Ptolemy and Copernicus are equally right. Both viewpoints furnish the same physical laws .... What point of view is chosen is not decided by principles but is a matter of expedience. For the mechanics of the planetary system the view of Copernicus is certainly the more convenient."[65] Against this, Werner Heisenberg, in his Goethe Lecture in Weimar in 1967, asserted the enduring realism of the Copernican theory as a result of its physical interpretability, and rejected, as a misunderstanding, the objection that described it as arbitrary: "It is true enough that Einstein's theory of relativity leaves open the possibility of considering the Earth to be stationary and the Sun as moving around it. But nothing whatever is thereby altered in regard to the crucial claim of the Newtonian theory, that the Sun, with its powerful gravitational field, determines the paths of the planets, and that the planetary system can therefore be really understood only by starting from the Sun as midpoint or center of gravitational forces." [e]

It is probably not an accident that this was said in the Weimar of 1967. The thesis of the equivalence of the world-systems has to be a scandal for every epistemological realism, but also for every political system with a claim to exclusiveness. Thus it is not illogical that the rejection of Einstein's relativity theory by official Dialectical Materialism was aimed above all at this conclusion of equivalence. Even if it is only a case of analogy, still the pluralism of cosmological systems is a paradigm of the way ultimate truths are not available to us. In 1952 D. I. Blokhintsev, in a critique of the non-'realistic' positions of Western physics, underlined this as an identical motive—that if all frames of reference, in the general theory of relativity, have equal rights, then Ptolemy's world-system would be assigned the same rights, retrospectively, as Copernicus's system. This is clearly intolerable for a view of history that sees in every present situation, and thus also in the contemporary one, the logical outcome of the total process of its past. The mere "historicism" that is involved in ascribing to Copernicus and his claim to truth an effect that was admittedly revolutionary for his age, but then relativizing this judgment to the situation at the time, cannot be accepted by a philosophy of history.[f] Here the Russian critic follows a historical interpretation of Copernicus as a revolutionary by which the metaphors that are based on the changes that Copernicus executed are converted into a picture of what those changes presupposed: Copernicus did away with the (as it were) feudal preeminence of the Earth in the solar system and thereby advanced a sort of democratic model; any historicization of this role of the theorist as perpetrator would level off the exemplary binding force of the event for the elimination of the geocentric world-system.[66]

Nowhere was posterity's indignation against the author of the anonymous preface noticeably directed against his motive of contradicting astronomy's claim to truth in the interests of the absolutism of a different truth. Only superficial consideration of his leveling off of astronomical hypotheses into equivalent fictions can overlook the fact he is not in the right here even in terms of the history of science. For in keeping with the consciousness of scientific progress, in the concept of hypothesis the point of view of (at least) corrigibility remains dominant, whether it is corrigibility through a new hypothesis's provisional verification, which is considered to be possible, or corrigibility through the mere falsification of the old one. Both cases were excluded in Osiander's 'philosophy of science' of equivalence because he had ready in the background a concept of final definitiveness, a sort of eschatological verification. Since this astronomical truth was not 'contained' in the unique divine revelation that is presupposed by Christian theology, it could only be contained in the other, and final, revelation, of which the first one contains a promise—though it would then be a truth that has ceased to have an object, as a result of the end of the world.

Science, too, has worked (though in its case only in thought experiments) on this kind of limiting case of certainty—for example, in the fiction of the Laplacean intelligence—in order at least to construct the outermost possibility of the knowledge that could be integrated by its means. The Laplacean demon is not a spirit that could make revelations, if only because the world-equation that it would have to convey could not be received and processed into individual items of knowledge by anyone—but also because the kind of totality of knowledge that it would contain would not even conform to man's interest in knowledge.[67] In Copernicus's case, this cognitive interest was characterized by a congruence between the possible and the accessible truth about the world. For the truth that is granted to the desire for theory ("quatenus id a Deo rationi humanae permissum est") is directly dependent on the idea that the world was created for man's sake.

In the history of science, the attempt to grasp the truth of the world as a whole, its *ratio*, its *forma* or *figura*, is only an episode, which was to be deprived of its basis by the consequences of this undertaking itself as soon as the delimitation of the 'topic' vanished along with the delimitation of its object. But for the modern age Copernicus is less a figure in the history of science than a figure in the history of consciousness. An emanation goes out from this figure and this work, an emanation that becomes evident at the beginning of the seventeenth century, and the connection of which with the sober reality of the cathedral canon in Frauenburg, *in hoc remotissimo angulo terrae* [in this remote corner of the Earth], can no longer be described as his "exercising an influence" in history. A Copernican enthusiasm [*Pathos*] develops that reaches another high point at the beginning of the nineteenth century, when Lichtenberg, in his concise biography of Copernicus, defines his hero's chief merit in terms of his having, "armed with reason and geometry, in the great struggle with truth that error, supported by all the power of sense-appearances, had successfully endured for about two thousand years, finally turned the victory, by one decisive blow, to the side of truth." [68]

Without the shining forth of a living conviction, without its rootedness in the certainty of the new world-concept and in the unconditionality of its claim to truth, there would be no way to comprehend the historical power of this work, which "exercised influence" only in a metaphorical sense, while what it really did was to offer itself as a central theme for the self-definition of a new consciousness. As a mere hypothesis, as a model of calculation, subject to the test of methodological convenience, the *De revolutionibus* could not accomplish that. It was in relation to this Copernicus, the Copernicus who was suited to the myth that formed part of a historical self-comprehension, that Osiander's power-depleting move had been an offense. By putting Copernicus's work back into the astronomical tradition and enlisting it in the service of the theological proviso of divine sovereignty, he made it questionable as the signature of the epoch-making break with the Middle Ages. For it was only in terms of the inevitability of a new truth that the epoch could understand its origin as a new beginning against the Middle Ages. Lichtenberg stylizes his Copernicus analogously to the other founder figure of the epoch, Descartes, when he ascribes to him the desire "to throw out all the rubbish for once and make a new start."

Descartes had defined his program in this way almost word for word, in the first of his *Meditations: funditus omnia semel in vita evertenda* [everything must be thoroughly overthrown for once in my life]. Of course Descartes himself had seen the foundation and the security of the beginning only in the protection provided by his method; he did not see them in Copernicus. For him, the rational construction of the system is independent of the real structure of the world; the ideas of the Creation are inaccessible, and the *ens perfectissimum* [most perfect being] owes man nothing beyond not allowing the world to become the great deception of his cognitive faculty. Everything else is the business of caution governed by method. This holds in particular for the conflict of the great astronomical systems, to decide which would violate the principle that man must not presume to participate in God's plans.[69] God, as infinite power, can only be the principle from which to derive a world-construction in the most general sense (*generalis totius huius mundi constructio*), and that means a universe of what is possible, into which the reality that is met with in a particular case can be subsumed as a part.

This principle of understanding of the world does not permit man too high an opinion of himself. The maxim "Ne nimis superbe de nobis ipsis sentiamus" [We should beware lest we think too highly of ourselves] is violated if man imagines the Creation as a whole as referred to himself and his cognitive capacity: "si res omnes propter nos solos ab illo creatas esse fingeremus."[70] On the assumption of what is still a medieval God, that is a renunciation of Copernicus's anthropocentric world-formula and, as a result, an exclusion of its rational use in connection with cosmology. Regarding the three world-models, those of Ptolemy, Copernicus, and Tycho Brahe, Descartes has to say—even without remembering the persecution of Galileo—that they are to be regarded as assumptions, which should not be judged by the criterion of their truth but only by that of their suitability for representing the phenomena.[71] From this point of view, he regards Ptolemy's hypothesis as unsuitable, and Copernicus's as equivalent to Tycho's, though also a little simpler and clearer, so that the only reason Tycho could have for making changes in it was that he thought that he could advance from hypothesis to objective truth.[72] Descartes also rejects Tycho's system because it was only verbally that he asserted that the Earth was at rest.[73] So there are after all other criteria than that of constructive simplicity by means of which to arrive at a rejection of all three world-systems and to make room for one's own.

For his own cosmology, Descartes claims that it is the simplest and the most suitable (*accommodatissima*) not only for representing the appearances but also for investigating their causes (*ad eorum causas naturales investigandas*). Nevertheless, even with this claim, which goes beyond the astronomical type of hypothesis, Descartes's proposal is supposed to remain only the outline of a possibility.[74] Here the idea of the origin of the world as the merely factual decree of an absolute power serves as the basis of inquiry into the rational unity of what is possible, as the object that is adequate to reason. Reason asserts itself against the contingency of the world by means of its own universe of possibility, for which it possesses the method of a *mathesis universalis* [universal science]. This universe necessarily includes, as a concrete instance, the world of what actually exists, but relativizes it by means of a knowledge that is supposed to be valid for all possible worlds, so that it need not become involved with empirical knowledge of the contingently given, actual world. Descartes demonstrated this in the fifth part of the *Discours de la méthode* [*Discourse on Method*] and from a certain point onward in the third book of the *Principia philosophiae* [*Principles of Philosophy*], where he carries out rational experiments in an unreal "new world," located in an imaginary space.[75] The question of the laws of a "world as such" is already in the background here as the autonomous theme of a reason that by that theme defines itself, for the first time, in an original manner, and that no longer seems to have need of any anthropocentric teleology. This nonneediness is the real difference between the modern age's rationality and Copernicus, a difference that makes his claim to truth, and what that claim presupposed, obsolete, and excludes it, as a transitional episode, from being part of the epoch.

3

The Reformation and Copernicanism

Anyone who is inclined to see history as the result of the 'strategies' of luminous and dark powers, of the cunning of reason and the craftiness of unreason, will not be surprised at the suspicion that that "infamous act" of the author of the anonymous preface to Copernicus's work could have been more than an isolated and private act, that is, that it could have been part of a conspiracy between Wittenberg and Nuremberg, between Melanchthon and Osiander, to deprive the new truth—the 'other' new truth—of its claim precisely to be that. For this is the important point here: It is not schools, terminologies, or systems that are in competition but concepts of truth. Here conspiracies are easy to imagine, because the supposed possession of truth has always inclined people to instrumentalize ethics in its service.

A historian has exclaimed indignantly, in regard to this suspicion that Melanchthon and Osiander could have conspired together, "But this is mixing fire and water together!"[76] This is not yet meant to allude to the future conflict, in the "Osiander controversy," but it is meant to represent Melanchthon, the Aristotelian, as incapable of plotting with the admirer of Copernicus. Of course this makes it too simple. Osiander was certainly well-disposed toward Copernicus's work, but his greater interest had to do with local dignity, because in Nuremberg the Reformation could not afford to take an obscurantist position in astronomy. This was where Regiomontanus had been active, where Johann Schöner was in high esteem, and where the technique of producing astronomical instruments was highly developed, so that the city's standing was closely tied to this discipline. Osiander was not a "critical" philosopher, in opposition to the "realist," Copernicus;[77] rather, he combined the Reformation's fideism with the respect that would inevitably be felt, in the stronghold of astronomical competence, for the astronomical reformation. Melanchthon, on the other hand, is characterized too narrowly and one-sidedly when he is described as an "Aristotelian." On the contrary, in his underlying philosophical assumptions he stands very close to the Humanistic anthropocentrism that made Copernicus believe in the possibility of truth in astronomy. It is only this proximity that makes his opposition instructive.

If one wants to stick to the realm of facts for which evidence can be presented, there is only one very indirect way in which Melanchthon participated in Osiander's offense: through the pressure that the University of Wittenberg exerted on Joachim Rheticus finally to fulfill his duties as master and begin lecturing. After his almost three-year-long stay in Frauenburg, Rheticus was not able to remain in Nuremberg for the duration of the work of the printers, which would have enabled him to be sure of the authenticity of the printed *De revolutionibus*. My imagination does not extend to entertaining the hypothesis that Melanchthon removed Rheticus from Nuremberg so as to give Osiander a free hand there. His apprehension about Rheticus's infection by the Copernican philosophy is, of course, well-known; he wanted to have Rheticus near him again, and to see him married, so as to make Rheticus's enthusiasm (which seemed to him pathological) die away. Thus it came about that while events took their course, behind his back, in Nuremberg, Rheticus lectured in Wittenberg on the *Almagest*, a subject that involved not even a hint of rebellion, and could satisfy Melanchthon.

It should not be forgotten that Rheticus's journey to Copernicus in Frauenburg, to obtain reliable information about his astronomical reform, had taken place with the consent, if not on the advice, of Melanchthon. But still more important is the fact that after he had a direct knowledge of the new system, Rheticus was able to believe that he would bring Melanchthon, too, over to its side, and with the authority of the *Praeceptor Germaniae*[a] would achieve a breakthrough in, at least, the Lutheran university. That is why Rheticus had not even waited for the printing of the *Narratio prima* to be completed, but on 14 February, 1540 had the first three printed sheets delivered to Melanchthon by Andreas Aurifaber.[78] The fact that Melanchthon, as it turned out, was not open to the new cause's conclusive evidence was a critical setback for the alliance of Rheticus and Copernicus. It must also have given Rheticus food for thought, precisely because he had thought Melanchthon capable of agreeing with him. This is the really important point: In his intellectual constitution, Melanchthon was not a 'born' opponent of Copernicanism.

Questions like what would have happened if Rheticus's expectations had turned out to be correct have only a limited heuristic value, but they have an attraction that methodological scruples cannot eliminate. The deeper reason for this attraction is that with such hypothetical reflections in relation to what actually happened we appeal to the undemonstrable presence of freedom in history, and at the same time seek to trace the blindness to possibilities that is always prevalent. So, what would have happened if Melanchthon had felt the same conclusiveness in Rheticus's report that its emissary felt? The decisive fact is certainly that around the middle of the sixteenth century only Melanchthon could have done the Copernican doctrine the service of providing it with a far-reaching success and establishment in the educational institutions. He was in a position to understand Copernicus's intention appropriately and, starting from Copernicus's hints, to render it intelligible to the age; and, no less important, he was also in a position to stand up for it with an authority that was a match for the forces of inertia. His determining influence on the intellectual life of his age and of his contemporaries simply cannot be overestimated. His judgment had a canonizing effect. Around the middle of the century there was hardly any scholarly authority left that could not be traced back to his. He was to such an extent one of the 'inescapable' intellects of the age that even an Ignatius Loyola, who in a letter to Peter Canisius of 13 August, 1554 had strictly forbidden the use, in the schools, of philological writings by the Reformers, had to concede, in a later letter (of 4 January, 1556) to Francis Scipioni, permission to use Melanchthon's writings on grammar, rhetoric, and dialectic for private study— though only after the author's name had first been blotted out. For this potency, the external preconditions for helping a new idea to triumph were unique. What, though, was the position with regard to inner preconditions, in relation to the Copernican idea?

At first glance it is amazing that the Reformation did not make the Copernican 'sign' its own. Did this not provide the most 'pregnant' authority to invoke in the conflict with the medieval and Catholic claim of harmony between the visible and the invisible—of congruence between the natural order and the order of salvation? Man was banished to his [literally] eccentric position and was thus definitively deprived of any pagan/cosmic security in the center of the world. The radical intensification of people's concern about the certainty of salvation had directly and exactly to accord with and spring from this new insight into the world's adverseness to salvation. In having faith 'nevertheless'—in the *sola fides* [faith alone] and *sola gratia* [grace alone]—man's physical lostness in the world could find its final fixed point, its absolute minimum of anthropocentrism. To appropriate Copernicanism and interpret it, as a sign, in this way would not have been in accordance with Copernicus's intention, but it would have presupposed admitting his claim to truth—though with the difficulty that it would not have shared the premises on which that claim was based. Here lay the resistance that made it impossible for Osiander to assume, like Copernicus, that the astronomical reform could reestablish harmony between man and his God, after the confusion created by Nominalism—that the *contemplator caeli* [observer of the heavens] would gain, by means of theory, the standpoint to which the *conditor caeli* [creator of the heavens] had related the order and intelligibility of his product. However, this intention of the act of reason did not get expressed in its result, insofar as that result was taken as a 'sign'—otherwise, the Counterreformation could not have understood Copernicus as an opponent.

As a result of the Reformation, man's interpretation of the world and his understanding of himself had become dissociated: The doctrine of justification as a forensic imputation left man's 'reality' untouched—it denied to the redemption in which one had faith the things corresponding to it in the cosmos, and understood the world in a way that bracketed out the Providence ruling over it. The world of nature and the world of faith no longer made any difference to one another. Melanchthon had systematized Luther's ideas in the *Loci communes* [*Commonplaces*] in 1521. Here, for the first time in the occidental church—Adolph von Harnack wrote in 1897 about Melanchthon—the Christian religion had been presented not "in the schema of a drama of God and the world and a sacred 'physics,' but rather as the awakening and the process of a new inner life." Such a turning could have found no less confirmation and support in the Copernican universe—a universe that seemed to deprive man of the indications of his being chosen and taken care of by nature—than a century later (in view of Copernicus) fell to the lot of the self-consciousness of man as one who was thrown back upon himself and now could master the world only on his own. It was the Reformers' "No," and above all Melanchthon's "No," that left the Copernican 'sign' entirely to the next century—and not only "left" it but made the next century the first to discover its latent power as a sign.

The discussion was not conducted with the kind of concern for fundamentals that, in retrospect, one could wish for. The casualness of Luther's and Melanchthon's remarks in relation to Copernicus permits only the impression that this was not, initially, a topic of primary importance. In the well-known passage of Luther's *Table Talk*, he could refer to an unnamed person as the "new astrologer" who, as a "fool . . . [wants] to overturn the whole art of astronomy."[79] The offense given by the vague information available at this early date, in June 1539, is not so much the content of the innovation as, rather, the mere fact that one is proposed; Luther's words in the original transcript[b] clearly describe a violation of Humanism's condemnation of all passion for innovation and of the attention that a person wants to attract by disregarding what is accepted. To be sure, Luther appeals to Joshua's command to the Sun to stand still, and thus, already at this early date, brings the principle of the sole authority of Scripture into play. One cannot impute to Luther any awareness that it is "not a comfortable feeling to be expected to assert ourselves as persons within an infinite universe in which our Earth only has the value of a particle,"[80] because no implication of this sort was perceptible as yet.

The superficiality and casualness of Luther's remark may veil a deeper uneasiness, if there could be any perception at all, as early as 1539, of the importance of the issue. A statement of Luther's in a different passage of the *Table Talk*, in a similar context, which puts in question not only a particular but any kind of cosmology, leads us to the heart of it: "No reason can grasp or understand even the natural works of God's creation . . . ."[81] That comes very close to the position represented by Osiander; it deprives every cosmological system of its claim to truth and leaves the "art of astronomy" no role but that of looking after the calendar. That man should be able, by his natural means, to fathom the reason in the Creation would be a seduction analogous to the rejected possibility that he could submit *ex suis naturalibus* [by his natural means] to the *dictamen rectae rationis* [com-

mand of right reason].[82] And for Luther the heavens that the Copernican reform thinks it can put in order or at least comprehend in their
order are, rather, the heavens that the fearful conscience pictures as
collapsing upon it.[83]

All of this cannot be carried over to Melanchthon without further
ado. He was, in his century, the greatest promoter of astronomy
and of its importance. He rediscovered what we would call, with
an outmoded word, the "educational value" ["*Bildungswert*"] of this
discipline. He untiringly employed his authority, in prefaces to astronomical publications, to promote old and new writings. In his Humanism man is still, or is again called to be, the observer of the
heavens. A philosophy that contained no understanding of the heavens, he writes, would only be a piecemeal job.[84] Opponents of this
conception, which carries the imprint of Stoicism, he calls—with
the tradition's most effective and most randomly applied term of
abuse—"Epicureans," because they behave as though they are indifferent to the Creation's worthiness to be admired; what is more,
they do not deserve the name of men because they are at war with
human nature, which after all was created to observe these divine
objects.[85]

If, following Harnack's thesis, the ancient and pagan drama of God
and the world was abolished by Melanchthon in the *Loci communes* in
1521, then at the latest in 1535, in his preface to Peurbach's *New
Theory of the Planets*, it is back again, or at least approaching: In the
heavens, God's will, which governs the world, and man's upward-
looking desire for knowledge confront one another (probably not
without an implication of astrology),[86] and a rejection of investigation of the heavens cannot be any kind of obedience to God's will.[87]
The need for knowledge of events in the heavens no longer appears as
curiosity, which could divert one from one's sole serious concern,
because God offers man this object as a legitimate one. When he
insists on its accessibility by investigation, he takes advantage of a
possibility of secure knowledge that God has assigned to him (*certam
rationem divinitus ostensam cognoscere*).[88]

Melanchthon's prefaces to important astronomical works show how
he circles around the problem that was posed for him by Nominalism
and by the Reformation, the problem of again, in the first place,
legitimizing astronomy's object and its claim to knowledge. One does
not perceive any of this in Copernicus; he has already gone a step

further and no longer sees the problem in the justification of the *object*, but in showing how the *subject* can prove itself a match for that object. Melanchthon repeatedly assures us of the comprehensible rationality of astronomy's object, but he reveals no awareness of the problem that is inherent in the fact that he does this in prefaces to books whose theoretical norm and whose application of that norm seem to demonstrate the opposite. What gave Copernicus the impetus for his reform evidently cannot be perceived on the level of the theoretical authority that Melanchthon exercises. This backwardness in his knowledge of the actual state of the discipline is made intelligible by the urgency of justifying the discipline. Melanchthon has already traveled a distance in order to be able to write these prefaces in the language of Humanism. And he is evidently not yet at the end of this path.

Melanchthon deals with Copernicus in the first book of his *Initia doctrinae physicae* [*Introduction to Physics*], whose first (of many) editions appeared in 1549.[89] In the chapter with the concise title "Quis est motus mundi?" ["What Is the Motion of the Earth?"], Melanchthon links up, one might think, with Luther's rebuke of innovation: There are people who, either from their *amor novitatis* [love of novelty] or from their need for recognition, deny the motions of the heaven of the fixed stars and the Sun and instead make the Earth move. What Melanchthon immediately brings forward as a telling objection against this is already aimed at the triumphal Copernican assertion, which was formulated earlier by Nicholas of Cusa and was later taken up by Giordano Bruno and Galileo, that the Earth has become a heavenly body. Here, formulated as a reproach, it reads, "Terram etiam inter sidera collocant." [And they place the Earth among the heavenly bodies.] Such pastimes, Melanchthon continues, are not (and he refers to Archimedes's report about Aristarchus) a new invention. Publically advocating such absurd theses is irresponsible (*non honestum*) and harmful as an example. Although Melanchthon reminds us of Aristarchus, it is remarkable that he does not bring his heaviest gun—which must have been known to him from Plutarch's account of Cleanthes's accusation—into action: the reproach of impiety. A comparison between the contents of the two reproaches is nevertheless instructive. Cleanthes regards it as impious to expel the Earth from the center of the cosmos, while Melanchthon only takes exception to its being elevated to the rank of a heavenly body.

Now Melanchthon makes use of an ambiguous formula in order to free astronomy from the suspicion that it cannot rely on the phenomena but instead must seek reality *behind* them. A phenomenon here is not only what *shows itself* but also what *is shown*. God shows the truth about astronomy's objects so openly, Melanchthon says, so manifestly, that one only has to take hold of them, and one does not have to run after complicated ideas.[90] The talk of a truth that is "shown" is devious in this context. It is not only meant to tie theory to the appearances and deter it from mistrusting what is immediately accessible, but also to form a transition, in the very next step, to the other process by which God "shows" truths—the process of the Bible. Of course there are people, Melanchthon admits, who deride a natural scientist who adduces the testimony of revelation. Just as Melanchthon had earlier said that he regarded it as *non honestum* to engage in the new astronomy, here he uses this term (which is specific to Stoic ethics) to say that he considers it *honestum* to direct philosophy's attention to heaven's communications and, in the obscurity of the human spirit, to bring divine authority into play wherever possible.[91] This is followed by Bible quotations, but also by reflections in which other positions of the Earth are tried out and the consequences resulting from them are critically discussed. The overall result is, as might be expected, negative, and an objection such as the one based on the uniform annual periodicity of the lengths of day and night is characteristic of the level of the argumentation.[92]

In conclusion a warning is again issued against those who indulge themselves by bringing the "liberal arts" into confusion (*conturbare artes*), and who only want to separate us from the truth. "Confusion of the arts" here undoubtedly also means asserting the claim to detach astronomy from the propaedeutic bundle of school subjects to which it had belonged since the Middle Ages, and to elevate it to a "science," in the strict sense, as Copernicus had (inevitably) done.

Melanchthon's own *Introduction to Physics* gives a gloomier picture of man's intellectual power, and especially of its competence in relation to astronomical questions, than do his propagandizing forewords to other people's books. It is also an official text, a book of instruction, to which Melanchthon could not give all the Humanistic distance from Reformation theology that he had increasingly gained since the crisis of 1523/1525. What he writes to Duke Albert of Prussia on 18 October 1544, in order to thank him for the stipend he had granted to Erasmus

Reinhold for the completion of the *Prutenic Tables*, sounds different: "After all, this human life has to have arithmetic, the art of measurement, calendars, and cosmographies; moreover, the beautiful order of the Sun's annual course is clear evidence from God that this nature was made by a wise and orderly master builder, so that, in praise of the Lord, we should all love these arts, by which this order is explained."[93] Since Reinhold had in any case used Copernicus's *De revolutionibus* as the basis for his tables, Melanchthon could have formulated this with his eye on Copernicus's dedicatory preface, it comes so close to the latter's world-formula—though without carrying the argument beyond a justification of astronomy in its technical aspect, as satisfying the needs of practical life. The divine promise, as Melanchthon interprets it, does not extend to the satisfaction of the need not to have to live in an unknown and untransparent world.

The public reception that this work had is shown by the fact that the second edition came out only a year after its first appearance. How seriously Melanchthon took the book's public effect is shown by the other fact that it was already a revised version. Emil Wohlwill was the first to point out that it exhibits rather radical changes in the remarks on Copernicus.[94] This would already be noteworthy in itself, if it were not also the case that, in an otherwise unchanged text, the only revisions relate to the attitude Melanchthon takes to the new astronomy. So this is evidence not only that there had been some change in Melanchthon's opinion but also of the amazing shift, within the decade since the first casual remarks, in the urgency of dealing with this question and of pronouncing on it.

We would be perplexed to adduce any motive, in the short time between the two editions (between 1549 and 1550), that could have influenced Melanchthon's basic attitude in this way, if there were not good reasons to think that the manuscript of the first version of the *Initia doctrinae physicae* must already have been completed by 1545. In that case the important passage moves closer to the letter in which Melanchthon declared himself most strongly and clearly, when he wrote (to Burkard Mithobius on 16 October 1541) that some regarded it as an outstanding accomplishment to do something as absurd as this Prussian astronomer, who sets the Earth in motion and stops the Sun, something that should induce wise rulers to intervene.[95] Still it seems possible that when he composed the book, Melanchthon did not yet possess a copy of the *De revolutionibus*, which had appeared in 1543. In that case his knowledge of the new thesis would no doubt have been based only on the *Narratio prima* or on other communications by Rheticus.

Just as Rheticus had misjudged Melanchthon's readiness to receive the cosmological reform, notwithstanding his assumed affinity for it, so Melanchthon did not understand that Rheticus's enthusiasm for what he had become acquainted with in Frauenburg was based on the Humanistic premises of anthropocentric teleology to which Melanchthon himself had continually drawn nearer and which prohibited the importance of astronomy from being exhausted by its function as a technique that is serviceable in life. Thus Rheticus had gotten into a difference of opinion with Melanchthon, but it was the kind of difference that results from having progressed further in the same direction. Melanchthon failed to realize this when he complained to Joachim Camerarius on 25 July 1542, about Rheticus's having become Copernicus's student.[96] But the disapproval never went so far that it came to a quarrel; Melanchthon assisted Rheticus with letters of recommendation when he went to Nuremberg to supervise the printing of the *De revolutionibus*,[97] and no doubt also encouraged his appointment as a professor in Leipzig in 1542.

At this time the public's knowledge, not indeed of the content of the Copernican reform, but of the great claims made for it was still entirely based on Rheticus's proclamation. The Basel printing of the *Narratio prima* is preceded by a letter from Achilles Gasser to Georg Vögelin in which we are told that the author "unquestionably brings to light the reestablishment, indeed the rebirth of a new and entirely true astronomy, regarding the principles of which he gives us extremely clear information."

If Melanchthon had initially based his remarks about Copernicus on Rheticus's testimony, then when Copernicus's chief work was in his hands it had to have the effect of a correction. Here we cannot exclude the possibility that the anonymous preface exerted its softening effect, since—if we do not follow the conspiracy thesis—Melanchthon, after Rheticus's departure, did not necessarily know who the preface's author was. If we suppose that it was a reading of the *Narratio prima* that provoked Melanchthon's polemic, then we also see the only possible occasion for the change of opinion that followed immediately: With its sober, nonkerygmatic manner, the *De revolutionibus*, which had become available in the meantime, hardly seemed to justify the original form of the polemic. Whether Copernicus's appeal to Cicero made an impression, we have no way of judging. Cursorily, as the busy man read, the first page after the title is most likely to have been the one that appeased him, however little that may satisfy our expectations of the "history of [a book's] influence."

For Melanchthon's having acquired a clearer knowledge and a changed evaluation of Copernicus's work between the first writing of the *Introduction to Physics* and its reissue, we have the important evidence of Melanchthon's memorial address for Caspar Cruciger in 1549. With the death of Cruciger, the interests of astronomy and natural science had lost their most important promoter in Wittenberg. Through Erasmus Reinhold, who read the address, Melanchthon reminds the listeners of the observations they had undertaken together, by means of which the geographical breadth of Wittenberg was supposed to be more accurately established, and which also led to a determination of the deviation of the equinoxes from the "usual calculations." In connection with these and other similar observations they had begun, he says, to admire and to love Copernicus more, and also to follow his procedure for comparing new data with the traditional ones.[98] In doing so they had established that the "true length" of the year should be changed from the medieval figure to the figure given by Ptolemy. Even if Melanchthon wrote this way only out of respect for the deceased, who had never expressed opposition to Copernicus, still the productive practical activities around him that were based on Copernicus must have made it impossible for Melanchthon to repeat his earlier judgments.

An indispensable step in order to be able to characterize the changes that Melanchthon made in the second edition of his *Introduction to Physics* is to compare the two versions of the critique of Copernicus, of 1549 and of 1550, with one another.[99] Above all, the imputations of unsuitable motives have been removed, such as the—for a Humanist—especially serious ones of a passion for innovation and a craving for intellectual recognition. The wording has also been neutralized in other ways. The characterization of the renunciation of geocentrism as "games" has been removed. The whole critique has acquired a didactic premise: In an outline intended for beginners, such as the *Initia* was meant to be, love of what is traditional and uncontested must be the guiding principle—because a truth that is shown by God can only be of this kind—whereas advanced students are free to engage in unusual reflections. The change in the sentence that, in the first edition, begins with the words *Bonae mentis* is subtly and painstakingly thought out—not so much because the moralizing element of intellectual well-meaningness is entirely excised, but because the change puts the 'God-given truth,' which on this subject had previously been assumed to be unproblematic, under a subjective condition, namely, that the way in which this truth is offered must be understood. In other words, it is no longer assumed as a matter of course that God must also have expressed himself unambiguously, and in a way that everyone is bound to recognize, in the astronomical realm. What suggests the conjectured influence of the anonymous preface to the *De revolutionibus* is the reservation that the advocates of the new doctrine do not regard it, with full seriousness, as true. All in all, Wohlwill seems to me to go too far when he speaks of a "revocation of the censure that had been pronounced." Melanchthon does not abandon the assumption (and was evidently strengthened in it by Osiander's addition to Copernicus's work) that the question at issue is one that essentially surpasses the capacity of the human intellect, which can expect clarification of it only through the assistance of divine authority. It is only in his confidence that this theological consultation will yield a result that can be treated as objective that he seems to have begun to waver. In the final analysis, this remark could have been made with the intention of conceding, again, the *bona fides* [good faith] of the authors of the new theory.

So in the substance of the matter not even a small step has been taken toward Copernicanism. On the contrary, an important deviation from Ptolemy and the tradition, which was contained in the first edition, in 1550 has already been omitted again and 'reptolemized' in favor of the 'didactic' fidelity to tradition that I mentioned. In the *editio princeps* [first edition] Melanchthon, submitting to the extremely clear evidence, made the two planets Venus and Mercury circle the Sun immediately and circle the Earth only indirectly, together with the Sun. It is true that in doing so he had only made use of the freedom that is supposed to belong to the astronomical *artifex* [artificer], but one can see that he did so with the concession that, in fact, not just everything is possible, but we must expect to find a comprehensible mechanism, if the sentence from the letter to Duke Albert that "this nature was made by a wise and orderly master builder" is not to be an empty phrase. Melanchthon had grasped a little piece of the logic that is inherent in the immanent history of the Copernican theory. But this step toward heterocentrism is already withdrawn, and the *Almagest* vindicated again, in the second edition. Melanchthon must have seen—and this is an impressive example—that when the consensus had been abandoned in one case, the restraint appropriate to an "introduction" (*prima institutio*) could no longer be preserved. Slightly alarmed, no doubt, by a presentiment of the inevitability of further steps, he draws back from the path he had just entered, of an immanent 'Copernicanization' of his picture of the world.

Even in the fifth and sixth decades of Melanchthon's life, his combination of Reformation theology and Humanist reception, of Scholastic Aristotelianism and Ciceronian neo-Stoicism, is much more complex than the schematic conception of a process of development allows us to perceive. The consulting of revelation and the principle of the sole authority of Scripture are brought into play on the basis of and with the help of the pagan idea of the divinity of the heavenly bodies, which by their nature can be assimilated to the idea of a *deus absconditus* [hidden God] and can thus take part in His sole mode of manifestation, through revelation, or can remain definitively inaccessible as a result of the silence of such revelation. Nevertheless, the aesthetic illuminating power that could justify Dilthey's enthusiasm for the founder (whom he had discovered) of the "natural system" of the history of ideas is missing: "Melanchthon's entire physics is permeated, from beginning to end, by Aristotle's feeling about the universe's affinity to thought. A ray of the radiance of the cosmic contemplation that follows the shining paths of the heavenly bodies shone into Melanchthon's spirit. This spirit never lost its delight in astronomical and physical studies. That delight cheered him in the midst of the *rabies theologorum* [madness of theologians] and the wretchedness of the political world of that time." [100] This relationship can no longer be seen in this way. Was it not precisely the Aristotelian theorem of the complete heterogeneity of the supralunar world that set a limit to the fulfillment of the universe's "affinity to thought," and at the same time offered fideism the vehicle by which to gain entrance to cosmology? The object's transcendence and the artificiality of the means with which it was treated were in the end mutually supporting, and insurmountable inhibitions prevented Melanchthon from interfering with this reciprocal relationship. One of the reasons for this, which also becomes manifest in the second version of the *Introduction to Physics*, was his consciousness of the need to provide the Reformation with a capacity for culture (which Luther disdained, but which was inevitable after the "eschatological phase"), and for that purpose to respect the school framework of education as it was handed down. However remarkable it may sound, the preservation of the canon of the "liberal arts" was one of the motives for not breaking astronomy out of the Faculty of Arts (to which, for example, a person like Caspar Cruciger in Wittenberg had belonged) by changing its epistemological status.

The relationship of forces, as between institutional and theological motives, in Melanchthon's cautiously keeping his distance from Copernicus is something that we can hardly judge. Harnack described the liberation of Christian teaching from the schema of a "sacred physics" as the core of Melanchthon's reforming accomplishment, as it had presented itself for the first time in 1521 in the *Loci communes*. Here Paul's Letter to the Romans appeared as the comprehensive presentation of Christian doctrine.[101] But was there, alongside the institutional one, also a theological motivation for the fact that in the last third of Melanchthon's life that "sacred physics" largely regained its old importance? In order to answer this question, we must put the *Introduction to Physics* back into the continuity of Melanchthon's life, from which we initially removed it in order to determine the changes in one of its chapters. As Harnack writes, one notices "beginning in 1522 and 1523 how Melanchthon becomes uncertain whether one can build Christendom on the basis of Paul and theology alone," and that he "returned, as well, to his first love, to the ideal of his youth." No one returns "as well" to the ideal of his youth, and whether "return" is a means of comprehension at all, in an intellectual biography, is doubtful. Melanchthon has an equivalent of "sacred physics" in this last third of his life, but this time it enters through the gap of Romans 1:20.[c]

Even in the version of 1540, Melanchthon's commentary on the Letter to the Romans still exhibits this breakthrough point clearly. Here there is nothing left of the original severity of the context composed of the three verses 1:18–20. In the New Testament document that forms the foundation of Reformation theology, the Stoicism of verse 20 is entirely in the service of the eschatological threat of wrathful judgment, in the face of which the pagans are meant to be deprived of even the final excuse that they could not have known of God's existence and His law outside of revelation. The use of biblical texts in isolation from their context easily extracted from this digression a jumping-off point for 'natural theology,' which in the Letter to the Romans would appear unusually friendly to the world. In Melanchthon, the talk of the pagans having to answer before the judgment of wrath suddenly took on the importance of confirming the ancient world's conception of man as destined to know divinity through the cosmos, and making the world be created only in order to help fulfill this destiny. Here physics is not the final content of this destiny, but only the instrument by which to arrive at knowledge of God.[102] Not even the Fall was able to frustrate this radically. Otherwise the whole of nature would be condemned to being in vain, as far as the purpose invested in it by its Creator is concerned. But this cannot be permitted, because nature does nothing in vain and does not exist in vain, as Aristotle said. From this admittedly teleological, but not anthropocentric, premise one can then deduce that knowledge of nature must be possible and can be awakened and kindled. The author of an introduction to physics, an *eruditus physicus* [erudite scientist], is able to contribute a great deal to that end.[103] In order to prevent the Creator's production of His world from being in vain, the "sacred physics" has returned, and has done so on the strength of the Letter to the Romans itself.

It is a process that has a significant similarity to the integration of ancient philosophy into the biblical proclamation of the Gospel, in the first centuries after Christ. In both cases the most important feature is that the idea of the Creation pushes back the idea of redemption and justification and itself becomes the central residence of faith and the subject of extensive commentary. Melanchthon's commentary on Genesis[104] is, from a systematic point of view, the connecting link between the commentary on the Letter to the Romans and the *Introduction to Physics*. The interpretation of the account of the Creation is governed by the opposition between the concepts of "nature" and "God's will." What is "natural" here is preestablished by the ancient world's codification, but is not automatically accepted. Thus Melanchthon resists the Aristotelian doctrine of the "natural places" of the elements, which had contributed so importantly to sanctioning geocentrism. For him the Earth has no essential nature (*natura terrae*) fixing the body that is composed of this element with necessity to a particular position in the universe. The association of a heavenly body with a particular location in the universe is a contingent disposition decreed by the divine will.[105]

To be sure, Melanchthon's pressing interest, in this passage, is not directed at the question of the Earth's cosmic location, but at the explanation of the phenomenon of falling, which is bound up with the Aristotelian doctrine of the elements. But even if the situation had been different and Melanchthon had meant at least to prepare the way here for a statement about cosmology, bringing in the divine will would not have permitted any variant model, because this will *communicates* itself in the Creation, so that the phenomenon and the reality cannot diverge. At the moment in which voluntarism is no longer linked to the terrors of God's hiddenness, it gives intuition precedence over reason. Only if the will also manifests itself can everything in nature be a "gift" and nature as a whole the permanent monument reminding us of the "thanks" to be offered for it.[106]

Facts become ordained—and, by the same token, supposed facts as well. This is also implied by Melanchthon's explanation of the individual steps of the six days' work. Earth and heavens are created on the first day as a receptacle for the remainder of the work (*reliqui opificii receptaculum*), but a relation of spatial order between the heavens and the Earth does not yet come about on this day, so that the Earth's "natural place" simply cannot be part of its essential definition.

Melanchthon's rejection of the "Platonic" requirement belongs in this context.[d] It is in its astronomical use that the term "nature" receives its sharpest profile as a concept of antitheological contrast: To ascribe to the heavenly bodies the characteristic motions that belong to them by nature (*nature cursus*), so as to deduce from this the limits to what is astronomically possible, belongs among the trains of thought that are contrary to faith (*impiae cogitationes*), because the heavens move not by virtue of their essential nature but by virtue of the divine word.[107] If one considers that for Copernicus adherence to the "Platonic" requirement is an essential precondition of the epistemological realism of his astronomy, then Melanchthon's rejection of it has to bring us close to the core of his disagreement with the new astronomy. Copernicus's claim to truth is not so much, as it was for Ptolemy, something unattainable and therefore to be abandoned in favor of a possible minimum claim, but rather something inadmissible and almost sacrilegious.

Conceived as science, astronomy has to separate appearance from reality, has to assume an unambiguous relationship of foundation between them—that is, it must always consider the phenomena with a view to the realities that produce them. Melanchthon revives what was contained in the ancient world's concept, allied with "theory," of the *contemplator caeli* [observer of the heavens]. The starry heavens are not primarily a mechanism having 'discrete consequences,' but rather a prospect for intuition. Melanchthon sees the heavens as full of signs, as a field of ciphers that carry significance for men, and for the reading and interpretation of which astronomy has to provide the alphabet. His Christianized astrology fits into this scheme. The cosmic order is, as such, secondary, insofar as it does not have the function of a decree relating to life and salvation—insofar as *ordo* is not understood as *ordinatio*. Astronomy and ethics both find their final authority in the will of God: Just as astronomy is a knowledge of the celestial motions insofar as they are due to divine commands, so moral philosophy is knowledge of actions in their relation as causes and effects insofar as God has deposited them, as something ordained, in man's spirit.[108] How it is constructed is unimportant in comparison to what it signifies. Reason cannot be competent—and it cannot be the epitome of the world's teleology for man (as Copernicus assumed)—to go beyond the foreground of appearance in the will to truth.

A late document of this worldview of Melanchthon's and of its anti-Copernican implication can still be found in his commentary of 1556 on the Old Testament book of Ecclesiastes. Melanchthon comments on the moving complaint at the beginning that everything is totally in vain, man's labor under the Sun is futile, the generations come and go meaninglessly, and only the cosmic scenery, behind the human drama, is unaffected and invariable: "But the Earth abides forever. The Sun rises and the Sun goes down; back it returns to its place, and rises there again." What Melanchthon sees in this text is primarily its 'human meaning,' and secondarily, and incidentally, a 'revelation' about cosmology. The talk of the Sun's course and of its unceasing return is part of the plaintive description of the human miseries that are repeated in a similar way. But then, in conclusion and entirely incidentally (*obiter*), Melanchthon admonishes scholars to take note that the text also contains the statement that the Earth stands still and the Sun moves on in its path.[109] In view of the phenomenon's explicability, he is not principally concerned with this communication, so he can pass over the discussions that are affected by it. But, he points out, it is in fact there, and it sets a clear boundary for disputations. This is the mild Melanchthon, distant from his former polemical severity.

I have spoken of Melanchthon's "backwardness" in comparison to Copernicus, and I mean this in the sense of the chronological synchronousness of things that are historically nonsynchronous. In the logical development of the Humanistic world-formula, Melanchthon halted one step short of Copernicus's position. Both of them extracted from the same sources the fundamental Stoic idea of anthropocentric teleology: the world that was created for man's sake. It was not exactly easy to arrive at this idea of the world from the starting point of the Reformation. When one is under the spell of the terrors of Nominalism's capricious God, with His acts of election and rejection, then the newly apprehended offer of grace to man, the absolution and justification that are granted to him in advance, are bound to put in the shade and render insignificant what could be presumed to have occurred for man's benefit in the Creation.

There has never been, in the Christian tradition, a balanced relation between the doctrine of the Creation and the dogma of redemption. This insoluble conflict lay at the root of Gnosticism, which thought that it could solve the problem only by means of two gods. The more radically the idea of redemption was to be applied, the more negative the result was bound to be for cosmology, in which the competing comforts of the world and of man's opportunities in it could only reduce, if they did not level off altogether, the distinctiveness of the expectation of salvation. In relation to Copernicus's world-formula this meant that the "for us" that it contained could not be granted the same level of importance as the Christological *propter nos* had for the Reformation. To put it in extreme form: Christ would not have had to die for man's benefit if the world had really been created for man's benefit. Therefore the first action meant to benefit man can only be weakly confirmed in comparison to the second: Man's dignity in the cosmos can only be deducible from the dignity of the event of God's incarnation as man in the cosmos. So the Renaissance formula of the *dignitas hominis* [dignity of man] does not take the teleological form according to which everything is there for man's sake, but rather the imperial and genuinely biblical form according to which everything is made subject to him.[110]

Where Melanchthon nevertheless emphasizes the teleology on man's behalf, he turns it back toward its origin in such a way that the ancient answer to the question of what the world is is only made tolerable by the corollary that while it is true that everything in the nature of things is produced for man, man in his turn was created for God, who wants to be known and to be contemplated in the world.[111] In this voluntaristic formulation, the theological amplification not only carries out what was begun cosmologically; it simultaneously restricts it to the economy of the happiness of knowing God.[112] The teleological idea becomes a restrictive principle even for astronomy, because only where usefulness in human life is affected can we count on it to be effective.[113]

The backwardness in the rational interpretation of the teleological principle is perhaps even more significant, among the people around Melanchthon, when the individual in question is the 'Copernican,' Erasmus Reinhold, the author of the *Prutenic Tables* (which were so useful in promoting the kind of fame that Copernicus did not desire). In 1541, Reinhold explained the Earth's deviation, through its mountains and deeps, from the pure form of a sphere by saying that God had abandoned rationality in His work in favor of usefulness for man.[114] Leonardo da Vinci had already written down a gloomy variant of this idea, one that was less friendly to man, when he described the possibility of life on the Earth as being bought at the expense of violating the spherical form through the uneven distribution of seas and land masses—but only as an episode, which would one day be brought to an end by weathering away, so as to make room for a uniform inundation of the whole surface and to restore the impossibility of man upon it. For Leonardo this interruption in nature's self-presentation by means of its pure form embodied an antiteleology only the clearing away of which was supposed to restore the *telos* of a Platonic eschatology. As Erasmus Reinhold expresses his teleological idea, it could very well already be directed against the Copernican system as a cosmological one: If the Earth is not a sphere, it cannot be a star either; but precisely its incapacity to be a star makes it man's home—pure rationality would make his life impossible.

Such objections, derived no less from the teleological view of the world (and an anthropocentric one, at that), must have been familiar to Copernicus when he wrote the exceptionally long third chapter of the first book of the *De revolutionibus* on the question of how earth with water could form a perfect sphere. He has to harmonize his demonstration of the Earth's spherical form with the irregularities on its surface from which man benefits, if the purity of the Earth's shadow image during eclipses of the Moon is to be a valid argument. Only as a sphere can the Earth be a star, and only as a star can it produce the astronomical phenomena. Only by producing what has to be explained can it satisfy the claim of reason.

Copernicus and Melanchthon differed regarding the human needs that had to be capable of being satisfied before one could speak of the world's having been produced on man's behalf by a maker who was not only well-meaning but also maximally intent on regularity (*regularissimo omnium opifice*).

4

Perplexities of Copernicus's Sole Student: Joachim Rheticus

The name of Joachim Rheticus extends, like a red thread, through all the phases of the early history of Copernicus's impact. The whole typology of Copernicus's reception is displayed here. Rheticus is the reconnoiterer, he is the convert, he is the herald, he is the heir, he is the doubter, and he is not far from being the apostate.

For there is no question about it, Rheticus struggles with the master's legacy in an extremely questionable way. This is not the case with most of what he has been blamed for. He was under pressure when he left Copernicus's work uncompleted at the printer in Nuremberg. He was following accepted academic procedure when he announced his Wittenberg lectures on Ptolemy. He did not have much alternative when he did not comply with Tiedemann Giese's request that he should bring a complaint before the Nuremberg City Council.

The later story of Copernicus's only student, after his death, deserves attention if only because it is an amazing case of the failure of the enlightenment that the subsequent centuries expect from science and as the prototype of which they were to celebrate Copernicus's accomplishment and his impact. That makes this case of the most direct influence, of emphatic conviction, of an inheritance that was offered and was accepted, an occasion for thought about the following question: How is it that enlightenments miscarry?

Enlightenments appeal to reason as their ultimate authority. The critics of the enlightenments—what else should they criticize them with except, once again, with reason?—assert that it was not reason [*Vernunft*] at all that made the enlightenment, but rather the understanding [or 'common sense': *Verstand*], which, bereft of all reason, put itself in reason's place. However transparent the pattern of such assertions is,[115] they do nevertheless inform us of a deficiency in what this reproach is aimed at. Could it be due to the fact that in their historical function, though not in their self-understanding, enlightenments, as aggregates of critical actions, do indeed clear away the rubbish of prejudices, but in the zeal of this procedure do not bear in mind that the needs that these prejudices served are still there, and that they must be satisfied in a new and different way? That is why, to the surprise of the exhausted enlighteners, after them the still vigorous dogmatists go to work on the new believers.

Joachim Rheticus, a man for whom life is a serious business, searches all his life for contents that will be convincing and fulfilling. When he meets Copernicus, the world for him becomes simple, clear, familiar, dependable. A truth is offered in a place where it could not have been expected. A truth, and no more than that? Not also 'significance,'[a] assistance in life, such as Melanchthon wanted to hold to in a sort of Christianized astrology? One of the commonplaces about the Copernican reform is that it made the superstition of astrology impossible once and for all. The unmoved heavens are also heavens that exert no influence, and the eccentric and unimportant Earth is scarcely any longer the focal point of all cosmic dispositions and references.

It is not even certain whether Copernicus himself had drawn this conclusion. A small, not entirely dependable indication is the already mentioned letter of John Apel, in Nuremberg, to Duke Albert of Prussia, a letter that he wrote to accompany the delivery of a horoscope and in which he mentioned that for assistance in interpreting the constellation there was still "an old canon in Frauenburg, if it cannot be found elsewhere."[116] Can we conclude from this that Copernicus cast horoscopes and interpreted them? Probably not. Such a skill would automatically be imputed to someone who had a reputation as an astronomer—how else should he live, if he did not have an ecclesiastical living, like the canon in Frauenburg?

This argument has to benefit Rheticus too. The way in which he distinguished his inclinations from his means of livelihood can be seen from his remark that astronomy and chemistry were on the side of his pleasures, and medicine on the side of his needs: "Chemicis delector et astronomicis, sed ex medicina vivo."[117] In his dedicatory preface to the posthumous astronomical treatises of John Werner he writes, regarding astrology, that it can predict not only the weather but also great political changes on the scale of the passage of power from one people to another, from the East to the West and from there to the North.[118] He concedes that the Bible attributes upheavals of this sort to sin, because a curse lies on the Earth on account of it; but the 'salvation story' seems, in fact, to be coordinated with the cosmic process only as a moderating or an intensifying factor. No prayer can withstand the celestial destinies, and the evils for which men are responsible are only piled on top of those that nature ordains. Man, insofar as he determines man's fate, has a moderating or an intensifying effect.

So it continues to be necessary to give sovereigns astronomical assistance, and it continues to be in their interest to furnish the necessary means to refine this assistance. To justify astrology in relation to Christianity it is enough to say that the dominion of the heavenly bodies over the inferior world is instrumentally related to God's dominion—that constellations depend on the divine will, as the example of Joshua shows.[119] It is characteristic of the state of the development of the concept of laws of nature that Rheticus cannot distinguish between the "ordinary" course of celestial events and an "extraordinary" intervention in their process. All of this could be the practice corresponding to Melanchthon's theory.

It is world destinies more than personal ones that Rheticus's astrology undertakes to deal with. On the basis of the anomaly of the apparent motion of the heaven of the fixed stars, he promises to the Hapsburg prince, who is just arming for war with the Turks, the sudden downfall of the Turkish Empire. By means of the astronomical art (if not yet in this art itself) the heavens become communicative: "Coelum per astronomiam loquitur nobiscum ...." The heavens and the Earth have been, from the beginning, like books about the change of empires, if only people had viewed them correctly, rather than with blind eyes, and had not passed them by with deaf ears. What historical books contain is only, Rheticus says, a transcript of what the heavens foretold and commanded. So this is a language that is directed chiefly at the world's rulers, and has to be made intelligible to them. The book metaphor gives astrology a political office. It enables Rheticus to take the dependence of terrestrial events on stellar constellations as the carrying out of a preexisting pattern, and history as the reading off of an already formulated text, which 'returns' *post factum* [after the fact] to the overall status of a text in the works of historians—which, Rheticus remarks, Caesar even provided for himself, rather than leaving to others, as his prototype, Alexander, did. God himself turned Julius Caesar's eyes to the heavens, so that he could read there the writing by which the Author of nature speaks visibly.

The book metaphors bring astronomy even closer to the model of the theological concept of revelation than would be necessary in order to justify astrology. This 'revelatory' conception of the possibility of knowledge of nature has important consequences for Rheticus. To begin with, the metaphor relieves astrology of the offensiveness of the determinism that is indispensable to it: The heavens are the oracle, given by the wisest seer of all, of the events of the future, of which he is at the same time the author. The "book of nature," only episodically translated into the fleeting language of action, gains its other permanent form, and an intelligibility that is less in need of an intervening interpreter, in the book of historiography.[b] Time lies symmetrically on both sides of the axis of action: One can read the latter beforehand, in one book, and afterward, in the other, as a chronologically definable 'event.'

Rheticus's concept of an event is pushed to apocalyptic extremes. The heavens, through their interpreter, Rheticus, do announce the downfall of the Turks to the Hapsburg prince, but they also announce even more threatening and far-reaching downfalls: When that precession of the heaven of the fixed stars reaches its third terminus, God will carry out His sentences of judgment and the powerful will be dethroned and the weak raised up.

Rheticus aims still higher when he makes available, for the completion of the *Chronica* of John Cario, which Melanchthon published, a time reckoning of his own that, being based on Copernicus's astronomy and his theory of precession, is supposed to reach from the creation of the world to its end. In other words, it would have functionally unified the chronological work based on biblical statements and that contained in astrology.[120] Here we find explicit confirmation, in the connection with Melanchthon's name and work, for the conjecture that Rheticus took over his supposedly Christian interpretation of astrology from his teacher, though certainly not without intensifying and taking to an extreme its adaptation to the concept of revelation.

Rheticus's departure from Copernicus and his blindness to the 'enlightening' claim of the latter's astronomical reform becomes almost anecdotally tangible when he announces precisely to his deceased teacher's closest friend, the bishop of Kulm, Tiedemann Giese, a calendar for the year 1550, which he intends to publish in an edition of 5,000 copies, and mentions, to stimulate Giese's interest, that with its use (though not as a result of it!) the plague will cease to rage in Prussia, since as early as January Saturn passes from the sign of Aries, which is unfavorable for Prussia, to that of Aquarius.[121] Finally, we still have a late astrological opinion rendered by Rheticus on the succession in Poland after the death of King Sigismund II, a document that would date from the year 1571.[122]

Rheticus takes a less specific position regarding the effectiveness of astrology when he writes, in his dedicatory preface to Euclid's *Elements*, that this art, which Ptolemy called "prediction by means of astronomy," is a great help in protecting health and property, and indicates the confluence and excess of the humors in the body as well as changes in the weather and the climate—but is controversial and still needs a new treatment, which he himself, at least vaguely, proposes to provide.[123] Even in his programmatic letter to Ramus in 1568, in which he responds to the challenge of an "astronomy without hypotheses"—that is, an astronomy that would meet the criteria for being a science—Rheticus quite spontaneously mentions his astrological efforts. Here, indeed, one can also see the conclusion of a development. Rheticus claims to have founded a new discipline of his own, which is supposed to concern itself with the oldest foundations of this "art." [124] Thus he claims for astrology, too, what he thinks he has made possible in astronomy: that it can offer an access to the origins and the beginnings, to the ages that lie closest to the original revelation of wisdom. From this point of view, too, the erection of the Egyptian obelisk in Cracow marks the culmination of his departure from Copernicus.

One of the earliest stations in this development, perhaps a symptom of a first experience of being at a loss after the death of his teacher of astronomy, is Rheticus's visit to Girolamo Cardano in Milan in the fall of 1545. He could be linked to this important doctor, mathematician, and astrologer merely by this combination of interests, which was his own. It is not without irony that Cardano had had his famous *Ars magna* published at the beginning of this year by the same printer, Petreius, in Nuremberg, by whom the *De revolutionibus* had also been published, and what is more, he had dedicated it to Andreas Osiander. Cardano mentions Rheticus's visit in his widely read autobiography.[125] The paths of these two men through life share the pattern that resulted from their universal and never-to-be-satisfied interests and led them into the obscure border regions of science. In 1570 Cardano wound up in the prison of the Inquisition on charges of heresy, a fate that Rheticus was spared probably only because deviations from the offical faith were part of Cracow's character as a city. Cardano was a magician, and Rheticus would have liked to become one.

There is at least one contact with the realm of the demons. People around Rheticus connected a serious psychic crisis in his life to demons. When, ignoring all admonitions to return, he withdrew from the Leipzig faculty of arts and taught mathematics in Constance, rumors of Rheticus's being possessed went around (*quaedam sparsa de daemonio*) through the usual channels, by way of traveling merchants.[126] In any case a good friend of his, the schoolmaster in Lindau, Kaspar Brusch, did not want to deny that such rumors had a kernel of truth ("quae etsi non sint vana aut falsa"), though he also had to say that after five months of serious illness Rheticus had feared for his reputation and had heard with bitterness that such stories had been circulated. Brusch also reports that Rheticus's parents, in nearby Bregenz, had pressed him to make a pilgrimage to a Catholic shrine in Alsace at which possessed people were healed ("in quo papistae multos daemoniacos liberant"), but Rheticus had hoped for "liberation" only by Christ.

The crisis is described by this eyewitness as a religious one. As Rheticus's friend he had visited him daily and talked with him, gave him biblical and theological materials to read, and also writings of Luther, Melanchthon, and Cruciger, which Brusch, remarkably, characterizes as "prophetic." Rheticus, he says, sometimes cried to the Son of God—from Whom alone he hoped for that "liberation"—out of a full heart and with passionate vows, and often in tears. This *liberatio* seems not to have been a matter either of "redemption" in the theological sense or of "relief" from pain in the medical sense. After his illness, we are told, Rheticus was changed and seemed like a different person. He devoted himself entirely to his studies and to reading from Holy Scriptures. Once he even preached to his friends with an earnestness that can probably only be described as "eschatological": against the unprecedented self-confidence of the world and of men, with which they live for the moment and do not believe in the demonic realm or in those dark avengers of their misdeeds. Melanchthon's student has turned away from the confidence in the world and the joy in it that had grown in this teacher over the years. He will also interpret the signs in the heavens differently—more threateningly and apocalyptically.

It was natural to ascribe a 'Faustian' quality to this crisis, and we know from the preface to Johannes Kepler's *New Astronomy* of 1609 that this was done. At the distance of half a century, all that remains from the times of the fathers is a rumor. Rheticus's crisis has become the prologue to Kepler's own: a fruitless struggle to describe the orbit of Mars, which for Kepler was fulfilled and solved only in his own work, by the abandonment of the Platonic prescription. The desperation of theory becomes conflict with the demon.

According to Kepler, during Rheticus's observations of the planet Mars he finally "took refuge with the guardian spirit of his kind," whether it was in order to avail himself of the demon's knowledge or "from impotent longing for the truth." But the demon was angry and attacked him, dragged him by the hair and knocked his head against the vaulted ceiling of the room, and finally smashed his whole body on the floor with the scornful exclamation, "That is the motion of Mars." For Kepler the kernel of truth in the story is the difficulty, driving one to the edge of spiritual disorder, of the problem of this planet's orbit; Rheticus had been beating his head against a brick wall, and that had had the appearance of a violent assault by a demon. But it was the resistance of the object itself: "One can see that Rheticus had to undergo this, because he had irritated Mars." It seems almost manifest how history presses toward making the refractory demon the partner of the desire for knowledge. Rheticus was not far from the configuration of the Faustian pact.

The time that he spent with Copernicus, the imprint of Copernican rationality, did not penetrate deeply into the intellectual structure of this man. It is the pattern of the intuitive *coup de main* that he interprets as a divine intervention. This was already evident in the *Narratio prima*, where he characterized the conclusiveness of the Copernican doctrine by the fact that its spiritual comprehension is produced more quickly than its proofs can be expounded in words. But people who have received illumination are easily disappointed.

His teacher, Copernicus, had appeared to him as one who was inspired. This imago of personal charisma soon has the other fundamental idea superimposed on it that the astronomy that mankind possesses must be based on a primeval endowment. It is not yet evident that these two ideas are not mutually compatible. As in theology, it remains difficult to understand why the possession of truth as a result of an original revelation has been less convincing and less durably effective than the truth that derives or is expected from a new revelation or illumination. For this construction, something or someone must always be found to blame; for this role, Rheticus, diverging sharply from his teacher, was to select Ptolemy. In fact, by a roundabout route via his intended target, Ptolemy, Rheticus's critique hits the figure (which no doubt was sacrosanct for him) of Copernicus. Copernicus's role is inevitably only made more puzzling by the fact that he claims to have found, at such a late moment in the history of the world as measured by Rheticus's astrological chronology, a truth that had been imparted to mankind from the beginning. When Rheticus was staying with Petreius in Nuremberg in 1542 he wrote in the dedication of his *Orationes duae* [*Two Talks*] that there were many proofs that the "liberal arts," with which God not only adorned this life but also in numerous ways made it easier, were of divine origin (*divinitus ostensa*). Since the beginning of the world they had been handed down by divinely inspired people to posterity, and their obvious usefulness for the whole of life was a result of this.[127] This idea of an original scientific revelation already implies the question whether original traces and remnants of it can be found, and also what instruments are proper to it. Admittedly, the conviction that 'higher' theoretical means are possible, and the striving to find them, cannot be understood unless one assumes that Copernicus's 'truth' had, after all, finally disappointed its adept's expectations, just as it had not finally been able to satisfy its author either in his earliest claims and expectations.[128] Wanting to know more of the prehistorical astronomical truth, Rheticus unexpectedly moved closer and closer to the position of Osiander, who had wanted astronomical knowledge to be reserved for a divine communication—with the difference, though, that Rheticus took this literally and began to hunt for it, as a given primal fact. When he did not confront Osiander and did not pursue the complaint before the council in Nuremberg, could he already have been aware of this approach to an alternative that did not lead to resignation, but did not lead to reason's truth either, but rather to participation in higher initiations?

Even if Rheticus assumed that God meant to give man the writing of His Creation, in great clarity, to read, nevertheless as one who was familiar with astronomical technique he could not ascribe that 'clarity' to the mere intuition of the starry heavens. Every evident proposition in this discipline could only be based on the instruments that ensured the accuracy of its observations. That is why for Rheticus a decade and a half later the original revelation of astronomy became the divine imparting of its most important instrument. When Rheticus had settled in Cracow—expressly because it lies on the same meridian as Frauenburg—he set up the Egyptian obelisk and simultaneously (and logically) rejected the authority of Ptolemy.

The obelisk, as something shaped by nature, is for him like an *acheiropoieton* ['something not made by hands'] that fell from heaven.[129] According to Pliny's account, the first obelisk was set up in Egypt by King Mitres on the strength of a vision in a dream: "Deum autem velle, ut suam geometriam in natura consideremus . . . . " [God wanted us to consider his geometry in nature.] The connection between the obelisk and the cult of the Sun was important because it pointed to heliocentrism and thus to Copernicus: "Thus, too, the Sun is the queen and the ruler of the celestial state; all the other stars move according to her rhythm and her stimulus. She is also the eye of the world, by the light of which everything is illuminated. Thus only by means of the obelisk can all the laws of the celestial state be precisely apprehended and described."[130]

The decisive difference from Copernicus becomes tangible: Rheticus does not believe human reason to be capable, when left to itself, of getting to the bottom of the mysteries of the heavens; it has to do something that Copernicus would not have thought of, namely, to have recourse to a tool that has a higher origin and an effect that instruments invented by human beings do not produce. By believing that he can provide the Copernican theory with an archaic guarantee that has the dignity of the sacred, Rheticus relapses behind Copernicus's rationality and his own agreement with Copernicus's standpoint. What moves the astronomer is no longer common human curiosity but instead the legitimacy of something that helped to fulfil divine instructions given to Pharoah; the initiate's investigations are characterized by ease and exactitude.[131] The obelisk is an intervening agency, an original document; it is not only an interpreter of nature but is itself its interpretation (*ipsius rerum naturae interpretationem*). So this expression, *interpretatio naturae* (which was to become central in Bacon), points to the background of the primeval and the magical, but also to that of the authentic legitimation of the desire for knowledge. It was appropriate that the Egyptian teachers of Pythagoras and of Plato had been the first to use an instrument the forgottenness of which was to make possible Ptolemy's epistemological skepticism.

But now the almost banal counterquestion: Do obelisks actually look like instruments that have a divine origin? "But who, when he saw such a mass of stone, would imagine that this mass was the interpretation—or even an interpreter—of nature? No one, I would think, unless it were one of those whose traces the shipwrecked and hopeless Aristippus found on the beach"—that is, a mathematician. What is meant is, no doubt, that the sign of the obelisk's divine origin dawns only on those who are able at their first sight of it to infer, at least hypothetically, the principle of its production and its function. So what has become stone here is not a revelation for everyone, but is, again, a revelation that requires its priests in order to be recognized as such at all, and then to be expounded.

What truth was Rheticus seeking when he set up the obelisk? It was undoubtedly still the truth of the Copernican theory. He was no longer satisfied by the argumentation that Copernicus had offered in his book. Of what could the further proof of its truth consist? To begin with, it is clear that Rheticus wanted to use the obelisk for determining the positions of fixed stars. This is evident already from the earliest mention of the obelisk, in a letter of 1554 to the Breslau doctor Hans Crato, in which Rheticus announces a new survey of the heaven of the fixed stars with the aid of the obelisk.[132] In that case the connection with the question of the truth of Copernicanism could only be in the expectation that he would be able, by means of increased accuracy in determining positions, to demonstrate the parallactic image of the Earth's annual motion, the apparent absence of which Copernicus had explained by the excuse that the sphere of the fixed stars was immensely distant.

So Rheticus understood his teacher's last charge as requiring him to take over the obligation of proving his system without the anthropocentric premise, by mere observation—by increases in technical accuracy and by demonstration to intuition. We do not know whether even the formulation of this asserted legacy has a basis in fact. But Rheticus makes it public, and asserts it vis-à-vis the highest authority, as the bequest of the man whom he honored not only as his teacher but as his father. After he had passed three years in Prussia, he says, the noble old man had enjoined him, at their farewell, to apply himself to the completion of what he himself, because of his age and the inevitability of his end, was no longer able to finish.[133]

If, however, one looks more closely at what can have constituted the incompleteness of Copernicus's work, with which the bequest is connected, it becomes clear that for Copernicus it evidently cannot have had to do, any longer, with proofs that were still needed for the first three motions of the Earth—that is, with diurnal rotation, the annual orbit, and the supposed change in the position of the Earth's axis—but rather, evidently, with the problem (which had been smoldering since his early controversy with John Werner) of the precession, and its speculatively assumed periodicity. It was most probably the problems of the heaven of the fixed stars, which are discussed in the third book of the *De revolutionibus*, that Copernicus bequeathed to his student.[134] This commission would not, indeed, be unconnected to Rheticus's astrological and chronological interests. By the technical linkage of the planets' positions to the positions of the fixed stars, all the inaccuracies of the astronomy of the fixed stars were transplanted into planetary astronomy. Here the Copernican reform had not yet produced the anticipated increase in accuracy that was necessary for the work of the astrologer to be reliable, and this may have contributed to Rheticus's growing reserve with regard to his Copernican inheritance.

That is why the reports in ancient writings that the Egyptians had already possessed the heliocentric system were invaluable for him. His train of thought must have been something like the following: How was it possible that right at the beginning of astronomy this truth, which in later times has been so difficult to establish successfully and which is so repugnant to the human spirit, was present? Either it was, as such, a component content of that original revelation—an idea that had to be repugnant to the theoretician and that could not be much use to him as argumentation in dealing with his colleagues and contemporaries—or else the secret of that early discovery had to be hidden in the instrument of observation. But what could the instrument of the obelisk have proved? The only compelling proof that Rheticus could have known of that did not need to depend on the results of the astronomical tradition since Ptolemy was the demonstration of the parallax of the fixed stars resulting from the Earth's annual motion. Rheticus must have believed that he could observe it with the obelisk, even though Copernicus had already bracketed it out as a proof. Of course, if he was not disappointed in this expectation it must have been because he gave himself up to an illusion. The obelisk could not come close to producing the necessary accuracy of determinations of position, which was first achieved by Bessel in the nineteenth century.

Much more important is the distancing from Copernicus that is involved in this train of thought. For, after all, Rheticus's seeking with much effort and with almost magical hopes for an observable proof was, at the same time, manifestly incompatible with the idea that Copernicus himself had already secured this truth beyond all doubt. This finally allows Rheticus to acquiesce in Petrus Ramus's formula that an astronomy without hypotheses was still to be found. One can formulate it still more sharply: In his search for an observable proof, Rheticus leaves the astronomical tradition by the means and according to the criteria of which Copernicus believed that he had found his truth, in that, for himself and for his theory, he had recognized the satisfaction of the "Platonic" requirement as sufficient and conclusive. If one can put it this way, very briefly, Rheticus wants to 'see' the motion of the Earth in space. His distancing of himself from Copernicus, in his fulfillment of Copernicus's bequest to him, consists in his departure from the "Platonic" requirement as the sufficient ground for astronomical verification.

Now it becomes understandable that at the moment when he hopes that he can open a new path, Rheticus leaves the tradition of astronomy that was handed down from antiquity and turns against Ptolemy as the corrupter of the Egyptian primeval wisdom; and at the same time it becomes clear that this turning was not possible without indirect offense to the memory of Copernicus. The teacher's and the follower's argumentations did not complement one another, did not strengthen each other reciprocally. The mere assertion of the possibility of the one reduced the weight and power of the other and with them one's confidence in its author's power of judgment. Copernicus, who has just been apostrophized as a father figure, is immediately consigned to the obsolescence of the fathers. This is not altered at all by the fact that Rheticus was not able to reproduce the new, and supposedly ancient, proof.

Only now does the famous exchange of letters between Petrus Ramus and Joachim Rheticus become recognizable as the magnificent finale of the first generation of Copernicans. In 1563, the prospect of being called to Paris as royal mathematician was held out to Rheticus. Petrus Ramus had the chair of rhetoric there, and no doubt it was also rhetoric when he offered this chair as a prize for the discoverer of an "astronomy without hypotheses." What Ramus meant by such an astronomy, though, was a deductive procedure. What he had in mind was to deduce all the *artes liberales* [liberal arts] in an exclusively logical manner, as he thought he had already successfully done with geometry. He did not succeeded in doing this with astronomy, and he communicated this result to Rheticus in 1563.[135] The current state of astronomy was so difficult to comprehend, he said, and so confused by hypotheses that neither in logic nor in literature nor in men was there help to be found for the reformation that he had in mind. Ramus appeals to Plato: He was the inventer of the requirement of an astronomy without hypotheses, according to which the stars would have to take their course without any unreliability (*moveri non ulla infirmitate*) and, in their permance, would have to have no need at all of the construction of hypotheses (*hypothesium machinis*).

It is surprising to find that the illustrious Paris 'artist' comes closer, with his advertisement for the reformation of astronomy, to Copernicus (about whom he seems to know nothing) than to his addressee, whom he is courting. It looks as though Ramus understands by an "astronomy without hypotheses" nothing but the satisfaction of the "Platonic" requirement in its strict form, in which only homocentric orbits around the center of the universe are allowed. There is no doubt that that was also the pattern of the astronomy of which Copernicus had dreamed when he was taking his first steps into the new territory of heliocentrism, and for which he had already had to lower his expectations somewhat in the *Commentariolus*. Ramus's expectations of astronomy also result from the essence of the heavenly bodies as the most extreme purity and simplicity of the idea of circularity.[136] Hypotheses, his letter goes on, appeared in astronomy only at a late date, having been introduced, according to Proclus's account, by the Pythagoreans, as the deformation of concentric circles by epicycles and eccentrics. Thus they were nothing that could compete with the great age of the true astronomy, from the Babylonians and Egyptians up to Plato.

It looks as though someone is seeking Rheticus's agreement who has heard only casual accounts of the latter's Egyptian inclinations and can understand these precisely in terms of the astronomical tradition. All of this leaves no doubt that under the title of his "astronomy without hypotheses" Ramus did not have in mind any science using physical arguments, or even a kinematic theory that diverged significantly from the classical model. This is made sufficiently clear by the example that he choses, which he takes from the dedicatory preface to the *De revolutionibus*, which he erroneously ascribes to Rheticus: The hypotheses from which astronomy has to be freed are, he writes, as monstrous as the epicycle of Venus; they are inventions that have nothing to do with the nature of the object.

But Joachim Rheticus's view of things had long since ceased to correspond to these expectations—and in fact precisely because he was no longer the Copernican he had been at the very beginning. If we have not lost parts of the correspondence, Rheticus answers Ramus's great summons only five years later. He will set about producing an astronomical work, he says, that will meet Ramus's demand that astronomy should be freed from hypotheses. But the addressee must have been surprised by what Rheticus added to this promise when he went on to say that the goal of freeing astronomy would be reached by founding astronomy exclusively on observations: "... ut hypothesibus artem astronomicam liberarem, solis contentus observationibus." [137] A decade and a half after its erection, the Egyptian obelisk is still the key to the accuracy of observation that can remove heliocentrism from the status of a hypothesis. The Egyptians, Rheticus writes, had possessed this knowledge, and only Ptolemy robbed us of it because his *magnae constructiones* [great constructions] had, instead, the effect of *maximae destructiones* [the greatest destructions] on the contents of that archaic tradition. The outsider, who no longer had to show consideration for the school and its canonical texts, breaks off all connection to Ptolemy. It was not without reason that Rheticus went no further with the commentary on Copernicus that he himself announced and that the people around him expected—because indirectly this would inevitably have become still another commentary on Ptolemy.

In the way in which Rheticus observes the bequest that he accepted from Copernicus, two motives intersect: the motive of obtaining a proof for heliocentrism that goes back to the sources of Egyptian astronomy and that gains access, by means of increased observational accuracy, to the parallactic reflection of the Earth's annual path, and the motive of pursuing the open problems of the long-term changes in the length of the year and in the positions of the seasonal points [the solstices and equinoxes]. Although the first intention had only been made possible by the fact that Rheticus had freed himself from Copernicus's premises and from his consciousness of success, and the second intention required a long-term series of observations that far exceeded the interval marked out by the lives of the two men, there was nevertheless something that the two intentions had in common, methodologically, and to which the future belonged: increasing the accuracy of observations. What Rheticus correctly recognized, and what finds a symbolic representation in the figure of the obelisk, is the importance of instruments for all the future possibilities of astronomy. The consciousness that with Copernicus astronomy had essentially arrived at the end of what was required, theoretically, would have blocked progress precisely at the only point where it was still possible. However obscure its secondary meanings, in connection with Rheticus's 'Faustian scene,' may have been, the obelisk nevertheless signalizes the gap through which an unanticipated future disclosed itself for astronomy alone.

Copernicus's student with the obelisk, who transforms the demand for an astronomy without hypotheses into his Faustian claim, is an unexpected embodiment of the effect of science in the form of its prototypical early protagonist. Rheticus is the first significant example of the difficulty of combining the rational claim to be making a new beginning and establishing science in its definitive form with the appeal to reason as man's essential nature and to its teleological claim vis-à-vis the universe of its objects. The difficulty consists in admitting and thinking through the possibility that centuries of error, of resignation, of stagnation and of folly had held back the scientific truth that was possible at any time.

Copernicus himself, and then the early Copernicans, flirt with the answer that it was the time span required to construct the observational basis of astronomy—a time span that could not be abridged, could not be forced, but could only be waited out—that had not been completed earlier. But that was true, after all, only for certain questions in the astronomy of the fixed stars. Every rationalism needs an answer to the question of what the "self-incurred tutelage" that was to be spoken of at the culminating point of the Enlightenment really means.[c] It needs this because after all it can only be reason itself that can incur the blame for this—that leads, out of reason, to unreason. In Kant's formula about self-incurred tutelage the early modern age's organic metaphors for history—the comparison to the ages of man—are parried: One cannot have oneself to blame for being under the tutelage to which children are subject, but (at bottom) only for the tutelage that results from giving up [the opposite]. But how can reason have given itself up?

Rationality becomes irrational where it seeks to give reasons for its own failure and thus evades the suspicion that that failure could ever be repeated. Rheticus, too, faces this problem. It is the problem of Copernicus's lateness and of the provisionality that, ultimately, could still be detected in this lateness. Included in the eschatological consciousness that the time allotted for the world is coming to an end is the consciousness of urgency, that one cannot wait any longer, and also the remedy of establishing a symmetry between beginning and end—of going back to the beginning and discovering the pure substance that must not be lost at the end. This picture of the degeneration and rediscovery of the truth requires a guilty party, and the irrationality of the schema concentrates itself in him. For this role Rheticus finally nominated Ptolemy, who had destroyed the astronomical tradition beginning in ancient Egypt and had suppressed omnipresent reason. The fact that not everything is possible for reason at every time cannot be part of reason's 'enlightenment' self-understanding, because reason claims—precisely in regard to its actual present situation—to tolerate no historical restrictions on what is possible for it (because such restrictions would also hold for each of its future situations).

This is where 'Romanticism' arises. Rheticus's Romanticism is fixed in the distant unhistorical exclave of Egyptian astronomy, to which one only needs to return in order to see reason's autochthonous accomplishment plainly before one—petrified, in the obelisk, and destroyed and unappreciated by Ptolemy. It is on him that the disturbing and demonic qualities are projected and to him that the paratheory addresses itself that describes the powers that managed to break in and interrupt and that were able to embody their effectiveness in a whole epoch, the Middle Ages. Here one no longer needs the idea of having brought it on oneself. Rheticus saw himself in the image of a Hercules who vanquishes the monsters lying in the way of the truth.[138] From his case one can already learn most of what subsequent centuries, insofar as they also understand themselves as the heirs of Copernicus, will have to work out in the way of difficulties in self-comprehension.

5

Not a Martyr for
Copernicanism:
Giordano Bruno

Seen in terms of the whole history of his influence, Copernicus triumphed in the end not so much *over* as *through* his opponents. No doubt he would have entered the history of natural science as a figure of the first rank even without them, but for the self-consciousness of modern man he would hardly have achieved more importance than, for example, Boyle or Lavoisier, to whom chemistry owes everything but to whom a world that is dependent on chemistry is no longer able to attribute anything.

The adversaries of Copernicanism formulated, in the form of rebuke, everything in the astronomical reform that could become potent for consciousness. This elementary state of affairs—of action through counteraction, the generating of potential energy as a response to contradiction—is concentrated in the figure of the unfortunate Giordano Bruno, in his fate and in his works. He gave it exaggerated, satyrical form in the dialogue *La Cena de le Ceneri, The Ash Wednesday Supper*. Measured by its historical influence, this dialogue is not his most important work, but it reveals how Bruno the metaphysician, for whom the death at the stake was intended and whose influence extended deep into German Idealism, and Bruno the Copernican, who provided the cosmological orientation for the metaphysician, are connected.[139]

The nexus between the Copernican and the metaphysician arises as a result of one of the most important objections to the annual motion of the Earth, the objection that is based on the absence of any parallax in observations of the fixed stars. Since the great orbit around the Sun is not reflected in shifts in the heaven of the fixed stars, the universe must be immensely large, and the stars, which in that case are extremely distant from us, must be huge luminary bodies if their light is still to be visible to us. With Giordano Bruno, the "immensity" that Copernicus had to concede becomes "infinity." What had been intended as an objection to heliocentrism is laid down as what the reform actually makes it possible to demonstrate. Bruno justifies Copernicanism on the basis of its implications, rather than its inevitability. The new cosmology has elevated the universe itself and as a whole to metaphysical dignity.

The infinite universe is the appropriate interpretation of a world-concept that no longer functions as a proof of the existence of a transcendent divinity, but instead functions as a demonstration of divine self-communication: "We know for certain that this space, being the effect and the product of an infinite cause and an infinite principle, must be ... infinitely infinite." The anti-Copernicans appear as fainthearted people who cannot keep up with the ascent into this dimension. Bruno's strength, which is also his predisposition for conflict with the Counterreformation, consists in the fact that he regards the argument about Copernicanism—and represents it in his writings—as only the foreground manifestation of a disagreement in which the issue is Christian theism itself. Bruno needed Copernicanism as truth.

The Ash Wednesday Supper presents the whole range of makeshifts that the lack of both the principle of inertia and the law of gravity had made indispensable for someone who wanted to be able to hold to the 'truth' of the new system. Where the new physics was still missing, metaphorical ideas of the organic coherence of the heavenly bodies veiled the explanatory accomplishment that was still owed to the reader. The anticipatory grasp at a new world-scheme uses ridicule of its opponents to cover up the uncertainties that it is not yet a match for, and seems to promise that there will be no need for another thousand years' work. But the new expansion of the universe is not an expedient of this kind—it is no longer a mere refuge from objections.

If one wants to see what is involved, behind the linguistic formulas, one has to realize what order of magnitude of "immensity" Copernicus can have had in mind when he tried to remove the objection to the annual orbit around the Sun. By means of a procedure that was already discovered by Hipparchus—estimating the angle of diminution of the Earth's shadow, in eclipses of the Moon, by reference to the duration of the Moon's transit through the shadow—Ptolemy had calculated the average distance between the Earth and the Sun as 1,210 Earth radiuses. Copernicus corrected this figure only by an insignificant amount—to 1,179 Earth radiuses—as Tycho Brahe and Galileo also did. The difference between this and the actual figure (which is about twenty times greater) remained unsuspected in Giordano Bruno's time, and despite all his joyousness in dealing with infinities it would have confronted him with an abyss of enormity in the distances even just to the nearest heavenly bodies, whose parallaxes would have had to remain definitively below any accuracy in observation.

Only the discovery of the finite speed of light created a new standard for the distances in the universe, one that seemed to be manageable in what at first were small multiples. Here theoretical explosions have taken place in our own century that cannot even be compared to the presentiments to which Copernicus's theory gave rise: In 1907, Bohlin still estimated the distance of the nebula in Andromeda as 20 light-years; up to 1952, one said 800,000 light-years; and in two more decades it has become 2.25 million light-years. For comparison: From Ptolemy until Copernicus the diameter of the whole cosmos had been calculated as 20,000 times the diameter of the Earth. Assuming these classic proportions, the new annual orbit of the Earth around the Sun would have had to be noticeable even to the naked eye.

Talk of the immeasurable and the infinite is always hasty. This can only be perceived in retrospect. When Kepler writes to Herwart on 28 March 1605 regarding the logical consequences of the Copernican theory, "If the Sun stands still and the Earth moves, then the sphere of the fixed stars must, in effect, be infinitely great," then we know how little the term "infinite" means, when it is used merely to explain the nonappearance of parallaxes for such a 'small' motion as the annual one. Since Kepler sticks to the idea of a solid outermost spherical shell of the cosmos, for him the distance of any fixed star is at the same time that of all the others. So the minimum distance necessary to explain the absence of parallaxes is at the same time the minimum figure for the radius of the whole universe. Kelper was not to stick to the closed sphere out of timidity; more important for him was realism in interpreting the concepts of rest and motion. For them, only the finite cosmos offered a real reference system, as long as one did not need— as a consequence of the change in the concept of motion to one that gave priority to straight-lined inertial motion—to venture the leap into the paradox of the ideal reality of absolute space.

This, in particular, Giordano Bruno did not need, because he did not possess the idea of inertia. If one considers what importance the reality of rest and motion had for the metaphysics and natural philosophy that had been accepted since antiquity, one will be able to realize again the proportions of the threat that had to be felt in the transition from Copernicus's "immensity" to Bruno's "infinity." The Nolan's[a] infinite space stands between the infinite universe and Newton's absolute space insofar as, lacking a reference system, he no longer admitted a realist interpretation of rest and motion. Such renunciations are not carried out without a prospect of what appear to be greater gains: For Bruno, the metaphysics that he had in mind, the metaphysics of the absolute world, easily made up for the loss of a real difference between rest and motion, which is something in which, above all, the tradition of proofs of God's existence had been interested. But we should not overlook the fact that the possibility of formulating the Copernican claim to truth cosmologically continued to depend on the real differentiability of rest and motion; in the case of Newton this is still discernible as a motive for his introduction of absolute space, despite the fact that nothing that could verify it empirically could be deduced from it.

Giordano Bruno was an expert in the art of 'fleeing forward': He took what filled his contemporaries with uneasiness or alarm—namely, the expansion of the cosmic housing to an unimaginable magnitude—and drove it one step further, into the pure negation of conceivability. After his two years' stay in England and the failure of his Paris disputation in 1586, Bruno tried to establish himself at the universities of Wittenberg and Helmstedt. But his supposition that German Protestantism could be won over to an alliance with Copernican metaphysics was mistaken. I have already tried to show why this was the case. In its adherence to the traditional cosmology the Reformation left standing the strongest support of the alliance, which it so bitterly attacked, between Christianity and Aristotelianism. But the self-consciousness of modern man was to be based on the negation of the possibility that he could read off from nature, as the Creation, whether any role at all had been 'intended' for him and what role it was. When the Reformation—like the Counterreformation—took its stand against Copernicus, it denied itself a share in this self-determination and did its part to make the conflict between theology and science a distinguishing mark of the modern age. In Germany, Copernicanism was not to be accepted and undisputed until around 1760.

For Giordano Bruno it was two centuries too early to be entering the German philosophical scene. On 8 March 1588 he took his leave of Wittenberg with a moving speech, which makes two things comprehensible: first of all why, despite all the tolerance he was shown, he left this place of refuge in a state of resignation and without being understood, and second why his influence on those who came after the Reformation could not be terminated by this departure. Nowhere else did the forgotten Giordano Bruno emerge again from obscurity in such a conquering manner as he did in the country where his canvassing for support had been in vain.

Bruno's sixteen years of wandering are framed by two dramatic conflicts. When he left his monastery in 1576, as a 28-year-old, this enabled him to escape an action on charges of heresy brought against him by his order, an action that included no less than 130 articles of accusation. In 1592 he fell into the hands of the Inquisition in Venice and was extradited the year after that to Rome. In this period of time only the two years 1583 and 1584, which he spent in the retinue of the French ambassador in England, and thus under diplomatic immunity, constitute a solitary point of rest, and at the same time—with the philosophical dialogues in Italian—the zenith of his productive power. For the high-flying expectations of Bruno, who was always wrought up and always demanding, this period too was still disappointing, still embittering. Of what he composed after the years in England, only the challenge of the 120 disputation theses that he caused to be defended in Paris in 1586, and which were at the same time to mean the end of his stay in France, is exciting. The Latin treatises that he finished and had printed in Frankfurt am Main no longer reveal much of the speculative intensification and the delight in attack that characterize the works he wrote in England.

These six dialogues in Italian constitute a loosely unified composition the foundation of which is laid by the *Cena* and its Copernicanism. This first dialogue is most clearly related to the situation that Bruno had happened into in England, where through his protector's influence he quickly got access to the Court and to the university at Oxford, but immediately also experienced the disappointment of his

Copernican ambition—not entirely unjustly, if the dialogue gives us an accurate pictue of Bruno's technique of argumentation, which may have entertained an audience that enjoyed seeing people discomfitted, but can hardly have helped to persuade the experts to whom it was addressed. Queen Elizabeth, at any rate, is supposed to have listened to Bruno with pleasure.

The circumstances in which he was living, and his audience, are hardly enough, by themselves, to explain the explosion of his literary activity during the English period. The fact that Bruno wrote in Italian and that he had the dialogues published in Venice makes it very clear that he wrote, in England, for a different audience, and that this was probably already entirely a result of resignation deriving from his failures. He has already turned away from the stage on which he is still acting, and feels drawn to the supposed freedoms of Venice, which were to be his doom.

But why then does this productivity break out only after his arrival in England? If there is any explanation at all, it can only be that it was only in England that Bruno hit upon the central idea of his speculation about the universe. The electrifying connection between his inclination, which was already present early on, toward the Copernican "paradox," and his philosophy, which was (no doubt) principally stamped by Epicureanism, could have been produced by an acquaintance with Thomas Digges's work, which had appeared in 1576 and was talked about by those who were familiar with the novelties of the time.[b] Digges probably only wanted to remove a contradiction that had remained in Copernicus's argumentation when he maintained, in accordance with ancient tradition, that the spherical form of the Earth was the "natural" reason for its rotation around its axis. But if there was such a connection between spherical shape and motion, then the form of action that was assigned to the Earth could not be taken away from the outermost sphere, the heaven of the fixed stars, which could just as well be a perfect sphere. So with the expression *mobilitas terrae* [mobility of the Earth], Copernicus only arrived at a possibility, which would have to be decided, one way or the other, on other grounds. For this argumentational dilemma, Digges found the appropriate remedy that one could bring the outermost sphere to a standstill by making it cease to exist as a body supporting the fixed stars. The uniform distance of the fixed stars from the observer's standpoint became an optical illusion and the postulate of visibility was abandoned. So Digges not only deduced a logical consequence of Copernicus's magnification of the heaven of the fixed stars but at bottom he even sharpened Copernicus's argumentation, to the extent that it remained within the horizon of the Scholastic relation between essential form and qualities. In fact for Digges the outermost sphere is retained, as such, but as far as appearance is concerned it is made to disappear, inasmuch as the fixed stars are freely arranged in space behind it and can be seen through it.

What for Digges had been the removal of a difficulty in the Copernican theory became for Bruno the reason for a new enthusiasm when he realized how the heliocentric "paradox" and the unlimited space filled with atoms and worlds built of atoms were connected. This forerunner must have made England appear to him as a space predestined for Copernicanism. The *Cena* serves to deride those who did not live up to this vocation and did not recognize its messenger.

Bruno's Copernican dialogue is a piece of Mannerism. The traditional didactic dialogue is severely deformed here by aspects of its staging, and the doctrinal content is pushed to the side by the drama. In this way the form announces how far removed Bruno feels from the 'schools.' The dedicatory letter to the French ambassador also reveals the fact that he writes as a house jester who has to amuse his lord and protector. The *Cena* is described by its own author as a mixture of dialogue, comedy, tragedy, poetry, rhetoric, and satire. Original elements of the Platonic dialogue are still visible: The central reference to a festive meal stands in the tradition of the *Symposium*. The fact that it is an Ash Wednesday supper, in particular, is most nearly explained by the fact that in ecclesiastical tradition the first day of Lent was the day on which public penitents were expelled from the community, into sackcloth and ashes. Bruno could have picked it out as a characteristic day for his role. He likes to see himself as the new Diogenes. But if, in such allusions, he celebrates his conflict with church and society as a liberating one, he remains, in the London garret of the French ambassador, a Diogenes whose lot is not to be honored by being visited by 500 men of the stamp of Alexander, who could stand between him and the Sun. This deprives his mockery of liberality and makes his lack of real freedom vis-à-vis the circumstances that he pretends to scorn just as noticeable as his lack of human independence. He suffers in the role of the favorite of great lords, which, however, he had sought, as his only opportunity not to have to be in flight for once. This discord determines the register of pitches in which the dialogue is conducted.

The scene begins with Humanistic cultural skirmishing, dialectical finger exercises, a parade of quotations, and rhetorical shadowboxing. Prudenzio is a typical pedant, such as Bruno had already mocked in a juvenile comedy; he is characterized by his objection that one cannot properly carry on a conversation among four people without first disposing of the erroneous idea that this could be called a "dialogue," because it is actually a "tetralogue."

We discover with surprise that Bruno, the Copernican, has already historicized the head of his school. Copernicus is not his authority but his 'prehistory.' He appears to Bruno as the still groping beginner who had not yet recognized the emerging process of a great clarification, but instead, as the last in a long series of astronomical laborers, had only surveyed the basis for a bold extrapolation without daring to make the leap into speculative totality. For himself, Bruno claims "to see neither with the eyes of Copernicus nor with those of Ptolemy, but with his own eyes, where judgment and conclusions are concerned." He must, it is true, put trust in the specialists' observations, but not in their interpretation. They are only the translators of a text, not its expositors—only the messengers, not the strategists.

What he grants to Copernicus is *giudicio naturale*, a natural power of judgment, and, connected to that, the courage to take a stand against the current of the tradition, almost without the protection of new evidence. This is characteristic of the Nolan's consciousness of himself: He historicizes Copernicus, by reducing him to the courage to contradict, and gives himself credit for having for the first time really provided this contradiction with its arsenal of arguments and embedded it in the consistency of a world-system.

This historicization of Copernicus will have to be understood as a projection from Bruno's own biography. What, after all, could have moved him, during the early years when he was in the monastery, to accept Copernicanism? Scarcely reasons of an astronomical or cosmological kind. What may have attracted him, in Copernicus, and established an affinity, was disagreement perceived as daring contradiction, as an incentive to leaving behind the grinding mill of the schools. In Bruno we can see how Copernicus, for his century, fit most readily into the thought-form of paradox, as someone who turned the world upside down—and in doing so certified that he had for the first time turned it right side up and set it going—and who announced the termination of the familiar propositions about the order of reality. There is a thoroughgoing stylistic correlation between the Nolan's mode of thought in many other matters and his Copernicanism, if one sees in the latter above all the scandalizing of consciousness. Paradox is a characteristic element of Mannerist style.

Bruno from the very beginning looks out for opportunities to have a powerful impact. He favors paradox because it is potent, even if its effect is superficial. That being the case, Copernicus may have become part of his equipment pretty accidentally, in the way in which such things have always been reached for by people who wanted to startle those who were (or even only appeared to be) comfortable and complacent. He himself says in the *Cena* that people at court had summoned him to arrange a discussion in which the topic was to be "his Copernicus, and other paradoxes of his new philosophy."[140] Here Bruno himself associated Copernicanism with the style of his philosophy and characterized the expectations to which he addressed himself by the example of the curiosity at court. Infinity enters the modern age's cosmology under the title of "paradox" and is eliminated from it, later, under the title of "antinomy."

Giordano Bruno was the first person to attach to the event of the Copernican reform the metaphor, which was used so much in the subsequent centuries, of the dawning of a new light. Of course, Copernicus had been only the first light of this dawn, the bright daylight of which had only broken by means of him, the Nolan. Admittedly, this day is new only in comparison to the darkness of the past night of the Middle Ages, because it is the Sun of antiquity and of its "old, true philosophy" that now returns.[e] Under the regime of this cyclical periodicity of history the absence of light becomes just as 'natural' an event as its return. It remains obscure how this is compatible with the simile (modeled on Lucretius) of the liberation of the human spirit from the prison walls of the finite universe, which Bruno propounds at the beginning of his Latin didactic poem, *De immenso et innumerabilibus* [*On the Immeasurable and the Uncountable*], as well as in the *Ash Wednesday Supper*. How is his enthusiasm about the power of reason compatible with the resigned form of historical consciousness that is inevitably contained in a cyclical picture, which always commands us to think that reason has already once, or more frequently, given up its own dominion? Thus in the language of a renewal of antiquity a claim is formulated that implies the irrevocability of what is now to be accomplished and that in that respect is not compatible with granting that all of this was already here before and nevertheless disappeared. How could the appearance of Copernicus have meant a definitive beginning of reason if the Pythagoreans and Aristarchus had already possessed his insight, with its prerequisites, and had nevertheless remained an episode?

The combination of metaphors of light and the fundamental cyclical pattern is especially well-defined in the *Excubitor* [*Watchman*] that Bruno puts at the beginning of his daring Paris theses of 1586. He compares the natural alternation of daylight and nocturnal darkness to that of truth and error in the world of the spirit. The antinomy of localizing oneself, historically, within such a fateful periodicity lies in the fact that one must inevitably admit that in the epochs of darkness one could not be acutely aware of the lack of the light of reason. For this very awareness would be one of the principal accomplishments of reason itself, its *docta ignorantia* [learned ignorance], which Bruno was familiar with from his acquaintance with Nicholas of Cusa. But how can knowing ignorance distinguish itself from ignorant knowledge?

In the *Cena*, at any rate, Bruno has no alternative, in the face of his opponent's incomprehension—even though he compares this incomprehension to his own pre-Copernican juvenile foolishness—except finally to pray to God that if He does not want to make the man capable of seeing, He should at least enable him to believe that he is blind. Here we meet the fundamental problem of all of those who see themselves as confronting deep eclipses and delusion systems[d] of mankind: How should people who are deluded in that way ever become aware that something rational has dawned on or been offered to them? What else is left but to give oneself the role, and finally the attitude, of the bringer of salvation? But the bringer of salvation himself guarantees nothing unless the cyclical process can be replaced by a linear one in which the "point of no return"[e] could perhaps exist.

However much Bruno would like to push Copernicus into the background, in favor of his own self-apotheosis, he does not succeed any more in creating a qualitative contrast between his own role and Copernicus's, or in attributing to himself more than the renewal of that "old, true philosophy." If Copernicus had only liberated one of the old truths from the "dark caves of blind, malignant, arrogant, and envious ignorance," there remained the uneasy question what powers they were that had been able to imprison reason in that way, and that, when the new day also comes to its evening, might perhaps imprison it again. In regard to this anxiety, Bruno made only the one step forward of observing that the distance of time between Copernicus's new beginning and the previous day of ancient astronomy had at any rate not been in vain, because a tradition of the data and observations had been carried across this interval.[141] If it was to be characteristic of the Enlightenment that it postulated for itself a beginning of rationality, and of its linear process, which was both new and definitive, Bruno cannot even be brought forward to testify to the clarity of the problem, as his fellow Neapolitan Giambattista Vico can be, a century later.

In the *Cena*, Bruno makes his Teofilo say that the important thing "is whether we find ourselves in day, and the light of the truth is above our horizon, or whether this light shines in the horizon of our antipodes, our adversaries—whether, that is, we or they are in darkness and consequently whether we, who have begun the renewal of the old philosophy, stand in the morning, which terminates the night, or in the evening, which ends the day." Just this question of one's own location in the course of history is, on any assumption that does not see increasing comprehension as a precondition of judging the preceding phase, a simply insoluble problem. For with the periodic disappearance of reason, the position from which a state of eclipse could be recognized as such, or light, as something past, could even merely be remembered or anticipated as something in the future, would also be obscured. All of this is included in the descriptions—which the Enlightenment itself was to give, in relation to the "dark Middle Ages," and which it was able to give because this past was seen as definitively past—of the absence of any felt need for rationality. That is why Copernicus is more than a reviver of Pythagorean ideas. In Bruno's picture of Copernicus we can see, sketched out in negative form, how the historical consciousness that is to come will define itself with the help of a different picture of Copernicus.

The destruction of the view of the world that determines the Middle Ages is still bound to the rules of the game that this view of the world prescribed. The metaphorical presentation of the new truth is intoxicated by breaking through the walls of the celestial spheres, by expanding space and multiplying the one world into the universe of infinite worlds. Reason as the overstepping of boundaries that were previously drawn, recognized, and finally hardly perceived any longer—this schema, to which Francis Bacon, above all, will give new vividness, stamps the self-understanding of the modern age as it gets under way. But at the same time it marks reason's inability to take small steps and to overcome things gradually, which can be gathered in Bruno's case from the difference between his own use of Copernicus as his point of departure and the distance he establishes from him by a leap. This process has an ecstatic character. Reason is bearable only for the few who are able to bear the painful consequences of its violence. Perhaps it is even only meant as the reason of the one person who takes it upon himself to shatter himself with it against the world: "For this burden is not suited to the shoulders of just anyone, but only of those who can carry it, like the Nolan."

That is why a major stylistic means of the *Cena* is derision and scorn of those who are ignorant and those who imagine themselves to be knowledgeable. Argumentation seems to be what least of all needs to be brought to bear in order to achieve acceptance for the new truth. "If someone, for lack of mental capacity, is not able to agree to the facts, he must subscribe to the Nolan's doctrine at least on many larger and essential points, and admit that what he is not able to recognize as true is certainly more verisimilar." This declaration by Teofilo, in the dialogue, shows to what an extent reason is seen as an uncommon and unexpected substance, which cannot be counted on and by which one can only allow oneself to be surprised.

In the *Cena* Bruno has a character say of himself that he has discovered "how one mounts to the heavens, passes through the outermost circle of the stars, and leaves the convex surface of the firmament behind him." In the midst of the age of discoveries and in the capital city of a nation that is devoted to them, this splendid gesture is well calculated. Bruno compares himself to the mythical inventor of the first ship, Tiphys, in the tale of the Argonauts, and in doing so takes credit for the fact that his expansion of the world takes place in a direction in which it affects no one injuriously. He refers back to the old topos of culture criticism according to which the first leaving behind of terra firma had been an act of overstepping what is given by nature, an act that combined hubris, curiosity, and a passion for luxury. The cosmic salvation-bringer does not belong among the discoverers whom Bruno, in their own country, reproaches for having "found out how to disturb the peace of others, to profane the guardian spirits of their countries, to mix what prudent nature separated, to redouble men's wants by commerce, to add the vices of one people to those of the other, to propagate new follies by force and to set up unheard-of lunacies where they did not exist before, and finally to give out the stronger as the wiser. They have shown men new ways, new instruments, and new arts by which to tyrannize over and assassinate one another. Thanks to such deeds, a time will come when the other peoples, having learned from the injuries they have suffered, will know how and will be able, as circumstances change, to pay back to us, in similar forms or worse ones, the consequences of these pernicious inventions." To this prophetic critique of the age of discovery and of the beginning exploitation of new continents Bruno contrasts the opening of the universe, as the alternative of a freedom that has now become limitless. "Our reason is no longer imprisoned in the shackles of the eight, nine, or ten imaginary spheres and their movers." The reason represented by this salvation-bringer is not that of the discoveries and inventions as access to new resources, but rather that of an ecstasy of theory, an infinity-mysticism that has a religious character.

According to his own testimony, Bruno already made the acquaintance of the Copernican thesis as "a small boy," and regarded it as a "sophistical snare," contrived "by those idle intellects who want to debate for fun and who make a profession of proving and defending the proposition that white is black." He judges his opponents in the light of this, his own experience. But his refusal to debate on the basis of astronomical arguments makes one wonder whether he himself has gone beyond a childish delight in paradoxes. The dialogue that he presents is, in its construction, not dialogical; it is not of the type that was originally counterposed to the Sophists' rhetoric. When his opponent, Dr. Torquato, impatiently exclaims, "Ad rem, ad rem, ad rem" [To the facts, to the facts . . .], and the Nolan only bursts out laughing and responds that he does not have to give an accounting, he only makes assertions, and "ista sunt res, res, res" [these are the facts, the facts . . .]—when he does this, it is both arrogance and embarrassment, but it is also a demonstration that the Copernican theory that he presupposes does not matter to him.

Bruno insists on the evidentness of the totality that he outlines. For it, Copernicanism is a digression, to which one cannot go back. This is a proclamation, to which the appropriate response is listening, and putting it in the genre of the dialogue was simply a mistake. "According to the custom of the Pythagorean school, which we follow, the students are permitted neither to ask questions nor to make objections before they have heard the philosophy lecture all the way to the end; for if the teaching is complete in itself and properly understaood, it removes all doubt automatically and eliminates any contradiction." This fits perfectly the metaphors of organism, in which Bruno expresses the unity of his universe.

The question to what extent, and up to what breaking point, Giordano Bruno really was the Copernican that he gives himself out as, and as which he was elevated to the role of martyr for the new scientific truth—this question must be answered in accordance with this finding, that Bruno merely passed through Copernicanism on the way to metaphysics. Then the Copernican reform became, in comparison to what Bruno grasps and projects as its logical consequence, a provincial event. In the center of the world, or near this center, an exchange of places is undertaken: The Earth is moved a little bit away from this center and placed on a circular orbit and the Sun is put in its place. Bruno has no appreciation of the fact that this description describes the degree of the change from the point of view of its results and that it trivializes the magnitude of the difficulties involved. Consequently he is also not interested in the lack of foundation from which the Copernican theory in fact suffered; he considers it to be in need of a new metaphysics, when what it actually lacked was a new physics.

Copernicus himself, in keeping with his system, had covered up this lack by appealing to the "natural" character of circular motion for a spherical body, and of the self-integration of the spherical bodies in the universe. Bruno makes it disappear in his notion of the world-organism. An organism is characterized by its self-preservation, with which all partial processes like metabolism and reproduction are coordinated. Following this schema, taken from the Stoic tradition, Bruno interprets the processes in the universe in accordance with the funtions of self-preservation, reintegration, and reproduction. But that means that only from a partial point of view is change real; the whole is unchangeable unity. In this regard, Bruno interprets Copernicus with the aid of Parmenides. The result is that he subsumes the partial physical truth of Copernicanism in a metaphysical identity that assigns disagreements like the one centering on the heliocentric change to the side of appearance, the side of the *doxa* ['opinions', as opposed to knowledge]. If one has this truth, there are no "truths" any longer.

Copernicus is also a "forerunner" ["*Vorläufer*"] in the sense that his truth could only be a provisional [*vorläufige*] one. To be burned at the stake for this would have been an inexcusable error. And Bruno did not commit it, either.

Bruno's daring speculations have been admired for their anticipations of later astronomical knowledge and problems. But with the path that he took starting from Copernicus, he shuts off every other path at the same time. He gave away the Copernican key to orientation in the universe, in order to soar to his metaphysics of 'absolute creation,' within the horizon of which the solar system paled into insignificance. In fact, enthusiasm for organic totality led to the atomistic character of all of its elements.

This becomes clear when one looks ahead to the modern age's incipient cosmology, which instead of merely assuming the existence of world-'islands' in space proceeds to the attempt to construct higher-level systems analogous to the Copernican system, with its center. It will imagine in the literal sense a world of worlds, in which the Copernican principle of perspectival analysis is applied again and again; so that while the existence of a center of the world is not denied, that center is passed on, in each case, to the next level of construction, as (hypothetically) the last one. Compared, for example, to Lambert's cosmology, Bruno's universe is without coherence and structure. The need that he feels is not to make the construction perspicuous but rather to spread out the prospect of the whole, in a comprehensive intuition, before himself, and to perceive in it the driving claim of each of its members to realize all of its potential and to enter into the universal participation of everything in everything. Motion is then the expression of this impulse, the continual metamorphosis of the world's substance from one formation to another.

But for that very reason this whole can also be regarded as a chaos of metabolism, which, without an immanent constructive principle, must continually destroy itself, on the way to the unattainable, so as not to congeal in contingent givenness. Bruno expressed this, as a *profetico dogma* [prophetic dogma], in what I think is for him a central myth, the myth of the universe as a potter's wheel. According to it "the whole is in the hand of the universal maker like a single lump of clay in the hand of a single potter, to be made and unmade—on the wheel of this spinning of the heavenly bodies—in accordance with the vicissitudes of the generation and the corruption of things, now as a good vessel and now as a bad one, from the same material." [142] In this world-metaphor, cosmological speculation and the cyclical idea of history are united once again. But the circle is not only an expression of the universe's self-referentiality and self-sufficiency as an imitation of an exclusiveness that, in the traditional metaphysics, only belongs to God; it also has some of the futility of a process in which every outcome that it ever arrives at is always and inevitably the negation of the possibilities that the substrate that was consumed in it had previously possessed.

The metaphysician easily describes the endless expenditure of suns and worlds; less easily, he tries to find a result that could justify it. In Bruno, this gesture of futility always stands directly alongside the gesture of great pomp: He is despised, at the same time that he is the bringer of salvation. As literature, the sequence of the action in the *Cena*, also, combines cyclicalness and futility. The second dialogue portrays the procession to the nocturnal Ash Wednesday supper as a grotesque odyssey of the Nolan and his group through a London that has been demonized into Hades and the Labyrinth—a procession in which the marchers finally find themselves back at their starting point again. Not by chance, the narrator, after describing this tortuous journey, breaks into a bitter appeal against the things that resist reason, which are also the means by which it confuses itself: "Oh variable dialectics, tortuous doubts, importunate sophisms, subtle snares, obscure enigmas, intricate labyrinths, bedeviled sphinxes—dissolve, or let yourselves be dissolved!" Even Bruno's God of the absolute creation is still Aristotle's reflexively self-related God, One Who in this case, it is true, can no longer only *think* nothing but Himself, but still cannot *create* anything but Himself in the form of the infinite universe.

When they finally sit down to their Ash Wednesday supper, an amusing incident interrupts the ceremony. A member of the Nolan's entourage, who is supposed to be given the last place at table, misunderstands the order of seating, takes the last place for the first and the best, refuses to accept it, and betakes himself, out of modesty, to what had been intended as the first place. The to and fro that now sets in is not without a concealed allusion to the Copernican situation.

It should not be forgotten that the opposition to the new system was fought out less with biblical verses than with the coupling of cosmic topography and metaphysical ranking. For man, the loss of his position in the center of the world was supposed to mean the forfeiting of a special distinction, which the Stoic tradition had based on the possession of the privileged standpoint of the world-onlooker. This argument did not enter the common consciousness of a distinctive position implied by geocentrism as much as did the vague background idea of a ceremonial in which being in the center and being the point of reference of something that happens in a circular fashion around that point has some importance. The little grotesque on the occasion of the distribution of the symposiasts' seats may be intended to make the reader aware of how difficult, and at the same time how unprofitable, it is to orient oneself, in a finite world, by the first or the last, by central or eccentric positions, instead of understanding this as a metaphorical convention that comes to nothing at the moment when the Nolan, in the presence of the party, smashes the cosmic framework by means of which alone a rank order of realities could still be represented. Of course Bruno boasts, as Galileo did later, of having elevated the Earth to the rank of the stars, but in an infinite universe that now consists of nothing but stars and that admits no 'higher' and 'lower,' all this means is that he eliminated the only exception to and the last proviso against the homogeneity of nature.

The reason for the disquiet that Copernicanism aroused went deeper than the mere fact that a cosmic ceremonial that was flattering to man had been violated. When Bruno identifies the visible stars with solar systems, the fact does not become at all clear, and he does not even find it worth explaining away, that not only has he assumed the existence of invisible realities in the world but those realities have become preponderant and have reduced the world that is accessible to man to a provincial enclave. This is not the place to describe the extraordinariness and the painfulness of this process. First of all, the important thing is to comprehend the pressure for compensation that weighs on Bruno's metaphysics. Invisibility is no longer a marginal condition, once Digges had already made the disconcerting declaration that the greatest part of the world remained invisible for us. But what was the meaning of a world that no longer existed for the admiring intuition of the supposedly privileged one among its creatures? Digges had not yet decided to give this question an answer that would have wagered everything on the correspondence between the infinitude of the Creator and the infinitude of the creation. Instead he had suggested that man practice a sort of pious renunciation of his share in and claim to the world. Bruno, on the contrary, combines with his enthusiasm for the infinity of the universe and the infinite number of worlds a pagan rebelliousness against any self-withholding on the part of theology's God, against every saving blessing that competes with the universe—and perhaps also against any redeemer who competes with his own pretensions as salvation-bringer.

The Copernican event disappears behind the new infinity as the mere stimulus that led to the conceiving of the world's true magnitude. For what, then, was Giordano Bruno burned to death on that 17 February 1600 in the Roman Campo di Fiore, if indeed it was *for* something and not merely as a result of something? Heine, commenting on Schelling's *Bruno oder über das göttliche oder natürliche Prinzip der Dinge* [*Bruno, or: On the Divine or Natural Principle of Things*], called him the "noblest martyr of our doctrine." But which doctrine is it for which Bruno bore the witness of death? Even if we disregard Kierkegaard's penetrating question, which is as relevant as ever, as to whether one is permitted to die for the truth, there remains the other question, which is worrisome in every historical case, as to whether the death was really for any truth at all, and if so, for which one.

It is certainly one of the bitter discoveries resulting from the demythologizing of the self-consciousness of modern science from the heroisms of its beginnings that Bruno was not one of its champions. In Bertolt Brecht's *Life of Galileo* he has Galileo say to the Curator of the University of Padua, "Mr. Giordano Bruno has been extradited from here to Rome. Because he spread the teachings of Copernicus." The Curator denies this. Bruno was extradited, he says, "because he was not a Venetian." His reply is, legally, perfectly accurate. Beyond that, however, he conceals, with his sobriety, the fact that Giordano Bruno's offense was a greater one than being a Copernican would have been. The truth is that Copernicus had given him a new God, or, more precisely, a new Son of God.

The documents of the extradition proceedings, after Bruno's imprisonment in Venice, have survived; those of the later trial in Rome, which led to the sentence of death, have to be considered as lost. They were destroyed during the transportation back to Rome of the papal archives, which Napoleon had had brought from Rome to Paris in

1810. The director of the Vatican archives, Mario Marini, to whom Pope Pius VII had entrusted the return transport of the archives between 1815 and 1817, allegedly regarded the records of the Holy Office as worthless and, with the permission of Cardinal Consalvi, had them torn up so they could be sold as scrap paper to a Paris factory. The proceeds, it is reported, amounted to 4,300 francs.

However, we are not entirely uninformed about the trial in Rome. In 1887, Gregorio Palmieri, then the assistant director of the Vatican's confidential archives, found some notes in which the state of the documents of the entire proceedings—that is, of both the Venetian and the Roman proceedings—had been summarized. The occasion for the making of this synopsis was a new member's joining the tribunal in the summer of 1597: the assessor Marcello Filonardi. When Leo XIII was informed of the discovery of this document, he ordered that it should not be made available to anyone. Thus a second "discovery" became necessary, which was made by Angelo Mercati on 15 November 1940; the document was placed in Pius IX's personal archive, which this time did not prevent it from being published.

In the thirty pages of this memorandum the course of the proceedings is reproduced in part by word-for-word quotations from the documents, in part by abstracts. Notes in the margins refer to the original documents, so that there is some basis for an estimate of their extent and their arrangement. In this connection it is instructive to observe that the Venetian and Roman examinations of Bruno, of which there are seventeen, altogether, are numbered continuously— which shows that, as far as the record keeping is concerned, the Roman proceedings were treated as a continuation of the Venetian ones. To one who is acquainted with the Venetian minutes, it immediately becomes clear that thorough work had already been done there: In the Roman examinations and in the statements of fellow prisoners from Venice hardly any new aspects emerge.

The most surprising thing about the investigation that was conducted in Venice is essentially confirmed by the Roman document, namely, that Bruno's Copernicanism was not part of the substance of the charges that were brought against him. Among the theses he is accused of having held, the thesis of the eternal existence of the world appears; but it is one of the basic elements of the medieval condemnations of orthodox Aristotelianism, and has nothing to do with

Copernicus. That also holds for the thesis of the plurality of worlds; it has to be seen here exclusively in a theological context, in which what is at stake is the basic constituents of Christian dogma—the Creation, the unity of mankind in guilt, God's incarnation and redemption of man—all of which presuppose the uniqueness of a world to which God's saving care relates and in which the 'salvation story' takes place. Only in the final article of the summary does a weak support appear for the common idea that Bruno was a victim above all of his Copernican enthusiasm. Here written statements of Bruno's are summarized, in which he defends himself against accusations that were taken from his published works. In this connection we learn that the thesis of the motion of the Earth had at least attracted attention, without, however, becoming the focus of an accusation.

According to the evidence of the Venetian records, it was Bruno himself who broached the subject of the Copernican thesis. In his defense of his *Ash Wednesday Supper*, Bruno said that he had meant, with this book, to ridicule the pre-Copernican backwardness of some doctors. The Venetian tribunal did not take up this offer of what evidently seemed to Bruno himself to be a harmless secondary topic, but instead went on to the next 'hard' question: Whether Bruno had ever praised heretical princes. So that was more important. Bruno had already had occasion to speak of the plurality of worlds as well, in Venice, and had had no 'success' with that topic with the tribunal. They did not want to allow topics to be forced upon them, and they may even already not have wanted to let the martyr of science—rather than the rebel against the central elements of Christianity—play his self-chosen role. When it finally came to extradition proceedings, there was no mention of Copernicanism. It was also only in passing that the proceedings before the Roman Inquisition gave Bruno the opportunity to express his Copernican conviction and to insist on the compatibility of the Earth's motion with the Bible.

If that is the case, the question still remains as to whether the Inquisition's proceedings ever, and at any point, penetrated to the core of the Nolan's spiritual constitution and of its internal coherence. The *Sommario* [*Summary*] shows the tribunal to have a pedantic blindness to the common center of the questions that it poses and to what the answers that it receives disclose. What happens to it is what happens to heresy tribunals in all ages and in all dogmatic systems: It learns nothing about the structure of the opposition by which it sees itself as confronted. Hidden among the random contents of the last collective rubric of the synopsis of the documents is the supreme metaphysical principle of all of Bruno's mental maneuvers. It says, God's nature would be finite if He did not in fact produce something infinite, and infinitely many things.[143] It is the future axiom of every religion of reason—the elimination of every uncertainty created by the hidden and capricious God—if God's creation is, at the same time, the exhaustion of His potential, the complete expenditure of His wisdom and goodness.

In this connection, the question to what degree of pantheism this leads and whether in the end one can still distinguish at all between the Creator and the creation is not very important. The crucial thing is the securing of the world as the self-exhibition of absolute reason, which at any time can flip over into the idealist solution, and is implicitly already identical with it. Christian voluntarism had already anticipated this solution, in its reverse form: If God had created the world by virtue of His nature, not by virtue of His will, there would always be new worlds, and infinitely many of them.[144] Lessing integrated the argument into his rational interpretation of Christianity; the God of reason thinks only Himself, and by thinking Himself He creates the world: "With God, every thought is a creation"—but if what He thinks and what He creates are identical, He can create nothing but Himself, nothing but "infinitely many worlds" or "the infinity of the world."[145] In the medieval Christian system the position of such an unreserved, essentially self-exhausting self-manifestation was occupied by the Son of God, whose origin, as "generation," had a status that no creation was permitted to approach.

Bruno's assertion of the infinitude of the world and of the number of worlds was a challenge aimed at this center of the system of Christian dogma: According to Bruno's great premise, no individual contingent fact, no person, no saving event, not even an individual world could claim to represent, to contain, to exhaust the power and the will, the fullness and the prodigal self-expenditure of the divinity. But if the Creation, as the natural reduplication of the divinity, was already an insurpassable reality, the incarnation of God within it could not even be an 'interruption.' To believe in it would already be an absurd mistake, a demonstrative disregard for what this universe already was and what it did not need to become, for the first time, through any sort of addition.

In this connection the record contains one brief flash of attention, by the Roman inquisitors, to the central point of the conflict. On 24 March 1597, the prisoners of the Inquisition are visited, in their prison, by the Congregation of the Holy Office. In the entry in the records on this visit, the only substantive matter relating to Bruno is the admonition that he should desist from such crazy notions as the idea of the plurality of worlds.[146] But even at the stake, the basic pattern of Christianity's self-understanding according to which God became involved with the singularity of man in the universe, and with man's salvation, against the world, becomes for Bruno the scandal that causes him to turn away from the crucifix that is held before him. It was the final consistent step on the path of the dismantling of images that, according to the Venetian documents, the young monk in the monastery in Naples had started on by removing the pictures of the saints from his cell, though admittedly he had still tolerated the crucifix there. This had brought the first suspicion of heresy upon the man who in the end, after all, could only express the imagelessness of the infinite by means of metaphors of thought becoming incarnate; who knew how to 'reify' even empty space; and who pictured the contest of the world systems in the fools' scene of the *Cena*.

Bruno did not slide into conflict as Galileo did; he was a "born" heretic who tried to strike the center of the dogmatic system and to achieve the most resolute distance from it. As it had already done for Gnosticism, for him too Christianity broke down because of the incompatibility of creation and redemption. A God who has to redeem cannot have created. But in view of this contradiction, Bruno decided—differently from Gnosticism—in favor of the Creation. He posited it as absolute, not because it came out of nothing, but because it ended in everything. Bruno really was an *eretico obstinatissimo* [very obstinate heretic], as the *Avvisi di Roma* called him shortly before his *solennissima iustizia* [very solemn execution].[147]

Of course not without simplification, one could characterize the epochs of antiquity, the Middle Ages, and the modern age by their respective definitions of the horizon of human consciousness of the world. In antiquity the horizon of reality and the horizon of visibility coincide, just as the horizon of reality and the horizon of possibility do. The Middle Ages held to the congruence, for nature, of reality and visibility, but separated the horizon of reality from that of possibility by using the concept of creation and that of its contingency to make the real world appear as only a selection from what is possible for omnipotence, and furthermore emphasized the backwardness of actuality, in comparison to potentiality, ever more intensely in their speculation, thus making that backwardness an offense to reason. This offense is still evident in the case of Leibniz, in his momentous rationalization that justifies the actual world, in relation to the unused possibilities, by the fact that it is precisely the best of all possible worlds.

Giordano Bruno solves this dilemma of reason in the other way, by making reality itself into the sum of what is possible, thus indeed reestablishing the ancient congruence under the new conditions produced by the Middle Ages, but at the heavy cost, anthropologically, of the final dissociation of the horizons of reality and visibility. Man, after all, could not be decisively important if the point was to justify the whole of what exists, and not to allow it to be put in question any longer from the perspective of the surplus of possibilities. Aristotle had had the disarming objection to the atomists' infinite world that this *apeiron* [boundless thing] could not possibly rotate in a circle.[148] This argumentation, based on the congruence between appearance and reality, had been eliminated by Copernicus. To that extent, the "first motion" of the Earth already opens up a passage to Giordano Bruno: When the diurnal motion of the first heaven is abandoned, the primary argument against the infinitude of the cosmos also collapses.

For the Enlightenment, the plurality of the actual worlds inhabited by rational beings was also one of the answers to reality's slipping away beyond the terrestrial and human horizon of visibility. What man's eyes could no longer behold had to be able to become an appearance within the horizon of other creatures. Thus for a while yet, by way of a detour through speculation, the *contemplator caeli* [observer of the heavens] is retained as the allotted purpose of the world-performance, though with a relativization of man's status as its sole addressee. The Nolan's cosmological misanthropy leaves no room for this idea. His enthusiasm for the plurality of worlds is directed against the provincial frame of reference precisely of the story of God and salvation that is centered on man and degrades the universe to the role of mere scenery. The disproportion between man and the universe was not his problem. The principle of the inexhaustibility of creation does not allow any fear of the emptiness of space to arise; space is as real as anything else. Bruno already anticipated the "réponse au vide" [response to the void] that Valéry describes as the elementary attitude of modern man to space—he already saw and overcame the anxieties [*Ängste*] that the abyss of the infinities was to inspire. When, in the third dialogue of the *Ash Wednesday Supper*, Bruno formulates the basis of his thesis of infinitude most pregnantly as the infinite operation, in an infinite manner, of the infinite cause, the reaction of his partner in the conversation is described as complete discomposure: "He was nonplussed and speechless, as though he suddenly saw a ghost before him."

On his departure from Wittenberg Bruno had said bitterly, "I came to you us a foreigner, an exile and a refugee, a plaything of fate, unprepossessing, with few possessions and no patron, burdened with the hatred of the multitude and consequently despised by fools and the vulgar." And he had to continue his journey in the same condition. But he had at least stated and had printed, in the home of the German Reformation's resistance to Copernicanism, as the formula of his triumph, that the Earth, our mother, had now become one of the stars.[149] In his cosmic disdain for man, Bruno even projects his disregard of the herald for himself (in view of the news that he has to bring) into the Earth, which, by becoming a star, makes its inhabitants disappear all the more into insignificance and become immaterial for the attention of a god.

One would have thought that after this departure from Wittenberg there would have been no return to the German scene for Bruno. He spoke a language that was completely unintelligible to Protestant Scholasticism. But he did return. It is a sharply etched combination of circumstances when the works of the heretic who ended at the stake come forward again in Germany on the occasion of another case of 'heresy,' for which, admittedly, it was too late for the stake. Herder could find it "amusing" that Goethe "on this occasion ends up sharing a stake with Lessing," as Goethe himself wrote to Jacobi on 11 September 1785. Jacobi had, in a devious way, implicated Goethe as well in his posthumous disclosure of Lessing's Spinozism, when he described Lessing's confidential confession of his most private convictions as having been produced by Goethe's "Prometheus" ode, which he, Jacobi, had long kept, in manuscript, as a gift of the poet, and which he now published, without the author's knowledge, in his

Spinoza book. At this moment an undercurrent of the modern age and especially of German Classicism came to the surface in a way that was surprising, disconcerting, and dangerously powerful.

Even in the mouth of the Weimar General Superintendent, Herder, the stake was now only a metaphorical joke. Nevertheless, Goethe could write, about this event, in the retrospect of *Dichtung und Wahrheit* [*Poetry and Truth*], that his "Prometheus" had served as the "priming powder for an explosion that revealed and brought into discussion the most secret relations of estimable men: relations that, unknown to those men themselves, slumber in an otherwise highly enlightened society." What Goethe calls an "explosion" and what historians were to call the "pantheism dispute" also stirs up, from the bottom layer of hitherto excluded ideas, the works of Giordano Bruno. Even if Jacobi, when he described Lessing's disclosures in those morning hours of 6 July 1780 as "Spinozism," had misinterpreted or misrepresented them—and even more if the removal of Lessing's inhibitions in this regard by "Prometheus" had been an intentional way of implicating Goethe—this did not make much difference to the aftereffects. For Jacobi sees himself as called upon by Lessing's bequest to him to go in search of the sources and the ancestors of the unbelief that had been unmasked in this way. In doing so, he hits upon Giordano Bruno's dialogue *On the Cause, the Principle, and the One*, and translates, for the first time, selections from it, which he appends to the second edition of his controversial book, in 1789.[150]

It is amazing how little real tradition and reception is involved in the exercise of a great historical influence. Perhaps Socrates is the first and most important example of the fact that the greatest influence can be based on the smallest quantities of information. Jacobi presented a few pages from Bruno's speculative dialogue, and they became one of the most influential texts of German Idealism.

Goethe did not exactly take pleasure in this indirect consequence of his "Prometheus." He fluctuates between rejection and astonishment. In 1812 he notes, "Jordanus Brunus. I see more clearly his complete unserviceability, indeed harmfulness, for our times." In 1829 he once again looks into Bruno's works, "to my astonishment as always, realizing for the first time that he was a contemporary of Bacon of Verulam." If one thinks back to a first excerpt that the young Goethe made from Bayle's *Dictionnaire* article "Brunus," his interest extends across a period of six decades, the expression of an unmistakeable but vague fascination, which is not fulfilled.

Apart from Bayle, Goethe knew the most influential early history of philosophy, Jakob Brucker's *Kurtze Fragen aus der Philosophischen Historie* [*Brief Questions from the History of Philosophy*], which was published in seven volumes in Ulm between 1731 and 1736, and of which an even longer Latin edition appeared as *Historia critica* in 1742–1744. The picture of Giordano Bruno in this work is unthinkable without the influence of the "critical" method, bearing the imprint of Bayle, which had defined historiographical truth as what is left over after the elimination of all the errors and contradictions in the sources. In the case of Bayle's article about Bruno this meant that while Bruno was indeed brought within the horizon of knowledge of educated people, he was also put on the level of the doubtful phenomena, anecdotes and curiosities that were piled up in this work. Bayle had gone furthest of all in doubting whether Bruno died at the stake. For this, he said, we had only the eyewitness report of the German pedant Caspar Schoppe. That was too little, in view of the close proximity between unreliability and falsification in testimony of this sort.[151] Bayle's skepticism reaches its climax when it is a matter of supposed facts in the sphere of religion. Above all, he finds the catalogue of the Nolan's writings to be full of inconsistencies. It is of some importance for his pride as an antiquarian that he can say explicitly of at least some works that he has seen them himself. Bayle's remark that Descartes took some of his ideas from the Nolan is not without importance for the latter's secret reputation. With Bayle, historiographical judgment of the Nolan begins to turn away from an exclusive focus on his role as a heretic.

For Brucker, the difficulty persists that Bruno's writings were as good as untraceable, "without which one cannot, after all, say anything detailed about him." A certain weakness for curiosities on the part of his contemporaries makes Brucker conjecture that it is the exceptional rarity of an especially notorious treatise, the *Spaccio della bestia triomfante* [*Expulsion of the Triumphant Beast*], that brought it about "that people became curious to find other writings of Bruno as well, and to investigate the circumstances of his life and his teaching more closely." This thinker of "almost insuperable obscurity" keeps the predicate that applies to everything medieval, that he was "darker than Thomas, Scotus, Ockham, and all the Scholastics." The undigested ideas of his predecessors had overpowered him, Brucker says; poetic furor carried him away; and finally, he was not entirely free of the intention of amazing his age by his inaccessibility—so that all in all he was "a veritable Scotist and a dark *philosophus*." All of this can only mean and no doubt is only meant to mean that this man not only contributed nothing to the age of rational thought but actually hindered its emergence. "He was a man with a very fiery and excited faculty of imagination, who had a great number of strange and singular, exotic- and curious-sounding ideas . . . . But at the same time he lacked a healthy power of judgment, and though he no doubt hit upon all sorts of truths, especially in the doctrine of nature and the mathematical sciences . . . still they were only accidentally discovered grains, which lie hidden under a frightful desert of extravagant, badly connected, and sometimes meaningless ideas and words." If we turn, instead, to the substantive characterization, which in this early historian of philosophy follows the pattern of assignment to specific "sects," then it is contained in the "chief paradox" that Bruno wanted to combine Pythagoras and Epicurus in one system.

This procedure cannot be expected to yield much justice, nor even a little insight, when it is applied to one who was an outsider to all systems and schools. But the attempt to locate Bruno in the topography of the philosophical systems was nevertheless also an unmistakeable rejection of an undisguised investigation of the "Spinozism" in his enthusiasm for the infinite universe. This was the bugaboo word of the century, the most terrible suspicion, and a description whose effect was the intellectual annihilation of what it was applied to. Both Bayle and Brucker were only able to protect Bruno from this verdict by pushing him back among the figures of the Middle Ages. Only Jacobi, by connecting Bruno to Spinozism, definitively altered this situation in the textbooks, a situation that still made it difficult for Goethe to perceive Bruno as a contemporary of the recognized protagonist of the modern age, Francis Bacon.

What was the aim of Jacobi, who (as Goethe attested) "was given metaphysics as a punishment," when he threw his selection of texts from Giordano Bruno into the dispute about Lessing's concept of God? In contrast to German Classicism and Idealism, Jacobi regarded pantheism as a systematically completed and historically insurpassable dogmatic formation whose line of descent extended from Cabalistic metaphysics, by way of Bruno, to Spinoza. With the latter, its definitive form had been attained. Spinozism, such as Lessing was supposed to have confessed to Jacobi, was both the exhaustion of a possibility and a blind alley. It was just this quality of being off the historical track, and irretrievably obsolescent, that that supplement to the second edition of Jacobi's polemic against Mendelssohn's *Morgenstunden* [*Morning Hours* (1785)] was meant to demonstrate. The effect of the supplement was not the one he had intended. It consisted, rather, in provoking proofs that the opposite was the case.

Spinoza's metaphysics was shown to be surpassable and to be capable of logical development precisely in that it had been unable to provide history with any reality. Jacobi's simple fundamental idea, that Spinoza had gone beyond Giordano Bruno and had literally brought him to an end, could now also be turned around, yielding the insight that Spinoza's systematic monism had reduced Bruno's dynamic scheme of the identity of creation and generation to a world of the mere modalities of the one substance. In so doing, one could say, it had brought to a forcible standstill the great cosmic unrest that could now be understood as the history of the self-realization of spirit.

Jacobi thought that by adding the Bruno text he had presented Spinozism as the final form of pantheism, and thus had demonstrated its intolerability for a changed consciousness. The text's role in initiating German Idealism proved the opposite. With it, Jacobi had had (as Heinrich Scholz put it) "here, as almost everywhere, a primarily antithetical effect; he made a not insignificant contribution to the magnificent efforts of the age of Goethe to transcend Spinozism, in that, by his relentless criticism, he provoked the summoning up of the most vigorous efforts against himself." The text's effect—or, perhaps better, the presentiment of its author's potency that it conveyed— turned against Jacobi's purpose and his thesis. Jacobi believed in the God behind nature, the absolute spiritual principle, Who was endowed with the highest dignity by His hiddenness and His inaccessibility for man. Goethe, excited and provoked by the "undivine book *About Divine Things*,"ᶠ was to place himself on the side of the defenders of the picture of Diana of Ephesus that fell from the sky, and of the craftsmen who copied it, and who did not want to have "another god, and a formless one, at that," forced upon them— "As though there was a god sort of in the brain ...." For Jacobi, pantheism was equivalent to atheism, because he saw it as absorbing, through its infinite world, the transcendence needed by his God, and making impossible the setting of limits that is presupposed in the notion of a Beyond. Strictly speaking, this thinking was still, or once again, medieval: Only a finite world could leave the 'space' vacant for a spirit and a will that were superior to it.

Hegel sums up Jacobi's relation to Bruno by saying that he "caused great attention to be paid to him, by his assertion that the sum of Bruno's teaching was Spinoza's One and All, or really pantheism." But Bruno avoided just this consequence, by not conceiving the world as an expression of the divine attributes, but instead making its infinitude express the idea of creation as the self-exhaustion of its author. He had wanted forever to avoid and to eliminate the question that was posed in Marcellus Stellatus Palingenius's widely circulated *Zodiacus vitae*: "Why hath not God created more?" [152] Bruno found in this formula of his fellow heretic (who was still persecuted after his death) the whole offensiveness of theological absolutism, and opposed it with a deduction of the 'completeness' of the world. This rationality is abstract; it does not favour man, except perhaps by sparing him worrisome questions.

If Copernicus had still attempted to save the old anthropocentrism with the means provided by a new cosmology, Bruno's infinity is characterized much more by the fact that with it he holds on to the last and perhaps the only possibility of tracing the new world back to the old God. Of course this could not be the Christian God, because a world as the fulfillment of God's entire potential had to preempt the position of the "only-begotten" one. When God's holding nothing back—which the theological tradition had held to by rejecting all deviations from the principle of the identity of essence between the divine persons—now relates to the world, it becomes the principle of the latter's intelligibility as well.

Looking precisely at the line that Jacobi draws from Bruno through Spinoza to Lessing enables us to see the unique function, in the formation of the modern age's self-consciousness, of the thought-figure that is (more or less accurately) described as pantheism. When the divinity was committed to a relation of necessity to the world, and finally to an identity with it, it could no longer continue to be the withholding of the mysteries of this world from man. In Bruno's universe the refuges of ordained or even meritorious ignorance have lost their accommodation. When he boasts of having broken through the walls of the finite cosmos, this is not primarily an expansion of the reality that can be experienced, but rather an elimination of underlying realities that cannot be experienced: Neither is there any longer a heaven, beyond the outermost sphere of the medieval cosmos, as the space of what is inaccessible, nor—as Bruno explicitly boasts—is there a place for an underworld. Brucker expresses this accomplishment in the words: "By this system one is freed from the fear of Hell, which poisons the most agreeable enjoyment of life." That is the point where Bruno links up with Epicurus and his neutralization of the world for the human heart that is intent on consolation and is plagued by uncertainty. The spaces outside the world to which both hope and fear refer are shown to be empty (or full, depending how one looks at it). In this homogeneous world that was acceptable to reason there could no longer be a favor shown to man; the whole was already the most-favored thing.

In his *Theory of the Heavens*, in 1755, Kant formulated as follows the connection between infinity in cosmology and the elimination of anthropocentric prejudice in his concept of nature: "The infinity of the Creation comprehends in itself with equal necessity all the natures that its superabundant richness produces . . . . Because, in its proceedings, it produces unmixed propriety and order, no particular purpose is permitted to disturb and interrupt its effects." [153] It is a different matter that—in spite of the misanthropic context, in which even the dreadful comparison to lice plays a part—this was not the only text in which Kant was not able to comply with his requirement of anthropological indifference, and that he was less and less concerned to obey it. But the return of anthropocentric and teleological tendencies is evidently connected to the fact that as a result of the transcendental dialectic, Newton's absolute space became a mere form of appearances and thus lost its function in relation to questions of a 'higher' order. To that extent, Kant's epistemological dualism was also a conscious way of blocking that which, as a threat, was called "Spinozism." The failure of this precaution (a failure that to us, as the heirs of Neokantianism, is hardly recognizable as such any longer) takes the form of German Idealism.

Kant and Jacobi, each with his 'counteroffer' to Spinozism, were both unable to arrest Idealism—or that which came after it and, in its turn, against it, but under conditions prescribed by it. Heinrich Heine was to explain this with his assertion that pantheism was "Germany's secret religion," and that the people who had already zealously combatted Spinoza a half a century earlier had foreseen this. At their head was Jacobi, "who is occasionally honored by being ranked among German philosophers," but who was nothing "but a quarrelsome intriguer, disguised in the mantle of philosophy," or finally, and most harshly, a "mole" who was unable to perceive the Sun of reason.[154] Heine denies that pantheism levels off humanity and leads self-consciousness to indifference: "On the contrary, the consciousness of his divinity will also inspire man to manifest it .... " But above all he sees in what Jacobi detested and what Heine describes as "the open secret in Germany" the conclusive countermove against the mechanistic conception of the world, and the Deism that went with it. "We have, in fact, outgrown deism. We are free, and we want no thundering tyrants; we have reached majority and can dispense with paternal care. Neither are we the work of a great mechanician. Deism is a religion for slaves, for children, for Genevans, for watchmakers."

Here we can see the underground inheritance of Bruno and his world-organicism. In this basic linguistic decision Heine also takes his stand against Hegel's critique of Bruno and on the side of the Romantic philosophy of nature. For all his sympathy for the "grand enthusiasm of a noble soul, which has a sense of indwelling spirit, and knows the unity of its own being and all being," Hegel had registered his reservations against regarding Bruno as the patron saint of Idealism.[g] It is true that he just had access to the first edition of Bruno's works, produced by Adolf Wagner in 1830; but besides some Latin treatises, he too only made use of the dialogue *On the Cause, the Principle, and the One*, which had become well-known through Jacobi's selections. These "disorganized riches" presented to him "often a dreamy, confused, allegorical appearance—mystical enthusiasm." On the other hand, though, it was "a great beginning of the effort to think unity." This sought-for unity lies in the elimination of the Aristotelian dualism of form and matter, through the thesis that matter "has life in itself," and is thus capable of endless transformations. "Form is immanent in matter—it is identical with it—so that matter itself brings about these changes and transformations, and runs through them all .... Thus it is in fact the presupposition of all corporeality, and consequently is itself intelligible, something general, or indeed it is the understandable again itself, its own final cause; it is the cause and the final cause of everything." But what Bruno had gathered from Copernicus's multiple motion of the Earth, which for him absolutely contradicted Aristotle's physics of the simple motions of the elementary bodies, has to remain intelligible, conceptually representable, and must not unintentionally go over into the organic indefiniteness of 'animation.' That is why Hegel regards Bruno's effort to think unity—to present "the logical system of the inner artist," of the productive idea, in such a way that the configurations of external nature correspond to it—as unsuccessful. What emerged, he says, was only a universe that dissolved in the all-devouring metaphorics of the organic, which had evaded the exertion of the concept: "The universe is an infinite animal, in which everything lives and moves in the most diverse ways." The moments of the world-process are only "collected," only "enumerated," and not, in a stricter sense, "developed"—in other words, the *coincidentia oppositorum* [coincidence of opposites][h] is not the dialectic. The verdict: Through Jacobi, Bruno "acquired a fame that exceeds his deserts."

Idealism shares Bruno's incomprehension of the historical facticity of the divine Incarnation as it is conceived by Christianity. But Idealism in the meantime has acquired less risky equipment with which to relate the central dogma of the religion allegorically to the process of spirit and to absolute knowledge as "the substantial aspect of the unity of the divine and human natures," so that it would no longer have to demonstrate such a conviction by turning away from the crucifix. The hypostatic union was integrated into the totality of history and the world; the world had not only *taken the place of* the Son of God, as in Bruno, but had itself *become* Him, as in Schelling's *Bruno*. The Nolan could not even have dreamed of a reconciliation of such a kind. For him, the meaning of God becoming man and that of His infinite power becoming the world had been in absolute competition.

Ludwig Feuerbach expressed this in one of the most penetrating analyses of the Christian combination of the Incarnation and eschatology: "The incarnation of the species with all of its plenitude into one individuality would be an absolute miracle, a violent suspension of all the laws and principles of reality; it would, indeed, be the end of the world. Obviously, therefore, the belief of the Apostles and early Christians in the approaching end of the world was intimately linked with their belief in incarnation. Time and space are actually already abolished with the manifestation of the divinity in a particular time and form, and hence there is nothing more to expect but the actual end of the world. It is no longer possible to conceive the possibility of history .... But if history nevertheless continues in the same way as before, then the theory of incarnation is in reality nullified by history itself." [155] So the continuation of history had not only disappointed the eschatological expectations; it also had to refute, in its interpretability as a divine person, the figure on which those expectations had been founded. Seen in the light of this insight, Bruno's conflict had discharged the internal contradiction of Christianity as a whole, and his death at the stake was still anything but an outcome of Copernicanism. However, the convergence, at the vanishing point, of the lines of development of the two contexts is anything but accidental.

6

Experiences with the Truth: Galileo

Even if, like Albert Camus, one says that this truth was not worth going to the stake for—so that Galileo, in contrast to Bruno, was right not to run that risk—one still has to see the "Galileo case" in the perspective of the Copernican claim to truth, because in that connection Galileo's capacity for asserting that claim and for advocating it with the incomparable new means and arguments that he had uncovered has to come up for discussion. The problem is not really how convinced Galileo was of the Copernican theory—which would determine whether his position either permitted or forbade him to evade the consequences of his conviction. For probably even the most fully realized evidentness of a proposition in natural science cannot seriously justify dying for the truth—unless this were endured, implicitly, for an entirely different proposition, as in Bruno's case, or by substitution for the principle that one should be able, in general, to advocate openly and with impunity propositions of this kind as well as of other kinds. There is no trace, in Galileo, of this consciousness of doing one thing intentionally for the sake of another thing. This constitutes his weakness, as well as the lack of hesitation with which he makes his turnings. It may be that he would have died for the metaphysical dignity of the universe or for the human spirit's right to enlightenment, if he had been able to see them as implicated in his affair.

Johannes Kepler had already reminded Galileo of the truth as a quality of Copernicanism 13 October 1597, when he urgently appealed to him finally to abandon his cautious reserve and to stand up for the Copernican world-system in public. If he did so its success could no longer be in doubt: "tanta vis est veritatis" [so great is the power of truth]. Galileo evidently needed this allusion to the power of truth, because he, for his part, had inaugurated the correspondence with Kepler on 4 August 1597 with the confidential admission that he had already for many years been an adherent of Copernicus's doctrine, that he had even already put the arguments in favor of his system in writing, but had not dared to publish them. What had prevented him? It was not, at any rate, fear of the Inquisition, whose anti-Copernican disposition at this point had not even been disclosed by the false optics of the Bruno case. What he had feared, Galileo says, was the fate of Copernicus himself: to meet with agreement in a few people and only derision in endlessly many. Nothing more, then, than that.

This first letter of Galileo's uses the great word "truth" repeatedly to conjure up the highest obligation, which he and Kepler had in common. His addressee certainly had a right to revert to that and to summon Galileo now to make his contribution to an undertaking that, while it was enormous, had been begun long since and had come a long way: "Confide, Galilaee, et progredere." [Have confidence, Galileo, and go forward.] Galileo never replied to this appeal.

The correspondence between Kepler and him is revived again only in the year 1610. Kepler has been through his most important experience with the "power of the truth": the abandonment of the "Platonic" requirement that all celestial orbits be circular. This was the weightiest concession involved in his first law of planetary motion, which the second law merely followed by sacrificing, in the uniformity of motion, the remainder of the Platonic requirement. Kepler's experience with the power of the truth was the experience of a break with a rule that was still absolutely binding for Copernicus—an experience, then, of submission to an overwhelming compulsion by the facts. Almost simultaneously, Galileo, with his new telescope to the sky, hit upon the 'intuitive' evidence of the Copernican analogy in the Jupiter system. His experience with the truth was that of a readily comprehensible realization [of his expectations]. This ease of acquisition did not allow him to understand or to deal with the fact that others were able to refuse to accept such self-evidence. Galileo did not forgive the truth for having proved to be so weak. Two decades after the first discoveries with the telescope, we find the topos of the power of the truth again, fatefully, in the mouth of Simplicio, the figure in the *Dialogue on the Two Chief World Systems* who embodied, for Galileo, everything that was opposed to what he perceived to be true: Truth must after all be strangely constituted, Simplicio says, if it does not radiate enough light to make it stand out clearly from the darkness of the errors surrounding it. Coming from the mouth of Simplicio, and directed against what Galileo considered his best Copernican argument—the explanation of the tides—that is pure cynicism.

Galileo also did not forgive Kepler for having held the power of the truth up to him. He regarded the phenomena of the ebb and flow of the tides as a direct proof of the Earth's motion, whereas Kepler, in the introduction to his *Astronomia nova*, had attributed them to the influence of the Moon. In view of this thwarting of his favorite proof of the Copernican theory, Galileo so unmistakably shut his eyes to the accomplishments of Kepler's book that he mentions him, in the whole *Dialogo* [*Dialogue on the Two Chief World Systems*], only parenthetically. In the question of the tides, he makes Salviati express ironical surprise at Kepler, who after all was in possession of Copernicus's teaching on the doctrine of the motion of the Earth and who, it had to be admitted, had an open and acute mind (*d'ingegno libero ed acuto*), and who nevertheless had introduced such a thing as the Moon's dominion over the waters—in other words, occult qualities and similar foolishness (*e simili fanciulezze*). Galileo did not insist so much on their shared Copernicanism as rather on his own, and supposedly best, proof for it. Kepler's assertion of an effect operating across empty space encountered, as it still did in the later case of Newton, all the suspicion appropriate to a relapse into medieval thinking.

Why was it, exactly, that Galileo took no notice of Kepler's results in connection with the theory of the planets? The difference between them over the question of the tides is comparatively superficial, but it is symptomatic of a deeper and more comprehensive lack of understanding. Galileo's rejection of action at a distance across space is, it is true, fear of the return of those "occult qualities," but at the same time it is also adherence to a convincing basic idea of mechanics—to a "paradigm," in T. S. Kuhn's sense. Galileo's "mechanics" does not allow him to free himself to attempt a "dynamics" of the planetary system. This becomes clear from Salviati's last great instructional lecture, which was no doubt added at a late date, in the Fourth Day of the *Dialogue on the Two Chief World Systems*. Galileo starts from the observation, which becomes manifest in the system of Jupiter and its moons, that there is a relation between distance from the central body and orbital velocity, both for a satellite and for a planet. He relates this fact to the law, which he had recognized shortly before the turn of the century, of the relation between the radial length and the period of oscillation of a pendulum. Looking at the matter in this way leads to an infringement of the requirement of uniformity of orbital motions, at least in the case of the Earth's annual course, if the Earth together with its Moon is assumed to be a rigidly connected partial system, which produces, in relation to the Sun, a greater or lesser length of the pendulum depending on the location of the Moon, at aphelion or perihelion. In accordance with this, the orbital velocity of the Earth/Moon pendulum, which is thought of as being 'attached' to the Sun, would have to become greater in portions of the orbit corresponding to the New Moon and smaller in those corresponding to the Full Moon. The only reason why this periodicity has not yet been observed by astronomers, Galileo says, is that theory first had to draw their attention to this empirical topic, so that it was from the future that confirmation was to be expected.

One notes that Galileo took no notice of Kepler's first law, according to which the Earth itself traverses its annual path at varying distances from the Sun; and consequently he cannot take notice of the second law, of the dependence of its orbital velocity on its distance from the Sun, either. The important thing here is not that Galileo, with a mistaken explanation, accidentally hit upon the fact, which also makes its appearance in the theory of gravitation, that the Earth and the Moon, insofar as they form a system with a common center of gravity, revolve around the Sun at a speed that varies with the Moon's orbital periodicity. The crucial thing is, rather, that Galileo's attention was fixed on the 'intuitive' model that he had before him in the analogy of the pendulum, in such a way that not only did he fail to gain access to the rudiments of a celestial mechanics but that access remained blocked for him precisely by his affinity for the intuitive model.

By the means that he thought would be most consistent with the overcoming of Scholasticism, he deprived himself of an anticipatory appreciation of the thought-paths, pregnant with the future, that Kepler had entered on. This should not be read as an assignment of blame, because the kind of consideration of models that Galileo undertook here was entirely in keeping with the tradition of representing and studying the celestial motions with the aid of mechanical models. The point here is only to describe the inner logic by which Galileo ignored Kepler, and the type of relation to the truth that it manifests.

Galileo was not able to forget the saying about the power of the truth. Perhaps he also did not forget, along with it, that he had confessed his own weakness and timidity to Kepler. Not accidentally, Galileo makes use of the same idea on 30 April 1633, in his second examination by the Inquisition. In defending his *Dialogue on the Two Chief World Systems*, to which objections were being raised, against the reproach that in it he had advocated Copernicanism as the truth, he uses the excuse that the dialectical pros and cons of the arguments had come out as though they were intended to force one, by their power, to a conviction that in fact they were supposed to make easier to refute. The power of the truth has become its apparent availability, at will, for rhetorical misuse. One should keep in mind what else Galileo could have said. He could have said that even with the restraint that had been imposed on him (under threat of punishment) in the evaluation of arguments presented in hypothetical form, the truth simply asserted itself, on its own account, in such a way that it inevitably created the impression that the author was giving it special treatment.

This was not yet Galileo's last indirect response to Kepler's appeal. In the *Discorsi* [*Discourses on Two New Sciences* (1638)] he appeals once again to the power of the truth, and again in a fundamentally changed sense, namely, as the power of reason against intuition [*Anschauung*]. That is then, after all, a late lesson from his failure both with the telescope and with the tides. So the ineffectiveness precisely of what he had considered more than a match for any opposition remains the significant fundamental experience in relation to which the pattern of this life and thought becomes intelligible for us. Through the offer of the telescope, the fact that one can refuse to see, and deny that one has seen, and dispute the possibility of ever seeing, and wish that one did not have to see, became, as a human and a historical possibility, the acute incentive to expect less from bringing forward immediate intuitive evidence than from the analysis of elementary and basic physical facts.

Galileo may have found the weakness of the truth, which he had discovered through other people's turning away from its evidence, confirmed in his own case in, at the latest, 1633, when the "territio levis" [slight fright]—the mention, merely as a formality, of torture, in the Inquisition's proceedings—was enough to move the old man's timidity to offer to add to the four days of conversation in his *Dialogo* two additional ones in which his intention of refuting the Copernican system should, "with God's blessing," become clear beyond all doubt.

Galileo's defeat by the "slight fright" will have been more of a burden to him than it would have been to someone who appeared to have succeeded in outwitting force. Insofar as the truth as such does not already seem worthy to him of the sacrifice of human life, the historical observer will find Galileo's greater failure in the area of establishing his own credibility. In his letter to Kepler of 19 August 1610, he reaches, without reflection, for an animal metaphor, the "stubbornness of an adder," to describe the resistance of the philosophers at the gymnasium in Padua to the light of reason (*contra veritatis lucem*). We shall discuss this resistance later. At the moment the important thing is that Galileo, for his part, and in his own way, shut his eyes *contra veritatis lucem*, by not pressing his observations of Jupiter's moons to the point where he could display the regularities governing their motions. Kepler was to formulate his third law of planetary motion, about the relation between the planets' distances from the central body and their orbital periods, only in May 1618. Irrespective of the fact that Galileo will take no notice of Kepler's discovery then, either, one only has to imagine that he himself could have discovered and verified this law in the case of Jupiter's moons. Then Galileo would have deprived his opponents of the excuse of calling them optical illusions or undefinable phenomena, by being able to determine the configurations of the Jovian system at any given time, in advance. Only then would a look through the telescope have been enough to put the reality of the satellites circling around Jupiter beyond doubt.

Galileo simply put too much faith in the self-evidence of intuition at the moment when it occurs, and did not do for the heavens what he was to accomplish for terrestrial mechanics, namely, to ensure the lawfulness of the phenomena by connecting them to mathematics. Only Cassini, in 1668, with his ephemerides for the "Medicean stars," supplied the foundation for a further step in cosmology that was to be just as momentous once again for the change in our consciousness of the world as the Copernican reform had been—namely, for the discovery of the finite speed of light, by Olaf Römer, a few years later.

Galileo's adversaries had certainly not been empirical realists; but he himself left them an escape route, where he could have forced them to become realists. If we look at the matter closely, Galileo behaved toward the new possibilities that Kepler had opened up, possibilities for comprehending the planetary system as a physical unit and thus implementing the Copernican truth, with no less indifference and no more understanding than his adversaries from the gymnasium showed toward the configurational analogy that he offered them with the view of the Jovian system. What Galileo may have anticipated was something like the effect that the view of Jupiter's moons in the telescope had on the aged Christopher Clavius, according to Leibniz's account: Overwhelmed by the *vis analogiae* [power of analogy], he cried out that the old astronomy was done for (*actum esse de recepta Astronomia*).

But it was not by means of the telescope but by means of terrestrial physics that Galileo sharpened the problem of the truth of Copernicanism. It is not the categorical way in which the Copernican system is presented that creates the explosive power of the *Dialogue on the Two Chief World Systems*—as the inquisitors, with their attention fixed on modalities, believed—but rather the amount of physics that had gone into this presentation. Here too Galileo was looking in the wrong direction: He was searching for a physical proof for Copernicus, such as the argument from the tides was supposed to present, and in so doing he failed to recognize the more pressing need to free Copernicanism from the obstacles, in the area of physics, that stood in the way of its prevailing in astronomy. Another way of formulating this state of affairs is to say that Galileo gave the most succor to Copernicanism where he himself least noticed it, namely, in the approach to the principle of inertia that took place in the progressive development of his mechanics. With this principle one could not prove the truth of Copernicanism, but one could strike the center of the resistance to it that was theoretically motivated. Galileo wanted and believed it to be the case that if one could not grasp the evidentness of Copernicanism, one could *see* it, even if it was on a seashore, watching the alternate ebb and flow of the tides.

The central distinctive characteristic of Aristotelian and Scholastic physics had been that it had assumed a real and a realizable difference between rest and motion—that is, that it did not recognize the principle of the relativity of motion and of the equivalence of the states of rest and unaccelerated motion. For every motion—that is, not only for a change in a motion—it required that there should be a force proportioned to it. This defined the expectations that follow from the assumption of a motion of the Earth and that would have had to be confirmed—for example, that the path of a falling body should be displaced with respect to the perpendicular. Tycho Brahe had still used this argument in his *Epistolae astronomicae* [*Astronomical Letters*], in 1596: "How is it possible that a lead ball, when it is dropped in the proper way from a very high tower, hits with perfect exactitude the point on the Earth that is perpendicularly beneath it?" A single such phenomenon seemed to have enough power of falsification to make one accept, in astronomical theory, a complexity that was monstrous and that could not be interrogated further in regard to its mechanical possibility.

Against this failure of physics to mesh with astronomy, even the apparent plausibility of the argument from the tides could accomplish nothing. The Copernican system's claim to truth was condemned to failure if the small but decisive quantity of Aristotelian physics in the geocentric system could not be replaced by a new physics. Here what was at issue was not primarily the means by which to prove the motion of the Earth, but rather the securing of its mere possibility by eliminating the classical objections—an accomplishment that could clear the way, for the first time, for a decision between the systems based on the criterion of their rationality. Galileo did not perceive the Copernican theory's need for this type of help.

Certainly he had long since begun—at least since the treatise *On Motion*, completed in 1590—to investigate the physical problems of free-fall, of projectile motion, and of other mechanical phenomena with a new impartiality with respect to the tradition. But there is no basis for concluding that he perceived the connection between this interest and the Copernican system's need for proof. The telescope and the tides appear to him as the argumentational levers with which, by a *coup de main*, to help the Copernican truth to achieve its breakthrough. Now one can certainly not say that his expectations of the telescope were mistaken; it was to become one of astronomy's most important instruments. But at this moment, and in this direction, he was wrong to expect what he did. What Copernicanism needed was an extremely nonintuitive element, and one that was empirically 'absent'—the concept of an abstract state of affairs, whose introduction into mechanics really could free Copernicanism from its opponents' cunning insistence: It needed the concept of inertia, the fundamental assumption of a physics that would make a radical break with Aristotle, and the classic case of an assertion that, as such, could never be demonstrated by observation or experiment, with whatever equipment.

There is hardly another event in the history of science about which such mutually exclusive theses have been set up as the transformation of physics into a form that suited—or, to begin with, a form that did not block—the Copernican system. One of the commonest assertions is that the Aristotelian philosophy of nature was a collection of speculative propositions about nature, remote from experience, which were replaced by a system of statements whose origin was empirical through and through. But the downfall of Aristotelian physics is well described by the lapidary sentence of Heinrich Scholz: "It perished as a result of its positivism." In fact this physics describes reasonably accurately precisely what presents itself to immediate, everyday experience. It has turned out to be by no means impossible for certain basic assertions of the Scholastic physics that was dominated by Aristotle to continue to be used, in spite of the wide dissemination of knowledge of elementary physics, right up to the present. A statement such as the one that everything that moves is moved by a force expresses the experience, not subjected to reflection, that we can have in the region of our life that is close at hand. It is simply not correct that the critical remedy for this philosophy of nature was more experience. What led to the break with the traditional physics was a different kind of experience, an experience that was already directed toward specific premises—selected and arranged in accordance with them—and placed under definite conditions: in other words, *experimental* experience.

This type of experience never presents itself immediately, and is not exhausted in intuitive givenness. It confirms or disproves assumptions in regard to a definite and, at least in principle, measurable aspect of a total phenomenon. Experience that is controlled—not to say prepared or dissected—in this way cannot stand at the beginning of a radical theoretical change. Instead, what stands at this beginning is a distancing from the immediacy of the life-world. The suspicion arises, then, that the everyday experience that supplies us with what 'goes without saying' for us represents neither the 'norm' of physical reality nor its totality: in other words, that it could be partial and provincial, because in it simple law-governed regularities are superimposed on each other or are concealed by additional factors. This most familiar world of experience is too complicated for it to have been able to give us even a presentiment of the axiom that every cause always has the same effect. At the same time, the Aristotelian cosmology had excluded the possibility of outflanking this complexity of experience in the life-world by regarding the processes in the starry heavens, which are recognized as simple, as representing generally valid facts about nature, and coming back, from their perspective, to terrestrial experience. To that extent Francis Bacon is right when he says that the metaphysical doctrine of the hiatus between the heavens and Earth made human ignorance durable, by providing one of the strongest arguments for it.

A physics that is based on experience in the life-world excludes, above all, two conditions of its objects' appearing as simple: It can neither display bodies on which Earth's gravity does not operate nor observe motions in a space that is not filled by a medium. For a historical analysis one must separate the two conditions. For Aristotle, the resistance of the medium was constitutive for the phenomenon of local motion as a possible experience. He assumed that a motion without resistance would be executed by the moving force with infinite rapidity, so that it would have to take place instantaneously and imperceptibly. This argument reduces the assumption of the possibility of a vacuum to absurdity 'ontologically,' that is, by arguing from the concept of motion as a temporal process.

One would think that the impossibility of a motion without resistance would already have been a sufficient basis for the principle that every motion requires the continuous action of a mover. For "natural" motions, that is indeed the case, but for "violent" motions there is also an internal resistance of the body against the change that is forced upon it. For the heavenly bodies, the awkwardness of further inquiry was headed off by the introduction of the special qualities of their matter and of the intelligences that transmit the moving power of the unmoved mover to them. So Aristotle had not been able to grasp even the "natural" motions as enduring conditions. For in this concept motion, as a predicate of a body, remained a sort of 'exceptional' state, a transition toward a final location and a state of rest. The heavenly bodies are distinguished from sublunar bodies only in that they are not able to reach their goal, in the status of the unmoved mover—or they are always en route to that goal. Thus the meaning of a motion, as a phenomenon, lies in the fact that it connects a body's initial condition with its final condition, and that it is not able to overcome this difference for a moment.

Galileo was only partly able to free himself from these assumptions. He was most successful in connection with the assumption of a normalized motion that would be possible in a vacuum, although the production of a space that contained no air lay beyond what was possible for him experimentally. He did not even assume that a vacuum existed in interstellar space—that is, that it is the predominant condition in the universe. Nor did this question matter at all, as long as one had not eliminated resistance, as a constitutive element, from the concept of motion, so that its magnitude could be bracketed out as one of the supplementary factors. What was required was to think that even in absolutely empty space, motion would not be carried out instantaneously—which initially meant above all that the concept of empty space could be taken into a physical system without contradiction. With his axiom of essentially finite motion that is independent of resistance, Galileo defends himself against the abstract notion of a transition to infinitely rapid motion. His strength, in relation to Aristotle, is the fact that he can work with fictions; his weakness is that he finds this so difficult.

Of the two fictions—the vacuum and inertia—that were needed to regulate physical experience and thus to release it from its bond to the life-world, Galileo accomplished only the first one. He only partly succeeded in formulating the concept of a motion that is not influenced by forces. He appropriated the late Scholastic idea of a force that is 'communicated' to the moved body and he generalized the Aristotelian naturalness of circular motion in such a way that now it could appear not only in heavenly bodies but also in the realm of the sublunar elements—though admittedly as a factor that we cannot experience directly. Thus Galileo had held to the distinction between natural and violent motions, but he had expanded the range of "natural" ones in such a way that they could be understood as playing a part in complex phenomena. For the theory of free-fall, this had the advantage that Galileo did not need to speculate about the external violence of a force with which he was not acquainted. The lawfulness of the acceleration of a falling object resulted from its teleological "inclination" toward the mass of the Earth. Here Aristotelianism lent its support to the quantitative formulation, from which the "force," as a constant proportional contribution, could be read off. The fact that Galileo holds fast to the *inclinatio* reveals, no less than does his disposition to see conclusive evidence for Copernicanism in his view of the Jovian system, how his structure of thinking is tied to analogy.

Galileo gave the phenomena of human close-at-hand experience cosmic equivalence. The section of the world contained in this experience proved to be too small, and the Earth's perspective on the universe proved to be too parochial, for the dominance of the principle of circularity to be perceivable in the examples of motion with which we are familiar. Experience is more than looking closely. As was traditional in astronomy, now in physics too the phenomena are "saved" by means of what is not phenomenal.

If one attempts, in a thought experiment, to imagine the motion of a perfect ball on a perfect plane, then—for the prehistory of gravitation—the geometric and the physical constructions of the plane always diverge. Galileo arrives at a result that is equivalent to the principle of inertia only by basing his physical definition of the plane on the equal distance of all of its points from the center of the Earth. On the assumptions of Aristotelian physics, but also, indeed, by the evidence of the mechanical phenomena that would be found on it, a geometrically regular plane—that is, a plane constructed as a tangent to the Earth's surface—would slope 'upward' as one moved outward, in any direction, from its point of contact with the Earth. On the assumptions of the *inclinatio* physics, since the plane's distance from the center of the Earth would be greater at every point than it would be at its point of contact with the Earth's surface, a ball placed on it at any point would roll toward the point of contact. So the physical concept of a perfect plane produces a surface that is parallel, all over, to the surface of the Earth. Bodies moving in straight lines on it would have to describe circular paths and would have, for a sufficiently distant observer, complete equivalence to the motion of a heavenly body like the Moon. That is why Galileo, still under the conditions of late Scholastic physics, can assume that in the absence of resistance an impetus that was conveyed to this body at one time would endow it with a lasting motion. This has been described as a "principle of circular inertia."

It is very closely connected not only to the Scholastic remnants in Galileo's physics but also to what he himself felt to be his greatest triumph: that he had made the Earth a star among the other stars. He had gone far beyond Copernicus in that he had not only given the Earth itself the motion and the position of a planet, but also had awarded to the objects of human close-at-hand experience in and in regard to this heavenly body equivalence with the phenomena that are perceivable in the heavens. Even if the circularity of his principle of inertia was an atavism, still it was indispensable, temporarily, in order to define at all the normal conditions of states of motion in the region of gravity, as long as gravity is still unknown.

It is in this context that we have to see the often smiled-at, but scarcely understood attempt, in the *Dialogue on the Two Chief World Systems*, to demonstrate that a falling body has a straight line of fall only in appearance, while in reality—that is, in relation to an extra-terrestrial observer—it has a curved path. The treatment of this subject is carefully separated out from the second day of the *Dialogo* as an episode, as *contar favole* [telling stories], as a "play within a play." Salviati is asked whether he has thought about the natural line described by a body falling from a tower; he answers that this line arises from the combination of the straight line that is natural to falling and the body's participation in the Earth's diurnal rotation. Experience is correct to take only the simple line of fall as certain, since the second component enters equally into the system of the observer, the experimental equipment, and the falling body, so that it does not belong to the object that is observed but to the conditions under which it is observed. "We never see anything but the simple downward motion, since this other circular one, common to the Earth, the tower, and ourselves, remains imperceptible and as if nonexistent." [a] Only the part of the stone's fall that we do not share remains perceptible for us: " ... e solo ci resta notabile quello della pietra, non participato da noi." However clearly Galileo may have emphasized the unseriousness of this episode—for example, as late as the letter to Carcaville of 5 June 1637—still it is evidently a serious enough matter for him to offer such unseriousness at all. And this is for good reasons, because the thought experiment opposes one of the most troublesome principles of Aristotelian physics, the principle according to which a body can only have *one* natural motion. In his playful digression Galileo has openly violated this principle.

One cannot simply assume that Galileo was *unable* to free himself from the traditional assumptions that he did not attack. A historical author has to be seen not only, nor even primarily, in the horizon of what he himself was capable of, and in the scope that his progressive emancipations achieved; for him, insofar as he is concerned about the success of a new idea, fitting it into the horizon of what his audience is capable of is of primary importance. One simply does not dare to imagine what would have happened to Copernicus's work if it had gone beyond its laconic argumentation and had also immediately offered the physics, as well, in terms of which the asserted motions of the Earth could become integratable. With Galileo, too, in spite of his incomparably greater loquaciousness, the standing methodological doubt remains as to whether the surviving rudiments of traditional ideas in his works are not evidence of an effort to keep the argument, as much as possible, within the horizon of his audience.

There are also serious objections to such an analysis—above all, his behavior toward the doubters in the scene with the telescope. It is certainly true that Galileo gives us more cause to attribute the theoretical retardations to himself than would be permissible in the case of Copernicus's economical procedure. But for Galileo, too, the thesis will hold that the Scholastic dogmatics that he conserves is used, functionally, primarily as a way of acquiring licenses. If he had wanted to introduce straight-line inertial motion immediately as the normal case of the state of motion, he would have been faced with the extremely uncomfortable and momentous inevitability of unlimited space. Only the constant deformation of the straight-line inertial motion by the attraction of central bodies will permit physics to abandon the 'naturalness' of circular motion. Gravitation domesticates the straight-line quality of the principle of inertia.

Galileo would have had to admit, as elementary, a form of motion for which no possible teleological explanation was available, and in doing so would at the same time have deprived himself of teleology as the still quite productive effective instrument for the theory of the acceleration of falling bodies. True, it seems only a difference of language, without substantive importance, when one attributes the motion of falling to the influence of the central mass rather than to the falling body's inclination; but the historically decisive difference still lies in the systematic differentiation between natural and violent motions. The inclination toward the "natural place" presupposes a *vis motrix* [moving force] that is immanent in the falling body, whereas gravity operates on it from outside, hence violently. Galileo does not see that the lawfulness he finds in the free-fall makes possible an interpretation that combines both types of motion. Then, of course, the assumption of the constancy of the acceleration of the fall at all distances from the center of the Earth, which has a positive effect for the constitution of the theory, would immediately have become problematic.

The systematic framework into which Galileo had to fit his physics if he wanted to count on an audience that was capable of understanding him did not allow him to avail himself of the advantage that lay in admitting falling motions that never reach their supposed 'goal.' This, of course, is what Newton will require in order to turn terrestrial mechanics into a universal celestial mechanics. By means of it, the previously 'natural' circular motions are explained as resultants from inertial motion, which is now the only 'natural' one, and falling motion, which has now become 'violent.' The new violence of all falling motions made it appear (if the paradox is permitted) entirely natural that it could miss its goal (as that goal was traditionally defined) and could be destined to circle it forever. Of course this process no longer had the metaphysical dignity that had been awarded by Aristotle to the outermost sphere's eternal rotation in its loving, but unfulfillable devotion to the unmoved mover.

Galileo's physics remains one that is valid *on* stars, whereas Newton's was to become one that had validity *for* stars. Galileo could have found the step beyond his position in the introduction to Kepler's *Motions of the Planet Mars* [*Astronomia nova* (1609)], if he had not inexcusably disregarded this work, as he did. Kepler defines heaviness (*gravitas*) as the mutual influence of bodies toward their being united. The phenomenon of falling specifies the general definition one-sidedly only because of the disproportion between the masses of the bodies in question. So the principle here is more that the Earth attracts the stone than that the stone strives toward the Earth: " . . . ut multo magis terra trahat lapidem, quam lapis petit terram." From the point of view of the interests of the progress of theory, Galileo's neglecting to give his theory of motion the quality of a universal cosmic mechanics appears as a reproach. But contrary to the classical canon of the historiography of science, we also have to register Galileo's instinctive self-protection against demands that were bound to create a subversive pressure on the temporary constructions of his theoretical process. The conditions under which Galileo could satisfy his governing interest were met by his ability to give Copernicanism a physical foundation that made sure that phenomena on the moving Earth agreed with those on an Earth that was at rest. It was still the procedure of equivalence, with which Copernicus, in the first chapters of his *De revolutionibus*, had compared the *mobilitas* of the Earth to that of the heaven of the fixed stars.

In this way Galileo subordinates everything that he observes and discovers to the service of the Copernican intention that he had already formed early on. He never had the experience of a truth that forced itself upon him against his will and his convictions, nor, probably, was he predisposed for that kind of experience. Through the telescope at that great moment in 1610 he had found embodied in front of him, in the image of the satellites circling around Jupiter, the basic physical idea that the normal motion of a ball on a plane, uninfluenced by forces, would have to present itself to a cosmic onlooker as a circular orbit around the central body, the Earth. His reflexive optics then made him see the Earth's Moon as a realization of the perfect ball rolling around the Earth on the ideal physical plane. The fact that the objects in the space that we experience in the life-world, in the atmosphere surrounding the Earth, participate in the motions of the heavenly body with which they are associated was ultimately due to the fact that each of them could be understood as a satellite, however tiny, of the Earth-star.

The unique guiding function of the view of Jupiter's moons was bound to strengthen Galileo's feeling that the evidence of intuition was willingly at his service. For precisely that reason, the weaknesses of his position remained hidden from him: The concept of circular inertia seemed to be so perfectly adequate to the needs of Copernicanism, and at the same time to converge so perfectly with the traditional requirement that one should employ uniform circular motions in astronomy, that the radical character of the astronomical problematic and of the new physics that was necessary to eliminate it escaped Galileo.

The relation to truth that underlies his intellectual intentions and accomplishments is also the reason for the fundamentally conservative pattern of his conflicts with his opponents, and above all of his succumbing to the medievalness of the Inquisition. He was not undismayed enough to sympathize with Kepler's dismay at the strangeness of a new truth; consequently he succumbed, even before the "territio levis" [slight fright], to untruth.

Galileo's relation to truth has a historical dimension. That dimension is subjected to criticism by a philosophy for which, at the beginning of our century, rebellion against the fascination exercised by modern natural science had become constitutive, and for which coming to terms with that fascination remained a task that it could not shirk. Phenomenology had been founded, around the turn of the century, against psychologism, and it culminated, in the last years of its founder Edmund Husserl (years in which he was prevented from exercising an influence), in a critique of physicalism. In his *Crisis of the European Sciences*, of 1935/36, Husserl made the charge, which at first sight seems paradoxical, that Galileo had deprived our knowledge of nature of the rank and the claims of science and had debased it into a technology. Mastery, Husserl said, had taken the place of intuition, and functional making use of things had taken the place of the theoretical attitude. Galileo had abandoned the intuition of phenomena both as the initial basis for and as the norm of accomplished knowledge of nature. Thus he had made the decisive step toward modern science's relinquishing and forgetting the "meaning-fundament" of all theoretical processes in the intuitional sphere of the "life-world." "The natural science of the modern period, establishing itself as physics, has its roots in the consistent abstraction through which it *wants* to see, in the life-world, only corporeity .... Through such an abstraction, carried out with universal consistency, the world is reduced to abstract-universal nature, the subject matter of pure natural science. It is here alone that geometrical idealization, first of all, and then all further mathematizing theorization, has found its possible meaning." [156]

Now, the difference between 'technique' and 'science' already constitutes the core of the Copernican problem of truth. It is against the concept of astronomy as a mere procedure for commanding the phenomena, while how they are constituted is a matter of strict indifference, that Copernicus had tried to reestablish the inseparability of the ideas of science and truth. But in this disagreement Galileo has to be seen on Copernicus's side, against astronomy's 'technical' tradition as part of the faculty of arts. Not only does he demand that astronomical constructions should not be contrary to physical assumptions, but he also considers it to be possible to present physical proofs for astronomical theories. But it is just this relation of foundation between astronomy and physics that induces Husserl to attach little value to Galileo's delight in the use of intuition in astronomy, in view of the abstractions with which he reformed physics.

This reform seems to Husserl to have played a central role in the modern age's process of technicization. Galileo detached physics, as a discipline that becomes mathematically formalized, from the intuitive world-experience that supports it. In doing so he carries out a "surreptitious substitution of the mathematically substructed world of idealities for the only real world, the one that is actually given through perception, that is ever experienced and experienceable— our everyday life-world." [157] It was not first of all by going into the arsenal of Venice and reading off from the hands of the artisans of weapons the laws of nature that were unknown to them and the calculating procedures that they did not comprehend that Galileo made the technicizability of knowledge of nature one of the assumptions of the new physics. He had already done this by adapting the objects to the method by means of which he meant to understand them. Husserl presented this idealization of the world of intuition, which prepared it for the theoretical grasp of the new physics, in the graphic metaphor of the "garb of ideas" that veils the "things themselves" in such a way "that we take for *true being* what is actually a *method* . . . ."

Here too, in the eyes of his late critic, Galileo becomes the precipitate intellect who sacrificed everything to the swift impact of evidence and who no longer noticed the compromises he had to make and things he had to forfeit for his gains. Especially the author of the *Discorsi*, the founder of modern dynamics, appears as the one who is responsible for the deviation of modern intellectual history in which the successful manipulation and provision of methods at the same time excludes consciousness of what was contained or has been missed in the theoretical and practical output of their modes of procedure. For Husserl the lack of a consciousness of foundations is at the same time a lack of responsibility—ethics requires one to know what one is doing. The new physics thoughtlessly left behind it the painstaking construction of its concepts through intuition and the laborious intermediate steps by which it should be established—in favor of the flying start of its advances.

So the factor that causes crises, according to this outline of a philosophy of history, is not progress itself but rather the lack of a grounding for its beginnings. Galileo's "fateful omission" was that when he was founding physics he did not go back to the "original meaning-giving achievement." Husserl moves him into the ambivalent position of "at once a discovering and a concealing genius" ["ein zugleich entdeckender und verdeckender Genius"].

These formulations of Husserl's contain a penetrating diagnosis of the genetic structure of modern science. That is true even if the therapeutic conclusions that he draws may have involved an indigestible prescription. That prescription was that we should find our way back to that suppressed and forgotten intuitiveness of nature as it is experienced in the life-world. Like other culture critics, Husserl too comes close to the thesis of an original fault behind the modern age. As a philosopher of history, he remained the Cartesian he had always been: He sees the beginning of the modern age as the zero point of a process, a point at which everything was still possible and the task was to carry out a radical act of foundation. But any glance at the case of Galileo teaches us that this was anything but an absolute beginning. None of his achievements would have had any prospect of meeting with agreement or even attention from his contemporaries if they had been put forward in such a way as to provide a rationally conclusive and historically unbiased proof. Husserl's insight that in Galileo discovery and concealment were related in a definable way is not made less significant because it evokes doubt as to whether this was really an avoidable lapse on the part of reason.

It is just not the case that Galileo founded a 'new' physics, in the strict sense—in doing which all the opportunities of a radical beginning would have been open to him. Instead, he undertook reoccupations in an existing view of the world that was developed and was embodied in the languages used in the schools of the time. He could not "discover" without "concealing." Husserl could not have recommended the "infinite work" of phenomenology[b] as a means of curing the crisis of European science—and, beyond that, of European consciousness—if he had not assumed that that original loss of intuitiveness was a more or less frivolous abandonment. The validity of this assumption is questionable. It is also doubtful on account of the implied function of 'bringing salvation' that is connected to it. Husserl sets up, against Galileo, the demand that the idea of science should be carried out, and indeed intensified to its highest degree, without relinquishing the things that were, historically, relinquished. To that end he had to deny, decidedly, that there is such a thing as a law governing the realization of new possibilities, according to which each of them can only be gained at the cost of conceding a narrowing of the horizon of what is possible. In the combination of discovery and concealment he saw only a contingent linkage, only a yielding to the temptation of the shortest path and the most obvious functional usefulness. Then a renewal of the original situation and a compensatory catching up with what had been neglected could be both considered possible and offered in the method of phenomenological intuition. History appears to be reparable: We hear about concealment, not about destruction [*Zerstörung*]. Husserl choses an almost idyllic image for that divergence between the foundation in the life-world and the unbound accomplishment when he says that science "hovers, as though in empty space, above the life-world." [158]

Analysis of Galileo's concept and expectation of truth yields a different finding from Husserl's, which charges him with "surreptitious substitution of idealized nature for prescientifically intuited nature." On the contrary, Galileo's cognitive procedure is endangered by the way it is tied to intuitiveness and to the analogies in vision, and by the precipitancy of its consciousness of evidentness. The astronomer and mathematician who searches the heavens for eidetic confirmations of the Copernican diagram; who thinks that he can set the changes of the sea's motion, in its terrestrial bowl, before us, for us to witness, as the tides; and who did not find his way to the basic assumption of inertia because it evades intuition, in our experiential world, and required a step into the nonintuitiveness of fiction—this Galileo could not have been the protagonist of science whom Husserl brands as phenomenology's antagonist.

We seem to be able to see how difficult he finds it to free himself from, for example, the intuitiveness of the Aristotelian concept of force, even though his findings compellingly urged him to do so: He moved only tentatively from the supposed proportionality of weight and rate of fall to the conjecture, which he expressed in the *Discorsi*, that the rate of fall is entirely independent of both the absolute and the specific weights of the object. Is it too daring to say that in his theoretical constitution Galileo stands closer to Goethe than to Newton?

Was it then only the old man's increasing blindness that imposed on him (more than it granted him) the freedom for the abstract imaginings of the *Discorsi?* In his moving letter to Diodati, Galileo described this final experience of the truth's withdrawing itself as the shrinking, to the residual reality of his own body, of that same universe that he had expanded many times over by his discoveries. His biographer, Viviani, also pauses over this extinguishing of the light of those eyes, "which within less than a year had discovered, observed, and taught others to observe much more, in the universe, than all mankind had been permitted to see in all past ages." [159]

If we consider, in its genesis, Galileo's most famous discovery in dynamics—his description of free-fall—then what was discoverable for him and what remained concealed from him becomes evident. This accomplishment, in particular, early became a quasi-myth intended to make Galileo appear as the prototypical experimenter who laid down for the first time, and above all for himself, the program of empirical investigation: By collecting a great mass of experiential data, to trace, dissect out, and secure the natural law that is hidden in them. But what is already the case with his predecessor in the theory of experiment, Nicholas of Cusa, still holds, more or less, for Galileo: that the experiments that are described and are called for were mostly only thought experiments, even if the description of them is followed by the toposlike formula, "I have often verified that."

The little mouse with an interest in history who could have watched Galileo's investigative activity from his hiding place in a corner would certainly have been disappointed by the reality that corresponded to the "hundred-fold testing" of which Galileo assures us in connection with his law of falling bodies, by the scantiness of his store of data, and by the way in which, as a result of unsatisfactory experimental conditions and methods of measurement, the regularities that he postulates disappear in his margins of error. The legend of Galileo's experimenting with free-fall from the leaning tower of Pisa is instructive not because it stubbornly propagates itself, since Viviani's *Vita*, even though it happens to be possible to refute it, historically, but because it makes it easy for us not to believe in its truth's having been evident to the senses: With the duration of fall of $3\frac{1}{3}$ seconds, which corresponds to the height of the tower (55 meters), Galileo simply could not have decided anything, analytically, with the methods of time measurement that he mentioned. But the calculations that he gives in the Second Day of the *Dialogue on the Two Chief World Systems* in connection with the mental attempt to determine the time involved in falling from the Moon to the center of the Earth reveal that the margins of error in his observations were still set much wider than would have been permissible with the scrupulousness that he claims to practice. Behind his discoveries we see not the sober industry of the devoted experimenter, who keeps on increasing his body of data until it surrenders its hidden formula to him, but rather the speculative anticipation that places its trust in nature's use of absolutely simple means to achieve its effects.

At the beginning of the Third Day of the *Discorsi* Galileo says that in his discovery that in free-fall the series of odd numbers determines the relation between the distances traversed by a body in equal intervals of time, he was "led, by hand as it were (*nos quasi manu duxit*), in following the habit and custom of nature herself, in all her various other processes, to employ only those means which are most common, simple and easy." But it is not attentiveness in observation but his metaphysical postulate that leads him to ask, "When, therefore, I observe a stone initially at rest falling from an elevated position and continually acquiring new increments of speed, why should I not believe that such increases take place in a manner which is exceedingly simple and rather obvious to everybody (*omnibus magis obvia ratione*)?" The principle that nature proceeds as everybody would have to expect allows us to conclude that "we find no addition or increment more simple than that which repeats itself always in the same manner."

Here, by turning his abstract attention in a different direction, Galileo could have perceived that the principle of the acceleration of a falling body becomes demonstrable in experience only if the velocities that are arrived at in each case are maintained permanently; then he would have had before him the principle of inertia, and would have had to arrive at a concept of force that can only be proportional to the change in the motion, that is, to the acceleration of fall. What stood in the way of this was the fact that free-fall was not representative of the phenomenon of motion in general, but only of "natural" motion in the sense defined by the distinction created by Aristotle. Consequently Galileo will not be able to gain, from the phenomenon of free-fall, a universal dynamic principle of constancy that could also hold for "violent" motions.

By assuming that the factor generating the acceleration of a falling body was its "inclination" toward the center of the Earth, Galileo also missed the obvious question whether the body's distance from this center of the terrestrial body affects the magnitude of the acceleration. That would inevitably have taken on importance in relation to a thought experiment with orders of magnitude like those in the case already mentioned of a body falling from the Moon to the Earth, and would at least have introduced Newton's problem. Galileo initially formulates his law of falling motion as the relation between the distances traversed in successive equal periods of time. The ratio between the duration of the fall and the distance covered in each case—the quadratic proportion—only emerges from the number speculation about the correspondence between the increments to the distance traveled and the odd numbers. There is no question but that this whole-number relationship, with its eidetic clarity, was capable of only approximate confirmation by the accuracy of the experiments that were possible for him—and that that was all it needed. Nor do the fictional partners in the *Discorsi* insist on witnessing such experiments. The question that is put to nature in the language that is metaphysically suited to it requires one to pay only the most fleeting attention to its answer. It is characteristic that even the Scholastic, Simplicio, immediately dismisses the Aristotelian assertion that the speed at which a body falls depends on its weight—so well does the procedure that Galileo applies accord with Simplicio's idea of how knowledge is to be gained.

In the context of the discussions about free-fall, in Galileo's last work, the metaphor of the power of the truth appears again. The requirement that a body falling from a state of rest must pass through every velocity up to the terminal one, which is constant on account of the growing resistance of the medium, appears as a demand that our faculty of imagination (*immaginazione*) can obey only with difficulty. The imagination is constrained, we are told, by the impression of the senses that the falling body immediately reaches a great velocity. A rational analysis of what we observe leads to the opposite conclusion. The initial velocity must be assumed to be very small (*piú che minima*). Regarding this transgression, by thought, of what is intuited—this emergence of the true state of affairs, as against its *primo aspetto* [first appearance]—Salviati, as Galileo's representative in the dialogue, says, "See now the power of truth (*la forza della verità*); the same experience which at first glance seemed to show one thing, when more carefully examined, assures us of the contrary."

If one investigates the connection more precisely, it is not in fact the same experience that at first suggested the thesis and then the antithesis. It is the abstract execution of a thought experiment that refutes the empirically recordable datum. It is penetrating into the border region of the non-intuitive that puts the most readily accessible intuitive data in the wrong.

The "power of truth," of which Galileo once again speaks, is the resigned inversion of his expectation, in 1610, that he would achieve the victory of Copernicanism with his telescope. It may have been the old man's answer, now only ironical, to Kepler's early appeal to him to place his confidence in the power of the truth. But it is also still a misunderstanding of his German contemporary, who for his part surrendered the pure form of geometrical speculation, which he had employed in his *Harmony of the Universe*, when the obstinate inaccuracy of between eight and nine minutes in the orbit of Mars forced him to change the *New Astronomy* over to the eidetically unsatisfying orbital pattern of the ellipse and to permit the orbital velocities to be non-uniform.

Grown blind, and living under the eyes of the Inquisition, Galileo, from his situation of enforced lack of [public] impact, launched the *Discorsi* as a final test of the power of the truth. Perplexed and discouraged by the ill-fortune of his previous books (*confuso e sbigottito da i mal fortunati successi di altre mie opere*), he planned to leave this last one in manuscript form.[c] Galileo did not need to blind himself, as Democritus, according to the legend, is supposed to have done; he was blind enough no longer to expect intuition to be able to gain possession of conclusive evidence of the truths about nature. Only after the problematic character of his reliance on such evidence had become manifest, as a result of the failure of his Copernican mission, does he now entrust the fate of his truth to a different language. According to his own words in the *Saggiatore* [*The Assayer*], in 1623, the "book of nature" is written in this language: "Philosophy is written in that vast book which stands forever open before our eyes, I mean the universe; but it cannot be read until we have learned the language and become familiar with the characters in which it is written. It is written in mathematical language, and the letters are triangles, circles and other geometrical figures, without which means it is humanly impossible to comprehend a single word."[d]

However far back the mechanical investigations and discoveries of the *Discorsi* may reach, even to the beginning of the century, it is only in Arcetri that the picture of Galileo the 'Platonist' is completed.ᵉ In the *Dialogo* Salviati had always appealed to anamnesis and maieutics when fundamental questions of mechanics were at stake. His predilection for this basic Platonic idea is more than an ornament; it is an interpretation of the experience of the investigator, whose inquiries had the character above all of thought experiments and of the anticipations that they generate. But the appeal to Plato is also one of the rhetorical means employed by a kind of science that, by its own self-assessment, is no longer Aristotelian—that is, is post-Scholastic—and that credits itself with a different art of translating the language of nature than that of conceptual abstraction. It was scarcely two centuries before the longing for a comprehension of nature that modern science left unsatisfied also saw in Plato the upholder of intuition and unity. Goethe writes, "In order to escape into simplicity again from the boundless manifoldness, compartmentalization, and complexity of modern natural science, one must always ask oneself the question: How would Plato have acted toward nature as it may now appear to us, in, with all its fundamental unity, a greater multiplicity?" [160] But how does it stand with the Platonicity of the origin of the modern kind of science, if one could so soon appeal to Plato against it as well?

In 1937, Ernst Cassirer defined the substantive contents of Galileo's Platonism as follows: "Motion, the fundamental phenomenon of nature, had now been taken into the realm of the 'pure forms'; knowledge of it appeared as being of the same rank, in principle, as arithmetical and geometrical knowledge, the knowledge of numbers and figures." [161] Whether that is already sufficient as a specific opposition to Aristotelianism's concept of nature seems to me to be questionable. The route to this result can after all also be described as an extension of the method that was always approved in astronomy. Because Galileo had dissolved the metaphysical dualism of the sublunar and the supralunar worlds, he obtained, as part of the general formulation of his concept of nature, the Platonic residues that had been 'absorbed' by the spheres of the heavenly bodies in the Aristotelian cosmology. Before he could idealize terrestrial mechanics in such a way that it was finally suited to serve as the mechanics of the universe as a whole, he had to carry out the mental operation of making the imagined bodies comply with the conditions that Aristotelianism had earlier assumed to apply to the spheres of the heavenly bodies. These conditions can be reduced to the formula that the materiality of the astronomical objects could be treated as a zero quantity. The Aristotelians' trick was to assume a "fifth elementary nature" in order to neutralize the problem of matter. The result was that Aristotelianism could accomplish the same things as a Platonism that imagined the Ideas themselves as physical realities. The operation of idealization, rather than the assumption of ideality, is the basic process from which Galileo starts in his thought experiments. The residues of Aristotelianism in him are the Platonisms, again, that Aristotle had already integrated into his system.

I would like to clarify the end point of this path by reference to the latest result of Galileo's physics that we can grasp. It becomes decisive, for the accomplishments of the *Discorsi*, that matter is no longer a zero quantity. It has itself taken on the determinability that is characteristic of what the tradition of hylomorphism had described as "form." Right at the beginning of the First Day the problem arises of what importance has to be attributed to matter in the production of the simplest machines. It would have been in accordance with the assumptions of the metaphysical tradition to make the characteristics of an object depend exclusively on its form, while matter, as the substratum of the form's realization, contributes no additional intelligible features. Instead, it is a sort of undefinable disturbing factor in the process of accomplishing the characteristics of the form. For the functioning of simple mechanisms, that meant that maintaining the same proportions had to be the sole stipulation in changing the size of a whole structure.

This was contradicted by the practical experience that Galileo had learned about in the arsenal in Venice. The engineers there knew that machines of certain dimensions functioned faultlessly, while if their scale was altered they became incapable of functioning. It seems that in order to analyze this state of affairs it is not sufficient to base mechanics on geometry, nor is it any help to refer to the imperfection and indeterminability of matter. Now Galileo's thesis is that hidden in this problem is an independent theory of matter. It cannot rely on the assumptions of a construction following the rules of traditional mechanics. The dogma of the imperfection of matter is not sufficient to explain "the disobedience of real machines in relation to the abstract and ideal ones." Even if one could assume matter to be perfect and unalterable, one could still show that "the mere fact that it is matter makes the larger machine, built of the same material and in the same proportion as the smaller, correspond with exactness to the smaller in every respect except that it will not be so strong or so resistant against violent treatment; the larger the machine, the greater its weakness." From this it follows, for Galileo, that "assuming matter to be unalterable and always the same ... purely mathematical demonstrations can be produced that are no less rigorous than any others." Translated into the language of the Platonism that is supposedly his, this train of thought means that the material limit to the realization of ideal machines is, in its turn, ideally determined, and hence can be made mathematically determinable.

Such determinability opens up, in general, access to the possibility of drawing the material substratum of nature, which antiquity had assumed to be chaotic and irrational, into the rationality of physics. As one factor among other determining and determinable factors, matter is nothing else but (using Aristotelian language, now) a "form" itself. But this outcome is neither Platonic nor Aristotelian, although in its systematic place it could be better prepared for by the Aristotelian cosmology than by the Platonic doctrine of Ideas. And Galileo's language, in this context, too, cannot be unambiguously associated with either of the two traditions. He contrasts "real machines" (*machine in concreto*) to the "abstract and ideal machines" (*astratte ed ideali*), but he is just as likely to say that one has to "abstract" from all the imperfections of matter (*astraendo tutte l'imperfezzioni della materia*).[162]

From the point of view of this end point of the development of Galileo's physics, the intermediate stage that he had reached six years earlier in the *Dialogue on the Two Chief World Systems* becomes more intelligible and more instructive. There the function of matter becomes thematic in the discussion of the question whether, in the experiment with the ball rolling on a plane, the imperfection of the material of which it is composed must be taken into account. Salviati, in the dialogue, solves this problem as follows: "Whenever you apply a material sphere to a material plane in the concrete, you apply a sphere which is not perfect to a plane which is not perfect, and you say that these do not touch each other in one point. But I tell you that even in the abstract, an immaterial sphere which is not a perfect sphere can touch an immaterial plane which is not perfectly flat in not one point, but over a part of its surface, so that what happens in the concrete up to this point happens the same way in the abstract." [163] After all, calculators in commerce also have to include in their calculations imprecisions of measurement, shrinkage, moisture content, and packing materials. In a similar way, the *filosofo geometra* [mathematical scientist], "when he wants to recognize in the concrete the effects which he has proved in the abstract, must deduct the material hindrances." If he knows how to do this, he can make sure that the real states of affairs behave no differently than the arithmetical calculations: "... che le cose si riscontreranno non meno aggiustatamente che i computi aritmetici." Thus the errors lie neither in abstractness nor in concreteness, neither in geometry nor in physics, but in the *calcolatore* [calculator] who does not know how to make a correct accounting ("che non sa fare i conti giusti").

What is asserted here is that the result, in a real experiment, does not necessarily fail to correspond to the mental construction on which it is based, to the extent that the material character of the apparatus was taken into account, mathematically, in the mental preconception. What is decisive is not the deviation of the real ball from the ideal one (which in this text is always described as the ball *in abstracto*), but the determinateness, and determinability, of the magnitude of this deviation. This is not altered at all by the fact that the influence of matter here is still seen entirely as a "disturbing" one (*gli impedimenti della materia*). The 'hyletic' [material] component is no longer the element of irrationality, but has itself become capable of precision, which means, here, of precision in the determination of the deviation from geometrical precision. The example from commerce shows that the calculability of the material substratum still has the character of the subtraction of what is fundamentally an 'alien' addition to the 'substance' of the commodity, so that it resembles, for example, the procedure by which, in medieval physical treatises, the resistance of the air had been taken into account. What prepares the way for the more precise ideas in the *Discorsi* is the requirement that matter should be calculable.

In the *Dialogo* Galileo makes Salviati say explicitly that the perfect ball on the perfect plane does not exist, so that cognition, in mechanics, does not refer to this ideality either. Only if he assumed this would he be a Platonist. Galileo does not need to be a Platonist, because he does not lose the possibility of discussing the case scientifically by speaking of empirical states of affairs. Admittedly, these realities are distinguished by the fact that they derive from the sphere of the arsenal, which is already technical. So one could say that the Platonism in Galileo's assumptions consists in the fact that the 'phenomena' of which he speaks have already gone through a process of approximation to ideal values, when they were prepared, artificially. The new Platonist goes into the arsenal because what he meets there is no longer nature but artificiality, governed by norms.

Galileo does not get the problematic of the characteristics of the motion of a ball on a plane from the Platonic tradition; for him, this topic comes from the eighth chapter of the treatise on *Mechanical Problems*, which was still ascribed to Aristotle. The author, who was at any rate a Peripatetic, was denied the expedient of describing the motion of a ball as "natural" and deducing its persistence from that. Aristotle had not been able to award natural motion in a circle to a *solid* spherical body because its rotation, on account of the different velocities of its parts, cannot be a perfectly constant motion.[164] Only the stellar spheres, as hollow bodies, could fulfill the condition of constant motion (disregarding the fact that they were not in the purview of physics in any case). This premise would already have excluded the Copernican solution of introducing the Earth's axial rotation as the equivalent of the diurnal rotation of the heaven of the fixed stars.

The marvelous things with which the ancient treatise on mechanics deals are based precisely on the artificiality of its products and mechanisms. In them, a process that still has an almost 'mythical' quality presents itself: the outwitting of nature, extorting accomplishments from it that it would not deliver 'of its own accord.' Ancient mechanics is described most precisely, in its attitude, by one of its component disciplines (in the classification that is handed down by Proclus)—by *Thaumatopoiike* [Working wonders]: "We wonder at things that happen in accordance with nature when we do not know their causes, but we wonder at things that are contrary to nature when they occur through art, for man's benefit . . . . If something that is contrary to nature is supposed to be done, it creates perplexity on account of its difficulty, and stands in need of art. That is why we call the part of art that remedies such perplexity mechanics."[165] The text quotes the verse of Antiphon: "By art we master what by nature conquers us." The preparation of simple tools like the lever, scales, windlass, pulley, and block and tackle brings them close to the pure conditions that are required in order for something to be representable by mathematics. The pattern of such cases of extreme artificiality is that the smaller thing is master of the larger; that is, for example, a small weight moves a large load. It is explicitly said that questions of this kind neither belong to physics nor can they be entirely separated from it; rather, they occupy a territory where mathematical and physical discussions overlap. The method, we are told, is mathematical, and the realm of objects is physical.

Now this is also the context in which the question belongs as to why round bodies—in the three basic types of the wagon wheel, the pulley, and the potter's wheel—are most easily mobile, or confer the easiest mobility. This is due, we are told, to the fact that in all three cases the supporting plane is touched by the smallest possible surface.[166] The author can no longer take the next step—the, as it were, *metabasis* [passing over] into the Platonic—which would be to infer that the rotation of the ball would be 'absolutely' easy, and thus continuing, if it touched the plane at only one point, or (which in this context would mean the same thing) if it were suspended in space without touching anything.

For mechanics, the only important thing is that the ball and the plane can be brought to ever more precise realization. Thus the mathematical idea becomes the norm governing efforts to perfect things through art. It is not nature, but the mastering of nature through cunning, which stands under the norm—a norm that is both compelling and necessarily fallen short of—of precision. Accordingly, ancient mechanics could not provide any help in the explanation of "natural" processes. Its overlapping with physics did not become effective as a stimulus for theory. For the classical philology that came into existence in the eighteenth century, the author of the *Physics* could not be identical with the author of the *Mechanics*.

The ball freely rotating in space, which Copernicus had seen the Earth as, was not an object that could be displayed experimentally. In the case of the ball rolling on a plane, Galileo did not fully resolve the difficulties of the traditions that converged in this experiment. It is not, in any case, an instance of Platonism, but is, rather, the adaptation of Galileo's physical assumptions to the Aristotelian cosmology. The thought experiment 'legitimizes' the tendency of motion to continue precisely by the fact that that continuance represents a circular orbit around the center of the Earth. But the decisive premise is missing, that the center of the Earth could still be the center of the universe. Aristotle had described the location of the Earth at the center of the universe as a secondary occupation of what was in principle an abstract position, to which it had to be possible to relate even the homocentric revolutions of the celestial spheres, since they could not have objective reality as their absolute space. The abandonment of this abstract cosmic orientation system created a need, of which Galileo took no notice, to trace the operation of gravity back to the *bodies* in the universe, rather than to its *space*. The center of the universe now was only an imagined point, a nothing without any effective power (*un punto imaginario ed un niente senza veruna facultà*).[167] But the center of the terrestrial body, moving in space, now also became a mere abstract point of orientation. At the stage represented by the *Dialogo*, Galileo does not yet see that in constructing an 'ideal' plane he has to abstract from its relation to the center of the Earth, so as to represent the motion of the ball on the plane as tending to continue in space.

Only at the beginning of the Fourth Day of the *Discorsi* did he start from the idea of rendering the tendency of motion to continue comprehensible by making it equivalent to the "natural" motions of the heavenly bodies. Only here does he construct, instead of the imaginary sphere parallel to the surface of the Earth, a tangential plane on which the motion of the ball would have the form of a straight line. As a result, however, a real experiment on the surface of the Earth is changed not only by material 'imprecisions' vis-à-vis the ideal case in the thought experiment but also by a combination of regularities of which each one can in principle be displayed completely in the concrete case. The 'impurity' of the real appearance is due not to its difference from the Idea, but to the complexity of the elements of which it is constituted.

If Platonism, as for example in Nicholas of Cusa, could take on the function of justifying the imprecision of empirical science in view of the status, in reality, of its objects, then Galileo, at any rate, judging by the norm of his concept of cognition and its object, is definitely not a Platonist. For him, the "appearance" can comply with a rule to any degree of accuracy that is desired, and the margins of error in our knowledge result only from the interference, in its objects, of law-governed regularities that are in principle determinable.

When Ernst Cassirer dealt for the last time, shortly before his death, with the problem (which he had circled around again and again) of Galileo's Platonism, he availed himself of the formula, which by now hardly adds to our understanding, that this Platonism was "of a new and very paradoxical nature. Never before had such a Platonism been maintained in the history of philosophy and science." [168] This was because it was not a metaphysical but a physical Platonism, and such a Platonism "was a thing unheard of." But in what was the unheard-of quality of this Platonism supposed to consist? Precisely in the fact that states of affairs that were no longer geometrical could nevertheless still be represented as being ideal. Cassirer writes, correctly, that Galileo would have lost the fruits of his entire scientific effort if he had actually accepted the Platonic theory of the perceptible world. It is true that he starts from the Platonic concept of science, but the objects of this science are ones that, according to its concept, it would have had to deny to itself.

The thesis that Galileo was a Platonist is always documented with a formulation that had also become useful in characterizing him as a heretic. When the indictment of Galileo was being prepared, in 1632, a catalogue of the incriminating theses in the *Dialogue on the Two Chief World Systems* was produced, which contains, as the sixth item, the assertion that there is a certain similarity between the divine and the human knowledge of mathematical truths. Galileo had in fact said that while in its extent God's knowledge was indeed infinitely greater than man's, in the conclusiveness of its insight into mathematical propositions it had no advantage over man's. There just could not be a higher degree of certainty than being certain that a state of affairs is necessary. It is the shared quality of conclusive evidence, which Plato's doctrine of Ideas presupposes when it assumes the existence of objective originals as the identical origin both of the appearances in nature and of the possibility of human knowledge of them. For Galileo's supposedly Platonic statement we shall have to be satisfied with the strict interpretation, in which it says that the conclusive evidence possessed by the divine cognition of mathematical objects—if it has such evidence—cannot be greater than the conclusive evidence possessed by man's understanding, if it gains insight into mathematical propositions. The special character of the discursivity of human understanding is, for him, only one of degree. He conceives of the divine intuition as an 'instantaneous' discursivity.

"Our method proceeds with reasoning by steps from one conclusion to another, while His is one of simple intuition .... These advances, which our intellect makes laboriously and step by step, run through the divine mind like light in an instant; which is the same as saying that everything is always present to it."[f] The fact that man needs time to arrive at his certainties no longer seems like an 'essential' defect of his intelligence, and no longer hopelessly separates its insight into the lawfulness of the world from that of its lawgiver.

The problem of Galileo's Platonism cannot be considered without its connection to the provocative question of whether, at least within their own system, the inquisitors proceeded consistently when they took offense at the absolutism in Galileo's theory of knowledge, or whether, ironically, the defendant stood on the side of the logical consistency of the Christian system. Here one should not forget that in its epistemology, Platonism tried to detour around the real world, so as to trace its certainties back directly to the intelligible cosmos of the Ideas. The susceptibility of every Platonism to a demonization of the real world, in favor of immediate, even ecstatic, relations to the ideal one, results from the inevitableness of admitting the impotence of the Idea over against reality. This is precisely the point at which the concept of creation had obliged the Christian tradition to oppose Platonisms. Absolute power cannot let it be said of itself that it was not able to accomplish what it intended in the Creation. But Galileo's inclusion of matter in the concept of the complete law-governed determinateness of all appearances satisfies precisely this requirement. The difference between this inclusion and Platonism (and Aristotelianism) is precisely its conformity to one of the essential premises of the Christian system. It is the definitive overcoming not only of every dualism but also of any disposition toward it, if we permit no unresolveable element of 'disturbance' or of 'opposition' in nature. Of course the theological price for this completely consistent deduction from a theologoumenon had to be the concession, in the theory of knowledge, that the Inquisition incriminated as restricting divine abundance to the criteria of our power of imagination.[169]

Galileo is a Copernican also, and especially, in that he cannot admit a constitutive insufficiency of human cognition vis-à-vis the totality of nature. In Platonism even astronomy could not have the rank of a sufficiently grounded knowledge. Plato had explicitly remarked, in the seventh book of the *Republic*, that the heavenly bodies that could be observed deviate, in their positions and motions, from those of Ideality, precisely because they are real. As in all dogmatic systems, the Roman Inquisition too does not spell everything out; otherwise the exceptional position of the censorship that was decreed against Galileo in 1616—as an explicit *censura in Philosophia* [censorship of philosophy]—would not remain so mysterious. Does the conflict at its heart not have to do at all with one world-system taking the place of another, one must ask, but rather with the claim to give expression, in the cosmic reform, to a hitherto unheard-of claim to truth, a claim to a knowledge of the world that is equivalent to God's knowledge of it?

If one can give credence to some of Galileo's recent biographers, his conflict with the Inquisition originated in the historical incident of a capricious vanity on the part of Pope Urban VIII, the former Cardinal Barberini. In that case it would not have reached into the deeper dimension of colliding systems. As is well-known, the pullers of strings in the collision of 1633 are supposed to have succeeded in putting the pope into highest dudgeon over the conclusion of the *Dialogue on the Two Chief World Systems*, and eliciting from his offended vanity what would have remained foreign to his intellectual character. Such psychologizing makes the conflict into a sort of natural phenomenon, whose devastations and relics we, then, could only register. As an external factor in producing the event, the offense taken by the pope would not be worth the trouble of analyzing. In fact, however, the key to an understanding that goes beyond the contingent psychological facts is located here.

The attempts to relieve the Curia of its responsibility for the trial of Galileo by making the Aristotelians in the universities appear as the actual inciters, because they saw themselves as deceived in their possession of knowledge or at least as threatened in its enjoyment, is indeed initially seductive, but immediately becomes less convincing when one considers that orthodox Aristotelianism—of the type represented, among Galileo's opponents, by, for example, Cesare Cremonini, who is so frequently quoted—had to maintain its own existence against two threats, one being Copernicanism and the other being the Counterreformation Church. There was hardly a community of interest there—even if it is true that Galileo's polemical temperament had inspired hostility in enough people to set in motion efforts of this kind to influence the pope, irrespective of where they came from. In the case of a figure like Urban VIII, who cannot be seen as a mere instrument of insinuators, 'influences' require a pre-existing disposition on the part of the one who is influenced—in this case, a profoundly altered disposition.

Maffeo Barberini had celebrated Galileo's discoveries with the telescope in a poem of his own,[170] and as late as 1623, when he had just become pope, had accepted the dedication of the *Saggiatore*. There can be no question about his ability to appreciate Galileo's importance. It is just as certain that Galileo cannot have meant the figure of Simplicio, in the *Dialogo*, to refer to the pope, even as a covert allusion. For such a hypothesis, even the fact that he put the anti-Copernican proviso that was imposed on him into the mouth of the foolish Simplicio (as an objection, at the end) does not provide a valid piece of evidence, because for an idea that was required of him at such a late date there was no other figure that was still possible, and this one at least made the pope's wisdom appear as a completely foreign body, in the configuration of the dialogue, because it did not even fit in the mouth of an Aristotelian.

What happened is well enough known. After the permission to print had already been granted, in Florence, the Roman censor, Riccardi, gave Galileo two orders. He required the work to have a preface prescribed by Rome and he required the addition of a train of thought that one could probably describe not so much as the pope's 'own' theory about the cause of the tides as a sort of theological proviso, which for the sake of brevity I shall call the "omnipotence proviso." In the decree accompanying the obligatory preface, dated 19 July 1631, Riccardi wrote, "The conclusion of the work must be made to agree with this preface, by Galileo's adding the reasons from divine omnipotence dictated to him by His Holiness, which must quiet the intellect [namely, of the reader], even if it were impossible to get away from the Pythagorean [that is, Copernican] doctrine."

This was more than the instruction issued by Bellarmine in 1616, that Galileo should speak of the Copernican doctrine only in the manner appropriate to a hypothesis. The cardinal could quite well have understood his instruction as protecting the interests of Copernicus himself, since it is very unlikely that he knew that the preface to the *De revolutionibus* had a different author. So Bellarmine was still relying on what he regarded as the author's own intention—not at all, of course, without seeing this as protecting the interests of theology, which are so clearly expressed in Osiander's text. In any case, Bellarmine did not need to be conscious of carrying out an act of violence against Galileo; he brought Galileo back to the intention of the authority to whom he appealed. The situation with the omnipotence proviso of 1631 is entirely different.

With his proof from the phenomenon of the tides, Galileo consciously and intentionally departed from the obligatory hypothetical form of presentation of Copernicanism. It is from this that the suggestion arose that the omnipotence proviso of 1631 was supposed to present a specific theory about the tides that had occurred to Barberini, whose offended vanity was then a reaction to the way Galileo treated his idea. But the real function of the omnipotence proviso in regard to the theory of the tides does not permit this assumption at all. It does not explain anything; instead, it has to do with the dignity of explanations. The specific order to speak of the Copernican theory only hypothetically is generalized by saying that even the supposedly best proofs of it can only have a hypothetical provisionality, because all knowledge of nature has to reckon with the divinity's unpredictable ways.

Galileo knew that among his physical arguments this was the only positive one. The main topic of the Fourth Day did not fit into the pattern of the book's argumentation, which in its form obeyed the instruction of 1616. Galileo may have succumbed to the suggestiveness of his own demonstration. He thought that the phenomenon of the tides gave such an overwhelming intuitiveness to the motion of the Earth in space—to which he attributed the periodic splashing motion of the masses of water in their basins—that in view of it the regulation of what he could say, which in any case was already fifteen years old, could be forgotten. After all, Bellarmine, too, had already been dead for a decade. But it is precisely against the special position of the theory of the tides that the new order is directed. One can still not exclude the possibility that the benevolent intention of the pope, who was familiar with Galileo's favorite proof from many conversations, was, with the omnipotence proviso, to take the edge off of the argument that was rendered prominent in this way, and thus, on behalf of the author, to make his book publishable after all.

Any demonization of the partners in this tragedy is a mistake. A letter addressed to the inquisitor in Florence by Riccardi, already on 24 May 1631, reminds him that the pope's intention is that "the book's title and subject should not be on the flux and reflux but absolutely on the mathematical presentation of the Copernican doctrine of the motion of the Earth." The term "hypothetical" is avoided. In spite of the mention of the argument from the tides, one can still say that the pope, like Bellarmine before him, is only applying Copernicus's supposed foreword to his successor.

But what may have been intended by the pope as a theological commonplace, and only meant to protect the book, had to appear in Galileo's eyes as an extreme provocation, demanding the self-destruction of the new theory of science. The omnipotence proviso contains the substance of the incompatibility of the thought of the Middle Ages and the new pretension of a scientific explanation of nature. Just for that reason, it was possible to exploit, against Galileo, the casual (but not disrespectful) way in which the papal proviso was provided for [in the book]. And indeed the congregation that prepared the Inquisition's case listed Galileo's treatment of the proviso that was imposed on him as the second point in the catalogue of things of which he was guilty.

Nothing could have been harder for Galileo than at the end of his dialogue to disarm his single and supposedly strongest proof—to have to offer the objection, himself, by which this as well as every other new truth was robbed of its power. The discussion of the theory of the tides is not, as is sometimes conjectured, an older piece that was attached to the body of the book; instead, it serves as a final intensification of the book's means. In the original plan this argument was meant to be given the privilege of having the whole book entitled *Dialogo de flusso e reflusso* [*Dialogue on the Tides*]. Galileo sets up a correspondence between the difficulty of his long cogitation over this problem and his admiration of the finesse of nature, which "performs with the utmost ease and simplicity things which are even infinitely puzzling to our minds; and what is very difficult for us to comprehend is quite easy for her to perform." [g]

This is the context in which Kepler's thesis of the tides' being caused by the influence of the Moon can be pushed aside contemptuously, with an aside about his open mind and his penetrating acuteness. The phenomenon of the tides would have lost all relation to the truth of the Copernican system, if Kepler could not be put in the wrong. At the same time, the assumption of an action at a distance between the Moon and the Earth had to appear to Galileo as a monstrosity of the kind typified by the Scholastic occult qualities, as long as he had taken no notice of Kepler's theory of the motions of the planets.

One has to read precisely how Galileo incorporates the argumentative foreign body into his train of thought and encysts it within it. He himself, the text says, refuses his assent to the fantastic opinion (*questa fantasia*) of the production of the tides by the motion of the Earth, and would not object if it was described as a futile hallucination (*vanissima chimera*) or as an extraordinary paradox (*solennissimo paradosso*). He has Signor Simplicio be assured, in the dialogue, of his sympathy and esteem for the perseverance with which he stood up for Aristotle's point of view. Simplicio is asked for his pardon in case he may have been offended by an excessively vigorous self-assurance in Galileo's arguments, and he behaves like a good loser, who, however, appealing to a "very eminent person," has in fact one further idea that he has to put forward. Even if they were already of one mind about the explanation, they had to ask whether God with His infinite power could not have produced the phenomena of the tides in any other way He chose, rather than by means of the motion of the Earth that Salviati takes to be the cause. If one were not able to exclude that possibility, then one would certainly have to describe it as inadmissable daring (*soverchia arditezza*) if, by a definite assertion, the indefinite multiplicity of the possibilities open to the divine power and wisdom were narrowed down to the range of the ingenuity of the human power of imagination.

We know from the *Praeludium philosophiae* [*Prelude to Philosophy*] of Cardinal Oregio that Urban VIII had advocated the omnipotence proviso in conversation even before he became pope. In accordance with Nominalism's way of thinking, the law of contradiction counted for him as the only means by which human reason could circumscribe the means available to divine power for the realization of phenomena in nature. Human cognition's burden of proof lay in this demarcation of possibilities that are free of contradiction: "Probare debes implicare contradictionem, posse haec aliter fieri quam excogitasti ... " [You must prove that it involves a contradiction for this to be able to come about in a different way than the way you thought up.][171] Looked at from the point of view of this proviso, not only Galileo's reasoning, but even Copernicus's premise that the simplest and most perspicuous natural order is the principle by which to decide between systems, appears pagan. Now, Copernicus did not at all want to say that the simplest procedure for realizing the phenomena is the most consistent with God's power, but rather, I daresay, that it accords with the goodness of His will for Him to make the truth accessible and intelligible for the creatures for whose sake He intended the world. But if man had to reckon with and reconcile himself to the powerlessness of the truth, precisely because he could not expect the will of the divinity to grant satisfaction, through the world, of the natural need for knowledge, then he had to be prepared for every sort of contorted deviousness in nature, and for hopelessness where penetrating the machinery of the universe, behind the phenomena, is concerned. Forgoing the reliable assumption of laws of nature meant, for every theoretical effort, that it was subject to the threat of the revocation of the condition of its being capable of anything. Miracles, which were no longer wondrous, became a continually present threat to every step of theory. Leibniz—in a very comparable controversy, at the conclusion of his second letter in his controversy with Samuel Clarke—was to express this in a single sentence: "Car avec les miracles on peut rendre raison de tout sans peine." [For everything may easily be accounted for by miracles.]

For Barberini, the theologian, God is a factor of uncertainty that can be borne only because—or precisely because—he has offered man a single, but saving certainty in the form of His Revelation, and because He makes man conscious of the unavoidability of this offer through the theological proviso that no desire to achieve certainty independently, through theory, has any ground to stand on. What had been done to Copernicus's work by Osiander's preface—putting its claim to truth in brackets—appears in Galileo's work as the finale, which is likewise forced upon him and likewise deprives his claim to truth of its power. It is clear that the omnipotence proviso did not simply attach itself to the *Dialogo* as a pious and comforting final thought that would calm people's minds and take the edge off of their conflict with each other. Instead, it made the effort expended in the whole argument appear in a light of grotesque futility. In his letter to the Grand Duchess Christina, Galileo had already called a spade a spade in characterizing the despotism of theology over natural science. The right to make demands, which theology awarded to itself over against the other disciplines, created a situation that was exactly as if "an absolute despot, being neither a physician nor an architect but knowing himself free to command, should undertake to administer medicines and erect buildings according to his whim— at grave peril of his poor patients' lives, and the speedy collapse of his edifices." [h]

All the more palpable is the discipline with which, in the dialogue, Salviati replies to the general proviso in a way that conveys perhaps only to the modern ear the impression of irony: *mirabile e veramente angelica dottrina* ... [an admirable and truly angelic doctrine]. It is quite in accord with such a truly heavenly doctrine to say that we are no doubt permitted to carry out inquiries regarding the construction of the universe (*disputare intorno alla costituzione del mondo*), but not actually to discover the manner of operation of the divine Creation (*ritrovare l'opera fabbricata dalle Sue mani* [discover the work of His hands]). But what would a mystery be—something like this is the way Salviati tries to make use of the proviso—if it did not call forth the curious activity of the person for whom it is a mystery? And how could an activity that leads the human spirit to admit its limitation and its weakness in the face of the depths of the divine wisdom not be permitted? While this is said very cautiously and without being explicitly directed against the consequences of the omnipotence proviso, nevertheless it does indicate what a possible resistance would be like: Only if the striving for knowledge has a prospect of arriving at truth, or at least of progress in the degree of verisimilitude, can finite knowledge become conscious of its finitude in relation to the infinitude of what is still reserved for it or even withheld from it. In contrast to this, merely invoking the infinitude of the internally noncontradictory possibilities available to omnipotence destroys any perception of a relation between what has already been apprehended and what still can be apprehended. It thrusts inquiring reason back into the passivity of its total resignation.

Perhaps this way of putting the conclusion would not have given the momentous offense that it did if it had not been impossible to avoid a conspicuous inconsistency with the text of the Fourth Day. When Galileo received the order from the Roman censor, the dialogue was not only complete, it was—in virtue of the local imprimatur—already being printed. We do not know whether Galileo consciously risked a situation in which Barberini's idea, which was familiar to him from conversation, had already been played out in another passage in the last part of the *Dialogue*. There Simplicio had already adduced, against the tidal theory, the equivalence of various possible explanations, and had finally stated his preference for the belief that "the tide is a supernatural effect, and accordingly miraculous and inscrutable to the human mind—as are so many others which depend directly upon the omnipotent hand of God."[i] Anyone to whom that seems hopelessly medieval and obviously ridiculous should remember that Newton was still to prefer this same immediate operation to the action at a distance of gravitation. But Galileo's Salviati bestows defamatory praise on the Scholastic, Simplicio; with this view, he is entirely in agreement with the doctrine of Aristotle at the beginning of his *Mechanical Problems*, where everything whose causes are hidden from us is declared to be a miracle. Of course, this is followed by a turning that corresponds in its pattern to the condemnation of Aristotle's natural philosophy that was carried out in the thirteenth century.

The omnipotence argument could still be used, at this point, in favor of Copernicanism. The classical objections to the motion of the Earth are now, in their turn, put under the proviso that God must also be able to act contrary to what philosophy finds admissible and sanctions. Thus on the assumption of God's unrestricted power (*assoluta potenza*) the motion of the Earth is drawn into the realm of permissible assumptions. But now—and this is Salviati's surprising move—if two possible miracles that violate natural philosophy are permissible, namely, on the one hand the motion of the oceans in their basins without any corresponding motion of the Earth, and on the other hand a motion of the Earth, produced by higher authority, that causes the tides, then this second possibility is still the simpler one and "so to speak, more natural among things miraculous." Otherwise, on account of the complex periodicity of the tides, even a multiple miracle would have to be assumed—which, for reasons that are not clear, is supposed to be troublesome.

Whatever the situation may be in regard to the rhetorical quality of this consideration, it is in any case a much less respectful application of the omnipotence proviso, if one compares it to the one that was imposed on Galileo for the end of the dialogue—much less respectful because it misuses it. It perverts the theological intention of the idea—that it should demonstrate the abundance of divine action—into its opposite. It subjects omnipotence to the criterion of economy, which, however, was supposed precisely to be delegitimized, for questions on the metaphysical level, as an inadmissible expedient reflecting the indigent condition of human thought. The "more natural" possibility in the domain of miracles—this comparative is a devious attempt of Galileo's, by a rhetorical reversal, to apply the weapons that are aimed at him against the center of his opponents' position.[172] One gets a taste of what Galileo *could* have made of the conclusion that was dictated to him.

The principle of omnipotence had played a decisive role, as a factor creating freedom for mental variation, in the decay of medieval Aristotelianism. Against a solid dogmatic structure, such a principle is useful only as long as its license, according to which everything is possible, is not put fully into practice, but instead only what was previously declared impossible is admitted into the process of thought. The fact that a certain framework of argumentation of the traditional dogmatic system remains in place establishes in advance, at the same time, what it has become meaningful to oppose. The more successfully the destruction of the dogmatic prior 'givens' proceeds, the more doubtful does the functional capability of the principle that propels this destruction become. One only needs to compare what had been accomplished in 1277 by the Paris edict of Bishop Tempier against Aristotelianism and what Urban VIII caused to be accomplished against Galileo, in 1632, with the same basic principle, but without the accuracy of aim of the first instance. One can also say that Bruno's deduction of the infinite world from the self-reproduction of infinite power brought to light the principle's unfitness for purposes of explanation (that is, that it impeded scientific theory) no less than the still Nominalistic variant of the principle that was imposed on Galileo did. The difference consists in the fact that the Nolan had reasoned that because everything is possible, everything is real, whereas Galileo was required to reason that because everything is possible, one could not know what is real.

Before the censor made clear to him what the general proviso meant, Galileo tried, rhetorician that he was, to turn the principle of omnipotence into a formula of humility with a guarantee of progress. Reality was always supposed to be greater than what man can grasp, but at the same time this was not supposed to make it hopelessly unreachable—so that its basic pattern remained mathematically comprehensible, and knowledge, in that respect, remained equivalent to God's thinking. Of course it never occurred to Galileo that the principle could also be turned in such a way as to mean that the region of human experience always remained too small for nature's lawfulness, because it was restricted to the terrestrial life-world and to the contingency of the piece of the universe that was optically accessible to it. The fact escapes Galileo that it is the terrestrial standardization of our experience that deprives us of the homogeneity of the universe and the oneness of its physics. This is what prevents him from understanding Kepler. The suspicion that the segment of the universe that is available to man could be constitutionally too small to allow even the nature of cosmic space, alone, to be related to the norms of Euclidean space (as the space of the life-world continues to be) is far in the future.

When, in the Third Day, Galileo sets up a discussion of the Copernican implication of a universe that is far greater than any appearance suggests, he forces the mind (*discorso*) as well as the imagination (*immaginazione*) to forgo forming an idea (*formar concetto*) of the actual dimensions and distances. Intuition is simply incapable of distinguishing large distances from the largest ones (*distinguiere le distanze grandi dalle grandissime*). But Galileo seems to be sure that the whole is made according to just the same pattern as each of its parts. The system of Jupiter and its moons, once again, in comparison to the solar system, instructed him about this. The idea of omnipotence commands us to transcend intuition without surrendering it. Galileo always shrank from the implication of infinitude. He was content to assume a universe that exceeded the capacity of intuition. Its rational construction only needed to be far enough ahead of the progress of knowledge so that the next step of theory would, as it were, still find room in what was assumed to be there.

Man could no longer demand, as he could in Copernicus's view, that the totality of appearances should be coordinated with him and his power of comprehension alone. That was now only one, and not the only, principle of the *sapienza* [wisdom] and *potenza* [power] that are realized in the creation. "It seems to me that we take too much upon ourselves, Simplicio, when we will have it that merely taking care of us is the adequate work of Divine wisdom and power, and the limit beyond which it creates and disposes of nothing. I should not like to have us tie its hands so. We should be quite content in the knowledge that God and Nature are so occupied with the government of human affairs that they could not apply themselves more to us even if they had no other cares to attend to than those of the human race alone." [173] Man is well provided for, but he is not in a position to be certain that he is the purpose for which the world is arranged.

An epistemology of partial capability, which promises that understanding will progress by forgoing insight into the essence of things and restricting itself to the aspect that can be formulated mathematically, accords with this view of the divinity's arrangements. In this way, still long before the censor's orders, Galileo arrives at formulas that seem already to satisfy the general proviso: "The substance of spots might even be any of a thousand things unknown and unimaginable to us . . . . Therefore I see nothing discreditable to any philosopher in confessing that he does not know, and cannot know, what the material of the solar spots may be." [j]

In a letter to Marin Mersenne, Descartes was to criticize in Galileo this self-restriction of the claims of theory, and to connect it to the erraticness of his scientific style. "His mistake is that he continually digresses and never stops to expound his material thoroughly, which shows that he never examined it in an orderly fashion and that without considering the primary causes of nature, he merely sought the causes of some particular effects, so that he built without a foundation." Precisely on account of this difference it was Galileo, and not Descartes, who became the founder of modern natural science. It is true that Descartes was to have the success of founding a large school that was dominant for a long time, but by that very fact he exposed himself to a historical demonstration of the sterility of his unrestricted claim. Restriction of his claim is what makes possible the universality of Galileo's range. In his third letter to Markus Welser he writes, "But in my opinion we need not entirely give up contemplating things just because they are very remote from us, unless we have indeed determined that it is best to defer every act of reflection in favor of other occupations. For in our speculating we either seek to penetrate the true and internal essence of natural substances, or content ourselves with a knowledge of some of their properties. The former I hold to be as impossible an undertaking with regard to the closest elemental substances as with more remote celestial things." [k]

The renunciation that Galileo advises us to make is not theologically motivated. But he exploits every opportunity to characterize the price that has to be paid for scientific success as, at the same time, a piece of Christian humility, or as expressing hope directed at the fulfillment of all human anticipations, which will take place only in the next world. Thus he says also, in regard to the old claim that cognition has to have to do with the essence of things, "Such knowledge ... is withheld from us, and is not to be understood until we reach the state of blessedness. But if what we wish to fix in our minds is the apprehension of some properties of things, then it seems to me that we need not despair of our ability to acquire this respecting distant bodies just as well as those close at hand—and perhaps in some cases even more precisely in the former than in the latter."[1] This is, at least for epistemological purposes, the final surrender of the assumption that the natural arrangement of the world contains at the same time a metaphysical index of degrees of accessibility—that is, that spatial distance also represents theoretical distance. Galileo's anticipatory move is that his method levels off the objects to the same effective distance. His balance between making claims and engaging in renunciation—his turning away both from metaphysical definitiveness and from theological resignation—consisted in his conscious acceptance of what Husserl, three centuries later, was to reproach him with, namely, being a discovering and a concealing genius at the same time. But only this made it possible for him to turn the bitter experience of the impotence of truth into the at least partial empowerment of reason.

The frenzy with which Galileo, in his in many ways unfree situation in Arcetri, went to work to write the *Discorsi*, can be understood as reflecting the intolerable possibility that the enforced formula at the end of the *Dialogue on the Two Chief World Systems* could have been left as his last word. Precisely because his life's passion, his enthusiasm for Copernicanism, could not once be expressed in the *Discorsi*, he succeeded in the most lasting demonstration of nature's accessibility for theory, as a more effective preparation for the undisputed acceptance of the Copernican system. The gesture of defiance has become more subdued. For the *Dialogo* Galileo had still chosen the device of—fictionally—moving the conversation back in time, so that it was put before the Index decree of 1616. As a historically identifiable person, the partner in the dialogue who is distinguished, for the most part, as the leader—Salviati—had already died in 1614. Now, in his last work, the only kind of defiance that was left to Galileo was the formal one of identification, of making the persons of the Copernican dialogue, Salviati, Sagredo, and Simplicio, appear once again and, in eloquent silence regarding their initial Copernican question, attend to the supposedly narrower business of dynamics.

Part IV

The Heavens Stand Still and Time Goes On

Introduction

The statement that Copernicus played a decisive role in changing the concept of space in the modern age stands already as an almost incontestable theme in the history of science. Newton's absolute infinite space appears as a consequence of the expansion of the space of the cosmos, an expansion that had to be undertaken as a result of the Copernican premises. Equally plain is the critical reformation of this logic that came about via the detour of the idealization of space, beginning with Leibniz.

Now, my thesis is that the intensification, which is characteristic of the modern age, of the problematic of time can also not be understood apart from the alteration that Copernicus had set about making in the model of the world. Of course the problematic of time is less spectacular and has less affective impact than that of space; this may explain the fact that Copernicanism's consequences for the concept of time have not hitherto been described.

With this topic, one must also be sufficiently cautious regarding what we can assert as historical "effect." For the mere possibility of the Copernican reform depends on certain changes in the Scholastic system, in regard to the concept of time, that did not need to be required in the case of the concept of space because here it was initially only a matter of necessary quantitative changes. Just because I have to show, in what follows, that the implications of the concept of time in ancient thought would not have provided the latitude for the Copernican reform, it becomes crucial to exhibit, already in the movement of medieval thought, certain widenings and variabilities in the concept of time without which Copernicus's basic idea could not have been carried out or could not have been defended. From this point of view, what one might describe as an "effect" of the Copernican reform appears as only the final step in the logical development of what the reform presupposed. Copernicus did not need to integrate the idea of absolute space and absolute time into his system, but without a certain amount of change in the direction of Newton's absolute concepts, in the ancient and High Scholastic positions in regard to this problem, the Copernican reform would not have been feasible. The new system's success as a theory pushes forward, sanctions and brings to full explicitness the process that is already presupposed in it. What may be called the "effect" of the reform consists in this.

At this point it is important to observe at what a late date Copernicus was still being explicitly blamed for not having taken into account, in his astronomical reform, what was cosmologically possible. This question is undoubtedly at work beneath the surface of most of the reactions to Copernicus's work, though they were unable to articulate it adequately. I intend to show here, for the special problem of time, that Copernicus took into account—parsimoniously, it is true, but still very painstakingly—systematic elements that were required by natural philosophy. In addition, of course, before anything else I have to show that the problem of time did in fact have importance for the Copernican change in the world-system—because just this is not obvious without further ado.

Now, it could be objected that for Copernicus and the tradition with which he was familiar the concept of astronomy was defined precisely by the fact that it not only *could* not raise or take into account questions of natural philosophy but was not even *allowed* to do this. For astronomy simply was not a physical discipline that had to clarify questions about causality, about the reality of space and time, about the actual nature of the heavenly bodies and their spheres, and about their actual distances and sizes.

But the special position of astronomy and of its methodology—or, for that matter, of the Earth and of the physics that are valid in its region—could no longer be maintained at the moment when Copernicus made the Earth itself into a heavenly body and thus into a subject for astronomy. That was not only a formula for a new dignity but also a change in the theoretical situation. If the Earth was a star among stars, it had to be possible to make general statements in this new homogeneous universe. Of course the fact that the insurmount-

ableness of the limits of terrestrial physics had originally been felt as
a lack had fallen into oblivion; as so often happens, the lack of
methodical feasibility had been transformed, in the course of time,
into a canon of material inaccessibility.

Now it is evident that Copernicus made it a rule for himself to
violate the canon of the traditional discipline of astronomy as seldom
as possible. Precisely this is why he does not explicitly discuss the
possibility of his astronomical construction in terms of natural philosophy, so far as such discussion lies outside the argumentation that the
precedent set by Ptolemy required him to engage in. But it was
obvious that he had to be prepared for his readers' examining his
thesis from points of view deriving from the accepted Scholastic
system of natural philosophy. These two prior assumptions determine
the characteristic way in which he proceeds in his thinking.

Copernicus's argumentation requires above all that the motions
that are ascribed to the Earth could in fact take over the functions
that had traditionally been accomplished by the motions of the
heavens that he pronounces to be apparent, rather than real. Now,
for the tradition of natural philosophy since antiquity, the diurnal
rotation of the heaven of the fixed stars, in particular, which was now
declared to be the phenomenal equivalent of a real axial rotation of
the Earth, was very closely connected to the problem of the foundation of the reality and the measurability of time.

The most serious consequence of this exchange of functions between the entities that stood at opposite cosmic extremes in the
Aristotelian-Scholastic system consisted in the fact that the Earth had
to take over what, in the tradition, it had seemed least of all equipped
for, namely, the most rapid and the most uniform cosmic motion, the
only motion that could satisfy the requirements of manifesting and
measuring time.

From no other point of view does the radicalness of the Copernican
reform become as manifest as it does from that of the problem of time.
From no other point of view does it become so clear why the Earth not
only *could* become a star among stars but *had to* become one. It is only
this necessity that allows us to understand the substance of certain
argumentative efforts and formulations of Copernicus's that otherwise seem more ornamental than anything else. These include above
all his assertions about the equivalence between the form of the world
as a whole (*forma mundi*) and that of the Earth (*figura terrae*), with

which two special chapters of the first book of the *De revolutionibus* are concerned. If the Earth is to be able to take over the function of realizing time—a function that, assuming the Aristotelian-Scholastic definition of time, it has to take over—it has to satisfy the requirements of uniformity that Aristotle set up for the outermost heaven. It is just this that Copernicus insists on, by making the two forms equivalent, as a precondition of the interchangeability of their functions. The Earth had to become a star, in the traditional sense, so that it could accomplish, in the new system, what only the celestial bodies, or indeed only one of these, had been credited with in the old system—what in Aristotle's metaphysics even the celestial bodies had not been credited with without the metaphysically demonstrable propelling effect of the unmoved mover.

Thus the fact that we know almost nothing about what one could call Copernicus's "concept of time" is not very important. The crucial thing is that we can see how, in regard to this concept, Copernicus takes the old assumptions into account in his new situation.

1

How the Movement of the Heavens Was Indispensable for the Ancient Concept of Time

In the strict sense, we have no concept of time. We comprehend what we mean when we use the term "time" by means of spatial metaphors, and we use them not only as clarifying illustrations but as an intuitive foundation [*fundierende Anschauung*]. The simplest form of this consists in imagining a point that moves on a straight line in such a way that for every subdivision of the line into equal parts, differences in the point's motion on the different sections cannot be established. But this statement only relates to the "concept of time" if we are not in a position to specify another straight line, and a point moving on it, with the help of which differences and irregularities in the motion of the first point could be determined. If we are left with no other possibility of gauging the regularity of a periodic motion against another motion, this motion manifests time, since other organs for determining time can be calibrated according to it. So we would always have only a relatively final gauge of time, or (putting it differently) a gauge that cannot be shown with certainty to be the final one.

The philosophical discussion of the subject of time did not begin exactly in this way. It took it as a premise that we have a concept of time, and asked how it is possible that real circumstances can be grasped by means of this concept. Aristotle arrives at a kind of ontological argument: If we possess the concept of time, the conditions of the possibility of this concept must exist. That is the core of the argumentation that he presents in the sixth chapter of the twelfth book of the *Metaphysics*. If one asks oneself what theoretical interest, overall, is the guiding one for Aristotle in these discussions, it seems to me to be that of demythologizing the concept of God, rather than that of proving God's existence. His radical solution of this problem relates to two points of departure of every mythology: the beginning and the end of the world. The discussion of the problematic of time can clearly be associated with this intention. The first sentence of the seventh chapter establishes, with a sigh of relief, that the reduction of metaphysics to the problem of motion abolishes the mythical categories of origin from night, from the indistinguishable, or from nothing. The absolute constancy of the final circular motion of the celestial sphere permits a completely demythologized concept of God: the concept of the unmoved mover as the pure thought that thinks itself.

The path to this result necessarily goes by way of the concept of time. It does so as follows: There is no effective sense in speaking of a beginning and an end of the world as a whole, because these expressions already assume that points are determined, for which it must be possible to specify a before and an after. Beginning and end are always already located *in* time. To conceive of a beginning or an end of time itself implies, for Aristotle, a contradiction. So from the point of view of its concept, time would have to be entirely unaffected by whether the world does not yet exist, already exists, or no longer exists. But for Aristotle the concept of time implies not only absolute constancy but also that time is determinability. Constancy itself already implies, after all, not only that its beginning and ending are inconceivable, but also that in the course of time definable periods can be compared with one another. Intervals of time and points in time must be ascertainable and localizable, in respect to each other, in relations of being before, at the same time as, and after. The core of this argument is that time is not only measured, in an accessory way, by clocks, but is in fact 'produced' by them in the first place. The concept of time implies that there is at least *one* absolutely public clock, however elementary it may be—which cannot mean: however imprecise it may be. On the contrary, imprecision is absolutely impermissible, because it presupposes the concept of *one* time, and thus of *one* clock, by which it would have to be possible to establish such an imprecision. So the one required clock must not only be public: it must also be an ultimate clock, whose precision neither needs nor is capable of further testing.

The point of this argumentation is that there must be a motion that is absolutely constant, without beginning or end, publically manifest, periodic, and with a suitably short phase in relation to the processes in the world. With a motion like the diurnal rotation of the heaven of the fixed stars, which satisfies all the other requirements, there is no *empirical* confirmation in regard to the requirement of constancy; to demand this would involve the contradiction of measuring the ultimate criterion against a further one. But the concept of time demands something like an absolute guarantee. Exactly that, then, is a subject having the character that will henceforth be called "metaphysical." Only a god—for Aristotle, the unmoved mover—can be the reliable agency to provide this guarantee. Thus, for the first time, a function is found for the God of the philosophers that indelibly separates Him from all mythical categories. At the same time, Aristotle has sketched out a procedure that was to find its culminating point—which was still insurpassable for Hegel—in Anselm of Canterbury's ontological argument: the inference from a concept to existence. It is the concept of time that implies the existence of absolutely constant motion and of its guarantor. It is not yet *the* ontological argument, but it is *an* ontological argument.

One sees in this context that it is not anthropological interests that commit Aristotle to the geocentric world-model. Of course a geocentric cosmology would also still allow the apparent diurnal motion of the heaven of the fixed stars to result from the rotation of the Earth. This was the Pythagorean solution. This is where the obstacle lay for Aristotelianism. It consists in the time concept's dependence on the real motion of the heavens. If the motion of the heaven of the fixed stars was a mere appearance arising from the rotation of the Earth, then there would be nothing that could satisfy the requirement of constancy that is implied in the concept of time. Everything that Aristotle has to say about the Earth as a body in the universe contradicts the conditions that could give it a *real* motion that is also *absolutely constant*. Even if he had wanted to, on account of astronomical insight, it would not have been possible for him to assert the Earth's diurnal rotation and to make it answer for the requirements of the constitution of time. His realism of concepts would not have allowed him to make the concept of time relative to the most convenient public motion that is accessible in a particular case. The Aristotelian theory of concepts does not permit one to take them as raising something to a higher degree—as an extrapolation, a limiting case. They are the result of an abstractive process that exposes the essential character that is inherent in the thing. Deviations are overlays obscuring access, rather than relativizations.

Hardened despisers of what they call "historicism" will say at this point that what Copernicus was able to do, Aristotle, too, who was so much more potent, philosophically, could already have accomplished: to invent the protective assertion of [the Earth's being] a perfect globe, which thus has a perfectly constant motion. But the context constraint that affects a particular, apparently dogmatic principle has to be related not only to the framework of the system to which it belongs but also, going beyond that, to the assumptions to which systems, in their turn, are subject. Then, for all the differences between schools, there will always only be a certain latitude that remains for the articulation of systems, a latitude that banishes to the marginality of the sectarian, the episodic, and the ineffective everything that tries to go beyond it.

In Aristotle, the geocentric principle and the theory of time are connected in a way that is brought about by antiquity's concept of reality. An aspect of this concept of reality is that it has as its correlate the model of the onlooker. The onlooker conceives of himself as at rest, as the mere point to which events are presented. The world acts for him. Antiquity's ideal of theory codifies the relation between rest and motion as the relation between the subject and reality. It follows, and is only a specific and prominent instance of this fundamental state of affairs, that it is the heavens that move—and not man, who contemplates them, and not the Earth, from which he contemplates them. In the same way, the concept of time that Aristotle formulated philosophically and made into a 'given' for the tradition is not based on man's self-experience as an actor and on his needs in regard to the measurement of time, but instead articulates the way time is experienced by the onlooker, to whom the world presents itself as the most public event.

The privileged position of the human viewpoint with respect to the world is implied in the linguistic datum that "phenomena," in Greek, do not in fact merely "appear"—they "show themselves," they present themselves of their own accord. 'Theory' as merely 'letting it happen' presupposes that the world shows itself precisely to the person who lets it go on as it will—not to the person who intervenes in it with experimentation and dissection. The special position of the objects of astronomy consists, then, in the fact that they satisfy this prerequisite of the theoretical attitude not only de facto but necessarily and insurmountably. It is the reality that satisfies the onlooker because it is sufficiently well-defined not to be able to be enhanced in its definiteness by any effort he might make. This complex—the onlooker at rest, and the reality that shows itself—is irreducible.

The Greeks, who were no doubt aware of the relativity of motion but to whom, in any case, it never seemed to be something remarkable or bothersome, decided it in favor of what offers itself as appearance. Reflection on one's standpoint—suspicion of the effects of perspective—would have been necessary in order to reverse this relation. That one covers a distance means that the goal that can be reached by this route comes closer, that it shows itself more clearly, but not that one's distance from this object is dealt with, or is mastered and disarmed as a positional handicap from which the subject suffered. The appearance is always changed and unfolded. Action, accordingly, means that the accomplishment of the imagined object is its self-realization. The priority of the aspect over the temporal implication of the Greek verb characterizes the agent's self-experience as that of someone who even in action persists in the theoretical attitude toward change in the object. The perfect is the completed stage of the object of action, not of the experienced course of the action as a path that has been traversed. That is why the 'moment,' the 'now' that is, as it were, taken out of time, does not exist. Every part of time is itself still time, and the concept of time cannot escape the problematic of the continuum.

Perhaps here, so as not to remain too dependent on Aristotle, we should glance at the Pyrrhonian type of skepticism, as one of the more extreme formations of Greek thought. Looked at closely, this skepticism is a radicalization of the onlooker position, and thus of the Greek concept of reality. It sharpens to the utmost what is implied in the concept of the 'phenomenon.' That leads to the maxim that one should let the world show itself and let the potential for man's orientation and assurance in the world present itself to him from that direction. The skeptic's obedience toward nature is still entirely the obedience of 'intuition,'[a] whereas the modern age's renewal of skepticism unintentionally adopts the biblical schema of obedience vis-à-vis a voice, even if it has now become the voice of nature. The ancient skeptic's mistrust is not directed at what shows itself but at the judgment that wants to lay hold of what shows itself. The source of all errors is to be sought in going beyond immediacy. Retracting the idea of theory, ancient Skepticism ends up with the model of the onlooker that was that idea's point of departure. To abandon oneself to the world, renouncing judgment's effort to grasp it, presupposes the intact Greek cosmos as the background of one's consciousness of reality, and in it the central position of the secure onlooker. The concept of time that proceeds from Greek philosophy and determines the tradition until after Copernicus is governed by the 'phenomenon's' implication of self-bestowal. To put it differently: It reproduces an aspect of the relation to the world that is comprehended in the concept of the phenomenon. Time is itself a phenomenal structure, a phenomenal background, against which processes and events are apprehended in an unambiguous positional order.

According to Aristotle's definition, time is the number of motion in regard to what is earlier and what is later. Because the number is not a mere projection onto the continuum, it finds its 'applicability' in the periodicity of the rotation of the heaven of the fixed stars. The accent is not on the act of counting as a subjective performance, but rather on the structure of countability that presents itself, in spite of continuity, in what appears. If one bears this in mind, then the cosmological distribution of roles between the Earth, at rest, and the rotating sphere is not first of all the result of specific theories, but is rather the realization of a disposition that was already present in the prior givens of language. I do not mean to say by this that such a disposition is out of the reach of thought, although it does produce conditions that form thought's point of departure.

In order to comprehend this state of affairs still more clearly, it must be remembered that the first and the second Copernican motions of the Earth are not necessarily combined with one another. The annual motion of the Earth around the Sun would be conceivable without giving up the diurnal motion of the heaven of the fixed stars; and, vice versa, one could maintain that the Earth rotates daily without giving up the orbital motions of the Sun, the Moon, and the planets around the central Earth. If it is the case that the second motion was the one with which Copernicus began, then it becomes clear that the problematic of time could appear only at a systematically secondary point and that removing it could have the character of protecting something that was already achieved. The activity of the Earth produces the phenomena, and these have already, at an earlier stage, ceased to be what 'shows itself.' Here one must bring to mind the formulas that Copernicus found for the Earth's relationship of domination vis-à-vis the astronomical phenomena. What *appears* in the utmost distance of the fixed stars is *produced* in the utmost proximity of the motions of the observer's standpoint.

The Copernican reform moved within a latitude that had been opened up by destruction of the ancient world's concept of reality with its correlation between the phenomenon that shows itself and the onlooker who is at rest. In a certain way, Newton's concept of absolute time was a relapse in relation to the possibilities that were created by this destruction, because absolute time is at the same time a pure onlooker's time. Its protagonist is the Laplacean universal intelligence. The compelling consequence of Copernicanism could only be to interpret the relativization of the Earth's position as implying the impossibility of an onlooker who is the focus to which a simultaneous cosmic reality relates. In this respect, the discovery of the finite speed of light, and the increasing minimalization of its magnitude in comparison to that of the universe, was of decisive importance.

Copernicus appealed to the fact that the Pythagoreans had already called the Earth a star. But he did not mention that in the same passage in Simplicius's commentary on Aristotle's treatise *On the Heavens* the Pythagoreans were also reported to have argued that the Earth is a star because it too is an instrument of time, insofar as it produces the alternation of day and night by its motion.[1]

The Pythagoreans do not occupy any privileged position among the ancient authorities to whom Copernicus refers. Nevertheless, very soon after its appearance the Copernican theory was described as Pythagorean. Jacob Brucker, in his widely read history of philosophy, still described Copernicus as having taken two central propositions of his reform from the Pythagoreans: the heliocentric thesis and the thesis of the Earth's motion around the Sun.[2] Here Brucker seems to have understood the Pythagoreans as maintaining that the Earth's motion around the Sun is connected to an axial rotation that causes the alternation of day and night: "The Earth is one of the planets, which are moved around the Sun; and as a result of this motion day and night are produced, when it is moved about its axis." While this interpretation is poorly suited to proving that the Pythagoreans were forerunners of Copernicus, it is not so far off the track if, instead of identifying the Pythagoreans' central fire with the Sun, one keeps the two separate. In that case the Earth's diurnal motion, bound to a rigid sphere, around the central fire can very well explain its axial rotation with respect to the Sun, and thus the alternation of day and night. For the rest, Copernicus's relation to Pythagoras is, for Brucker, only an example of the historical fact that the history of philosophy, which he undertakes to write, is able to turn paradoxes into truths.[3]

We do not need to discuss here the old question, which in recent years has been decided very much against the Pythagoreans, of the extent to which they advocated cosmological views that were in harmony with one another, and whether these views were even justified in a manner that was to some extent clear and comparable to science in its form. While the confusion of the central fire with the Sun, which may have been due to obscurities not only in the transmission of the school's doctrine but in that doctrine itself, did establish the Pythagoreans' post-Copernican reputation, it completely obstructed access to comprehension of this setup. Evidently they were concerned with the Earth's diurnal rotation, which made it the instrument of time, and with how to explain it. For the great difficulty with the assumption of a special motion belonging to the Earth, in antiquity, was that for this body in the universe, with which we are so well acquainted, loopholes like Aristotle's celestial matter were not possible. Someone who wanted to assert that the Earth had a motion of its own also had to be able to explain it by earthly means. For ancient thought, the prototype of the possibility of something having a motion of its own is fire. If one was inclined, for astronomical reasons, to assume an axial motion of the Earth, it was natural to use the information that volcanic phenomena provided about the fire in the Earth's interior to explain its motive power. If for Hicetas and Ecphantus, but perhaps for other Pythagoreans as well, the central fire has no independent position at all, but is located in the center of the terrestrial ball, and thus also at the center of the universe, then this is not a late misunderstanding but rather a correct conservation of the original function of the thesis of the central fire. A transition from the assertion of a terrestrial motive power to a cosmic one could explain why it was possible to loosen the connection between the central fire and the Earth. The Earth then was no longer thought of as an envelope surrounding the central fire, but instead as a body independently revolving around it. One cannot talk about an abandonment of geocentrism, here; but the central fire was to become the principle of the motion of all the celestial bodies, in connection with which the motive power was perhaps imagined as proceeding by way of a central axis. In the process by which the central fire was rendered independent, an adaptation of the theory to the requirement that the number of the celestial bodies should be the ideal number, ten, may also have been important.

If it is correct to understand the explanatory function of the central fire in this way, then it also becomes obvious that the motion of the Earth around the central fire could not have anything to do with its annual motion around the Sun. The mysterious "counter-Earth," of which the Pythagoreans are supposed to have talked, would merely be the antipodal side of the Earth, which had been separated off when the central fire was released from its envelopment by the Earth and rendered independent. This is consistent with the fact that Aristotle discusses the whole problem of the central fire and the Earth from the point of view of his doctrine of the elements. Here the question is only whether, in the sublunar interior of the cosmos, fire could appear again underneath the earth, and thus inside the Earth, as though only this was its most suitable place. Aristotle rejects this speculation with the argument that here "center" is used not only in a spatial sense but also in a metaphorical sense. But then one should not value the center so highly, because that which contains has a greater dignity than that which is contained. The discussion shows that Aristotle, too, can hardly have conceived of the Pythagoreans' central fire as an independent heavenly body.

What is crucial, however, is that in its astronomical aspect (as distinct from the aspect of the doctrine of the elements), the Pythagoreans' cosmology, for Aristotle also, is still geocentric. The distance of the Earth from the center of the universe would only be so slight that the phenomena of the visible heavens run their course just as though the Earth were located exactly in the center. The astronomical observer is in any case separated from the center of the universe by the radius of the Earth, and the Pythagoreans merely made the Earth distant from the central fire by that much more again.[4] However unclear its contours may be, the whole discussion shows that it cannot by any means have been a question of the problems of the system of the planets, and thus of something that would rank as an anticipation of Copernicanism. The absurdity of the Pythagorean doctrine consists, for Aristotle, not in its systematizing a different center, but rather in the simple assertion of the motion of the Earth: If the Earth moved, it would be a star, which, in view of all our experience of the heaviness and inertness of the element of which it is composed, it cannot be.

This makes clear at least what the Pythagoreans could not have meant when they spoke of the Earth as a star and as an organ of time. At the same time we see in sharper relief how Aristotle formulated the problem of time anew and compellingly, for those who came after him, when he said that the requirement that its phenomenal foundation should be absolutely uniform was satisfied only by the diurnal motion of the heaven of the fixed stars. That is why, while Copernicus could go back to the Pythagoreans for formulas to use as part of the complex of his protective material, there were demands placed on his defenses in natural philosophy that could neither be satisfied nor even understood by means of the Pythagorean theory.

The Pythagoreans spoke neither of the Earth's annual motion around the Sun nor of the possibility that the apparent diurnal motion of the heaven of the fixed stars was produced by the Earth's axial rotation, but solely of the diurnal periodicity of the Sun's apparent travels. If it is correct that with the theory of the central fire the Pythagoreans meant to indicate a motive force for cosmic revolutions, then this already contains the problematic of the reliability of their "organ of time": Fire is indeed the element of the greatest agility and speed, but it is not that of the most precise and insurpassable regularity. As the motive force of the axis of the universe, fire might meet all the requirements for describing the phenomena, but it certainly could not be adequate to the implications of the concept of time. The Aristotelian theory of time could have emerged directly from a critique of the weaknesses of the Pythagorean theory: Aristotle could see here that the problem of the ultimate motive force of the cosmic processes and the problem of time were closely connected. They could only be mastered by a *single* metaphysical solution. This solution demanded, above all, that one should hold to the reality of the motion of the heaven of the fixed stars.

The misunderstanding of the Pythagorean central fire as a heliocentric first step on the way to Copernicanism was to prove fruitful, as misunderstandings are said to be. Kepler made a thorough study of Aristotle's account of the Pythagoreans' cosmology; he translated and commented on the thirteenth chapter of the second book *On the Heavens*. It is true that the difference between the central fire and the Sun escaped him, but instead of that the function of the center of the universe being occupied by the supposed Sun, as a way of propelling the motions of the planetary system, dawned on him. In connection with the difficulties in the text, Kepler makes use of the traditional suggestion that the Pythagoreans had a secret doctrine, which Aristotle, too, can only have known approximately, so that a great latitude remains for conjectures. Kepler writes, "They spoke figuratively, so that by the world 'fire' they meant the Sun, and I too share their view that the Sun stands in the center of the universe and is never displaced from this position, and that the Earth rushes once each year around the Sun, that is, around the central place of the universe . . . . "[5] Copernicus "brought forth [this doctrine] of his own accord again, since he knew nothing of how the ancients had meant it." Copernicus was not influenced by the Pythagoreans; instead, it was his original conception that for the first time made it possible to understand what the Pythagoreans' cryptic statements must have meant. The interpretation of the ancient writings would have continued to be fruitless "if he had not discovered, out of his own head, how matters stood with this revolution of the Earth." Aristotle, at any rate, had not understood what the Pythagoreans meant. That the meaning of the "counter-Earth" could only have been the Moon had been confirmed by optical means in the meantime. The four moons of Jupiter and the two of Saturn had shown it to be a 'normal' systematic finding—"That is how far we have been able to go, up to the present, by means of the long telescope."

Much more important, now, is the fact that the identification of the central fire with the Sun allows Kepler to translate the original propelling effect of fire, as the self-moving element, into his physical explanation of the motion of the planets as produced by the Sun. Not only does the Sun illuminate, warm, and animate the Earth; it "carries it around in the universe, with its light rays, as a river carries a ship with it." The possibility of this explanation results, for Kepler, from the fact that the Sun, as the observation of sunspots certified, rotates on its axis in the direction the planets travel on their orbits, so that evidently the planets "follow the light rays that go ahead of them (which spring from the rotating Sun, and rotate with it), but they do not reach their speed—the most distant, Saturn, being farthest from reaching it, and the closest, Mercury, being closest." Kepler believes that he has uncovered the secret meaning of the Pythagoreans' doctrine for the first time by seeing in their dogma of the central fire a solution of the physical problem of the propulsion of the planets, a solution that Copernicus evidently had not found (and, in accordance with the tradition of astronomy, did not have to seek).

Kepler knows very well what he is doing when he brings the mover of the spheres, which Aristotle had shifted to the unmoved mover's transcendence, into the center of the universe. Soothingly, he gives his configuration a detailed analogy to the theological doctrine of the divine Trinity, whose identification of the center with the Father deciphers, in its turn, the Pythagorean designation of the central fire as the "guardhouse of Zeus." It was only with restraint that Copernicus had attached himself to the Sun metaphysics of the Renaissance; Kepler, in contrast, sees in the Sun a living center of power, a dynamic agent, which belongs, like the heart, in the center of the organism, "so that it can distribute its power through the whole universe, subtly, uniformly, uninterruptedly, and without any excess in one period over another." Here the element of a guarantee of uniformity, which is decisive for the problematic of time, is at least suggested, though only for the annual motions around the Sun and not for the diurnal rotations, which remain unexplained.

According to a report that is transmitted to us (admittedly only) by Plutarch, who is drawing on Theophrastus, Plato in his old age is supposed to have regretted very much having left the position in the center of the universe to the Earth, to which it did not really belong. This account cannot be interpreted as implying a precocious Copernican turning on Plato's part, after the *Timaeus*. If one understands the unadulterated Pythagorean position as being that the central fire was not supposed to be an independent heavenly body, but something that, enclosed by the Earth and located at the center of the universe, served to propel the axial rotations, then only a minor modification of the cosmology of the *Timaeus* is required, which may very well have occurred to Plato. It is not primarily a matter of the arrangement of the heavenly bodies, but rather of the sequential ranking of the elements. But in that case the conclusion that in his old age Plato gave up the Timaeus's geocentric cosmology would be mistaken. Since in this dialogue he assumes a motion of the Earth around the axis of the universe, one will be able to conjecture that the

Chapter 1

attractiveness for Plato of the Pythagorean thesis of the central fire
was that it seemed to clear up the problem of propulsion, which he
had not solved.

Now, here too, looking at it from a post-Aristotelian standpoint,
one will have to see the unsatisfactoriness of the Platonic cosmology
in connection with the question of the constitution of time. However,
Plato excluded the requirement of absolute regularity in the real or
apparent motion of the heavens, which was implied by the concept of
time, by denying to empirical astronomy in general the strict precision that follows from the claim of the concept or of mathematical
construction, and he ascribed this precision, in the strict sense, to a
second astronomical science, which relates to the invisible ideal
realm. If one takes the position that the report of Plato's late turning
in favor of the central fire is at least something that would follow from
the requirements of his cosmology, then the objection that fire, by its
nature, cannot be an absolutely regular propelling force and thus
cannot comply with the norm of the concept of time is in fact consistent with Plato's astronomical dualism. How should time in the empirical world and founded on the phenomena of the visible heavens be
able to fulfill the absolute norm that can only be described in the form
of ideality? However, there is no Idea of time. And therefore there is
also no concept of it, from which norms for its foundation in reality
could be inferred, as Aristotle was to do later. Instead, time is the
image of an Idea that we try to translate with the world "eternity." As
an image, it could neither contain a norm nor satisfy a norm—it
would be, to put it (anachronistically) in the language of Nicholas of
Cusa, essentially "imprecise." In regard to the Aristotelian unfolding
of the problem of time, that is not a preliminary stage or a dogmatic
disagreement, but a blind alley for the problem.

Plato's cosmology can be characterized as presenting a compromise between the intolerability of chaos and the inaccessibility of the
Idea. In the myth of the demiurge one can gather this from the world-
craftsman's double motivation: On the one hand, the reason why he
becomes active at all is the intolerability, for him, of seeing the
preexisting, disordered primal matter; on the other hand, the reason
why he chooses the way of ordering that matter that Plato describes,
and not a different one, is the pregivenness of the Ideas. The reason
for this master craftsman's effort to come close to this pregiven model
lies in his 'moral' character and his 'technical' proficiency. If a world

is indeed to be produced against and out of chaos, it should be the best possible one, which, so long before Leibniz, can only mean the most faithful imitation of the one cosmos that is preformed in the Ideas. One has to keep this beginning in mind if one wants to understand the decisive difference between Plato's unfolding of the problem of time and Aristotle's. With Aristotle, the norm of absolute regularity that is required in the concept must be realized absolutely in the phenomenal motion of the heaven of the fixed stars, in connection with which, a claim is made to a metaphysical guarantee of this phenomenal reality—a guarantee that takes the form of the unmoved mover—in order to overcome the dualism of ideality and phenomenality, leaving no gap between them, at this one point.

In Plato, the problem (which has never been completely resolved) of the foundational sequence in which space and time stand in relation to each other becomes visible for the first time. In the cosmology of the *Timaeus*, space is prior to the world and indifferent to the existence or nonexistence of the cosmos, while time is intraworldly and dependent on a specific structure of the world. Space is not, as it becomes in Aristotle, the interior space of the cosmic shell—it is not the container space of physical objects and processes. As a completely indeterminate magnitude, it takes into itself everything that it is not, without as a result becoming determinate in its turn. Since the cosmos that the demiurge produces is, as an imitation of the Ideas, an expression of reason, what precedes it, without being absorbed into it, cannot itself be a rational quantity. To that extent Plato's space anticipates Newton's absolute space, which, when it crops up in the first formulations of natural laws and then has to be eliminated from the empirical concepts that these make possible, is a necessary presupposition, but nevertheless cannot be included as a rational magnitude. Although it is *logically* necessary, absolute space is *physically* unreal.

Platonic time, on the other hand, comes into existence together with the cosmos and by means of the cosmos. It is the mode of appearance of nature's regularity, and thus of the reason that is striven for, between chaos and the Ideas, in the demiurge's work. Time as appearance is time that can be read off [from the phenomena]. The meaning of this capacity of time to be read off from the cosmic appearances becomes a pressing question where this capacity seems to be problematic, namely, in connection with the maximal periodicity, the cosmic Great Year. If the demiurge means to produce a distinct measure (*metron enargēs*) in time's manifestation—to make it conspicuous, as it were—and if he has already actualized time in the alternation of day and night, as the smallest natural division of time, then is man intended as an addressee—as one who will read off what can be read off—in connection with the Great Year as well, or are there (as in the medieval conception) other subjects who may be onlookers at the world process? It remains unclear, in Plato, who is supposed to be the observer and the beneficiary of the cosmic clock.

On this very point the transmission of the text also leaves us in the lurch. The smallest unit of time is introduced, in the *Timaeus*, before there is any talk of man as a possible addressee. How this must be understood depends solely on whether one sticks to *kai ta* [and the], which the manuscripts give—in which case the Sun's role is restricted to providing the measure of motion, and illumination of their paths, for the other heavenly bodies, and not for men—or whether one follows Archer-Hind's emendation and reads this as *kath' ha* [according to which], so as to include man in the teleological relationship. If the well-attested reading should be corrupt, it would still testify to a suggestive way of understanding the text, in which Plato would not be assumed to regard the whole cosmos—and not only terrestrial nature—as anthropocentrically organized.[6]

If the greatest and most perfect unit of time, the unit of the cosmic Great Year, should disregard man's power of comprehension, then that too would be an expression of the transcendence of the ideality that is imitated in cosmic time. The utmost approximation to that ideality is at the same time the utmost remoteness from the conditions of empirical ascertainability. Thus we can see even in details what is said with regard to Plato's concept of time within his outline of a cosmology: "The nature of the highest living being was eternal, and this character it was impossible to confer in full completeness on the generated thing. But he [that is, the demiurge] took thought to make, as it were, a moving likeness of eternity; and, at the same time that he ordered the heaven, he made, of eternity that abides in unity, an everlasting likeness moving according to number—that to which we have given the name 'time.' For there were no days and nights, months and years, before the heaven came into being; but he planned that they should now come to be at the same time that the heaven was framed."[b]

Talk of the cosmic clock in Plato does not yet have the metaphorical content that the metaphor of the world as a clock was to acquire in the late Middle Ages and the modern age. It was the introduction of the mechanical clock that for the first time gave this set of metaphors its decisive defining element, namely, the combination of accuracy and *internal* propulsion. The Pythagoreans' central fire then became the central conversion of the power of suspended weights or, finally, of the tensile power of a spring; the clock's working no longer seemed to be substantially and constantly bound to the influence of an externally supervening power. Here, for the first time, the ambiguity of the idea of the world as an internally driven mechanism becomes palpable. The self-propulsion that is guaranteed in its regularity leaves behind it the problems of the doctrine of the elements and the intelligences that move the heavenly bodies; at the same time, it becomes doubtful whether a self-propelling thing does not mean also an end in itself. Only the clock's hand continues to be a teleological element, but at the cost of becoming an external and superimposed one. It is the question posed by the great practitioner of world metaphors, Jean Paul, in his *Hesperus*: "Does the working of the world-clock reveal as much goal-directedness as its construction does, and does it have a face and a hand?"[7]

Fascination with the mechanical clock is combined, early on, with self-assertion's theme of immanence, of shifting the place where power is applied and where the phenomena are produced into the central realm of physical reality. Seen as a configuration of relations, that anticipates Copernicanism as, perhaps, the decisive transformation of the concept of reality that was arrived at in antiquity and that was based on the elementary idea of the nature that 'shows itself.' Reality does not show itself—it is made to appear: The star that is beneath our feet produces the stellar phenomena that are over our heads.

2

How Antiquity's Concept of
Time Did Not Fit in the
Middle Ages

The difference between the ancient world's concept of time and the Christian one has been described, in a formula that has already succeeded in gaining general acceptance, as the difference between a cyclical and a linear schematization of its passage. The most important evidence of this is probably still the fact that the Stoics' cyclical idea of world-periods was emphatically corrected by the early Christian authors, who replaced it with the proposition that the world must, indeed, perish, but only once; for this implies, we are told, that this occurrence is not senseless, but serves the purpose of the judgment that follows it.[8] However, this is more the idea of the unity and uniqueness of the historical process, which was linearized as a result of the reduction of Stoicism's world-repetitions to one phase. It is doubtful whether this is a sufficient differentiation of the epochal concepts of time.

For that purpose we must ask, more precisely, how the schema of the circle characterizes the ancient world's concept of time. When Aristotle says that time is the number of motion, he means the aspect of motion that is countable, and with regard not only to the distinguishable sections of the motion but also to their position in respect to one another in a relation of before and after, earlier and later. Of course these sections are determined—that is, what is countable is determined—by means of the assumed absolute uniformity of a *cyclical* motion. The circle is distinguished by the fact that it is possible to establish a boundary anywhere on it, in relation to which (as beginning and end) a period can be counted, as a unit, and which determines what is before and what is after, what is earlier and what is later. But each revolution in this circle—that is, each cosmic unit of time, the multiplication or subdivision of which makes it possible to measure time—has, in its turn, a position in an ordered pattern of sequential units, which cannot be interchanged. And just this pattern of sequential units, as what can be counted, is *linear*. It is only in the ordinal pattern, which permits one to conceive of time as the number of motion, that the other pattern—the pattern of the circle, the cosmic-intuitive elementary unit of motion—stands behind each unit.

If one bears in mind this intermeshing of cyclical and linear patterns in the ancient concept of time, an explanation follows for the fact that it was impossible for Greek thought to trace the concept of time back to and found it in internal time consciousness. Inner experience seems to be bound to the linear metaphors of the point moving in a straight line, or the stream. It would not have been possible to establish how units of inner experience could be constituted as units of measurable time, or (especially) how such experience could guarantee the required uniformity. The connection between the concept of time and the act of counting can only be established if we can find something countable that is given in intuition and that does not have to be standardized by reference to another given, as far as the uniformity of its motion is concerned.

That is why, in Aristotle, the concept of time is the only one that does not permit the indefiniteness that occurs, in the relation of form to matter, in the constitution of concrete objects, but instead lays claim to absolute precision—the unconditional identity of the concept and its concrete embodiment—in the concrete phenomenon. Of course, this precision is only possible because the motion of the outermost sphere is not self-motion. Here it could appear as though Aristotle's interest in the satisfaction of the requirement of regularity that is contained in the concept of time is already theological. But we must not read Aristotle with the eyes of the tradition that is dependent on him, the tradition that inverted his interest in the foundation of the concept of time and turned the topic of time into the lever for its cosmological proof of God's existence.

Seen from the point of view of the Middle Ages, the existence of a God was not very important to Aristotle. God had neither created the world (so that He would be responsible for it), nor did He have, directly, to produce the changes in this world (and to answer for them), nor was He, in His behavior, anything like a morally exemplary God, or One Who was capable of entering into covenants affecting human history. The unmoved mover was a God Who was bound to a purpose; the definition of His characteristics was exclusively oriented to the function of providing an absolute foundation for the possibility of time.

If one considers the concept of time in connection with the possibility of the Copernican reform, one must keep in mind above all the extent to which the outcome of the ancient world's constitution of the concept of time blocked change in the basic conception of the cosmos. If the Middle Ages had only been a continuation of antiquity by other means, then Copernicus could not have been an event that was still medieval in type. Of course there is an important continuity extending across the connection between the concept of time and the proof of God's existence. Only the release of the topic of time from the constraint of the tasks it had to perform in connection with the proof of God's existence expands, at least, the scope for heterogeneous thought-motives to have an effect. But the difficulties with antiquity's concept of time had become clear a great deal earlier, and had only been pushed to the rear by the importance of the problem of proving God's existence, in High Scholasticism.

The central difficulty was the connection between the required uniformity of the motion of the heavens and the world's lack of a beginning and an end. If uniformity is only another term for the functioning of a standard of measure that is not measurable in its turn by something else, then the logic of the concept of time prevents it from being possible to ask about what comes before and after time itself, since after all it is nothing but the sum of all possible befores and afters. It is important, and had important consequences, that in this regard Aristotle proceeds differently than he does with space, which could likewise be proved infinite on the ground that it is itself the measure of all possible insides and outsides with regard to any boundary. So the unity and uniqueness of the world are not secured, on the Aristotelian assumptions, from the point of view of the concept of space, whereas they are required from the point of view of the concept of time. For if there were several times associated with several worlds, then again one could imagine a time against which the different world-times could be measured and compared. To put it differently: The plurality of worlds would require a super-world, in which there would have to be a phenomenon like the rotation of the heaven of the fixed stars to provide the foundation in reality for the ultimate, unified concept of time. The concept of time makes the idea of the plurality of worlds contradictory. That, in particular, was a gain that the Middle Ages could not reap for themselves, because the unity *and* the eternity of the universe were equivalent implications of the Aristotelian concept of time.

One can also express this by saying that the Middle Ages could not fully take over the conceptual realism that was the point of departure of the Aristotelian theory of time. The concept of time could not 'seek' the cosmology appropriate to it. Instead, pregiven assumptions about the world marked out the framework for the concept of time that would be possible, and thus for the alteration of the received one. As soon as the concept of time can no longer be used in an argument like the "ontological" one, it tends increasingly to adapt itself to the theory of concept formation and to present time as the outcome of a process of abstraction. If the saying of Augustine that is quoted again and again in this connection, that even with the standstill of the heaven of the fixed stars time would continue as long as even a single potter's wheel was still turning—if this was supposed to be valid, then the unity and uniformity of time could no longer be anything more than the assumption that all time-establishing and time-measuring motions are comparable. In that case there would only be an ascending hierarchy of more comprehensive periodic motions, with, in each case, only a relatively—never an absolutely—ultimate member. This relativization leads one in the same direction as the realization that time measurement can never be more than the comparison of periodicities and that it was an error to think that in that comparison the greater uniformity must always measure the smaller one. For example, comparison can benefit from a higher frequency beat, even if it is less precise, as when Galileo would verify the isochronism of the motion of the pendulum in the cathedral at Pisa by means of his pulse.

Once the statement has been made that time could continue as long as merely one potter's wheel continued to rotate in the world, the logic of this reduction presses even further, beyond this statement. The next step would be to separate the concept of time from any physical motion at all and to conceive of time as a quantity of an ongoing condition, independent of motion or rest. Duration would then be the character that belongs to everything by virtue of its identity. In this connection, it is true that it seems to make no difference whether something preserves itself in its existence or whether it is preserved in it; but undoubtedly the specific character of 'duration' as such first became comprehensible after the idea of creation had brought the idea of contingency in its train. What has been extracted from nothing does not have duration and durability on its own account, but endures only by virtue of a sort of additional input. Aristotle's world depended for its motion on something outside it, but it was a matter of course, and unproblematic, that that which was moved could 'endure' by itself. For the medieval world, this became questionable, and thus duration became a possible topic in connection with the problem of time. Time, as something created at the Creation, is based on the specific characteristic of all creatures by which their radical and continual dependence on the divine will is grasped.

In Augustine's exegesis, the importance of the seventh day after the work of creation is seen in the fact that the completed creation rests while the Creator rests, and it rests "the more durably and securely, inasmuch as it needs Him, rather than Him needing it, for its rest." [9] This God is not primarily the mover of the world, but rather the guarantor of its repose as its enduringness. The key word in Augustine is *manere* [to remain, to endure]; it means being ensured against the consequences of being something that came from nothing: "Because —despite all the change in the created universe—nothingness never returns, this created universe will always endure in its Creator (in creatore suo semper manebit)." That, he says, is the meaning of the fact, in the biblical text, that the seventh day of the week of the Creation, unlike the others, does not have an evening. Only toward the end of the Middle Ages will this same seventh day be interpreted in terms of its relevance to the problem of cosmic motion. [a]

Through Augustine, time becomes the form of contingency— hence the emphasis on its independence from the existence and the motion of the heavenly bodies. That independence is an unavoidable consequence even of the mere fact that in the biblical Genesis the ancient world's time indicators are only created on the fourth day. So they come into being at a point that is already within time, and even after the passage of a period of time that is already articulated into periods. Augustine has recourse here to distinguishing between an indefinite time, before the creation of the heavenly bodies, and a definite time, after their creation, and in doing so appeals to

Aristotle's differentiation between undefined matter and defining form.[10] In that case temporality is an essential aspect of the Creation and of its creatures, quite independently of whether this time is measurable or not. Here too the paradoxical problem of the seventh day is an issue: The seventh day must be created, but cannot be created after the sixth day, since after it the Creation is finished and the Creator rests. So it has to proceed from the character of the completed creation: "Sed etiam temporis spatium creaturae temporali concreatum est; ac per hoc et ipsum sine dubio creatura est." [But the course of time is created together with temporal creation, and for that reason it too is undoubtedly a created thing.][11]

Time has no 'fatality'—it is not something that impinges and is decreed from outside; instead, it belongs to the inner structure of things. But time is also the index of a completed world, whose duration and processes are only two aspects of one and the same state of affairs, just as the God Who preserves the world and the God Who moves it are identical: "explicat saecula ... tanquam plicita" [He unfolds the ages ... as if they were folded up].[12] This internalization of time dissociates from the concept of time the element of measurability, which is now supposed to behave as form does to matter. It is true that motion as a periodic process continues to be necessary in order to be able to measure duration, but duration is autonomous, in relation to this comparison, because the moving thing must also already be—with regard to its identity—something that endures, in order to be able to exhibit periodicity at all. This concept of time, which attempts to stick to the objectivity of time, will find its most sharply defined formulation in the second half of the thirteen century in the Franciscan Peter John Olivi. He too will invoke the priority of rest over motion in regard to the *veritas essendi et existentiae* [truth of essence and existence], and in doing so he will in fact use the Augustinian term *manentia* [enduring things].[13]

This dissociation of temporality from measurability anticipates, in an important respect, the concept of absolute time—because for this concept it is just as constitutive as it is problematic that it does not satisfy the requirement of metrical determinability, though in its function it does not need to satisfy that requirement since it is needed only as a definitional element.

A further motive in the medieval development, in addition to this separation of the concept of time from the concept of a measure of time, is a tendency toward subjectivization. However, this tendency had an impassable limit in the epoch's need for time to be unitary and public. The controversies about the calendar, and the calendar reforms that were realized or were only called for, maintained a lively consciousness of the public character of time. The heart of this interest in the calendar was the consciousness of the historical character of the saving events of Christianity, and the necessity of continually demonstrating that consciousness by strict administration of the feast days.

In the interest of the guaranteed uniformity of time, Aristotle had introduced its metaphysical connection to the concept of God. The Middle Ages were not able to retain this connection because the assumption, by dogma, that time had a beginning and an end made the ontological insistence on absolute uniformity impossible. Something that has, and must have, a beginning and an end cannot have an absolute quality, and therefore needs neither an eternal motion nor a substance that is immune to change. But precisely because time lost its metaphysical foundation, it had to retain its connection to cosmology. The limit to the subjectivization of time that is set by its publicness and unity becomes clear when it is transgressed and this transgression is immediately retracted.

This can be gathered from Olivi's late attempt to infer from the premise that each created thing has its own time the conclusion that there is a plurality of times—and then to make the unity of those times come about only as an abstractive accomplishment of the intellect. This Scholastic, whose speculation always teeters on the brink of heresy, nevertheless finally shrank back from the solution of plurality: None of his important contemporaries had come to this particular conclusion, he says, and there was a lack of the shared premises that would be needed in order to give his arguments any weight. But this outcome, and this resignation, are clearly related to a specific situation, and the decision to adhere to tradition is characterized clearly enough as bound to its time, and as temporary: "Et ideo ad praesens nobis tenendum est tempus esse unum numero ...." [And therefore for the present we are to hold that time is single in number.] Nevertheless, Olivi's position characterizes the central strand of the medieval logical development.[14] The ideas of things having their own times and of the plurality of times point toward the new latitude that is necessary if the diurnal rotation of the Earth is to make the motion of the heaven of the fixed stars, which Aristotle had nominated as the normative motion for time, into an appearance, with the result that any stellar standpoint in the universe that moved in that way would have its own time.

Among the ingredients that seem to tend toward detaching the concept of time from its cosmological foundation, the Neoplatonic element cannot be overlooked, if only because it contributed, by way of the School of Cambridge, to what Newton was to conclude from the Scholastic results and from the requirements of Copernicanism. At first glance it looks as though Neoplatonism initiated the subjectivization of the concept of time, since it makes time into a kind of character or accomplishment of the Soul. But where Neoplatonic texts speak of the "Soul," the reference is primarily to the third hypostasis, after the One and the Mind [*Nous*]: the World-Soul, only the secondary multiplication of which produces the anthropological meaning of "soul."

The systematic ambiguity or interchangeability of cosmological and anthropological systematics was never eliminated from Neoplatonism. It still served to equip it for the speculative projections that were characteristic of its late influence on modern idealism. An indication of how far the concept of time had been decosmicized in the Neoplatonist Plotinus, around the middle of the third century A.D., is that he was the first to be able to pose the question of whether time would continue to exist if the heavens came to a standstill. This is the same question that, in Augustine's formulation, traverses the medieval involvement with the topic of time. For the context that culminates in Copernicus, it defines the decisive consideration.

In Plotinus, the origin of cosmic time was directly bound up with the World-Soul's proceeding from the Mind. Here the concept of time accords with the kind of mutability that is characteristic of the Soul's distance from its origin. Time is the expression of the fact that the Soul is a restless being that is, of itself, unsatisfied. This dissatisfied constitution is a condition of the fact that, instead of relying on its origin, it comes to terms with matter. If the Soul ever came to rest again, time, too, would cease—but this would be the self-contradiction that would put an end to the existence of the Soul. In relation to the metaphysical system of the procession that arises from the One, both time and the Soul are an exceptional condition, an episode with all the marks of temporariness and of the necessity of retrogression. Here the ambiguous significance of the Platonic "appearance," vis-à-vis the Idea, becomes a process: On the one hand, an appearance is an image that follows the example set by the original, and is thus provoked by it; on the other hand, it is a negative falling short of what is obligatory, an irreparable failure with respect to the pure origin. In the Neoplatonic schema, the emanations arise from the One as a result of its 'overflow,' its self-unfolding and self-exhibition, but at the same time they are only deficient forms of their origin, representing both a falling away from it and the need to turn back to it. The Soul and time are characterized, systematically, by this double description, as the unfolding of restlessness and the unsatisfiability of their demand. The Mind, as a new version of the "unmoved mover," moves only that which has separated itself from the Mind and which longs to be the Mind, itself, again.

Accordingly, in the process of the issuing forth of the World-Soul and the material cosmos, the motion of the heavens is secondary. It cannot be the foundation of the concept of time, if that concept relates to the 'motion' of the Soul itself. The question of the possible continued existence of time if the heavens stood still is based on this state of affairs. It is obvious that for Plotinus not even the potter's wheel is needed in order to permit this continued existence of time as the constitutive spontaneity of the Soul—its specific difference from the Mind. It is true that the revolution of the heavens makes time manifest, but it, in its turn, takes its course "in time." If the sphere came to a standstill, it would still be possible to determine the duration of the standstill. A scenario of movements and processes is described here in which time represents a state of alienation, of externality, of expatriation. If the Soul returned from this state to its origin and was absorbed in it once again, "even the sphere of the universe would not exist, which does not exist originally, because it too is and moves in time. Even if it stood still, but the soul continued in activity, we could measure the duration of the heavens' standstill, as long as the soul's action is outside eternity." [15]

Plotinus has already gone a step further in the process of rendering time autonomous than Augustine will go, and specifically in the direction of Newton's absolute time, with which he shares the idea that the visible world exists "in time." For Augustine, time is a created thing like the material world and along with it. That dependence is what matters to him, and he heightens it even further when he refers to the rotating potter's wheel as the minimal condition of time. For in this way time becomes the creature of a creature, dependent on man's set of tools, and far removed from the danger of itself becoming an absolute.

Of course, with the potter's wheel Augustine is reaching back to a metaphor that was already classical in astronomy. In the Alexandrian compendiums the planetary system is repeatedly compared to a rotating potter's wheel, on which the planets, like ants, are distributed between the center and the periphery and participate in the motion of the wheel on the one hand, while on the other hand they carry out their own motions as well.[16] Augustine's use of the potter's wheel as an example of the periodic motion that is needed in connection with the problem of time would be a sort of retraction of the cosmological simile. The independence of time from the motion of the heavens does not, it is true, lead to considerations that anticipate Copernicus, but it certainly does lead to the celestial motion losing its metaphysical importance.

Evidently some Christians had unprofitably concerned themselves with the question whether the heavens move or stand still, specifically in view of the biblical description of the heavens as the "firmament" ("'fixed,'" "firm"). Augustine answers them that many subtle and difficult deliberations would be needed in order to establish whether the case was thus or otherwise. He lacked, and they should also lack, the time to engage in those deliberations and to carry them out—time that they and he should rather expend on the true doctrine, for their salvation and for the good of the Church.[17] Since he nevertheless does enter into many another "superfluous question" like this one, the passage shows that evidently there were no weighty theological or metaphysical grounds on which to base a decision regarding those doubts about the motion of the heavens.

An important circumstance in the medieval formation of the problem of time was that for early Scholasticism, up to the beginning of the thirteenth century, Aristotle's theory of time was not known directly from his writings, but was accessible only in roundabout fashion, by way of the Roman and Patristic authors, so that its contours and reasoning were obscured. A typical occurrence in the history of the influence of a doctrine is that the loss of the argumentation on which a particular thesis was based can enhance its effectiveness. As long as, for lack of the presence of the Aristotelian physics and metaphysics in their original form, the systematic information still remained indefinite and nonobligatory, this body of doctrine was capable of extreme deformation by the new autochthonous needs arising from its being combined with theological premises.

In the early Scholasticism of the eleventh and twelfth centuries, the Aristotelian conception of time is known vaguely. The unmoved mover is familiar from Boethius's statement, which is quoted everywhere, from the *Consolatio philosophiae* [*Consolation of Philosophy*]: "Stabilisque manens das cuncta moveri." [And remaining still you give motion to everything.] But the connection that Aristotle constructs between the unmovedness of the mover and the necessity of a uniform basis for time remains inaccessible. The connection between the concept of time and the Aristotelian assertion of the eternal existence of the world (which, being cited from Saint Ambrose, was now entering the foreground of scandalous propositions) was also unrecognizable.

In his *Opusculum de opere sex dierum* [*Essay on the Six Days' Work*], Thierry (of the School of Chartres, before the middle of the twelfth century) cited, as an explanation of how the biblical "first day" was possible even before the creation of the Sun, the rotation of the heaven of the fixed stars. The problem of time is governed by the constitution of the diurnal unit before the existence of the Sun, and the answer necessarily lies in the priority of the stellar day. Thierry explains the motion of the starry heavens not by a motivating influence from God but by the special fineness of the created celestial substance. A creation of such great lightness as the starry heavens cannot stand still: "... summae levitatis est et stare non potest." [18] The motion of the outermost cosmic body cannot be a motion "in space" because it itself encloses the entire world-space, as the sum of all possible changes of place. These two premises, that of a motion that is necessary for the celestial substance and that of the impossibility of its moving in space, together imply circular motion, which then also begins from the first moment of the Creation and produces its first complete rotation in the first day of that first week.[19]

Since the heavens and the Earth make up the contents of the first act of creation, it would also have been possible to explain the first day, as a stellar day, by asserting an axial rotation of the Earth. However, too many Aristotelian presuppositions had already been absorbed to allow this, although that "maximal lightness" of the celestial material is still the lightness of fire and not of the Aristotelian celestial substance. The decisive step that blocks all nongeocentric reflections is the assumption of a celestial motion that is required by the substance involved. However, the meaning of circular motion as a real predicate of the outermost sphere was apparently not clear to Thierry, because he immediately connects to the theory of the necessary motion of the heavens the necessity of an unmoved center of the world. It is not the Earth's elemental sluggishness that makes it rest at the center of the universe, but rather the necessity, deduced from the motion of the heavens, that its circular motion should have a reference point that is at rest.[20]

Here Thierry quite evidently is describing a world-model with whose original reasons he is not conversant. For of course the production of time by an absolute motion requires that its uniformity should manifest itself for at least *one* observer who is absolutely at rest (unless he is moved in a way that is just as uniform). But that is not the salient point of Thierry's argumentation; for him, even the *reality* of circular motion, lacking any external space or body to refer it to, depends on the center being at rest.

What we have before us here is no doubt a reflection of the introduction of the first translations of Arab astronomical writings into the Latin school activity of Chartres. Thierry was not able to assimilate consistently the concept of relative motion from astronomy and the concept of real motion from the sources that informed him about Aristotelianism. For in the School of Chartres the discussion of the motion of the heavens leads, under the pressure of these difficulties, to a result that was to be significant for the tendency, in the course of the medieval development, toward unburdening the center of the world of a functionally or substantially necessary state of rest: It leads to the invention of additional celestial spheres beyond the sphere of the fixed stars. With William of Conches, a contemporary and fellow scholar of Thierry, the relativity of the motion of the heavens is no longer related to the Earth, at rest, as its *suppositum* [thing put beneath], but rather to an unmoved heaven (or one that is slowly moving, or moving in the opposite direction) beyond the heaven of the fixed stars, as its *superpositum* [thing put above]. For it was only by means of something unmoving or something that moved less that a motion could be discerned at all.[21]

In order for time to be possible, the absolutely uniform motion not only has to be *real*; it also has to *appear* as such. Since, however, the newly introduced spheres beyond the sphere of the fixed stars are not apparent to the observer on Earth, the point of reference of this construction is manifestly no longer man. As already became evident in connection with Plato's cosmic clock, every transcendence leads to uncertainty in regard to the question to whom the cosmic manifestation is addressed. The Scholastic concept of the *veritas ontologica* ['ontological truth'] makes it clear that the Middle Ages had a special way of relating the (in the strictest sense) self-presentation of its realities—a side turned toward a theoretical contemplation that is not that of man.

The introduction of spheres beyond the sphere of the fixed stars accorded with the medieval need for mediating agencies between the world and an unmoved mover whose world-transcendence is defined with a precision that is always increasing. But it also accorded with the problematic of the Aristotelian concept of space, a problematic that resulted from the assumption of a sphere that was supposed to move precisely on the border between space, within the world, and nonspace, outside it. This difficulty could have been an occasion for thinking through the possibility of denying motion to the heaven of the fixed stars itself. But for the Middle Ages, complications in the region of the inaccessible were a small price to pay for the continued existence of the basic cosmological pattern of the tradition. Accordingly, this provocation was not strong enough.

Another motivating force was the objection, which had already been directed against Aristotle in antiquity, by Theophrastus, that he had not allowed the divine influence on the world to reach all the way down to the Earth, and thus he had limited nature's metaphysical administration to the spheres of the heavenly bodies, ending with that of the Moon. In that case the Earth's immobility in its central position would be a consequence not only of its elemental sluggishness and its being fixed in the center of the universe but also of the absence of moving energy coming from the sole origin of such energy, the unmoved mover. Augustine had declared it to be inadmissible to say that divine Providence does not extend past the region of the heavenly bodies to the Earth, because this would mean that the sublunary things could also endure on their own, without divine assistance.[22] It is a matter not only of the perfect administration and order of the world, from which the Earth must not be excepted, but also of its sheer continued existence. Here the intensification of the idea of universal Providence into that of the universal preservation of the world—and finally to the idea of continual creation—already introduces a factor of homogeneity into cosmology: If the Earth owes not only its *origin*, in the Creation, but also its *continued existence* at each moment to a special act of preservation on the part of the Divinity, then it too must participate in the rationality of all divine acts. In that case it must be, in principle, 'cosmic,' in the original sense of the word—and that means, logically, that there cannot be any radical difference between it and the heavenly bodies.

When the Earth has thus become equally an object of the divine influence in the world, it is also at least inclined to participate in the regularity of world-processes. There is no longer, at least, any compelling theological exclusion of the assumption that the Earth could be a cosmic organ and could have an absolutely uniform motion. Once again in the School of Chartres, John of Salisbury, in his didactic poem *Entheticus* [*Treasure House*], written in 1155, summarized a fundamental objection to Aristotle in this verse:

Sed tamen erravit, dum sublunaria casu credidit et fatis ulteriora geri.
[But still he was mistaken when he thought that sublunary things are governed by chance and remoter things by destiny.]

As long as medieval cosmology takes place in the framework of the doctrine of elements—that is, as long as it assumes that fire is the element of the heavenly bodies—the prime mover's reaching the lower elements does not present any difficulties. Only the reception of Aristotle's doctrine of the special nature of the fifth substance puts before one the dilemma of, on the one hand, being expected to accept such a clear physical dualism of the world's matter, and, on the other hand, not being permitted to restrict the universality of a divine effectiveness. For Aristotle this had been no problem, if only because his unmoved mover simply does not 'act,' vis-à-vis the world, but, in an absolutely one-sided relationship, is set as a goal of action, and transposed into motion, by the outermost sphere.

Since the heavenly bodies' matter had been invented specifically in order to equip its processes with a quality that was as empirically unknown as absolute uniformity, after the reception of this special doctrine it was an extremely doubtful proceeding to claim such characteristics for the elements beneath the lunar sphere as well.

But exactly this had to be done, and Thomas Aquinas does it in a

Chapter 2

way that explicitly renounces the original definition of first matter as
absolute undefinedness, with this universal principle: "Diversarum
rerum diversae sunt materiae." [The matter of diverse things is
diverse.] This principle is stated with a certain relief. It excludes any
consideration of a dualism of a Gnostic stamp between the demiurge
and matter. There is no matter that is adequate to omnipotence—
no *materia proportionata* to every possibility.[23] The splitting up of first
matter into specific special matters for minds and bodies, stars and
terrestrial things, stabilizes the metaphysical ascendancy of the one
Cause of all things. The motivations behind the unhesitating acceptance of Aristotelian positions are not always as clearly evident as
they are here. In such propositions, too, Scholasticism shows itself to
be a sum of positions established against the threat of Gnosticism. The
ready acceptance of intracosmic dualism makes up for the forcible
exclusion of any point of departure for a metaphysical dualism.

This aspect needed to be at least briefly illuminated because it
makes the connection between the Scholastic system and geocentric
cosmology intelligible in its stability and in its later immune reactions. Scholasticism's repetition of the Aristotelian division of the
world into sublunary and superlunary realms no longer has any of the
original function of preserving the explanatory convenience of the
Platonic *chorismos* ['separation'—of the Forms from matter] in the
context of the Aristotelian philosophy of nature. The important thing
is no longer the special characteristics of the substance of the heavens;
now it is the particularization of matter. The specificity of the celestial
matter resides in the fact that it has to be a substratum only for change
of position (*innovatio situs*), but not for change in constitution (*innovatio
essendi*), as in terrestrial objects. Every Aristotelian would have had to
ask, at this point, whether this difference does not belong on the side of
form, and whether one should not therefore make a renewed search
for a unitary substratum for both specific forms.

As long as inertia of state and gravitation of masses, as explanations
of cosmic motions, have not been discovered, only the ubiquity and
permanence of the divine action on the world can open up and keep
open the scope for reflection as to whether the definition of time
according to the Scholastic short formula *tempus motus caeli* [time is the
motion of the heavens] cannot also be satisfied by an *apparent* motion of
the heavens combined with a *real* motion of the Earth. But the requirement of ubiquity does not allow one to differentiate the matter of the

universe, which is the universal disposition to accept action upon it, in such a radically dualistic way as had been done in the Aristotelian cosmology. No more than a cosmic underground that is not affected by God's action can be permitted in the medieval pattern of thought —no more does the assertion of an immediacy of omnipotence in relation to the terrestrial nadir of the scale of cosmic agencies, an immediacy that leaps over all those realities, accord with that pattern of thought. The destruction of the Scholastic obstructions is not yet the stimulation of a motivation to exploit the latitude that that destruction brings about.

The sole attempt to found time, in the context of the Aristotelian philosophy, on a sort of endogenous motion of first matter was made by the Franciscan Bonaventura. The advantage of this experiment lies in the fact that Aristotle's first matter is the identical principle of all physical processes and objects, so that it can 'effortlessly' satisfy the requirement that the basis of time should be unitary and ubiquitous. A consequence of theology's concept of creation was that one could not assign to any ingredient of the created world an exclusively negative undefinedness, sheer passivity and an absolute lack of definite purpose. It must be possible to assign some positive characteristic to anything that originates in God's will. Now Bonaventura sees this positivity of first matter in its immanent disposition to have form invested in it—in its tendency toward the realization of an objective form. So the form is not a pure additive to matter, something that is imprinted, externally, onto a matter that is indifferent to it; instead, it is something that is suitable to matter's disposition. It is something that one could characterize, with a later term, as matter's conatus. Now, it is on this elementary 'transitional' character of the primal stuff that Bonaventura, in his *Commentary on the Sentences* of 1253–1254, founds both the unity and also the uniformity of time as the homogeneous characteristic of all processes: "Tempus autem habet esse ex hoc, quod materia tendit ad formam .... " [But time derives its being from the fact that matter tends toward form.] It is easy to see that with this theory Bonaventura gets rid of several difficulties in the harmonization of Aristotelianism and Christian premises, in regard not even primarily to the theory of time but to the definition of a unitary 'aim' that from now on is necessarily invested in matter.

The elevation of matter to a rational element also has cosmological consequences, inasmuch as it has to level off the dualism of types of matter in the Aristotelian universe. But in the School of Padua, immediately before Galileo's activity, this process, with its epistemological implications, had still not reached its conclusion. When Galileo took up his professorship in Padua in 1592, his predecessor there, Jacob Zabarella, had died just three years earlier. Zabarella had summed up once again, in the thirty books of his Scholastic summa *De rebus naturalibus* [*On Nature*], the yield of Scholasticism for the philosophy of nature, and in doing so had also dealt with the obligatory topic *De natura caeli* [*On the Nature of the Heavens*]. In the question of the matter of the heavens Zabarella largely follows what "the Commentator," Averroes, had arrived at as the most extreme leveling off of Aristotle's position, namely, that the matter of the heavens is homogeneous with that of all other physical objects. Its special character is now derived only from the fact that no *forma contraria* [opposite form] exists for any form that is imprinted on this matter. So the celestial sphere's dependability for the concept of time now consists only in the fact that no other form of existence is 'offered' to it. Its actual state is the satisfaction of its potential: "satiatur enim ab illa forma totus illius materiae appetitus" [for all the appetite of that matter is satisfied by that form].[24]

This theory avoids the paradox of assigning a specific determination to something that by its definition is indeterminate, and reduces the cosmological dualism to its minimum. In the process, the Aristotelian precepts are followed strictly: If the concept of first matter is defined by the *mutatio substantiae* [alteration of substance], then it cannot be due to the matter itself if a change of form does not take place. But in that case is there still any sense in retaining the distinction between matter and form? According to Zabarella, this distinction in the heavenly bodies is conditioned by the double character of the human cognitive faculty: The heavenly bodies are objects of both perception and thought. The term "matter" stands for the circumstance that what is thought about exists outside the thought: *pro substantia ipsa corporea extra animam existente* [for the actual corporeal substance existing outside the soul]. This epistemological premise allows him to speak of both a *materia formalis* [formal matter] and also a *forma materialis* [material form]—that is, of both a form that is such as to require no matter for its realization and also a matter that arrives on its own at concrete reality: *materia quaedam actu per seipsam*. In view of this restriction, astronomy, as a purely kinematic science, has its basis in the fact that the heavens' matter is no longer characterized by a special sublimity and purity, but by its reduction to the concept of a *materia unde quo id est mobilis de loco ad locum* [matter that is of such a kind that it is mobile from place to place].[25]

So shortly before Galileo, the stellar object is objectively, and for any possible observer (i.e., not only for one in man's deficient position), absolutely not cut out for a physical theory.[26] In the strict sense, Zabarella says, the motion of the heavens is not a natural motion at all, because it is produced by a principle that lies outside nature; it has only an analogous structure to the natural motions of the other bodies in the universe, because its moving principle, the intelligence that dwells in it, is not the form of the heavenly body. To that extent astronomy, if it should expect itself to consider questions of causation, would not belong to the realm of natural philosophy. Only to the extent that it renounces causal questions—that is, to the extent that it considers effects instead of causes—is it part of the science that is possible for man: . . . *ita pertinens ad scientiam naturalem.*[27] The dualism of substance, within natural philosophy, has been dismantled, but the epistemological situation, only a few years before Galileo's reaching out into the universal physics of nature, is still unchanged. It is precisely in the reduction of the hylic dualism that the guarantees for the Aristotelian concept of time have been held to firmly: The heavens 'accomplish' nothing but their motion, in the most dependable fashion.

The breadth of the Scholastic discussion of time can be described as resulting from the disintegration of the original unity of the Aristotelian definition of time into the two elements, *numerus motus* [number of motion] and *motus caeli* [motion of the heavens].

The more general definition, "number of motion," is cosmologically indifferent. It leaves open the question whether only *one* motion underlies the concept of time—and whether this is an arbitrarily chosen motion, or the set-apart and guaranteed motion of the heavens, the *motus regularis et certus et nobis notissimus* [motion that is regular and dependable and best known to us]—or whether, in the last analysis, all motions equally make the concept of time possible, according to the formula *tot tempora quot motus* [there are as many times as there are motions], where the unity of the concept of time would be nothing but the abstraction of what all 'times' have in common. With the rudimentary definition of time as *numerus motus*, the concept of time and the measure of time no longer necessarily coincide. Where the *measure* of time receives a nominalistic interpretation, it becomes possible to trace the *concept* of time back to inner experience.

At first glance, the other residual definition of time, as "the motion of the heavens," looks more conservative. But in the Scholastic contexts it is no longer the old definition, because the motion of the heavens that provides the foundation for time is now interpreted causally rather than phenomenally. The point here is not to discuss once again the materials constituting the Scholastic discussion of time;[28] rather, what I want to do is to evaluate the typical features of those materials in regard to the way in which the motion of the heavens is involved in the concept of time, up to the point where the doctrine of the motion of the heavens is set free from the theory of time.

Avicenna, one of the two Arab commentators who were authoritative for Scholasticism, had provided, in his commentary on the *Physics*, the most often quoted catalogue of basic theses about the nature of time (*de quidditate temporis*), with the aid of which each Scholastic author could establish his own thesis, by means of the often used procedure of exclusion, as the one that was (in each case) left over. The first of his opinions (*opiniones*) asserts that time is a sequence of now-points, an accumulation of atomic moments (*aggregatio momentorum*); the second one defines time as an eternal substance (*substantia aeterna*); the third defines it as an arbitrarily chosen motion (*motus loquendo indifferenter de quocumque motu*); the fourth defines it as the motion of the heavens (*motus caeli*); the fifth defines it as the rotation of the heavens (*recursus caeli, id est una revolutio*); and the sixth, finally, identifies time with the heavens (*tempus est ipsum caelum*). The decisive thing for Avicenna was that the motion of the heavens is not only the *phenomenon* of an absolutely regular, permanent, and public motion, but the *cause* of all the other processes within the universe.

This basic thesis involves the great difficulty that the thought experiment of the heavens standing still can no longer be made, because in accordance with the assumption, the potter's wheel too would then immediately come to a standstill. Precisely the production of this difficulty, which is intolerable in relation to sundry requirements of theology, leads all the more compellingly to the separation of the motion of the heavens from the concept of time—for example, in the formula of Roger Bacon that no specific motion

whatever is the basis of time. Any motion can accomplish this: "Nullus motus est praecisum subiectum temporis ... sed omnis motus
indifferenter." Of course this is only an important assertion if the
universal causality of the outermost heaven is restricted, and motion
is held to be possible in the world even if the heavens stand still—as,
for example, the motion of a mechanical clock.

The other Arab commentator, the great authority of orthodox
Aristotelianism, Averroes, also initially joins in Avicenna's causal
reflections, but then gives the matter a new turn of his own by setting
up a connection between the concept of time and *inner* experience.
The result of this connection is finally that the early great question as
to whether time could persist if the heavens stood still is joined by the
other, late one as to whether time would exist if no soul capable of
cognition existed.

Of course in the "Commentators"'s [Averroes's] theory of time,
self-experience is not objectively set apart. We perceive the motion of
the outermost heaven not only directly, as a phenomenon, but also
indirectly, in any physical process we care to choose, as, according
to the doctrine of "formal" causality, the cause can be perceived in
its effect. Since our own mutability depends on the motion of the
heavens, we have, in the experience of it, a direct access (*per se*) to the
essence and the unity of time. So the preference for inner experience is
a methodical one, without the Augustinian emphasis on its special
dignity. In the relation between one's own mutability and the rotation of the outermost heaven, the element of the finitude of time as a
created thing has to drop out, since for the Commentator the heavens
are eternal.

It is important to observe how in this conception of Averroes the
importance that antiquity ascribed to the intuition of the public
phenomenon of the moving heavens is reduced. That too has become
plausible to the Scholastic authors: Their idea of man is no longer that
of the *contemplator caeli* [contemplator of the heavens]. This comes out
most clearly, in Averroes, in his reference to Plato's allegory of the
cave. Time exists not only for those for whom intuition of the heavens is always possible but also for those who are blind or for whom
the view of the heavens is blocked. Plato, Averroes says, treated as
fundamental a state that is completely closed off from the heavens
("... ut dicit Plato de incarceratis a pueritia sub terra ..."). The
allegory of the cave no longer serves primarily to present and to lead

one on the path out of the cave, but rather to probe the possibility of life in the cave. The cave scene has become an argumentational device in a situation in which philosophical issues have become intensified—hence the incomparable privilege of self-experience as an indirect access to the motion of the heavens, and hence also the fact that it does not matter whether the directly perceived motion is real; it can be real or imagined (*sive imaginatus sive existens*), since as an idea it is an alteration in the intellectual organ and thus an effect of the ultimate origin of all alteration.[29] Here the allegory of the cave is used in a way that strikes one as an early form of Poincaré's thought experiment with the heavens that are permanently hidden. In elementary reflections of this sort, man's way of possessing the world begins to disconnect itself from intuition.

This extension of the Aristotelian theory of time by the addition of the element of the ubiquity, not, indeed, of a public phenomenon, but of a causality that is universally present, is in the spirit of Aristotle's thinking at least insofar as the shared "formal" quality of the cause of both inner motion and outer motion explains the epistemological connection between the former and the latter. Insofar as the motion of the intellectual soul itself originates in the motion of the outermost heaven, it is, according to the principle of Aristotelian psychology, not only potentially everything that it knows, but also equipped to serve as the concept of time. Although such claims are uncommon in the tradition of commentaries on the ancient philosopher, Averroes claimed this systematically consistent extension of Aristotle's theory as his original accomplishment: "Quicquid scripsi de tempore secutus sum expositores, sed hic non." [In what I have written about time I have followed the expositors, but in this case I have not.] Surprisingly, in spite of the threat of an exclusion of all spontaneity from the human soul, this was not able to deter Christian Scholasticism from adopting the doctrine.

Albert the Great [Albertus Magnus] and Thomas Aquinas "accepted [the formula] and made it their own without hesitation or criticism." [30] In his *Commentary on the Sentences*, Thomas summarized it as follows: "Sentimus tempus secundum quod percipimus nos esse in esse variabili ex motu caeli." [We are aware of time in that we perceive ourselves to exist in an existence that changes, depending on the motion of the heavens.][31] If one understands this as an indicator of the relation of tension, in Scholasticism, between cosmology and anthropology, it makes the absorption of what were originally human definitions by a cosmologically oriented metaphysics especially conspicuous. Inner experience simply becomes again one of the constituent parts of outer experience—a more painstaking auxiliary way of safeguarding it. This orientation toward cosmology makes one doubt whether one should concur with Anneliese Maier's statement that the systematic supplement to Aristotle that Albert and Thomas took over from Averroes already "from the beginning introduces a certain admixture of relativism and subjectivism" into the subsequent Scholastic discussion.

But the important question in the context of our present subject is what was the importance of this modification of the problem of time in the approach to Copernicus. It is clear that Averroes's extension implies a consolidation of Aristotelianism's systematic state. The reality of the motion of the heavens becomes still more indispensable—insofar as this was still possible in the system—for the reality of the consciousness of time and for the constitution of the concept of time. This is demonstrated in the "Commentator"'s own text immediately following the principle we have quoted. Here he, too, asks himself what his thesis entails in regard to the assumption of the heavens' standing still. If it were possible for the heavens to stand still, it would also be possible that we should find ourselves in an existential state of unchangeability; but this is impossible, and consequently it is necessary that even a person who does not perceive the motion of the heavens through intuition should sense change.[32] The hypothesis of the heavens' standing still is reduced to absurdity by appeal to self-experience, according to which we never find ourselves to be unchangeable. If we are constitutionally changeable, then the heavens, as the condition of the possibility of that, must also be the same.

Only at first glance is this argument a reversal of the relation between anthropology and cosmology. In Aristotle's terms, the self-experience of changeability is the better-known thing only "for us," as a consequence of our deficient state (as far as method is concerned) in relation to the knowledge—which according to the nature of things would take precedence—of the reality of the motion of the heavens.

It has ceased to be possible to carry out the Augustinian thought experiment with the potter's wheel. As long as the potter's wheel turns, this too—even if it is by way of the organism of the person who operates it—is an effect of the motion of the heavens. But Averroes's doctrine also accomplishes something else that the imagined case of the potter's wheel accomplishes: It separates the constitution of time from its measurement. When it took the appearance of the moving heavens as its basis, the Aristotelian theory had seen in it both time's foundation and its measure. Our reflexive perception of our human changeableness does still lead us to the foundation of time, but it no longer leads us to its measure. If our intuition of the heavens were blocked, this doctrine would produce only the possibility of, in the Aristotelian manner, deducing from the concept of time the necessity of an (admittedly unknown) absolutely uniform and endless motion—in connection with which the periodicity, as a measure of this motion, would be a matter of course insofar as, according to Aristotle, a uniform motion could only be a circular one. The image of the potter's wheel also presupposed that time does not come into existence by being measured, but rather that the thing whose time is measured also exists in time independently of this measurement, just as objects exist in space independently of whether the quantity of their spatial extensions and distances is in fact measured. So time is, on the one hand, a measure by which processes are measured, and, on the other hand, it is itself what is measured. Averroes's doctrine makes the measurement of time by means of the *tempus proprium* of [the time proper to] the primary motion secondary in comparison to the constitution of time and of the consciousness of time. At the same time, however, the causally transmitted 'form' is the basis of the 'applicability' of the celestial motion's standard for the measurement of time to all intracosmic processes. That relation follows from the Scholastic axiom "Ratio mensurae respectu mensurati importat causalitatem." [The systematic relation of the measure to the measured implies causality.][33]

The integrating power of Aristotelianism, which produced the 'classical' formation of High Scholasticism, is decisive for the historical state of affairs that one can Copernicus's "delayed arrival," when one has become aware of his specific medievalness. Of course opposing forces to this integrative accomplishment sprang up very quickly, forces that were induced primarily by the 'objectionableness' of the intrasystematic implications. An exception to this generalization was a technical event that was bound to reshape the concept of time and to detach it from its cosmological mooring. This was the appearance of the mechanical clock.

Since antiquity, regarding the world as a clock meant primarily perceiving in it the establishment and the indication of the universal measure of time. The new medieval metaphors of the world-clock place the accent not on the clock's indicating front but rather on the mechanism that drives it. The mechanical clock is the paradigm of internal propulsion. Here it is really a secondary factor, of slight importance, that this driving mechanism only stores the energy that it was supplied with, that it has to be "wound up," and that this is done with motions that, in the Scholastic system, can be traced back to the outermost heaven.

In the "Lodgebook" of the Picard Villard de Honnecourt, around 1235, the idea of the *perpetuum mobile* [perpetual mover] emerges for the first time, probably not without the influence of a tradition that came from Alexandria and had been trasmitted by way of Arab architects. It is true that this tradition reaches the Latin West at the same time as the full reception of Aristotle from the Arab lands, but this does not license us to conclude that the idea of a purely immanent, technical self-movement was formulated for the first time in the middle of the thirteenth century as something corresponding to the eternal revolutions of the heavens. A sufficient reason why this simultaneity is a mere contingent coincidence of the sources [of the respective ideas] is that the mechanical idea of perpetual motion could by no means find in Aristotelianism the theory corresponding to it, but on the contrary had to find the contradictory principle that everything that moves is moved by something else. We have to reckon, instead, with a connection between magic and mechanics. Here mechanical thinking is still, and will continue to be for a long time, under the power of the idea of outwitting nature. The drawing in Villard de Honnecourt's "Lodgebook" allows us to infer only a thought experiment, not a public demonstration of the new ability, such as the great tower clocks will present in the next century. The caption under the drawing does, however, also show that it was not a matter of an eccentric toying with ideas, but rather of a much-discussed problem for craftsmanship: "Maint jour se sunt maistre desputé de faire torner une ruée par li seule ...." [Many a day masters have argued about making a wheel turn by itself ....][34]

This is followed in 1269 by the *Letter on the Magnet* that was written during a Crusade by Petrus Peregrinus de Maricourt (also a Picard). This is one of the few texts of the Christian Middle Ages that give evidence of empirical and experimental inquiry. The author tried to establish the connection between the description of the magnet and cosmology, a conception that was to be taken up again, in a thorough-going manner, by William Gilbert, at the end of the sixteenth century. As a by-product of this theoretical project, Petrus Peregrinus hits upon the proposal of a magnetically driven *perpetuum mobile*. He writes to his friend Siger de Foucaucourt that he wants to reveal to him the construction of a wheel "that, miraculously, moves continually." This mechanism differs from Villard's in that it introduces into the construction a power source that, according to the theory of the magnet, becomes operative only through a direct connection with the motion of the heavens. Petrus Peregrinus has discovered the bipolarity of the magnet and related it, causally, to the celestial poles, from which (he says) the magnet receives its power, which therefore can claim the designation of a divine power (*virtus dei*). Thus the perpetual motive power of this mechanism is directly related to the motive power of all motions within the world, the diurnal rotation of the outermost sphere around the world-axis. The *perpetuum mobile* complies with the idea of immanence only as a display, a speculation on the public's expectation of marvelous things. Petrus Peregrinus then goes one step further and connects the magnetic mechanism with the temporal function of the outermost heaven: If one gives the magnet an approximation to the spherical form of the heavens, he says, and arranges the axis that is represented by its poles so that it is parallel to the axis of the heaven of the fixed stars, then one obtains a clock that is driven by the cosmic time indicator and is synchronized with it. The "genuine" cosmic time presents itself by means of the artificial mechanism. Logically, the Crusader's testament recommends to the one who is to put the secret invention into practice that he should not only make himself familiar with the nature of things but also acquire knowledge of the motions of the heavens.[35] It is easy to see that the mythology of this clock is entirely based on the assumption of the reality of the motion of the heavens.

The heavens, which since antiquity had been the epitome of what was inaccessible to man, move, intellectually, into the function of an inexhaustible source of energy for man. The fact that this power is designated as *virtus dei* does not by any means remove it from being laid hold of and utilized; one senses that the ancient cosmic hierarchy no longer sanctions insurpassable limits. At the same time all of this still remains entirely consistent with the [inherited] system, because only what possesses a divine nature is exempted from the elementary characteristic, common to all processes in the world, that makes its appearance in the Scholastic language as the "tendency toward a state of rest" (*inclinatio ad quietem*). It is not only something like a generalization of organic tendencies to exhaustion and decay, but also an expression of the fact that all motions and processes are regulated by goals—with the exception of circular motion, which thus is also best equipped to be excepted from the universal fate of exhaustion. This primacy of rest succumbs to the critical destruction of that "tendency toward a state of rest," and is finally dissolved in a rational conception for which rest and motion are both states that change their identity only as a result of forces impinging on them from outside. For High Scholasticism the rule continues to hold that nature cannot possibly aim at motion for motion's sake.[36] The *perpetuum mobile* is projected from the beginning as an outwitting of nature that runs counter to nature's principle.

The decree of the Paris Bishop Tempier, in 1277—the decisive turning of Scholasticism against its Aristotelian integration—affects the topics of time and the motion of the heavens in two respects that at first glance are contradictory: On the one hand, it insists on the reality of time, against the tendencies to subjectivize it, while, on the other hand, it does not permit the interpretation of this reality as unconditional dependence on the motion of the outermost heaven. The delimitation in favor of a realism of time is contained in the condemnation of proposition 200, which certainly does not represent an Aristotelian thesis: "Quod aevum et tempus nihil sunt in re, sed solum in apprehensione." [That eternity and time have no reality in fact, but only in thought.] Since Augustine, time had been regarded as a constituent part of creation and thus as sanctioned in its reality. It remains unclear whether this sanction also excludes the view that only the human intellect adds, to the relations of earlier and later that are given (as countability) in the objective motion, the completion represented by the actual counting—in which case the definition of time as *numerus motus* [number of motion] does after all have a subjective complement.[37] In the theory that makes the *numerus motus* definition into a doctrine of subjective and objective complements, we should not underestimate as an objectionable feature the fact that the objective element of motion appears as the material substratum, which only receives its "formal," and thus essential, definiteness thanks to the subjective factor of number. This ontological state of affairs makes it probable that the condemnation of the proposition that I cited was also meant to affect the Thomistic thesis of the complementary role of subjectivity.

In regard to the connection between the problematic of time and the situation in the approach to Copernicus, very important in the decree is proposition 190, which strikes at the heart of the 'topographic' consequences of Aristotelian philosophy for Scholasticism. To assert that the first cause (*prima causa*) is the most remote of all causes (*causa remotissima*) contains an error, we are told, if it is understood as meaning that in that case the first cause cannot at the same time be the closest one (*propinquissima*). Perhaps it is an exaggeration to say that this condemnation of the proposition of the necessity of cosmic mediation signifies the beginning of the modern age. But what we are dealing with here is unquestionably the beginning of the prelude to it. The condemnation of a central principle of the Aristotelian systematics does not simply restore the situation that had existed before the consolidation of this system. Now, for the first time, one can be confronted with the question of what possibilities for the Earth and for man had been excluded or restricted by the cosmological 'channels' of Aristotelian metaphysics, which had the result that the cause that was most remote in terms of spatial distance could never be the most proximate cause in the sequence of factors. Only the first and outermost heaven could receive the regular and undiminished influence of the unmoved mover and display it in its circular motion. An equivalent assertion could not possibly be made for another cosmic body. But if the principle of the necessity of mediation is eliminated and the absolute cause becomes the most proximate cause, in the sequence of effects, for every point and every object in the world, then the outermost heaven, as a unitary and moving reality, has become superfluous in this system, even if it is not yet contested. The Aristotelian guarantee of the possibility of time, which consisted in the effective connection between the unmoved mover and the first, moved sphere, can now, in principle, be claimed (under conditions that are, in other respects, the same) for another cosmic motion—for example, for the terrestrial globe.[38]

The Paris edict of 1277 does not yet contain anything that anticipates Copernicus. Nor is it by any means an act of liberalization vis-à-vis a system that is growing increasingly dogmatic. Instead, it is restorationist in its intention; but in that intention it is one of the many illustrations of the fact that in history the *intentio auctoris* [author's intention] is not very important. In regard to the concept of time, the edict marks the new latitude for thought processes that make the Earth participate so directly in the relation to the first cause that it could carry out the effects of that cause with the same precise regularity as the outermost heaven. Furthermore, it is no longer necessary that there should be a connection between the Earth's state of rest and the circular motion of the outermost heaven, because the edict expressly condemns the assertion that God could put the world in motion only in a circle, and not in a straight line.[39] A century later, Nicole Oresme will appeal to this statement in order to reject the cosmological dogma of the necessity of there being a body at rest in the center of the universe. If the universe as a whole could be moved in a straight line in a space that is external to it, none of its bodies, including the Earth, would have the required state of rest any longer. Likewise time, insofar as it is based on the circular motion of the heavens, would only be the measure of intracosmic motions, and not of motion of the world as a whole and in the space that is external to it.

In its long-term effect, the Paris decree was understood and appealed to less as a prohibition of particular propositions than as a license for assertions whose extravagance was system-contrary. The author of the decree only wants to reject philosophy's presumption in asserting certain necessities, by which divine omnipotence could be restricted. The philosopher, for his part, employs this situation in order to make use of the very same divine omnipotence for his hypothetical construction. He effortlessly alters the Aristotelian concept of space, which had recognized one world-space only inside the outer sphere, so that it had to exclude a motion in space on the part of that sphere. In a space in which arbitrary motions are possible, the concept of time can no longer be tied to the rotating sphere. It is then quite logical that the Augustinian consideration of the heavens' standing still, which had been excluded in the integration of the Aristotelian system, is admitted again by the decree of 1277, which in two places condemns the assertion that it is impossible for the outermost heaven to stand still.[40] If one may not assert that the heavens' standing still would necessarily have as a consequence, in the world, that a candle could not even catch fire, i.e., that no change would be possible, then neither does Averroes's reverse reflection hold any longer, that any internal change in man could serve as a basis for time because it would be conceivable only as a causal reflection of the motion of the heavens and its regularity.

Averroes's extension of the system had made clear where the sensitive spot lay in the conflict between cosmology and theology, and what a vigilant defense of omnipotence had to accomplish. Paradoxically, along with omnipotence and its options in the world, man's freedom was also to be defended against the universal determinism of Averroist orthodoxy. There must be motions that result from actions that are endogenous to the world, if freedom in the world is supposed possible. Consequently, the potter's wheel, as an instrument of an acting person, must remain possible even under a heaven that is standing still. This is just where the importance of the Sun's standstill in Joshua lay for the Middle Ages—an event that, if it was to have its effect in deciding the outcome of the battle, precisely could not cause time, within which the battle could be won, to stand still as well. This excludes a universal and mediated cosmic causality and requires that not every motion in the world below should depend on the primary motion [of the outermost heaven].[41] It is considered quite possible that other motions could satisfy the requirement of uniformity. Likewise it is insisted, against Averroes, that the time that a process takes can be measured by means of the motion of the heavens even if it is not caused by that motion. The end point of this development is a reversal of the original conception, in that a lowly terrestrial motion can serve to measure the time involved in a process of stellar motion.

Nominalism separated the "number of motion" completely from the "motion of the heavens." The argumentation that William of Ockham gives for this is characteristic: God could create an additional heaven that would have a more rapid motion than the outermost heaven that is real at present, without changing the latter at all, and in that case the motion of the heaven that is outermost at present would no longer be the real foundation of time, even though it would not have been deprived of any real predicate—so that being the foundation of time cannot be a defining characteristic of the heavens. Thus the designation of the primary heaven as the cosmic clock is setting something up—it is the choice of an instrument with which to determine time for other processes: "Tempus est motus quo anima cognoscit quantus est alius motus . . . . " [Time is the motion whereby the soul knows how great another motion is.][42]

The argumentation is similar to the argumentation against the reality of motion. What we are given is the location of a physical body in various positions in space at various times; the term "motion" designates what we infer from that. God could create this body in one position in space and then annihilate it and create it anew in a different position in space; the perception would be identical, and the inferred motion would not be real. Thus the name "motion" designates something for which there is no legitimizing perception. Motion *and* time are terms for an identical reality, the constitutive defining elements of which are only physical bodies and space. Time is a point of view on this reality and thus implies an agency that chooses; the subjective element is constitutive. The advantage of the prime motion [the motion of the outermost heaven] consists, for the early Ockham, in the fact that among all the motions in the cosmos it is the most uniform and the fastest, and the quantity of other motions can be determined with the greatest certainty by means of it. Of course, the idea that this advantage could be founded in reality, that is, that the outermost heaven could receive the exclusive metaphysical guarantee of the Aristotelian unmoved mover, is made questionable by the remark in Ockham's *Commentary on the Sentences* that the most rapid motion could be not only that of the eighth or ninth sphere but also that of the Sun or the Moon, which are equally available and thus equally eligible as temporal parameters. But if the reference motion is not objectively prescribed as such, the possibility moves into the realm of discussion that one could also make any other suitable motion into the measure of time, and thus finally identify 'time' and 'measure of time.' [43] It is true that in this passage Ockham explicitly appeals to the "Commentator"; but he is wrong to do so, because for him the decisive causal connection between all subordinate motions and the prime motion is by no means the reason why each of them can provide the basis for the consciousness of time. Instead, the model of the efficiency of mechanical clocks already clearly stands in the background.

Consequently the metaphysical objectionableness of the fact that a motion that is low in the cosmic hierarchy can become the temporal measure of a motion that stands higher in that hierarchy is no longer acute. The fact that one can measure all motions adequately with clocks follows from the fact that clocks, in general, are parameters that are chosen and set up. Thus time is not measured as something that has been present all long; instead, it is produced, for the first time, by measurement, just as the term "motion" represents the interpretation of a connection that consists of nothing but the appearance of a body and the difference between its positions in space.

A crucial consequence for the historical process that is our topic here is that no necessary connection is assumed to exist any longer between the possibility of the concept of time and the reality of motion. In this respect, an *apparent* motion can accomplish exactly the same thing as a *real* one. From this viewpoint, an astronomical reform that declared the rotation of the outermost heaven to be a mere appearance resulting from the axial rotation of the Earth would no longer have been open to criticism based on the demand for a real foundation for time.

The original Aristotelian direction of the process of the generation and the manifestation of time in the universe had become reversible. The events in the heavens could be measured with the aid of the *motus inferior* [inferior motion], just as, to begin with, the motion of the outermost sphere could also be measured with the aid of the Sun's motion and the Moon's motion and fractions of these. This was especially true if it was a matter of the motion of a heaven that is not perceptible by us, such as the ninth or even the tenth heaven, both of which were added in the Middle Ages (and which in any case had remained mere assertions, as far as the problem of time was concerned).[44] If the motion of the heavens was to become a mere appearance—through the Copernican development that was in store—then the consequences of this for the problem of time were softened in advance by Nominalism's founding of the concept of time on something that in the strict sense cannot be 'given' at all, namely, on motion, as such: "Nec aliquis motus cuicunque apparet sensui." [Nor is the motion of anything evident to the senses.][45]

Strictly analogous to the logic of this conclusion that time is intraworldly is the logic of the conclusion that it is supraworldly. For the Aristotelians, this possibility had not existed, because for them a conceptual realist understanding of time had been one of the chief arguments in favor of the uniqueness of the world: Since there can only be *one* time, there can only be *one* world, and its outermost envelope, as the basis of this one time. It was again one of the statements of the Paris decree of 1277 with which, now as well, the problem of time had to be confronted: The first cause must not be tied down to being able to create only one world.[46] According to William of Ockham, if the outermost heaven were the organ of time for every world, it would have to follow that there were several outermost heavens, and consequently several times. Now, for Nominalistic thought, the consequence of a plurality of world-times that could be compared with one another or with one that was arbitrarily given priority (whoever it might be that would have to do this) is no more alarming than the plurality of times within a world, times that cannot be unified as a world-time in any way but according to the principle *unum per aequivalentiam* [one by equivalence]. Here the supraworldliness of time is related to an imagined concept-constructing intellect. But the step to a new ontological interpretation of this time that overarches worlds was natural as soon as the Paris decree's prohibition of thinking of the world as necessarily one was taken seriously.

An opponent of Ockam's theory of time, the Franciscan Gerardus Ordonis, appears to have carried this train of thought further and made the reality of time independent of the existence of any world at all.[47] This would then have been an anticipation of absolute time, for the reality of which it does not matter whether a world already exists or still exists and whether *one* world or many worlds exist. The path to absolute time would not have been taken, then, without the provocation of a development in the other direction, toward the removal of time's cosmic character, and its instrumental relativization. However, this reciprocal intensification of extreme opposite positions that had resulted from the difficulties in Aristotle's doctrine of time (which was their common root) has the result that, as Copernicus draws near, the blockage by the problem of time of consideration of the heavens' standing still is removed. Where the rotation of a potter's wheel, again, or even merely its imagined rotation, is sufficient in order to measure time and (thus) to produce it, the Earth's rotation, assumed to be extremely regular, would certainly be enough to satisfy this requirement. The same Joshua who, after Copernicus, was so often to be cited against him, serves here to support—up to the extreme of the derealization of time—the separation of the motion of the heavens from the constitution of time, on behalf of an expanded latitude for cosmology.[48]

Chapter 2

Although Ockham appeals to the "Commentator," his conception of the possible founding of time on a motion that is only imagined has nothing to do with the latter's doctrine of the self-experience of changeability. In Averroes this changeability continued to be bound up, causally, with the real motion of the outermost heaven, and it was only because of that connection that it could become the ontic foundation of the consciousness of time. This connection does not on any account allow us to picture the outermost heaven as standing still. Ockham, on the other hand, is not thinking about self-experience but about an imaginary motion that can be normatively defined as regular. If we assume that the world stands still, this fiction takes the place of the motion of the outermost heaven. Only an imagined motion can be idealized in such a way that its regularity could not be surpassed even by a divine creation.[49] Thus, in Ockham, Averroes's first step beyond Aristotle has developed into the further step of pushing the anthropological foundation of time all the way to the pure artificiality of the inner fiction and its idealization.

The epistemological situation in which Nominalism thinks that it finds man is one in which he is constitutionally beset by difficulties. If man becomes "creative," as in this case, where he invents a motion in case the world stands still or he loses contact with it because it is destroyed or he goes blind, this is not an index of his special dignity but rather of his making do as best he can in his *status viae* [transient condition]. But what was discovered at the beginning of the fourteenth century on this assumption of an anthropology of dire necessity is integrated, a century later, into an anthropology of creative surplus. Nicholas of Cusa expressed the discovery of man's status as the measure of the world with Protagoras's much-maligned saying, and applied it not only to logic and mathematics but also to the measurement of time. Time itself is an instrument of the soul. Years, months, and hours are measures of time that were created by man, the Cusan writes in his dialogue *On The Globe Game*. So is time itself, since it is the measure of motion, and an instrument of the soul that measures. Thus the soul's rationality does not depend on time, but rather the rationality of the measure of motion that is called time depends on the rational soul. Consequently this soul is not subject to time, either, but rather is prior to time in the way that sight is to the eye: It is true that one cannot see without eyes, but sight does not get its essence from the eye, since the eye is, after all, its instrument. Thus the rational soul, too, although it is unable to measure a motion without time, is not therefore dependent on time. Instead, the reverse is the case, since it makes use of time as an organ for distinguishing between motions.[50]

This anthropological idealization of the concept of time is significant, in the century before Copernicus, because it is systematically connected with the process that I described as the "idealization of the center of the world": The naturalism that gathers man's position in the world from the fact that the Earth, as his home, occupies the center of the world, is pushed aside by the reflection that the 'center' of the world, as the reference point of its rationality, is wherever man finds himself. This suspension of the indicative role of cosmology for man's self-consciousness is the fundamental idea of the *Oration on the Dignity of Man* that Pico della Mirandola had drafted for the opening of a gathering in Rome to which he invited the philosophers of the Renaissance in 1486.

Pico imagines the Creator addressing His final creature. The Architect of the world, having completed His work, feels a desire for a being that can judge, love, and admire the reason, the beauty, and the magnitude of this product. Consequently, the final creature does not stand inside this work, as a part of it, but rather stands over against it—over against an already completed totality. Therefore, the original 'world-program' contains no prototype for this observer, no assigned standpoint. As a creature, he is just as superfluous for nature as he is necessary for nature's Author. Accordingly, we read in the Creator's speech of investiture to His creature that He has given Adam no established abode, no fixed form, no definite task, so that he himself can choose and adopt an abode, an appearance, and a task. Man is not so much placed in this idealized center of the world as he is exposed there. The position of Stoicism's observer of the world, at rest, has become the initial position of a self-shaping and world-transforming agent. All other creatures, we are told, have a definite nature, within firmly established laws, imposed upon them. Man is supposed to determine his nature himself on the basis of his freedom; which now means to see the world as serving his self-definition. Even before Copernicus, the shift of the accent from cosmology to anthropology, a process on which the Middle Ages as a whole had worked—under the terms set by their prior assumptions—in tiny steps, and with setbacks, is pushed so far forward that the loss of the real center of the universe, as a consequence of a relation to the world that could no longer be characterized cosmologically, could be accepted. The removal of the cosmic character of time is not only included in this process; it is an element of the process that reacts, in turn, on cosmology.

3

The Perfection of the Earth as a New Precondition for the Old Concept of Time

Galileo was the first person who was able to state that Copernicus had made the Earth not only a star among stars but *il primo mobile* [the first mover] itself,[51] that is to say, the equivalent of the heaven of the fixed stars, which had previously had the fastest motion in the cosmos. In order to protect his reform from questions deriving from natural philosophy, Copernicus could have availed himself of the licenses resulting from the medieval alteration of the concept of time and asserted its independence from the motions of the heavens. But just this—not to confront the strongest opposed position and demonstrate that he could deal with it—contradicted his principles of argumentation. So I would not even regard it as proved that, in regard to his own position on the concept of time, Copernicus was an Aristotelian and a realist. But it undoubtly seemed to him that a way of providing for the subject of time that was adequate to an Aristotelian/realist position would remove all other possible objections a fortiori. Just as he does not violate the Aristotelian concept of space as a container (although after the outermost sphere has come to a standstill that concept becomes, at least, superfluous), neither does he violate the concept of time that is based on the motion of the heavens. On the contrary, he tries to satisfy its maximum requirements. If, in doing so, the phenomenal rotation was supposed to be sufficient, instead of the real one, the guarantee of absolute regularity had to be localized in the real cause of the appearances. So what Copernicus needs is for the functions, in the universe, of the outermost heaven and the terrestrial body to be interchangeably equivalent. The new star has to be equipped to reoccupy the position of the *primum mobile*.

The founding of the concept of time on the phenomenal diurnal

motion of the heaven of the fixed stars clearly suits the new conception of anthropocentric teleology that is in the background of the
Copernican reform. For this principle of understanding the world it
does not matter on which end of the world's radius the real motion
takes place, as long as the apparent rotation offers the absolute
foundation that the Aristotelian deduction of the absolute regularity
of the first motion from the concept of time had produced. But only
the diurnal rotation could be deduced, in its absolute reliability, from
the perfectly spherical form that the Earth was to be shown to have.
In the case of the annual motion, no equivalence with the outermost
heaven could be set up. But this was also in accordance with the
requirements of man's relation to time. Thus it is not by accident that
Copernicus emphasizes the agreement between the cosmological
foundation of time and the circumstance that human beings measure
time chiefly by means of days. The most favorable public perceptibility and the dimensions of the life-world converge for the deepest of all
the reasons recognized in this view of the world.[52]

At the beginning of the second book of the *De revolutionibus*, where
Copernicus sums up the results of the first book, introduces the thesis
of the Earth's threefold real motion, and announces that he will
demonstrate how these motions explain the appearances in the
heavens, he also explains why he will begin this presentation with the
Earth's diurnal rotation. Once it has grasped the real origin of the
phenomena, human reason is in a privileged position in regard to the
self-presentation of its object, because the motion with the greatest
manifestness to the senses and the greatest publicness is at the same
time the motion that belongs to the terrestrial body primarily and
directly (*maxime ac sine medio*). With this formula, Copernicus declares
a difference between the Earth and all other heavenly bodies, since
the traditional theory of the transmission of cosmic motion admits no
instance of a solid body having an axial rotation. The Moon's synchro-
nous turning around the Earth was regarded as a consequence of its
being firmly bound to its geocentric sphere. So the Earth's diurnal
rotation—in contrast to its annual motion around the Sun, which
it shares with the other planets and in which it merely takes over
what had previously been the Sun's position—is a novelty in the
Copernican construction, which does not appear in the classical
cosmology and would be inexplicable in it. The formulation according
to which the Earth possesses the diurnal rotation not only principally

and by its nature but also "without the intervention" of other factors is based on this singularity.

The derivation of the concept of time from the diurnal motion is governed by the analogy of the relation between the unit and number. Copernicus sees the constitution of the greater units of the month and the year as a nominal operation, while the real unit by which time is determined is objectivized by the diurnal motion: "quoniam ab ipsa menses, anni et alia tempora multis nominibus exurgunt, tanquam ab unitate numerus" [inasumuch as from it the months, years, and other periods of time under many names are derived, as the numbers are from unity]. It is characteristic of the Platonism of this passage that of course the temporal magnitudes are mentioned that can be constructed from the magnitude of the day, as an ultimate unit, but the magnitudes that are produced by dividing it up—such as hours, minutes, and seconds—are not mentioned. More important is the fact that precisely in this connection Copernicus formulates explicitly his principle (which was developed in the first book) of cosmological equivalence, that is, that the astronomical change does not affect natural philosophy: However the question of the relative motions of the heavens and the Earth may, on other grounds, be decided, this has no consequences for the concept of time.[53]

The connection between the principle of equivalence and the concept of time makes clear, once again, why Copernicus begins his work with a discussion of the question of the form of the universe as a whole and of the Earth in particular, and that it had to be important for him for the spherical form to be equally perfect in both cases.

Copernicus deduces the perfect sphere's mobility from the fact there is no difference, for it, between all particular positions. The peculiarity of this train of thought becomes comprehensible when one remembers that Aristotle had, in reverse and as though incidentally, made the Earth's spherical shape a consequence of the tendency of every part of the element earth toward its "natural place," the center of the universe. Then the Earth's situation at rest in the center of the universe followed precisely from the fact that all of its parts are equally unable to reach their "natural place,"[54] so that the terrestrial body's spherical form and its state of rest were results of stable nondifferentiation. In a world with no beginning, the actual standstill was a logical result of a contradictory process. The spherical form is a compromise between the antinomic 'ideality' of this natural place, which would have had to combine the entire mass of the element, earth, in one point, and the need to hold to the ordering principle of natural places in order to explain the phenomena of falling. Of course the Parisian Ockhamists of the fourteenth century drew from the same premise of the nondifferentiation of all parts of the Earth's mass with respect to the center of the universe the opposite conclusion, namely, the instability and continual unrest of the terrestrial body, such as becomes noticeable in earthquakes. With Copernicus, the Earth's spherical form cannot be the result of imagined physical processes. Instead, it is already a precondition of essential characteristics of this heavenly body, especially of its natural mobility.

Deriving characteristics of elementary bodies from their geometric forms was originally the atomists' idea. They had assigned the spherical form to the atoms of fire, as the most easily mobile element, and as the next step they had deduced from this the atomistic fine structure of the souls, as the principle of organic motion. Copernicus's way of speaking of the Earth's *mobilitas* [mobility] keeps dextrously remote from an unequivocal assertion of (already) an actual state of motion, instead of the greatest mobility. The obstacle, likewise deriving from atomism, consisted in the fact that according to its geometric fine structure, the element earth should be disposed toward extreme sluggishness and immobility. Thus in the perfectly spherical form of the Earth two opposite characterizations collided, one coming from its matter and the other from its form. So its possible motion is not yet actual motion. Nor can it be that, because otherwise the parallel inference from the *figura mundi* [shape of the universe] to the predicate of real motion would have to be just as compelling as in the case of the *figura terrae* [shape of the Earth]. Then Copernicus would have been in the same dilemma that Plato probably already had to resolve when he in fact made both of them—the heavens and the Earth—move, so that the phenomenal diurnal rotation could only be understood as the difference between the two rotations.

All that the *forma caeli* [shape of the heavens] guarantees, for Scholasticism, is the sphere's natural nonresistance to its motion—it is the sphere's *naturalis aptitudo, ut sic moveatur* [natural aptitude to be moved in this way]. If the form had been accepted as itself the principle of motion, the world would have been released from the contingency of divine preservation. If the heavens' motion were the realization of their form, then according to Richard of Mediavilla, if God wanted to bring the heavens to a standstill (as revelation, in the story of Joshua, assures us that He did want), He would have to destroy them, because they cannot assume a state that contradicts their nature.[55]

The occasion for Copernicus's handling of the question of the shape of the Earth is, on the surface, only the way in which he models his work strictly on the arrangement of the *Almagest*. There Ptolemy had treated the problems of the forms of the vault of the heavens and the Earth in the third and fourth chapters of the first book. Given this strict correspondence, the differences are all the more instructive.

Ptolemy relies on the methodical rule of taking the phenomena as his point of departure and doing justice only to them. Therefore he does not emphasize the inferred form of the entire sphere of the fixed stars, but only the circularity of the orbits of the stars as they are observed. The globe shape of the sphere is inferred from the circularity and parallelism of the apparent motions of the fixed stars. Copernicus's procedure in regard to this question is more speculative and deductive. The world, he says, is globe shaped because this represents the simply perfect form (*forma perfectissima*) and a totality that needs neither to be added to nor to be subtracted from (*tota integritas*). Thanks to it, the whole of the world has at the same time the greatest capaciousness and the maximum stability. That it must be the ideal form for physical bodies can be seen merely from the fact that all objects in the universe either already present themselves in this form—like the Sun and the Moon—or strive to fill themselves out to match it. An example is a drop of water: "Hac universa appetunt terminari ... dum per se terminari cupiunt." [Everything tends to take on this shape ... when it seeks its natural shape.]

Copernicus, quite logically, lags behind Ptolemy in a decisive point. Toward the end of the chapter Ptolemy had said, about the rotation of the heavens, that it proceeds with the greatest ease, because the circle and the globe guarantee the easiest motion. Here, at the latest, Copernicus must have had to run into the doubtfulness of the argument on behalf of the Earth's motion—which for him, after all, inevitably was reflected in the heavens' standing still. Only dissolving the outermost sphere, as a bodily unit, would have allowed him to escape the argument that the connection between perfectly spherical form and motion had to hold for the heavens just as much as for the Earth. If he had really been a Platonist, perhaps he would have looked for the solution in a motion of both of the bodies, the Earth and the heavens. This solution was even mentioned by Ptolemy, as a rejected alternative, alongside the thesis of the immobility of the vault of the heavens.[56] With this argument, then, one could not go beyond the mere equivalence [of the two spherical bodies].

Between Ptolemy and Copernicus, the interest behind the proofs has shifted. While the ancient astronomer still produces proofs for the Earth's globe shape that are based on perceptual data, Copernicus starts immediately from the globe shape as something commonly taken for granted, in order to concentrate entirely on refuting criticisms to the effect that it is imperfectly realized.[57] Significantly, Ptolemy does not mention the most important piece of evidence, although it was already known to him and had to suggest itself: The shadow of the Earth, which becomes visible in eclipses of the Moon, has a perfectly circular border.[58] It would have been contrary to his system to introduce an astronomical argument for an object that is supposed to belong exclusively to physics. For Copernicus, there could no longer be any considerations of system in regard to this point, if the Earth itself was supposed to have become an astronomical object.

Of course when Copernicus, at the beginning of the *De revolutionibus*, qualifies the Earth—merely by the way he thematizes its form—as a heavenly body, he has not yet reached the goal of his argumentation—the goal of suspending the priority of the sphere of the fixed stars as the basis of time. The principle of equivalence did not allow him to notice the possibility, as a next step, of at least considering that the outermost sphere might not be real. The phenomenalization of the diurnal motion would have followed automatically from that. But only equivalence accomplishes the cosmological minimum that allows one, in other respects, to pursue the argument on a purely 'astronomical' basis and thus to approach the construction of the system of the planets as a whole. The principle of equivalence and the conservative inclination to make only a minimal departure from what Ptolemy had established saved Copernicus from getting stuck in the geocentric explanation of the diurnal motion. As long as the thesis of the Earth's spherical form and the continuance of its natural motion remained tied to its occupying the position in the center of the universe, the assertion of its (diurnal) rotation had to exclude that of its (annual) revolution. In this context, too, everything points to the course of the argumentation as it is presented not reproducing the route by which the theory was originally produced.

It is precisely the fact that the perfection of the spherical form is supposed to satisfy the condition of uniformity only for the Earth's diurnal motion and not for its annual revolution around the Sun that makes it evident how this argument is connected to the problematic of the constitution of time. When Copernicus presented his first outline of his system, he introduced his critique of the astronomical tradition by affirming as an axiom for any system that a heavenly body that is perfectly round in shape can only have a uniform motion.[59] As it stood, this proposition did not need to disturb an Aristotelian either, because among his 'heavenly bodies' there were no solid rotating globes, but only the hollow spheres nested inside one another. Aristotle had explicitly refused to grant absolute homogeneity to the rotation of a solid globe, because, on account of their different speeds, the motions of its parts cannot constitute a single process of motion.[60] Since even the different, special matter of the heavenly bodies could not alter this circumstance, on strict Aristotelian premises even the elevation of the Earth to a star was not sufficient to produce the homogeneity in its diurnal rotation that could meet the requirements of the concept of time. Here Copernicus had to count on a certain forgetfulness on the part of his strict readers. He was able to do so because that thesis from Aristotle's *Physics* had never had to be applied to a solid heavenly body. Or does the perfect globe point to a Platonic rather than an Aristotelian background?

Now, for a strictly methodical history of science the derivation of an intellectual element may be of no interest, or only minor interest, if that element did not serve to substantiate a theory, but only to shield it from criticism. In regard to this distinction, however, one should bear in mind that that which is able to carry out the task of shielding from criticism at the same time contains, in the horizon of its derivation, an answer to the question of the effectiveness of what it (in that way) shields. For only if the shielding idea does not appear too far-fetched and exotic to people at the time can it accomplish what is always a precondition of the addressees' ability to assent. Of course this does not yet constitute a theory's historical impact, but it does respond to the immediately previous question: the equipping of a field of operation with potential plausibilities. It is not possible to infer much from this regarding the influences affecting the author himself and his theoretical development.

Regarding the specificity of an influence, then—or, beyond that, of a fundamental determining doctrine—the fact that Copernicus may have found the doctrine of the mobility of a precisely spherical figure in Nicholas of Cusa is not, in itself, very significant. It may be all the more important, then, that at least under the heading of "time," the rebuke that the work lacked consistency with natural philosophy could not throw it back into the obscurity where it would have no impact. D. Mahnke, L. A. Birkenmaier, and R. Klibansky have pointed to actual, probable, and possible contacts between Copernicus and the Cusan's thought. A possible transmitter is Carolus Bovillus [Charles Bouillé], a copy of whose book, which was in Copernicus's library and is furnished with original marginal comments, is preserved in Uppsala.

The characteristics of the perfect globe are the subject of the Cusan's dialogue *On the Globe Game*. This game, which the Cusan's speculation exploits, is played by throwing a wooden ball, which, on account of a concave asymmetry, can traverse a curved path, and is supposed to come to rest in the midst of a field of nine concentric circles.[61] The ball's concave 'defect' leads the Cusan to the idea of a faultless globe form. It looks like good Platonism when he asserts, here, that a geometrically precise globe cannot take on any material appearance at all. But the reason given for this does not follow the line of Plato's consideration of the relation of the Idea to the appearance, but instead pursues the reverse path of suspending the appearance by intensifying its precision. Thus the roundness of a real body can always be imagined and made more precise than it is actually given. But the limiting case of this intensification would be a body that is no longer perceptible. The suspension of the visibility of the perfect globe can be understood as meaning that in the strict sense, then, a straight line or a surface could only touch it at one point, in which case its *extremitas rotundi* [extreme of roundness] would become identical with the point on the surface of the globe; but a curved surface cannot be composed of points, which cannot, in their turn, be curved.[62]

Of course the late medieval thinker wants to demonstrate that the world cannot in and by itself attain what, according to the prescription of its essence, it is meant to become, but can only do so by suspending and going beyond itself. The formula for this demonstration is to intensify a characteristic to its absolute purity, whereby this intensification explodes the object on which the imagination carries it out. This procedure can also be applied to the world as a whole, insofar as it is enclosed in the outermost sphere as a round shell. The perfection of this sphere would have to deprive the world of any perceptibility from outside. That could not be unpleasant even to the Aristotelian, because it would only be another argument for the impossibility of an outside. This train of thought could become dangerous for the Copernican to come, because it would have likewise to deny the Earth any perceptibility, on account of its strict equivalence to the outermost heaven.

The Cusan explicitly characterizes the special position of the perfect globe as an exception to the Platonic duality of Idea and appearance. This duality exists only for the individual objects that constitute the world, whereas the totality that they make up, for its part, can no longer be comprehended as an appearance. This brings the cardinal, in the dialogue, to express himself as follows to his partner, the Duke of Bavaria: "Although from the Platonic point of view you speak the truth, there is nevertheless a difference between roundness and other forms. Even if it were possible to realize roundness in matter, it would nevertheless be invisible."[a] The material reality of the perfect globe would have a characteristic that cannot appear within the physical world, and that nevertheless belongs to that world itself. The predicates of the perfect globe allow us to except the body of the world from all the restrictions of the traditional physics and cosmology. They culminate in the assertion that the perfect globe would never lose a motion that had once been given to it. That is said, here, under the full protection of the impossibility (which was secured previously) of such a globe within the world. But the train of thought can only too easily be turned around: If the premise of the impossibility of a perfect globe within the world should not prove to be correct, in any case everything else that was said about it would remain. And especially what was said about its motion. What is in motion never comes to rest unless it can behave differently at different points in time. But the perfect globe cannot do that, since the defined conditions for every moment of its motion are completely identical.[63]

The requirements for inexhaustibility relate exclusively to the body's form, to which the inexhaustibility is supposed to belong, and not to its matter. That had to be of interest to Copernicus, since he could no longer give to a stellarized Earth the special matter that Aristotle gave to the celestial world. The Cusan, too, requires the inexhaustible motion to belong to the body "by nature"; if this has to be required explicitly in the case of the perfect globe, it can only mean that it must have been set in motion immediately in the act of its creation.[64] The Scholastic principle of mediatedness remains intact; all motions, insofar as they are "natural," come about through participation in the motion of the outermost sphere: "quem motum omnia naturalem motum habentia participant" [in which motion everything that has natural motion participates].

What Copernicus will do—namely, to transfer the predicates of the outermost sphere to the Earth—had to remain impossible for the Cusan. In relation to the heavens, all intraworldly realities still have the Platonic deficiency due to the appearance's mere participation in the Idea. This relation includes a gradient of degrees of definiteness, just as much as in the case of the Aristotelian causal chain. The Earth cannot exhibit any "precision"; it cannot possess perfect roundness. It is imprecise in regard to the definiteness both of its form and also of its centering on the center of the world, and of its state of rest. The Cusan is not a forerunner of Copernicus. He invents the motion of the Earth as an expression of that imprecision. For that very reason, that motion is as far as possible from giving the Earth astronomical dignity in the strict sense by which it could explain astronomical phenomena and provide the concept of time with something equivalent to absolutely uniform motion. The Cusan's beautiful speculation that the Earth also is a *stella nobilis* [noble star] does not alter this at all.[b]

If Copernicus should have found the idea of the perfect globe in the Cusan, nevertheless this discovery could only have helped him with a formulation, which for him was a means to satisfy a totally heterogeneous constructive function. He needed an Earth that did not fall short of the heavens in any way, because it had to be the organ of time.

The decisive connection to the tradition represented by the Cusan consists in the exclusive thematization of form. It must be made plausible to the inhabitants of the surface of the Earth, to whom it presents itself as anything but an *absolutus orbis* [perfectly round thing], that seen on the scale of the cosmos, such deviations have no importance: "quae tamen universam terrae rotunditatem minime varient" [that nevertheless there is little variation in the Earth's general roundness].[65] Under the burden of proof that is incumbent on him here, Copernicus even breaks his parallel to the structure of the *Almagest*; he inserts a chapter to which there is no corresponding one there, in order to safeguard his assertion about the form of the terrestrial body against the Aristotelian doctrine of the elements as well. This third chapter of the first book of the *De revolutionibus* has to be read above all from the viewpoint of the uniformity of the Earth's motion, if one does not want to suppose that the proof of the Earth's globe shape, which was already furnished in the second chapter, was, in the author's eyes, still insufficient. In spite of the elementary heterogeneity of the parts played by earth and water in the construction of the telluric heavenly body, its homogeneity has to be 'saved.'

The difficulty consists in the fact that while an 'ordered' state of these two elements—i.e., their uniform distribution around the center of gravity—would indeed guarantee the outcome of a spherical shape, in return for that a concentric layer of water would be laid over the Earth. The perfect result of balancing the elements would at the same time be a sterile result. In one of his visions of destruction—the satire on the wedding of the virgins—Leonardo da Vinci had celebrated the Earth's return, through erosion by water and wind, from its deformity (which was produced by volcanic action) to its ideal original form—to the smooth roundness of a globe that is entirely covered, again, by water. He celebrated it as carrying out nature's will to render impossible its disturbance by man, whose species would be washed away, with all its traces, as though it had never existed. For Copernicus the perfect globe form cannot be a fact in, and at the end of, the Earth's history, attainable only at the cost of the destruction of mankind—for whose benefit the world is after all supposed to be arranged.

The fact that this new planet is composed of the elements earth and water, and that in the nearby region of our experience it is so uneven, needs to be 'smoothed over,' theoretically. For Copernicus, water serves the purpose of preserving the form of the Earth as a whole by compensating for the irregular distribution of the masses of earth. While it is doing so, the difference in the shares of the two elements must not move the center of gravity (*centrum gravitatis*) out of the center of volume (*centrum magnitudinis*). This makes it comprehensible that a polemic is conducted, in a special chapter, against certain Peripatetics who had asserted that the share of water in the mass of the terrestrial globe was ten times as great as that of earth. On the one hand, Copernicus cannot admit that the [literally] 'philanthropic' elevation of the land masses above the seas threatens his system, by deforming the globe shape and thus the uniformity of its rotation; and on the other hand, the depth of the ocean basins cannot be supposed to be so great that the shapelessness of the terrestrial body would lead to an eccentric position of the center of gravity. The thesis of the disproportion of water and earth, which was stated by Aristotelians, violates the other Aristotelian regularity of the concentric surrounding of the center by the total mass of earth. Of course Copernicus can no longer speak of the center of the universe as the center around which the terrestrial body is organized, and he has quietly replaced it with the center of gravity. If the mass of water had ten times the volume of the mass of earth, the latter would no longer be large enough to constitute both the center of the Earth and the continents, projecting above the ocean. Water, then, cannot be the predominant element of this heavenly body; its teleological role is associated with Earth's qualification as a star. In spite of the still unknown extent of America, the discovery of the new continent becomes important as a confirmation of the heavenly body, Earth's homogeneous construction, on which the cosmic motion's disposition for uniformity, and thus for the constitution of time, depends.

The outcome of this excursus would remain entirely unintelligible to the reader of the *De revolutionibus* if it had not removed, for the subsequent chapters—which have the task of securing the equivalence of the heavenly bodies and the Earth, in regard to their suitability for motion—the problematization (deriving from the doctrine of the elements) of form as the exclusive criterion for this suitability. It has now finally become clear—so Copernicus concludes his excursus—that earth and water have a common center of gravity, and that there is no separate center of the Earth's volume, although the element earth is heavier. The appearance of a preponderant share belonging to water arises through the fact that as the space between the parts of the earth that split apart is filled with water, a greater part of the Earth's surface is composed of water. It is precisely in this context that Copernicus inserts the proof that the cosmic shadow of the terrestrial body confirms the cooperative effect of its two elements in producing the unity of the perfect form: *rotunditate absoluta, ut philosophi sentiunt* [perfectly round, as the philosophers maintain].[66]

Ptolemy had dealt with the Earth only in order to exclude it from astronomical theory. He was able to explain, by its shape and size, that the starry heavens are divided by the horizon into two hemispheres. With that, the observer's standpoint did not need to be added in, astronomically. Copernicus, on the other hand, has to make the Earth a topic for his theory, as the chief factor in the apparent motions in the heavens. The protection provided by the equivalence of the spheres allows him to distinguish between the apparent circular motion (*motus circularis*) and the globe's disposition for motion (*mobilitas sphaerae*). If he did not distinguish between mobility and actual motion, he could not keep the argumentational advantage that he has gained for the Earth. He would find himself still in the situation of Plato, who had to make both bodies rotate around the same axis in such a way that only the difference between the two motions produced the phenomenal day. At this point, then, the process of speculation has to terminate and the specifically astronomical argumentation has to begin, and provide a decision. Thus the formula comes about with which Copernicus introduces the fourth chapter on the revolutions of the heavenly bodies: The mobility of a globe consists in its capacity to be moved in a circle and thus to realize its essential form as that of the simplest body; and here the motion that continually returns to itself does not allow a distinction between beginning and end or between different parts.[67]

The goal of Copernicus's argumentation has to be to show the 'naturalness' of the Earth's motion in space. This is a necessary condition of its uniformity, although it is not, by itself, a sufficient condition, because natural motions could also be accelerated. The advantage to be derived from this predicate follows at once from the arguments that Ptolemy had urged against the motion of the Earth, arguments that Copernicus, in exact parallel, discusses in the seventh chapter of his first book. Ptolemy had considered both a motion of the Earth on a trajectory through space and also a stationary rotation. In both cases he arrives at "ridiculous" consequences. If the Earth had a motion like the fall of heavy bodies, "it is obvious that it would be carried down faster than all of them because of its much greater size: Living things and individual heavy objects would be left behind, riding on the air, and the Earth itself would very soon have fallen completely out of the heavens." But in the case of a stationary rotation, merely to explain the diurnal motion, the result would also be, according to Ptolemy, "that the revolving motion of the Earth must be the most violent of all motions associated with it, seeing that it makes one revolution in such a short time; the result would be that all objects not actually standing on the Earth would appear to have the same motion, opposite to that of the Earth."[68]

Copernicus's defense against this argumentation is extremely laconic. Of course one should not proceed on the assumption that a cosmic motion of the Earth would be "violent." Once one admitted that the motion under consideration belonged to the Earth as a "natural" motion, all those unpleasant consequences would disappear. Remember how much easier it is for Copernicus to speak of a "natural" motion because all he needs for that purpose, besides the form of the object, is an appeal to the act of creation. What occurs, in that way, "by nature," behaves suitably and cannot fail to preserve itself in its perfect arrangement: "Quae vero a natura fiunt recte se habent et conservantur in optima sua compositione."[69] It is easy to imagine how disappointed, exasperated, or charmed the contemporary reader who was familiar with the possible arguments—whom we always have to keep in view as the addressee of every invitation to entertain a theory—may have felt when confronted with the economical expediency of this argumentation.

Copernicus has been criticized for having adhered unreflectingly to the distinction between natural motion and violent motion, just as he did to the obligation to base his constructions on circular and uniform motions. Such historical admonitions, to the effect that the protagonists of science would have done better to be more revolutionary, originate in a disregard of the trivial circumstance that every act in the realm of theory has an audience and, even before it can count on approbation, has to get above the threshold of an initial plausibility of argumentation. The conservative disposition that adhered to the requirements of the circle and the sphere, as the edietic evidence of rationality, and to the differentiation between nature and violence, enabled Copernicus to declare that the Earth satisfied the conditions for being one of the heavenly bodies, and thus enabled him to make the difficulties relating to the phenomena of inertia—difficulties that he could not parry—appear negligible. After all, the reformer's burden of proof is not only, and not originally, that the changes that he proposes can be carried out, but also that his criteria for the need to change the existing system can be satisfied. In relation to the addressee of his work, the intolerability of the traditional theory can only be demonstrated by showing that it does not fulfill its own requirements and claims. The realism that interprets the "Platonic" requirement^c as a regularity following from the nature of the heavenly bodies enables Copernicus both to demonstrate the defect of the old system and to establish the reliability of the new one: "... fieri nequit ut caeleste corpus simplex uno orbe inaqualiter moveatur" [a simple celestial body cannot move irregularly in a single sphere].[70]

Inevitably, the connection that both Aristotelianism and Ptolemy had set up between the requirement of circularity and the special matter of the heavenly bodies is dissolved. One can show, in this passage, that for Copernicus the metaphysical principle of the anthropocentric teleology makes its entrance precisely in the position of that "fifth essence." What is the basis of the anticipatory confidence that one can now work with the "Platonic" requirement in connection with the Earth as well? Copernicus considers that two factors could bring uncertainty into a system of perfect circles and globes: a variability in the moving power or an unevenness in the body that is moved (*revoluti corporis disparitas*). It is plainly not enough that reason shrinks from both possibilities, because they would deprive cognition of its basis; instead, this is the point at which an appeal to the world-formula in the dedicatory preface becomes necessary, that is, to say that it would be shameful to assume the existence of variability and unevenness in a world that has to be credited with being arranged in accordance with the best order.[71]

It becomes evident that this principle affects the Earth in two ways: It secures for it the perfection of uniform diurnal rotation, freed from empirical objections, but at the same time it burdens it with the perspectival distortions involved in man's view of the world, the disentanglement of which is supposed to allow us to restore simplicity to all the other motions of the system. Consequently, the summons to investigate the Earth's relation to the heavens carefully, which concludes the fourth chapter of the first book of the *De revolutionibus*, cannot refer primarily to the question of the diurnal motion. The apparent unevenness of celestial motions must become a perspectival phenomenon, and that is only possible if the Earth not only rotates in place but also travels on an orbit and produces a succession of eccentric points of view. The warning not to ascribe to the heavens what belongs to the Earth cannot restrict itself to rotation as the equivalent of the diurnal turning. By natural sequence the next chapter, the fifth, begins with the summary statement that it has already been shown that the Earth, too, has the form of a globe, and it is now necessary to investigate whether its motion corresponds to its shape and what position in the universe it occupies. Only in this way can there be a reliable account of the appearances in the heavens.[72]

Once again the connection between form and motion is brought to mind, and thus the turning is accomplished—in Copernicus's tactics of proof—that imposes the burden of proof on the person who wants to deny the real predicate of motion to the Earth from the outset. The observer's standpoint cannot be neutralized any longer. It is itself the primary topic of astronomical theory, because it represents the most important condition affecting what appears to the observer: "Terra autem est, unde caelestis ille circuitus aspicitur et visui reproducitur nostro." [Now it is Earth from which the rotation of the heavens is observed and brought into our view.]

4

The Deformation of the Earth
and Absolute Time

Even in the history of science, ironies are not unknown. The same perfect globe form of the Earth by means of which Copernicus had protected his reform from objections from natural philosophy had to be surrendered, scarcely two centuries later, so that a physical proof of the first motion of the Earth, the diurnal rotation, could be produced. It was carried out on this Earth, by demonstrating its deformation. This was in keeping with the Copernican precept of making the Earth itself the primary object of theory. While it was only in the nineteenth century that the demonstration of parallax in the case of the fixed stars led to the expected astronomical confirmation of the annual motion, Maupertuis was able in 1736, in one of the first research undertakings organized on a worldwide scale, to verify the oblateness of the Earth at the poles, and thus to demonstrate almost visibly its motion in relation to absolute space. It is not by chance that by means of this proof of Copernicanism he provided Newton's physics with its breakthrough on the Continent.

Copernicanism's undisputed success was ultimately based on the fact that one of its essential prerequisites had been destroyed. The Earth had fulfilled the expectations not of Copernicus but of Aristotle. To that extent Newton, although he wanted to fulfill definitively Copernicus's claim to truth, stands, in terms of the outcome, closer to Aristotelianism, inasmuch as, for dynamic reasons—by virtue of the very motion that Copernicus wanted to vindicate on the basis of its globe form—the Earth could no longer be a perfect globe. Neither, therefore, would it fulfill the function that Copernicus had assigned to it, of satisfying Aristotle's conditions for the reality of time. I have shown that Aristotle, in a sort of ontological argument, deduces the characteristics of the motion of the heavens, and the guarantee of that motion, from the concept of time. In regard to this procedure, too, Newton's "absolute time," with the unconditional derivation of its reality from the conceptual system, stands closer to Aristotle than to Copernicus's attempted solution.

The first and only Copernican to recognize and articulate the fact that the Aristotelian condition for the foundation of time could not be fulfilled by the motion of the Earth, and that consequently time would ultimately have to be 'decosmicized,' as a merely 'logical' relation, was the postulator of an "astronomy without hypotheses," Petrus Ramus. Ramus says that while the heavens are undoubtedly our clock, they are not a condition of the reality of time. Otherwise Copernicus would not have been able to take motion from the heavens and confer it on the Earth, and to measure time more exactly, by this motion, than any astronomer before him.[73] In Newton, the process of theory will be more roundabout, being geared to the central Copernican need to present motion as a 'natural' state: both absolute time and absolute space as systematic presuppositions of the definition of inertia.

Newton's program for his theory, reduced to its simplest formula, is still Copernicus's: to press forward through the apparent motions to the true ones. As in Copernicus's case, for Newton also this intention cannot be carried out without affecting the concept of time.

In spite of his key role in the mathematizing of natural science, Newton perfects the epistemological realism that Copernicus had given to astronomy; his motto, *Merus mathematicus, merus idiota* [A pure mathematician is a pure ignoramus], holds also, and especially, for Newton. His preface to the *Principia*, of 1686, makes it clear what his claim to truth aims at: The question of the difference between apparent and real motions ultimately leads one to the "forces of nature." But their reality is exempted from the relativity of appearances only because one can define a state of physical bodies in which they are free from the operation of forces, a state that is in part the successor to "naturalness" in the Scholastic conceptual system. Here Newton enters the neighborhood of the question that Copernicus had asked himself in regard to the Earth's motion and that he thought he had answered with his demonstration of its 'form.'

If that is the case, the result for the concept of time is that it would have to depend on the possibility of a motion that is uninfluenced by forces, as the possibility of an absolute uniformity. For Newton, this cannot consist in the rotation of a perfect globe, because circular motion is no longer the ideal of "natural" motion. It has become, in the Aristotelian sense, "violent." If circular motion is accelerated motion, the perfection of a globe would prove not that it has real motion, but the opposite. Since Newton, the demonstration of the truth of Copernicanism required precisely that the Earth should deviate from spherical perfection.

Newton's concept of force requires the same application of the principle of sufficient reason as Copernicus's moving globe does. The globe's absolute homogeneity had to mean that there was in it no reason for a state that it had once taken on to change—that is, that it conserved both rest and motion equally. It was only in its form that a preference was given to motion. For Newton, body that is not influenced by forces has no reason to change its state, that is, for example, to grow tired in its motion or not to keep to a straight line. Aristotelian physics had not permitted either of these interpretations of being in a state; it required every motion to have a moving force, as its "accompanying causality."

The state of uniform motion could not be defined without the concept of time. What else could uniformity mean except that the body traverses equal distances in equal times? But how could units of time be determined to be equal? Newton could have said that the gauge of time would be a body whose motion is not acted on by any force. But this, in particular, could not be unambiguously established, empirically, since "being acted on" now had to be admitted not only in the form of contact or inherence, but also as *actio per distans* [action at a distance] (no matter how that was to be explained). Disregarding the question of how one was supposed to be able to make sure that a motion was not acted on by forces, the prescription of such a gauge of time was in any case rendered uninteresting by the presumption that in empirically accessible physical reality there is no state whatever that is not acted on by forces. So the Aristotelian procedure of inferring from the concept of time the existence of an absolutely uniform motion could no longer be applied. However, the argument for a time that was thought of as absolute, and only inferred, without any empirical foundation that could be displayed, was not therefore any less "ontological." Newton could not fix what *time* is if he wanted, by preference, to define what *force* is. For time is already implicated in the concept of uniformity, which he needed in order to introduce change as the only determinable feature of force. In change, alone, lay the prospect of an empirically quantifiable magnitude. So the interest in the definition of concepts led in a different direction from the interest in the measurement of magnitudes.

Absolute space and absolute time are just as much, and just as little, "Platonic" as they may be "Epicurean." In no variety of Platonism had space and time themselves ever been Ideas. They are 'Ideas' now only in the sense of realities that are withdrawn from, but necessarily presupposed in appearance, realities without which the basic *concepts* of physics cannot be realized, although they are not what makes its basic *accomplishments* possible. The insight that had led Newton to the absolutism of time is expressed in one sentence in his *Principia*: "It may be, that there is no such thing as an equable motion, whereby time may be accurately measured. All motions may be accelerated or retarded"—that is, it may always and everywhere be the case that bodies have forces operating on them— "but the flowing of absolute time is not liable to any change." [74]

The use of the metaphor of the "stream" of time shows that the concept is not made to comply with the requirement of determinability that is asserted against it. It is, on its side, an ultimate defining—and thus indefineable—element of the theoretical system in which the concept of force appears. One could show, starting with the topic of gravitation, that the preference given to the reality of "force" is due to its leaving open to Newton an underlying equivalence, which cannot be decided either way by theory, between assertions about physical "force" and assertions about divine "power." That is why the ontological dignity of the assumption of an absolute time can be reduced to that of a condition of the possibility of the concept of force.

The deformation of the Earth, which destroys the Copernican basis for the assertion of the "naturalness" of its motion, forces us to separate the concept of time from its basis in standardized cosmic motions; both "absolute time" and the transcendental idealization[a] of the concept of time are consequences of this. Once Kepler had concluded that the planetary orbits were not circular and the motions on them were not uniform, the next step was already inevitable: that the bodies in the cosmos, too, could no longer display the supposed rationality of the pure globe shape. After Newton, rotations and orbital revolutions, as the basic phenomena of the starry heavens, are no longer the basic states and normal processes of an 'embodied' rationality, but instead are the outcomes of states and forces that are superimposed on each other in them. Just as there is no pure inertial motion, as an empirical phenomenon, it is equally improbable that the figures of the circle or globe, which are mere limiting cases of complex operations, should come about. It is the deformations that allow us to infer the true reality of forces and that make the Copernican claim to truth realizable.

That there was a Platonic influence behind the absolutism of space and time in Newton's conceptual construction is very dubious. But if one wants to see the Neoplatonism in the School of Cambridge as already significant for the new physics, its possible role in the abandonment of homogeneity in the dynamics of circular motion should not be disregarded. This is something that one expected least of all under the heading of "Platonism." But it is due to the *Enneads* of Plotinus, which had just become influential in England, and in whose heterogeneous late presentation of ancient philosophy the doctrine of circular motion is most likely traceable to the *Mechanical Problems* (a work that was handed down under the name of Aristotle). On the Platonic and Aristotelian premise that the circular motion of the heavens realizes reason, the theory of the composition of circular motion had, for Plotinus, the inestimable advantage of representing, metaphorically, a feature of the mind that had not previously been thematized, namely, self-consciouness.

Plotinus had answered the question as to the cause of the circular motion of the heavens with the brief formula: "Because it imitates the mind." [75] The mimesis of the mind by the sphere consists in the fact that the latter, by moving around itself, remains in its place. The mind "is both at rest and in motion; for it moves around itself. So, then, the universe, too, both moves in its circle and is at rest." In the motion of the heavens the world-soul displays the mind's intimacy with itself, its "movement to itself, a movement of self-perception and self-consciousness and life, which never goes outward or to something else, because it has to encompass everything." The material representation of self-consciousness is identical with that of the characteristics of an organism, which can be defined as a unity that remains while all its parts are changing. Spirit and matter are, in Plotinus's system, incompatible, like unconditional identity and pure multiplicity; the soul forces the spheres to move in a circle as a compromise: "Body is naturally transported in a straight line, but the soul holds it back, and the combination of the two produces something that is both motion and rest." One could say that in this Neoplatonic conception the cosmology of ancient atomism, which was based on the straight line, was synthesized with the metaphysics of the universe that was founded, beginning with Xenophanes and Parmenides, on the circle and the globe.

Precisely this same interlocking combination is repeated in Newton's theoretical construction, when he makes an initial situation of objects falling in a straight line, such as atomism postulated, into a precondition of their acquiring circular motions around a central mass. The arrangement of the planets' orbits around the Sun, in which they all travel in the same direction and deviate only slightly from the plane of the ecliptic, makes it possible for him to introduce his God as the One Who determines the initial situation of the bodies with respect to one another, and thus the ratio of height of fall and gravitation that accords with the goal state: "I do not know of any power in nature which could cause this transverse motion without ye divine arm." [76] Newton's assertion to Bentley that this was an idea of Plato is instructive in regard to the continuity of tradition that he would like to establish. The motions of the planets proceed as though they had been created by God in a region far removed from our system and then allowed to fall, from there, toward the Sun, in which process the diversion into a circular motion, for Plato, could only have been produced by the bodies of the spheres. One sees immediately the advantage that the abandonment of the rationality of circular motions has for Newton: It allows him to introduce an element of theological facticity [i.e., brute 'givenness'] into the system, since it is only by an act of will that the planets' falling could have been directed sideward, and the Sun's power of attraction altered, in such a way as to give rise to a stable planetary system. [77]

When Newton refers, in support of this primal cosmological situation, to Plato, he does so at second hand. [78] It is remarkable that he does not directly cite the authority (with which he was familiar) on which his authority relies, which, since the latter is a treatise on ballistics, is certainly Galileo's *Discorsi*. In the Fourth Day of that work, after presenting the third theorem, about composite parabolic trajectories, Sagredo expounds an idea that is supposedly taken from

Plato. The demiurge had to accelerate the heavenly bodies that he produced up to their final velocity for their circular paths in the cosmos; he did this on rectilinear paths, until the intended degree of velocity was arrived at, and then converted the path into a circular one ("converti il moto loro retto in circolare").[79]

Galileo had already made one of his figures expound a similar idea long before, at the beginning of the *Dialogue on the Two Chief World Systems*. There the issue had been the incompatibility of rectilinear motion and a perfect world-order: Rectilinear motion was by its nature aimless and infinite, and no body can strive by its nature to reach a place at which it is impossible to arrive, and where there cannot be a goal for it.[80] If one wanted to engage in a cosmogonic fantasy, one could say that nature could have used rectilinear motion, too, to overcome chaos and to collect the materials from which to construct the bodies in the universe, while allowing only perfect rest and pure circular motion after the work was complete. One could go even a step further, if one wanted to appeal to Plato. Even after their completion, the bodies in the universe could still move in a straight line for a certain length of time, only to wheel around into circular motion, one after the other, as they reach the places intended for them, and to remain there forever. Although Galileo's enthusiasm for this conception only becomes comprehensible in the context of the doctrine of complex parabolic trajectories, in this passage he already calls the idea sublime and entirely worthy of Plato.[81]

About Galileo's supposed Platonisms, what needs to be said has already been said. If no appropriate passage can be found to substantiate Galileo's and Newton's reference to Plato's theory of an initial condition of the planets' paths, nevertheless one should not get over this difficulty by suggesting that they invented an authority. For what is ascribed to Plato here is something that, given the interpretation of the consistency of his system at the time, seemed completely indispensable. The key to this is in the requirement that only uniform motions should be allowed on the planets' orbits—a requirement whose origin in Plato was assumed as a matter of course. But in that case Plato's demiurge would have violated this primary rule of the system by putting the bodies he had made into motion on their final orbits, and thus moving them in a nonuniform way. The places where the planets were formed and received their initial acceleration could not be in their final orbits. They could wheel into those orbits only after arriving at their final velocities. If this assessment is correct, what we are dealing with is a retrospective application of the "Platonic" requirement to the *Timaeus*.

In the form in which we find it in Galileo, this touching up of Plato's cosmogony cannot yet have anything in common with its later use in Newton. All the causal preconditions that make the idea important for Newton are still missing in Galileo. For Galileo, the rule is that under the Aristotelian conditions of "natural" motion, only rectilinear motion can be accelerated; both the demiurge and the creator God, too—since their actions are only permitted to cause "natural" motion—could only use a rectilinear path for the initial acceleration of the heavenly bodies.

It is true that with his laws of planetary motion Kepler had overturned both the "Platonic" requirement and also the principle of the "naturalness" of such motion, but in regard to the problem of the initial state nothing had changed. His second law had 'saved' the ideal of uniformity of motion in a higher sense, by separating the two parts of the "Platonic" requirement and making uniformity independent of the shape of the orbit. Consequently the historical circumstance of the chronological order of his discoveries is also well justified by their logic, in that the first law's daring, by which it gives up the divine circles, was only set free by the sheltering effect of the second law. The Platonic regularity of the planets' motions in their circular orbits emerged in retrospect as a special case of the proportions that were regulated by the second law for ellipses. The regularity now consisted in the identity not of the portions of the orbit that are traversed in units of time, but rather of the areas that are swept over by the radius vector. This is a typically modern process: A principle of constancy that was eidetic is replaced by one that is quantitative and formal, and thus transformable. The third law only applies to different bodies what the second law asserts for an identical body, namely, the dependence of velocity on distance from the central body. For Newton's problem of the beginning of the world, what followed from Kepler was that every acceleration of a heavenly body had to change its orbit itself, so that the body could be assigned to its final place in the system of orbits only when it had arrived at its definitive velocity.

Here there could not be a free play of forces, because for Newton the arrangement of the bodies in the system was an expression of the highest wisdom, an eidetic pattern that was binding a priori, and that therefore could not be the outcome of a contingent process. On the contrary, it was a highly ingenious calculation that intended, by means of a rectilinear fall from a certain distance, a deflection to one side, and a "doubling" of the gravitation of the central body, to cause the planet-to-be to wheel around into the orbit provided for it—that is, into the orbit established according to Kepler's third law.

Kepler's deformation of the circle, or more precisely his demotion of the circle to a limiting case of the deformation of an ellipse, produced for the first time a need for an explanation of the shape of the planets' orbits; and this explanation, in its turn, left no other assumption open except that the rotating globe, too, will be deformed. The circle and the globe had been self-evident, needing no explanation: Ptolemy had founded the "Platonic" postulate simply on its appropriateness to the nature of divine beings[82]—which is more like a refusal to provide a foundation. The ellipse was not so much an offense against something that could no longer itself be divine as it was, rather, a challenge to explain something that till then had not needed to be explained.

Kepler himself articulated the theoretical program that leads to Newton as a consequence of his planetary laws: "You ask why I teach that the equations are to be computed from the triangular surfaces? ... The answer is, Because I have made up my mind to derive the calculation from the hypothesis of natural causes, just as the ancients derived it from the hypothesis of perfect circles and of uniform motions on these."[83] Now, one can transfer this idea from the shape of the orbit to the shape of the terrestrial body by saying that through the hypothesis and the demonstration of its deformation, the terrestrial body, which Copernicus had made into an astronomical object, has once again become a physical object in the traditional sense. But for that very reason it could no longer satisfy the conditions of the constitution of time. The concept of absolute time expresses the fact that a condition of the possibility of physics cannot itself be given in the form of a physical object.

Here I do not wish to follow the concept, proposed by Thomas S. Kuhn, of a "paradigm change." My doubts about it relate to its neglect of the role of continuity as a precondition of every possible discontinuity. So I prefer the idea of the "reoccupation" of a framework of positions that remains intact, that is presupposed, functionally, and that makes partial changes not only 'bearable' but, above all, 'plausible.' The only reason why Aristotle was able to succeed so well and so soon with his critique of the Platonic Academy was that he did not, in a revolutionary manner, eliminate the difference between reality and ideality, but conserved it within the reality of his cosmos itself, when he used the attributes of his celestial matter to reoccupy the position of ideality. He gave the real heavens the attributes of the ideal cosmos. A reoccupation of this kind was repeated in the reverse replacement of Scholastic Aristotelianism by the mostly self-styled "Platonism" of early modern natural science.

Copernicus promoted the Earth as well into 'ideality,' and produced once again, by means of its perfect motion, the 'ideal' time that according to Aristotle was conceptually necessary. Newton could no longer guarantee this a priori, with a planet that travels on an elliptical orbit with a nonuniform velocity and has an irregular shape. He reoccupied the position of that 'ideality' once again, with absolute space and absolute time. The perfection that was in the immediate vicinity, in what man experiences, had remained a functionally unavoidable episode.

Methodological idealizations must not be confused with idealities here. When Newton attacked the problem of the Moon's orbit, in 1666, he took up the traditional idea of a sphere and considered the pressure that the Moon's body exerts on the inner surface of that sphere when it slings the Moon around the Earth. There would have to be a compensatory force corresponding to this pressure, if one thought of the sphere's imaginary steadfastness being suspended. This force could only proceed from the Earth, by its influencing the Moon in the same way that it influences a stone that falls to the ground. The outcome of this initial elementary reflection was that the orbits of the heavenly bodies could be figures that result from the action of at least two forces, and that no longer had anything to do with the realization of a geometrical norm. Mechanics determined the selection of the geometrical apparatus that complied with it, rather than the reverse.

The ideality of absolute space also accorded with a different geometry from the eidetic one of the tradition. The model to which absolute space corresponds in 'type' is analytic geometry, which describes each of its objects as a relation between an 'external,' fixed coordinate system and the states of affairs that present themselves 'within' this frame of reference. Here the situation of the objects in their frame of reference is always only contingently given, not rationally explicable. Leibniz was to object to the transference of this facticity of the object's situation with respect to the frame of reference from geometry to the universe. His objection was not only metaphysical, in regard to the lack of a "sufficient reason," as in his correspondence with Clarke, but also related to the geometry that was presupposed, in connection with which he developed his *analysis situs* as a geometry of the 'inner' determination of objects, for which the 'outer' frame of reference makes no difference (which, when the geometry is carried over to physics, makes that frame of reference unreal).[84] Leibniz changes the mathematical orientation of his model of the universe in order to make it possible to carry out his principle of rationality. By doing so he gains, against any conceptual realism regarding space and time, the inner explicability of all objects in a universe that can be interrogated in regard to qualities—or at least he does not give it away in his initial model.

Leibniz's opposition to Newton in the name of the principle of sufficient reason cannot make us forget that when Newton made the generation of time entirely independent of its connection to the motion of a body in the universe, and locked its reality out of reach, in the reservation of its absolutism, he saved the concept of time from the abyss of a catastrophe that Copernicus had accidentally prepared for it. Copernicus had been unable to anticipate that the Earth's perfect globe shape and its real motion with respect to absolute space would be mutually incompatible. A half a century after Copernicus, the dismantling of what have been called the "cosmological perfections" was already beginning. The first conjecture about the Earth's deviating from the pure globe form was uttered by William Gilbert, when he attributed the compass needle's magnetic declination to a corresponding deformity of the "magnet," Earth, a deformity that he thought was constituted by the irregular distribution of the land masses and mountain ranges. The next step was a step in reflexive optics: In 1666, Cassini discerned in Jupiter the combination of a rapid axial rotation and flattening at the poles. Newton used this observation as the basis of the generalization that the polar axes of the planets had to be shorter than their diameters at their equators.[85] If the planets did not rotate, he writes, they would have a *figura sphaerica* [spherical shape]. The occurrence of centrifugal forces established not only the reality of motion in relation to absolute space, but also the end of the "naturalness" of (or the assumption of an inertia appropriate to) rotation.

On behalf of his generalization, Newton speaks cautiously of the parts of the planets' bodies "striving" (*conentur*) to heap themselves up at the equator. He lacked an appropriate cosmogony that would have allowed him to assume a fluid matter that would lead to the deformation of rotating bodies in the universe. For the Earth, though, a teleology on behalf of man once again makes possible an easy application of his theory. The Earth, Newton says, is at least in part a fluid body, insofar as its oceans can very well yield to the forces that press away from the globe shape. Consequently, at the equator all solid land would have to be submerged by the raised level of the ocean, if this was not prevented by a providential lifting of the land masses. Thus the deviation of the solid part of the Earth from the globe shape is not a consequence of a history of the Earth that began with an overall condition of elasticity, but instead is from the very beginning an adaptation of the Earth's form to the necessary consequences of its rotation for the relation between sea and land. That can only have been said with an eye to the possibility of human life, although Newton does not say a word about the connection. The result of this roundabout deliberation is the same as if he had developed a cosmogonic theory about the early deformation of the terrestrial spheroid.

The dismantling of the "cosmic perfections" was like an eidetic symptom of a new contingency of the universe, which seemed to contradict the claims of rationality that had been shaped in the course of hisotry. That makes other things intelligible besides Leibniz's turning against the absolutism of space and time. After Newton, the real universe tended toward the status of an event that is insular in the ocean of space and episodic in that of time. The expansion of space already by Copernicus, and after him by the telescope, by the finite speed of light, and by the first hypotheses about the paths of comets, made the material systems appear more and more as rare and sporadic phenomena. The solar system did not remain, as it still was for Copernicus, the essential and central part of the universe. Bruno's speculation was realized, in that the cosmic totality filled itself with an increased number of what were called, with a hyperbolic plural, "worlds," and at the same time emptied itself, in a rivalry between matter and space. Only after the middle of the eighteenth century does the new unity of a world of worlds delineate itself both systematically and cosmogonically, a unity in which solar systems were not the ultimate systematic magnitudes but instead were parts of higher-level systems of motion and force, from which Lambert ultimately wanted to construct one single system organized around a central body.

If the world is an episode in absolute time and an island in absolute space, then its incidentalness in relation to the 'salvation story'—an incidentalness that, for Newton, was essential—is accented more sharply than it ever was in a medieval text. Its creation and its destruction remain systematically appropriate data that are familiar and imminent in terms of recollection and expectation. The creation is inserted into the *receptacula* [receptacles] of space and time as a lost enclave, and the apocalyptic events will eradicate it from them again, as a disturbance of pure ideality. As presuppositions of the world's contingency, space and time inevitably move to the side of the divinity—if not of his attributes, at least of his organs. In the case of space, Newton made this explicit with his formula of the *sensorium divinum* [divine organ of sensation].[b] If space and time contain the event of the universe as a contingent fact that does not 'fill them up,' then any falling back on the point of departure in antiquity, which conceives of time as depending on the motion of the cosmos, and which (following Augustine's reasoning) includes it in the Creation and ties it to the latter's perishableness, is cut off. If Augustine had said that time would persist even if the heavens stood still, still there remained the minimum requirement, as a substitute, of a potter's wheel that still turned. For Newton, time remains what it was and is, without regard to any real motion: "The duration of perseverance of the existence of things remain the same, whether the motions are swift or slow, or none at all: and therefore this duration ought to be distinguished from what are only sensible measures thereof . . . . "[86]

In Newton, the concept of time is involved in the definition of absolute space. He defines the concept of real motion by its relation to "unmoved places," which are associated with immovable space. Their identity is definable only by means of the reference to absolute time: "Now no other places are immovable but those that, from infinity to infinity, do all retain the same given position one to another; and upon this account must ever remain unmoved; and do thereby constitute immovable space."[87] The fact that absolute time has to be logically prior to absolute space in this way becomes noticeable even in the way in which it functions as a condition of experience. There is indirect experience of absolute space insofar as the rotation of objects represents a distinctive reality. The forces that appear in connection with it, and their effect in terms of shape, objectivize that relation to absolute space. There is no comparable access to the reality of absolute time.

Absolute time is distinguished from relative time in astronomy by the equalization of time (*per aequationem temporis vulgi*), because "natural days" do not possess the expected obedience to measure.[88] Astronomers had a "more correct" time for the measurement of the celestial motions than they could "read off" from them. The suspicion that the uniform motion by which time could be measured exactly is not given in nature at all drives Newton to a concept of time as a limit that for its part can no longer be measured or used for measurement. It is precisely the idea of natural parameters' continual need for correction.

Even if one assumed, as a thought experiment, that it had not been possible up to the present day to establish empirically an irregularity in the Earth's diurnal rotation, this would not in principle alter at all the theoretical situation, which is that since Newton one would have had to surmise, expect, and thus continue to search for such an irregularity. The problematic of time, too, is a consequence of Copernican realism. Newton says explicitly that he wrote the *Principia* in order to demonstrate physically the reality of the Earth's motion. The completion of this proof at the same time expels the reality of time from the reality of nature.

The Earth's irregularity obtrudes itself on the eighteenth century as something with increased immediate significance. In November 1755, the news of the earthquake of Lisbon, with its 30,000 fatalities, shocked Europe. In Königsberg, Kant took up his pen three times to explain the uncanny phenomenon of the unsettled ground to his fellow citizens and to integrate the event, once again, into the German Enlightenment's theodicy. Nevertheless, the great earthquake became the symptom, which was passionately taken up, of piercing doubt about nature's friendliness to man, as a specific outcome of its general uniformity. It was not accidental that in 1752 the Prussian Academy of Sciences, in Berlin, had offered a prize for an essay on the question whether or not the Earth's diurnal motion had always had the same velocity; by what means one could make sure of this; and, if it had been irregular, what might be the causes. In 1754 the deadline for submission had to be postponed for two years, and on 3 June 1756 the prize was awarded to Father Frisi, in Pisa.[89]

Kant did not wait this long, but instead published his answer, in June 1754, in the *Wöchentliche Königsbergische Frag- und Anzeigungs-Nachrichten* [*Königsberg Weekly Inquirer and Advertiser*]. We can assume that he counted on a wider interest for his thesis than that of an Academy—that he had in mind the same audience to which, a year later, he meant to present the *Universal Natural History and Theory of the Heavens*. Both dates are deceptive in connection with any reflections concerning the history of the influence of the respective works, because the larger work perished when its publisher went bankrupt, and the smaller one was so far from reaching the scientific public that its thesis of the slowing of the Earth's axial rotation had to be discovered anew, a century later, by Robert Mayer, Delaunay, and Airy. Kant's theory, too, is based on the state of affairs that had already troubled Copernicus, at least briefly, in regard to the Earth's perfection—namely, the different characteristics of masses of earth and of water.

How did the Berlin Academy arrive at the topic of their prize question? It is natural to think once again of Maupertuis, who had been president of the Academy since 1746 and who had been driven to Lapland, two decades before, by the question of the consequences of Newton's physics for the terrestrial body.[90] Maupertuis was the embodiment, at the midpoint of this century, of the faculty for imagining problems. It was natural that after the empirical confirmation of the irregularity of the Earth's shape, he now wanted to promote a search for irregularity in its rotation too—an undertaking for which, at that time, there was no possibility of an empirical technique, and which therefore was bound to excite the theoretical imagination of a Kant. For him it was, at the same time, the first sample that he offered, in a special case, of the productivity of the cosmology that he had in store—as he himself says, "a test piece of a natural history of the heavens."

The conjecture that the Earth's rotation is changeable had been stated for the first time—as a pure speculative paradox—by Giordano Bruno, in his Paris *Articles* of 1588.[91] Even if one says that the thesis as such only carries the Cusan's "imprecision" over into the Copernican context, nevertheless having set up a connection between it and the concept of time remains the Nolan's own accomplishment. For him, the Copernican problem of the constitution of time was radically intensified by his assumption of an infinity of worlds. All the advantages that Copernicus thought he had obtained by means of the equivalence of the Earth and the heaven of the fixed stars were lost again as a result of the necessary admission that there had to be as many times as there were worlds, so that the priority of the apparent motion of the heavens could be merely contingently terrestrial and thus subjective. If there could no longer be a homogeneous cosmic time, any conceptually necessary restriction had lost its force in connection with the special terrestrial time. The loss of that highest perfection of uniformity is the price of the riches of the organic fullness of motion that, for Bruno, makes up the individuality of the cosmic bodies and thereby protects the new infinitude from the tedium of the mere reproduction of constant possibilities. Imprecision, which for Platonism was a defect, has become a precondition of the positiveness of the individual.

The conjecture that there is a regular alteration in the velocity of the Earth's rotation—that is, that it grows continually slower—was proposed for the first time in 1668, before the Royal Society in London, by Robert Hooke, who was (not accidentally) the man who had achieved one of the most important improvements in mechanical clocks, by introducing the spiral spring as the balance. The motives of his conjecture were still extremely medieval, though they were not far removed from the type of biblical speculation that Newton also engaged in. While the idea of lengthening human life by means of medical knowledge was involved in the motivation of the new idea of science, Hooke was occupied instead with the problem of the great ages reached by the Old Testament patriarchs. He did not derive from this any hope for a macrobiotics, but was entirely concerned, instead, to relieve the Bible of the appearance of being untrustworthy. For this purpose he recommended considering whether the Earth's rotation might not have slowed down, as a result of friction, in such a way that the high figures given for ages in the early period could in reality be related simply to correspondingly shorter units of time—in which case the lengths of modern lives would indeed have to be counted as shorter numerically, but not in their absolute temporal magnitude.[92]

Kant's answer to the Berlin Academy's prize essay question took the tides in the oceans as its point of departure. The tidal wave traveling in the direction opposite to the Earth's rotation and breaking on the East coasts of the continental land masses would gradually have to slow the Earth's turning. This effect would only be suspended if the time required for the Earth's rotation and for the Moon's revolution around the Earth had become identical, i.e., if the length of the day and that of the month had been equalized. Then the Moon's influence would always operate on the same side of the Earth's surface, just as, conversely, the Earth's influence on the Moon has already achieved coordination with it. Kant regards the partial system composed of the Earth and the Moon as a late result of the process of cosmogony. If the Earth had already had its satellite during the phase when it was in a fluid state, its speed of rotation would long since have been assimilated to the Moon's period of revolution.

None of the expectations that Copernicus had connected to the perfection of the Earth's form was fulfilled. Apart from the irregularity of the axial rotation, inconstancies emerged both in the supposed rotation of the Earth's axis with respect to the heaven of the fixed stars, during the annual revolution, and in the position of the axis of rotation within the Earth itself. To ensure the identity of the constitution and the measurement of time, these three things would have had to be given as constants. Then people tried to standardize to mean values the periodic variations that were caused by various kinds of factors. But given the need for ever greater accuracies, the Earth's unreliability was bound to lead to separating the definition of the unit interval of a second (with its reference, by way of the hour and the minute, to the astronomical day) entirely from its cosmic basis and making it independent of the Earth's motion.[93]

In 1936, in the Imperial Institute of Physics in Berlin, A. Scheibe replaced the last stellar time constant with the terrestrial instrument of the quartz clock. The irregularity of the Earth's rotation turned out to have an order of magnitude of 10^{-8} [1 part in 100,000,000]. Today this very slight variation is traced both to tidal friction and to processes of thermal expansion and contraction in the body of the Earth. It is not only a matter of reliable periodicities: As late as the beginning of 1974, the Earth's rotation accelerated and days were thereby shortened by a thousandth of a second—the greatest acceleration determined up to that time. If climatic influences should play a role in this, and thus bring meteorology into the calculation, a new stage of insecurity would be arrived at in the long sequence of confirmations of distrust in the reliability of the Earth as the organ of time—if the process of rendering time autonomous were not concluded.

In 1952, the International Astronomical Union (IAU) also proposed separating the definition of time from the Earth's rotation. Something was accomplished in which the earlier reform of the measurement of length had failed. When, in the attempt to find a norm based on the "reason in nature," the unit of measure of 1 part in 10,000,000 of the Earth's quadrant was introduced,^c the suggestions of Huyghens and La Condamine that the norm should be the length of a one-second pendulum were passed over. To be sure, the astronomers in the IAU stuck to the position that the second had to be defined as a fraction of the Earth's annual revolution around the Sun. It was the requirements of physical precision that this "ephemeris second" too did not satisfy and that thereby frustrated the attempt to save a definition of time that was tied to the cosmos. In 1964, the International Committee for Weights and Measures, in Paris, for the first time laid down a definition of the second in terms of the atomic physical constant of the "cesium frequency." In 1967 it was adopted as part of the International System of measurement (SI), and since 1969 it has been the legal unit of time in the Federal Republic of Germany.

The fact that no astronomical phenomenon can satisfy the requirements of a precision metrology in connection with time determination is part of the consequences of Copernicanism for the concept of time. The difference between the astronomical second based on mean solar time and the atomic second entered public consciousness—hardly perceptibly, it is true, but at least it was noted in the press—as a result of a 61st second that was added to the last minute of the month of June 1972, through a worldwide manipulation of clocks.

In comparison with the calendar controversies of earlier centuries and their effect on people's consciousness of the world's state of order, there was no noticeable public perplexity in connection with this event.

Part V

The Copernican Comparative

Introduction

According to later legend, Copernicus is supposed to have predicted the discovery of Venus's changing phases, as a directly perceptible confirmation of his system. Of course he had no way whatever of knowing that this phenomenon would ever become visible for human eyes. Less than a century passed, and Galileo saw it through his telescope. Johann Heinrich Lambert was to have less good fortune with Venus when he forecast that this planet too must have a satellite and that people would see it at the next opportunity. This was to present itself on 1 June 1777, when Venus would pass close to the Sun's disk. In an essay "Vom Trabanten der Venus" ["On Venus's Satellite], in the *Astronomisches Jahrbuch für das Jahr 1777* [*Astronomical Yearbook for 1777*], published in Berlin in 1775 under the supervision of the Academy of Sciences, Lambert gave all the data and conjectures from the earlier, doubtful observations of Cassini, Short, and Montaigne. Above all, he refuted the assumption that the satellite would have had to be seen during Venus's two transits of the Sun in this century, in 1761 and 1769: Both times, he maintained, the position of Venus's moon on its orbit had been such that it had remained outside the Sun's disk. As a consequence of this possibility's not being considered, false resignation—assuming a supposedly definitive result—had set in. That is the point on which Lambert focusses his attack. The possibilities of testing earlier observations and conjectures had not been exhausted. "Perhaps the appetite for investigation has also diminished. For admittedly if one sees no trace of a satellite for several years, one easily grows tired of tracking it further." Even the most widely used text book of natural science of the period, Erxleben's

Anfangsgründe der Naturlehre [*Elements of Natural Science*], contains a reference to the date of 1 June 1777.

Lichtenberg, who edited the *Anfangsgründe* after its author's death, inserted one of his pointed supplementary remarks here, on the subject of Lambert's conjecture of a moon of Venus: "... but it was not observed. So the author, who died on August 18 of that year, lived to learn the lesson of that day."[1] For his contemporaries, Lambert typified the courage to use one's imagination. But when he turned out to be wrong, they did not hesitate to take note, with some malicious pleasure, of the absence of corroboration. For Lichtenberg, an aversion to "teleological philosophers" ["Endzweckspilosophen"], as he apostrophizes them, plays a role here as well. Lambert's soft spot— which he presented cautiously enough—for Venus's moon was not, in fact, accidental: How could such a great planet do without the equipment that is normal in the solar system; how could it not itself represent this system in a smaller form? The thought-form of his early *Cosmological Letters* still stamps this late interest of Lambert's in the question of the existence of a satellite of Venus, and thus the last empirical result to which his speculative courage led him. It is the thought-form of the "Copernican comparative," which can be reduced to the formula, which Lambert himself gave it, that more than two centuries after Copernicus people "had still not become nearly Copernican enough."

What could that mean? It could imply that the fundamental Copernican principle of putting in question and penetrating the illusion of the observer's being located in the center of the universe had not yet been applied sufficiently radically to the problems that present themselves in astronomy. But it could also imply applying the Copernican pattern of the solar system, with the body with the greatest mass at the center of the motion of the other bodies, over and over again, as the structural principle for solving all the problems in the construction of the universe. Lambert resolutely followed the first path, but he was not able to forgo the intuitive evidentness of the second. That led him not only to the last mistaken speculation of his life, but also, much earlier, to the most momentous one, of succumbing to the temptation of the extreme consequence: an absolute central body of the cosmos.

It is probable that the anonymous biography of Lambert that appeared in September 1778, in Wieland's *Teutscher Merkur* [*German Mercury*], was written by Lichtenberg.[2] This presentation is in accordance with the pattern that Fontenelle had coined almost a century earlier in his obituary notices for the members of the Academy of Sciences in Paris. As though it were equipped with a guarantee from nature—which, the century is convinced, knows how to achieve its ends—the talent forces a passage for itself through every unpropitious circumstance and every lack of understanding on the part of those around it. The mathematical *puer autodidactus* [self-taught child] is the darling of Enlightenment biographers because he demonstrates the epoch's ideal of thinking for oneself. Thus the description of Lambert's life is that of a man who "had himself to thank for almost everything"; it shows "how an extraordinary genius breaks through all obstacles and prepares for itself the path on which it was destined to shine." True, the boy does not once again, like Blaise Pascal as a child, discover a good stretch of Euclid for himself, but he has the good fortune (which permits us just as well to infer that he was favored by nature) to find "accidentally an old book of mathematics"; and since he found it among the Latin books that he had eagerly read, he immediately had an opportunity to discover his "decided inclination" when faced with the choice between humanistic tradition and mathematical science. What he immediately recognized in the old book of mathematics were the mistakes that it contained, which, however, he was unable to correct. During repairs to his father's house, he immediately proves himself to be an applied mathematician, when "with his book in his hand he puts various questions to the builders about the practical application of some propositions in it"; this causes one of the builders, in turn, to give him a book of mathematics. And now nature again carries out a bit of theodicy by bringing together what belongs together. The biographer writes, "What a satisfaction for his appetite for knowledge, when he discovered that this book had been written precisely to correct the errors in his own book! Now, without any further instruction, he learned from both books, without trouble, the elements of arithmetic and geometry." However low an opinion the Enlightenment might have of the old miracles, it was glad to let smaller ones occur on its side, when it was a matter of promoting the goals of this same Enlightenment, and awakening confidence in the ways of reason.

If we can believe the recollections of Sulzer, who was director of the philosophical division of the Berlin Academy when Lambert arrived in Berlin in January 1764, Lambert narrated the story of his conversion to mathematics to the Prussian king in the classical form. The king wanted to see immediately the philosopher whom Sulzer had recommended to him for appointment to the Academy. He was warned that this scholar did not exactly recommend himself by external dignity or adroitness. The king's response is, once again, concocted too well in the style of the epoch: "We shall extinguish the lights; bring me the man at night; I do not want to see him, but to hear him." Whatever illumination may have been provided, the impression made was a poor one. The king began the conversation by asking Lambert what sciences, in particular he had learned, and the answer was, "All of them." He must be a capable mathematician, then; Lambert answered in the affirmative. "And what professor instructed you in mathematics?" Answer: "I myself." To which the king responded, "You are a second Pascal, then?" Answer: "Yes, your majesty." Frederick had to revise immediately his first negative impression that "he had had the greatest idiot proposed for his Academy," because the Russian ambassador tried to promote Lambert's membership in the Academy in St. Petersburg. The king is supposed to have justified Lambert's appointment to the Berlin Academy with the words, "With this man, one has to look at the immensity of his insights, and not at trifling matters."

So it was not only his environment and his circumstances, but his own figure, that stood in the way of this potency. Lichtenberg—if it was indeed he—wrote, "It is true that his early education had left behind it ineffaceable traces of his original low station, which expressed themselves continually in his shy, embarrassed manner, his unmatched and sometimes quaint clothes, the miserable furnishings of his rooms, his loud laughter, often silly jokes and comical gestures, and his taste for bright, unmixed colors, coarse foods, and sweet, bad wines, and sometimes induced him to mix with common citizens at coffee klatsches, to engage in their political discussions and to laugh heartily at their remarkable notions. But under this bizarre exterior the finest qualities of the heart and the intellect lay hidden . . . . " Here we see traces of the eighteenth century's admiration of Socrates. But the thinker for himself was also endangered by himself. Constructive rationality could itself turn into prejudice. "Since, however, he had as it were derived all of his knowledge from himself, it was difficult to make clear and (so to speak) instill in him something that he did not hit upon of his own accord or could not at least think through and, as it were, make his own; so that for that very reason he could invent things much more easily and confidently than he could judge them, and he often looked at something entirely backward, without being able to be set straight easily. He himself used sometimes to call his genius a machine . . . . " Thus we arrive at a disproportion, reminiscent of the anecdote about Thales, between higher contemplation and a sense of reality: "In his own sphere, where he understood everything with the greatest clarity and certainty, he did usually judge correctly, but outside it, and where the talk was of men and affairs, his views were so unbelievably wrong, and lacking in the most ordinary common sense . . . . "

The power of the imagination, which Lambert embodied, corresponds to the priority that the Enlightenment tried to give to possibility over reality. The imaginary as an exotic, utopian, or cosmological thought experiment is always at the same time a perspectival device that intensifies the features of what actually exists, to the point of caricature. An episode in Lambert's correspondence with M. Holland illustrates how easily the imaginary can be set up as a substitute for the real. Lambert complains that the practice of only reading reviews of books, and dispensing with the books themselves, has become an epidemic. In this connection "the best journals seem to have the worst effect." From this point of view it has ceased to matter whether the book in question exists or not. This led him to think that "one could write a useful library of books that do not exist, but that nevertheless could exist." [3] Holland's answer is that one should make the best of the fashion by putting on the "mask of a journalist," and in that field no one had yet thought of a "journal of possible books. If I have in hand only fragmentary and unconnected ideas about a subject, then I shall present them as selected passages from a work to be published later. Learned news from the ideal world will have a relation to the real one. I shall announce improvements in a school or at a university as having already occurred, when I wish that they might be true . . . . " [4] Lambert comes back to the idea by return of post, and pushes it one step further yet: "Books that have actually been published could also provide an occasion for reviewing better ones that have not been published, without mentioning the former ones. A nonexistent journal could also be reviewed, with all the characteristics that it should have." [5]

Lambert's universe emerges as the kind of expansion of reality in which even Leibniz's possible worlds could still become real.

1

Perspective as the Guide for Cosmological Expansion

The scientific situation in which Lambert wrote his *Cosmological Letters* was characterized by the transition from the problems associated with the construction of the solar system to those associated with the construction of the universe of the fixed stars. It is true that the outermost planets, Uranus, Neptune, and Pluto, had not yet been discovered. But the question of the construction of the planetary system, which seemed to be closed with the victory of Copernicanism, had surprisingly arisen again, when it became possible to determine the orbits of comets. The knowledge that these so long feared celestial phenomena were identifiable periodically returning bodies and that their paths could be represented as (it is true) very eccentric, but still similar to the paths of the planets, was a result that was prototypical for the Enlightenment.

Lambert was an important participant in this integration of the comets, as celestial bodies moving in accordance with laws, into the solar system. But precisely the widening and enrichment of the solar system brought to consciousness the way in which the classical questions of astronomy were limited to the region influenced by the Sun. Man's prospects of penetrating the depth of the space of the fixed stars with his theory and grasping it constructively, as he had done with the solar system, had not even been touched yet.

What had happened shortly after Copernicus meant nothing more than the dissolution of the outermost sphere of the fixed stars, the existence of which had been assumed since antiquity. Once its motion had been shown to be an illusion, it was natural to think of the stars not as mounted on the inside of a globe that encloses the world, but rather as distributed freely and irregularly in space and as having distances from the Earth that correspond to their apparent magnitudes. Space had gained depth but had lost structure. Lambert stands between the two great accomplishments of modern astronomy, the constructive description of the solar system and the disclosure of relations of order in the universe of the fixed stars. This situation in the history of science, which he not only marks but reflects, is described in a single sentence as the point of departure of his *Cosmological Letters*: "While more recent astronomers are very anxious to put in order the sphere of our solar system and to assign each comet its orbit, and to determine in advance its eventual return, hardly any attempt has been made to find out something probable on the arrangement and position of the fixed stars."[a]

The two centuries since the appearance of Copernicus's chief work now draw together into a unit: In this period, in spite of the discoveries by means of the telescope, nothing more has occurred that is critical for the whole picture of the universe. The year 1750 marks the turning, with the *Original Theory or New Hypothesis of the Universe*, by Thomas Wright. In this work the thesis is advocated that the phenomenon of the Milky Way is produced as a perspectival effect of a standpoint inside this vast, flattened circular aggregation of stars. If Kepler had had, against his will, to abandon the ideal circular form for the orbits of the planets, here it returns again, realized speculatively, and reconciled with "our ideas of a circular creation, and the known laws of orbicular motion."[b] Wright indicates that he sees the distribution of the planetary orbits, which deviates only a little from a plane, being repeated on the large scale of motions within the Milky Way. The crucial thing for him is that the irregularity in the visible distribution of the stars is only the false appearance of a true order, which is invisible from our eccentric standpoint.

The guiding principle, for Wright, is manifestly the metaphysical need for a well-ordered arrangement of the world that is all-encompassing, even if it is painstakingly concealed from man. A breakthrough to new methods could not be achieved in that way. Nor did Wright get much of a response from professional astronomers. William Herschel was the first to approach the problem with sober, quantitative methods. In 1784, in his first publication on the construction of the heavens, he was still able to write that people had pictured the starry heavens as the inner surface of a hollow globe, with the observer in the center, and that only now, with the help of his

statistical method of counting fields of stars, was that illusion beginning to dissolve. "In future, therefore, we shall look upon those regions
into which we may now penetrate by means of such large telescopes,
as a naturalist regards a rich extent of ground or chain of mountains,
containing strata variously inclined and directed, as well as consisting
of very different materials." [c]

Lambert had neither Herschel's instruments nor his methods. But
for him, too, the first phase of the Copernican epoch draws together
under the characterization that, despite all explicit denials, the spherical form of the heaven of the fixed stars had been held to in the
resignation that did not seek a theoretical grasp of the phenomena.
Beyond the solar system lay the space of the starry heavens as a given
fact that could not be further analyzed and could not be penetrated.
For the heaven of the fixed stars, the Copernican turning had remained without effect: "With respect to the whole world-edifice we
seem to be much in the position in which Pythagoras, Philolaus,
Aristarchus, Nicetas, and other Greek sages were in respect to our
solar system." Lambert concludes his description of cosmology's pre-
Copernican situation with the significant sentence, "We are still
waiting for the Copernicuses, Keplers, and Newtons of the whole
world-edifice." [d]

Thomas Wright had pursued a perspectival train of thought. If our
standpoint is located inside a lensshaped system of uniformly distributed stars, then the result for us, in the direction of the plane of this
lens, must be the view of accumulated stars that we have before us, as
a phenomenon, in the band of the Milky Way. It must have been just
as difficult to arrive at this reflection, then, as it seems simple to us
now. One has to keep in mind the fact that for the traditional
classification of the celestial phenomena, the belt of the Milky Way
was something like a special kind of constellation, alongside other
constellations—that is, it was something that we observe from afar
and from outside, as a foreign region of shimmering realities. The step
that Wright had taken installed the solar system as part of the system
of this reality; it created unimagined relationships. However strong
the metaphysical impulse to order the world once again and to put the
Castilian King Alfonso in the wrong may have been, the core of
Wright's accomplishment is a piece of perspective. Thus he also says
that in the Milky Way the stars are "continually crowding behind one
another, as all other objects do towards the horizon point of their

perspective, which ends but with infinity."ᵉ For the comprehension of Lambert's cosmology, too, it is indispensable to pay attention to his early and lifelong occupation with the problems of perspective.

The twenty-four-year-old autodidact wrote in 1752 the *Anlage zur Perspektive* [*Draft on Perspective*], an unpublished preliminary step on the way to his *Freye Perspektive* [*Free Perspective*], published in 1759. The manuscript begins with the sentence, "Visible things often present themselves to our eyes in a very different form than they actually have."⁶ This early interest of the author of the *Cosmological Letters* in the problems of perspective prepares the way for the work of 1761 in a very specific manner. The development of the faculty of spatial imagination and representation is one of the elementary prerequisites of the modern age's sense of cosmological construction. More than a form of thought, perspectivism becomes a form of life, if the passion for reflecting on one's own standpoint can be styled in this way.⁷ During his studies in Cracow, the young Copernicus had applied "special diligence to perspective," which according to the statutes of this university, from 1449, had to be regularly presented by the occupant of the chair of astronomy, and which was in fact lectured on almost every semester.⁸ And it is certainly not an accident that on the occasion of his visit to Frauenburg, Joachim Rheticus presented to his future teacher, as a guest's gift, not only the Greek folio edition of the *Almagest* and the *Elements* of Euclid but especially the *Optics* (printed for the first time shortly before then, in Nuremberg) of (pseudo-) Witelo. Rheticus's gift is still contained in the list of the remains from Copernicus's library that are preserved in Uppsala (since the Frauenburg cathedral library was removed to Sweden during the Thirty Years' War), and marginalia in Copernicus's hand show that he used it.⁹ Then in the foreword to his *Dioptrics*, in 1611, Kepler gave what one might call free variations on a speculative type of perspective, such as the view of the universe from the Moon or from Jupiter. This is already something like the exotic uses of perspective with which the Enlightenment was to try to see its own taken-for-granted civilizing standpoint from a distance.

The convergence between the theory of perspective and the 'decentralizing' of the Enlightenment's point of view also constitutes the homogeneous impetus in Lambert's two works—only two years apart—on perspective and on cosmology. While in the theory of perspective the thing is to grasp the mechanism of illusions so that one can produce them, cosmology has to reverse the process by using this knowledge for the purpose of discovering illusions to which we could be subject as a result of our standpoint and our motions in the universe. The technical producibility of appearances—the inversion of optics in perspective—provided more ingenious means of penetrating appearances. Lambert describes the difference between optics and perspective as follows: "The art of seeing concerns itself with the grounds by which we distinguish the appearance of things from their true form, and on the basis of which we want to infer the latter from the former. The art of perspective leaves the true form behind and only endeavors to project the apparent form." Now one can say that the elementary task of cosmology consists precisely in specifying the location of an observer's standpoint that corresponds to a perspective that is given in advance. Not accidentally, this is the point at which Lambert goes beyond his first outline, in his *Anlage zur Perspektive*, towards his *Freye Perspektive* of 1759. The first entry in his *Monatsbuch* [*Month's Book*] of November 1756, on this chief work, reads *Artis Perspectivae problemata inversa* [The problems of the art of perspective inverted]. He carried out this program in section 8 of the first part of the *Freye Perspektive*.

If one takes perspective not only in the traditional sense, as linear perspective, but also as the so-called "atmospheric perspective"—that is, as the theory of changes in colors and brightnesses at different distances and in passing through the intervening medium—then Lambert's early work leads on to his investigations in the theory of color, but especially to his foundation of photometry, in 1760. His *Photometria* belongs among the untiring efforts of Lambert and his contemporaries to make measurable what was still unmeasurable. Determinations of brightness were to become an important tool of astronomy in spatially differentiating the plane prospect of the heavens. The initially merely verbal explosion of infinitude in early speculative cosmology gives way to the process of securing, theoretically, what one could call the "yield" of astronomical progress in terms of space. However weak this methodological basis may still have been, Lambert possessed what he himself calls "sufficient audacity" to bring his inferences "as close to certainty as can be managed."

The literary form of a correspondence, which Lambert chose for his cosmology, also has its function, according to his own words, in displaying something like the scales on which the arguments can be weighed, "so as to see how much they still fall short of full weight." Thus an art of conjectural argumentation is developed, which provides a collection of examples of both sufficient and insufficient proof—an exemplary "rational doctrine of the probable." The moment in the history of science where one is feeling one's way beyond the borders of secured results is reflected in the attempt to represent the part that one possesses as already the whole. Lichtenberg, again, described this type of thought—which is just as indispensable in the history of science as it is dangerous—in a contribution of his own, in his Göttingen notebook, entitled "On the Universe as a Structure": "In this exalted part of natural philosophy, surmises, as long as they do not go entirely outside the bounds of a reasonable analogy and are not hazarded without any check, are often instructive, usually pleasant, and never harmful. As long as the power of imagination does not trespass against the dignity of the object, it can soar without interference; for what observation actually teaches us about the construction of the universe safeguards the imagination in the highest flight that seemliness allows it."

Lambert himself provided a literary affiliation for his *Cosmological Letters*. He would have liked to see them as the second part of Fontenelle's *Conversations on the Plurality of Worlds*. True, he had not been able to match his predecessor's liveliness and wealth of notions; but he insists that he has given Fontenelle's conception a new and much more important content, since "world," for him, has become a term for a system as part of a higher system.

Lambert's reference to Fontenelle has to be measured against another standard as well, that of the impact of their works, and of the impact to which they lay claim. Since their first appearance in 1686, and still more as a result of Gottsched's translation of 1726, Fontenelle's cosmological conversations were one of the most successful books of the Enlightenment, the prototype of an accomplishment that Goethe was to describe as that of the "popularizer" ["Gemeinmacher"]. What was depreciatingly called "ladies' philosophy" ["Frauenzimmerphilosophie"], what Casanova at least begins to read aloud to Clementina, what was widely imitated as a model of the "courtly style," and what Kant contrasts, as a model of "affected popularity," to true popularity, was taken very seriously by Lambert as exemplifying an ideal of Enlightenment style—not only on account of the object they had in common, but also on account of the importance that is given to the object. Fontenelle preforms the eighteenth century's effort to accomplish enlightenment by finding, for man's eccentric position in the universe, an adequate—which is to say, once again, an eccentric—position from which to represent it. Reason meant that things that were taken for granted became mere facts, but also that the facts had to set in motion an apparently endless process of accounting for them.

Of course Lambert's relation to Fontenelle also has to be seen in the context of his critique of the outcomes, which were becoming visible, of the eighteenth century. What Fontenelle had provided, and his procedure—a courtly presentation of a cross section of the results arrived at by a science, with speculative prospects, and without noting what was provisional in this, and what could still be questionable—characterizes the beginning of a decay of the scientific attitude, and of the presentation of science, into something aesthetically pleasing and nonbinding. Lambert writes in a letter, "If I compare the discoveries of this century with those of the previous one, I have to conclude that the zeal that animated learned men in those days has been transformed into frivolities, and instead of seeking something really new and true people have spent 60 years disputing about Leibniz's principles of dynamics and metaphysics, without producing anything at all thereby. When people began to get bored with this, they didn't turn to the remaining traces of the previous century, but instead to belles lettres . . . . " [10] To write the second part of Fontenelle's *Entretiens* [*Conversations*], then, also had to mean to make up, in relation to the model author, for what the decay of astronomy through its aesthetic mode of treatment had, if not caused, at least concealed. Lambert did not write the continuation of the *Cosmological Letters*, which according to his own testimony he had already begun; it was supposed to be *Letters on the Course of Events on the Earth's Surface*. Later on, he thinks that the epistolary form was not suitable for this topic.[11] But his cultural criticism, and doubts about the literary form he had chosen for his presentation of cosmology, seem to converge in this increasing distance.

In 1767, at the Berlin Academy, Lambert reads an essay on the benefits to each other of serious science and belles lettres. He opposes the "ever more prevalent requirement that one should present the most abstract matters *à la portée de tout le monde et dans un beau style*" [so as to make them comprehensible by anyone, and in an elegant style.].[12]

He especially criticizes the idea that science has to exert its influence directly on the public; what the scientist accomplishes for himself and for the other members of his guild, or for a statesman, has "its important influence on the public only through its more remote consequences." Lambert describes the pains he took to present these principles in exactly fitting images and examples, "because the eyesight of the *beaux esprits* [wits] is much too weak to see it right away and recognize it in the distance." Perhaps from the standpoint of 1767 Lambert after all no longer regarded the effort of the *Cosmological Letters* as sufficient to bring out the refractions that impede a forthright transmission of knowledge and the disappointments of an all-too-quick assumption that one has been instructed.

Or did Lambert, after such a brief membership in the Berlin Academy, already see his fellow members as so far removed from the position that he had idealized his partner, in the *Cosmological Letters*, as occupying? He goes on to report, in regard to the essay that he read, that it exhibited 'philosophy and what is still to be done, in a dignity and loftiness that are usually quite unknown to the *beaux esprits*; and I did this so that what they do not want to learn, they should at least learn to respect." Now Lambert no longer believes in the public that his *Cosmological Letters* had presupposed, as real or possible. His doubts are founded in a diagnosis. He sees the progress of science failing to prevail in the consciousness of the age. The decline of the epoch's philosophical interest and philosophical character takes place just at the moment when he himself believes that philosophy, as science, is no longer very far from being perfected. Thus he writes to Gottfried Ploucquet, on 1 May 1767, "In fact the philosophical age seems to be ending, at which, for several reasons, I am not surprised. But it does seem peculiar to me that this is happening at a time when we are only a few steps away from making logic very useful and making philosophy quite methodically well-founded and useful . . . ." [13]

What effect had Lambert expected from his *Cosmological Letters*? As far as I can see, we have only one piece of evidence in which he expresses himself on this subject to a colleague. It is a letter of 24 March 1761, to the Göttingen mathematician Kästner. Kästner is definitely one of the "belletristic" scholars; he did not have the scruples that caused Lambert to shrink from the combination of mathematics and rococo. It is characteristic of him that after Lambert's death he composed, with a light touch, an epigrammatic verse on the absent satellite of Venus, and furthermore that he versified the astronomical scene as the rococo game of an erotic calumny.[14] Lambert had had his *Cosmological Letters* sent to Kästner, and now he inquires about the pleasure and the surprise to which they may have given rise in him. He expects Kästner to have been surprised at the daring with which he depicted the construction of the universe ("la structure de l'univers"). In the long run he thinks his system has a chance of gaining dominant acceptance, though he also doubts that the effort of observations and calculations necessary to test it will be seriously undertaken.[15] Lambert seems to place more trust in the procedure of teleological deduction and analogies than in the process of empirical verification. The effect that Lambert expects and the means by which he tries to achieve it do not entirely correspond to what, in accordance with his theory of science, he should have set himself as his goal.

Who read the *Cosmological Letters?* Where their effect is perceptible, it is specific: Speculative and constructive cosmology, which abandons the realm of intuition [*Anschaulichkeit*], supplants the aesthetics of the "starry heavens" spanning the Earth and presenting themselves to intuition. Precisely this border line, where intuition is forfeited and the price becomes evident that will have to be paid for the gains of modern science, and where Lambert himself hesitated, marks the parting of the ways for his readers. Goethe's aversion to Lambert's work appears natural and logical. On 19 February 1781, Goethe writes to Lavater that Knebel is occupying himself with more general natural philosophy, "considering the lines that can be drawn from the way this great whole is bound together"; and in this connection Lambert's *Cosmological Letters* played a role in the circle of his male and female friends, where one conversed "about a beautiful harmony of the spheres" and where Goethe remained at a distance, "since this huge clockwork interests me, myself, only as a source of the most obscure misgivings . . . ." A few days earlier he had written to Charlotte von Stein, apostrophizing Knebel, that "the terrestrial harmony is after all mightier than the celestial."

We know that Jean Paul read the *Cosmological Letters*; he writes about them on 22 June 1799 to Herder.[16] It is true that in *Hesperus* Flamin advises Clothilde to regard the topography of the heavens as a piece of religion: " . . . . a woman should learn the catechism and Fontenelle by heart"; but Jean Paul's great cosmological metaphors cannot derive from Fontenelle, because the universe, in this language, is present in metaphors less and less as what one imagines, but rather as what one knows and can only recite. The "ranks of suns," the "waves of suns," the "great, falling universe of worlds," the "roaring, dashing sea of worlds and suns"—all of this is more than and different from what the treatise on the plurality of worlds could yield. Greater cosmic motions have come into view than the Copernican motions of the Earth; man's position in the universe is defined as an "eternal chase through the universe," and the "mute spirit" has found its way into "the wild, gigantic mill of the universe." [17]

Herder too read the *Cosmological Letters* and called them "a glory of the human intelligence." In the *Adrastea* he sees Herschel's telescopic discoveries as fulfilling what Lambert had only been able to conjecture. When we are told that the new telescopes "display the immeasurable universe itself in reflection, and will display it still more clearly as they are improved," and that with them one sees "the heavens like an immense garden," the return to the language of intuition and the expectation that what previously was only inferred will present itself directly to the eye is unmistakable. "In the true sense, then, new worlds and stars dawn on us, and perhaps we glimpse the aurora of the Creation." On the way to such evidentness, Lambert's work was only a preliminary stage, a phase of transition and of means that are heterogeneous with respect to the goal—of instruments that become eliminable once again. "Modest Lambert," Herder exclaims, "how high Herschel's reflector has elevated your fame, even beyond your own thoughts!" True, Herder says, the telescope has discovered systems of stars without visible central bodies (systems whose possibility Lambert had, after all, left open), but the cause of the huge central body of the cosmos is still not lost. Laplace has shown by calculation that a luminous body of the same density as our Earth and with a diameter 250 times greater than that of the Sun could not send us any light, on account of its gravitational power, so that if there should be bodies so great in the universe, they would have to remain invisible to us. Here, in honor of Lambert, the invisible has found its (for the moment) final refuge. "How high Lambert's calm spirit soared, which postulated dark bodies as central bodies of world-systems, for us." [18]

2

The Copernican System as a
Prototypical Supersystem

The imaginary correspondence begins by Lambert's partner's expressing his desire for instruction, a desire that has not been satisfied by reading. In the language of the century, he cannot speak of his "curiosity" without awakening negative associations. But he appeals to the fact that just in the preceding year it had been possible, for the first time, to identify a comet as a recurring phenomenon. "I hoped to satisfy in full my curiosity about the orbit of planets and comets, and I congratulated myself in advance on my good fortune, that I soon shall know in a reasoned way how it happens that comets return at appointed times."[a] However reassuring the incorporation of the comets into the solar system may seem to the writer of the letter, he is just as disturbed by the remainder of what still awaits clarification in the universe, and which might indeed remain beyond the reach of clarification.

This last is the point of a naively stylized objection, which notes the existence of an uneliminable disproportion between the world's size and its age—a disproportion that inevitably ties astronomy's progress, as a theory, to long-term waiting. The distances of the fixed stars from the Earth, we are told, are light-years greater than the presumed age of the universe since the Creation. On the basis of this disproportion between the universe's age and its size, every day new, hitherto unseen stars would have to make their appearance in the heavens. In view of the long time required for new data to arrive, astronomy's tendency to speculation would be its insurmountable fate. True, the new light that continually appears in the heavens can become an image of enlightenment itself: "The night should become continually brighter for me, and each night I rejoice at the newly arriving light from other stars." [b] But one who bases astronomy on the rigor of observation exposes it to the accidental character and the uncertainty of an experience that will never be completed but is, as it were, always waiting.

This objection is extremely characteristic of what one could call the burdens resulting from the Copernican turning. The expansion of the temporal horizon had not kept up with the expansion of space that the theory had made necessary. The propositions of geology, which would require such an expansion of time, still lay in the future. The beginning of Lambert's cosmological correspondence makes this distortion in the structure of the comprehension of the world [at the time] perceptible. In the middle of the eighteenth century the difference between the level of the things that are decisively altered by our knowledge of the world and the level of those that are determined by our immediate experience suddenly becomes much more dramatic. The entrance of the magnitude of the speed of light into cosmological discussions is a prize example of what science must demand of man in the modern age, on an abstract plane, and withhold from him on the plane of intuition—that is, of what was definitively to tear asunder our life-world and our models of the world.

It is true that the discussion about light's finite or infinite speed is an old one, but Francis Bacon was probably, nevertheless, the first participant in the discussion who was seized by alarm at the idea of the consequences of light's having a finite speed. It seemed to him shocking (*dubitatio plane monstrosa*) that in looking at the starry heavens we could catch sight only of something past, something that might long since have ceased to be real—that the *tempus visum* and *tempus verum* [apparent time and true time] could split apart arbitrarily.[19] It would be necessary—Bacon's train of thought went—to assume that the speed of light was finite if one had to ascribe a corporeal nature to light. But as a material emission it would involve an immense loss of mass on the part of the bodies emitting it. That would not be readily compatible with the nature of the heavenly bodies, which, since Aristotle, represent imperishability and immutability—unless the light-matter that was emitted was so fine that the loss of substance would remain imperceptible. In the case of such a subtle matter, again, one could assume a very great, almost instantaneous speed of propagation. The reassurance that Bacon obtained in this way was all the easier for him as his rejection of Copernicanism required him, in any case, to let the heaven of the fixed stars rotate around its axis with an immense speed; in which case, why should light not be capable of just such an *immensa velocitas*? Bacon, the anti-Copernican, thought that he had once again warded off the consequence that the presentation of the starry heavens, which man had so long related to himself as the one who was called to observe it, could be a mere appearance also in that it was not the homogeneous total reality that it presents itself as in intuition, but instead was nothing but an accidental section through the many layered depths of huge temporal differences.

In the moment in which the telescope had broken through the limits of natural optics, a new boundary limiting access to reality began to emerge here: things that are invisible on account of time. For them, even a mankind that possessed the most powerful instruments would always remain too young—if it was unable, after bursting open the constriction of space, to do the same thing to the constriction of time and (putting it briefly) to make the world correspondingly older. Even then, the preservation of the reality of our perception of the heavens as the full presence of what presently exists had to fail, as will become apparent.

In 1728, the year of Lambert's birth, Bradley makes public his discovery of the aberration of the fixed stars, in a letter to Halley. The annual periodicity of this deviation yields the first empirical confirmation of the Earth's motion around the Sun, assuming that the speed of light is finite. Bradley had been looking for an entirely different confirmation of the Copernican theory, one based on the optical parallax of the fixed stars, as a reflection of the Earth's annual motion around the Sun. Instead of the intuitive demonstration of this perspectival phenomenon—for which his instruments and his methods were still not precise enough—Bradley succeeded only in giving an indirect proof, one that could be calculated by way of the connection between the annual orbit and the speed of light. But the phenomenon of aberration did nevertheless enable him to specify the value below which the optical parallax of the most favorable stars had to lie, and what minimum distances to fixed stars consequently had to be reckoned with. Bradley wrote that it now had to be admitted that the parallax of the stars was much smaller than had been asserted by all those who claimed to have deduced it from their observations. It was very probable that for a star like Gamma Draconis it did not amount even to a second of arc, in which case the star would have to be more than 400,000 times more distant from us than the Sun. This letter of Bradley to Halley marks one of the moments in the history of modern astronomy in which the changed orders of magnitude with which one would have to deal in the future suddenly become clear.

Only with the introduction of the magnitude of the speed of light into cosmology does it become evident how thoughtlessly people had hitherto used the term "immeasurable," and how little it had actually meant, beyond its protective function for Copernicanism. However clearly Lambert (in the interest of his intended construction) holds to the finitude of the universe, he has to struggle no less hard with the reproach that the nesting systems that he projects require a space of hypertrophic dimensions. Generous reckoning with millions of light-years has not yet become the everyday language of astronomy. Bradley still easily calculated with multiples of the (up to then) by no means accurately determinable elementary cosmic distance from the Earth to the Sun; but he had already pushed forward, by conversion, into the dimension of light-years. What Lambert had in view—the circling of ever more complex systems of stars around center points that themselves were supposed to be connected, through orbits, to a comprehensive central system—meant entering into an entirely new order of magnitude.

One can also put it this way: To bring systematic order into the universe of the fixed stars required, above all, space. In his twelfth letter Lambert exhorts his partner in correspondence to think generously, where spatial requirements are concerned. There is no reason at all "to be so thrifty in respect to space in the world." If what mattered had been to fill empty space maximally with bodies, then it would have happened at the expense of the gravitation of the central bodies and the possibility of ordered motions in the universe. In other words, in that case Copernicanism itself, as the first step into the new space requirements, would have had to be mistaken. "Our Earth would then have had to stay still in one place, and its orbit would have been filled with other planets. But in this way the most beautiful and the most varied element would disappear from the world. For its perfection, motion is much too necessary: where motion is, however, there space must remain. And it is clear in itself that the central forces, the orbits, the solar systems around each fixed star demand motion and space." [c]

Lambert writes this in the age of late Cartesianism and of Leibniz's successors, which is to say, of a philosophical tradition that had pledged itself to the principle of a universe that was optimally filled and justifiable in every one of its parts. That made empty space expensive and in need of strong arguments to establish it. One can see clearly what resistance Lambert finds in himself and expects in his reader when, on behalf of his scheme's constructive perspicuity, he intensifies his quantitative requirements beyond anything that can be imagined. Something is sacrificed, or at any rate a sacrifice that was carried out long since is now consciously implemented.

The problem of parallax—of perspectival displacements of position in the heaven of the fixed stars—presented itself in an even more acute form for Lambert than for early Copernicanism. For now it was no longer a matter only of the Earth's motion around the Sun, but also of the vast revolutions of the solar system itself within the system, the Milky Way, of which it was a part. This need, raised to a higher power, for such very long-term changes in the heaven of the fixed stars finally to be demonstrated, inclined Lambert to lend credence to the premature confirmations of parallaxes that Bradley had rejected. Thus in 1718 Halley had compared his determinations of the positions of some bright fixed stars with Hipparchus's catalogue of stars and confirmed that changes had occurred. But here the change, in the meantime, in tolerances of precision in measurement played a role whose importance was difficult to determine—insofar as those changes could not be explained, in accordance with a perspectival model, as consequences of the motion of the observer's standpoint.

The extent to which Lambert waited, as it were, from one day to the next for news of this kind, can be seen from the announcement in the *Cosmological Letters* that while they were being written—in fact, during the pause that Lambert had inserted after the fourteenth letter, with which he had actually meant to conclude the collection— the first observations of changes of location in the heaven of the fixed stars had been made. Whether this is a dramatic caesura that Lambert intentionally inserted in the correspondence, we do not know. We can only learn, from what is, admittedly, a much later communication, in a letter, that he had interrupted the composition of the letters (which he had begun in June 1760) in August and September of that year, and only continued it in October.[20] In any case, the fictional correspondence uses that news to explain the fact that the venture is, after all, made of augmenting the conjectures still further and extending the sequence of letters: "Since this verification came sooner than I myself expected, it led me to develop in greater detail my conclusions, which until then had naturally followed one another, to see what else they could still disclose when I considered them in a similar generality and carried the analogy even further." [d]

Lambert justifies this step, which goes even further afield—these "daring ideas," whose "boldness should make up for what is lacking in the precision of the proof" [e]—precisely by the fact that his anticipatory sketch, more than satisfying explanatory needs, serves to direct the observer toward new possible experiences, whether they be confirming or condemning.

Using the image of the scales, again, Lambert admits that the propositions of the last part of his *Cosmological Letters*, written after the interruption, are of unequal weight. He considers the thesis of the motion of the fixed stars around a center to be "as good as completely demonstrated." [f] It follows from two rational postulates: that of the motion of all celestial bodies, and that of the self-preservation of the whole. The postulate of the motion of all celestial bodies is still based on the teleological principle that space should be filled (*principium plenitudinis*). The abandonment of this principle, in the strict sense in which it entails the *horror vacui* [abhorrence of a vacuum], is compensated for by a dynamic interpretation of the filling of space by the orbits of the celestial bodies and systems of bodies. Examination of the whole at one moment does not show it to be a dense reality; only the addition of the dimension of time, as the summation of all processes of motion, does this. But if all bodies are continually in motion, then the preservation of the whole as an ordered system means that these motions cannot be rectilinear, but must periodically return to where they started.

To Lambert himself, the introduction of a principle of preservation [or "conservation": *Erhaltung*] into cosmology will hardly have seemed like another piece of teleology, considering how closely assertions about preservation were bound up with modern rationality.[21] "The maintenance of the world-edifice eliminates the linear motion, and the fundamental laws of mechanics, together with the law of gravitation, make that motion completely central." One has to bear this universal principle of preservation in mind when one takes exception to the most serious defect in Lambert's cosmology, which is his preference for the thesis of a central body. It is true that it is above all the analogy of the Copernican solar system that leads him astray. But he had, nevertheless, kept open the possibility of a difference between the partial systems in the universe and the total system, when he did not take it for granted that Newton's physics must also be the final principle of a cosmology too: "The mechanical grounds rest on the law of gravity, but the cosmological grounds rest on the preservation and simple order of the whole." So cosmology needs to be derived directly from rational premises. This is so if only because it would never be possible to demonstrate, for any empirical system, that it must be the ultimate framework of order.

As with every other teleology, Lambert's raises the question of what background idea provides its orientation and thus determines the 'quality' of this world-contruction. He himself occasionally displays a set of political metaphors that guide his thinking. In the eighteenth letter he speaks of the subordination that rules in the world, and consequently of the requirement that this principle should display itself all the more simply, the closer the construction comes to the whole of the system. The central bodies of the individual systems— the "regents" of the planet/satellite systems and solar systems, and of the Milky Ways and the systems that subsume them—seem to Lambert to be guarantors of the world-order, which are necessary in accordance with the principle of preservation, even if, as dark bodies, they may definitively escape observation. Hardly doubting, but more as a challenge, he asks his partner in the correspondence whether he should do away with these central bodies once more and "introduce democracy into the world-edifice?" The answer, in which Lambert expresses his whole predilection for the idea of central bodies, reads, "The world-edifice is too big for that, and if every part is not to do what it likes, without order, there must be more powerful forces present to keep the edifice in a definite order and to make it into a whole that is well arranged and harmonious through many levels." [g] One must realize, in this connection, that Lambert does not yet have at his disposal the solution of the many-body problem, which Laplace was to be the first to give for the theory of gravitation, and which would have appeared to Lambert, in his system of metaphor, as the introduction of democracy into the world-edifice. But (not without a historical and political reference) this solution was provided for the first time a half-century later, in the *Exposition du système du monde*, in 1796.[22]

The cosmological expansion of space and the principle of teleology are closely connected for Lambert. The construction, which is secured by the highest teleological suitability, justifies the space to which it lays claim. If the unity of the universe can be systematically implemented, the "immensity" of the new order is legitimized. Once again, one must not lose sight of the situation in which Lambert finds himself in the history of science. Newton's world, after the forfeiture of the Aristotelian and Scholastic unity of the cosmos as a housing, was not yet by any means *one* world again in any physically fulfillable sense. Gravitation might be effective in the solar system alone, or might be limited to the space around such systems. So there was no force that 'filled' space. Only the ubiquity of gravitation leads to a physical basis for the unity of the world, as a real quality. Lambert's position in relation to this problem is characterized by the fact that empirical certainty of the validity of the law of gravitation outside the solar system was only to be furnished after his death, by William Herschel's discovery of double stars. The double stars also demonstrated for the first time systems of bodies that gravitated around their common center of gravity, without having any central body. This discovery, too, still gives no definitive evidence in regard to the propagation of gravity outside such systems, i.e., between the "worlds" in the sense used up to this point. So Lambert was left to found unity, as a principle of his cosmological construction, on a fundamental teleological assumption.

Basing one's arguments on the "aims of the creation" was admittedly no longer as much an unobjectionable matter of course as it had been in the first half of the century. Lambert even doubted the applicability of this principle for the sciences that deal with the terrestrial phenomena of nature. For cosmology, as a science of "simple" states of affairs, the application of the principle of pure teleological suitability seemed to him to be indispensable. He writes, around 1762, in his response to a letter of Jacob Wegelin to [J. J.] Bodmer, on the subject of the *Cosmological Letters*, "I thought that an examination of the world-edifice, where everything is simple, had the most immediate claim to be made as a test. Things on the Earth's surface are far too complicated, and discovering the general laws is much more difficult. I tried to make this test in the *Cosmological Letters*—and not, as I hope, with the intention of forcing my pet preconceived opinions on readers, but rather of leaving them entirely free to judge; in short, [I tried] to see how far teleology, when compared to the real world, so far as it is known at present, could promote discovery, or at least successful conjectures." Lambert stated clearly in this letter that even an astronomy that works only with rigorous proofs cannot "exist without cosmological principles." Such principles, such as the principle of homogeneity or that of the conservation of substance and force, must, moreover, simply be granted, without being capable of an empirical proof. To someone who is not prepared to go along with this, "one will not be able to prove that a new Sun, Moon, and stars do not rise every day." For Lambert, there was no critical leap involved in proceeding from the assumption of the homogeneity of the rule of natural laws in the cosmos to the conclusion that the habitability of *one* body in the cosmos requires us to assume the habitability of all cosmic bodies. So the principle of the universal habitability of the world, which he exploits so much, does not occupy any special position among his assumptions. At any rate, he does not think that with it he has crossed the border that he wonders about again and again, at which "the cosmological principles begin to be arbitrary in their application."

So Lambert's procedure consists in systematically investigating the productiveness of teleological premises. Teleology "makes effects into intentions and causes into means, but it assumes the effects and the causes as known. In my opinion every science should also be useful for invention, and teleology still lacks this advantage most of all." The carrying over of the Copernican model to all the higher-level problems of the construction of the universe is also an application of this principle. The suggestion that one has found the construction plan of creation makes the smallest units of the total system into an absolutely valid orientation for the larger units and the largest ones. It is no longer evident to us, as a matter of course, that this was a consistent way of Copernicanizing one's understanding of the universe. But Copernicus himself had proceeded no differently when he had 'applied' his solution for a problem of part of the planetary system— namely, for the configuration of the Sun, Venus, and Mercury—to the whole planetary system. Lambert only thought that once the construction principle had been perceived, its application could be made completely consistent.

Today we would describe consistent Copernicanism from another point of view, which is certainly quite present in Lambert, but nevertheless is secondary to the one I have been describing. To us, consistent Copernicanism seems to be the carrying out of the elementary insight that man's point of view and his optics, in relation to the universe, are arbitrarily eccentric, or, in the least favorable case, extremely unsuitable. Lichtenberg insistently pursued the consequences of Copernicanism in this direction: We touch only the surface of nature, which we are placed in part too remote from and in part too close to be able to see it sufficiently accurately. We should also remember Karl Ernst von Baer's attempt to demonstrate the Copernican eccentricity of our consciousness of time, in relation to nature, and thus the "dependence of our picture of the world on the length of our moment."

Lambert did not take the easy route in introducing teleological propositions into cosmology. Also the term "introducing" would have seemed entirely inappropriate to him; he thought he was only making explicit assumptions that had always entered into inferences from observations. "They creep in tacitly, and a rigorous investigator sees that one presupposes them as being granted, and, as it were, obtains them surreptitiously."[23] In the course of his extensive scientific correspondence, this position was repeatedly confronted with objections. Holland writes on 19 March 1769 that the derivation of natural laws from the idea of the most perfect being was subject to such great difficulties because our concepts of perfection were so indefinite. Thus Descartes had derived his law of the constant quantity of motion from a mistaken concept of perfection, because he counted immutability as one of the outstanding attributes of the divinity. Newton's followers, on the other hand, maintained that this supposed natural law contradicted not only experience but also the divine attributes. Conclusion: "And thus, if the expression is permitted, every philosopher has a different God."[24]

Lambert sees that the derivation of cosmological propositions from attributes ascribed to God by philosophy leads one into the dilemma of characterizations that are adequate to those attributes. Such a deduction demands not only the best of the possible worlds, but also an infinite world. Leibniz avoided this metaphysical symmetry, which Giordano Bruno had implemented, by eliminating the reality of space: If infinite space existed, there could not be a world in this space. The undifferentiatedness of infinite space would not allow the establishment of a world as a rational act. Since there is a world, the space in which it exists cannot have an absolute reality.

Lambert could not agree to either of the two solutions; it will be most nearly correct to say that he evades the problem. His postulate of the world's systematic construction in finite steps on the analogy of the solar system requires him to conceive of it as finite, "and as a result half of the infinity of God's intentions comes to nothing." Finitude restricts the validity of every teleological conclusion: "The whole has its limits here, which impair the universality of the teleological propositions, insofar as one must always add the proviso, 'As far as the world structure reaches.'"

Lambert also considers this restriction in connection with the question of the habitability of the bodies in the universe, which appears to him as the "chief purpose of the Creation." If the world is factually finite, then the assertion that the habitable part of it, in turn, is factually restricted becomes consistent. If one can no longer say that the fitting implementation of divine omnipotence is the infinite universe, then neither can it be required any longer that the world-structure must be inhabited as far as it extends. Here Lambert anticipates the long-term disappointments that the Enlightenment's pet idea of a universe that is everywhere inhabited and rationally superintended was bound to suffer. The most radical motive of this desire cherished by the Enlightenment for a rational community extending into the cosmic realm is expressed in Hermann Samuel Reimarus's rhetorical question, "Why should we alone be worthy of reality?" [25]

The plurality of inhabited worlds becomes a necessary element of theodicy when man alone is not a sufficient foundation for the right of this universe to exist, but on the contrary his uniqueness in the cosmos makes his inadequacy for that purpose a concrete demonstration of insufficient reason.[h] Adorno took up this Enlightenment idea in the following way: "If indeed the Earth alone among all the heavenly bodies were inhabited by rational beings, the idiocy of such a metaphysical phenomenon would amount to a denunciation of metaphysics; in the end, men would really be gods . . . only under a spell that prevents them from knowing it . . . ."[26] When the terrestrial history of reason becomes an only partial, eccentric realization of cosmic rationality, a cosmological consideration of it can make do with the assurance that this is not the whole or the final history of reason.

Lambert shares his century's desire for the whole universe to be habitable, and inhabited, by living creatures (perhaps not all of them rational ones). He argues by analogy: "On our Earth, which since the invention of magnifying glasses we can now observe even in its smallest parts, we find everything so full of inhabitants that we can no longer be in doubt about seeing the population and animation of all parts of the universe as an intention of the Creation that admits no exceptions." In addition, our experience with the improvement of optical instruments teaches us that we are still far from having discovered the smallest order of magnitude of living creatures. Should we not also be able to extend to the other bodies in the universe this experience of the inexhaustibility of what is real?

The limiting case here, too, are the comets, which seem to represent the extreme conditions for the assumption of living creatures. Lambert considers it to be possible that comets are not always members of one and the same central-body system. Their parabolic orbits could pass over into hyperbolic ones, and thus a comet would change over from one system into another one. Together with the idea of their habitability, this idea gives the comets and the variety of their orbits the appearance of a methodically contrived device intended to produce an inexhaustible variation of perspectives on the cosmos from this wanderer. Its possible inhabitants, in Lambert's imagination, become astronauts, who would have been "created for the purpose of viewing the edifice of the heavens, the position of each sun, the plane and course of their planets, satellites, and comets in all their interconnectedness."[i] Here the Enlightenment's domestication of the frightful comets seems to be complete: They are integrated into the functional system of the cosmos in the highest form, that of usefulness for theory. Of course, the contemplators of the heavens who were favored in this way would have to have long lives; centuries would have to pass for them as hours do for us; "immortality ought to be their heritage."

The invention of the lightning rod and the determination of the orbits of comets were the prototypical accomplishments of the Enlightenment: The objects or symptoms of man's fear of the natural phenomena that are inaccessible, incalculable, or uncontrollable for him turn out to be within the reach of the instruments of cognition.

When Herschel, in 1781, initially announced under the title *Account of a Comet* the discovery of Uranus, which was later recognized as a planet, this confusion palpably set aside the oldest and sharpest differentiation among celestial phenomena. At the same time, however, it showed that in this question too what mattered was not eidetic marks, but rather the parameters by which the object's orbit is determined. Lambert made significant contributions to the study of comets in several works, spread over his whole life. In his *Insigniores orbitae cometarum proprietates* [*Some Notable Qualities of the Orbits of Comets*], which was published at the same time as the *Cosmological Letters*, in 1761, he established what was later called "Lambert's theorem," which enabled him to manage with a minimum of observational data. The proposition reads, "In every parabolic trajectory the time in which any given arc is described depends only on the corresponding chord and the sum of the radius vectors of the extremities of the arc." If one thinks of the dark background constituted by the centuries that were ruled by fears about comets, this piece of geometrical elegance has almost an 'end-in-itself,' aesthetic quality. Lagrange will say of the theorem that "in this case geometry remained the victor over analysis." That, too, is an insight into Lambert's character as a theorist.

Lambert himself stylized the careless attitude with which he brought the formerly frightful phenomenon into the comfortable environment of his house. In a letter of 1 May 1770, he writes about his observation of the comet of 1769: "I have contented myself with observing it here from my window, with a three-foot-long telescope, a lorgnette, and my pocket watch. However, this was nevertheless sufficient to determine its path reasonably accurately." [27] In another letter Lambert was able to indicate with pride, regarding his home observation of this comet, that on the basis of eight observations he had achieved a tolerance of between one and two minutes and had thus far surpassed the accuracies of other observers, who had corrected themselves in the meantime. [28]

However, only very long-term observations could bring complete mastery of the phenomenon of the comets. Lambert learns of the comet that had been sighted in January 1771; only the nightwatchman of a little town had noticed it. Lambert is able to determine the elements of its path and concludes from them that this comet is identical with none of the known ones. Whether his thoughts went

back at this moment to the idea that comets could change from one
central system to another, we do not know. A more obvious response
for him was the idea of institutionalizing the surveillance of the
heavens for comets, and entrusting it to the nightwatchmen. It is clear
that he preferred to attribute the unknown identity of the comet that
was sighted to the lack of observational data rather than to the
speculative hypothesis of its having changed systems.[29]

How does Lambert see the relation between the intended and the
achieved changes in the consciousness of the universe in a person who
submits to the enlightening effects of science? The literary form of the
correspondence gives him an opportunity to reflect on the imagined
success or failure and to work it over each time in a fresh beginning.
He makes his correspondent insist on knowing whether it is not possible
"to reach certainty in another way than through new doubts and new
uneasiness." The process of science, which with its solutions always
immediately produces other problems, gives one the impression that
the difficulties are only continued on a higher level, because "they are
simpler at the start and become more erudite, when one has solved the
earlier ones."[j] This suspicion is indeed terrible; it touches the heart of
the question of what science can accomplish for man.

Once again, the distressing problem is clarified by reference to the
example of the comets. Science, as enlightenment, is not already
successful simply by virtue of having turned the signs in the heavens
into phenomena. Epicurus's dream of neutralizing the phenomena
has been dreamed to its end. What can no longer be a sign becomes a
possible agent. And is the secret expectation that we might after all
receive some message—obtain signs—from nature really refuted by
that change? Is causality more familiar and more reliable than the
world of signs? To Lambert's correspondent, at any rate, Copernicanism seems to increase the risk of living in the world. "How unconcerned was the good Ptolemy, with his resting Earth, and how
tranquil his followers remained until Copernicus came and began to
lead the Earth around the Sun! But now we have more excitement,
and Copernicus would not be too proud of his triumph if he knew that
we must now be concerned that a comet might come and drag the
Earth with it beyond the fixed stars. . . . I soon wished that comets had
their old significance again and presaged wars and all kinds of calam-
ities, which befall us anyhow and which are less universal than effects
of a kind that threaten not only single countries but the whole Earth,

and that come in a much more unexpected manner than any war does."[k]

The answer that Lambert allots to his correspondent's doubts is instructive in regard to the difficulty of relating the profit of scientific truth to the expectations of life. The fact that new knowledge leads to new questions and doubts is after all simply "the common road along which we come from one truth to another." The only thing one can regret is "that all this takes somewhat longer than we might wish." But no acceleration of the process of cognition would change at all the basic structure by which solutions produce problems in their turn. Posterity is no consolation for the person to whom the truth is not promised and for whom his own life span—with the results, at that time, of mankind's great effort—becomes a mere episode in a process that seems never to be able to guarantee that a present will ever be anything but an episode in relation to the truth. The troubled questioner will no doubt get used to a language in which essential needs have to be referred to future achievements. "Be satisfied always in what we know for certain and leave future possibilities to be resolved by posterity if they cannot be tackled yet."[1]

In comparison to these difficulties with the truth, difficulties that science presents to the age of enlightenment, dealing with comets is child's play. If prediction based on calculation cannot entirely eliminate fear, it can nevertheless set in motion anticipatory precautions. Theory's restlessness at least yields practical reassurance. But the seriousness of this argument is evidently not very important to Lambert. He pushes it to the point of absurdity. If the astronomers were able to predict the flooding of the Earth as the result of the influence of a comet, or even the abduction of the Earth from the vicinity of the Sun into the orbit of a comet, they would have the merit of giving everyone the opportunity to look around, in good time, for a ship, or to prepare himself for the Siberian cold of places far from the Sun. But such an answer assigns too much importance to man. In the great economy, man is not as important as he himself may believe: "The preservation of entire celestial bodies appears to me, at least, to be more important than the preservation of creatures that propagate their species and are reborn year after year."

Precautions made possible by predictions are only a side effect. Prediction begins the process of confirming the theory from which it is derived. Prediction integrates the event into a reality that is still, or again, a cosmos. The prediction confirms the event as in conformity with order, rather than the event confirming the prediction as in keeping with its system. This functionalizing of the phenomenon, as the "appearance of an appearance," has radically changed astronomy's interests. Lambert put the accomplishment of his *Cosmological Letters* in this functional context and measured the right to speculate against this criterion. "Analogy is not sufficient to determine truth with certainty, but it provides an inducement for conjecturing the truth and finding it out completely by means of experiments and investigations serving that purpose. This is where its use is distinguished from its misuse. The extent to which this distinction was precisely observed in the *Cosmological Letters*, I leave to people of keener vision to judge, as long as my intention of using analogy in moderation and correctly is granted."[30] Lambert points out in his defense—in detail, and giving page references—that he proposed concrete observations to the astronomers, in the *Cosmological Letters*, observations "that, if they are continued for many years, can prove the system astronomically."

Lambert's correspondent had complained that since Copernicus the Earth had not only been put in motion but had also been exposed, by its motion, to accidents and collisions. Lambert's response reverses the direction of the argument: "I hope that you will now find less reason to be angry with the excellent Copernicus for having dislocated the Earth from its rest, because you see that any comet can do that, and that perhaps various of them have already produced changes of this kind."[m] Copernicus is not justified by the fact that he made the Earth into a star, but by the realism of his inclusion of the Earth in the complex of the causal factors of nature. And just this gives Lambert an opportunity to pronounce his central formula for the possibility of a Copernican comparative: "You can also understand from this that we are perhaps not yet Copernican enough, although I am not at all of the opinion that we could reach that state by assuming that the Earth will in time become the satellite of a comet."

Lambert's fantasy of having the comets inhabited by creatures who practice astronomy and whose life span approaches immortality now no longer has the appearance of sheer playing with ideas, but instead of a way of expressing the requirement of the Copernican comparative. For all that these comet inhabitants do is to avail themselves of the extreme advantages for theory that follow from their eccentric orbits and from the possible exchange of systems. The position at rest in the center of the universe, as the specially favored position from which to observe it, has long since been forgotten. The alarming conclusion that they are speeding, lost, through space is outbidded by the encouragement for theory that their continual change of position provides. The truth of this universe is not experienced by a god at rest, but by an observer who is enmeshed in the complexity of its motions and deciphers them by means of his changing perspectives. Lambert makes his partner in the correspondence also join in the idea of the comet inhabitants: "Each entry into a new solar system is their spring [season], and they celebrate the fall when they leave it again.... It gives me no small pleasure that there are such creatures who take in at a glance the world-edifice in its entirety, and I would like some day to journey with these astronomers around millions of suns."[n] It is a sketch—before Laplace—of a new intelligence that would be adequate to science's cosmic dimension. Changing from one system to another is its form of existence.

This century of the Enlightenment does not defend itself only against the oppressive suspicion that reason's fate could be absolutely bound to the lowlands represented by human history, and delivered up to it as the ultimate authority. It also agitates the other problem of whether reason has to be tied to human nature, contingently equipped, as it is, with this particular organism. Lambert has his partner in correspondence write that "the general concept that we should form about the inhabitants of the world is still much too narrow, because we have not seen diversities other than those we find around us on the Earth.... How difficult it is for us to imagine a thinking being, an intelligent creature, without also giving him two hands, two feet, a head, and other parts resembling a man. By stretching our imaginations, we add, perhaps, wings because we feel that we lack the ability to fly."[o] It is remarkable that Lambert nowhere connects the idea of the inhabitants of other worlds with reflections on their morality. The comet inhabitants would have to be brilliant astronomical theoreticians, and in addition also approximately immortal—but what effect might these dealings with the starry heavens have had on their morality? This question obtrudes itself if one thinks of the use Voltaire made, at about the same time, of the idea of inhabited worlds, in which his axiom is that light is the same for Sirius and for us, and morality must also be the same. The superiority of the inhabitants of other worlds in the sense organs with which they are equipped, as in *Micromegas,* is combined with a superior morality, though in its norms it is the same as that of mankind.

In the idea that in view of the riches of reality, the organs that man has at his disposal are limited, Lambert also sees a problem of language. The captivity of our faculty of imagination in a type of perception that is governed by the preeminence of sight allows us to speak only indirectly about realities that cannot become optically present to us: "The nature of light, its effects, the partnership between soul and body either are not known to us as yet, or are known only insofar as we receive the impressions that light makes upon us, and we seek to fill in the rest through inferences for which we have as yet no other principles than the ones we arrive at through seeing!" [P] One should call to mind, in connection with this passage, to what a great extent our technical world has grown into the relativizations that were anticipated in speculations like these. A world in which a measuring apparatus not only surpasses perception in precision and range, but also can largely replace it, and at the same time offer its data at will, by means of pointers, values on a graduated scale, diagrams, curves, punched tape—such a world has moved far away from specifically optical structures of objects as the familiar and ingrained ways in which a life-world is given. The variation in what is possible as a means of access to realities has long since left the sphere of the imagination, and also of mere assertions of contingent facts about man, and has become the technical task of procuring information for which no "organ" is competent, of storing it and of processing it up to the point of a plan of action that is appropriate to the facts. All of this has become a sphere of observations and languages that is separate from man.

As was Leibniz at the beginning of the century, Lambert was struck that the Copernican reform had left everyday language unchanged. With this language we continue to interpret our sense-experience in a Ptolemaic manner. Seen from the higher standpoint of the cosmology of the Copernican comparative, a new justification of this behavior now emerges. The Copernican system's claim to realism proves to be temporary. A Copernican language would have overcome the backwardness of everyday language in relation to cosmic reality only to the extent of one single step. The cosmological system of worlds nested inside one another, and of their revolutions within revolutions (which are in fact oddly reminiscent of Ptolemy's epicycles), is reflected in a whole nested system of possible languages. In this system, any language that refers to a partial system can be corrected with the aid of the higher-level system, in the same way in which Copernicus had corrected the everyday language that is associated with the world of the senses. The complexity of these corrections, one on top of the other, is greatest in the lowest-level system. The language that would be valid for an observer on the Moon is a degree more complicated still than the one that is valid on the Earth, if this observer does not want to waive a complete description of the reality that presents itself to him, and to take into account in his language (as one does in a prepared sample) only the configuration "Earth and Moon," or "Sun, Earth, and Moon." An astronomically adequate language, which would have to regard the solar system, too, as only a lower-order system within one new and integrative relation after another—a fully 'corrected' language, that is—would be not only unobtainable but also impracticable.

Tellurian astronomy is forced to approach the total cosmic system from, as it were, the wrong direction. Copernicus had thought of the observer's eye as transferred into the Sun, and immediately the intertwined planetary appearances disentangled themselves. This perspectival transposition was comparatively easy to carry out; true, the privileged standpoint was lost for the Earth, but it could be regained with one single leap. The Copernican comparative changed this situation entirely. Lambert describes the change in his eighteenth letter: "So inexhaustible is astronomy that we must necessarily be satisfied with hypotheses and understand only how much these hypotheses must gradually be extended and made more complex. They go through uncounted degrees closer to the truth, and with each degree a new yardstick of space and time begins to be used. In the Copernican hypothesis our yardstick for space is the radius of the Earth's orbit, for time a year or the duration of the Earth's revolution on its orbit." [q]

At least in astronomy, Lambert writes, we have completely accustomed ourselves to the Copernican language. In this language we proceed from the closed orbits of the planets around the Sun, and only by a special effort can we speak of the cycloids in terms of which these paths have to be described if the motion of the Sun, in the next higher-level system, is to be taken into account as well. In the Copernican system we forgo putting the Sun, in the midst of the elliptical orbits of its planets, in motion, "and use the so-far-customary language, because we can never use the true language in a more definitive way." Now this is another point to which Lambert can make his partner in correspondence react with shock and confusion. The expectation that stands at the beginning of the modern age and of all enlightenment— that with a fundamental reform the prejudices of language, of the senses, of custom, and of the past can be eliminated, and the truth gained by a *coup de main*—finds itself disappointed and discouraged, on the edge of a kind of satiation with Copernicanism. "Now we shall be Copernican enough for once, or we shall never be, or we never should have been. . . . " [r]

Astronomy's classical expectation that simplicity will finally prove to be the principle on which the universe is constructed has once again become an illusion. We shall never know the true path—the cycloid constructed out of one superimposed revolution after another—that the Earth traverses in absolute space, and consequently we shall never command the true language for describing reality, as Lambert admits in this letter. In order to be able to pursue astronomy nevertheless, one has to decide in favor of a language in each case. Thus the history of astronomy already presents itself as a history of language, whose overall syntactical principle is that of Copernican correction, and whose test remains the question whether this principle has been exhausted. "Ptolemy stayed with the popular common language. Copernicus began to make the first step and taught us the alphabet of the first astronomical language. But he did not know that the language was merely hypothetical and would lead us to the true language only by way of many steps. Tycho Brahe misjudged the matter and confused its vowels with its consonants, but people soon found that the pronunciation suffered thereby and was not at all fluent. Kepler and Newton have completely removed this confusion and found the means to raise the still somewhat rough language to its true refinement and to contrive their epic poems in accordance with the mythologies of their times. But times change, and our immediate descendents will allow this language only in poems, or they will use it to abbreviate terms, where it won't matter. But wherever something important is under way, they will speak in a higher tone." [s]

But what, then, are we to conclude about the principle of teleology? Copernicus had insisted that even though man was displaced from the center of the world, he was still the reference point of the universe's inherent purpose—indeed, that this purpose is only fulfilled by man's proving himself a match for the difficulty of grasping the way the world is constructed. Lambert sticks to the principle—which he calls the "great principle"—that disorder in the world can never be anything but apparent disorder. But in the meantime the true relations of order have lost their comprehensibility for human reason. Here a seemingly theological model is at work. The hiddenness of order behind the prospect of disorder not only describes the goal of an effort that must be summoned up unceasingly but also serves to elevate and glorify that order. In Lambert's cosmology man's position, if we ignore for the moment the possibility of living on the Moon, is, for the purpose of perceiving the cosmic order, the least favorable one that is conceivable. Could he have understood just this as the teleological point? What else can it mean when he writes that where the disorder "seems to be greatest, the true order is far more magnificent yet, but only more hidden from us." The point where the order reaches its greatest magnificence is also where reason undergoes its greatest difficulty in perceiving this order. This has altered the view of the starry heavens in relation to history: "I contemplated the starry heavens for this purpose as recently as yesterday evening, because I had never actually been able to find a certain symmetry in their apparent form. Once again I sought it in vain."[1] The idea of order has become a sheer rational postulate, to which all realizations in intuition continue to be denied.

This unfitness of man for the cosmos is due, to use an expression of Karl Ernst von Baer's, to the insufficient length of his moment of life. The perspectival changes in the heaven of the fixed stars in which the motion of the solar system through the universe, as a part of the rotating Milky Way, and the latter's motion (in turn) around the center of its system could be reflected would have to be so minor that, even with the highest accuracies that were imaginable in Lambert's time, they would be ascertainable only across centuries and millennia. "But in what millennia to come," Lambert makes his imaginary partner ask resignedly, "will our descendents live who will discover a change in the locations of the fixed stars, and thus discover the laws of the whole firmament?"

Here we can find a very subtle reason for Lambert's interest in systems containing central bodies. Lambert may know that a system of bodies can rotate around their common center of gravity; but the periods of rotation around central bodies can be significantly shorter. That gives such systems the advantage of providing more favorable conditions for observation. A universe that was equipped with great masses all the way up to the final great systematic center would be a universe all of whose revolutions would have shorter periods. So here a teleological consideration relating to the possibility of knowledge (though admittedly still of knowledge by longer-lived creatures than man) turns the scale in the direction of the thesis of central bodies— which may represent Lambert's intentions more accurately than if we interpreted him as shrinking from the idea of an 'empty center,' or as needing the reassuring solidity of a completed system. "Inasmuch, however, as we know that the last step will end at the body that directs the whole creation around itself, I believe our thinking is sufficiently Copernican; or do you want, Sir, to go even further?" To be sure, that, in the nineteenth letter, is an utterance of the partner; in his own role in the dialogue (as late as the tenth letter) Lambert's formulations are more cautious: "I leave it therefore undecided whether a system of fixed stars revolves only around a central point, or whether at that point there is in fact a body of enormous mass toward which the fixed stars gravitate." [u] In the sixteenth letter, Lambert still represented himself as undecided as to "whether I should leave empty the common center of the system, or place there an enormously great dark body whose mass should be heavy enough to keep each fixed star in its orbit just as our Sun does this with respect to the planets and comets." [v] Only in the eighteenth letter is his inclination very marked to equip "the center of the whole realm of reality," as "the first driving wheel of each movement of the world-systems," with that inestimably great central mass. Lambert is "enthralled by the splendor, size, majesty, and beauty of that position, which spreads order and laws across the whole creation." [w] If this position were a mere center of gravity, in empty space, then the order of the system would be nothing but the "continual mixture of millions of individual effects ...," approximately the way it works in the course of events here on Earth. "But I do not regard the world-edifice as set up in that way."

The discussion of the problem of order ends, in the nineteenth letter, with the conclusion that the choice of a language decides between order and disorder. Everyday experience on Earth provides us with the superficial appearance of disorder, which, however, turns out to be a combination of patterns governed by universal laws, the discovery of which brings order to the fore. "In the firmament, the reverse is true. The simple order is the apparent one."[x] The planet with a satellite revolving around it is the simplest constructive component; but if the planet in its turn is regarded as the satellite of a sun, the circular motion of its moon loses its initial appearance of beautiful regularity. It is true that Copernicus also seemed to have repaired what had initially presented itself as disorder, and to have established simple relations in the solar system—but this was only because though his thinking was Copernican, it was not sufficiently so. As soon as science's mightily patient subject had found even the first trace of a motion of the Sun in relation to the fixed stars, this provided a reason "to regard the order of the planets, also, as fictitious, and to conclude once and for all that the true order is the most complicated one and will never be reached by us." The paradox of this state of affairs is that the lowest-level system can be brought to a high degree of order, in the language that is appropriate to it, but that at the same time this order is the most deceptive in relation to the goal of describing the whole.

This paradox also underlies the situation in the history of science that is reflected in Lambert's thinking: The supposed termination of astronomy's problematic state by the Copernican description of the solar system seemed to leave only an empirical demonstration of the parallax of the fixed stars to be desired; but precisely the demand for this concluding evidence unexpectedly and unintentionally led beyond the narrow Copernican problematic. The requirement of the Copernican comparative originates in the experience of the price that is paid for the reassurance produced by a gain in order. After all, it was not yet certain whether in the comparisons between contemporary observations of the fixed stars and Hipparchus's catalogue changes had been found, or whether a disagreement between the ancient and the modern observations was only due to the increased accuracy of instruments and to characteristics of light that were not yet known to the ancient astronomers. Lambert admits this uncertain basis as late as his tenth letter. But what he is concerned with here is not that at all, but the peculiar way in which the solution of the problem of the structure of the world of the planets conceals the problem of the structure of the world of the fixed stars. As was soon to become evident in connection with the discovery of further planets beyond Saturn, as well, what was decisive for astronomical progress was no longer the scandal of the great inaccuracies but rather the instructive value of very small irregularities. "The world-edifice, considered as a whole, has a thousand driving wheels mutually so oriented that each bigger one locks into the nearest smaller one and this into the next one and causes in the smallest one countless wobbings."

Although Lambert mistrusts the Copernican language, as one that suggests arrival at a definitive order, in his procedure of "analogy that progresses from system to system," he clings to it out of desperation. Describing every system as an analogue of the solar system is an expedient adopted, so as not to lose control of the uncertainty that has just been opened up, by thinking that is still far from sufficiently Copernican.

The repetition of the Copernican prototype on every level of the cosmological construction gives this diagram a sort of eidetic validity. The teleologically deduced simplicity and clarity consist in the fact that there is only *one* basic pattern. But this revives a problem that had already been involved in the ancient idea of the relation between form and matter: The validity of the form makes the diversity of its instantiations in concrete cases arbitrary and unimportant. Since it would never be possible to demonstrate empirically that the final cosmic central body was absolutely at rest, the system simply cannot be protected from the suspicion of always having more superimposed layers: "But if instead of these three levels there were innumerable levels, if these three levels did not make complete enough the subordination of systems, if there was no proportion between the number of planets and comets around a sun and the number of all the suns, then we should undoubtedly have to admit still more levels." [y] That is in the eleventh letter, and in the nineteenth this fundamental question about the attained or attainable output of the system in terms of order has still not been settled. On the contrary, resignation in this regard seems inescapably final: " . . . I am even more astonished that in all this only the simplest order is apparent, and the true order is as complicated as it could ever be. We shall never be able to survey the true one entirely, and consequently it would never serve us, even if it were the simplest of them all." [z]

The economy of the claim to theory and the deduced rationality of the cosmic object are radically divorced. Only forgoing surpassing the language of the level of construction on which one finds oneself guarantees the appearance of a perspicuous order—of reality's making some concession to the needs of a finite creature. Only then and only in this way does this order accommodate itself to our use, as Lambert says, and "present itself to us under the simplest form insofar as we need it. Perhaps this appearance is so essential that it was merely in order to maintain it that the true order had to be so very complicated." The totality of this universe has sublimated itself into an idea—and no longer in the sense in which "idea" was understood in antiquity. The Copernican step is no longer the step from illusion to reality [*vom Schein zum Sein*] but rather from illusion to appearance [*vom Schein zur Erscheinung*]. The correct choice of one's perspective, and thus of one's language, has brought about an evidence relative to time, an evidence that is relativized by the lengthening of the observational basis and the enhancement of observational accuracies in the interest of a greater system. Only what will come to light again and again in such steps, as the basic pattern, is supposed to be established from the beginning. Like all the scientific exertions of the modern age, this too develops into an "endless task." But that, in turn, restores to the point of departure—to the despised dependence of the senses on immediate appearances (a dependence that is subject to correction by theory)—a sort of legitimacy in the life-world. "The more levels came to be added, the more complicated became the path of the original celestial globes; only their appearances remained simple, as they had to be if the firmament was to present to each planet the most perfect clockwork." It is time—Lambert thinks, as Copernicus did before him—that remains and has to remain unaffected by the reconstructions of the world.

3

A Retrospect on Lambert's Universe, from the Twentieth Century

After the two centuries since Lambert's *Cosmological Letters*, the question of the quality of the universe in regard to order may sound antiquated. In fact it has never fallen silent, but has only been formulated differently. Its continuing connection with Lambert's name is documented by Karl Schwarzschild's address to the Göttingen Academy in 1907 about the *Cosmological Letters*.[31] This important astronomer was in more than one respect a successor to Lambert, above all in that he gave a new stimulus to photometry—the discipline founded by Lambert—by exactly determining the laws governing the darkening of photographic plates. It now became possible, by considering the relation between apparent and absolute brightnesses, to determine the spatial distribution of stars, which Lambert could only anticipate as a result of establishing perspectival changes in the firmament of the fixed stars. Photometry is also the point of departure for Schwarzschild's critism of Lambert, that a consistent application of this method would have put him in a position to make a realistic estimate of the degree of emptiness in the universe, and to cease all teleological speculation to the effect that the dynamic filling of space must be the principle of cosmology. This universe is empty to a greater extent than the space needed by rotating systems could ever require or justify. Prodigality rather than economy turns out to be the main heading in the bookkeeping of the cosmic household.

That does not leave the twentieth-century theoretician unmoved either. He admits that what Lambert would have called completely inappropriate, teleologically, also makes us uneasy and makes us feel with painful acuteness "how remote we must still be from knowledge of the idea that governs the whole." But something that Lambert had not touched in any thought now seems more alarming than the emptiness of space: the "indescribable waste of the Sun's energy." The emptiness of the planetary system in which our Earth is located has as a consequence, Schwarzschild says, that only a millionth of what the Sun radiates strikes the surface of bodies.

As Lambert had spoken the language of his age, so the astronomer of the beginning of the twentieth century speaks the language of his. If Lambert saw the teleological nature of the universe in its elimination of the risk of collisions despite the greatest possible filling of space with moving bodies and systems of bodies, for Schwarzchild the minimization of this risk is the outcome of a process that can most readily be described in the language of Darwinism: "Living creatures were not created in accordance with purpose; instead, the struggle for existence selects out those that are adapted to a purpose. The planetary system was not originally constructed in such a way that no collisions come about; instead the collisions themselves have produced an arrangement in which they are excluded." The modern astronomer has ceased looking for cosmic harmony and has "laid down absolute disorder, instead of order, as the guiding principle." People have begun to picture the stars as the molecules of a gas which whiz about in a completely accidental accidental manner and for which at any point in space one particular direction is just as probable as any other. Purposelessness can also be a great reassurance—namely, the reassurance that the end is not pursued at the expense of the means and is not directed against an element of the system.

One only needs to remember that in Lambert's system in the end, despite all the teleology of the whole, man has to approach its rationality from the wrong direction. At least the surrender of teleology spares him any consciousness of being the intended object of such meanness. Schwarzschild is correct when he says of the absolute disorder that he has proclaimed that a view of this kind also contains philosophy. This built-in philosophical component cannot be got rid of, because natural scientists, who are occupied "with concrete things," have still always "attached themselves with remarkable obligingness to the ideas that are regarded in each epoch as progressive." In one sentence Schwarzschild passes in review the "philosophical fashions among natural scientists," beginning with "Kepler's cheerful deism," and proceeding by way of "Newton's melancholy mysticism" and "Laplace's religion of reason" to "Mach's critical empiricism," which "nowadays constitutes the natural scientist's public costume." Such labels simplify the problem of the scientist's implicit philosophy.

What philosophy is involved in Schwarzschild's confrontation with Lambert? In any case it is not that of an interest in historiography, but more nearly that of a self-presentation, which still claims, with the impetus of the Enlightenment, to be able to announce the definitive end of all of mankind's prejudices. Rationality becomes the negation of what historically had been premises of rational deductions. But the guiding principle of absolute disorder in fact relates only relatively to what was once regarded as order. Lambert's *principium plenitudinis* [principle of plenitude] required complicated provisions against catastrophes and collisions, i.e., a fussy and solicitous reason in the background. He did not ask himself what this reason would in fact have achieved. Was an increase in the matter in space really the same as an increase in purposefulness or in meaning—was it a sufficient reason for the whole? In accordance with his 'philosophy,' Schwarzschild is able to point out that selective evolution produces progress only where a surplus and an emptiness complement one another. This universe of absolute disorder functions precisely because it is, on the one hand, so empty and, on the other hand, so old. The primacy of chaos is justified by the fact that its terrors are long since past.

But if we imagine Lambert's fictional correspondent as a member of the audience at Schwarzschild's Academy lecture, we shall nevertheless see him shrink from the idea that the universe could be nothing but a late stage of chaos. Of course for this current state there is no problem about laying claim to the millions and billions of years that Lambert was prevented from assuming. The time factor accomplishes what can no longer be attributed to a reason in the background. On account of its age, and by means of its age, the relative chaos has already completed a good part of its catastrophes, which have become less common (and promise to become still less common) as a result of the agglomeration of matter and the corresponding emptying of space. The hidden consolation of the theory that operates with absolute disorder lies in this asymmetry of time in relation to every present moment. Of course, its argument is different. It asks by what right it is that the idea that the cosmos is chaos itself, grown old, has an alarming effect.

Schwarzschild points out that in the realm of micro-reality we do not feel such an alarm. Or are we frightened by the idea that the molecules of air in any given space move accidentally and purposelessly—that is, chaotically? But if that can and does leave us unmoved, how then do we justify regarding universal space differently from a randomly chosen volume of a gas? "As men, we do not want to assume that the importance or the inner value of any event depends upon the absolute magnitude of the things that are involved in it. Such a view of things would overwhelm us. But if absolute magnitude does not make a difference, then neither is there a difference between molecules and stars." Schwarzschild does not mention the relative magnitude of what is visible and tangible for man, or can become fateful for him. Even if he is right to omit this in the language of science, there is also a language in which one can discuss this omission. Had Lambert failed, in his scientific language, to take account of this absence of a difference between absolute magnitudes?

There had still been, in his thinking, a remnant of the traditional metaphysical boundary between the Earth and the heavenly bodies. This boundary separates things that are essentially different. Only cosmology can present the rationality of the world in pure form, while the disciplines that deal with terrestrial objects and processes have to be satisfied with a lower level of rationality. To give only one example, Lambert was not disturbed by the fact that he was unable to bring his investigations in the area of hydraulics to the point of a precise mathematical presentation. Here he was willing to accept the chaotic element: "It is after all admitted on all sides that a geometrical rigor is not to be hoped for here, as long as one does not calculate the course of every little bit of water, and that in the meantime one must be satisfied if the calculation does not deviate from experience all too markedly." [32] He encourages his addressee not to demand more than such approximation, and to "cite the still reasonably tolerable agreement between the calculation and experience as the basis for the acceptability of the calculation." Here, then, on the terrestrial 'darker side' of the universe, the metaphysical demand to see things in strict harmony with the plan of a calculating reason is no help. Here philosophy becomes the irksome admonisher who is inclined to demand things that are unrealizable and who only hinders or delays practical results in the zone of a reality that is at our disposition. "For after all, no amount of philosophizing will make the formulas any more correct than they are." It is important to see how with this object, which the traditional scale of values had characterized as an inferior reality, Lambert speaks an entirely different language than he speaks when it is a matter of the heavenly bodies. This also already marks the deep gulf that separates Lambert's cosmology from Kant's almost simultaneous cosmogony. Lambert could not be disposed toward describing the universe from a genetic point of view.

The questions of order and chaos, fullness and void, the size of space and the length of time also have their historical perspective. That is yet another Copernican comparative, in another dimension. What are we to make of alarm at the void? To an age that has begun to have nightmares of overcrowding, the perfection of a well-filled universe—even if it were capable of theoretical verification—is no longer plausible. It finds, instead, a source of confidence in the objection that Schwarzschild raises against Lambert's teleology: that in spite of the immense speed with which our Sun has traversed space for tens of millions of years, it has never come (nor will it come) significantly closer to another star. It was not that Lambert needed such a large housing in order to accommodate his clockwork; rather, it was because his universe was still, after all, laid out on much too small a plan, that it needed such a clockworklike order of nesting systems and provisions against catastrophe. Only when the universe is conceived as sufficiently large do the stretches of time become available that can satisfy the requirements of evolutionism. There is another aspect to this too: Lambert regarded bodies as what stands opposed to the void, what counterbalances it and fills it up. What he could not know was that this is only an illusion relating to our senses—that the void reaches deep into the structure of solid bodies themselves, and the proportions of the universe are only mirrored there.

This allusion to the change in our concept of the inner structure of bodies is closely connected to the way in which Lambert's problem of central bodies definitively disappears from cosmological discussions. Full of admiration for "Lambert's magnificent scheme," Schwarzschild not only examined the idea of the central bodies from the point of view of their mechanical superfluity but also tested its possibility by the means developed in astrophysics. He conjectures that there will be a natural upper limit for dense stellar masses, beyond which their atomic structure would have to collapse on account of the pressure inside them. Even the central body postulated by Lambert for the Milky Way would have to have the mass of 130 million of our Suns, and with the same density would be far beyond the natural limit that Schwarzschild supposes. So our cosmological imagination has to forgo these "fantastic monsters," which it can do all the more easily since the object that Lambert wanted to achieve by means of them can also be achieved without them.

Here Schwarzschild reaches directly back to Lambert's political metaphors. He makes mathematical mechanics guarantee "that order can also prevail in a republican association of stars." Still, he expounded this opinion in a meeting of a "Royal Society."

Schwarzschild describes the universe of the simulated correspondence as a "beautiful cosmological myth that Lambert composed." But this friendly derogation relates above all to Lambert's excessive application of the principle of teleology. However, Schwarzschild is candid enough to say that the philosophy of continual triumphs over prejudice has always only vanquished what was already dead, and that it has found it all the more difficult, in viewing what it has conquered, to recognize its own constraints. He has to admit, in the end, that at the beginning of his century, too, teleology is still not entirely dispensable. It is always only in retrospect that the use of the principle of teleology seems questionable. Bacon had described it, in view of the excesses of Scholasticism, as an intact virgin that gives birth to nothing. But if, instead of the universe's disposition for habitability and for astronomical knowledge, one posits its disposition to be stable across very long periods of time—a principle, then, of the preservation not so much of the human race as of the "arrangement of the whole visible universe"—then it very quickly turns out that in so doing one hits upon a necessary implication of astronomical statements. It cannot be correct to say that the aspect that the universe offers to its observer is a merely contingently given cross section of a chaos at an arbitrary point in time. Schwarzschild writes that a teleological inference forces us "not to satisfy our minds with the idea of the great universal gas, but rather to look for laws of a very definite, specific kind in the motions of the fixed stars." It is the same consideration that had caused Lambert—like Aristotle, before him—to exclude rectilinear motions of stars and systems of stars a priori, because they are contrary to the principle of the preservation of any formed whole. Schwarzschild's reflection relates to the peculiar fact that a particular view of the Milky Way presents itself to a viewer at a contingently given time at which the observed state is extremely improbable. If, in the ocean of time, in the atom of a moment, we perceive an ordered state of affairs, we can hardly regard this as a short-lived transitional stage in an accidental process. "That is to say, we cannot believe that the Milky Way is a transient phenomenon and that we have accidentally been born into the particular million years in which the stars have just come together into a ring or a lens; instead, we have to assume that the arrangement of the fixed stars in a somewhat level, flat formation also holds firm in cosmic time periods."

If one follows this line of thought, one must exclude the possibility that the stars move with the same frequency in arbitrary directions. Our solar system would require a hundred million years to traverse the diameter of the Milky Way; but in that amount of time, if the motion of the system's elements were irregular, it would have flown apart. "So we have to assume that the fixed stars, similarly to the planets, by nature move more or less parallel to the plane of the Milky Way and always led back into this plane by the gravitation of the whole."

Here Schwarzschild does something that is very similar to what Lambert had done: He initiates a teleological reflection, and then asks whether reality provides indications that it might prove to be correct. "But what we might wish in this way for the sake of our principle, for the sake of the order of the cosmos—that is reality." It is correct, Schwarzschild says, that the fixed stars' motions mainly run parallel to the Milky Way. In spite of the vast material of available observations of the fixed stars, astronomers had succeeded in discovering this only in recent years. Now it is important to see how Schwarzschild explains this delay, as he perceives it to be. The explanation leads directly back to the problems associated with Lambert's eidetic attachment to the Copernican schema.

Schwarzschild says that "where people have abandoned the image of a gas, at all, they have mostly pushed the analogy with the planetary system too far." Premature recourse to a system that was graspable as an image—picturing "the Milky Way as like a great wheel, which slowly turns"—obstructed penetration to definite empirical results. What one will find does not appear to Schwarzschild as something derivable from the principle of absolute disorder. The metaphor that he chooses so as to define the overall view of the universe that is emerging is no longer the metaphor of clockwork. If more order prevails in this world system "than one dared to surmise only a few years ago," then what comes forward here, "as though through a dissolving fog, are the outlines of a huge organic form."

Is this another "beautiful cosmological myth"? Or is a teleological principle again leading, as with Copernicus, into a new epoch of astronomy? As it turns out, the century and a half since Lambert have not enabled Schwarzschild to gain a definitive distance from the philosophy that motivated him. The astronomer at the beginning of our century is also able to define his situation in the history of science only by seeing it as the overcoming of a disorder that had previously seemed impenetrable. Schwarzschild formulates it like this: "There seems to me to be no doubt that with this we are standing at the beginning of a new epoch. The chaos begins to get clearer."

How do matters stand, in this new epoch, with "man's position in the cosmos"?[a] The answer can no longer be Lambert's. Man is not the addressee of the cosmic performance; not even the almost immortal inhabitants of comets could be that. How can the outlook for man in this universe be formulated? Perhaps precisely thus: that the scale of his life and his fate lie below the threshold of cosmic relevance. As a cosmological phenomenon, man does not exist. But just for that reason it is possible for him to exist beneath its orders of magnitude. The Earth's habitability is not affected by the postulated disorder. The Sun, on its path through the universe, would be able to repeat the history of life on Earth a hundred thousand times before it once came close enough to another sun to make organic life impossible in this planetary system. Even the principle of absolute disorder gives Lambert's desire for a habitable universe a certain prospect of fulfillment. Schwarzschild defines it like this: "Thus in spite of all the disorder on a large scale, for each molecule of the universal 'gas' and for each of its Earth 'atoms' a colossal free time en route—a prodigious stretch of undisturbed potential development—would continue; and what holds for our Earth will also hold for thousands of other satellites of fixed stars." This comforting reassurance preserves the premise, inherited from the Enlightenment, that life and reason in the universe should not be exposed to the fragile singularity of their contingent existence on Earth. Here, however, the "new epoch" still has its disappointments in store.

4
Competing Proposals for the System of Systems: Kant and Lambert

Lambert once said of Kant that it was "this philosopher, among all of them, to whose type of thinking mine is the most similar .... "[33] In 1766, when Lambert wrote this, this similarity was already attested, astonishingly, by the appearance, only six years apart, of the principle cosmological works of both of them: Kant's *Universal Natural History and Theory of the Heavens*, in 1755, and of course Lambert's *Cosmological Letters*, in 1761. The reproach was natural, and could not fail to be raised, that the similarity that Lambert gladly remarked upon was in fact an unadmitted dependence.

While Schelling, on the occasion of Kant's death, said that Lambert had repeated and made famous Kant's daring ideas about the systems of the fixed stars, the Milky Ways, and the nebulas, "without mentioning Kant,"[34] Schopenhauer was the first to suggest that Kant himself, in his *Der einzig mögliche Beweisgrund zu einer Demonstration des Daseins Gottes* [*The One Possible Basis for a Demonstration of the Existence of God*], in 1763, had intimated "that Lambert in his *Cosmological Letters* (1761) silently borrowed that theory [namely, of the origin of the planetary system] from him."[35] Schopenhauer even goes a step further, making Kant, indirectly, the author of Laplace's theory as well. Laplace could have known the French translation of the *Cosmological Letters* (which had appeared in Berlin in 1770 under the title *Système du monde*), and could have got from it the supposed borrowings from Kant's work. This illegitimate descent of every important cosmological accomplishment from Kant provokes Schopenhauer into a vigorous moral sally: "It must greatly distress us when we find minds of the first order suspected of dishonesty, a thing that is a disgrace even to those of the lowest rank. For we feel that theft is even less excusable in a rich man than in a poor one. But we dare not be silent about this, for here we are posterity and must be just, as we hope that one day posterity will be just to us."

Let us leave the issue of dependence aside for the moment and only examine whether Kant actually asserted what Schopenhauer attributes to him. Kant added to the preface of his essay on the demonstration of God's existence, in 1763, a note in which he refers to his *Theory of the Heavens*, of 1755. He says explicitly that it "did not become well known," and he grants that it must also "not have come to the notice of the famous Mr. J. H. Lambert," who "proposed the very same theory of the systematic constitution of the world-edifice as a whole." Kant infers precisely from the agreement between their theories that Lambert must have been ignorant of his forerunner, on the assumption, which goes without saying, that otherwise he would have named his source. But even if one assumes that up to this point Kant's note is too weak in the way in which it grants Lambert's independence, it is only on the assumption of such independence that the last sentence of the note retains its meaning. This sentence describes the coincidence of the ideas in the two works, down to their "smaller features," as favoring the expectation that "in what follows, this outline will receive further confirmation." Only if both authors had arrived independently of each other at such a congruent result could one infer that their agreement was grounded in reality, so that it legitimated their theory. That Kant could have expressed in this context the accusation against Lambert that Schopenhauer imputes to him is out of the question. The opposite is the case.

This makes all the more interesting the question of how the agreement between the basic conceptions of the two authors could have come about. Kant himself expressed his opinion on this matter in 1781, in a letter to Johann Erich Biester. He had to correct a misunderstanding that actually reversed the relationship of dependence that was suggested later. On the basis of information from J. F. Gensichen from Königsberg, Goldbeck had remarked in his *Literarische Nachrichten von Preussen* [*Literary News from Prussia*] that the *Universal Natural History and Theory of the Heavens* had often been mistakenly regarded "as a product of Lambert's spirit [*des Lambertschen Geistes*]." This notice caused Kant to ask Biester to publish in his *Deutsche Bibliothek* [*German Library*] a correction, for which he encloses, with his letter, a draft of a text. In it he insists on the priority of his own work, but compares it to a faint silhouette in relation to the "masterly synopsis, borrowed from no one, of the cosmological system." In this connection, now, Kant explains the "agreement of [their] conjectures" by the fact that both authors had been led to common ideas by the analogy between their greater world-systems and the planetary system. Lambert himself had remarked on this to him in 1765 in a letter, "when he had accidentally become aware of this agreement between our conjectures."

But Kant also points, in this connection, to an essential difference between Lambert and himself. For him, he says, the confirmation of the elliptical shape of the cosmic nebulas had been essential in reaching his hypothesis, whereas Lambert had not gone into this form of the nebulas, with its similarity to the planetary orbits, at all. One can add that he did not go into it although this circumstance, which favors his conception, should not have escaped him; and one could also add, going even further, that it would not have escaped him, either, if he had actually been acquainted with Kant's *Theory of the Heavens*. What Kant had himself described as his "competition" with Lambert, he now decisively softens in the last sentence of his draft correction: "But the correction of the shares [of different people] in conjectures that will probably always remain conjectures is of only minor importance."[36] Written in 1781, this concluding sentence also indicates his distance from his own early cosmological treatise of 1755, a distance that will occupy us later on.

In 1782, Kant published a review of Lambert's correspondence, which Bernoulli had edited. He took the opportunity to pay his respects to "the great man's comprehensive spirit and his prodigious activity." Kant does not do this without deploring the "almost extinguished zeal of men of learning in the dissemination of useful and fundamental knowledge." In that connection he mentions the remedy that Lambert had already recommended for this cultural crisis, "namely, the setting up of a confederation that would work with united powers against the barbarism that is gaining ground and, partly by improving certain still faulty methods, would again introduce thoroughness in the sciences." If one relates this remark of Kant's back to the beginning of the epistolary relations between Lambert and himself, it sounds like a self-correction that came too late, since it could not reach its deceased addressee.

In their very first exchange of letters, despite the declared similarity in their type of thinking, an instructive difference came to light. Kant had rejected the pessimistic tone with which Lambert, in his letter, had complained about the failure of his *Architektonik* [*Architectonics*]: It would have had to be a novel in order to be well received by publishers and readers. "Around here people philosophize only about the so-called belles lettres." Kant is not prepared to join in this lamentation. It is true that his addressee complains justly about "the endless trifling of the 'wits' and the wearisome loquacity of the scribblers who set the tone nowadays, and who have no better taste than to discourse about taste." But this very state of affairs is a positive symptom for him after all: It manifests the "euthanasia of the false philosophy .... when it expires in ridiculous games." Because for the invigoration of the true philosophy, nothing was more necessary than "that the old one should destroy itself." That is an idea that (here as elsewhere) readily connects itself with the word "revolution": " . . . thus the crisis of learning, in a time when there is nevertheless no lack of good minds, constitutes for me the best hope that the so long wished-for great revolution of the sciences is no longer far away." [37]

The disagreement with which the epistolary intercourse between Lambert and Kant commences seems more instructive to me than the question (which is already almost settled) of the possible dependence of Lambert's cosmology on Kant. For this disagreement also allows us at least to surmise where the roots of the differences between their cosmological systems could be located. If in Kant's colossal picture the coming into being of worlds, and their coming into being through the perishing of other worlds, is the leading theme, while in Lambert it is the (one almost wants to say "fated") circling of circles within circles—the perfect functioning of a clocklike apparatus—then the cosmological views of the two thinkers look like metaphors for the disagreement (which, while it did not speak of worlds, nevertheless implied them) in their first exchange of letters.

In his first letter to Kant, Lambert also goes into the relation between his *Cosmological Letters* and the *Universal Natural History and Theory of the Heavens*.[38] He does not see this question in the perspective of someone who is defending himself against a suspicion or vindicating himself, but on the contrary in that of a gratified pride in what he proffers to Kant in the formula about the similarity in their type of thinking. He speaks of testifying to Kant "the pleasure that I find in the fact that we have fallen into the same path in very many new ideas and investigations." It is in this context, of their common possession of a scientific attitude and a critical thoroughness that (for Lambert) had been hopelessly pushed aside by the spirit of the age, and of the resulting convergence of their philosophical efforts, that Lambert places his statement that his *Cosmological Letters* originated independently of any knowledge of Kant's early treatise: "I can assure you, Sir, that I had not yet come across your ideas about the world-edifice . . . at that time." He had first become acquainted with Kant's cosmological theory through his *Der einzig mögliche Beweisgrund . . .*, of 1763. It was only from the note to the preface that he had learned that the cosmogony in the seventh "reflection" of the second section of this treatise was a summary of the speculations of the earlier work.

Lambert's description of the primal scene of the *Cosmological Letters*, which he had given in the twelfth letter and which he now (in his letter to Kant) relieves of any appearance of being a literary fiction, by dating it precisely, belongs in the same context—a context not indeed of demonstrating, but of emphasizing his own originality. What Lambert describes had happened as long ago as 1749. After his evening meal, and contrary to his habit at the time, he had contemplated the starry heavens from the window of his room, and while he was doing so it had occurred to him "to regard the Milky Way as an ecliptic of the fixed stars." He had immediately written this idea on a piece of paper—of which he states explicitly that it was of quarto dimensions. The detail gives a documentary character to the anecdote. More than ten years later, when Lambert began composing the *Cosmological Letters*, the paper was all that Lambert had in front of him. This statement is corroborated in still another letter, which Lambert had already written on 7 March 1773, to Johann Lorenz Böckmann, and in which he describes his manner of working in such a way that in the fictional correspondence each letter did in fact give rise to the next one, in a production procedure based on an artificially engendered dialogue situation. In this letter too he describes as the fundamental idea of his work that the Milky Way was, "in regard to the system of the fixed stars, what the ecliptic is in regard to the solar system." To Kant, Lambert states explicitly that his discovery of the idea of explaining the phenomenon of the Milky Way as a perspective associated with a standpoint inside it had been independent of Wright's theory of the Milky Way. It was only in 1761, in Nuremberg, that he had heard of Wright and the incomplete translation of his (also epistolary) treatise. This information is instructive because in composing his *Theory of the Heavens* Kant, after all, was already acquainted with Wright's theory and also mentions him repeatedly, along with Bradley and Buffon. Thus Lambert's primal scene, isolated from possible influences—the situation of a 'lone wolf'—has remarkable authentic power.

At the same time it is anachronistic. It is not an astronomer from the age of telescopes that are growing ever more powerful and measuring instruments that are growing ever more accurate who applies himself, here, to his object. The bourgeois scene at the window is still the scene of the ancient onlooker, around whose location, which is at rest, the starry heavens turn. Accordingly, it is appropriate that this is the hour of a "sudden idea" ["Einfall"], in the strict sense of the expression—that is, not a moment, for example, in which the outcome of a long-term series of observations falls due. The scene resembles, in its layout, the one we are already acquainted with of Lambert's observation of the comet of 1769, which he had also carried out from his window. A further element of nonsimultaneity lies in the fact that he imports into the static situation of lingering before the view of the starry heavens the dynamic idea of an advancing motion. He speaks of the fact that more and more unusual things disclose themselves to him in the reality of the starry heavens and that their variety is inexhaustible, so that "an astronomical eye never becomes tired, because it finds a continual *plus ultra* [further beyond]." This language could never have been used for the *contemplator caeli* [contemplator of the heavens]: the world of the stars as a reality that continually widens in front of us, into which we penetrate and which nevertheless withdraws itself from us, as a horizon that remains beyond our reach, the farther we have pressed forward into it. "I never get far enough," Lambert had written in the twelfth *Cosmological Letter*, and "the desire always grows to go still farther." [a]

Thus the original scene cannot be used up in the sudden idea that it produces. Eleven years later, when Lambert had written the *Cosmological Letters* and Wright was now brought to his attention, he diminished the importance of the idea and its effects. The *Cosmological Letters* would not attract attention until an astronomer should discover something in the heavens that could only be explained on the assumptions of this system. If matters should ever reach that point, Lambert continues in his letter to Kant, of course the priority—and, still more, the apriority—of his cosmology would be called in question. He presents us with one of his caustically cutting attacks on belles lettres, and on the typical education of his time, when he goes on to say that at the moment his anticipations were empirically confirmed, "admirers of Greek literature" would present themselves and would not rest until they had proved that this system had already been known to Philolaus, Anaximandor, or some other Greek sage, and that he had only brought it forward again and "dressed it up better"—for these are "people who find everything in the ancients, as soon as they are told what they should look for."

Lambert does not leave this polemic, which he has already carried into the future, without a point that can be turned against his own stylization of the original scene. In the last sentence, with which he concludes his commentary on what he has in common with Kant in cosmology, he elevates the fundamental systematic idea that they share to an overdue logical next step in the history of science. He expresses his amazement that "Newton did not already hit upon this," since after all he was aware of "the mutual gravitation of the fixed stars." Thus he reminds Kant of the decisive factor—the immanent logic of the process of science—that had brought them, independently of each other, to such similar positions; and in this way he banishes the whole problem of a possible dependence to the same level where the original scene and the quarto page already find themselves.

Lambert's ignorance of the *Universal Natural History and Theory of the Heavens* verifies the lack of impact of this early work of Kant. It had appeared anonymously in Königsberg in 1755 and had immediately disappeared in its publisher's bankrupt estate. We know of only one review, which appeared in 1755 in the Hamburg *Freye Urtheilen und Nachrichten* [*Free Opinions and News*]. But the book's lack of impact is not only the result of a commercial misfortune, but also of what one could call Kant's free acquiescence in this fate of his book. Although, with his growing fame, the public wanted to see the early treatise again, Kant could not agree to a new edition. According to Borowski's account, it was only Herschel's discoveries, toward the end of the century—discoveries that in many respects confirmed Kant's theory—that led him to consent to at least an abridgement of the early work, which was assembled by Johann Friedrich Gensichen and appeared in 1791.

For the rest, the lost book surfaced again toward the end of the '90s in the first editions of Kant's collected works and in unauthorized reprints, but only the great influence of the two most important popularizers of science in the nineteenth century made Kant's work generally known: Alexander von Humboldt's *Kosmos*, which was omnipresent around the middle of the nineteenth century, and Hermann von Helmholtz's memorial address in Königsberg on the fiftieth anniversary of Kant's death. "Über die Wechselwirkung der Naturkräfte und die darauf bezüglichen neuesten Ermittlungen der Physik" ["On the Reciprocal Action of Forces in Nature, and the Most Recent Research in Physics on This Subject"]. "It was Kant," Helmholtz says, "who, being strongly interested in the physical description of the Earth and of the world-edifice, had undertaken the laborious study of Newton's works, and as evidence of how deeply he had penetrated into Newton's fundamental ideas had formulated the ingenious idea that the same attractive power of all ponderable matter that now maintains the course of the planets must also once have been capable of constructing the planetary system from matter that was loosely scattered in space." Here, to be sure, as a result of an all too large-scale treatment of history, the fact is ignored that Kant took from Buffon the inference from the shared orbital plane and direction of the planets to their having a common principle of development, and from Wright the analogy between the planetary system and the Milky Way.[39]

The public impact of Kant's cosmology came too late to be able still to provide stimuli to science. Laplace, with his *Exposition du système du monde* [*Exposition of the System of the Universe*], had taken over this function, forty-one years after Kant, and still largely in response to the same suggestions. Apart from accidental circumstances, the incomparably greater effectiveness of this work was due precisely to the fact that it was only very much in passing that it advocated the idea of evolution, in cosmology, while for the rest it provided this idea with the vehicle of the further development of celestial mechanics that was indispensable for specialists in the field.

Even if Schopenhauer were right, with his subterranean genealogy tracing all evolutionary cosmologies back to Kant's *Theory of the Heavens*, this would not accomplish much toward the comprehension of the connections in the history of science—apart from the question of how far, in Kant's own view, the early work from 1755 represents that by means of which he would have wanted to become an ancestor.

In this regard, his behavior toward his work is certainly not insignificant—especially the fact that he agreed to an "Authentischer Auszug" ["Authentic Summary"] only in 1791. This was meant to appear as an appendix to the translation of Herschel's treatise *On the Construction of the Heavens*. The few corrections that the editor undertook at Kant's instigation or with Kant's approval also raise once again the question of Kant's priority vis-à-vis Lambert in regard to the theory of the Milky Way, and go beyond what had previously been said about this. To Kant, we read, belongs "the right of the one who first took possession of something that previously belonged to no one." But above all Kant makes the text insist that important differences separated the two cosmological theories and that in this connection Lambert had the worst of it—for instance, in that "he divided the Milky Way into innumerable smaller parts and assumed that our planetary system was located in one of those parts, to which all the stars outside the Milky Way were also supposed to belong." [40] Lambert is criticized for his uncertainty regarding the nature of the cosmic nebulas. He had once regarded them as remote Milky Ways, but in another place he had seen them as light from suns reflected by the dark central bodies that he hypothesized—something that he conjectured specifically in the case of the nebula in Orion. Admittedly, this critique overlooks the fact that with the theory of the Milky Way as shaped like a lens, the pattern of cosmic systems was, initially, laid down. Spiral nebulas with an appearance like that of the nebula in Andromeda could still suggest the conjecture that they were perspectival views of other Milky Ways. The confused and unspecific form of the nebula in Orion was an entirely different matter. The assertion in the second of the notes authorized by Kant that while Lambert had conjectured the presence of other systems like the Milky Way, he had not succeeded in identifying known nebulas with such hypothetical systems—this assertion will hardly be capable of being defended.

How are we to understand Kant's late consent to the preparation of the "Authentic Summary"? There is an instructive remark, which is the only one explicitly said to have come word for word from Kant. The fourth supplement deals with the "very probable correctness" of Kant's theory of the genesis of Saturn's rings, and in this connection reference is made to Lichtenberg's "important approval." For Kant, the explanation of Saturn's system of rings had become a sort of test case for his "theory of the genesis of the great celestial bodies," and thus Lichtenberg's approval on this point could be related to the whole—though, as will be seen, with doubtful justification. In his lectures on physical geography, Lichtenberg provided a collection that he had made of fifty theories of the genesis of the Earth; among these is found Kant's geogony in the form he had given it in his short essay "On the Volcanoes in the Moon" in the *Berlinischer Monatsschrift* [*Berlin Monthly*] in 1785. Lichtenberg's sole comment on it is, "If what one wants is a dream, I doubt if one could find a better and more profound one." [41]

Like the theory of Saturn's rings, the volcanism on the Moon that was supposedly verified by Herschel's observations of 1783 was above all, for Kant, a confirmation of his "daring conjectures" about the homogeneous origin of the solar system. It was true that comparisons between the ring-shaped mountain ranges on the Moon and volcanic craters were intuitively plausible, but they were entirely out of keeping with the relative sizes of known volcanoes on the Earth: "For the shape, alone, does not decide the matter; the huge difference in their sizes must also be taken into account." Kant's short sentence enables us to see the extent to which the discipline of thought depends on our not yielding to *eidetic* evidence without the control of *quantitative* criteria. It is just on this point that Kant receives Lichtenberg's approval,[42] which must also have been important to him because in the editions of Erxleben's *Anfangsgründe der Naturlehre* [*Elements of Natural Science*] that Lichtenberg edited he had left unaltered the remark with which the new conception of the Milky Way was ascribed to Lambert: "What a great conception of the universe and its Creator Lambert ventured!" It may have been precisely this ascription, in the widely disseminated textbook, that led Kant to authorize the second note in the "Authentic Summary," in which doubt was expressed as to Lambert's having identified the nebulas with the Milky Way systems that he postulated. "So one cannot really describe this idea as a conception that Lambert ventured . . . ." Lichtenberg finally took this complaint into account only in the sixth edition of the *Anfangsgründe der Naturlehre*, in 1794, in which he did not touch Erxleben's sentence, but merely added, "Actually, six years before Lambert Herr Kant already . . . ."

Of course the parenthetical character of the correction also manifests the way in which in the meantime the competition between the two great cosmologists from the middle of the century has been overshadowed by a third name, who had brought the idea of the cosmic system of systems "much closer to being established" by his discoveries. Lichtenberg, in 1785, had tried to characterize William Herschel's accomplishment, in his *Construction of the Heavens*, as a universal orientation, and in doing so he had also typified the impulse that was already behind Lambert's thinking. The situation here was "perhaps as though someone who was enclosed in a wood in which the trees stood somewhat apart, but evenly distributed, so that he was nowhere entirely unable to see into the open, but saw it now close at hand and now far away, were able to draw an outline of the wood without leaving his place."[43]

Not only the supplements but also the omissions make the "Authentic Summary" instructive. Perhaps they allow us at least to approach the answer to the question of why Kant resisted so persistently the renewed—or, better, the first effective—publication of his early cosmological work. Here it would certainly be a projection based on an uncritical faith in science to think that Kant did not allow the work to be reprinted because he would have had to rework it on account of the results of subsequent research.[44] We shall have to imagine how Kant himself, as the years went by, may have seen his early work, in whose preface the audacity was already emphasized of the attempt to turn over the responsibility for the formation of the cosmos from chaos to a nature that is left to itself. Throughout Kant's book one encounters a bold tone that may have made him uneasy later. The author appears, in daring first-person sentences, as the Prometheus of his world-artwork: "I assume the matter of the whole universe to be in a state of general dispersion, and make of it a complete chaos. I see material forming itself in accordance with the established laws of attraction .... I enjoy the pleasure .... I annihilate this objection by clearly showing .... I have assured myself...."[b]

It cannot be an accident that at the very point at which the "Authentic Summary" breaks off, Kant speaks, in connection with the explanations of Saturn's rings, of the "pleasure of having comprehended one of the rarest peculiarities of the heavens, in the whole extent of its nature and production."[c] He invites the reader to pursue this pleasure "to the point of extravagance," only returning to cautious dealings with the truth when one has "given oneself up to imaginative ideas without constraint." Not to take these ideas too rigorously "if we are to satisfy our inclination for what is singular"[d] must have become irritating to Kant later, when he decided to leave out the speculation (even though it had been equipped with every reservation) according to which the Earth, too, could once have had a ring like Saturn—a clever digression that was aimed in an aesthetic direction, since the beauty of this sight would have been intended for those "who were created to inhabit the Earth as a paradise." Once reason's critique of reason as the organ of frivolous extensions of our knowledge had become a model for Kant, he had to feel reserve in relation to the speculative agility of his early work.

Lambert could not have written down the challenging sentence with which Kant sums up his cosmogonic purpose and of which he assures us that it can be pronounced without audacity: "Give me matter, and I will construct a world out of it!" [e] Lambert's fundamental idea of the maximal perfection of nature at every time and in every place did not allow him to dare what Kant had dared, namely, to leave this nature to its own chaos, as its initial state, and to show how "the blind mechanics of the natural forces can evolve so glorious a product out of chaos." [f] Lambert's need for a construction is so intensively and exclusively at work that he cannot give the idea of development any space. Evolution implies that at no time is the final systematic state arrived at and realized as the epitome of possible perfection. Remarkably, Leibniz's idea that the most perfect of all possible worlds can only be the one that is able, at every time during its existence, to become still more perfect than it is—so that it satisfies the claim of rationality not only in its spatial construction at any one time but also in the dimension of time—is lost for Lambert.

It is just for that reason that for Kant the claim to make order arise from chaos does not mean the same thing as actually to "derive reason from the irrational." [g] This he had already said in the preface to the *Theory of the Heavens* in order to set himself apart from Epicurus and atomism. For Kant, nature is lawful even in chaos; that is why chaos is already a part of nature's history, whereas Epicurus could only make cosmogony begin with a *deviation* from the initial state. But the price of the rationality of the initial state, in Kant, is that his universe can never cross over entirely and finally into the state of order.

The way Kant differs from Lambert is perhaps best expressed in the formulation "Creation is not the work of a moment." [h] Here the early Kant cast his first glance at the dimension of his late philosophy of history, in which he regains, as the plane on which history is accomplished, the infinity of which cosmology was deprived by the dialectical critique of reason. In Kant's early cosmological work, Newton's physics is only the system of rules by which, in the infinity of absolute time, the infinity of absolute space is traversed by a concentric and centrifugal sequence of waves of world-formation. From the one center constituted by the universe's beginning and its central body, the conquest of the void by existing things radiates. If the world could ever be completed, it would be the transformation of all matter into definitive form. But it is precisely this precision of systematic fulfillment that is never arrived at.

Viewed genetically, Lambert's perfection has become the extreme of improbability, because "nothing in nature is weighed out as exactly as possible." Conformity between an ideal model and the specific characteristics of reality is not found in the objects of nature "because in general the multiplicity of the circumstances that play a part in every natural state does not permit a precise regularity." Producing itself, the *natura naturans*[i] pursues and at the same time impedes its own goals. But this very idea justifies the genetic way of thinking and puts in the wrong a cosmology that would like to trace the present state of the world directly back to, and regard it as expressing, the divine creative will. The idea, for example, of seeing a factor of rational order in the agreement between the rotational directions of the planets is immediately paralyzed by the counterargument that there simply cannot be any reason for preferring this direction to the opposite one. Also, the planets' keeping their orbits in a common plane has no justification in rational usefulness for a purpose, since this state of affairs intensifies their gravitational influences on each other and thus prevents their motions from being exactly uniform. Thus the "clearest mark of the hand of nature" becomes precisely the "lack of the most exact definiteness in the circumstances that it endeavored to attain."

Inexactness is the stigma of nature's immanent self-constitution. Reason, which sees the world it confronts as beautiful and guided by purpose, "rightly becomes indignant at the audacious folly that can venture to impute all of this to accident and a lucky chance." But such indignation leads it astray into a "lazy philosophy"—into the mistaken assumption that the universe is perfect. Teleology, for Kant, becomes a large-scale structure that, in the infinite history of the cosmos, can no longer be demonstrated in the details. "One cannot deny, except out of obstinacy, that the admired mode of explanation that assigns reasons to attributes of nature by adducing their usefulness does not pass the test it was expected to pass here." The anticipatory antithesis to Lambert becomes obvious in the case of the comets, whose extravagant orbits are least of all suited to fit the "doctrine that the world has immediately the highest degree of organization." So Kant reckons them among "nature's defects"— something with which Lambert's omnipotent reason could not be charged. Kant's concept of nature is more elastic, embracing "all kinds, from perfection to nothingness; and its defects are themselves a sign of superabundance, by which its essence is not exhausted."

These statements, which are so instructive in regard to Kant's difference from Lambert, are in the eighth chapter of the second part [of the *Universal Natural History and Theory of the Heavens*], with which Kant advised those to begin reading who might become indignant at the "daring of this undertaking"—but just this part of the text is no longer contained in the "Authentic Summary." As Gensichen reports, Kant would not let himself be persuaded to put forward any more from that work. It is not convincing when he gives as the reason for this that the rest contained "mere hypotheses, to too great an extent for him to be able to approve of it entirely any longer." So he refused to include in the late summary precisely what he had recommended, in his preface in 1755, as a means of removing possible indignation in view of the work's daring contents, and what more than anything else could have illuminated, at its roots, his competition with Lambert.

Lambert's cosmological correspondence justifies its epistolary format by the continual inquiries of his partner in the correspondence about what his theoretical assertions concerning the universe could mean for man and for his self-understanding. It is natural to suppose that the disagreement between Lambert and Kant that is expressed in the thesis, on Kant's side, that the Creation is never complete also had to have consequences for his thinking about man. But the only thing that the proposition does for Kant is to increase "the appearances that nature makes"—to ascribe to it an industrious "production of new objects and new worlds."[j] Man is, in a peculiar way, removed from this process. The regions of the universe in which man is at home have already, for their part, arrived at that "proper perfection" the concept of which still involves, in spite of all the superabundance of endless processes, the basic idea of a completable and completed cosmos. That means that the universe as a whole is the endless forming of endless matter, but that inside this motion islands of perfection continually arise, which, after the state they have attained, cannot experience anything but losses in their content of order.

The existence of man, "who seems to be the masterpiece of the creation,"[k] at a particular point in the universe is in fact the sign that this part of the whole, which he occupies, represents a "world-edifice that has been brought to perfection," in the intermediate realm between the worlds that are *still* coming into being and those that are *already* falling apart. Man's relation to this endless process of nature's self-production is still, as it is for Lambert, essentially contemplative. But if, for Lambert, man is, as it were, brought to the contemplation of nature from the wrong side, Kant here has not yet deepened, in terms of anthropology or the critique of reason, the perspectival lessons of the difficulties in describing the solar system and the Milky Way. It still looks as though human reason, alongside the Deity, receives the eternal demonstration of the worlds: "All nature, which involves a universal harmonious relationship to the satisfaction of the Deity, cannot but fill the rational creature with an everlasting satisfaction, when it finds itself united with this primary source of all perfection. Nature, seen from this center, will show on all sides utter security, complete adaptation."[1] One sees how Kant at this time lags behind Lambert's Copernican comparative, because, although the anthropocentrism of his teleology is complicated by the idea of development, it still makes up the concealed point of his fabric of thought.

For Kant, it was only the idealization (induced by the antinomy of pure reason) of the concept of the world as a totality that is in principle not given for any experience, though it is certainly set for all experience as a task—it was only this idealization that led his thinking to its logical Copernican consequence. One only needs to compare to the early cosmological work the note that the *Critique of Pure Reason* contains, in the chapter of the Dialectic on "The Transcendental Ideal": "The observations and calculations of astronomers have taught as much that is wonderful; but the most important lesson that they have taught us has been by revealing the abyss of our ignorance, which otherwise we could never have conceived to be so great. Reflection upon the ignorance thus disclosed must produce a great change in our estimate of the purposes for which our reason should be employed."[m]

In Kant, the infinitude of the uncompleted world is compensated, for man, by the narrow horizon of an episodic completion relating to the conditions of his life. Man is a limiting case, like the transitory order of nature on which his life depends. Infinity is not yet the dimension that he considers himself capable of; that occurs only in the mediation between the absolute claim of practical reason and the factual conditions of physical reality—a mediation that presents itself in the idea of history as progress and in the Postulate of immortality.[n] An additional reason why Kant did not allow the whole conclusion of part one, in which he provides a kind of initial summing up of the method of cosmological analogies, to be included in the "Authentic Summary" is that the enthusiasm for infinity—in the "infinite multitude of worlds and systems," and the "abyss of a real immensity, in the presence of which all the capability of human conception sinks exhausted"[o]—cannot please him any longer. But just this is repeatedly secured, in the text of the early work, by the idea of a symmetry between Creator and creation, divine infinitude and cosmic infinitude. The power that manifests itself in nature is "infinite, and likewise infinitely fruitful and active"; and the instrument by which it reveals itself "must therefore, like it, be infinite and without bounds." Such an alarming "fertility of nature," which is nothing less than the "exercise of divine Omnipotence itself," is domesticated precisely by the narrowness of man's power of theoretical comprehension. It is not terrifying but more nearly comforting that the piece of nature that we are acquainted with is only an atom in relation to "what remains hidden above or below our mental horizon." Man's optics converge with a structure of reality that is favorable to him in the zone on which he has to draw to satisfy his needs, while the whole is based on a proportion that at this point is still administered exclusively by a metaphysics whose destruction, by the transcendental dialectic and its antinomies, is in store.

In this early text, Kant already has the tools of the dialectic at his command.[p] The procedure that he adopts at the beginning of part two to establish the necessity of introducing the concept of development is dialectical. If, in the solar system, the plane, direction, and regularity of the rotation of the planets agree to a large extent, this compels us to infer the existence of a homogeneous material medium.

But the emptiness of the space between the planets gives no evidence of the existence of a material substratum of that unity. This circumstance induced Newton to see "the immediate hand of God" at work here, "without the intervention of the forces of nature." [q] The train of thought already has the structure of a dialectical antinomy. Kant himself says that the reasons "on the two sides are equally strong, and both are to be regarded as equally claiming complete certainty." The task of a philosophical solution of such an antinomy has to be to find a concept that denies the mutual exclusiveness of the conflicting arguments. Kant accomplishes this by relating the two positions—the necessity of a material cause, on the one hand, and its demonstrable absence, on the other—to different points of time in a genetic continuum. Just as, a quarter of a century later, in the *Critique of Pure Reason*, he will use the concept of "appearance" to overcome the antinomic positions in the dialectic, here he finds the concept of development to be necessary in order to produce the inevitable but nevertheless undiscoverable material causality. What is not given in the present condition of the cosmic system can nevertheless be the cause of its present state. It is just this schema for the solution of the antinomy that he encounters in the planetary system that causes Kant to postulate an initial state of space, uniformly filled with matter, which at the same time possesses, for him, the advantage of the highest rationality, namely, that it is "the very simplest [state] that can follow upon nothing." [r]

This homogeneous matter is "the state of nature which immediately bordered on the Creation." One can look at Kant's cosmogony from an entirely formal point of view as setting up a correlation between an initial state in which matter is uniformly distributed in space and a final state in which all matter is lumped together in one rigid body. The same power of attraction between masses that condensed the diffuse original matter and made it into regular bodies finally operates on them in a destructive and catastrophic manner—just as earlier the ancient Stoics' fire had first produced and then destroyed the cosmos in each period of the cyclical process. In Kant the place of the ancient cycle, which organizes and then disorganizes the same world-material over and over again, is taken by the process that spreads in concentric waves, letting everything that once was organized deteriorate again and always taking new matter into the grip of its increasing central powers. Just as in Kant's later philosophy of history the individual was to be made into the agent of a progress that goes through and beyond individuals, in his cosmogony what deserves the name of "cosmos" is always only a transitional state in a huge metabolism, which balances the inclination toward downfall against organizing power, so as "to make up for the loss with advantages." The whole is not 'cosmos'; only the crest of the wave is, at each point in time. "The inevitable tendency which every world that has been brought to completion gradually shows towards its destruction, may even be reckoned among the reasons which may establish the fact that the universe will again be fruitful of worlds in other regions to compensate for the loss which it has suffered in any one place."[s] The central forces grow in proportion to the enlargement of the mass of the central body by disintegrated worlds. The destructive component of gravitation is at the same time the basis of the progress of its constructive taking hold of new matter that is ever more distant in space.

It was not only Lambert who was to be occupied with the problem of the central body; Kant, too, sees himself compelled to assume "a universal center of the whole universe which holds all its parts together in a combined connection, and makes the totality of nature into only one system."[t] There is no need to object to Kant's theory, as Hugo von Seeliger objected to the hypothesis of an infinite universe filled with matter, that such a universe could not be stable—because Kant's universe is unstable in the highest degree. Its systematic unity depends on the continual balancing of downfall against new formation, of the strengthening of the central forces and the propagation of the shaping process. It is not an expanding universe in the sense that matter as such expands, initially, in space, or even that space itself expands; instead, it is the expansion of productive force in an infinite space that is filled with matter. It is only on the edge of downfall, in the ever widening vortex of the central force, that worlds unfold, shortly before they are swallowed up. Cosmos is only for a moment, between chaos and chaos. Chaos and cosmos, as dissimilarly differentiated states of the distribution of matter, are equally law governed. The essence of this lawfulness is that it tends toward "the state characterized by the smallest reciprocal action."

For Lambert, the universe, in its perfection, is definite and stable, because it accords with a plan calculated to meet certain ends. That excludes great catastrophes, because they would put in question precisely what he regards as the essential element of this plan, namely, the habitability of the world in all of its parts. His universe cannot be infinite, if only because it is a reality of calculated risk and of precision that is appropriate to the end. If in Kant's planetary system there is no less order and no more risk to the individual bodies, that is the late and episodic result of a long process in which the partial catastrophes have already been passed through and, each time, have become less probable for the future. In this theory habitability cannot be a lasting and ubiquitous characteristic of the bodies in the universe. It is only the result of the fact that the total reality, in analogy to the distribution of habitability on the Earth, also has its "temperate zones." In relation to the claims of teleology, Kant relies on the genetic distribution of functions over time: What is not yet or no longer habitable can have been or can become so. Here again the (not yet dialectically suspect) concept of infinity is helpful, which makes it unnecessary for every finite item to meet the criterion of teleological usefulness. "If the character of a heavenly body sets up natural obstacles to its being populated, it will be uninhabited, even though it would be more beautiful, in itself, if it had inhabitants. The excellence of the creation is not reduced by this; because among all magnitudes, infinity is the one that is not made smaller by the removal of a finite part."

Kant related to the whole of his cosmology the "important approval" that Lichtenberg had expressed in relation to his theory of the formation of Saturn's rings. From the point of view of Lichtenberg's discussion, Kant had a right to do this, since Lichtenberg—enveloping it in a playful parody of Lavater's physiognomics—had treated Saturn's system, with its rings and its five (already discovered) satellites, as an intuitive model of the solar system.[45] But Kant absolutely could not permit this generalization of Lichtenberg's approval. His theories of the origin of Saturn's system of rings, on the one hand, and of the origin of the planetary system, on the other hand, differ to an extent that he himself probably did not sufficiently appreciate. His theory of the origin of Saturn's rings most nearly corresponds to the theory that Laplace was to give for the origin of the solar system; they have in common the fact that they take as their point of departure an accumulation of matter that is already rotating. In this way they are protected from the most serious objection to which Kant's explanation of the solar system was to be open, namely, that they are not able to deduce the fact that all the bodies belonging to this system revolve in the same direction.

For Kant, Saturn is the planetary fossil remaining from what had, in the past, been a comet. After the comet wheeled from its highly eccentric orbit into a planetary orbit remote from the Sun, its formerly fiery comet's tail became a mass rotating around the core, and the preponderant part of this mass, because of its tangential velocity, remained in an orbit around the planet and condensed to form its rings. Kant was very proud of this explanation of the origin of Saturn's rings; here something "the cause of which no one has ever been able to entertain a probable hope of discovering, has been explained in any easy mechanical way that is free of any hypotheses." [u] But from this model, in particular, his theory of the origin of the planetary system deviates crucially. The solar system does not originate from a total mass that is rotating initially, but from a medium, uniformly distributed in space, in which a core mass that is formed by accident induces matter to fall toward it and continually grows stronger by accretion of mass. The initial situation of the system is more similar to the one in the cosmogony belonging to ancient atomism than Kant would have admitted. As in Lucretius, everything depends on the motion of the masses, which are falling in a straight line, being deflected by collisions. But the problem is that one simply cannot expect of these deviations that they should always, or very preponderantly, take place in only one direction.

For a cosmology whose ultimate principle was a *deus calculans* [calculating God], any deviation from the idealized geometrical orbits had to be irritating; for a mechanistic cosmology it had to be suspicious, in reverse, that the orbits of the planets were still so limited in their eccentricity, so close to being circular in form. Kant thought that the eccentricity of the planets' orbits would have to increase in proportion to their distance frm the Sun; but Mercury and Venus, with their eccentric orbits, do not fit this schema. When Herschel discovered Uranus, in 1781, this was a corroboration for Kant inasmuch as he had predicted the discovery of additional planets beyond Saturn, but at the same time, from a systematic point of view, it was a deep disappointment, because Uranus's orbit came closer to a circular shape than was permissible on Kant's assumptions. In any case, the discovery of the planet Uranus caused Kant to formulate his theory of the origin of Saturn's rings more cautiously in the "Authentic Summary," in 1791, and above all to forgo the story of the conversion of the former comet into a planet.

This bold hypothesis about Saturn's early history had been important to Kant for reasons other than its mechanical consequences. It
was also important as an exemplary case of enhanced order, successful
integration, and reassurance and rectification in the system's frontier
zone. The comet that was converted to a planet and that now carried
its former wild fiery tail around with it as petrified rings was supposed
to testify to progress and the strengthening of the system's rationality
as the direction of cosmic evolution. Because for Kant the comets
belong among the defects of nature, the form of their orbits and their
distribution mark them as imperfect members on the edge of creation,
possessing "lawless freedom" and unsuited to providing "rational
beings with comfortable places to live." Here the difference from
Lambert is manifest. For Lambert, the comets' distribution outside
the orbital plane of the planets is precisely an expression of planning
on the part of the will to order and to preserve, which in this way
protects the system from catastrophes. Lambert does not see reason as
endangered by going beyond the system's boundaries, by expanding
its realm; rather, he sees a danger coming from outside, from the
Rococo of a played-out, belletristic culture. It is not an accident,
then, that he expects the best opportunities for rational creatures to be
provided by the extravagant nature of the comets, or indeed by the
limiting case of a comet's being able to change over from one solar
system into another. From this point of view, he sees the planets more
as domesticated ways of filling up the scope provided by the cosmos,
and their inhabitants as the bourgeois species among the populations
of the universe. This perspective seems to justify the preponderance of
the number of comets over the number of planets. The traditional
preference that astronomy had given to the planets is ascribed to a
pedestrian perspective on the universe: "I would sooner ask why there
are still some *planets* around our Sun. If it is not merely because
between the ellipses that the comets describe, some space remains, I
would look for the reason simply in the fact there were also supposed
to be inhabitants around the Sun who needed an unchanging moderate temperature. In other words, as a favor to the delicate creatures
that we are, the Earth had to make its progress on an almost circular
orbit."

One may feel that this single sample, the dissimilarity between
Kant's and Lambert's theories of comets, makes the antithesis between their typical metaphors excessively sharp. Still, we find at least

an indication here that the disparity between their systems is founded on a difference between their concepts of reason itself. Noting this, it should not be forgotten that a draft of a dedication of the *Critique of Pure Reason* to Johann Heinrich Lambert was found among Kant's literary remains. Lambert's death had caused him to abandon this intention.

5

What Is "Copernican" in Kant's Turning?

In a comparison that is hardly surpassable in terms of the influence it has exercised, Kant—in the preface to the second edition of the *Critique of Pure Reason*, in 1787—occasioned the formula of the "Copernican turning," without using it himself. The effect that this formula has had becomes comprehensible only when one observes in it the need to find not only an adequate expression, but above all an intuitive, a graphic expression for the radicalness and the special character of this philosophy. As so frequently happens in similar cases, the history of its influence has equipped the original casual coinage with a weight of meaning that it did not originally possess. In the meantime so much sedimentary evidence for the appropriateness of talking of a Copernican turning in Kant's philosophy has accumulated that some professionally skeptical people have been surprised and disappointed, after rereading the passages in the text more carefully, by their comparative narrowness and unfruitfulness.[46]

The best way, in our context, to realize the importance of this preface, which was written six years after the first appearance of the work, is to imagine that Copernicus had been able to write another preface to the *De revolutionibus*, after the same interval of time. What Kant writes expresses his experience with the difficulties of his own work—with the first segment of the history of its influence, and with the difference between that and what its author had expected. Part of the caution we have to exercise is suggested by the fact that Kant used the term "hypothesis" in a sense that is not entirely remote from the sense in which it is used in Osiander's preface. We must assume that Kant knew nothing about the actual authorship of that preface and that he believes he is following Copernicus's own example when he suggests that the reader treat his thesis initially only as an experiment. At the end of the eighteenth century, and especially with this book, it is no longer possible to establish confidence by falling back on authorities; the place of that procedure is taken by appeal to the lawfulness of history—to its typical sequence as an external expression of its inner rationality.

For Kant, philosophy is not the oldest and the earliest of man's sciences. It is the last to arrive at the possibility of entering upon the sure path of a science. But for that very reason it can look back at the process undergone by the other sciences in order to make sure how the passage into that sure path announces itself and is carried out. So Kant expects that what will happen with philosophy will be similar to what happened with the other theoretical interests that have already become sciences before it. Only logic seems to have no need at all of a radical change in its procedure. It has been secure, as a science, since the earliest times, and since Aristotle it has neither progressed nor regressed. Of course, this advantage is due to the extreme restrictedness of its object.

It did not prove to be as easy for mathematics as it was for logic to become a science. After a long prehistory in the Orient, it very probably first had to pass through a "transformation by a revolution." Through a happy idea that occurred to a single man, Thales of Miletus, and through his demonstration with the isosceles triangle, mathematics entered "the secure path of a science for all time and in endless expansion." Natural science underwent a comparable "intellectual revolution" only much later, namely, at the beginning of the modern age. Kant had said earlier that a light had dawned on the mind of the founder of demonstrative mathematics, and he now says of this turning of a theoretical discipline toward science that by its first accomplishments a light dawned on all students of nature. It is the light of a formal precedent. It enabled those who came later to understand that "reason has insight only into that which it produces after a plan of its own." [a]

Now, Kant's text is constructed in such a way that one immediately thinks of applying these prototypical accounts from the history of science to metaphysics. In comparison to the objects mentioned up to this point, the object of metaphysics appears as that of theory's oldest interest, which would still "survive even if all the rest were swallowed up in the abyss of an all-destroying barbarism." Just this age and this intensity of the interest in it make one wonder whether it might not be unable to attain at all what it has so far not been able to reach, namely, the rank of a science. In order to remove this objection to the analogy between the first two transformations and the third "intellectual revolution," which is due, one would need to show that the radicalness and totality of this third revolution are unique, so that it could only be completed after the others. In that case, in accordance with the analysis of the prototypes, one would have to expect that such a turning would come about only very suddenly, in the manner of an illumination, and through a single person or a few people. If one should not be permitted to anticipate this, then the retrospective view of the historical path of metaphysics, which "has hither-to been a merely random groping, and, what is worst of all, a groping among mere concepts," would come into skeptical conflict with the (still teleological) premise that "in one of the most important domains in which man desires knowledge," he must be able to place trust in his reason. The long futility of history, where metaphysics is concerned, needs to be explained, just as much as the expectation that it will finally not be futile needs to be justified. In advance of any examination of the texts, one could conjecture that the paradigm of Copernicus would have to be the most useful one for Kant's particular needs at this point, because it also provides an explanation for its historical lateness, in its theory of the production of a universal illusion.

For Kant, too, does exactly this, when in the dialectical part of his work he answers the question why, in the realm of metaphysical problems, reason does not simply fail, but misleads itself by its own mode of functioning, maneuvering itself into the blind alleys of its history. Consequently, in the case of metaphysics we have to speak of a certain necessity of illusion, and thus of the long futility of its history. Just as much as it is a theory of knowledge, the *Critique of Pure Reason* is a theory of error, a theory of its obstinacy and plausibility, and of the demands that are made on reason when its own critique is carried out. The preface to the second edition of the *Critique* compensates for the fact that the reader will not retrace Kant's own process of thought leading through the dialectic of reason, but instead is forced into a systematic reversal that offers him the analysis of the self-generation of error only when it can reply on the evidentness of the analysis of the possibility of experience.

The linked sequence in which something is described as almost futile, but then as possible, nevertheless, is repeated as a pattern of Kantian thought in the philosophy of practical reason, and driven to the point of paradox in *Religion within the Limits of Reason Alone*, which was written six years, again, after the preface to the second edition of the first *Critique*. This text is the first to make fully clear how in Kant, the old system-position of "authority" is reoccupied by the function of the exemplary figure, which no longer represents anything whose contents are binding as such, but instead now only represents the feasibility of what seems, on the face of it, to be an intolerable demand. The religious-philosophical interpretation that Kant will give, in 1793, to the figure of Christ opposes any new foundation of morality through teaching or through example, and is not even satisfied with "the exhibition of the good principle as an example for everyone to imitate," but concentrates on the mere existence of a will that proves that what is almost inconceivable is possible: that a human will can be a holy will, and that consequently no appeal to what is "human," against the absolute demand of the moral law, is permissible. Thus religion too becomes an agency counteracting a resignation that finds corroboration in the total experience of human history. A paradigm is needed not in order to learn what one ought to do, but in order simply no longer to be able to appeal to the fact that under conditions of historical reality it cannot be implemented.[47] The exemplary figure is the assurance that one can enter on and take a path whose demands must seem monstrous.

Now, it is not difficult to rediscover the long historical experience of the futility of metaphysics in the short experience of the futility of changing it that Kant himself had had in the half-decade between the first and the second editions of the *Critique of Pure Reason*. One only has to compare the philosophical 'immediate expectation'[b] of the completion of his critical enlightenment, as he had outlined it (as "the History of Pure Reason") in the fourth chapter of the Doctrine of Method at the end of the work, with the caution that he shows in the new foreword. In those days his imagined audience had not been of the kind that one would have had to invite to try something out hypothetically. Kant had completely credited and counted on his effect on his readers "to achieve before the end of the present century what many centuries have not been able to accomplish; namely, to secure for human reason complete satisfaction in regard to that with which its appetite for knowledge has all along occupied itself, though hitherto in vain." Now he reminds his readers how mathematics and natural science "by a single and sudden revolution have become what they now are," so as to invite his readers to reflect on what was essential in that "changed way of thinking," and "at least by way of experiment, to imitate their procedure." It seems to have become advisable to let solicitation take the place of proclamation. This solicitation involves, among other things, the gesture toward the prototypical figure, Copernicus.

The passage from the recommendation that we should try imitating, for once, the exemplary conversions, to the following sentences that contain Kant's own critical turning, is not as smooth as the text, which has no paragraph break, would like to make it seem. If the traditional view that knowledge can and must conform to the objects has been shown to be untenable, then the suggestion of a new attempt can really only mean to see the objects, for once, as depending on our knowledge. That would make evident, above all, how it is possible to know something about objects a priori. Just this is the context in which Copernicus's name is introduced: "We should then be proceeding precisely on the lines of Copernicus's first thoughts. Failing of satisfactory progress in explaining the movements in the heavens on the supposition that the whole host of the heavenly bodies revolved around the spectator, he tried whether he might not have better success if he made the spectator revolve and the heavenly bodies remain at rest." What this means for metaphysics is "a similar experiment": to trace the intuition of the objects back to the character of the faculty of intuition, and to trace the priori concepts of experience back to the rules of the understanding.

Now is this transcendental inversion the same "new method of thought" that could have been gathered from the course of the history of science? Kant's footnote, which is inserted just at this point, speaks against this: "This method, modelled on that of the student of nature, consists, then, in looking for the elements of pure reason in *what admits of confirmation or refutation by experiment*." The footnotes to this preface constitute a special sequence of thought that emphasizes the experimental type of thinking, and about which there need be no doubt that it originated later than the main text. These supplements work in a direction that is noticeably contrary to the internal logical tendency of the preface. That is especially true of the footnote that once again takes up the name of Copernicus. It does not refer to the reversal that

Copernicus carried out in the relation of motion, but rather to the methodological relation between hypothesis and confirmation, exhibited historically in Copernicus and Newton. The way the footnotes hang together reinforces the preface's intention not to demand more from the reader, initially, than a provisional acquiescence in the transcendental thesis, and to let the experiment with it decide. The historical interval of time between Copernicus's hypothesis and Newton's mechanics also contains the exemplary point that the experiment could not produce certainty immediately, which, however, should not be interpreted as weighing against it. So Copernicus's prefiguration of the transcendental turning consists precisely in the fact that he had to present a hypothesis without being able to undertake to prove it; the experiment was one that did not tolerate impatience, but required the dimension of the history of modern physics (of which, for Kant, Copernicus himself was not yet a part). Of course for testing the propositions of pure reason there is no experimenting with their objects, but only reason's "experience," described in the dialectic, that in its pure use it inevitably ends up contradicting itself.

Copernicus figures in this preface precisely not as one of the revolutionaries of science who display, in pure form, the moment that Kant regards as prototypical. In mathematics, the Greek idea had been to construct objects and to find the quality of knowledge in the fact that nothing is contained in the object that was not put into it in accordance with its concept. Likewise with Galileo and his successors: They had grasped the fact that "reason has insight only into that which it produces after a plan of its own." But Copernicus does not support this conception of scientific knowledge. He changed nothing in astronomy's procedure for describing the phenomena, least of all if, like Kant, one attributes to him the concept of hypothesis that is contained in the anonymous preface to his work. He changed the model, not the method for setting up and handling models. That is why Copernicus is not, for Kant, one of those real transformers of science. In the text of Kant's preface, he is the one who produces a metaphor for the kind of changes that are realized by the revolutions in the history of science in which the lawfulness of the data originates in the influence of the knowing subject. In the pertinent footnote, Copernicus is the producer of a hypothesis for the truth of which he is not yet able to answer, and which is able to pass the test—that it does in fact bring about the historical transaction of bringing astronomy, also, to the kind of path that distinguishes a science—only when it becomes physically interpretable. Taken by itself, then, the name of Copernicus does not fit the typical pattern in the history of science that Kant developed previously.

Having established this, it becomes possible to understand the astonishing fact that in all of his works Kant never returned to the Copernican analogy. In the canonical repertory of his references to exemplary accomplishments in the history of science, Copernicus's name is always missing.[48] In the material for the work of his old age that he did not complete, the sequence of names "Galileo, Kepler, Huyghens and Newton" has consolidated itself for the founding accomplishments of modern physics.[49] Evidently the idea of that second preface, that what has to be accomplished in metaphysics can be carried out all at once and by a single person and in a unique turning, is now repugnant to the late Kant, whose *Critique of Pure Reason* remained ineffective in a way that he could recognize. In regard to what is exemplary in the history of science, he now prefers the idea that an ascending sequence of people—for example, the sequence ending in Newton—carried out not a single turning, but a continual advance of knowledge. In that case, too, it continues to be significant that this sequence never begins with Copernicus.

Only the bold late interpretations of the second preface in terms of a "Copernican turning" still make it seem surprising that the astronomical reformer no longer appears in the ranks of those who carried out the turning toward science. A more careful examination would have led to the observation that in the footnote that mentions Copernicus by name, Kant has already become uncertain in relation to this prototype and to the interpretation of him that he has just set up. The fact that only Copernicus *and* Newton, together, produce the prefiguration is the first step toward an ascending sequence, with an analogy entirely different from the one that is developed in the main text.

The passage to which the second mention of Copernicus is joined as a footnote is, initially, confusing. Perhaps the author and the reader have already forgotten, here, the role in which Copernicus had already been introduced once before. So above all the position of this footnote about Copernicus in relation to the text needs to be clarified.

Looked at closely, what Kant holds out the prospect of to his reader is not only a test of the truth of the transcendental experiment that he is suggesting—the test being the dialectic—but also, even beyond that, the fulfillment of the desire of every metaphysics to pass beyond the limits of all possible experience. He holds out this prospect in the form of "knowledge that is possible a priori, though only from a practical point of view." With regard to the evidence of this knowledge, the whole *Critique of Pure Reason* changes its function and its modality: It does make room for such an extension, but it has to leave the room empty. As soon as Kant has added this to his assurance that the transcendental experiment will enable metaphysics to take the secure path of a science, he inserts the sign referring the reader to the second Copernicus comparison, in the footnote. So the relation between Copernicus and Newton is no longer an analogy for the relation between hypothesis and verification (i.e., for the relation between the analytic and the dialectic in the transcendental logic), but rather for the relation between theoretical and practical reason. If Kant had said earlier that the elements of a metaphysics should only be sought in "what admits of confirmation or refutation by experiment," the comparison with Copernicus and Newton shows that the theory of practical reason also stands to the theory of theoretical reason in this sort of relation of experimental confirmation.

It is not difficult to make that precise. The new analogy makes it clear how the relation is to be conceived between (on the one hand) the restriction of theoretical reason to appearances and (on the other hand) the accessibility, for practical reason, of what cannot be an appearance. Kant would fail in this intention (an intention that follows compellingly from, if nothing else, the way the footnote is coordinated with the text) if he could only refer to the fact that Newton's "fundamental laws of the motions of the heavenly bodies gave established certainly to what Copernicus had at first assumed only as an hypothesis." At this point, that would be a trivial circumstance. Instead, what matters to Kant is to show how Copernicus's "hypothesis," by positing a new order in the realm of appearances, provided Newton with the topography and the clue by which to infer, from this order of bodies, a system of forces that could not, for its part, become an appearance. For all the Ptolemaic system's (later underestimated) effectiveness in the realm of kinematics, its continued acceptance would have made it impossible to assert that bodies operated on one another in proportion to their masses. Newton was sufficiently clearly aware that with his thesis of an attractive power between bodies he had gone beyond what experience could legitimate and that here "hypotheses" in the strict sense were no longer permissible. Kant, then, localizes the "hypothesis" on the side of Copernicus, but lets it be proved by "the invisible force (the Newtonian attraction) that holds the world together."

Kant was familiar with the preface that Roger Cotes had added to the second edition of Newton's *Principia*, in 1713, and in which the actual epistemological program of the work had been retrospectively explicated. This contained for the first time the formula that Newton—after the conjectures and fictions of others—was the first and only person who had been able to prove from the appearances ("ex apparentiis demonstrare potuit") the gravity ("gravitatis virtutem") that belongs to all bodies.[50] Now Kant asserts explicitly that that force that holds the universe together "would have remained forever undiscovered if Copernicus had not dared, in a manner contradictory to the senses, but yet true, to seek the observed movements not in the heavenly bodies but in the spectator." This, for the first time, is the *tertium comparationis* [the common aspect of the items being compared] for the way in which the accomplishment of the critique of theoretical reason makes that of practical reason possible (not without being retroactively confirmed, in its self-definition, by the latter). It is the restriction of theory's objectivity to appearances, as a result of the transcendental turning, that first makes the expansion of practical knowledge into the empirically inaccessible realm of freedom free of contradiction, because unaffected by the (deterministic) conditions of experience.

So the *Critique of Pure Reason*'s turning is also Copernican insofar as—following the path indicated by Newton's turning toward a principle that was no longer empirical—it opens up access to practical reason. Only if we understand the controversial footnote in this way are we spared the supposition that it was slipped into the wrong passage in the text, in an effort to clarify something that is no longer being talked about at all. In the text to which the note refers, the last statement, immediately before it, is that because of the space that was created by theoretical reason, but left open, we are not only at liberty but "indeed we are summoned ... to fill it up with practical data of reason." The footnote citation is right next to this expression, "to fill it up." [c] The issue in the analogy to the relation between Copernicus and Newton—which prefigures this kind of process, of the possible or even necessary filling up of a position in the system of the visible world by the concept of an invisible force—is precisely the meaning of that expression. The way to the misunderstanding of the footnote is paved by its last sentence, which returns to the relation between hypothesis and proof within the *Critique of Pure Reason*, for which the connection between Copernicus's astronomy (always seen from the point of view of the anonymous preface) and Newton's physics is certainly useful, but does not have the specificity that has been developed in the footnote.

In the last work that Kant himself published, the *Streit der Fakultäten* [*Contest of the Faculties*], of 1798, there is another, this time anonymous, allusion to Copernicus. It confirms what may have changed, for Kant, in this reference figure. The issue is human history. It is compared to the apparent motions of the planets, the irregularity of which could finally be ended by choosing a point of view that was different from the terrestrial one. The question then is whether there is a comparable special point of view—optically unattainable, but accessible to reason—for the ups and downs of history as well, a point of view from which the arbitrary motions of history would reveal themselves as a law-governed progress in one direction only. Kant's answer is that for historiographical reason, this systematic reference point cannot be constructed. But one cannot exclude the possibility, either, that some special standpoint for an observer of history would be possible. To human sight, at any rate, only isolated elements of a path present themselves. Any attempt to lay claim, nevertheless, to a view of history as a whole has to lead to a situation that would be analogous to the post-Copernican situation of Tycho Brahe, who was willing to surrender the rationality of the whole in return for the advantage of a geocentric construction of the planetary system, i.e., of man's being optically privileged.

Kant's conclusion in regard to history is that man, as its observer, can never be a theoretical onlooker, but only a moral one. Kant enlarged on this point in connection with the example of the French Revolution. The historical event does not make it possible to read off the progress of history, but only the motion of the onlooker, of which the event becomes an indicator—giving him information about himself. That is, unspokenly, the pattern of the Copernican turning once again: From the motion of the object, the subject apprehends his own motion. The systematic central point, however, from which the totality would have to become surveyable and transparent as a unified rational process, remains unoccupied and unattainable. Reason may be operative in history, but it is not operative in historical consciousness, in any case. We may supplement the anonymous Copernican analogy by saying that that consciousness remains hypothetical, in view of the sporadic character of the indications of an 'orbit' that are available to us.

To reduce Kant's "Copernican turning" to the laconic material in the text does not mean to undervalue the wealth of consequences that this material produced. Kant himself, in the second preface to the first *Critique*, only wanted to provide a piece of graphic encouragement toward the experiment that he was offering to his reader. So it is neither surprising nor reprehensible that a metaphor having the epochal importance that this one possesses sets the reader's constructive imagination in motion even before he has finished reading it. Kant is making use of a historical potential that seems, in the hindsight of historiography, to have been bound to become an independent force. This inevitability is, in turn, a symptom of the wealth of meaning that had become attached to Copernicus's name. In the meantime, there has been so much talk of "Copernican turnings" that the specificity of the metaphor does need to be uncovered. For inevitably a change in a long-accepted system of views is already something that has to be measured by the Copernican standard, and all the more so if in such a change the conflict between the intuitive content of the life-world and rational construction—the conflict for which Copernicus produced the prototype—also enters in. But not everything that one may have discovered, in repeated comparisons, as what is specific to the Copernican reform also holds, for that reason, for Kant's own reference to it, especially for the transcendental analogy.

An important factor that has contributed to the misunderstandings of that analogy has been the insufficient attention that has been paid to a single word in the text. Kant says that the situation with his proposal that for once we should experimentally assume the direction of determination in the relation between subject and object to be reversed is just the same as "with Copernicus's first thoughts" [*mit den ersten Gedanken des Kopernikus*].[d] If one takes that at its word, it appears to distinguish between Copernicus's finally resulting total system and his "first thoughts," so as to put itself in relation to the latter only. We must not assume that Kant ever looked into Copernicus's *De revolutionibus* itself in the way that he did with Newton's *Principia*. In the literature that he can have been familiar with—for example, in Jacob Brucker—it is commonly asserted that Copernicus, as a successor to the Pythagoreans, first hit upon the possibility of substituting the Earth's axial rotation for the diurnal rotation of the heaven of the fixed stars. It does not matter that historically these cannot have been Copernicus's "first thoughts." The crucial thing for understanding Kant is that he remained dependent on the popular assertions about Copernicus's point of departure and that he consequently saw the mere exchange of the real predicates of rest and motion between the Earth and the heaven of the fixed stars as the essential analogy to his own reform.

The restriction of the analogy to Copernicus's "first thoughts" excludes the possibility that the transcendental experiment that is suggested to the cautious reader also metaphorically implies the daring transposition of the optical standpoint of an observer who would then be located in the central body of the system. That is a speculative amplification. For what Kant says about Copernicus is only that "failing of satisfactory progress in explaining the movements in the heavens on the supposition that the whole host of the heavenly bodies revolved around the spectator, he tried whether he might not have better success if he made the spectator revolve and the heavenly bodies remain at rest." There is no mention whatever of heliocentrism and the Earth's annual movement around the Sun. No transposition of the optical standpoint takes place, but only the commutation of motion and rest between the starry heavens and the Earth. That suffices for everything that is needed here in order to affect the reader's perception.

It is truly not a misfortune to read more into this than it contains. Only one should not say that with the associations that press forward from our Copernican horizon we are still disclosing Kant's self-understanding and self-consciousness. To the Kant interpreter, the heliocentric revolution in Copernicus's work may seem more appropriate, as a means of picturing Kant's transcendental undertaking, than the diurnal rotation of the Earth, which does not affect the status of geocentrism; but he might at least say that he regards Kant's actual comparison as too weak, or historically incorrect, and that he regards only his own continuation of it as adequate. But then one can no longer describe Kant's solution of the problem of the objects' agreement with a subjective a priori in this way: "Kant's general answer to this is what he described as his 'Copernican turning': Before Copernicus, people had assumed that the Sun revolves around the Earth. That seemed self-evident, because it is what appearance teaches us. Copernicus taught, contrary to immediate appearance, that instead the reverse is the case: The Earth revolves around the Sun; and he carried out this revolution because he succeeded better, that way, in explaining the motions of the heavenly bodies . . . . Kant's revolutionary reversal teaches that, instead, objects conform to knowledge." [51] Apart from the fact that Kant himself nowhere speaks of his Copernican turning,[52] he also consistently leaves the Sun out of the discussion. For it is only in connection with the annual motion that it makes sense to speak of the Sun's revolving around the Earth being replaced by the Earth's revolving around the Sun; it is true that the Earth's rotation takes responsibility for the Sun's diurnal motion across the sky, but it does not allow us to say that now "the reverse is the case: The Earth revolves around the Sun."

Kant makes it unequivocally clear that by "Copernicus's first thoughts" he means the thesis of the Earth's diurnal motion, when he speaks only of the old conception's making "the whole host of the heavenly bodies revolve around the spectator," whereas Copernicus "made the spectator revolve and the heavenly bodies remain at rest." Thus the Copernican analogy is conceived in a consistently narrow fashion; it contains nothing from Copernicus's "second" and subsequent thoughts, and above all nothing about heliocentrism. The spectator who is prototypical for the transcendental turning does indeed revolve along with the Earth in place of all the heavenly bodies, but he does not revolve with it around the Sun. That only becomes thematic in the Copernicus-Newton analogy of the second footnote, with the entirely different object of clarifying the relation between theoretical and practical reason, and the relation between gravitation and freedom.

Hermann Cohen made the observation that in the second preface Kant does not speak as the author of the work, but here he has "himself become the reader again."[53] The commentator did not draw any conclusions from this for the Copernican metaphor. In fact, however, Cohen's observation suggests a different way of understanding the text than as a source of information about Kant's self-understanding. In its context, the reference to Copernicus is clearly a rhetorical move. Kant is courting his reader, by suggesting to him, prior to any apodictic argumentation, an easier, provisional approach to his work. The sentence that precedes the reference to Copernicus begins with this turn of expression: "Let us make trial, for once, then ['Man versuche es daher einmal'], whether we may not have more success in the tasks of metaphysics, if . . . . " It is the *reader* of whom Kant is speaking, and who immediately returns, in the reference to Copernicus, as the "spectator." He is the one who is supposed to remember Copernicus and, aided by the thought of Copernicus's daring, summon up courage. Also in the sentence that immediately follows the mention of Copernicus, Kant does not speak of himself but of the reader, who is supposed to be encouraged to take up his suggestion of the provisional experiment: "Now a similar experiment can be tried in metaphysics, as regards the intuition of objects." It cannot very well be assumed that in 1787 Kant still regards it as an experiment for himself to act similarly to Copernicus in his "first thoughts." It is just that he is evidently not as sure of the effect of deductions and proofs as he was six years earlier—and reason always has recourse to rhetoric when it has had this kind of disillusioning experience. Precisely because Kant did not present himself in his own self-consciousness in the text that led to the formula of his "Copernican turning," it is necessary to regard this text, which was pregnant with such future consequences, as a rhetorical instrument and a metaphor.[54]

Finally, it should not be overlooked, as a linguistic aid for the reader, that Kant opposes to the reader "the whole host of the heavenly bodies." To the public in Kant's time—as opposed to the public in Copernicus's time—it was self-evident that the fixed stars are individual bodies in space. Copernicus's audience would not have been able to understand at all what Kant wanted to demonstrate: that the fact that a priori determinate characteristics belong to all objects equally is due not to the objects' being determined, without exception, in themselves, but to the constitution of the subject of experience, just as the fully homogeneous motion of "the whole host of the heavenly bodies" is due not to their all having been rendered physically uniform, but rather to the single motion of the spectator. Copernicus's contemporary would not have been able to gather this sense from the analogy because he conceived of the heaven of the fixed stars as a rigid body and as physically connecting all the points of light that appear in it, so that it would have had to seem no more difficult to him to explain their simultaneous motion as real than to explain it as apparent. The only reason Kant was able to offer his reader the economical advantage, as a product of the transcendental thesis, that it made the whole manifold move in the same direction, was that after Copernicus (though of course as a consequence of his having brought the heavens to a standstill) the outermost heaven had long since been dissolved into "the whole host of the heavenly bodies." Thus the transcendental exchange of functions that makes the subject's legislative activity into the unifying ground of the lawfulness of the appearances is not typified clearly enough in the Copernican reform. This marks the most intelligible motive for Kant's having recourse only to Copernicus's "first thoughts," and directing no interest, here, to his heliocentric change: The phenomenal diurnal motion brings the greatest multiplicity into the strictest unity.

Kant's forgoing of possibilities with which his interpreters were to supply him is in the service of rhetorical terseness. Thus the experiment that he proposes to his reader cannot be interpreted as a case of the freedom to change one's standpont.[55] So one will not be permitted to extract from or read into his reference to Copernicus, as the *tertium* [the 'common aspect,' again], the "freedom of consciousness" to choose its standpoint: "Copernicus experiences the freedom to choose his standpoint, and makes use of it, as a physicist: Kant recognizes this freedom as such, and assigns reason the standpoint of freedom .... Just the fact that Copernicus avails himself of the freedom to distribute the roles differently, in his chosen model of the revolutions of bodies around one another, than Kant does much later, makes him a companion of Kant."[56] Entirely apart from the question of what Kant looked to in Copernicus, it has to be noted that for Copernicus there had been no freedom to choose a standpoint if only because he does not "choose" his model at all. Osiander would be vindicated if Copernicus had "regarded his interpretation as a possibility that is from the outset not 'truer' than Ptolemy's, but only has the advantage of greater simplicity." One already has to 'modernize' Copernicus severely in order to say of him, as a companion of Kant, that he "lays claim, for theoretical consciousness, to a standpoint that consciously includes the freedom of reason to 'experiment' .... " That conception relates neither to Copernicus's first thoughts nor to his last ones.

Kant, with his rhetorical figure, keeps completely within the framework of the equivalence that Copernicus developed between motions of the outermost sphere and of the Earth. Contrary to all 'standpoint' mysticism, Kant's reader can remain on Earth precisely because the transcendental turning that is offered to him is not, and is not meant to be, idealism.

Only Newton's physicalization of the Copernican system makes it possible to document a prefiguration of the Kantian problem of freedom—but just this [prefiguration] has nothing at all to do with a freedom to change one's standpoint experimentally, because from that point onward the arrangement of the solar system is no longer at our disposal. In any case, Kant's footnote in the second preface establishes unshakably that Copernicus alone does not yield anything for the problematic of freedom, because only Newton's introduction of gravitation developed the practical 'deeper meaning' of the Copernican reform. One may regard the liberation of modern rationality from the previous blind and unquestioned bondage to the accidental and parochial perspective of the Earth as one of the freedoms in this epoch's consciousness, but this has nothing to do with Kant's concept of freedom as a quantity that first becomes negatively possible as a result of the restriction of theoretical reason to appearances (in all their perspectival conditionedness), and localizable in the space thus left empty. Kant shows us just this in his footnote on the paradigmatic quality of the relation between Copernicus and Newton: The analogy between freedom and gravitation directly excludes the perspectival element, which still entirely governs the Copernican reform (as seen through the confusion caused by Osiander), but which no longer has any place in the realism of the relations of masses in Newton's cosmological physics. Only in this way was an exemplary counterpart created for the absolutism of freedom in Kant's concept of reality.

It remains a documentary demonstration of the intensity of Copernicus's historical influence that even painstaking commentators did not, in the strict sense, 'read' the two relevant passages in Kant's second preface, but rather, in a sort of anticipatory confidence based on a disposition (which was activated by mere contact) to engage in metaphors of this kind, 'presumed' them instead of 'consuming' them. This is the fate, at once splendid and miserable, of a rhetorical element in the context of theory—of its temptations to hermeneutical generosity and easy satisfaction. Naturally, when the universal and perhaps most widely read historian of modern philosophy could write that "Kant liked to compare his undertaking to Copernicus's,"[57] one cannot blame Bertrand Russell when he seizes the opportunity pithily to deny Kant the right to compare his accomplishment to Copernicus's. Russell, too, is guided by something he has never read when he says that Kant spoke of his "Copernican revolution." Then he pronounces this to be Kant's misunderstanding of the fact that he had produced, instead, a Ptolemaic counter-revolution, by restoring an anthropocentric conception, contrary to the immanent intention of modern science.[58] In this amazing statement the contradiction of the formulation (ascribed to Kant) of his "Corpernican revolution" converges with the other tradition of Copernican metaphors, in which the sequence of the humiliations of man by modern science begins, prototypically, with Copernicus.

Kant found no place in this sequence, nor would he have appealed to it. But he shares something with the members of this sequence that could have made Russell skeptical in regard to his negatively evaluative contradiction: Kant also, as the second preface shows, had to deal with an inner resistance of his audience to taking part in the transcendental turning. Explaining inner resistances to the demands made on us by scientific theories is, of course, the chief way in which the Copernican metaphor, in its role as a figuration of man's loss of value, is employed. Certainly the best-known instance is Freud's inclusion of himself in the sequence beginning with Copernicus and Darwin.[e] Still more impressive is the genealogy that Carnap gave to his psychological physicalism in order to integrate its "bitter demand" into the regularity of modern history: "By Copernicus, man was expelled from his exalted position in the center of the universe; by Darwin, he was deprived of the dignity of a special creature that was superior to the animals; by Marx, the factors by which the course of history is to be explained causally were brought down from the sphere of ideas to that of material events; by Nietzsche, the origins of morality were stripped of their nimbus; by Freud, the factors by which man's ideas and actions are to be explained causally were referred to obscure depths, to 'lower' regions .... Now psychology, which, as the theory of psychic or spiritual processes, has hitherto been clothed with a certain sublimity, is to be degraded to the status of a part of physics."[59] The fact that Kant does not appear in this catalogue of misanthropisms is in keeping with the tenor of Russell's charge that he usurped the role of the Copernican revolutionary.

Now, an adequate understanding of Kant can least of all consist in imputing to him, in his reference to Copernicus, the intention of giving a model of the universe the function of guiding man's understanding of himself. In that case, the separation of theoretical from practical reason would have broken down at a critical point in Kant's self-detachment from his early period. No knowledge of nature, no variety of 'physicalism' can have any consequence for man's assessment of himself. At any rate, this firm demarcation vis-à-vis the tradition of interdependence between cosmology and anthropology is the outcome of the critique of reason. To put it differently: On Kant's assumptions, Copernicus had said nothing, and could not have said anything, about man. Of course it is not unimportant—on the contrary, it is extremely instructive—that Kant found the restrictions of the critique of theoretical reason more of a burden than he could bear, and that he consequently summoned up enormous compensations for them in the further development of his philosophy. In the perspective of these compensations, the critique of reason can be seen as an instrument of man's moral and aesthetic self-assertion against his being physically determined—but it cannot be seen as evidence of metaphorical submission to such determinism. If one thinks of the section of the *Critique of Judgment* entitled "On Nature as a Power," it is true that one sees Kant facing the question of the position of reason in the whole of reality, and answering that reason is reality's "standard," which "has that infinity itself under it as a unit, and in comparison with which everything in nature is small," and which founds "in our minds a preeminence over nature even in its immeasurability." [60]

Once one has carried out the duty of looking to see what Kant's text really contains, what he really says, nothing prohibits the further hermeneutic steps of determining whether there is an additional affinity between Kant and Copernicus—or whether such an affinity even stands in the background of what is presented only in a superficial form, with a view to the hesitating reader, in the second preface to the first *Critique*. One can then go so far as to say what Kant could have said, if not what he should have said. In my opinion the most intelligent attempt in this direction consists in bringing out, in the analogy between the trnascendental turning and the diurnal rotation of the Earth, the phenomenal correspondance not in the heaven of the fixed stars but in the bodies that make up the solar system.[61] The argument for this cannot, of course, present itself as a historical one: that Copernicus had been so predominantly concerned with the problems of the planets that any reference to his turning would in its turn have to stick to the primacy of the planets. For the interpretation of Kant's text it does not matter at all what, historically, was primary or urgent for Copernicus, but only what was bound up, for Kant, with an appeal to Copernicus as a model. But this was reflected only in the expression "the whole host of the heavenly bodies." The way in which Kant aimed his rhetorical interpolation simply does not permit us to expect that he recognized the historical importance of the problem of the planets for Copernicus or that, if he recognized it, he took the least account of it. That would hardly have helped the reader who is imagined as needing encouragement in his first leap, either.

Seen in relation to Kant, the problem of the planets would have offered the most precise point of reference in Copernicus's theory, because in the phenomenal motions of the planets 'two sources' of real determining factors are interwoven: their own real motions on their orbits around the Sun, and the projection of the Earth's own real motion, in its annual revolution and in diurnal rotation, onto the former motions. Insofar as the planets' motions are not to be ascribed only to the onlooker (or, in the comparison, to the subject), but instead an objective share of motion that is independent of the onlooker (or the subject) contributes to the phenomenon, a comparison that related to the theory of the planets would have been least susceptible to interpretations assimilating it to transcendental idealism. Nothing would have been more important to Kant—especially before the second edition, in view of the first idealist misunderstandings—than to avoid such assimilation, if he had realized that his taking into account only those "first thoughts" of Copernicus (that is, the relation between the Earth's rotation and the phenomenal diurnal motion of the heaven of the fixed stars) in fact creates a more "idealist" impression than any possible theory of the planets could have produced. The method of such a theory certainly does justify a structural comparison with the method that a transcendental epistemology has to apply if it distinguishes between two irreducible 'sources' of all experience. Just as Copernicus could arrive at the real motion of the planets only by subtracting the additional motions that come from the observer's moving along with the Earth, so Kant had to attribute to the subject its role in producing the identical 'direction of motion' in the a priori determinateness of its objects, in order to obtain the 'value' of their own determinateness as a remainder. In that case it would not have been the outcome of the Copernican reform, but its 'method,' that Kant would have wanted to compare with his own.

Unfortunately, all that one can say is that Kant should have thought of that. But he did not.

Part VI

Vision in the Copernican World

Introduction

The telescope is, as it were, the index fossil in the Copernican formation. In the seventeenth century a canon of instruments emerges in which the new epoch typifies its way of taking hold of reality: the clock, the telescope, the microscope, the scales, the pneumatic pump, the thermometer.

This list is not homogeneous. The pneumatic pump embodies the victory over the error of the *horror vacui* [nature's 'horror of a vacuum'], and man's ability, to a degree, to produce cosmic conditions for physical processes on the Earth. The clock, the scales, and the thermometer are associated entirely with the demand for the quantification of scientific experience; their decisive accomplishments are still in the future, and depend on increases in their accuracy, as in the function of the scales for the "changed method of thought"[a] in chemistry in the second half of the eighteenth century. The telescope and the microscope continue to be bound up with the concept of reality as what is evident to intuition, a concept for which all natural processes are in principle clarifiable through vision, even if, as a matter of contingent fact, the coordination between the organ and the object may be interrupted.

The role as key instrument that the telescope takes on in the self-consciousness of the modern age's cognitive process is conditioned by the assumption that success for theory consists in the expansion of the field of objects, while their species remains in principle the same. From this perspective on reality, it would have been surprising to find that the use of the scales could explain a natural process in a case where microscopes of increased power produce nothing. It is only very late that the invisible turns out to be only a part of what is kept in reserve for theory to accomplish. The assumption, which is self-evident on the ancient world's premises, that optical totality would be definitive for theory, is still involved in Schopenhauer's idea that astronomy would be established "by a single empirical intuition, if we could freely pass through universal space, and had telescopic eyes."[1] The whole tradition of atomism had believed that one would only need to see the shapes of the atoms (which at present are inferred) in order to comprehend the characteristics of everything, with certainty, as their effects, and in order not to demand more than this. The metaphor of the mechanism whose performance one can grasp through intuition as soon as one has merely removed the housing that obscures one's vision and converted all of its parts to a scale appropriate to our optical faculty determines the expectation of the way final theories would appear. The forms and processes in the micro-realm have the same types of configuration that also appear in the macrorealm. One day one would see in the microscope something like the system of Jupiter with its satellites.

The invisible makes up a continuum of reality with the visible. The visible surrounds a core of the invisible that is too small, and is itself surrounded by a corona of the invisible that is too distant. On both sides, it is a reserve to be penetrated in proportion to our measureable accomplishment in magnification. The unbroken eidetic character of the object in the telescope and the microscope is in contrast, from the very beginning, to the quantifiability of the instrumental accomplishment. A very important fact, in connection with Galileo's unfortunate experience with the telescope, is that he did not think of carrying out measurements with the instrument. Of course, his telescope was not well suited to that either, because it did not produce a real optical image; Kepler was the first, with his astronomical telescope, to make it possible to use cross-hairs in the plane of the real image and thus to combine optics and measurement. But among the inimitable features of the history of modern science is the fact that Kepler, entirely devoted to the treasure of observations that he inherited from Tycho, neither recognized the importance of his own invention nor made any use of it at all. The stubborn geocentrist Christopher Scheiner turned Kepler's idea to good account and was the first to measure the speed of the Sun's rotation. Kepler 'recognized' the telescope, he decisively improved it, but he did not make use of it. He was like the people who, two centuries later, were to discover the planet Neptune by calculations, before—and so that—it could be optically 'discovered.' If the epitome of all of astronomy's theoretical intentions had been to calculate the location of a heavenly body at any moment with the same accuracy with which one can observe it,[2] then to accomplish the same thing for a still unknown heavenly body was a model instance of astronomy's perfection.

"The chief thing is always invisible." Lichtenberg illustrated this statement of what is now a specific contingency of our sensorium—a statement that is disillusioning, in a new way, far beyond the Middle Ages—with the elementary instance of the magnet: We perceive its characteristics without any difficulty, but all of these are not what makes up its attractive effect. Ancient astronomy is vindicated, in a surprising way, for having left the centers of its epicycles empty, because the suggestion that a central body makes comprehensible the motions that circle it concentrically is only superficial: "What our eye perceives in this revolution is not what constrains the planet."[3] The specific invisibility of the chief thing does not reduce Lichtenberg's admiration for the telescope, but it is no longer the scepterlike sign of human dominion over the world, which Kepler proclaimed it to be,[4] but rather an expedient for escaping the dilemma of cosmic eccentricity. Having discovered it compensates man for whatever the observation of the world can impose on him in the way of disproportions: "Without extracting humiliating reflections about man from this—reflections which, if he needs them, he can find close at hand—let us rather admire the spirit of the creature that was able to procure this knowledge by means of a bit of glass that it ground with dust . . . . "[5] With Lichtenberg, however—simultaneously with William Herschel's telescopic triumphs ("almost as though the telescope had just been discovered"[6])—the attitude to the world that is symbolized by the telescope begins to abate: The optical prolongation shares in the ambiguity of the heavens, denying man both the fulfillment of his centrifugal curiosity about the world and aid in his centripetal process of understanding himself. No one in the eighteenth century reflected as radically as Lichtenberg did on the consequences of Copernicanism; it is not an accident that the last thing he wrote was a biography of Copernicus.

The great moral anticipation that the Enlightenment had associated with visual penetration into the depths of the universe—as a preliminary form on the way to the mutual understanding and alliance of its rational natures—was fulfilled all the less as it became the motivation of expenditure for the instruments and for the organization involved in research. Utopian thought reflects the great gesture of those anticipations and of their moral implications: "The telescope is the moral cannon that has lain in ruins all those superstitions and phantoms that tormented the human race. It seems as if our reason has been enlarged in proportion to the immeasurable space that has been discovered and traversed by the sight."[7] The disconcerting awareness dawned, in people's minds, that once again it was not "the immeasurable universe itself" (as Herder hastily decided) that was revealed in Herschel's telescopes, but rather the advance knowledge—implied by the demonstration that everything observed previously was merely a sample—that what was now observed, and what would be observed at any given time, would also be surpassable. There were no experiences of fulfillment for this epoch that was so ready for them at every moment.

It is true that the telescope is the signature of a new anthropological situation and the paradigm of an accomplishment that grows out of that situation, but the way in which it represents this accomplishment is increasingly not direct and specific but rather metaphorical. Other instrumentations are measured against this prototype and its promise of definitiveness. When Leibniz announces his *scientia generalis characteristica* [universal characteristic], he writes that the establishment of the proposed sign system will give mankind a new kind of organ, which will increase the power of the understanding much more than optical lenses have been able to accomplish for the eyes, and it will be as much superior to microscopes and telescopes as reason is to the eye.[8]

Romanticism parries the disappointments of the Enlightenment with a system of precisely coordinated reoccupations. Scarcely acceptable tolerances on the part of rationality become model instances of the expansion of reality. The fact that scientific experience requires a reality that is adapted to (one almost has to say 'prepared,' in laboratory fashion, for) the ways in which it can seize it is simultaneously its strength and its weakness: It always also paves the way for the powerful impression that a new openness to the obscure and a new liberality with the setting of limits has to produce. For the Enlightenment, different worlds were worlds containing a superior rationality, while for Romanticism they were worlds containing things that were unapproachably different in kind. Accordingly, Romanticism brings with it a need for organs that transmit experience without denuding, domesticating, destroying their objects. The mere resigned suspicion that the senses with which man is equipped will turn out to be only a contingently given selection from what is possible easily turns into the admission of a hidden world of vague currents and forces in which the extensions of reality may be hidden, but cannot be identified. The Romantic idea emerges of a universal instrument, of which only the first stage would have announced itself in the telescope: "Could there be magnetic or electrical telescopes? Mirrors for magnetism and electricity? Lenses, telescopes, and mirrors for every type of force? In what sequence do you suppose bodies refract electricity or magnetism? For certainly there are sequences here, just as there are with light. There may also be achromatic lenses for every force, no differently than for light." [9]

The telescope is a magnifier not only in the optical sense, but also in an eminently historical sense: The preconditions on which its ascent is based are quantitatively intensified and qualitatively transmitted by its accomplishments. The break with the postulate of visibility is founded in an anthropology that makes it both comprehensible and manifest that man can only be a match for a world for which he produces his own organs. Every success of the telescope confirms what its possibility had caused people to expect: the backwardness of visibility in relation to reality, of the organ in relation to the challenges it faces. To the extent that the organ's medium multiplies, it becomes impossible not to see that the explosion of reality into which it makes it possible to press forward cannot be overtaken and made up for. The expansion of space and time is always ahead of the expansion of our instruments, even though it was first introduced and impelled by the latter. But here something sets in that has its parallels only in the Enlightenment's reflexive exoticism (with a faint prelude in the reflection on peripheral peoples that occurred in late antiquity): The subject that drives and is driven into distant and alien regions becomes noteworthy and awkward for itself to the same extent that it seems to get away from itself in space and time. Montesquieu, in the 59th of his *Lettres Persanes* [*Persian Letters*], gave the formula for what I would here like to call "implied reflection," which will find its unexpected articulation only in the optical geotropism of space travel: We have our relation to reality only by carrying out a secret turning back onto ourselves. [10]

1
How Horizons of Visibility Are
Conditioned by Views of Man

The Greeks did not invent the astronomical telescope. In contrast to the common explanation of their backwardness in technical inventions, one cannot blame this on the circumstance that they had the institution of slavery and did not need technical means of production. Consequently the telescope is one of the criteria by which the early modern age, to its own amazement, could decide in its own favor the doubtful question as to whether antiquity was superior to it.

The question *why* the ancients did not arrive at the invention of the telescope was raised in the Paris Academy of Sciences in 1708. Fontenelle reports on a lecture by de la Hire on the appearance of burning glasses in ancient texts; the lecturer had found this in Aristophanes. It is the reporter, Fontenelle, who poses the additional question of why, while one could burn things with lenses, one could not magnify objects optically. It was, he says, the mistaken theory of vision, in antiquity, that prevented people from establishing the connection "entre un verre qui brûle et la manière dont se fait la vision" [between a glass that burns and the way in which vision takes place]—entirely apart from the fact that the ancients' burning glasses were probably composed of glass balls, either solid or filled with water, with which an arrangement that would function as a telescope would hardly come about by accident.[11]

An echo of the great *querelle des anciens et des modernes* [quarrel of the ancients and the moderns], the effort to consolidate the historical self-consciousness of the modern age, can still be perceived in the discussion that Lessing—for totally different reasons—devotes to the problem of ancient optics in the 45th of his *Antiquarian Letters* of 1769. Here the issue is whether the ancient gem cutters had already used optical auxiliary instruments, as some authors, pointing to Seneca's glass spheres filled with water, maintained. Lessing tries to show that these glass spheres could not yield any optical progress because the cause of the effect was not seen in the spherical surface of the body but in a sort of *qualitas occulta* [hidden quality] of the sphere's contents. Here, too, then, the reference is to a technique to which the theory is not adequate and which it is not able to help along. As long as theory is lacking, Lessing's statement—which is important for the history of science—holds: "The easiest discoveries are not in fact always necessarily the earliest ones." However obvious and palpable it may be, what theory does not legitimate or indeed does not even permit evades our attention, our interest, our readiness to receive it. The path from the mirror to the lens, from catoptrics to dioptrics, continued to be barred because "they had undertaken no experiments with the artificial means of the glass, and it remained a deep secret from them how, by the different surfaces of this artificial means, refraction could be brought under our control."

If the requirement of precision in producing lenses could not be fulfilled, knowledge of their optical effects was useless, because the disqualifying characteristic of distortion was more conspicuous than the accomplishment of magnification. An additional important observation of Lessing relating to the history of science and technology is that a side effect can conceal the main effect and impede its recognition and utilization. An accidental discovery is so improbable even in the case of the simple magnifying lens because the conditions in which its optical effect would be unambiguously positive are improbable: "For a crystal that was ground by chance into a lens shape will also be only approximately lens shaped, so that while it would magnify the shape of a small body lying under it, it would also distort it. So what special use could the person who noticed the magnification expect from it if he was still so far from the conjecture that the distortion was a result of the inferior precision of the spherical surface, and could be remedied by correcting the latter?"

That is a masterpiece of reflection on the history of science, in its effort to draw the horizon of possibilities into the context of facts and demonstrable historical circumstances. What is possible is connected to what one is able to see, and in the artificial distortion of natural forms (to which the ancient consciousness of reality had to be especially sensitive), the regular limiting case of magnification was not even suspected, and was consequently not discovered, even *en passant*. Lessing knows that what is important is the structure of attention, not how closely to the organ of perception, or distantly from it, phenomena are deposited. He made graphic what had become clear to him in this subsidiary question in the gem-cutting controversy—namely, the historical finding that people can have a backward attitude to the light—by means of a delightful variant of the Enlightenment's metaphors of the sunrise. "And this, it seems to me, is what almost always happened in cases where we see the ancients halt in the neighborhood of a truth or an invention that we nevertheless cannot give them credit for finding. They did not take the last step to the goal, not because the last step is the most difficult one or because Providence directly arranges matters so that certain insights should not emerge before certain times; instead, they did not take it because they stood, so to speak, with their back toward the goal, and some prejudice misled them into looking for this goal in an entirely wrong direction. The day broke, for them, but they looked for the rising sun in the West."

That is not yet Lessing's final word on this paradigmatic piece of the *querelle*. At the end of the controversy about the ancient magnifying glass we find a relativization of quantitative optical progress. The antiquarian secondary topic of the art of gem cutting unexpectedly turns out to be instructive in regard to the question of the quality, or the intensity, of vision. The telescope is not mentioned, but it is implied in the concluding turning, which suspends the whole letter and its witty pugnacity, and which begins with the seemingly scholastic formula that the question whether the ancients or the modern had the greater keenness of vision requires us to draw a distinction. What follows then has to be taken together with Lessing's own philosophy of history and its seemingly pedantic schema of education: "We see more than the ancients; and yet perhaps our eyes may be worse than the eyes of the ancients. The ancients saw less than we do; but their eyes, speaking generally, may easily have been sharper than ours. I fear that this may be the upshot of the whole comparison between the ancients and the moderns."

The absence of optical magnifiers, and then their long incubation after the end of the thirteenth century, when the optical characteristics of glass lenses for eyeglasses were discovered, cannot be satisfactorily explained by either the economic milieu or the failure of theory alone. Just as little can (in reverse) the invention and the rapid progress of the telescope and the microscope be attributed to acute real needs, however early people may have recognized what allies military men and merchant sea captains could be for theoretical concerns. Preconditions will have to be sought, rather, in the possibility of reception—in people's capacity to be affected by an offer of new optical instruments. One who, like the Greeks, keeps it in his consciousness that he is the onlooker of the world and of its phenomena, which show themselves to him, and that he is directed at and adapted to this privilege by nature, will not be inclined to suspect that his eyes and the world could be so disproportioned to each other that essential regions and objects in reality would be withheld from man's access. Only where that kind of reflection finds support—where it is nourished and enriched by analogous experiences and unexpected displacements of the horizon—can the range of human optics be taken as a contingent fact and first be held to need, and then be held to be capable of technical expansion.

Fontenelle already saw that the philosophical preconditions of the telescope belong, in the widest sense, to the context of a skeptical anthropology. He expressed this in the most charming manner in his *Dialogues des Morts* [*Dialogues of the Dead*], in a conversation between Galileo and the Roman gormandizer Apicius. The Roman hedonist expects the inventor of the telescope to be able to strengthen and to refine all man's other sense organs as well: "I would have begged you to concern yourself with the sense of taste, and to invent an instrument that would increase the pleasure in eating." Galileo tries to explain to the Roman that man's sense organs are perfectly sufficient as long as reason does not demand something from them that it needs for its own expansion, to appease its dissatisfaction. Of course people who want to know whether the Sun has spots, whether the planets rotate around their axes, or whether the Milky Way is composed of little stars can never get enough from their organ of vision. "If all you want to do is to enjoy things, you do not lack anything you need for that purpose; but to know those same things, you lack everything. Les hommes n'ont besoin de rien, et les philosophes ont besoin de tout [Men do not need anything, and philosophers need everything]." If it turns out that reason can arrive at new truths, this speaks, at bottom, against it, because it exposes the imperfection of its nature. Galileo concludes the conversation with the thought that the bon vivant would be badly served by the carrying over of reason's capacity for expansion to all the sense organs, because every moment in history would be excrutiatingly relativized by a future that could bring ever new gratifications that had to remain denied to it at present. "If new pleasures should be discovered, could you ever be consoled for not having been reserved to live your life in the final times, when you could have had the benefit of the discoveries of all the centuries? As far as new knowledge is concerned, I know that you would not begrudge it to those who come after you."

Fontenelle lets us see the negative side, which the new historical consciousness of the modern age's superiority with respect to antiquity and of knowledge's capacity for expansion implies, and which was almost always to be overlooked in the triumph of the *querelle*: Progress burdens the individual with the consciousness of his factual restriction to the arbitrary standpoint of his finite existence in a history that seems to be able always to bring new things in response to reason's needs. The new subject that gains from this can only be an agency that spans individuals and generations and is indifferent to happiness, which always pertains only to individuals. Galileo warns the Roman's shade against imagining an expansion of human needs on the analogy of the explosion of theory that was initiated by the telescope. Apicius accepts this: "I see that there is no special privilege in knowledge, since it is left to those who want to trouble themselves with it; and nature has not taken the trouble to equalize the people of all the centuries in this regard. But pleasures are much more important. It would have been too unjust to bear, if one century had been allowed to enjoy more of them than another; and consequently they have been apportioned equally." [12]

The skeptical anthropology to which, as a context, the disposition toward the expectations that are realized by the optical magnifier belongs, itself finds support and confirmation in Copernicus's cosmological reform (even if it finds no such things in the intentions of the author of the reform). If it can no longer be taken for granted that man exists in the center of the universe as the creature for whom the creation provides, and that from this center of the universe he possesses the most favorable, comprehensive and, as it were, true-to-scale view of what makes up nature, then it is natural to see him as needing assistance in his theoretical efforts. So the telescope and the microscope fit into a ready-made 'slot.' Thus doubts about whether man is organically equipped for the role of observer of the universe coincide, historically, with the elementary discovery in astronomy that his cosmic standpoint must be as unsuitable as one could imagine for gaining an accurate idea of the construction of the system in which this standpoint is located.

In the *Apology for Raimond Sebond*, in 1580—that is, not many years before the emergence of the telescope—Montaigne provided a skeptical anthropology of this kind, and in doing so dealt, not accidentally, with the topic of the contingent givenness of the senses with which man is equipped. One could reduce Montaigne's theory of sense perception to the formula that while man does know everything through his senses alone, he cannot, by that means, know everything. The first reflection that Montaigne proposes ends in doubt as to whether man's equipment includes every sense that nature permits. There are animals that are capable of life although their outfit of senses lacks one or more of the faculties that man possesses. The analogy suggests itself that man, in his turn, could lack possible senses. If this were the case, our intellect would never discover this lack. Beyond the sense organs there is nothing that could serve to enable us to check our faculty of perception as such, in individual cases or as a whole; there is not even one of the five senses that can objectivize another one. One who was born blind could not possibly be made conscious of the fact that he did not see, because he would not understand the meaning of this statement, and therefore would not understand the deficiency that is imputed to him. Montaigne shows by this example that we cannot infer from the fact that our five senses are sufficient, or indeed that we are satisfied with what they accomplish, that because we cannot establish a need for further senses, they do not exist. It is the nature of every cognitive narrow-mindedness that it is narrow, in particular, in relation to itself. The perfection of the senses when they are kept apart from the demands of reason—the perfection that Galileo's shade uses as an argument against Apicius, in Fontenelle's dialogue of the dead—was already pushed aside, a century earlier, by the skeptical contemplator of the human world (though of course without his thereby granting reason a right to be alarmed).

The telescope, which still lies in the future, fits this skeptical anthropology so precisely that a later reader could think that he saw it being announced. Montaigne depicts astronomy as a prime instance of the disproportion between the capacity of our perception and the abundant claims of our cosmological theories, which he typifies as mere anthropomorphisms. "It is as though we had had coach-makers, carpenters, and painters who went up on high to set up contrivances with various movements, and to arrange the wheels and interlacings of the heavenly bodies in variegated colors around the spindle of Necessity, according to Plato . . . . These are all dreams and fanatical follies." [a] Is it possible at all to imagine such self-deception being dispelled by reality? For Montaigne this is, no doubt, only a thought experiment, the formulation of an 'optative' image, which reads like a piece of archaic futurology: If it should please nature, one day, to open its inner self to us and to let us see the mechanism of its conduct undisguised, with eyes equipped for the purpose—what abuses and what miscalculations we would then discover in our poor science. Probably not a single object would turn out to have been correctly understood. "Que ne plaist il un jour à nature nous ouvrir son sein, et nous faire veoir au propre les moyens et la conduicte des ses mouvements, et y preparer nos yeulx? Ô Dieu! quels abus, quels mescomptes nous trouverions en nostre pauvre science!" The tele-scope suggests itself so much here that the unsurpassed German translator of the *Essais*, Johann Joachim Christoph Bode, allows himself to be seduced into the anachronism of letting it already be in his author's hand.

If—so Montaigne's train of thought went—the physical dispro-portion between man and the universe is only a model of the meta-physical disproportion in which man imagines the divinity according to his own image and close to him, and thinks he can measure the ground of being by reference to nature, as its effect, then it turns out to be "above the order of Nature; its condition is too lofty, too remote, and too dominant to be bound and fettered by our conclusions. It is not through ourselves that we arrive at it; our way lies too low. We are no nearer heaven on the top of Mont Cenis than at the bottom of the sea . . . ." [b] And just at this point one reads in the German translation this startling advice: ". . . nehmt Fernröhre zu Hülfe, um euch davon mit euren eigenen Augen zu überzeugen" [use a telescope, to per-suade yourself of this with your own eyes]. The original text speaks of astronomy's classical model instrument, the astrolabe, with the aid of which the proportions of the cosmic system can be pictured: ". . . consultez en pour veoir avecques vostre astrolabe" [consult your astrolabe and see].[13] This example has only an anecdotal character, but it shows what I mean by the expression "a ready-made 'slot.'"

Montaigne, the skeptic, is conservative because he sees man's imagination as constrained by the limits of his experience; and for him, the realm of that experience is still a constant magnitude. He cannot know that any progress whatever would be possible here. For the discovery that he made is a resigned one, formulated in the shocking statement that for reason, the least advantageous position is the one that it in fact finds itself in, in man: " ... la pire place que nous puissions prendre, c'est en nous."[c]

That is already an anthropological iteration of the Copernican turning, whose initial model is itself affected—is historically relativized: What does it matter whether Copernicus was right, if in reality it does not matter whether he should have been wrong? We do not have an advantageous standpoint in ourselves—that reflects most plainly on our effort to determine this standpoint. The logic of the Copernican turning is reflected in its own futility: In that the new system was able to supplant its predecessor after so many centuries, it made the possibility of certainty regarding a system itself problematic.

Only an optimistic anthropology permits one to exclude from reality everything that is not directly accessible to man's faculty of perception. Even if it can also be rationalized as a principle of economy, the traditional astronomy's postulate of visibility corresponds to the assumptions of an anthropology in which man and cosmos are seen as coordinated in such a way that no essential incongruence can be assumed between man's organic equipment and the constituents of reality. The postulate of visibility follows from the symmetrical construction of the geocentric universe and man's central position in it. If all the points of the heaven of the fixed stars are equally distant from the Earth, the suspicion has to be excluded that there could be stars that, on account of their distance, remain invisible for men: If any fixed star at all can be perceived, then all of them can be. There is no conceivable need for an auxiliary means that could have promised to accomplish more here than the human eye can accomplish. The postulate of visibility has a pagan tendency. It denies any idea of disclosing to man only a limited view of the world and of conceiving of the world's sublimity as contained in an inaccessible space reserved to God. The idea, associated with metaphors of majesty, that not everything in the world concerns man—and especially not what is above him—is hard to harmonize with the other formula, that man is called to contemplate and to admire first the cosmos and then the creation—so that what his gaze cannot rest on at all has no meaning. We can gather in reverse, from the sarcasm of the later despiser of the world, what the anthropological significance of the postulate of visibility was. Mark Twain writes in the third of his *Letters from the Earth* that the sky in Paradise contained no stars, and the first one would have appeared after three and a half years, and most of them would never have been noticed—because the telescope did not yet exist.

What is Copernican is not only the restriction of the horizon of visibility or the determination that man's position is eccentric. Much more important is the fact that almost all the knowledge that astronomy was able to acquire after Copernicus about the construction of the universe and of the systems of which the solar system is a part is based on taking advantage of visual conditions associated with not being in the center. The disillusionment that could be associated with being expelled from the center was at the same time, potentially, the point at which actions having relevance to theory could be brought to bear. It is true that when man's standpoint had been displaced into its first motion through space—the annual orbit around the Sun—Copernicus himself had written off the possibility of confirming this empirically by reference to the heaven of the fixed stars, but he had not been able to prevent new instruments and new accuracies from keeping alive a search for parallax in the fixed stars. Even if what would have had to appear did not in fact do so for centuries, other things did.

If the validity of the traditional astronomy's postulate of visibility was based on the fixed stars having a finite and equal distance from the Earth, while light was propagated instantaneously, this assumption was not yet cancelled because Copernicus had to make the distance of the fixed stars "immense" and because this value became variable, also, by an amount equal to the diameter of the Earth's annual orbit. The statement still held good that the visibility of *one* fixed star guaranteed that of every other fixed star. What turned out to be more important is that the standstill of the heaven of the fixed stars made superfluous the cohesion and corporeality of the first sphere, which had been useful in explaining its supposed motion.

Chapter 1

Thomas Digges drew from this, as early as 1576, the conclusion that the fixed stars were distributed freely in space and that the closed spherical shape of the universe was only an illusion resulting from the limits of visibility. The whole unfriendliness of this idea to humans first becomes apparent in the statement, which could not be justified from Copernicus's own point of view, that the greatest part of the world is invisible to man. This has to do with favoring the principle of omnipotence over the principle of teleology. Galileo's early alliance with the telescope turns out, from this point of view, to be an anticipation of his late collision with the principle of omnipotence: The telescope is technology's contradiction of the concession of a sphere that is withheld from optical access.

This contradiction is put in question by the discovery of the finite speed of light by Olaf Römer in 1676. He did something that would never have occurred to Galileo, which was to compare the times of the eclipses of Jupiter's moons by the planet with the Earth's distance from it. The finite magnitude of the speed at which light is propagated, with a value that is very high in relation to the dimensions of the solar system, brought with it not only a sudden increase in the size of the universe but also a disparity between space and time that seemed to set an absolute limit on man's perception. One could not yet deal as generously with time as one could with space. The biblical chronology permitted the lapse of only a few millennia since the creation of the world. For the light of distant stars to be propagated as far as where man is stationed, greater lengths of time quickly became necessary than the canonic age of the world made available at all. The interest in the world's chronology becomes understandable when one takes into account the fact that the introduction of gravitation into cosmological thinking had as a consequence another sudden increase in spatial orders of magnitude. In the final remarks [the "General Scholium"] added to the second edition of the *Principia*, Newton assumes that the stars whose light reaches us represent, once again, suns, with associated systems of planets and comets, which are regulated, internally, by gravitational relationships analogous to those in our system. But what about the influence of these systems on one another? Here Newton seems to assume it as undoubted that gravity does not overcome the distance between the systems. Otherwise they would have to be in motion relative to each other, and ultimately to collapse into one another. So self-preservation, as the rational principle with which the creation is endowed, requires a space between the systems that exhausts gravitation, and the size of which is hinted at by the eccentricity (which is just becoming evident) of the comets' orbits, which are indicators of how far the field of the Sun's gravity extends. If the stellar universe was not permitted to be a unitary system of mass-forces, man's optical neediness could no longer be taken into account. It was not insignificant for the recognition of these orders of magnitude that the value that Römer had found for the velocity of light was confirmed only in the year of Newton's death, 1727, when Bradley discovered the aberration of the light of the fixed stars.

For the process that one could describe as the reflective assessment of vision,[d] it is decisive that what is invisible could become an ultimate threat. Not accidentally, Ludwig Feuerbach will pose the question, "Is there anything absolutely invisible?"[e]—a question that tortures a view of man for which intuition is his crucial capacity, and that had first proclaimed itself in a cosmological context in Lambert's work, in the form of the suspicion that the world could be greater in light-years than it is old in years. For the postulate of visibility, that is the critical consequence—even after the celebration of the triumph of the telescope—of the Copernican introduction of immeasurability. Only now does it become impossible to exclude every other suspicion as to whether man is empirically equipped for the world.

Man could no longer be the designated witness of the wonders of the creation if the time required for light to reach him from unknown stars and star systems was longer than the entire duration of the world. Retrospectively, Copernicus himself is supposed to have anticipated the damage to the role of the one who was called to contemplate the heavens. In two apocryphal anecdotes, the focus is on Mercury and Venus, the two key planets of the Copernican system.

In his *Traité d'astronomie* [*Treatise on Astronomy*], which first appeared in 1764 and was one of the most widely read books of the second half of the eighteenth century, Jerome de Lalande expressed doubt as to whether Copernicus had ever seen the planet Mercury. Lichtenberg corrected Lalande: In the passage to which Lalande refers, Lichtenberg tells us, Copernicus says only "that this planet gave him a lot of trouble and quod ei in gratiam ter locum mutaverit" [that in return it will have changed its location repeatedly].[14] The anecdote appears again, in a version in which it is given a tragic point, in Ludwig Feuerbach: "Copernicus is supposed to have mourned, on his very deathbed, that in his entire life he had never once seen Mercury, despite all his efforts to do so." [15] Feuerbach supplements the anecdote with an important commentary, which confers on the dying astronomer's sorrow the dignity of a faculty for sensing the future: "Today the astronomers, with their excellent telescopes, see Mercury at broad noon. Thus the future heals the pains of the past's unsatisfied drive for knowledge."

The importance of the Copernicus anecdote consists for Feuerbach in the fact that it defines the relation between reason and the senses not only anthropologically, and not only epistemologically, but also historically. Reason is always only an anticipation of intuition; vision and touch do not constitute the raw material, the substratum, of thought, but rather the essence of the fully realized relation to reality. To have reflected on the planet Mercury in terms of its importance for the construction of the solar system, but never to have seen it, must, for Feuerbach, have filled the dying Copernicus with the consciousness of a definitive lack. Mankind, in its history, could not leave this matter at that. If reason is not the perfection of sensuality but its anticipation, then the unsatisfied knowledge drive shows what can be in the future. That drive becomes a sort of temporal instinct, by which the individual becomes conscious of the interest of the species, but by which at the same time he lays claim, for himself, to a counterinterest, and experiences its painful futility. "What mankind desires in the youth that is the past, it possesses in abundance in the old age that is the future." This circumstance, the temporal dimension of the knowledge drive, becomes tangible in the figure of Copernicus. The triumph and the melancholy of pretension and resignation are right next to each other when Feuerbach makes Galileo call after Copernicus, "Oh, if you . . . could have lived to see the new additions to and confirmations of your system, what delight you would have derived from them! Thus speaks the man from the real 'next world,' the man of the future, to the man of the past."

The call, across time, that Feuerbach makes Galileo direct at Copernicus, is an almost word-for-word quotation from the Third Day of the *Dialogue on the Two Chief World Systems*, where Galileo has his Sagredo exclaim, "Oh Niccolò Copernico, qual gusto sarebbe stato il tuo nel veder con sí chiare esperienze confermata questa parte del tuo sistema!" [Oh Nicholas Copernicus, what a pleasure it would have been for you to see this part of your system confirmed by so clear an experiment!]^f The exclamation relates, in this passage, to the telescope's accomplishments in confirming the Copernican system by means of the observation of changes in the size of the planets depending on their position in relation to the Earth, and of Venus's changing phases. "These things can be comprehended only through the sense of sight, which nature has not granted so perfect to men that they can succeed in discerning such distinctions. Rather, the very instrument of seeing introduces a hindrance of its own"—i.e., irradiation. Copernicus's perspicacity was admirable, Salviati says in the dialogue, but still one must regret his misfortune, which prevented him from living in their time and viewing the confirmations of his theory through the telescope.

That is the germ of a further anecdote, the first trace of which is found in John Keill's *Introduction to the True Astronomy* (1718), and which also survived the doubt that Alexander von Humboldt was to register about it in his *Kosmos*. It attained its greatest publicity through d'Alembert's acceptance of it as a plain fact in his article "Copernic" in the fourth volume of the French *Encyclopedia*, which appeared in 1754.[16] If the telescope was the manifestation of a lack of anthropocentric teleology, this supposed fact moved it into the framework of a different teleology, which fascinated the Enlightenment, namely, the teleology of history. This function did not dawn even on the critic of the Enlightenment. Herder writes, "The firm, calm inventor did not live to see the consequences of his system. A few days before his death, the seventy-year-old man saw the first copy of his printed book on the revolutions of the heavenly bodies. But his discovery continued to have effects. With the help of new telescopes, invented after Copernicus's death, Galileo saw what Copernicus had inferred: the phases of Venus . . . . "[17]

Copernicus had not inferred anything of the sort. He would probably not have shared the high expectations that Galileo attached to the discovery of phases of Venus and to their power of proving the Copernican system. That Galileo should have been more skeptical at this point follows from the simple circumstance that the phases of this planet, up to "full," had to appear not only in the Copernican system's heliocentric construction but also, already, in the preliminary stage (which we can infer) of a partial system composed of the inner planets, Mercury and Venus, revolving around the Sun. Copernicus would have known that in Ptolemy's system, too, given adequate optics, Venus would have had to appear only in varying crescent phases, and never in the "full" phase, so that the power of demonstration against Ptolemy that Galileo sought could only consist in demonstrating that Venus has a "full" phase.

But the call that Galileo makes Sagredo address to Copernicus across the intervening century not only exaggerates the telescope's power of demonstration but, above all, fails to appreciate the position of the strict validity of the postulate of visibility, which for Copernicus is still unquestioned. He could have anticipated neither the telescope nor any positive results from it. Only as the consequences of Copernicanism are worked out is it also noticed that the assumption that everything real is visible has an all too pagan stamp, and does not fit into the pattern of the Christian tidings about man. One could say that this is part of a retrospective attentiveness to the potential of the Copernican turning to promote Reformation attitudes. The ancient world's resting contemplator of the heavens is recognized as a delusive literary prolongation of man's paradisaic rights and privileges. At the same time it becomes clear, as a basic modern pattern, that the loss of Paradise (the more profoundly it is felt, the more this is the case) makes the regaining of it, if not a longing, then a formula for a program of method. In this context, the telescope appears as the instrument by which to compensate for a status that man has lost.

Bacon's idea of making the regaining of Paradise into the goal of historical progress through science is not the secularization of an originally religious idea of the final promises, but rather—as a technique of argumentation—a way of avoiding an anthropology that makes man's organs constitutively, and thus insurpassably, insufficient in relation to the world. Something that was lost can be gained again; something that was denied in one's essential structure cannot be attained. It is an idea that determines the magnitude of expectations.

Consequently it is not surprising that Bacon is disappointed by the first results arrived at by means of the telescope. His metaphors for the optical instruments (*perspicilla*) are extravagant; on the basis of optical contact, they already anticipate astronomical commerce.[18] The process of seeing becomes an action, and is thus protected from the mistrust that is Bacon's attitude to the mere intuition of objects, the acceptance of their visibility as the mode of givenness that is suitable to them. Man cannot begin *ab initio contemplationis* [from a beginning of contemplation]; he is an inquisitor who presses the objects hard so that they will reveal themselves to him. The telescope represents the situation in which nothing is given as a present, but everything can be obtained by exertion. Hence the disappointment in view of what was disclosed by Galileo's telescope—things that Bacon enumerates with evident contempt. Indeed, pointing to the poverty of the results, he hesitates to concur in them, to accord his trust. The world has expanded, but experiences of extraordinary things have not appeared. They are important discoveries (*inventa nobilia*), but they are suspect, because there are so few of them: "Quae (sc. demonstrationes) nobis ob hoc maxime suspectae sunt, quod in istis paucis sistatur experimentum, neque alia complura investigatu aeque digna eadem ratione inventa sint." [Which demonstrations I regard with suspicion chiefly because the experiment stops with these few discoveries, and many other things equally worthy of investigation are not discovered by the same means.][19] In its instrumental demonstration in the telescope, the mechanistic idea of method—the production of knowledge *veluti per machinas* [as by machines]—seemed not to be confirmed by the extent of what it produces.

A half a century after Bacon, Joseph Glanvill took up the connection between the idea of the Paradise that was lost and to be regained and the idea of optical progress, and developed it further with a view to the balance sheet of the accomplishments of the new instruments. In *The Vanity of Dogmatizing* (1661), Glanvill made Adam in Paradise distinguished precisely by the fact that he did not yet need "Galileo's tube" in order to contemplate directly the whole of reality, the "celestial magnificence and bravery." He was not yet confronted by the reservation-boundary of the invisible, but was even able to perceive the influence of magnetism by means of vision. Clearly discernible (as already in the case of Bacon) is the implication that the punishment of labor in the sweat of one's brow, which was decreed at the expulsion from Paradise, is an expenditure that serves as restitution for the loss that was caused, and at the same time legitimizes what is regained. In 1668, in his famous essay *Plus ultra*, Glanvill presented a first balance sheet of human experience as augmented by instruments. For him, as for Bacon, the passage beyond the Pillars of Hercules and the discovery of new lands and continents was only the prelude to an expansion of the world in all dimensions. To Glanvill, the eschatological formula of the coming new heavens and new Earth is not excessive as a designation of the state in which Paradise has been regained, and as a definition of the preeminence of the telescope in this process: "We have besides, New Heavens as well as a New Earth a larger and truer prospect of the World above us. We have travell'd those upper Regions by the help of our Tubes . . . ." [20]

In Bacon, the feedback process was not yet in operation by which the results arrived at with the telescope retrospectively articulate people's expectations, determine the type of result expected in the future, and thus form a permanent propelling mechanism of development. The 'prehistory' of this invention and the history of its effects mesh together in such a way that the history of its effects again and again becomes the prehistory of new displacements of the horizon of experience. Glanvill still has to deal with opponents like Galileo's, who regard the telescope as an illusion-producing instrument; but the skeptical anthropology of the Adam who is exiled from Paradise localizes the inaccuracies and illusions in the natural eye. It is the products of human work and technique that must remove and correct the defect, as the thermometer, the barometer, and the pneumatic pump (Glanvill mentions these in one breath along with the telescope and the microscope) make possible for man perceptions that, with his unaided senses, would have much less precision than even his optical perceptions possess or that are not possible by organic means for him at all.

But not only the instruments but also the objects have to offer a basis for joyful optical anticipation. Glanvill believes that nature's modes of functioning are based, as a matter of principle, and without exception, on processes that are visible or can be made visible—that is, that they are mechanisms, in the sense of the miniaturization or magnification of familiar technical contrivances. This association of empiricism with what one could call optical optimism persists up to the nineteenth-century molecularists. One can look on at the way nature does things, once one has been put in a position to scrutinize it accurately enough—something of the ancient onlooker, with his presuppositions, is still involved in this conception, which led the Romantic Johann Wilhelm Ritter to the statement, "All chemistry is astronomy." [21]

That is also the basis of Glanvill's *Plus ultra*, with its expectation of unlimited progress in microscopic and macroscopic optics, and its innervating effect, in accordance with the principle, "Hope is the fuel of Activity and Endeavour." [22] But what are the great questions that motivate this expectation? What are the enigmas that optical progress is finally supposed to present, unveiled, to the human eye? It is very remarkable that for Glanvill it is explicitly only two questions to which a clear answer is anticipated: whether the Earth moves and whether the planets are inhabited. [23]

The regaining of Paradise reestablishes, by *artificial* means, the same kind of intuition that Adam is supposed to have had by *natural* means before the Fall. That anticipation is not a matter of course. One could also have deduced, from the same anthropological premises, that the science that is possible for postparadisaic man would have to be heterogeneous in type to the ideal of intuition (which would be the 'paradisaic' form of relation to the world), so that that science now could only be based on what can be done by indirect means.

What did the people do to whom the 'regaining of Paradise,' with the new optical instrument, had to appear as a delusive, if not a blasphemous, formula? If anthropological skepticism implied that human optics were deficient, this did not yet have to imply converting to a historical expectation of instrumental progress. Where the negative anthropology, with its epistemological skepticism, is simply the correlate of a human situation that can only be defined with the aid of theology, the terrestrial effect also only becomes an expression of man's expectation of compensation in the next world. It was in this sense that Montaigne's thought was continued by Pascal. From this point of view, the outcome of the (fundamentally disillusioning) displacement of limits was not the optical gains but the confirmation of a realm of the 'not-yet,' which cannot be caught up with. Pascal's infinity is more a definitive reservation than the space for the expansion of human theory, and is anything but an 'infinite task.' The telescope is integrated into the defense of Christianity. The Bible, Pascal writes, speaks of a greater number of stars than there are in the catalogue made, in antiquity, by Hipparchus, which knows only 1,022. "Combien les lunettes nous ont-elles découvert d'astres qui n'étaient point pour nos philosophes d'auparavant!" [How many stars the telescopes have revealed to us that did not exist for our philosophers in former times.] [24] Once again, as in Galileo, the telescope stands against the disbelief of the philosophers; but it must not be a triumph of human progress, but rather of the unheeded Scriptures.

Voltaire relentlessly exposed and attacked the weakness of this apologetic note. If one takes the Bible as seriously, he says, on the subject of the number of the stars, as Pascal does in order to credit it with a revelation that has been overlooked, one increases its defeat in every case where its 'natural science' has manifestly turned out to be mistaken—as in the question of the immobility of the Earth and the motion of the Sun. Anyone who wanted to lay claim to the Bible on behalf of natural science would destroy its binding force in matters of ethics. Almost imploringly, Voltaire points to this inevitable consequence of Pascal's integration of the telescope: "Voyez, je vous prie, quelle conséquence on tirerait du sentiment de Pascal. Si les auteurs de la Bible ont parlé du grand nombre des étoiles en connaissance de cause, ils étaient donc inspirés sur la physique." [Consider, I beg of you, what consequence will be drawn from Pascal's opinion. If the authors of the Bible spoke with full knowledge when they spoke of the great number of the stars, then their inspiration extends to matters of natural science.][25] For the rest, infinity and great numbers were more an expression of confused indeterminacy than of precise anticipation: One only needs to regard the heavens unwaveringly, and one's view becomes dazzled and confused, and finally one thinks one sees an infinity. Thus arises the prejudice (ce préjugé vulgaire) that was reduced by the ancient catalogue of stars to precise ideas, whereas what the Bible did was to encourage it. In any case, the contemporary catalogue of stars, by Flamsteed, to which Voltaire refers explicitly, contains only three times the number that Hipparchus had counted. By mentioning this comparatively modest number, Voltaire counters Pascal's (in his eyes) 'farfetched' alarm in the face of the threatening abyss of infinity.

Even five years before the turning, in 1653, that is marked by the *Memorial*, Pascal categorized the importance of the telescope differently. In the preface to the planned treatise on empty space, he had presented one of the earliest sketches of a philosophy of history. He connects to the distinction between memory and reason the distinction between two dimensions of the human interest: on the one hand, to hold onto what has already been gained in the way of humane manifestations, and on the other hand to gain what can still be sought and discovered in the way of hidden truths (vérités cachées). These two 'organs' seem to contain the anthropological ground plan of the double interest as, itself, an unhistorical definition. There is still a comfortable separation between the two things that will soon be connected in the *pensée* I quoted: the Bible as a divine institution of memory and the telescope as an instrument for finding hidden truths. "Elles ont leurs droits séparés . . . . " [They have their separate rights . . . .] The telescope expresses the fact that reason has an unrestricted freedom of expansion, is inexhaustibly fruitful, and advances in its discoveries endlessly and without interruption.[26]

The infinitude of science, as a correlate of the infinitude of its object, is still unconstrainedly accepted in this early fragment. The hiddenness of nature's secrets has no tendency to humiliate, but rather, in relation to the expenditure of time and energy, to challenge. That is why, for reason, there can be no admiration of antiquity. Whatever it may have demanded of itself, in any case it stood at the beginning of the expenditure of time and energy, and also of the possible gains in experience. What the ancient theoreticians of nature lacked was not vigor of reason but good fortune in experience. Thus the ancients could be excused for not having had a correct idea of the Milky Way. They were not able to use technology to remedy the weakness of their eyes: "la faiblesse de leurs yeux n'ayant pas encore recu le secours de l'artifice." However, it did not seem excusable to Pascal when his contemporaries held fast to the traditional conception of the Milky Way and scorned the aid of the telescope (*la lunette d'approche*), which had revealed an infinity of small stars behind this celestial appearance. Without the alarm of his later thoughts, Pascal sees the telescope enter into relations with infinity.

Here too, already, in the fragment on empty space, there is no congruence between man's cognitive faculty and its principal object, nature. But the disproportion can still be compensated for by means of the factors of time and labor. The discovery of progress is the discovery of the unity of theory's subject in its exertions across time: The stock of hidden truths, of which Pascal speaks, is reduced by untiring and consistent exertion. The telescope gives a unique example of how an optical horizon that was constant through millennia is transformed into a continually displaceable boundary. Pascal does not suspect that there might be an opposing process by which the stock of those 'hidden truths' increases in proportion to those that have been discovered at any point in time.

Hardly a decade later, the idea of the two infinities, and of man's being placed between them, has, instead, a threatening sound: Any effort to measure oneself against these infinities turns into an act of self-humiliation. Natural science has become the *grand sujet d'humiliation* [great cause of humiliation]. Because men did not contemplate these infinities, but provided themselves with the appearance of a central position in the universe, they foolhardily set about the investigation of nature, as though they culd arrive at an appropriate relation to it.[27] Concepts of the type represented by 'infinity' have not only a regulative function but also a resignative one: Progress, as an infinite process, implies that every phase that it arrives at produces disappointment by its element of surpassability.

To begin with, the new 'world-optics' still enter the service of the Baroque metaphors for the world: The human eye penetrates through one new backcloth and stage set after another, into backgrounds that in their turn are also backcloths and sets. In 1654, Pascal receives the letter of the Chevalier de Méré, which could have become at least the provocation of his reflection on the two infinities. Méré writes that the hundred thousandth part of a poppy seed contains not only this world, in miniature, but all the worlds that Epicurus dreamed of, the surface of the Earth and the sea, with their abysses and depths, the mountains and valleys, the cities, workers, some of whom build and some of whom tear down—and some even make telescopes, which the people in those microworlds have to use because their eyes and their other senses are proportioned to their dimensions.[28] The inadequacy of the senses for the world is repeated in the worlds of which a presentiment is just beginning to dawn on man, and since they are "worlds," rather than it being the case that their discovery can mean a reduction of the unknown, they multiply the stock of these "hidden truths" beyond all limits.

A new structure of the process of theory suggests itself, in which the stock of problems increases in proportion as problems are mastered. Georg Simmel described this basic circumstance of the reverse relationship between the objective accomplishment and the state of consciousness in the case of optical instruments: "It is true that infinite distances between ourselves and objects have been overcome by microscope and telescope alone; but we were first conscious of these distances only at the very same moment in which they were overcome."[29]

Copernicanism tore asunder the fit (which was already loosened, anthropologically) between the world and man's organs: the congruence between reality and visibility. It did not remain a question of the range of the organ that could be achieved by technical means, or of doubts about the completeness of the faculty of sensation in relation to the information that nature sent out and the subtlety that it had to offer; the internal unity of perception—the consistency of what the actual senses, each separately, could convey—was also problematized. Since Aristotle, there was a coordinating sense—almost unchallenged in the tradition—between the specific sense organs: the *sensus communis* ['common sense']; and with it the compatibility of information from the senses, as the correspondence of characteristics belonging to an identical object, was unquestioned. Molyneux had posed to Locke, as the *experimentum crucis* [crucial test] for empiricism, the problem of whether someone who had been born blind and had learned to distinguish a ball and a die by the sense of touch could also, on acquiring the faculty of vision, identify these bodies by this means. It is quite appropriate that this problem, viewed more as a mental than as an empirical exercise, occupied people's thoughts in the eighteenth century.

The question of the translatability of the data of each individual sense into those of the others—the problem of how we produce a 'common' object of all the senses—ended by raising questions about the unity of the concept of space. Was the structure of tactile space identical with that of visual space? Berkeley had asserted for the first time in 1709 that geometry does not have to do with the structure of optical space. The conclusion that Berkeley had drawn from that—that in that case there could not be any geometry at all for visual space—was declared by Thomas Reid, in 1764, to be at least not compelling.[30] It is not without rational charm to see that a line leads from the ramifying consequences of Copernicanism for optical self-consciousness into the history of the origins of non-Euclidean geometries, the most important precondition of which was the contingency of the idealizations that are carried out in Euclidean geometry.

The breakdown of the postulate of visibility—understood in the widest sense—is brought to a point by a kind of reversal: The visible world is not only a tiny section of physical reality, but is also, qualitatively, the mere foreground of this reality, its insignificant surface, on which the outcome of processes and forces is only symptomatically displayed. Visibility is itself an eccentric configuration, the accidental convergence of heterogeneous sequences of physical events. The morphology of organisms is only a model instance illustrating this statement. What it was once essential to know—the form and appearance of the organism, the sum of its classifiable characteristics, its typical colorings—all of this turns out to be the mere external aspect of chemical, metabolic processes, which are controlled by the core contents (which cannot be apprehended eidetically) of the gene. It is the ideal case for a Laplacean demon: the totality of life, gathered together as information. Optics has become an aspect of objects that does not make a difference: an appearance of appearances. What is accessible at all in this way has turned out to be an arbitrary section through reality, having little significance in relation to what can only be indirectly inferred from it. Even optical accessibility in the telescopic dimension is a side effect of physical realities in space, which became possible in a narrow bandwidth of conditions that exist nowhere in such restricted, clouded, and distorting forms (but in agreement, for those very reasons, with the vital requirements of the organic subject, man) as at the bottom of the terrestrial atmosphere. This final stage in the reversal of the postulate of visibility means that the invisible has occupied the position for which, in the metaphysical tradition, the visible seemed to possess the sanction of being the access to reality.

2

The Proclamation of the New Stars, and One Single Person's Reasons for Believing It

Even if one can speak of a disposition in its favor that was a precondition of the historical influence that the optical magnifier exercised, still the suddenness with which the telescope appeared and spread was an event that largely evaded the means of comprehension of people at the time, and took on irrational characteristics.

Among the classical means for domesticating such an act of invention was shifting it back into the Middle Ages, into the neighborhood of the suspected pacts with the devil of heretics like Roger Bacon—or even into antiquity, to which, despite the absence of evidence in the sources, a secret knowledge of the telescope was ascribed. It would have been the pinnacle of a technique of legitimation through anachronism to trace the invention of the telescope back to Aristotle himself. In the *Dialogue on the Two Chief World Systems*, Galileo makes Salviati report such an attempt. He knows certain gentlemen, he says, who are still living and active today, who were present when a doctor lecturing in a famous Academy, on hearing a description of a telescope—which he had not yet seen or used—said that the invention was taken from Aristotle. He had had a text fetched, and looked for a certain passage, where the reasons are discussed that make it possible to observe stars during the daytime from the bottom of very deep wells: "Here you have the well; it is the tube. Here are the thick vapors, which the invention of lenses imitates. And here finally you have the strengthening of the power of sight by the passage of the rays through a transparent medium that is denser and darker." [31]

Domesticating the telescope as part of the groundwork constituted by the authority of Aristotle is a device that serves not only to legitimate it but also to eliminate the facticity of the late and seemingly arbitrary temporal beginning with which the Enlightenment, with its own consciousness of the radical beginning of reason, was to collide again and again. Even Galileo did not dare to found the epoch-making incision that the invention of the telescope marked, after the horizon of cosmic experience had been constant since primeval times, on the spontaneity of human reason alone. He avails himself of the formula that it had pleased God, at just this point in time, to grant such a wonderful invention to the human spirit. The most important early historian of the telescope, Peter Borelli, defends the desire to go beyond the optical limits set for man by God, and thus to improve God's works, against the charge of arrogance. God, he says, wanted to withhold this art from us for many centuries on account of the Fall of the first human beings, but now, out of special paternal kindness, He deigned to open the eyes that had been darkened as a result of sin.[32] Of course that is not a satisfying explanation of the delay of an invention that is so clearly connected to the essence of reason, but it integrates the fact into the salvation story, which evidently holds in readiness some intraworldly concessions, in addition to redemption as it is understood by theology.

The domestication of the invention of the telescope involves the intervention both of accident and of mysterious importation. One encounters early on in the literature the legend of the foreigner, in Dutch costume, who arrives in Middelburg in 1609 and buys concave and convex lenses from the lens grinder Lippershey; holds them one behind the other, to test them; approves the merchandise and disappears with his purchase, never to be seen again. The lens grinder imitates what the foreigner had done with the lenses and notices, to his amazement, the effect of combining them. Of course the story leaves open the possibility that it could have been the Devil, but in any case he did not demand a pact, and the subsequent consequences were based only on the old concept of technique as imitation (though nature no longer sufficed as the object to be imitated). No less formulaic is the account of the lens grinder's children who, while playing in their father's workshop, accidentally hit upon the optical effect when, through two lenses, they suddenly see the cock on the church steeple standing before them close enough to touch. The historiography of the telescope has also taken an interest in a somewhat more serious milieu, which was found in the family of Jacob Metius, whose brother Adrian had distinguished himself not only by his knowledge of mathematics but also by being an assistant to Tycho Brahe on the island of Hven. None of the competing inventors was able to get the longed-for patent from the Dutch Estates General, because the invention had evidently become an open secret in the shortest possible time. In these artisans' circles, in spite of the brother who worked for Tycho, it did not occur to anyone to direct his gaze, with the telescope, at the starry heavens.

There is apparently no connection between the prehistory of telescopic astronomy on the Continent and the first appearance of the idea of a new celestial optics in England. There, Thomas Digges had gone even beyond the logical consequences of Copernicanism when he asserted the preponderance of the invisible part of the world over the visible part. It was this excess that gave rise to the conjecture that Digges had not only pursued Copernicus's theory but had already been involved with the first optical results arrived at by means of the telescope. In his *Pantometria* of 1571, five years *before* his exciting astronomical treatise, one finds both in the preface and also in the 21st chapter of the first book a description of an optical apparatus that is supposed to have been produced by his father, Leonard Digges. Such optical instruments were already known in England before 1580, but had evidently not been used for astronomical observations. The telescope, which was a sort of artisan's tool, did not fit into the systematic context of astronomy, the "liberal art." Only a year before Galileo's proclamation, that is, in June 1609, Thomas Harriot had undertaken celestial observations, and in the winter between 1609 and 1610 he had also involved friends and students in his observations and had equipped them with instruments of their own. It seems that the English regarded only the Moon as an appropriate object to bring closer by means of the telescope. It looks as though this may still have lain within the framework of the postulate of visibility: The new optics allowed one to see more clearly what one had already seen, but evidently did not yet allow one to anticipate that things that had never before been seen could be made visible. This could not lead to the initial success—spectacular, but incompatible with the tradition—of the "new stars," the success that Galileo will exploit, not least of all by being able to combine with it a celestial christening in honor of the house of Medici.

Above all, however, as the state of the sources shows, these amateurs lacked the faculty of linguistic proclamation of the new things they may have seen. Galileo's *Sidereus nuncius* [*Astronomical Announce-ment*], in March 1610, is also an event in the history of descriptive prose. As the first document of a reflective celestial optics, moreover, it is an exemplary illustration of the needs for familiarity that will be bound up with Copernican optics, and that culminate in the En-lightenment's expectations that the universe will be inhabited.

When on 24 August 1609, Galileo presented the Senate of Venice with a telescope having a ninefold power of magnification and pointed to the instrument's usefulness in warfare, he had still hardly tested its peaceful application. Might his aiming it at the heavens have first occurred after the resignation that resulted when it turned out not to be possible to maintain a monopoly of the invention on behalf of the power of Venice? In any case, Galileo had the tangible success that he was able to make plausible to the Republic the invention as his own, and its usefulness "on sea and on land," and to secure for himself, in return, a lifelong guarantee of his professorship in Padua, at triple his previous salary.

Galileo's first biographer, Vincenzio Viviani, describes the process: "In approximately April or May 1609, there arose in Venice, where Galileo was at the time, a rumor that a Dutchman had demonstrated to Maurice, the Count of Nassau, a sort of glass, through which distant objects seemed to be close at hand. This was all that was said. Having this information only, Galileo at once returned to Padua and set himself to investigate how such an instrument could be made. Having immediately rediscovered (*ritrovó*) this the following night, the day after that he put the instrument together in the way in which he had imagined it (*nel modo che se lo aveva immaginato*). Despite the imperfection of the lenses that were available, he saw the effect that he wanted, and quickly reported this to his friends in Venice." [33]

However little it is now possible to clarify who the original inventor of the telescope was, it is equally clear that all the candidates for this historical position are mute. The *idiota* [layman] whom Nicholas of Cusa had raised, in the form of a spoon carver, to the antitype of the scholar, and whom he had introduced into his dialogue with a specific antirhetoric, turns out to be helpless to assess the potential of his invention, and speechless when it comes to proclaiming it. The early historian of the telescope even compares the typology of this inventor to the choice of the apostles: In the same way, in this case, God had wanted to have the doors to His secrets opened by men who had been taken by Him *ex populi faece* [from the dregs of humanity].[34] Galileo might perhaps not have invented the telescope if he had not heard that the telescope existed; but he understood very quickly (even if it was not quite as quickly as his biographer would like it to have been) how it must have been made, without needing additional particulars. This process is just as characteristic as was Galileo's connection with the arsenal of Venice. There he learned not only to extract theoretical laws from the technicians' practical solutions, but also to think about feeding theoretical laws back into the possibilities of technical practice. For the telescope, this meant that Galileo arrived very quickly at improvements in the instrument's effectiveness, without which the astronomical breakthrough would not have succeeded. When Galileo beheld Jupiter's moons for the first time, early in 1610, this was for him not only an almost palpable confirmation of the Copernican system, but also a roundabout mental route to a new idea that was, once again, technical: to put the navigation practiced by the sea power, Venice, on a more reliable basis with the help of the short periodic motions of this system of satellites. Galileo's genius lay in the anticipations with which he caught on to the idea of the telescope. The telescope was a curiosity on display at the annual fair before, in Galileo's hands, it became an instrument of theory.

Pure visibility, as a quality that is accessible for everyone, does not exist. On the contrary, man continually unburdens himself from the overload[a] of what would be optically possible—he always 'directs' his gaze first at objects that are equipped with other qualities besides optical ones and that make claims on him. The telescope, as an instrument that depends on addressing oneself to a very narrow part of the visual field, was only an empty possibility. In spite of his bias in favor of the telescope's having originated in Zeeland, and against any other claim to priority, Borelli stated clearly enough in his history of this invention that for Galileo's intellect and acumen, the tiny opening represented by a foreigner's invention had been enough to enable him to gain access to a new world.[35]

The ambiguity that lies over the origin of the telescope, as to whether it was a magical implantation or a pure logical consequence of theory's expansion of the universe, reaches deep into the early phase of its astronomical employment as well. What Galileo says about this in the *Sidereus nuncius* lets this ambivalence show through. On the one hand, he declares emphatically that he himself devised the telescope "after being illuminated by divine grace," while on the other hand, a few lines later, he refers to the rumor that had reached his ears in May 1609. Of course the reader can render these consistent by observing that one does not need to impute to Galileo a claim to priority, but only, as the important aspect for him, his having independently found out the principle that had already been realized elsewhere, and having brought it to a new and increased effectiveness. Perhaps he wanted to promote that ambiguity when he gave to his proclamation of the telescope and of his first discoveries by means of it this one title, the language of which permitted the enthusiastic interpretation of *The Sidereal Messenger* as well as the soberer *Astronomical Announcement*. Of course, a closer examination leaves no doubt that one could not very well say of a messenger that he "contains" observations. Nevertheless, the most important among the first readers of the pamphlet, Johannes Kepler, at the emperor's court in Prague, promptly misunderstood the phrase, as we can see from the title of his response, *Conversation with the Sidereal Messenger* (*Dissertatio cum sidereo nuncio*).

In contrast to the Dutch legends of its origin, Galileo's self-narrated early history of the telescope combines the elements of work and grace, monetary expenditure and inspiration—i.e., the typical ingredients of the history of modern science and of biographies of scientists. The principle and the means—his involvement with the laws of the refraction of light, and the lead tube with the two lenses—represent a quick crossing of the threshold of accidental discoveries and artisanal testing, and measurable progress from ninefold, by way of sixtyfold, to one-thousand-fold magnification, as Galileo describes the three phases that he traversed in ten months.[36] He expressly passes over the fact that he had offered his invention, in its first stage, to the Venetian Senate, and had recommended it for terrestrial use: "But forsaking terrestrial observations, I turned to celestial ones." [37]

With the *Astronomical Announcement*, on 13 March 1610, the 46-year-old Galileo comes forward for the first time as a publicist. The edition is composed of 550 copies. The report, written down in a hurry, carries no marks of literary pretensions, as later works of Galileo's do, but as a descriptive accomplishment it is nevertheless something still largely unknown, and for which the way was not paved, linguistically, at the time. The time pressure that Galileo evidently feels also has to do with the fact that he expects new discoveries from one day to the next, so that the form of a definitive account would not have been in keeping with the actual state of affairs. This expectation was to prove to be correct, because in the *Sidereus nuncius* important phenomena that were already within the reach of the early instrument are absent, such as the unexpected shape of the planet Saturn, which Galileo was to discover only at the end of July 1610. The phases of Venus are likewise absent, since he saw them only in December 1610—which also allows us to infer that the telescope was not from the very beginning, and exclusively, in the service of the evidence of Copernicanism. It is true that Galileo was already a Copernican, before the telescope; but it provided him in a short time, as though of its own accord, with the verification that he needed, and fixed him on his Copernican mission with the evidence of intuition, for which he was more receptive than, for example, Kepler, with his inclination toward speculative construction.

This state of affairs contributed to Galileo's so expressive indignation at the refusal of his colleagues to look through the telescope or to accept what they saw. The *Sidereus nuncius* is written in Latin, which identifies it as a publication for fellow specialists; but the disappointment I referred to was to push Galileo toward a different audience, and into a different language and a different literary form. But in this we can see how he misunderstood his failure. For much more important than the fact that the *Astronomical Announcement* was not written in the Italian language is the fact that it was not written in the language in which Galileo thought the book of nature was written. No use is made of the language of mathematics. The approximately two hundred observations of Jupiter's satellites that he had carried out restricted themselves to estimating their various distances in terms of the unit represented by the diameter of the central body. He came no closer, in this way, to the (in the strict sense) astronomical goal of calculating their configurations with respect to one another—and especially to calculating the times of the eclipses of the individual moons, which would have been critical for his suggestion that they could be used for navigation.

What excited the discoverer was the miniature model of the solar system, not as an exact scientific exercise, in the sense in which that would be understood in the future, but as a configuration. In spite of his use of the telescope, and although he did not "celebrate the neglect of all artificial means as itself a method for extending his knowledge" (like, for example, Goethe, with his rejection of the telescope and the microscope), this Galileo would have been described by Kant as a "naturalist of pure reason." The belief that nature offers to our first artfully strengthened glance the pure patterns of its construction and its procedures contains something of the ancient concept of reality as momentary evidence,[b] in the form of the expectation of a rapid triumph for the new cosmology.

It is extremely instructive to observe that the only quick and great success that was given to Galileo's proclamation of the new stars confronted him with the crucial omission in his Copernican mission: He had not read Kepler's *Astronomia nova*, which had appeared in the previous year. For as early as the beginning of May 1610, Kepler responded publically to Galileo's essay, which he had received through the Tuscan ambassador in Prague on 8 April 1610.

One has to imagine the situation of the two (after the death of Tycho Brahe) most important astronomers. Galileo had broken off an earlier correspondence with Kepler in 1597, after Kepler had unsuccessfully called on him to stand up publically for the Copernican theory and had sent him his *Mysterium cosmographicum*. Galileo had never gone into it. Now Galileo wanted Kepler to ratify his telescopic discoveries in the heavens. But for his part, he had not taken note of Kepler's new work, which, by discovering the elliptical form of the orbit of Mars, broke through to a new science of the heavens. So Kepler begins his *Conversation with the Sidereal Messenger*, addressing Galileo directly, with the graceful reproach that he had just rested himself from his long years of work on his *Commentaria de motibus Martis* [*Commentary on the Motions of Mars*, = *Astronomia nova*], all the while anticipating Galileo's reaction to this work, and wishing for the resumption of their correspondence, which had been interrupted for twelve years—when, about 15th March, he had received the first news of the new stars. Instead of reading other people's books, Galileo had been making his own exciting discovery.

This discovery arouses Kepler's admiration and amazement—perhaps also something more and something different, for one must not overlook the fact that the book he had sent to Galileo twelve years earlier had provided a geometrical/speculative demonstration of the completeness of the Copernican system with six planets. Now, Galileo described the stars that circled Jupiter as planets, not as moons. This plural of "moon" simply did not exist yet, and in any case, the Copernican analogy to the larger system could be made more obvious by using the name "planet." Galileo did not hesitate for a moment before describing the motions of these satellites around Jupiter as circular orbits, i.e., holding to the "Platonic axiom." He had taken no notice of Kepler's new theory of the orbits of the planets. It is amazing to what an extent Galileo was indisposed, and remained indisposed his whole life long, to be the partner that Kepler desired in laying the foundation of the "new astronomy."

Compared with Kepler's break with tradition in giving up the planets' circular orbits, Galileo's looking through the telescope at the Jupiter system seems extremely conventional. It is an attempted *coup de main* of intuition, by means of the telescope, to carry the day for Copernicanism. Kepler's *Astronomia nova*, on the other hand, is a classic instance of theory based on warding off intuition. It is true that after Tycho Brahe's death Kepler had been able to get possession of his notes, with the most accurate astronomical data up to that time (data that he mentions in the title of the work), but Tycho's son-in-law got revenge for this by depriving him of Tycho's unique instruments, with which these data had been obtained. Kepler was not able to reach the level of these observations himself. He also suffered from eye trouble (polyopia).[38] It is a significant fact that the greatest achievement in laying the foundation of modern astronomy was accomplished by a nearsighted and diplopic person who was denied the best instrumental equipment.

The difference in the time expended by the two astronomers in achieving their successes—in the one case, many painful years, and in the other, an opportunity seized in a few weeks—is also eloquent. Against this background, the understanding that Kepler shows for the news of Galileo's innovations stands out all the more impressively. We should not forget, however, in our proper admiration for such greatness, that the mildness of his reproach and the generosity with which he absolved Galileo had a negative effect. It was made all too easy for Galileo to overlook the fact that he should have read Kepler now, instead of relying entirely on the telescope. But pride in no longer reading as much as people had done earlier, but doing more looking, instead, has entered even the historiography of science as a criterion of the separation of the modern age from the Middle Ages—and not, in every case, validly.

Chapter 2

Kepler describes very precisely the scene in front of his house in the middle of March 1610, when he meets the imperial councilor Wackher von Wackenfels and receives from him the first reliable report about Galileo's new stars. The scene is almost grotesque, since both of them lose their composure, the one as a result of delight, the other as a result of perplexity; confusion and laughter, in view of the surprising news, make it difficult for both of them to speak and to listen: "ille gaudio, ego rubore, risu uterque ob novitatem confusi . . . ."

When Wackher has left him, Kepler calls to mind Galileo's reputation, his honesty and his acuteness. But his immediate thought is of the conflict in which this report places him with his own *Mysterium cosmographicum*, in which he had deduced the number of the planets, six, from the fact that the spaces between their orbits would have to correspond to the five regular Euclidean solids. We have to assume that this first report did not enable him to distinguish clearly whether what Galileo called "planets" were phenomena that he had seen in the neighborhood of Jupiter, or ones that revolve around it, as was soon to be clearly stated on the title page of the *Sidereus nuncius*. Kepler himself, even before the publication of the *Mysterium cosmographicum*, had once considered whether one did not have to assume two more invisible planets between the orbits of Jupiter and Mars and those of Venus and Mercury. He had even thought that he could determine the particulars of their orbits. As a 'sightless' procedure, this anticipated what was to become reality in the discoveries of Uranus and Neptune. Kepler had then given up his reflection, in favor of the speculative demonstration of the solar system's geometrical completeness. In the second edition of the *Mysterium cosmographicum*, in 1621, he himself excludes any idea that his early assumption of an additional planet in the neighborhood of Jupiter could give him some priority in relation to Galileo's discoveries. The event shows to what an extent people had already 'reckoned with' invisible things in the heavens.[39] For Kepler would have had to accept permanent invisibility for his speculative planets, since before the *Mysterium cosmographicum*—i.e., before 1596—an artificial optical instrument was not to be thought of. Perhaps that would have struck Kepler as only a little 'blemish,' no more important than, for Leverrier, the absence of optical confirmation of the eighth planet, which he had established by calculation and which already had its full reality as the source of perturbation in the orbit of the seventh planet, and was only 'discovered' because telescopes also existed now.

The earlier history enables us to understand how Kepler immediately contrives an expedient that could save his demonstration of the mathematical plan of creation ("salvo meo mysterio cosmographico"). In doing so he does not allow himself to be deflected by Galileo's need to have discovered "planets." The new stars, he says, are not planets in the traditional sense at all, but system-elements that would correspond to the Earth's Moon. Here it should not be forgotten that the Earth's Moon had only just received its special position in the system from Copernicus. For him, the Moon was, as it were, left over, as the last instance of a celestial body with a geocentric orbit. To that extent, as a residue of the outdated error of the Earth's central position, it had become a disturbing peculiarity. It was entirely logical that Kepler's reflection, on Copernicus's behalf, now aimed at removing this last remnant of the special character of the Earth with the help of the news from Venice, by supposing that not only the Earth, but all the other planets too, had moons.

What distinguished Kepler from Galileo's doubters and opponents in Padua, Bologna, and Pisa was the fact that he immediately began to think astronomically, without dwelling long on the question of the reliability of the telescope. It certainly was helpful for him, at this first moment, that he did not yet know anything more precise about Galileo's new stars and that he had his defensive interest in protecting the old number of the planets. An important fact in this connection was that it was a matter of four new planets. Kepler relates this number to the still moonless main planets (and explains the absence of a moon for Mercury as a result of its being outshone by the nearby Sun—otherwise it would have had to be five new stars). This interpretation of the new discoveries in the heavens finally made all the planets into 'Earths,' and made the special case of the Earth/Moon system into the normal case. We still find a trace of the motive for this reflection in the Copernican formula for the Earth, and for the exceptionality of the satellite revolving around it, that Kepler uses in the *Dissertatio*: "... terra, unus ex Planetis (Copernico) Lunam suam habeat, extra ordinem sese circumcursitantem" [the Earth is the only planet (according to Copernicus) that has its own Moon, revolving around it, out of order].⁴⁰ Galileo's discoveries—Kepler realizes—could have invalidated this *extra ordinem* and established nature's specific procedure: Not only were these planets of a second order, but they were 'characteristic equipment' of the first-order planets. The discovery of the new stars appeared as a step toward establishing Copernican consistency.

When the emperor himself gave Kepler the first copy of the *Sidereus nuncius* that arrived in Prague and expected him to comment on it, he was well prepared, mentally, by his quick uptake on and digestion of the rumors that had preceded it. Not much later, the Tuscan ambassador at the imperial court delivered to Kepler the copy that Galileo had meant for him. It had to cause embarrassment when Kepler met Wackher again. The clarity that now existed required rethinking above all in regard to the four supposed planets in the neighborhood of Jupiter, which Kepler, in his first response, had wanted to distribute to the old planets, as moons. If there was no option left except to leave them to Jupiter, then one had to contrive a new ordering principle for the speculative assignment of moons to planets, because here too it was necessary to demonstrate proportions that were worthy of God. Above all, Kepler has to make sure to make it possible to avoid the designation "planet." In September 1610, he invents the term "satellite" for Jupiter's moons. Kepler's Copernican use of Galileo's discovery—his supposition that all of the planets are equipped with moons—still makes people pay attention to Venus's transits of the Sun a century and a half later, as we saw in connection with Lambert.

Kepler accurately assessed the situation in which Galileo had to find himself after the proud announcement of his discoveries. The news of the fateful night at Magini's house in Bologna had already reached Prague at the end of April 1610: A succession of learned witnesses had declared the appearances in the telescope to be delusive, and Galileo had responded with nothing but silence. Kepler had already offered his support before that, in his letter of 19 April , and he persisted in this attitude with his *Dissertatio* in the beginning of May. How solitary he was, standing by Galileo's side, was at first concealed from him, until Galileo wrote to him on 19 August 1610, that he had been the first and remained practically the only person to have placed complete confidence in his assertions. He would show his gratitude for this in a second edition of the *Sidereus nuncius*, to which he intended to add an answer to the *Dissertatio*.

A second edition of the *Sidereus nuncius* never came about, nor did any other form of public gratitude on Galileo's part toward his sole adherent, who would have been the greatest even if there had been many.

3

The Lack of a 'Paratheory' to Explain Resistance to the Telescope

"It is remarkable, indeed, how incorrectly most people act when they are obliged to form a judgement of their own on some new subject." Sigmund Freud wrote this sentence, in his "Autobiographical Study" of 1925, about his early difficulties in gaining acceptance for psychoanalysis; and he compared the way some of his opponents resisted his ideas to "the classical manoeuvre of not looking through the microscope so as to avoid seeing what they had denied."[a] Here Freud has (characteristically) projected the refusal of Galileo's opponents to look through the telescope onto the microscope, which evidently conformed better to the comparison with the view of the 'underworld' that psychoanalysis had opened up.

The strong point of psychoanalysis, in its own difficulties in gaining acceptance, was that Freud possessed a 'paratheory' that not only explained to him his opponents' resistance but actually made this resistance into an instrument that verified his theory. The advantages of such a paratheory are of incomparable value rhetorically (or, many people would like to say here, "strategically"): Internally, it protects one from doubts, and externally it destroys the resistance, by categorizing assent to the theory as a symptom of rationality and inner freedom, indeed as an ethical accomplishment, and putting the reasons for resistance to it beyond the pale, as extratheoretical and scientifically and morally indefensible. Freud (and he was not the first to do this, nor did he remain the only one) used psychoanalysis to explain both the interest in psychoanalysis and also the resistance against it. As early as 1914 he wrote, in "On the History of the Psychoanalytic Movement," that he had been protected from embitterment by "a circumstance which is not always present to help lonely discoverers." They, he says, had to torment themselves with the question of how to explain their contemporaries' rejection, and they had to guard themselves against letting this opposition affect their confidence in their convictions. "There was no need for me to feel so; for psychoanalytic theory enabled me to understand this attitude in my contemporaries and to see it as a necessary conse-quence of fundamental analytic premises." [b] When a theory provides itself with a specific paratheory explaining the weakness of its influ-ence, it determines at the same time where and how the decision about its success has to occur—namely, where the resistance shows most clearly the features of the prognosis, so that the overcoming of that resistance will satisfy the conditions of the theory. Thus Freud did not show great happiness at his success in America. In his view, only the outcome in the countries of old culture would be decisive: The resistances to his theory had to be overcome "on its native soil." Thus in 1925, in "The Resistances to Psychoanalysis," he could count both the resistance and its therapy as evidence for the theory.

With the aid of this paradigm, one can study how far Galileo was from being equal to his failure with the telescope. He was amazed at the refusal of his learned contemporaries to look through the tele-scope, and still more amazed at his lack of success with those who had not even refused to avail themselves of the offered look. The historio-graphy of science has made things too easy for itself by regarding the opponents of Galileo and his telescope as mere staffage in the picture of a triumph that was achieved in spite of them, and by deriding them in the same way that Galileo had already done when he wrote to Kepler on 19 August 1610, "What do you think of the chief philoso-phers of our gymnasium, who, with the stubbornness of a viper, did not want to see the planets, the Moon, or the telescope, even though I offered them the opportunity a thousand times? In truth, just as he [Odysseus] closed his ears, so they closed their eyes against the light of the truth (*contra veritatis lucem*). That is monstrous, but it does not astonish me. For men of this kind think that philosophy is a book, like the *Aeneid* or the *Odyssey*, and that the truth is to be sought not in the world and in nature but in the comparison of texts (as they call it)." Just this—that he had not been surprised—is a misrepresentation on Galileo's part; just as his explanation in terms of 'book philosophy' is a gross simplification.

For the forms of rejection of the telescope were so various that one would have to be inclined to see them as results of anecdote formation and mythicization, if the transmission of the facts were not testably reliable. There was the orthodox Aristotelian, Cesare Cremonini, with whom Galileo had already clashed in the dispute about the new star, in 1604, and who at that time had already had nothing else in his mind but "to defend the heaven of his Aristotle" ("mantenere il cielo del suo Aristotele"). Cremonini, who was himself to become entangled in proceedings of the Inquisition, and having been acquainted with whom later strengthened the suspicions directed at Galileo, refused throughout his life even to glance through the telescope, because this "could only confuse his head," as he wrote with charming candor in a letter to Paulo Gualdo on 6 May 1611. Libri, too, in Pisa, declared Galileo's observations to be impossible, because they contradicted Aristotle's definitions: "He wanted to tear the new planets forcibly from the heavens, and talk them out of existence by logical arguments, as though such arguments were magical incantations," Galileo writes to Kepler. The most distinguished and perhaps the most serious opponent, around whom the resistance to the innovations banded together, was Giovanni Antonio Magini, the mathematician at the University of Bologna, who wrote on 26 May 1610, to Kepler, of all people, that the four new attendants of Jupiter had to be eliminated and stamped out. In Magini's house in Bologna, on the night between the 24th and the 25th of April 1610,—i.e., in the month after the appearance of the *Sidereus nuncius*—an observation of the moons of Jupiter had taken place in which the participants asserted that they could not perceive the satellites. After this defeat Galileo had departed sadly early in the morning, Martin Horky, a student of Magini, writes to Kepler on 27 April. It was also Horky who, with a pamphlet, began the public polemic against Galileo's telescope. As late as 1618, Prince Cesi, the founder and leader of the Accademia dei Lincei, complained that the present-day philosophers would "rather close their eyes and bury themselves in the dark forest of the ancient writers than use their sense organs in the service of the truth." The recipient of this letter was the same Bellarmine who, on 26 February 1616, communicated to Galileo, in a conference that up to the present day has still not been fully clarified, the Inquisition's first anti-Copernican decision.

Even before the date of his conference with Bellarmine, Galileo wrote to the Grand Dutchess Christina of Tuscany the famous letter against his opponents' theological wing, which, however, begins with a complaint about the professors who had been as enraged by the new discoveries in the heavens as if he had fastened such phenomena onto the heavens with his own hands, in order to confuse nature and science: "... quasi che io di mia mano avessi tali cose collocate in cielo, per intorbidar la natura e le scienze." If these people, who denied the new not out of love of the truth but out of a preference for their own opinions, had only chosen to make observant use of their senses, they could have obtained certainty. Galileo would hardly have formulated this critique in this way if he had been capable of more rigorous reflection on the presuppositions of his own results in physics and if he could have inferred from these that for his opponents, too, a mere empirical willingness to look at the phenomena could not be sufficient to abrogate their philosophy.

Galileo had no theory of the opposition to the telescope, still less a theory that would have suggested to his adversaries to wonder whether they were not being prevented by timid backwardness, obstinate disingenousness, and suspect self-regard from seeing what had recently become visible. Thus he arrived at the conclusion that the whole crowd of the Aristotelians (*tutta la turba peripatetica*) had shown itself incapable of following even the simplest and easiest demonstration. His oppenents' resistance appeared to him as a symptom of an intellectual indolence that entrenched itself indoors in studies, to "look things up" in Aristotle instead of looking *at* the things themselves. Galileo did not reflect on the complicatedness of natural optics, which does not automatically conform to the technical increase in what it can accomplish.

Thus, his own visual failure in two important areas was escape his notice. In his *Saggiatore* he declared comets to be phenomena beneath the sphere of the Moon, as the tradition had done. In spite of having seen sunspots in good time, by means of the telescope, he recognized their reality and importance so late that he became embroiled in the momentous dispute with the Jesuit Scheiner, in which he unjustly laid claim, for himself, to the phenomenon, which he had no doubt already seen in the summer of 1610, but had overlooked. When three comets were observed in the heavens in 1618, there was a good reason for Galileo, in his turn, not to "look": The paths that he would have had to grant to them contradicted his stubborn adherence to the dogma of the circular form of the paths of all heavenly bodies—if, indeed, he had recognized them as heavenly bodies at all. Galileo's optics, too, were subordinate to his dogmatics, and he resorted to the same tactic of evasion that his opponents had employed against the moons of Jupiter: declaring inconvenient phenomena to be optical illusions.

If one keeps that in mind, one begins to wonder whether we have not accepted, in our histories of science, in the formulas that Galileo coined, an all too definite picture of inferiority on the part of those who denied the moons of Jupiter and those who refused to look through the telescope. The historical rightness of scientific knowledge and of its protagonist is reflected not only in the unambiguous wrongness, but also in the foolishness of those who denied it. Something that looks as though it may have become a stereotype needs to be tested; there is no other way to keep alive our ability to sense prejudices. Here it may be useful also, for once, to examine more closely what was thought in the historiographically eclipsed sphere of opposition to the telescope. An exemplary text is the pamphlet of the Florentine Francesco Sizzi, which appeared in Venice in 1611 as a direct reply to the *Sidereus nuncius*.[41] The dedicatory preface of the *Dianoia astronomica* [*Astronomical Demonstration*] goes back, according to its date, to August 1610, which shows how immediate the reaction was. The title page promises to show that the excitement generated by the four planets that Galileo had recently seen in his telescope is unfounded.

So among Galileo's discoveries it is only the satellites of Jupiter that provoked Sizzi's polemic. Like Kepler, Sizzi takes the *nuncius* as a messenger, an allegorical herald figure of the new, against whom he takes up his weapons and enters the arena.[42] It was the news of the increase in the number of the planets that had alarmed him, causing an extreme intellectual perturbation.[43] Should one take that literally? At first glance one might suppose that just what had filled Galileo with enthusiasm, at the sight of Jupiter's system of satellites, must have deeply alarmed Sizzi and other people. But on closer examination one finds, surprisingly, that what Sizzi sees himself as drawn into, and what he endeavors to settle, is not an early Copernican conflict. Galileo's confirmation of Copernicanism does not bother him.

What drove Sizzi to doubt and finally to dispute the reality of these new planets is the infringement of the canonical number of seven. Just as Galileo, looking at the Jupiter system, finds an eidetic authority in favor of Copernicus, Sizzi sees himself being deprived of a feature of the world-order that is equally eidetically guaranteed. For him, the supposed moons of Jupiter (if he had to admit their existence) would be abnormities of the stellar system—something like monsters and prodigies, which, as blunders on the part of form-giving nature, have no place in the rational purity of the heavenly bodies' spheres. What Sizzi carries out is more than argumentation—it is an apotropaic [ill-averting] gesture, with which he not only disputes and denies but drives away and shuts out what is demanded of him here. His polemic is a defensive action on behalf of the cosmos.[44] Appearances have no effect against a solidly founded preconception of the world-structure.

That there can only be seven planets is not a question of the empirical data. If it is already difficult to see what one does not expect, it is almost impossible to accept, as a result of mere optical experience, what is not admissible in the context of one's a priori [*präsumptiv*] understanding of the world. The opposition to Galileo's telescope is after all only a gross case of the subtler factors affecting vision, of which Goethe will say to Chancellor Müller in 1819, "One perceives only what one already knows and understands. Often one fails to see for many years what only a more mature state of knowledge and culture enables us to become aware of in an object that confronts us every day. Often only a paper-thin wall divides us from our most important goals—we could boldly break through it, and the job would be done. Education is just the art of teaching how one can go beyond imagined or in fact easily surmountable difficulties."[45] The fool's role that Galileo's opponents have long played in the historiography of natural science has rendered them harmless for us and obscured their significance as indicators of the difficulties in our relation to reality that are always present and become especially acute in historical situations where radical change is under way. The failure of their obstructed faculty of vision is only a correlate of the exaggerated expectations that Galileo himself had invested in his optical discoveries.

When Sizzi thinks that he has to save the canonical number of the planets, what is at stake for him is the system of analogies in a world in which the seven-armed candelabrum of Moses corresponds to the seven planets, and in which this number seems to possess a quality of lawfulness when even doctors have to agree that, in the course of an illness, the seventh crisis is the decisive one. In comparison to this, the difference between Ptolemy and Copernicus seems unimportant. It is true that Ptolemy is claimed as authority for the reliability of the classical number of the planets and for tracing this number back to the Egyptians and to Moses, Abraham, and Noah; but even the new school of astronomers (*recens astronomorum schola*), despite their joy in changing things, their elevation of the Earth to a planet and inclusion of the Sun among the fixed stars, had not changed the number of the planets.[46] In the spectrum of the possible characteristics and actions, there is no room for an increase in the number of planets; the astrological meanings have been assigned, and what could not or must not act, does not deserve to be actual either.

The establishment of expectations is an important factor in the history of science. On this occasion, at the beginning of the seventeenth century, the increase in the number of the planets was headed off, one more time, through Kepler's influence. But the problem recurred at the end of the eighteenth century, in 1781, when Herschel discovered Uranus (which Hegel, in turn, deduced as the last admissible expansion of the system). Herschel himself had not 'seen' a planet, because there simply could not be a new planet. What he saw was a comet, although none of its eidetic characteristics could fit this classification: It lacked a tail, was round in shape, had a perihelion that was extremely distant from the Sun, and a slow motion on its orbit, and that orbit lay in the plane of the ecliptic. The threshold that the evidence had to reach before one would admit that one had discovered a planet was very high. In such processes, in the history of science, obscure heuristic means sometimes play a positive role, like the so-called Bode formula[c] for the radii of the planets' orbits. Another thing that was not irrelevant to optical tolerances was the alchemistic linkage between the number of the planets and the series of metals. The discovery of a new metal could open up the canonical correlation. In a letter of Max Lamberg to Giacomo Casanova, dated 10 December 1790, he speaks of the suspected discovery of aluminum and adds that a new, unknown metal would have to be designated by the name of a planet, if one could discover an eighth one. There was a sort of historical invisibility that took place in vision itself, the overcoming of which was entirely comparable to the technical enhancement of vision. Lichtenberg may have been the first to formulate this pointedly when, according to Gamauf's account of his lectures on astronomy, he expressed surprise—in connection with the dispute about who was the first to discover sunspots—that they had not been perceived much earlier, and without auxiliary means: "But no doubt the reason why people did not notice these spots is that they did not want to notice them . . . ."

Of course, one should not make things too easy for oneself here with the word "want." If one does not want to let understanding and the lack of it stand as arbitrarily allocated historical fates, an attempt must be made to understand not only what progress and increasing knowledge are but also how the intellectual process functions that one could call "self-closure." This also holds for those who rejected the telescope. Among their reasons, the one that would be the most worthy of any rationality may be the desire to ward off the invisible part of the world as something supposedly contradicting the unity of the universe of possible operations—however close such a defense may still stand to astrological needs. The incapacity of the opponent of the telescope to comprehend it is definitely not incomprehensible.

What Sizzi cannot admit is the existence of bodies in the world that are not situated in a single universal interconnected field of operations, because, since they cannot be seen without magnifying instruments, they could not for their part exercise influence either without such instruments. If natural optical contact is not possible between us and them, then in the opposite direction as well, no influence can come about. Something whose existence would thus be in vain, because it would be beyond man's reach and unable to reach him, does not exist.[47] The traditional postulate of visibility turns out to be a special case of a more general postulate of influence. Whatever is real must show itself. If these little planets existed, a rumor about them would have had to surface at some time, as a manifestation of an influence that was below the threshold of distinctness, as in the cases of the nebulas, the stars of the Milky Way, and the surface of the Moon. The postulate of influence guarantees that the tradition since antiquity will be complete in its testimony to reality, insofar as it must have registered, at some time, even minimal impinging influences. The seemingly foolish global proposition that whatever is not mentioned in the inherited texts is not real either—this apparently caricatured extreme expression of Humanism—is inferred from the postulate of influence.[48]

In this context one would expect Sizzi to analyze another assertion in the *Sidereus nuncius*, namely, Galileo's explanation of the ashen light of the New Moon as a reflection of the light of the full Earth. Here

Galileo had been arguing above all on behalf of the Earth as a heavenly body that was not merely passive, but just as active as any other. Sizzi still takes it completely for granted that planets are heavenly bodies having their own luminosity. He makes the same claim for the Earth's Moon, which, as he admits, primarily reflects sunlight, but also has a luminosity of its own, as one can see during eclipses. What the assertion of the Moon's having its own light is meant to repel is the old view, described by Galileo as Pythagorean, that "the Moon is like a second Earth." This would not only deprive the Moon of its traditional character as a heavenly body, but would also make it possible to open up, for the supposed satellites of Jupiter, the expedient of viewing them not as additional planets but 'constituent parts' of a Jovian system similar to that of the Earth and its Moon.

One is almost tempted to say that this is the revenge for the ceremonial manner of Galileo's "celestial message" that he has discovered new stars and, in the exuberance of naming them "Medicean," has overlooked the 'provincial' small format of the newly seen objects. The Earth's Moon must remain a star so that Jupiter's moons cannot become stars. For Sizzi, the instrument stands in a laughable disproportion to the claim that it is supposed to help to prove—the creation is supposed to have failed in its purpose by such a little bit, and must thank the paltry accident of this *perspicillum* for attaining it again. For if these new planets need the telescope in order to be known and to be seen, they would necessarily need the same (or a similar) instrument in order not to fail in their existential purpose.[49] What Sizzi defends is not only the cosmos as the nature that surrounds and determines man, but also—if one wants to express it in one of the great propositions of the philosophical tradition—that man is the measure of all things: of the real things, that they exist, and of the unreal, that they do not.

Someone who looks in Sizzi's tract for materials relating to the impending Copernican conflict will prick up his ears at the argument that the new planets around Jupiter destroy the Sun's privileged position in the center of the world system. Might the dissolving of the new stars take precedence even over the defense of the geocentric world? On looking more closely one discovers that what reads like a good Copernican argument derives from a tradition with which Copernicus already was familiar. The Sun is the source of light and warmth, even of the power of motion; it is the vehicle of the first created light, and metaphorically the heart of the living world. It has to occupy its place in the center of the world: "medium in mundo locum ex necessitate requiret." But the position in the center of the world that is referred to here is one that is to be preserved, not to be gained. It is the position of the middle sphere of the seven homocentric spheres that rotate around the Earth. For Sizzi the topographic center is occupied, now as before, by the Earth; but he has the idea of a dynamic/energetic center as the most favorable position from which to operate on the remaining heavenly bodies. It is the same conception that Kepler, at the same time, is combining with heliocentrism—using, incidentally, the same linguistic material that Sizzi employs, which thus proves to have no exclusive connection to Copernicanism. Sizzi's sole concern is with this symmetry, which can only be preserved by the planets' numbering seven (with the Sun, as one of them, in the middle); if there were four more in the neighborhood of Jupiter, it would be destroyed.

Sizzi does not content himself with demonstrating that Galileo's planets cannot exist; he also wants to make it clear how the supposed discovery could come about. The second part of the little essay is devoted to showing that the perception of the satellites of Jupiter could be due to an illusion. Here the modality of possibility is sufficient. Sizzi appeals to the optics of Euclid, Alhazen, Witelo, and Porta. This tradition of optics contains a theory not so much of normal vision as rather of its pathology: a set of techniques for producing illusions, *visus fallaciae* [deceptions of vision]. Like traditional mechanics, optics too resulted from the intention of finding out nature's tricks, of tricking something out of her that she is not ready (or only ready in accidental cases) to give up. The reports of atmospheric mirror images, of mock suns and mock moons, of coronas and false comets are the material with whose aid one can study how, if the occasion arises, to make such things. The suspicion that the telescope is an instrument of sense deceptions, of hallucinatory reduplications, is not farfetched. It accords with the milieu from which the instrument had so unexpectedly emerged, and with the legends about its origin.

This passage makes another fact about the history of modern science paradigmatically tangible. Galileo writes at the beginning of the *Sidereus nuncius* that on account of the rumor of the invention of the telescope he had occupied himself for ten months "in inquiring into the principle and devising means by which I might arrive at the invention of a similar instrument." He had succeeded after he had adressed himself to the theory of the refraction of light. But here Galileo tells nothing about the results of these investigations, apart from the procedure that he had invented for determining the power of the instrument. Instead, he explicitly announces that he will "explain the entire theory of this instrument on some other occasion." The reaction to the *Sidereus nuncius* has to be seen in the context of this absence of theoretical 'ground' underneath it: By not publishing his preliminary optical investigations, Galileo left space open for doubts about the results of the first observations with the telescope, for what he had to earn for himself in the course of almost a year can by no means be taken for granted as knowledge that his learned contemporaries will have acquired in school. Besides, the introduction of a scientific instrument is different from that of, for example, the compass, where something that is produced by nature could be exploited without its theory having been furnished beforehand. But even among instruments the telescope has a special position because there not only were no methods for checking its astronomical use but, from everything that (in the horizon of the time) could be regarded as possible, there could not be such methods. In relation to stellar objects, the optics of visible light were the final authority, beyond which no appeal was possible.

Sizzi's arguments against the reliability of the telescope may seem to us to express a dogmatic animosity against what should not exist. But if one tries to put oneself within his contemporaries' horizon of possibility and to imagine what they had to take into account and what they could be prepared for, the probability becomes preponderant that they would expect nothing from the new visual apparatus but delusive things, deception, deceitful or unintended misrepresentations. The retrospective angle of vision of the history of science does not perceive the byways, detours, and blind alleys because it acknowledges as an 'event' only what led further; in this way the course of this history gets the appearance of a logical sequence, but each of its components loses the quality of action. But historiography has to understand events as actions, and it does so by trying to imagine the distribution of possibilities and probabilities in a situation at the time.

For this endeavor the obscure Sizzi, whose name marks a blind alley, is no less instructive than Galileo, whose name stands on the highway of scientific progress. Sizzi's doubts are an index of the telescope's ambiguity as an instrument of theory. The unsatisfied need for a substantiation of what it accomplishes puts it in the unfavorable light of one of those deceptions for which Galileo's contemporaries were metaphysically prepared. The program of epistemology as one of the central concerns of the new epoch originated in this kind of mistrust.

Sizzi is not a skeptic, but the observations through the telescope seem questionable to him in several respects. Galileo, he says, has not applied the criteria of astronomical science so as to ensure the objectivity of his discoveries. His observations are not even sufficient to support a claim to have demonstrated a regularity in these phenomena that is appropriate to the stellar region of the universe. What the imaginary celestial messenger brings, Sizzi says, does not adhere to the order and the law of motion of celestial realities, but rather reminds us, in its changeability, of the terrestrial processes that are described by Heraclitus's saying that one cannot bathe twice in the same river.[50] That speaks in favor of deception, of the complex mutability of conditions on the Earth on which the observer is situated, who has no opportunity to have the same view of Jupiter through the telescope twice.

At this point Sizzi leaves the alternative open as to whether the motion of the heavens or the motion of the Earth is part of the complex of the conditions of observation that are continually changing ("nam caelum vel terra assidue movetur"); neither here nor anywhere else does he express any opposition in principle to Copernicanism. For the argument that he is about to present, the motion of the Earth, as an additional irregularity, might be more welcome than otherwise.

If the defenders of the telescope argue that optical refractions can be eliminated as a possible explanation of the points of light in the neighborhood of Jupiter because such phenomena would have to appear with other planets as well, Sizzi replies that when the real objects are at different distances, the factors producing such duplications would be entirely different too. He adds, not without a hint of an accusation, that it happens that Galileo's telescope is constructed in such a way that it produces the peculiar refraction phenomenon only at the distance of Jupiter. Galileo himself had provided an occasion for this insinuation. In the *Sidereus nuncius*, talking of the moons of Jupiter, he had invited all astronomers "to examine them and determine their periodic times, something which has so far been quite impossible to complete, owing to the shortness of the time"; but he had also indicated that one needed not only a very accurate telescope but one "such as we have described at the beginning of this discourse." This remark may have struck Sizzi as an anticipatory excuse in case Galileo's colleagues should fail to observe the new stars.

Did the interaction of the instrument with individual organic characteristics of the observer also play a role in the optical illusion? This is the point at which Sizzi has difficulty in dealing with his own recollection of the telescopic observation that he carried out with Galileo. At this point he drops the allegorical interpolation of the "celestial messenger." He addresses Galileo himself and asks him to remember the night when they, together with learned and eminent men, had observed Jupiter. That night Galileo had at first seen only *one* "secondary image" of Jupiter; then someone else had declared to those who were present that he saw two, whereupon Galileo, after repeated observations, also saw two such *imagines* [images], as did others as well that night.[51] Then follows Sizzi's explicit declaration that he himself never saw Jupiter in anything but its unreflected and undoubled singleness: "Ego vero numquam Iovis imaginem nisi simplicem intuitus sum . . . ."

How can one explain this disagreement? Here, as in other cases, as well, of resistance to Galileo's telescope, one must bear in mind how poor the optical quality of the early telescopes was, as a consequence of chromatic aberration. Even without the dogmatic difficulties of the Aristotelians, what could be seen through such a telescope already gave the impression of an optical illusion and a mirage. Even granted his good will, it was probably not easy for a hasty observer either to admit or to deny what he had seen. Only in such an objective situation could Sizzi develop his explanation of the course of that critical night, which contains a sort of psychology of assurance produced by obligingness and of agreement casually uttered out of fatigue and weariness at being kept from dinner, and to avoid staying longer.

The strengths and the weaknesses of Sizzi's essay are instructive in relation to the theoretical situation. One of its weaknesses lies in the fact that he produces too many "good arguments" against the new planets, arguments that do not always mesh with one another without contradiction. Thus on the one hand he considers it certain that the *perspicillum* is a source of deception, which can produce nothing but mirages, while on the other hand he invokes the fact that this instrument has existed since antiquity, in order to pronounce spectacular new discoveries improbable. We can understand the illusion argument more easily than the tradition argument, which deprives the claimed progress, as such, of credibility. Sizzi depends on the *Magia naturalis [Natural Magic]* of Porta, in which all the indications of a little-known existence of the telescope were collected, beginning with the famous distant viewing installation on the pharos of Alexandria and the "telescope" of Ptolemy and continuing up to Albert the Great, Cornelius Agrippa, and Pope Leo X. If, in spite of his possession of a telescope, the moons of Jupiter had remained hidden even from the man who had founded the accepted astronomy, could one then accept their existence on the authority of a Galileo?[52] And the papal promoter of all the arts and sciences, possessor of every means and all power—would he not have had to see, if there were something to see? Sizzi has to admit that the magician Porta was the only person who named these telescope possessors—and that he did so only in obscure and ambiguous words; but one could explain this restraint by the fact that hitherto one had wanted to avoid making the instrument into a popular entertainment.[53] Perhaps Sizzi would not trust such uncertain testimony if the prior decision had not already been made for him, which makes the conflict over the telescope into a prelude to the later *querelle*—namely, that there is nothing essentially new in the world: "Nil dictum, quin dictum prius; nec factum, quin factum prius." [Nothing is said that was not said before; nothing is done that was not done before.] For optics, that is only a different formulation of the postulate of visibility.

The strength of Sizzi's essay lies in the fact that it exposes the methodological backwardness of Galileo's first publication. Just because Sizzi's resistance to the telescope has nothing to do with opposition to Copernicanism, Galileo's overestimation of the new instrument's power to demonstrate the truth of Copernicanism becomes all the more palpable. The paradoxicalness of the connection between the telescope and Copernicanism that Galileo set up becomes evident: Copernicus had directly contradicted the sensual evidence of the Earth's immobility and its central position, in favor of a rational construction; now Galileo called his contemporaries to the telescope in order that sense-appearances should convince them of the correctness of that contradiction. Can one credit the obscure opponent of the founder of modern physics with having noticed this intra-Copernican paradox?

One can, I think, see some indication of this. Sizzi goes back to Aristotelian Scholasticism's basic epistemological principle that all knowledge begins with the senses and that nothing becomes available to reason that has not previously passed through the organ of perception. For astronomy, however, he asserts that this axiom is least correct for it, because the stellar objects are known by means of reason itself.[54] What we know about the size of the fixed stars, which surpasses that of the Earth and the Moon, as well as about the greater remoteness of the fixed stars in comparison to the planets, and about the globe shape of the heavenly bodies in general, does not originate in our perception, but in a process of indirect trains of thought and inferences (*intervenientis ratiocinationis beneficio*). Sizzi explains the inapplicability of the epistemological axiom to astronomy as a result of the fact that astronomy excludes the establishment of causal connections, except for the operation of light on the faculty of vision. But this effect does not satisfy the condition of an unambiguous causal connection, since otherwise any given visual appearance could only have been produced by the reality of what is given in it. The possibility of deception is expressed here almost in the terminology of Cartesian doubt.[55] So the problematic character of Galileo's telescope consists, for Sizzi, precisely in the fact that it depends exclusively on light. The only way to confirm what one sees through the telescope is to look through it again. Thus is summed up in the statement that science cannot be gained through visual appearances alone: "sic per visum solum scientia haberi non potest."

To secure his discovery against his adversaries, Galileo would have had to satisfy the requirement of applying the standard of classical astronomical objectivity to the stellar phenomenon. He had said himself, at the end of the *Sidereus nuncius*, about his observations that "from these data the periods of revolution of these planets have not yet been reconstructed in numerical form." He had indicated a half-monthly period only for the outermost of Jupiter's four satellites. In his haste to bring to the public the news of the "excellent and very clear argument" for Copernicus, Galileo had not complied with the rules of the astronomical fraternity, for his "planets," but had only summoned them "to apply themselves to examine and determine their periodic times." It is not accidental that the modern age, the outcome of which was to be the forfeiture of 'intuitiveness' [*Anschaulichkeit*], begins with an enthusiastic effort to mobilize intuition, and with the ignominious disappointment of that effort.

Perhaps it has become clearer why Galileo had no theory by which to understand the resistance to the telescope. This lack also deprived the conversions that still ensued, and those that did not, of an effect whose meaning was assigned in advance. Consequently the legend according to which the anti-Copernican Clavius is supposed to have changed his mind on his very deathbed, at the mere report of the discoveries with the telescope, and prophesied the downfall of the old system, is a late product. I reproduce this legend—for the reader's pleasure, too—in the version given by John Wilkins: "'Tis reported of Clavius that when lying upon his Death-bed, he heard the first News of these Discoveries which were made by Galilaeus his Glass, he brake forth into these words: 'Videre Astronomos, quo pacto constituendi sunt orbes Coelestes, ut haec Phaenomena salvari possint': That it did behove Astronomers to consider of some other Hypothesis, beside that of Ptolomy, whereby they might salve all these new appearances. Intimating that this old one, which formerly he had defended, would not now serve the turn: And doubtless, if he had been informed how congruous all these might have been unto the Opinion of Copernicus, he would quickly have turned on that side." [56] Christoper Clavius, the "second Euclid" and the leading calendar specialist of his time, died on 6 February 1612—two years after the appearance of the *Sidereus nuncius*. More reliable, perhaps, than the legend of his conversion is the report of what he said about the moons of Jupiter: If these things are supposed to exist, a wonderful telescope must first be created, in order to produce them. [57]

The attempt to identify the telescope as an old stage property of occult knowledge and thus as a witness against Galileo's innovations was an unsuccessful piece of rhetoric, if only because it put the attack on the instrument's weaknesses in the wrong. It could have been more effective to accept the telescope and to present Galileo's innovations as a consistent extension of the Scholastic system. In any case, this alternative was adopted too late when, in 1630, the Jesuit Christopher Scheiner, in his work on the sunspots, *Rosa ursina*, described the telescope as the mature fruit of the Aristotelian tradition. He sees the instrument as the extension of a procedure that for centuries had allowed inferences from given states of affairs to causes that are not given: "Sed vera et solida philosophandi ratio ... ab effectu ante oculos posito ad ignotas rerum causas indagandas progreditur." [But the true and sound philosophical reasoning advances from the effect that is placed before our eyes to the unknown surrounding causes of things.][58] With celestial phenomena, however, it had not yet been possible to apply this procedure, because, on account of their all too great remoteness, it had not been possible to bring even the effects, from which one had to infer the causes, to a state of clear givenness. But now, by divine benefaction, access to the celestial spaces was opened up and an accurate view of these shining celestial bodies was made available, indeed, as though we could regard them from close by and could even touch them.[59]

In contrast to Sizzi, Scheiner exaggerates what the telescope accomplishes. He does this not only because he thinks he has integrated it into the supposed continuation of the Middle Ages, but also because in the priority dispute with Galileo about the discovery of the sunspots he had procured recognition for a phenomenon that Galileo was able to overlook. The sunspots forced people to abandon the special nature of the heavenly bodies. From now on, their heterogeneousness is only due to their spatial remoteness. A result arrived at by means of the telescope becomes an argument for the possibility of what the telescope accomplishes. It loses the questionable character of magical cunning and becomes the almost natural perfection of the organ of vision. Thus a hopeless situation for theory is replaced by a sort of right, if not a duty (*fas*), to experience the heavens in the full sense of bodily presence. Of course Scheiner's cautious attempt at domesticating the telescope is anything but a daring deed. The artificial avoidance of affront seldom enters history, but it is a part of 'histories of influence' like these, if one considers how many times greater the distribution of the *Rosa ursina* was than that of the *Sidereus nuncius*, and what had to be done to make the telescope plausible.

Jupiter's moons were given astronomical reality only in 1693, with Cassini's tables of their periods of revolution. When Fontenelle gave an appreciation of Cassini, after his death, in one of his famous eulogies in the Académie des Sciences, he found a formula for the scholarly resistance to Cassini's numerous telescopic discoveries (especially those of Jupiter's oblateness and the duration of its rotation), a formula whose importance as a source of illumination lies in the fact that it could not yet have been applied to Galileo's case: "Le refus de croire honore les découvertes fines." [It is an honor, for fine discoveries, when people refuse to believe them.] In the one century since Galileo's telescope, resistance to the demands of scientific claims had ceased to endanger cognition's freedom of motion. For the early spokesman of Enlightenment, the fact that discoveries exceed what his audience is ready to accept, and that they disturb people's minds, has become an honorable indication of their quality. For the Enlightenment, the effect of this quick inclination to take unwillingness and resistance as indices of the quality of the truth for which acceptance has to be won—practically to imitate the Galileo situation—was not only usefulness. The hectic intensification of its operations against a horizon of expectations that did not expand rapidly enough led to its means becoming worn out and exhausted. Having no theory of the resistance to it, but instead turning it into a means of self-confirmation, remained the mark of a specific disregard of consequences [*Rücksichtslosigkeit*] that was characteristic of theory's self-propulsion.

4
Reflexive Telescopics and Geotropic Astronautics

Copernicus had deduced from the shape of the Earth's shadow in eclipses of the Moon that the Earth is precisely spherical. Thus, without auxiliary optical means, he ushered in the practice of reflexive vision: In a star (in the classical sense) he found confirmation that the Earth too is a star. When Galileo aimed his telescope at the Moon, he saw a duplicate of the Earth: forests, seas, continents, and islands. But at the same time he deduced from the secondary moonlight that the Earth 'shone' like a star. Even before the postulate of physical homogeneity was complete, attitudes toward celestial phenomena had changed in such a way that man's interest in the Earth could no longer be excluded from them.

The Earth had become a star, with the consequence that stars now could only be multiple Earths. Reflexive vision first becomes differentiated by the idea of evolution, which allows us to perceive, at one time, different nonsynchronous stages of worlds. Thomas Burnet writes in his *Sacred Theory of the Earth* (an early work of modern geology, published in 1684) that in the telescope the Moon appears like a great ruin, and the Earth's appearance would be no different from a comparable distance; and this impression would be confirmed if we could descend below the Earth's surface. The Earth is a star, but as such it acquires a history, whose debris it contains, and the marks of whose age it carries on itself. Lichtenberg will express this in the idea that planets should be regarded as prototypes of a development, from which knowledge, prospective as well as retrospective, can be gained about the Earth. Ultimately one found in the stars things that could be looked for on Earth. In 1868 Janssen and Lockyer discovered in the spectrum of the Sun lines coming from an unknown element, which could only be identified in 1895 as the noble gas helium, which it was possible to isolate in a tellurian mineral. Nowadays we do not need to be told about the importance of astrophysical theories for the understanding of processes of atomic fusion. The starry heavens have become a textbook for the technology of what cannot naturally be found on Earth.

Exerting himself to look out into space, man did not descry something entirely different and alien; rather, what was held out to him was a cosmic mirror of his own world, of its history and its potential. Whether this had only been due to the anthropomorphic narrowness of language and the feebleness of description was something that was bound to emerge, at the latest, with the first steps in astronautics. It was just as much to be expected as it was to be feared that the first people on the Moon would describe the landscape, which had never been seen, with the means with which they were familiar, depending on which of the competing powers would be the first to set foot on this ground. Thus, in fact, in February 1969, the commander of *Apollo 8*, Frank Borman, told the Royal Society in London that the closest thing to a likeness of the surface of the Moon would be the Mojave Desert in California. Edward Sapir pointed out as early as 1912 that the language of the Paiutes, a desert tribe, contained the most detailed means for describing features of the landscape because their ability to find once-discovered watering places again depended on this. But the linguistic inadequacy of astronautics did not lead to the metaphorical event that seemed, in 1960, to be approaching, when it was still possible to underestimate the extent of the role that images and television transmission would play. Language, in these events, remained a paltry rudiment, a mere seasoning—in spite of uninterrupted verbal production—to the stream of transmitted pictures. Of course that is not an accident, since one cannot imagine the technical state of space travel arbitrarily separated from the perfection of the process of transporting images, which is only a sector of the business of exchanging information. One culd almost say that it would not have been worthwhile to send man to the Moon if we could have brought only his words back with him, though, on the other hand, perhaps it would not have been necessary to send people to the Moon at all if what was to be brought back was, above all, pictures. In any case, language hardly got a chance either to fail or to succeed. No classic travel narrative will be handed down about the first walks on the Moon. Man did not succeed, on this occasion, in showing that the only reason why the Earth is not a desert is that he exists, contemplates it, and can talk about it.

An instructive phenomenon is the doubts of the people who felt that even the transmitted images were not exotic enough for them to be able to believe in the reality of the event. The important thing, here, is not this curiosity, as such, because there are always people who doubt something that most people regard as proven. But the technical capacity for simulation, which had become so important in the preparations for space travel, now furnished the arguments for doubting the reality of its highest accomplishments. That is also why there was more active skepticism (about whether people had really landed on the Moon) in the country that could claim credit for the accomplishment. A fluttering American flag, on a Moon that lacks any atmosphere, appeared as an accidental failure of simulation, and there were plenty of people who thought they could identify the landing place in the American desert in the state of Arizona, by moonlight. People pointed out that the exposure time of the cameras, which made short streaks appear in the pictures of the sky, was too long, since short enough exposure times should be possible on the Moon to make the stars appear as points. All of this is significant only as illustrating the search for the authenticity of what was so surprisingly unalien. Still more interesting is the objections people had to the astronauts' dialogue. Shortly before Shepard and Mitchell landed in the Fra Mauro Range, there was a conversation about the weather and the clearness of the view, a conversation that should not have occurred on the Moon. "It seems to be a beautiful day in the Fra Mauro region," said one; "If only you could set down over there where the haze is, Al!" said the other. That was like a bait for doubters. Did it mean something that someone had seen all of this "with his own eyes"? Had man's horizon of reality been expanded?

The first footprints in the dust of the Moon: A deceiving demon, and a very small caliber one in comparison to Descartes's, could easily have produced them for the world as a theatrical illusion. During this decade of astronautics only one single picture could not have been invented, but simply went beyond anything the imagination could have anticipated: the picture of the Earth from space. If one tries to relate the centuries of imaginative effort and cosmic curiosity to the event, then the both unexpected and heart-stopping peripety of the gigantic departure from the Earth was this one thing, that in the sky above the Moon one sees the Earth. Kepler had described it in advance, but in this case knowledge was not the important thing. In August 1966, *Lunar Orbiter II* transmitted, from its orbit of the Moon, the first picture of Earth shining over a lunar landscape. The fixed orientation of the television camera during the first manned landing, in July 1969, deprived the terrestrial viewer of the opportunity to look back at the Earth, and suggested to him that what was important was the piece of the Moon that was not shown because it lay in the dead arc behind the camera. Only mobility and color in the transmitted picture allowed one to grasp the uniqueness of the moment of cosmic reflection that enabled man to experience—above the lifeless desert of something that had once had the unattainable quality of a star—the seemingly living star, the Earth. This roundabout view of the Earth exonerates Cardinal Francesco Barberini, the nephew of Urban VIII, who (as Castelli writes to Galileo on 6 February 1630) had blamed Copernicanism for degrading the Earth by making it a star.

Now perhaps all of this would only be a sentimental reminiscence if, with the backward view of the Earth, the expense of manned space travel had not become an episode. Something that no one had expected to come about so easily and so quickly, and that economic difficulties alone would not have brought about—namely, a weakening of cosmic curiosity and a turning of interest from the remote world to the proximate one, from a centrifugal direction to a centripetal one—took place rapidly and almost silently. There is a depression that follows the attainment of absolute goals, however imaginary this absoluteness may have been. This pattern is not sufficient for the comprehension of this historical event. The decisive thing for this purpose is a revision that brought to an end the Copernican trauma of the Earth's having the status of a mere point—of the annihilation of its importance by the enormity of the universe. Something that we do not yet fully understand has run its course: The successive increases in the disproportion between the Earth and the universe, between man and totality, have lost their significance—without its having been necessary to retract the theoretical effort. The astronautical success was a disproof, *in extremis*, of ancient expectations, and it also destroyed the Enlightenment's myth of reason's being compelled to compare itself with the cosmos. One can also put it this way: Equivalence is established between the microscopic and the telescopic sides of reality—absence of difference, in a sense that no longer has any tinge of Pascal's abysses of the infinities.

A decade of intensive attention to astronautics has produced a surprise that is, in an insidious way, pre-Copernican. The Earth has turned out to be a cosmic exception. Of course what man has taken is only a few steps in the region close to the Earth, even if we include the explorations we have undertaken, within the planetary system, by means of automatic equipment. Still, in relation to the expectations directed at these exotic bodies, the optical evidence of these thrusts into space is overwhelmingly in favor of our own planet. The monotonous quality of the photographs and data reported back to us, from Jupiter (on the outside) to Mercury (on the inside)—i.e., within the scope of what could ever be practical for travel by human beings—was the disappointment of the (open or secret) assumption that mankind still has some option other than the Earth.

What presented itself to human view was nothing but scarred and cratered worlds or stifling hot hells with no indications of potentialities for life. Nor is it to be expected that beyond Jupiter enough solar energy is intercepted to allow us to find anything but conditions hostile to life, if instrumental probes should ever reach Saturn, or even Uranus, Neptune, or Pluto, without being forgotten in the long years of travel thither. For of course the exploration of space requires not only technical apparatus that is continually capable of functioning but also the functional identity of what one has to call, in the broadest sense, the "ground station." How does the subject of this interest and this ability to wait constitute itself? On 29 February 1972, after a delay of one day, the planetary probe *Pioneer 10* was started on its way to Jupiter. Twenty-two months later, when it flew around the planet, at a respectful distance, the continuity of the undertaking, in the consciousness of the public, had already been lost, and only a sort of 'administrative identity' guaranteed the resumption of a contact in which the travel time required by radio messages of data regarding the planet's radiation belt had become so great that it was no longer possible to intervene "without delay" in the steering of the apparatus and its instruments.

The umbilical cord of terrestrial information and steering breaks as soon as the times involved in radio transmission to and from the probe become longer than the requirements for the intervention of 'reason' make appear advisable. In the case of an expedition to the nearest fixed star, the radio transmission of queries to the terrestrial base would already require years, and would soon be running up against the limit of what organization can sustain against its being forgotten.

However long radio contact can be maintained with the Jupiter probe of 1972, we shall never learn what was accomplished by the little aluminum plate with engraved, encoded information about the Earth and man, with which the probe was equipped. Even during the short history of space travel, the expectation that the apparatus might someday, after it has left the solar system, came into the hands of nonterrestrial intelligent beings has already taken on an anachronistic quality.

That aluminum plate is a late tribute to the Enlightenment's cosmic expectations. The Enlightenment had believed itself to be situated in a universe composed of inhabitable worlds and rational creatures. "How could we be the only ones worthy of real existence," Reimarus had asked—and that was still a relatively harmless formulation of the feelings for which reason's terrestrial reality was an unworthy one, one of the lowest of its realizations. The simultaneous existence of a higher rationality, in the cosmos, was supposed to guarantee, at the same time, the ascending future of terrestrial rationality. It was entirely in accordance with the logic of Copernicanism that the terrestrial conditions for the existence of reason could not be specially favored ones, since after all the Earth itself seemed to be more an out-of-the-way provincial location than anything else, a location where reason was excluded from palpable certainties to such an extent that a Copernicus had been able to be, historically, only one of its latest figures. By means of reason, this backwardness in relation to the cosmic standard could and should be made up for. Membership in the rational community of the universe could be earned through merit. Just as the Enlightenment's gaze was directed centrifugally at the purer cultural forms of ancient China, and at the natural stages of noble savagery in all the continents, so it was likewise oriented centrifugally toward these projections into the universe. Reason, in its terrestrial contingency, must not be solitary and abandoned to the factually given conditions of this one history only.

In this connection, hardly any role was played by the expectation that one would ever be able to make contact with the extraterrestrial intelligences. Visits from alien worlds or to them were novelistic exaggerations of the fundamental pattern. However, there is an unmistakeable connection between the still intact belief in immortality and the idea that the habitable heavenly bodies would offer higher forms of existence to man's better part. Such ideas were not set up with a view to their being confirmed or refuted. Nevertheless, reason manifests itself in them as the agency that compares reality with what it could be. This is why it induces imaginings, without being able to determine what their contents will be. The Enlightenment gave the universe's habitability for rational beings the rank—if one wants to put it in Kant's language—of a practical postulate.

Its strength was that it was not exposed to falsification by experience. Looked at abstractly, the astronautical decade has not changed that situation, since there cannot be a proof of the uninhabitability and uninhabitedness of distant planets. But now that the friendly heavenly bodies in our own (narrower) system have proved, when viewed close up, to be unsuited as objects for a daringly founded "exobiology," the insistence of intuition—of what has been seen—stands against what can now be conceived only in assessments of probability, and the boundary beyond which we can anticipate alien life or even intelligences that we could address has moved dramatically farther out into space—beyond practicable reach, at any rate.

The idea of intelligent civilizations in the universe has been brought so far beyond the stage of a rationalistic postulate that it already determines expensive research activities and still more expensive projects for research. In September 1971, a first conference of Russian and American astronomers agreed on the possibility of communication with extraterrestrial beings when it concluded that it would "now be possible and worthwhile to attempt to make contact with such extraterrestrial civilizations, which, however, would probably be vastly superior to mankind in their civilization and technology." This formula, from the Associated Press, may have processed the outcome, to put it in journalistic form, but it contains the motivations that, a scant four years later, already determine the enormous energy expenditure of the station at Arecibo in Puerto Rico. However alarming the improbability may be of finding precisely the frequency and the directional parameters on which such communications could be received or sent, the most surprising thing is the expectation that other intelligences not only exist, for this dialogue, and are sufficiently intelligent, but that they could also have a need comparable to ours to search for signals coming out of the universe and, at enormous expense, to send quanta of energy into the universe. Does the assumption that such civilizations could be decisively superior to ours already justify, as such, the expectation that they would have to have a similar need to ours for communication extending across their horizon into the realm of the uncertain? Here questions begin that are more radical than the question whether what we have to communicate would be likely to be understood, and whether we shall ever understand transmissions that we might receive. I think the issue needs to be radicalized in a dimension that perhaps could again be described as the "Copernican" one.

Why in fact is it assumed, or even considered possible, that cosmic intelligences should at the same time be of the same type as ours, and yet so much more "civilized"? On reflection, the probability that living creatures on other planets could be substantially superior to man turns out to be slight. For one has to assume that they too will be products of a process of development, an organic selection in the struggle for existence—and a selection that simply does not favor the qualities that we recognize as superior to ours and that we would like to have, but instead favors precisely the qualities that we have to suffer under in ourselves: a will to self-preservation, fascination by those who are stronger, rivalry for rank and position. If, however, they should have got somewhat further than we have in coping with the infantilisms of progress and of technicization, then their need to look around them in the universe for models or for onlookers at their own insufficiency will have diminished proportionately.

This is not yet the heart of the question either. It has been said that to grant life and reason only to the Earth is still a variant of the old anthropocentric prejudice. I am afraid that the burden of proving one's post-Copernican freedom from prejudice lies in the opposite direction. The unexamined premise of searchers for extraterrestrial communication lies in the fact that they regard the probability of life on planets of alien solar systems as already a basis for assuming the presence of reason there. Nothing would be more ill-founded than this, if reason turned out to be by no means a logical continuation of organic evolution as such. That the process of the development of life must always, or even only in one additional case, lead to reason (even if that reason were not embodied in a human form) seems to me to be something taken for granted that is brought into this problem complex, but that in fact only repeats again the old favor that man rendered to himself of taking the crown of creation into his safe keeping. All the anthropological indicators point to the conclusion that man on Earth is the exception in nature, insofar as he was able to preserve himself in existence only by means of an unexpectedly successful repair of a deviation, a repair that would be extremely inexpedient according to the rules of organic nature. Reason then would not be the summit of nature's accomplishments, nor even a logical continuation of them. Instead, it would be a risky way around a lack of adaptation; a substitute adaptation; a makeshift agency to deal with the failure of previously reassuring functional arrangements and long-term constant specializations for stable environments. One can emphasize this break in evolution more crassly or less, but in any case it does not permit us simply to take it for granted that the extraterrestrial existence of life (if that has the great probability that it is nowadays granted) already brings with it a corresponding likelihood that rationality will emerge from the genetic stream of organic substance.

The anthropological hypothesis that man might not be a logical result, consistent with the rest of organic evolution, does not imply any depreciation of reason. Whatever necessity of self-preservation this faculty of anticipatory adaptation may have originated in, its reaching out for and anticipation of possibilities consolidated itself—independently of all self-preservation—as the consciousness that conceives itself and lays hold of itself as possibility. That is to say that for reason, and only for it, the poverty of its origins is a matter of complete indifference, since its history consists in upward self-revaluation: turning a makeshift function into the better and the best, security into certainty, naked necessity into the modality of evidence, and consciousness into self-consciousness.

Reason can be a triumph over any origin and any eccentricity, but it can hardly be understood as the normal goal of the processes of matter, as Teilhard de Chardin wanted it to be, with his assumption of complete "noospheres." His thesis that while the appearance of man on the Earth is indeed merely contingent in place and time, it is something 'preformed,' normal in terms of a law governing its genesis, and generated from the universal characteristics of matter—this thesis is indeed distinguished by its irrefutability and by its dignity as a speculative gesture, but it deprives us of any prospect of understanding something of the history and reality of man. With its peculiar combination of elements specific to the Enlightenment and others belonging to the tradition of Christian metaphysics, Teilhard's speculation contributed a lot, in the decade before the astronautical episode, to bringing expectations of traces of cosmic life and cosmic intelligence to a high pitch. Now that the greater part of our solar system has been scouted out with distinct radioed images and instrumental soundings, the thesis of the preformation of matter for life and reason—which has fallen to the level of edifying philosophy—is more likely to encounter skepticism. Nothing that shines up there seems to tolerate life, let alone to hasten unremittingly toward reason.

Seeking the "consolation of philosophy" is not unreasonable, and is only thoughtlessly brought into disrepute. That also holds for the idea of the habitableness of the universe. But just at this point one falls all too quickly into new disconsolations. What would the distribution of those "noospheres" at distances of millions of light-years from one another mean, when any attempt to measure reason against reason and make it be the winner would have to founder on the mere fact that the answer to any signal could only be expected when no trace of the culture—or even the species—that needed an answer can be present any longer? The billions of solar systems in the universe may imply the probability of living creatures here or there, but the order of magnitude of the distances between them destroys, at the same time, what for metaphysical speculation could still be called the "meaning" in this state of affairs. The slowness of the speed of light makes the final decision as to what the superabundance of suns and the frequency of planets, the emptiness of space and the probability of living creatures add up to for reason. Contrary to its title, relativistic physics set the absolute limit against which all technical finesse becomes powerless and in view of which even our Milky Way already far exceeds the order of magnitude that would allow communicative 'simultaneity.' This reflection concludes, as the "dreadful cynicism" of nature, what had begun as reason's self-consolation. The insurmountable wall constituted by the speed of light for the first time denies man an existence that has cosmic importance.

Part of the euphoria of the astronautic departure and race is the metaphor of the "mothership Earth"—that is, of the mere foothold, in the universe, for centrifugal activities in space, activities that now brace themselves only episodically, and with one foot, on the ground from which they started. Only half of this is a metaphor of intimacy and security; the other half is one of mobility and transiency. The centrifugal impetus of astronautics is like a remnant of the special value assigned to the stellar reality by metaphysics, and of its corresponding degradation of the Earth as the dregs of the universe. When gravity seemed to be victorious, it became synonymous with a burden. A sufficient reason why the Earth is not the mothership of astronautics is that it is the solidity of its ground to which the spaceships so speedily return. The reflexiveness of Copernican vision is repeated in the movements by which seeing Earth was to be followed by walking on it. It is more than a triviality that the experience of returning to the Earth could not have been had except by leaving it. The cosmic oasis on which man lives—this miracle of an exception, our own blue planet in the midst of the disappointing celestial desert—is no longer "also a star," but rather the only one that seems to deserve this name.

It is only as an experience of turning back that we shall accept that for man there are no alternatives to the Earth, just as for reason there are no alternatives to human reason.

Translator's Notes

Translator's Introduction

a. From James R. Newman's review of several books about Copernicus in *Scientific American*, October 1957, p. 155.

b. *Die Genesis der kopernikanischen Welt* (Frankfurt: Suhrkamp, 1975) is a greatly expanded version of Blumenberg's *Die kopernikanische Wende*, edition suhrkamp 138 (Frankfurt: Suhrkamp, 1965). It is the third of Blumenberg's books to appear in English, the others being *The Legitimacy of the Modern Age* (Cambridge, MA: MIT Press, 1983), a translation of the revised edition (1973–1976) of *Die Legitimität der Neuzeit* (Frankfurt: Suhrkamp, 1966); and *Work on Myth* (Cambridge, MA: MIT Press, 1985), a translation of *Arbeit am Mythos* (Frankfurt: Suhrkamp, 1979). Blumenberg's other works include *Paradigmen zu einer Metaphorologie* (Bonn: Bouvier, 1960), reprinted from *Archiv für Begriffsgeschichte* 6 (1960): 7–142; *Schiffbruch mit Zuschauer. Paradigma einer Daseinsmetapher*, suhrkamp taschenbuch wissenschaft 289 (Frankfurt: Suhrkamp, 1979); *Die Lesbarkeit der Welt* (Frankfurt: Suhrkamp, 1981); and *Lebenszeit und Weltzeit* (Frankfurt: Suhrkamp, 1986). Some of his most important papers are reprinted, with a very useful introduction, in *Wirklichkeiten in denen wir leben* (Stuttgart: Reclam, 1981). One of these papers, "An Anthropological Approach to the Contemporary Significance of Rhetoric," has been translated in K. Baynes, J. Bohman, and T. McCarthy, eds., *After Philosophy* (Cambridge, MA: MIT Press, 1986), pp. 429–458.

c. Alfred North Whitehead, *Science and the Modern World* (New York: Macmillan, 1925), p. 19; quoted in Thomas S. Kuhn, *The Copernican Revolution. Planetary Astronomy in the Development of Western Thought* (Cambridge, MA: Harvard University Press, 1957), p. 122.

d. Nikolaus Kopernikus, *Gesamtausgabe*, ed. F. and C. Zeller, vol. 2 (Munich: Oldenbourgs, 1949), p. 5.

e. Part II, chapter 4, of this book, in the second paragraph. (In the original, *Die Genesis der kopernikanischen Welt*, p. 237.)

f. Kuhn, *The Copernican Revolution* (cited in note c), p. 126.

g. Part III, introduction, third paragraph from the end. (*Die Genesis*, p. 308.)

h. See p. 183. (*Die Genesis*, p. 216.)

i. E. A. Burtt, *The Metaphysical Foundations of Modern Physical Science* (first published 1924) (New York: Doubleday Anchor, 1954), pp. 52–56; Alexandre Koyré, *The Astronomical Revolution*.

Notes to Pages xvii–xxiii

Copernicus–Kepler–Borelli, trans. R. E. W. Maddison (Paris: Hermann; Ithaca: Cornell University Press, 1973), p. 65; T. S. Kuhn, *The Copernican Revolution*, pp. 127–130. Edward Rosen opposes this tradition of interpretation in his *Copernicus and the Scientific Revolution* (Malabar, Florida: Krieger, 1984), pp. 66–69. A major variation on the "Platonism" interpretation points to the glorification of the Sun by Renaissance Neoplatonists like Ficino, and to Copernicus's similar rhapsody in book 1, chapter 10, of the *De revolutionibus*, and argues that only a heliocentric model like Copernicus's could satisfy someone who felt this way. Koyré, in *The Astronomical Revolution* (same page), calls this "the motive—the real motive—which inspired the mind and soul of Copernicus." Kuhn (pp. 129–130) is sympathetic to this interpretation, as is Frances Yates, in her *Giordano Bruno and the Hermetic Tradition* (New York: Random House [Vintage edition], 1969), pp. 153–154. But as Blumenberg points out (in part II, chapter 6, paragraph 6, *Die Genesis*, pp. 274–275), "The traditional astronomy in fact already contained a heliocentrism that was perfectly in keeping with its system"—the heliocentrism that placed the Sun symmetrically "in the middle" between the orbits of Venus, Mercury, and the Moon, on the one hand, and Mars, Jupiter, and Saturn, on the other. A heliocentrism of this type was in fact explicitly embraced by the Sun enthusiast, Charles Bouillé (Bovillus): see part II, chapter 4, note 73, and the text thereto (*Die Genesis*, p. 245). Robert S. Westman makes the same general point, against Yates, in *Hermeticism and the Scientific Revolution* (Los Angeles: William Andrews Clark Memorial Library, 1977), p. 16.

j. E. Cassirer, "Galileo's Platonism," in *Studies . . . in the History of Science, Offered to George Sarton*, ed. M. F. Ashley Montagu (New York: Schuman, 1946), p. 280. Blumenberg quotes this remark on p. 417.

k. There are at least two other ways of dealing with the absence of Plato's central doctrine from those early modern "Platonists," Copernicus, Kepler, and Galileo. One is to talk of a Platonic "hypothetical method," as Cassirer (following Paul Natorp, and the Marburg School of Neokantianism) does in the paper cited in the previous note. It seems, though, that a separation of 'method' from ontology reflects modern assumptions and is likely to distort what Plato himself was up to. The second way is to postulate a "Christian Platonism" whose most influential early formulator was Augustine, and which transmitted Plato's mathematical rationalism, stripped of his ontological dualism by the homogenizing effect of the Christian idea of creation, to Copernicus, Galileo, and Kepler. (This analysis is proposed by Jürgen Mittelstrass in his *Die Rettung der Phänomene* [Berlin: DeGruyter, 1962], pp. 178–258.) This account does not explain why the implementation of "Christian Platonism" as natural science, which Copernicus, Galileo, and Kepler carried out, took place more than a thousand years after its foundations were laid by Origen, Augustine, et al.

l. *The Legitimacy of the Modern Age* (cited in note b), p. 179; *Die Legitimität der Neuzeit* (1966; cited in note b), p. 144; *Säkularisierung und Selbstbehauptung*, suhrkamp taschenbuch wissenschaft 79 (Frankfurt: Suhrkamp, 1974) (the revised edition of parts I and II of *Die Legitimität der Neuzeit*), p. 211.

m. *Legitimacy*, p. 135; *Legitimität*, p. 89; *Säkularisierung*, p. 156.

n. *Legitimacy*, pp. 135 and 126; *Legitimität*, pp. 88 and 78; *Säkularisierung*, pp. 156 and 144.

o. Part III, chapter 2, of this book, in the last paragraph. (*Die Genesis*, p. 370.)

p. Part II, chapter 4, second paragraph. (*Die Genesis*, p. 237.)

q. L. Moulinier, *De la Docta ignorantia* (Paris: Alcan, 1930), p. 32; cited in J. P. Dolan's introduction to Nicholas of Cusa, *Unity and Reform* (Notre Dame: University of Notre Dame Press, 1962), p. 5.

r. Part II, chapter 2, next to last paragraph. (*Die Genesis*, p. 199.)

s. Kuhn, *The Copernican Revolution*, p. 122.

t. See p. 158. (*Die Genesis*, p. 188.)

u. See p. 162. (*Die Genesis*, p. 193.) On the surprisingly meager fruitfulness of Nominalism for natural science, see *Legitimacy*, pp. 348–353; *Legitimität*, pp. 344–349; *Der Prozess der theoretischen Neugierde*, suhrkamp taschenbuch wissenschaft 24 (Frankfurt: Suhrkamp, 1973), pp. 151–157.

v. See p. 197. (*Die Genesis*, p. 233.) The next quotation is from the same page.

w. See p. 197. (*Die Genesis*, p. 234.)

x. "Exertion against the late Scholastic Middle Ages": same chapter, last paragraph. The idea that, between epochs, systematic "positions" that have been established in one epoch can be "reoccupied" by contents that do not originally belong in those positions is the central idea of *The Legitimacy of the Modern Age*. Blumenberg does not use the term "reoccupation" in this passage, but the process that he describes seems clearly to be an instance of that pattern. On the "most difficult rhetorical act" of man's self-comparison to God, see the last page of "An Anthropological Approach to the Contemporary Significance of Rhetoric" (cited in note b).

y. From the paragraph preceding the one in which John Leonicenus is mentioned—see index (emphasis added). (*Die Genesis*, p. 243.)

z. *Die kopernikanische Wende* (cited in note b), p. 9.

aa. Michael Heidelberger, "Some Intertheoretic Relations between Ptolemaean and Copernican Astronomy," *Erkenntnis* 10 (1976): 323–336; the quotation is from p. 323. Heidelberger cites Blumenberg's *Die kopernikanische Wende*, "p. 8ff.," from which my last quotation was taken. Blumenberg makes the same points that he does there in the last four paragraphs of part II, chapter 1, of this book.

bb. For this critique see, for example, the papers of Martin Hollis in *Rationality*, ed. B. Wilson (Oxford: Blackwell, 1970), and in *Rationality and Relativism*, ed. M. Hollis and S. Lukes (Cambridge, MA: MIT Press, 1982).

cc. See pp. 171–172. (*Die Genesis*, p. 202.) The first nine paragraphs of this chapter constitute almost a manifesto against 'judgemental' rationalism. An important related text is Blumenberg's "Ernst Cassirers gedenkend ...," in *Wirklichkeiten in denen wir leben* (cited in note b), pp. 163–172.

dd. See pp. 14, 15. (*Die Genesis*, pp. 23–24.)

ee. Thomas S. Kuhn, *The Structure of Scientific Revolutions* (Chicago and London: University of Chicago Press, 1962). Heidelberger's paper is primarily a critique of Kuhn's theory as it applies to the Copernican revolution.

ff. *The Legitimacy of the Modern Age*, p. 465; *Aspekte der Epochenschwelle*, suhrkamp taschenbuch wissenschaft 174 (Frankfurt: Suhrkamp, 1976), p. 16. The next quotation is from the same page (translation slightly revised).

gg. "An Anthropological Approach ..." (cited in note b), p. 436. Original: *Wirklichkeiten in denen wir leben*, pp. 111–112. On the inadequacy of explicit rationality as a guide to action, see my introduction to *Work on Myth* (cited in note b), pp. xxvi–xxix.

hh. W. Bouwsma, review of *Legitimacy* in *Journal of Modern History* 56 (1984): 698–701; M. Jay, review of *Legitimacy* in *History and Theory* 24 (1985): 183–196, p. 188 in particular. That Blumenberg is not unaware of this issue is shown by his reference, in *Legitimacy*, p. 464 (*Aspekte der Epochenschwelle*, p. 15), to "the admittedly outmoded ideal of a 'history of ideas' [*Geistesgeschichte*]."

ii. *Legitimacy*, p. 99; *Säkularisierung*, p. 114.

jj. On this interpretation of the point of departure of Greek philosophy, see *Legitimacy*, pp. 243–245 (*Legitimität*, pp. 211–213; *Der Prozess der theoretischen Neugierde* [cited in note u], pp. 23–25); and in the present volume, part IV, chapter 1, paragraphs 7ff. On the special role of vision in the "ancient concept of reality" see Blumenberg's "Wirklichkeitsbegriff und Möglichkeit des Romans," in *Nachahmung und Illusion*, Poetik und Hermeneutik 1, ed. H. R. Jauss (Munich: Fink, 1969), pp. 10–11, especially note 3. Both the Greek *theorein* and Cicero's translation of it, *contemplatio*, combined sensual (visual) and intellectual, 'aesthetic' and 'religious' aspects.

kk. See part VI, chapter 4, of this book; "An Anthropological Approach . . ."; and *Work on Myth*, part I, chapter 1 (and my introduction to that volume, especially pp. xiv–xvi).

ll. On *Anschauung* "intuition" see translator's note b to part I, chapter 3.

mm. Part I, chapter 5, paragraph following text to note 85 (emphasis added). Together with part II, chapter 5, of *Work on Myth* (which sketches the *motivation* of this far-fetched but uniquely revealing project), this chapter provides one of the most interesting historical/systematic 'diagnoses' of German Idealism with which I am acquainted.

nn. Part I, chapter 7, text to note 131. Compare, on the question of man's "responsibility," as articulated in "final myths," *Work on Myth*, pp. 288–291 (*Arbeit am Mythos*, pp. 319–323).

oo. On this, see "An Anthropological Approach . . . ," passim.

pp. Blumenberg describes how "the Copernican intention that Galileo had formed early on" (p. 401; *Die Genesis*, p. 469), which seemed to be well served by his notion of 'circular inertia,' prevented him from giving a hearing to Kepler's anticipations of Newton's concept of gravity. He discusses Galileo's faith in visual intuition—and the problems it got him into—in this chapter and especially in part VI, chapter 2. It is interesting to compare Blumenberg's account of Galileo's problems with truth, in these chapters, with Paul Feyerabend's in *Against Method* (London: Verso, 1978), in which Feyerabend picks out the same canny use of rhetoric by Galileo, but seems less aware of the (necessary, but) painful *difficulty* of Galileo's position.

qq. Other important texts of Blumenberg on the ambiguous roles of science and truth in our lives include part III of *Legitimacy*; "An Anthropological Approach . . ."; *Die Lesbarkeit der Welt* (chapter 1); and *Paradigmen zu einer Metaphorologie*, sections 1–4 (pp. 12–58), on metaphors relating to truth. It is also worth comparing these to part II, chapter 3, of *Work on Myth*, where he analyzes (sympathetically, as always!) the genesis of that aboriginal "antithesis" of rationality: dogma. Finally, *Lebenszeit und Weltzeit* spells out and studies the 'ontological' conflict between life (as experienced by individuals) and science (as a process whose obligation is to the world, as opposed to individuals).

rr. The sources of these quotations are Fontenelle, *Entretiens sur la pluralité des mondes* (1686), I; Goethe, *Werke*, ed. E. Beutler (Munich: Artemis, 1948–), vol. 16, p. 395 (I have translated Goethe's whole passage in translator's note a to the author's introduction to part III of this book); and Nietzsche, *Genealogy of Morals*, section 25, trans. Walter Kaufmann, in *Basic Writings of Nietzsche* (New York: Random House, 1966), p. 591. Blumenberg discusses these quotations in *Paradigmen zu einer Metaphorologie*, pp. 107 and 122; in *Die kopernikanische Wende*, pp. 158–160; and the Nietzsche passage, again, in part I, chapter 7, of the present book. The final chapter of *Die kopernikanische Wende*, entitled "Metaphorische Kosmologie—Kosmologische Metaphorik" (pp. 122–164) is especially valuable as a supplement, on these topics, to the present book.

ss. See the first entry for Karl Ernst von Baer, in the index, for the page on which this statement is made (*Die Genesis*, p. 639).

tt. See notes 58 and 59, and translator's note e, to part V, chapter 5.

uu. See text to translator's note k to part V, chapter 4 (*Die Genesis*, p. 682).

vv. Fourth paragraph after text to note 50, in part V, chapter 5 (*Die Genesis*, p. 702).

ww. Last paragraph of part VI, chapter 1 (*Die Genesis*, p. 746).

xx. Paragraph 7 of part VI, chapter 1 (*Die Genesis*, p. 726).

yy. Sixth paragraph from the end of part VI, chapter 4 (*Die Genesis*, p. 791). (All the quotations in this section are from this chapter.) This acount of the "break" in evolution that is represented by man, and by human 'reason,' can be found in (for example) Arnold Gehlen, *Der Mensch. Seine Natur und seine Stellung in der Welt* (1940; 4th ed. Bonn: Athenäum, 1950), and Paul Alsberg, *In Quest of Man: A Biological Approach to the Problem of Man's Place in Nature* (Oxford and New York: Pergamon Press, 1970). See Blumenberg's "Anthropological Approach ...," especially pp. 438–439, for more on this.

Part I

Introduction

a. Wernher von Braun, in *Proceedings, XIth International Astronautical Congress, Stockholm, 1960* (Vienna: Springer, 1961), vol. 1, p. 650.

b. *Höhle*. The reminder of Plato's allegory of the cave, in the *Republic*, is probably not unintentional.

c. Henri Poincaré, *Science and Hypothesis*, in his *The Foundations of Science* (New York: Science Press, 1921), p. 110. Original: *La Science et l'hypothèse* (Paris: Flammarion, n.d.), p. 141. The "difficulty" to which Poincaré refers here is the fact that "space is symmetric, and yet the laws of motion would not show any symmetry; they would have to distinguish between right and left. It would be seen for instance that cyclones turn always in the same sense ... " (p. 141).

Chapter 1

a. Sophocles, *The Plays and Fragments*, trans. R. C. Jebb (Cambridge: Cambridge University Press, 1908), vol. 5, p. 183.

b. "Interworlds" refers to the spaces between Epicurus's infinite worlds, in which he pictured the gods as living carefree lives with no knowledge of or concern for either human beings or their worlds.

c. *Anschauung*: 'viewing.' On the multiple significance of this key term, which I later consistently translate as "intuition," see translator's note b to part I, chapter 3.

d. *ruhend*, here, should be understood not only as "at rest," in the physical sense (that is, as not in the, for Aristotle at least, "unnatural" state of motion, and not actively involved and interfering with what one observes), but also as "calm," not disturbed by passion; and probably also as "idle," in keeping with the Greek idea that theory was a disinterested pursuit appropriate to those who were not burdened with practical responsibilities. On the connection of the idea of "rest" to antiquity's concept of reality as "momentary evidence," see the author's "Wirklichkeitsbegriff und Möglichkeit des Romans," in *Nachahmung und Illusion*, Poetik und Hermeneutik 1, ed. H. R. Jauss (Munich: Fink, 1964), p. 11n3. An important discussion of this subject in this book is in paragraphs 6–13 of part IV, chapter 1.

e. See note c.

f. For Plato's "second astronomy," see the *Republic*, 528E–530C; the "true Earth" is mentioned in the *Phaedo*, 110A.

g. *Ruhe*. See note d.

h. This account of the purpose of Epicurus's pursuit of natural science is expounded on pp. 157–158 and especially pp. 264–265 of *The Legitimacy of the Modern Age* (Cambridge, MA: MIT Press, 1983); *Die Legitimität der Neuzeit* (Frankfurt: Suhrkamp, 1966), pp. 116–118 and 238–239.

Chapter 2

a. Both *Vergötterung* and *Vergottung* can be translated as "deification," but the former, with *Götter* in the plural, implies polytheism, and consequently can be used figuratively for "idolatry" of the sort that is warned against in the lines quoted from Deuteronomy; whereas the latter, with *Gott* in the singular, would affect the whole world, including man, and would thus exclude idolatry (as a relation to something other than oneself) as well as Gnostic dualism and Christian orthodoxy.

b. See *The Legitimacy of the Modern Age*, p. 564 and note 26 to the same page; or *Die Legitimität der Neuzeit* (1966), p. 544 and note 27.

c. The medieval *artes liberales* were grammar, rhetoric, dialectic, arithmetic, geometry, and astronomy.

Chapter 3

a. On the "observer in repose," see note g to chapter 1, part I.

b. "Intuition" is the usual translation of *Anschauung*, in philosophical contexts, even though what is meant has little to do with our current use of the term to designate a nonrational mode of access to truths. The reason is that there is no colloquial English word that has the same breadth of meaning as *Anschauung*, covering the acts of "looking," "contemplation," "observation," and "perception"—all of which were included in the Latin *intueri*, and thus are at least implicit in "intuition." And it is precisely this breadth of meaning that is often crucial to passages in this book in particular, where the author is showing how the rich ancient idea (also inherent in Greek *theoria*—see note e—and in Latin *contemplatio*) of 'beholding,' with its indissoluble combination of aesthetic, moral, and scientific aspects, was narrowed down to the mere process of the collection of visual 'data,' which are instrumentalized for the nonvisual functioning of modern 'theory,' to such an extent that "intuition," in English, is free in the end to be transferred to the essentially antitheoretical (and equally nonvisual) realms of occult or instinctive processes or 'genius.' The change in meaning parallels the long transition from the ancient concept of reality as "momentary evidence" to the modern concept of reality as the "realization of a system that is consistent in itself" (Blumenberg, "Wirklichkeitsbegriff und Möglichkeit des Romans," in *Nachahmung und Illusion*, Poetik und Hermeneutik 1, ed. H. R. Jauss [Munich: Fink, 1969], pp. 10–13). Good accounts of the terminological histories of *Anschauung* and *Intuitio/Intuition/*intuition can be found in the articles under those headings in the *Historisches Wörterbuch der Philosophie*, ed. J. Ritter and K. Gründer (Basel: Schwabe, 1971–). So I must ask reader (and I have often used single-quotation-mark scare quotes to remind him) to have in mind the obsolete or technical sense, in which the visual orientation is still strong, whenever he sees the word "intuition" in this translation.

c. The story is told in Plato's *Theaetetus*, 174 A. For the history of the anecdote see the author's "Der Sturz des Protophilosophen. Zur Komik der reinen Theorie, anhand einer Rezeptionsgeschichte der Thales-Anekdote," in W. Preisendanz and R. Warning, ed., *Das Komische*, Poetik und Hermeneutik 7 (Munich: Fink, 1976), pp. 11–64.

d. This paragraph sums up the process described in note b, which makes it necessary for us to use "intuition" in an obsolete or quasi-technical sense.

e. *Theoria*, in Greek, means first, a looking at, viewing, beholding, observing, or being a spectator, and second, contemplation or reflection. It carries no suggestion whatsoever of the modern idea of 'developing' or 'devising' theories. (See note b for the context of this change.)

f. *Zu den genuinen Gegebenheiten*. I render *Gegebenheiten* as "phenomena" because the inappropriate associations of that term with Platonism seem less misleading, in this case, than the inappropriate associations of the alternative translation, "data," with modern methodical "handling." What the author has in mind is, of course, the contents of a direct, pretheoretical confrontation with reality.

g. See the author's *Die Lesbarkeit der Welt* (Frankfurt: Suhrkamp, 1981) for a history of the metaphor of the "book of nature."

h. On the relation between curiosity as a subjective state and as an institutionalized process (and for further discussion of Diderot's *Encyclopédie* article), see the introduction to Part III of *The Legitimacy of the Modern Age*, especially pp. 236–241, or *Der Prozess der theoretischen Neugierde*, suhrkamp taschenbuch wissenschaft 24 (Frankfurt: Suhrkamp, 1973), pp. 15–22.

Chapter 4

a. The author gives a more extensive summary of Kant's early cosmogony in *The Legitimacy of the Modern Age*, pp. 212–214; or *Die Legitimität der Neuzeit* (1966), pp. 182–184.

b. On Kant's later interest in the transmigration of souls, in connection with the possibility of moral progress, see the author's *Work on Myth* (Cambridge, MA: MIT Press, 1985), p. 292; or *Arbeit am Mythos* (Frankfurt: Suhrkamp, 1979), pp. 323–324.

c. See the *Republic*, 530 C.

d. "Two things fill the mind with ever new and increasing admiration and awe, the oftener and more steadily we reflect on them: the starry heavens above me and the moral law within me": Kant, *Critique of Practical Reason*, trans. L. W. Beck (Indianapolis: Bobbs-Merrill, 1956), p. 166; *Gesammelte Schriften*, Akademie ed. (Berlin: 1900–1942), vol. 5, p. 161.

e. This "Conclusion" [*Beschluss*] is attached to part 3 of the *Allgemeine Naturgeschichte und Theorie des Himmels*, a part that unfortunately is not included in the standard English translation by W. Hastie. The subject of part 3 is the inhabitants of the planets. Kant is not certain that they are all inhabited, but thinks that inhabitants of the outer planets would be less influenced by gross matter, and would therefore be more spiritual, and more virtuous, than us. And in the conclusion it is evident, despite ironical disavowals of knowledge, and a warning not to base hopes on such imaginings, that Kant is attracted to the idea that a cosmic transmigration of souls might allow us "to partake of a closer intuition of the remaining [i.e., nonterrestrial] wonders of the Creation."

f. According to Kant's *Critique of Practical Reason* (1788), our consciousness of the moral law presupposes, as unknowable but necessary "postulates," the reality of God, freedom, and the immortality of the soul. The last of these creates the possibility of the cosmic transmigration of souls, and thus the enhanced possibilities of 'intuition' of the cosmos that Blumenberg is discussing here. See the passage in *Work on Myth* cited in note b.

g. *Idee*, "Idea," is a technical term first used by Kant to designate "a necessary concept of reason to which no corresponding object can be given in sense-experience" (*Critique of Pure Reason*, A327/B383). The three such Ideas that were prominent in that critique were those of the unities of the subject, the world, and God—in each case, of a totality that cannot be given in experience and therefore can never be completely 'grasped' by empirical concepts. In the *Critique of Judgment* he sets alongside these "Ideas of reason" a number of others, including the "aesthetic Idea," a "representation of the imagination which induces much thought, yet without the possibility of any definite thought whatever, i.e., *concept*, being adequate to it" (section 49). So the Ideas of reason are concepts to which no intuition can be adequate, and aesthetic Ideas are intuitions to which no concept can be adequate. The perception of nature as "sublime" is a primary instance of an aesthetic Idea. Since according to Kant we are compelled to think (and experience) in ways conditioned by these Ideas, we are necessarily permanently confronted by the inadequacy of our concepts to the world, but are at the same time unable to reconcile ourselves to that inadequacy.

Chapter 5

a. "Contingent 'fact'" here translates *faktisch*, an adjective that was given prominence by Heidegger as characterizing Dasein's way of being (where it was closely related to *Geworfenheit*, "Thrownness"). As the text, here, illustrates, Blumenberg does not see 'facticity' [*Faktizität*] only in human existence, but he does retain the negative connotation, suggesting something that is not necessary—that is contingent, perhaps even random or arbitrary—and that is not its own source or master. To translate these terms, rather than resorting to "facticity" (and "factic"), which might suggest more terminological continuity with Heidegger than is actually present, I have used various paraphrases involving the word "contingent."

b. Kant, *Universal Natural History and Theory of the Heavens*, trans. W. Hastie (Ann Arbor: University of Michigan Press, 1969), pp. 145–146.

c. "Was in der universellen Materialisierung seine Unverfügbarkeit behauptet hat." "Materialization" refers to the modern reduction of the world to permutations of homogeneous 'matter,' which is the intellectual side of the project of making reality 'available' (*verfügbar*), subject to our disposition (*Verfügung*)—a project that the heavens, with their apparent *Unverfügbarkeit*, seem to put in question.

d. Putting into practice the increasing medievalism of his aesthetic sympathies, Schlegel converted to Catholicism in 1808.

e. Heine, *Confessions*, trans. (together with Tolstoy, *A Confession*) P. Heinegg (n.p.: Joseph Simon, 1981), pp. 45–46. The next quotations are from pp. 47 and 46.

f. See note a on *faktisch* and *Faktizität*.

g. The story of Goethe's "Prometheus" material (the ode and the fragment of a drama that he built around it) and its delayed impact is told in some detail in chapter 1 of part IV of *Work on Myth*.

h. The capital "I" in "Ideal," here, is to indicate that a more or less Platonic sense is intended— relating to "Ideas" (*eidoi*, also translated as "Forms") as supersensible entities. These terms should not be confused, however, with "Idealism" and "Idealist," which in this book refer to German Idealism (Fichte, Schelling, and Hegel), as distinguished from idealism in general *and* from Plato's doctrine of Ideas. All of these must also be distinguished from *Kant's* doctrine of "Ideas" as "necessary concepts of reason," which is described in note g to the previous chapter. (How productive those two little Greek syllables have been!)

Chapter 6

a. On *Anschauung*, 'intuition,' see, again, note b to Chapter 3.

b. "Intentionality" (a medieval concept that was revived by Franz Brentano and opopularized by Husserl) refers to the quality of mental acts such as believing, assuming, or perceiving, by which they 'point to' ('intend') a state of affairs (which may or may not exist in reality). In such a relation the object is thought of, primarily, as simply existing, rather than as a means to anything else.

c. On the "principle of insufficient reason"—which is to be contrasted to Leibniz's "principle of sufficient reason"—see the author's "An Anthropological Approach to the Contemporary Significance of Rhetoric" (cited in note b to the translator's introduction), p. 447, where it is described as the "axiom of all rhetoric" and "the correlate of the anthropology of a creature that is deficient in essential respects."

d. Anselm's formula for God, in his famous ontological argument for God's existence.

Chapter 7

a. Humboldt began work on *Kosmos* when he was 65 years old (in 1834), and worked on it until his death in 1859, his ninetieth year.

b. Blumenberg examines this aspect of the *Kosmos* phenomenon at greater length in chapter 18, "Ein Buch von der Natur wie ein Buch der Natur," of his *Die Lesbarkeit der Welt* (Frankfurt: Suhrkamp, 1981).

c. The great age of German literature, which marks the 'coming of age' of German as a literary language, is conventionally divided into "Classicism" (*die Klassik*: the mature work of, above all, Goethe and Schiller) and "Romanticism" (*die Romantik*: Hölderlin, Novalis, Kleist, et al.).

d. On Kant's "Ideas," see note g to chapter 4.

e. The reference is to the anecdote about Thales of Miletus's falling into a spring. See note c to chapter 3.

f. See chapter 5, text to note 83.

g. Nietzsche, *Genealogy of Morals*, third essay, section 25; *Basic Writings of Nietzsche*, trans. Walter Kaufmann (New York: Random House, 1966), p. 591. See *The Legitimacy of the Modern Age* (Cambridge, MA: MIT Press, 1983), p. 140, for another discussion of this passage.

Chapter 8

a. "Life-world" is a term coined by Edmund Husserl in the 1930s to designate the primary world of the individual's experience, as opposed to the object-world of science. Blumenberg discusses this concept in the introduction and the first paper in his *Wirklichkeiten in denen wir leben* (Stuttgart: Reclam, 1981), p. 4, and discusses it at length in part I, "Das Lebensweltmissverständnis," of *Lebenszeit und Weltzeit* (Frankfurt: Suhrkamp, 1986).

b. On Kant's "Ideas," which are not to be confused with Plato's "Ideas" (*eidoi*, also often translated as "Forms"), see note g to chapter 4 and note h to chapter 5.

c. "Cosmological antinomy" refers to the first of the four Antinomies in Kant's *Critique of Pure Reason* (A426–429, B454–457), which presents an undecidable conflict between the thesis that the world is limited in space and time and the antithesis that it is infinite in both respects. Thus the problem is "dialectical" in the Kantian sense of that term (the Antinomies are the centerpiece of what Kant calls the "Transcendental Dialectic"), rather than in the Hegelian or Marxist sense.

d. The author surveyed the tradition of metaphors of light in his "Licht als Metapher der Wahrheit," *Studium Generale* 10 (1957): 432–447.

e. *Verblendungszusammenhang*, "delusion [literally: blinding] system," is a term used by T. W. Adorno and M. Horkheimer in their *Dialektik der Aufklärung* (Frankfurt: Fischer, 1969), p. 40.

f. The author distinguished several "concepts of reality" that have been influential in our tradition—among them that of the "realization of an internally consistent context," a type of reality in which simulation presumably does not have to be discriminated against—in his "Wirklichkeitsbegriff und Möglichkeit des Romans," in *Nachahmung und Illusion*, Poetik und Hermeneutik 1, ed. H. R. Jauss (Munich: Fink, 1969), pp. 9–27 (quote on p. 12).

g. See the essay cited in note f for further discussion of Leibniz's new concept of reality. Additional comments on our "difficulties with the concept of reality" can be found in the author's *Wirklichkeiten in denen wir leben* (Stuttgart: Reclam, 1981), in the introduction (the quote is from p. 3) and elsewhere, and in "Gebärden des Wirklichkeitsverlustes," the first of three short essays published under the collective title "Über den Rand der Wirklichkeit hinaus," in *Akzente* (Munich), February 1983, pp. 16–27.

Part II

Chapter 1

a. *Wirkungsgeschichte* is sometimes translated as "effective-history," which conveys the compact terminological character of the original German at the cost of being nearly incomprehensible to those who do not happen to know the original. The title of this chapter, of which my paraphrase is so lengthy, is *Die Vorgeschichte als Bedingung der Wirkungsgeschichte*. To clarify what follows I should mention that *Wirkungsgeschichten* have especially been sought in literary and intellectual history, as histories of the influence exercised by particular texts or authors.

b. The struggle against such "absolutisms of the present," and against the "injustice in the dimension of time" that they entail, is the theme of the address that Blumenberg delivered when he was awarded the Kuno Fischer Prize of the University of Heidelberg, in July 1974: "Ernst Cassirers gedenkend," *Revue Internationale de Philosophie* 28 (1974): 456–463; reprinted in his

Wirklichkeiten in denen wir leben (Stuttgart: Reclam, 1981), pp. 163–172. Incidentally, the word *Destruktion* that he uses here to describe the business of the theory of history is the term used by Heidegger for his proposed treatment of the history of metaphysics, a term appropriately rendered by J. Derrida as *déconstruction* (which, however, has become such a trademark of his school that it would be misleading for us to use it here). So it should not be taken as implying 'blind,' uncomprehending destruction (as the normal German *Zerstörung* might). Finally, on the relation between "interests" and theoretical knowledge, which Blumenberg mentions in the next sentence, see *The Legitimacy of the Modern Age* (Cambridge, MA: MIT Press, 1983), pp. 235–236.

c. Osiander's unauthorized preface described Copernicus's doctrine of the Earth's motions as purely hypothetical, in the sense of the traditional procedure in astronomy of postulating motions for purposes of computation and prediction but making no claim to their existence in reality. For a full analysis of Osiander's contribution see part II, chapter 2.

d. No doubt the reference is to Goethe's remark in the last year of his life (which was quoted on the endpaper of the original, German edition of this book), that Copernicus's was "the greatest, noblest, most momentous discovery that man has ever made, more important in my eyes than the entire Bible" (*Werke*, ed. E. Beutler [Munich: Artemis, 1948–], vol. 23, p. 844).

e. Chapter 13 of *Das Kapital* corresponds to chapter 15 of the standard English translation by S. Moore and E. Aveling (London: 1887; New York: International Publishers, 1967).

Chapter 2

a. "Sich als indifferent zu ihrem Dasein zeigt"—in contrast to God, Whose concept (according to Anselm's famous "ontological argument") implies His existence.

b. J. Kepler, *Gesammelte Werke*, ed. M. Caspar (Munich: Beck, 1938–1955), vol. 13, p. 35, and vol. 15, p. 172.

c. "Orderly use of omnipotence" is presumably an allusion to William of Ockham's notion of God's *potentia ordinata* "ordered, or ordained power," which he contrasts with God's *potentia absoluta*, absolute power unconstrained by self-imposed limits, such as laws of nature.

d. Text in A. Maier, *Zwei Grundprobleme der scholastischen Naturphilosophie* (Rome: Edizioni di Storia e Letteratura, 1951; rpt. 1968), p. 169. My translation of this passage is partly based on M. Clagett's in his *The Science of Mechanics in the Middle Ages* (Madison: University of Wisconsin Press, 1959), p. 528.

e. Text in Maier, *Zwei Grundprobleme*, p. 173.

f. Celestial motions that continued forever would be "infinite" in the literal sense of "unending."

g. Text in Maier, *Zwei Grundprobleme*, pp. 179–180.

h. E. H. Haeckel (1834–1919) formulated the doctrine according to which "ontogeny recapitulates phylogeny"—the sequence of steps in the genesis of the individual corresponds to the sequence of steps in the genesis of his species. Or, by analogy: The steps in a systematic presentation of the Copernican doctrine correspond to the steps in its genesis.

i. The Prime Mover was the "final" cause, the cause "for which" things happen in the world, but not the "efficient" cause, the "agent whereby" they happen.

Notes to Pages 169–197

Chapter 3

a. *Destruktion*. See note b to part II, chapter 1.

b. *Historismus*, "historicism," in Germany, signifies primarily the attitude, advocated by L. Ranke and analyzed by F. Meinecke, E. Troeltsch, and others, that regards each historical period as having its own validity, which is to be interpreted without the application of external (supposedly universal) standards. (Thus it is very different from—indeed, antithetical to—what K. Popper has more recently dubbed "historicism.") Of course, as Blumenberg is pointing out, a "historicist" manner need not reflect a thorough-going historicism.

c. 'Nostrocentrism' would be a view of the world as centered on *us* ("temporal nostrocentrism": centered on our time). On the opposition between it and a thoroughgoing 'historicism' see Blumenberg's "Ernst Cassirers gedenkend . . . ," *Revue Internationale de Philosophie* 28 (1974): 456–463, reprinted in his *Wirklichkeiten in denen wir leben* (Stuttgart: Reclam, 1981), pp. 163–172.

d. *Idee*, "idea," here has connotations of "guiding idea" or "ideal," as in Kant's technical use of the word (see note g to part I, chapter 4).

e. In English in the original.

f. These two sentences paraphrase parts of the third sentence of Copernicus's preface to Paul III.

g. "Homocentrism" refers to the analysis of planetary (and stellar) motions in terms of spheres all of which have the same center—admitting no epicycles, eccentrics, or equants (all of which involve other centers, besides the Earth).

h. The "supposed Platonic requirement" was that the celestial phenomena should be explained by means of combinations of uniform circular motions. See note e to part II, chapter 5.

i. On "intentionality" see note b to part I, chapter 6. Here it might be paraphrased as "inherent tendency."

j. On the way in which Augustine's path prefigures that of medieval Scholasticism, see also *The Legitimacy of the Modern Age* (Cambridge, MA: MIT Press, 1983), p. 135.

k. The author expounds Leibniz's argument on this point in *The Legitimacy of the Modern Age*, pp. 149–150.

l. Throughout this passage, "evils" translates *das Übel*. I have avoided the singular, "evil," so as to avoid implying an identity between this Aristotelian discussion and the post-Gnostic and Augustinian "problem of evil," with its bias toward wickedness as the source of what is bad in the world. German avoids this implication with the aid of its (fairly sharp) distinction between *übel* (evil = bad) and *böse* (evil = wicked).

m. The quotation is from 2 Esdras 6:59, *New Oxford Annotated Bible with the Apocrypha* (New York: Oxford University Press, 1973), p. 36 (of Apocrypha). "2 Esdras" is the Protestant equivalent of "4 Esdras" in the Vulgate.

n. The story is told in Joshua 10:12–14.

o. Thomas Aquinas distinguishes between "ontological truth," which is the correspondence or conformity (*adaequatio*) between an existing thing and its idea as conceived by God, its Creator, and "logical truth," which is the *adaequatio* between a knower and the thing that he knows.

Chapter 4

a. Translated by H. Nachod in *The Renaissance Philosophy of Man*, ed. E. Cassirer, P. O. Kristeller, and J. H. Randall, Jr. (Chicago: University of Chicago Press, 1948), p. 86.

b. On "the radicalism of theological voluntarism," and what was needed in order to deal with it, see *The Legitimacy of the Modern Age*, part II, chapters 3 and 4.

c. Roman law and canon law.

d. Copernicus's *Commentariolus* was a brief sketch of his heliocentric system, which he wrote some years before the *De Revolutionibus* and did not publish. Some copies of it circulated in manuscript form. It was translated and annotated by Edward Rosen in *Three Copernican Treatises* (New York: Columbia University Press, 1939; rpt. New York: Dover, 1959).

Chapter 5

a. In the Scholastic literature Aristotle was referred to as "the Philosopher."

b. In the author's use of the term *Übertragung*, "carrying over," one is always aware of the parallelism with "metaphor," the Greek root of which is likewise "to carry over." I do not translate it as "metaphor," but reserve that translation for *Metapher*, so as to preserve the connotation that these latter terms have of belonging to a specialized terminology.

c. See Kepler's text in note 87 to this chapter.

d. Kepler's third law, announced in his *Harmonices mundi* [*Harmonies of the World*] in 1619, states that the proportion between the cube of each planet's mean distance from the Sun and the square of the time it takes to make a complete orbital journey around the Sun is the same for all the planets.

e. The conception of the task of astronomy as that of explaining the observed motions of the heavenly bodies in terms of combinations of *uniform circular* motions (which, for Copernicus, excluded Ptolemy's use of "equants"—centers of motion differing from the geometrical center of the circle in question) was ascribed to Plato by Simplicius in his famous (sixth century A.D.) commentary on Aristotle's *On the Heavens*. In its "strictest form" (see the paragraph corresponding to note 136 in part III, chapter 4) it would also have required all the circular motions to have the same center (as in the "homocentric" system developed by Plato's pupil Eudoxus), thus excluding eccentrics and epicycles as well. The "Platonic requirement" was first entirely abandoned by Kepler, as Blumenberg describes in part III, chapter 2, where he quotes a letter from Kepler to David Fabricius.

Chapter 6

a. *Versuch einer immanenten Entstehungsgeschichte der kopernikanischen Theorie*. I believe my rather free translation conveys the sense of this title, but readers may wish to be aware of the original German wording.

b. A "conjunction" of a planet with the Sun occurs when they share the same longitude vis-à-vis the Earth. An "inferior conjunction" would place the planet in question "below" the Sun, i.e., between it and the Earth; a "superior conjunction" would place the planet "above" the Sun, i.e., beyond it. With the observations available in antiquity, the motions of the "inner planets,"

Venus and Mercury, which never stray very far from the Sun, could be explained by a model of either kind.

c. That is, as if Mercury and Venus orbited the Sun while it orbited the Earth, just as planets, in one traditional construction, traveled on circular "epicycles" around a point (the "deferent") that simultaneously traced a circular path around the Earth.

d. Noting a possible misreading of this sentence, Professor Blumenberg asked me to explain that unlike the diameter of Venus's ("abnormal") epicycle, which was described in the paragraph before last, the "thickness" that is referred to here is a real distance determined by the maximum and minimum Venus-Earth-Sun angles and by the variations in Venus's brightness or magnitude (its "phases"), which have to be explained, rather than by the diameter of the (imaginary) epicycle.

e. On "deferents," see note c.

f. Here and in what follows, "gravity" translates *Schwere*. Ordinarily, in pre-Newtonian contexts, *Schwere* would simply be "weight," but because the accent here is on a relationship to a center (rather than on weight as an 'intrinsic' quality of an object, in the least sophisticated, 'commonsense' understanding of the concept), I have preferred the quasi-technical "gravity." Please bear in mind that of course no implication of "force," or of Newton's other concepts and laws, can be read into it yet.

g. See the text corresponding to note 79 in part II, chapter 5.

Part III

Introduction

a. In this well-known *Zwischenbetrachtung* Goethe wrote, "But among all the discoveries and new convictions nothing may have produced a greater effect on the human spirit than the doctrine of Copernicus. The world had scarcely been recognized as round and complete in itself when it was expected to relinquish the enormous privilege of being the center of the universe. A greater demand may never have been addressed to mankind. For think of all the things that went up in smoke as a result of accepting this: a second Paradise, a world of innocence, poetry and piety, the testimony of the senses, the conviction of a poetic-religious faith; it is no wonder that people did not want to give up all of this, and that they opposed such a doctrine in every way—a doctrine that justified those who accepted it in, and summoned them to a previously unknown, indeed unimagined freedom of thought and largeness of views" (*Materialen zur Geschichte der Farbenlehre,Werke*, ed. E. Beutler [Munich: Artemis, 1948–], vol. 16, p. 395).

b. This formula is paraphrased by Blumenberg in *The Legitimacy of the Modern Age* (Cambridge, MA: MIT Press, 1983), p. 394, as follows: "... that God can also produce a phenomenon in nature in a different way than is made to appear plausible by a particular explanation. Consequently it would be inadmissible daring (*soverchia arditezza*) to want to narrow down and commit the divine power and wisdom to a particular idea by asserting that single explanation to be true." cf. *Dialogue Concerning the Two Chief World Systems*, trans. S. Drake (Berkeley and Los Angeles: University of California Press, 1970), p. 464.

Chapter 1

a. "*Täter*" ['perpetrator'], in German, can mean either simply a 'doer,' or a 'culprit'—one who did something wrong. This ambiguity, with its suggestion that someone who 'intervenes' in the workings of nature (as Copernicus does by "moving the Earth" and "making the Sun and the heavens stand still") is ipso facto a wrongdoer, is exploited by the author throughout this chapter. Fortunately we have a rough equivalent in the similar ambiguity of the English "perpetrator."

b. Luther's famous remark in his *Table Talk* is cited in note 79 to part III, chapter 3.

c. The author introduced the concept of "absolute metaphor" in his *Paradigmen zu einer Metaphorologie* (Bonn: Bouvier, 1960; rpt. from *Archiv für Begriffsgeschichte* 6[1960]: 7–142), p. 9.

Chapter 2

a. In contrast to the dominant Lutheran conception of man's "justification," which was formulated by Melanchthon and focused on the forensic "imputation" of Christ's justice to man, Osiander described justification as the entrance of Christ's divine nature into man, which produces a renewal that renders him really, substantively just.

b. "Pregnance" [*Prägnanz*] refers to the kind of sharply defined outline of events that suggests mythical 'significance.' For the background of this translation see *Work on Myth*, p. 111, note i.

c. On the "Platonic" requirement see note e to part II, chapter 5.

d. Blumenberg discussed aspects of this argument in *The Legitimacy of the Modern Age*, pp. 149–150.

e. W. Heisenberg, *Across the Frontiers*, trans. P. Heath (New York: Harper and Row, 1974), pp. 128–129.

f. By a "philosophy of history" Blumenberg means here (as usual) one of the classical modern philosophies of history (e.g., Comte, Hegel, Marx), which claim knowledge of the pattern of history as a whole. On "historicism," which is not at all the same thing, see note b to part II, chapter 3.

Chapter 3

a. *Praeceptor Germaniae*, "teacher of the Germans," is an honorific title traditionally applied to Melanchthon because of his leadership in the Reformation era's reorganization and expansion of education in Germany.

b. The "original transcript" is quoted in note 79.

c. "His invisible attributes, that is to say his everlasting power and deity, have been visible, ever since the world began, to the eye of reason, in the things he has made" (*New English Bible*).

d. On the "Platonic" requirement, see note e to part II, chapter 5.

e. The dogma that Christ died "for us."

Chapter 4

a. 'Significance" (*Bedeutsamkeit*), as the distinctive quality of what myth offers, is discussed in part I, chapter 3, of *Work of Myth*, esp. pp. 67ff. (*Arbeit am Mythos*, pp. 77ff.).

b. On the "book of nature" and the "book of hisory" see the author's *Die Lesbarkeit der Welt* (Frankfurt: Suhrkamp, 1981), especially (in connection with Rheticus) chapter 7.

c. Kant writes of "self-incurred tutelage" [*selbstverschuldete Unmündigkeit*] in "An Answer to the Question: What Is Enlightenment?" (1784), in *Kant's Political Writings*, trans. H. B. Nisbet (Cambridge: Cambridge University Press, 1970), p. 54. For a discussion of Kant's analysis of this phenomenon see *The Legitimacy of the Modern Age*, pp. 429–434.

Chapter 5

a. Giordano Bruno was born in Nola near Naples, and hence is sometimes called "the Nolan."

b. Thomas Digges, *Perfit Description of the Caelestiall Orbes according to the most aunciente doctrine of the Pythagoreans lately revived by Copernicus and by Geometricall Demonstrations approved* (1576), ed. F. R. Johnson and S. V. Larkey, in *Huntington Library Bulletin* 5 (1934). See *The Legitimacy of the Modern Age* (Cambridge, MA: MIT Press, 1983), pp. 369–371.

c. See *The Legitimacy of the Modern Age*, pp. 551–552, and note 5, where the source of this quotation is given as *La Cena de le Ceneri* I, *Dialoghi italiani*, ed. G. Gentile and G. Aquilecchia (Florence: Sansoni, 1958), p. 29.

d. The distinctive term *Verblendungszusammenhang*, "delusion system," was used by T. W. Adorno and M. Horkheimer in their *Dialektik der Aufklärung* (2nd ed., Frankfurt: Fischer, 1969), p. 40.

e. This phrase is in English in the original.

f. In his *Von den göttlichen Dingen und ihrer Offenbarung* (Leipzig: 1811), Jacobi expanded on the 'emotivist' theism that he had advocated in his *Über die Lehre des Spinozas* (Breslau: 1785).

g. Hegel discusses Bruno in his lectures on *The History of Philosophy*, part two, section three, B, 3. In the translation by E. S. Haldane and F. H. Simson (London: 1896; rpt. Atlantic Highlands, N. J.: Humanities Press, 1974), the passage quoted is in vol. 3, pp. 121–122. The subsequent translations are my own.

h. The *coincidentia oppositorum* was a key notion of Nicholas of Cusa's, which Bruno took over. See *The Legitimacy of the Modern Age* (Cambridge, MA: MIT Press, 1983), pp. 490ff.

Chapter 6

a. Galileo, *Dialogue Concerning the Two Chief World Systems—Ptolemaic and Copernican*, trans. Stillman Drake (Berkeley and Los Angeles: University of California Press, 1967), p. 163 (slightly revised).

b. See "Lebenswelt und Technisierung," in the author's *Wirklichkeiten in denen wir leben* (Stuttgart: Reclam, 1981), pp. 7–54 (esp. p. 41), for references, and a more comprehensive account of Husserl's attitude to modern science.

c. This is from the dedication of the *Discourses* to the Count of Noailles.

d. Translation from A. C. Crombie, *Medieval and Early Modern Science* (Garden City, NY: Doubleday, 1959), p. 142.

e. Arcetri was the location of Galileo's house, near Florence, to which he was confined from December 1633 until his death in 1642.

f. Galileo, *Dialogue Concerning the Two Chief World Systems*, trans. S. Drake (Berkeley and Los Angeles: University of California Press, 1967), pp. 103–104.

g. Op. cit., p. 448.

h. *Discoveries and Opinions of Galileo*, trans. Stillman Drake (Garden City, NJ: Doubleday Anchor, 1957), p. 193.

i. *Dialogue* (cited in note f), p. 421.

j. *Discoveries and Opinions*, p. 98.

k. *Discoveries and Opinions*, p. 123.

l. *Discoveries and Opinions*, p. 124 (slightly revised).

Part IV

Chapter 1

a. On *Anschauung*, 'intuition,' see note b to part I, chapter 3.

b. *Timaeus* 37 D–E, translated by F. M. Cornford in *Plato's Cosmology* (London: Routledge and Kegan Paul, 1937; rpt. Indianapolis and New York: Bobbs-Merrill, n.d.), pp. 97–98 (slightly revised).

Chapter 2

a. On this late medieval interpretation of the seventh day, see part II, chapter 2, text to note 20.

Chapter 3

a. See Nicholas de Cusa, *De Ludo Globi. The Game of Spheres*, trans. P. M. Watts (New York: Abaris, 1986), p. 59.

b. For further details on the Cusan's cosmology and its relation to Copernicus, see *The Legitimacy of the Modern Age*, pp. 502–515.

c. The requirement that the astronomical phenomena should be explained in terms of uniform circular motions. On the ascription of this requirement to Plato, see note e to part II, chapter 5.

Chapter 4

a. The "transcendental idealization of the concept of time" refers to Kant's account of time, in the Transcendental Aesthetic of the *Critique of Pure Reason*.

b. On this formula see also *The Legitimacy of the Modern Age*, pp. 80–82.

c. 1 part in 10,000,000 of the length of the quadrant passing through Paris was the definition of the meter when it was introduced in 1790.

Part V

Chapter 1

a. J. H. Lambert, *Cosmological Letters on the Arrangement of the World-Edifice*, translated by S. L. Jaki (New York: Science History Publications, 1976), p. 45.

b. Thomas Wright, *An Original Theory or New Hypothesis of the Universe* (London: 1750), p. 63.

c. W. Herschel, "Account of Some Observations Tending to Investigate the Construction of the Heavens," *The Scientific Papers of Sir William Herschel* (London: Royal Society and Royal Astronomical Society, 1912), vol. 1, pp. 157–158.

d. Lambert, *Cosmological Letters*, trans. Jaki, p. 53.

e. Wright, *An Original Theory* (cited in note b), p. 63.

Chapter 2

a. Lambert, *Cosmological Letters*, trans. Jaki, p. 55. (I have amended this and some of the following translated passages.)

b. Op. cit., p. 56.

c. Op. cit., p. 124.

d. Op. cit., p. 50.

e. Op. cit., pp. 45–46.

f. Compare op. cit., p. 51. (The quotations in the next paragraph are on the same page.)

g. Op. cit., p. 170.

h. *Des unzureichenden Grundes*: the opposite of what is called for by Leibniz's famous "principle of sufficient reason." See note c to part I, chapter 6.

i. Lambert, *Cosmological Letters*, p. 73.

j. Op. cit., p. 56.

k. Op. cit., p. 57.

l. Op. cit., p. 58; the quotation in the next paragraph is from the same page.

m. Op. cit., p. 62; the next quotation is from the same page.

n. Op. cit., pp. 79–80.

o. Op. cit., p. 92.

p. Op. cit., p. 93.

q. Op. cit., p. 171. The next quotation is from the same page.

r. Op. cit., p. 174.

s. Op. cit., p. 175.

t. Op. cit., p. 175.

u. Op. cit., p. 110.

v. Op. cit., p. 152.

w. Op. cit., p. 166.

x. Op. cit., p. 178. The next two quotations are from the same page.

y. Op. cit., p. 115.

z. Op. cit., pp. 179–180. The remaining two quotations are from the same page.

Chapter 3

a. "Die Stellung des Menschen im Kosmos" is a common phrase in popular philosophy, and was the title of a book by Max Scheler (Darmstadt: O. Reichl, 1928).

Chapter 4

a. J. H. Lambert, *Cosmological Letters*, trans. S. L. Jaki (New York: Science History Publications, 1976), p. 121.

b. Kant, *Universal Natural History and Theory of the Heavens*, trans. W. Hastie (Ann Arbor: University of Michigan Press, 1969), pp. 23 and 28. (In some cases I have amended Hastie's translations.)

c. Op. cit., p. 129.

d. Op. cit., p. 130.

e. Op. cit., p. 29.

f. Op. cit., p. 18.

g. Op. cit., p. 25.

h. Op. cit., p. 145.

i. Spinoza's phrase for nature seen as its own active creator.

j. Kant, *Universal Natural History*, pp. 145–146.

k. Op. cit., p. 150.

l. Op. cit., pp. 155–156.

m. Kant, *Critique of Pure Reason*, trans. Norman Kemp Smith (London: Macmillan, New York: St. Martin's 1929), p. 488.

n. On Kant's "Postulates," see note e to part I, chapter 4.

o. *Universal Natural History*, p. 65. Subsequent quotes are from the same page.

p. Kant's transcendental "dialectic," as classically presented in the division of the *Critique of Pure Reason* that carries that name, must of course be carefully distinguished, at least initially, from the various ideas designated by the same term in Plato, Hegel, and Marx.

q. *Universal Natural History*, p. 72. The next quote is from the same page.

r. Op. cit., p. 74. The next quote is from the same page.

s. Op. cit., p. 148.

t. Op. cit., p. 147.

u. Op. cit., p. 116.

Chapter 5

a. Kant, *Critique of Pure Reason*, B. xiii, trans. Norman Kemp Smith (London: Macmillan, New York: St. Martin's, 1929), p. 20. Most of my quotations from the *Critique of Pure Reason* follow Kemp Smith's translation, but I have followed Blumenberg's example and omitted page references when the context makes them easy to locate.

b. *Naherwartung*, "immediate expectation," is an expression applied to the early Christian expectation of the imminent Second Coming of the Lord.

c. *Auszufüllen*, because of German grammar, is the last word of the sentence in the original.

d. *Critique of Pure Reason*, B xvi. Kemp Smith (p. 22) translates this as "on the lines of Copernicus's primary hypothesis," but gives the original German in a note.

e. See Freud, *Introductory Lectures on Psychoanalysis*, lecture 18; *Standard Edition of the Complete Psychological Works*, ed. J. Strachey (London: Hogarth Press, 1953–), vol. 16, pp. 284–285; *Gesammelte Werke* (London: Imago, Frankfurt: Fischer, 1941–), vol. 11. pp. 294–295.

Part VI

Introduction

a. "*Änderung der Denkungsart*" reminds us of Kant's discussion of scientific revolutions (discussed by Blumenberg in the previous chapter), in which Kant uses the same unusal terminology.

Chapter 1

a. Montaigne, *Essais* II, 12; ed. P. Villey (Paris: Presses Universitaires de France, 1965), p. 536; trans. J. Zeitlin (New York: Knopf, 1935), vol. 2, pp. 197–198 (slightly revised).

b. *Essais*, ed. Villey, p. 531; trans. Zeitlin, vol. 2, p. 192.

c. *Essais*, ed. Villey, p. 568.

Notes to Pages 632–663

d. In the original: *Reflektierung der Optik*. Since in German there is no verbal distinction between 'reflection' and 'reflexiveness,' the phrase suggests both vision's 'looking at [reflecting] itself,' and the viewer's reflexively considering the role of vision, as such, in his relation to the world.

e. Feuerbach, *Nachgelassene Aphorismen, Sämtliche Werke*, ed. W. Bolin and F. Jodl (2nd ed. Stuttgart: Frommann-Holzboog, 1960–1964), vol. 10, p. 298. See *The Legitimacy of the Modern Age*, p. 445.

f. Galileo, *Dialogue Concerning the Two Chief World Systems*, trans. Stillman Drake (2nd ed. Berkeley and Los Angeles: University of California Press, 1967), p. 339. The subsequent quotation is from p. 335.

Chapter 2

a. "... der Mensch entlastet sich ständig von der Überflutung ...." *Überflutung*—literally, "inundation"—reminds us of the psychological concept of *Reizüberflutung*, "stimulus overload." *Entlastung*, "unburdening," is a central concept in Arnold Gehlen's anthropology.

b. The ancient concept of reality as "momentary evidence" is one of several such concepts that are distinguished and discussed in the author's "Wirklichkeitsbegriff und Möglichkeit des Romans," in *Nachahmung und Illusion*, Poetik und Hermeneutik 1, ed. H. R. Jauss (Munich: Fink, 1969), pp. 9–27 (see especially pp. 10–11).

Chapter 3

a. Freud, "An Autobiographical Study," trans. J. Strachey, *Complete Psychological Works*, vol. 20, p. 50.

b. Freud, "On the History of the Psycho-Analytic Movement," trans. J. Riviere, *Complete Psychological Works*, vol. 14, p. 23.

c. J. E. Bode, astronomer at the Berlin Academy, "made public in 1772 the relation according to which the distances between the large planets are in a nearly regular geometric progression" (B. Sticker, article, "Bode, Johann Elert," *Dictionary of Scientific Biography*, vol. 2).

Author's Notes

Part I

Chapter 1

1. Sophocles, *Oedipus at Colonus*, 1225; trans. F. Storr, Loeb Classical Library (London: Heinemann; New York: Putnam's 1912), p. 261: "Not to be born at all is best . . . . " Aristotle, *Eudemian Ethics* I 5; 1216 a 11–15; trans. J. Solomon in *The Works of Aristotle*, ed. W. D. Ross, vol. 9 (London: Oxford University Press, 1915).

2. B. de Fontenelle, *De l'origine des fables*, ed. J. R. Carré (Paris: F. Alcan, 1932), p. 18.

3. C. Rosenbaum, *Die Philosophie Solomon Maimons in seinem hebraischen Kommentar gibath-hammoreh zum moreh-nebuchim des Maimonides* (Giessen: 1928), p. 51. The Old Testament root is Wisdom 1 : 14: "Creavit enim, ut essent omnia . . . " (Vulgate).

4. Jacob Burckhardt, *Griechische Kulturgeschichte* V: "Zur Gesamtbilanz des Griechischen Lebens," *Gesammelte Werke* (Basel: Schwabe, 1955–1959), vol. 6, p. 352.

5. Jacob Brucker, *Kurtze Fragen aus der Philosophischen Historie*, vol. 1 (Ulm: 1731), pp. 227–229.

6. Johnann Christoph Gottsched, *Gedächtnisrede auf den unsterblich verdienten Domherren in Frauenburg Nicolaus Coppernicus* (Leipzig: 1743), pp. 128–129.

7. Diogenes Laertius, *Lives of Eminent Philosophers*, II 10.

8. Plutarch, *De facie in orbe lunae* 923, trans. by H. Cherniss and W. C. Helmbold in *Plutarch's Moralia*, vol. 12, Loeb Classical Library (Cambridge, MA: Harvard University Press; London: Heinemann, 1957), p. 55; also in *Stoicorum veterum fragmenta*, ed. von Arnim (Leipzig: Teubner, 1921–1924), vol. 1, #500. The list of Cleanthes's lost writings is in Diogenes Laertius, VII 174.

9. *Stoicorum veterum fragmenta* (cited in note 8), vol. 1, #499; vol. 2, #642, 644.

10. *Stoicorum veterum fragmenta* (cited in note 8), vol. 1, #538.

11. B. Snell, "Pindars Hymnos auf Zeus," *Antike und Abendland* 2 (1946); also in his *Die Entdeckung des Geistes* (2nd ed., Hamburg: Claassen und Goverts, 1948), translated as *The Discovery of the Mind* (Cambridge, MA: Harvard University Press, 1953).

12. Seneca, *De otio* V 3–4, trans. J. W. Basore in *Moral Essays*, vol. 2, Loeb Classical Library (Cambridge, MA: Harvard University Press; London: Heinemann, 1932), p. 191: "Curiosum nobis ingenium natura dedit et artis sibi ac pulchritudinis suae conscia spectatores nos tantis rerum spectaculis genuit . . . ." Compare *Epist.* 64, 6, with its comparison between "contemplatio sapientiae" and "contemplatio mundi": "quem saepe tamquam spectator novus video."

13. Kant, *Kritik der Urteilskraft* [*Critique of Judgment*], section 86.

14. Seneca, *Naturales quaestiones* VII, 2: "illo quoque pertinebit haec excusisse, ut sciamus utrum mundus terra stante circumeat an mundo stante terra vertatur . . . digna res contemplatione, ut sciamus, in quo rerum statu simus, pigerrimam sortiti an velocissimam sedem, circa nos deus omnia an nos agat."

15. Cicero, *De natura deorum* II 6, 16 (= *Stoicorum veterum fragmenta*—cited in note 8—vol. 2, #1012), trans. H. Rackham, Loeb Classical Library (London: Heinemann; Cambridge, MA: Harvard University Press, 1933), p. 139. Lactantius (*De ira* X, 36) quotes Chrysippus in order to interpret him as implying that what man cannot make requires that man's capability should be enhanced, so that it can be made: "ars, consilium, prudentia, potestas."

16. Lucretius, *De rerum natura* I 62–79; trans. W. H. D. Rouse, Loeb Classical Library (London: Heinemann; New York: Putnam's, 1924). (The last translation is my own.)

17. Lucretius, *De rerum natura* VI 649–654.

18. Lucretius, *De rerum natura* VI 677–679.

Chapter 2

19. H. Jonas, *Gnosis und spätantiker Geist*, vol. 1 (2nd ed., Göttingen: Vandenhoeck und Ruprecht, 1954), pp. 163–164, note.

20. Arnobius, *Adversus gentes* II, 20ff. On this see H. Blumenberg, *Das dritte Höhlengleichnis*, Studi e Ricerche di Storia della Filosofia 39 (Turin: 1961) (rpt. from *Filosofia* 11 [1960], pp. 705–722). On the history of its influence: P. Krafft, *Beiträge zur Wirkungsgeschichte des älteren Arnobius* (Wiesbaden: Harrassowitz, 1966).

21. Examples of this are given in M. Werner, *Die Entstehung des christlichen Dogmas* (2nd ed., Bern: P. Haupt, 1953), p. 266–267.

22. Origenes, *Contra Celsum* IV, 23; trans. H. Chadwick (Cambridge: Cambridge University Press, 1953), vol. 1, p. 199.

23. Deuteronomy 4:19.

24. Lactantius, *De ira dei* 7, 5: "Homo autem recto statu, ore sublimi, ad contemplationem mundi excitatus confert cum deo vultum et rationem ratio cognoscit." Similarly in *Divinae institutiones* II 1, 17, and II 2, 18–20. As a late Stoic formula it appears in Manilus, *Astronomica* IV 913–918:

Atque ideo faciem coeli non invidet orbi
Ipse deus, vultusque suos corpusque recludit
Semper volvendo, seque ipsum inculcat et offert,
Ut bene cognosci possit, doceatque videndo
Qualis est, doceatque suos attendere leges.
Ipse vocat nostros animos ad sidera mundus . . .

This is quoted by Montaigne, *Essais* II, 12 (Paris: Didot, 1870), p. 223B.

Notes to Pages 28–32

25. Lactantius, *De ira dei* 14, 1–3: "... spectatorem operum rerumque caelestium ... ut rerum omnium dominaretur."

26. Lactantius, *Divinae institutiones* II 9, 25.

27. Irenaeus, *Adversus haereses* IV 20, 4; *Patrologia graeca*, ed. J. P. Migne (Paris: 1857–1894), vol. 7, col. 1034: "... qui et mundum hunc attribuit humano genere ...."

28. Irenaeus, *Adversus haereses* V 29, 1; *Patrologia graeca* (cited in note 27), vol. 7, col. 1201: "... omnia quae sunt talia, pro eo qui salvatur homine fata sunt .... Gentes autem quae et ipsae non allevaverunt oculos suos ad coelum ...."

29. Philo Alexandrinus, *De migratione Abrahami* 8ff.

30. Lactantius, *Divinae institutiones* III, 24.

31. On Satan's change of position in the course of the destruction of Gnosticism: M. Werner, *Die Entstehung des christlichen Dogmas* (2nd ed., Bern: P. Haupt, 1953), pp. 258–259.

32. Thomas Aquinas, *Expositio in Aristotelis De caelo* I, *lectio* 20, n. 199; ed. Spiazzi (Turin: Marietti, 1952), p. 98: "Consueverunt enim homines vocare caelum illud quod est extremum totius mundi, et quod maxime est sursum: non quidem secundum quod sursum accipitur in scientia naturali, prout scilicet est terminus motus levium (sic enim nihil magis est sursum quam locus in quem fertur ignis): sed sumitur hic sursum secundum communem modum loquendi, prout id quod est remotius a medio, vocatur sursum. Consuevit etiam vocari sursum id quod est locus omnium divinorum (ut tamen divina non dicantur hic corpora caelestia, quae non omnia sunt in suprema sphaera, sed secundum quod divina dicuntur substantiae immateriales et incorporeae): dictum est enim supra quod omnes homines locum qui est sursum attribuunt Deo."

33. Augustine, *De doctrina christiana* II 39, 59. Compare II 29, 45.

34. Augustine, *De doctrina christiana* II 39, 58: "In quibus omnibus tenendum est, Ne quid nimis; et maxime in iis quae ad corporis sensu pertinentia, volvuntur temporibus, et continentur locis."

35. Augustine, *De doctrina christiana* II 29, 46: "Quae per seipsam cognitio, quanquam superstitione non alliget, non multum tamen ac prope nihil adiuvat tractationem divinarum Scripturarum, et infructuosa intentione plus impedit; et quia familiaris est perniciosissimo errori fatua fata cantantium, commodius honestiusque contemnitur."

36. Augustine, *De doctrina christiana* II 40, 60: "... sed etiam liberales disciplinas usui veritatis aptiores, et quaedam morum praecepta utilissima continent, deque ipso uno Deo colendo nonnulla vera inveniuntur apud eos; quod eorum tanquam aurum et argentum, quod non ipsi instituerunt, sed de quibusdam quasi metallis divinae providentiae, quae ubique infusa est, eruerunt et quo perverse atque iniurose ad obsequia daemonum abutuntur, cum ab eorum misera societate sese animo separat, debet ab eis auferre christianus ad usum iustum praedicandi Evangelii."

37. Boethius, *De geometria*; *Patrologia latina*, ed. J. P. Migne (Paris, 1844–1864), vol. 63, col. 1353 AB: "Utilitas geometriae triplex est, ad facultatem, ad sanitatem, ad animam .... Ad animam ut philosophi. Quam artem si arcte et diligenti cura atque moderata mente perquirimus, hoc, quod praedictus divisionibus manifestum est, sensus nostros magna claritate dilucidat, et illud supra, quale est caelum, animo subire, totamque illam machinam supernam indagabili ratione aliter discutere et inspectiva mentis sublimitate ex aliqua parte colligere et agnoscere mundi factorem, qui tanta et talia arcana velavit."

38. Cassiodorus, *Institutiones divinarum et saecularium lectionum II. De artibus ac disciplinis liberalium litterarum; Patrologia latina* (cited in note 37), vol. 70, col. 1216 C: "... si casta ac moderata mente perquirimus (sc. astronomiam), sensus quoque nostros, ut veteres dicunt magna claritate perfundit, quale est enim ad caelos animo subire, totamque illam machinam supernam indagabili ratione discutere et inspectiva mentis sublimitate ex aliqua parte colligere, quod tantae magnitudinis arcana velaverunt."

39. Odo of Tournai, *Liber de restauratione monasterii S. Martini Tornacensis*, Monumenta Germaniae Historica, Scriptorum, vol. 14: 274 (reference due to O. Meyer): "... iam vero, si scolae appropriares, cerneres magistrum Odonem nun quidem Peripateticorum more cum discipulis docendo deambulantem, nun vero Stoicorum instar residentem et diversas quaestiones solventem, vespertinis quoque horis ante ianuas ecclesiae usque in profundam noctem disputantem et astrorum cursus digiti protensione discipulis ostendentem zodiacique seu lactei circuli diversitates demonstrantem ...."

40. William of Ockham, *In I. Sententiarum*, Prologus 1 HH: "... si videam intuitive stellam existentem in coelo ... illa visio potest manere stella destructa."

41. William of Ockham, *In I. Sententiarum* 17. 8 F: "... si deus conservaret corpora coelestia in eodem statu quo sunt, in infinitum continuaretur generatio illorum generabilium et corruptibilium."

Chapter 3

42. Giovanni Pico della Mirandola, *Heptaplus* VI, 4; ed. E. Garin (Florence: Vallechi, 1942), p. 318: "Similiter est finis secundarius principali appendens et annexus, quod sapienter significat (sc. Moyses) dicens posita sidera ut lucerent in caelo et illuminarent terram. Neque enim bonum inferiorum primarius finis est caelestium. Sed id primum intendunt, ut sibi luceant, tum postremo ut et nos illuminent."

43. Pico della Mirandola, *Heptaplus* II, 4; ed. Garin (see note 42), p. 236: "... in quem usum fundata et cui muneri delegata a Deo fuerint (sc. sidera) .... Caelestium corporum duae in universum manifestae operationes: motus et illuminatio ... statuta illa ut lucerunt in caelo et terram illuminarent."

44. Pico della Mirandola, *Heptaplus* II, 7; ed. Garin (see note 42), p. 242: "... videamus ne nos illorum servos velimus, quos nos fratres esse natura voluit ... sed animus est, sed intellectus, qui omnem ambitum caeli, omnem decursum temporis excedit."

45. Jacob Burckhardt, *Gesammelte Werke* (Basel: Schwabe, 1955–1959), vol. 3, p. 352.

46. *Briefwechsel der berühmtesten Gelehrten des Zeitalters der Reformation mit Herzog Albrecht von Preussen*, ed. J. Voigt (Königsberg: 1841), p. 112. An assessment of the 'suspicion' of astrological activity [on the part of Copernicus] that arises here will have to be given later.

47. Fontenelle, *Dialogues des morts* (1683): "Molière et Paracelse"; *Oeuvres complètes*, ed. G.-B. Depping (Paris: 1818; rpt. Geneva: Slatkine, 1968), vol. 2, p. 246: "Je suis persuadé que si la plupart des gens voyaient l'ordre de l'univers tel qu'il est, comme ils n'y remarqueraient ni vertus des nombres, ni propriétés des planètes, ni fatalités attachées à de certains temps ou à de certaines révolutions, ils ne pourraient pas s'empêcher de dire sur cet ordre admirable: Quoi! n'est-ce que cela?"

48. Copernicus, *De revolutionibus* I, prooemium: "Quid autem caelo pulcrius, nempe quod continet pulcra omnia?"

49. Copernicus, *De revolutionibus* I, 9: "quae omnia ratio ordinis, quo illa sibi invicem succedunt, et mundi totius harmonia nos docet, si modo rem ipsam ambobus, ut aiunt, oculis inspiciamus."

50. Galileo, "Lettere intorno il sistema Copernicano," no. 5, *Opere*, ed. E. Alberi (Florence: 1842–1856), vol. 2, pp. 45–46. [My translation differs from the long-standard English translation of Thomas Salusbury—printed as "Letter to the Grand Duchess Christina" in *Discoveries and Opinions of Galileo*, ed. Stillman Drake (Garden City, New York: Doubleday, 1957), p. 196— in details suggested by the German translation that the author used: Ernst Cassirer, *Das Erkenntnisproblem in der Philosophie und Wissenschaft der neueren Zeit* (Berlin: B. Cassirer, 1911), vol. 1, p. 272.]

51. Bacon, *De dignitate et augmentis scientiarum* V, 4; *Works*, ed. J. Spedding, R. Ellis, and D. Heath (London: 1857–1874; rpt. Stuttgart-Bad Cannstatt: Frommann-Holzboog, 1963), vol. 1, pp. 644–645: "Nam si summus ille opifex ad modum aedilis se gessisset, in pulchrum aliquem et elegantem ordinem stellas digerere debuisset ... cum e contra aegre quis ostendat in tam infinito stellarum numero figuram aliquam vel quadratam, vel triangularem, vel rectilinearem. Tanta est harmoniae discrepantia inter spiritum hominis et spiritum mundi."

52. Cicero, *De natura deorum* I 9, 22; trans. H. Rackham, Loeb Classical Library (London: Heinemann; Cambridge, MA: Harvard University Press, 1933), p. 25.

53. Bacon, *Novum organum* I, 129; *Works* (cited in note 51), vol. 1, p. 222: "... ut non imperium aliquod, non secta, non stella, majorem efficaciam et quasi influxum super res humanas exercuisse videatur, quam ista mechanica exercuerunt." Cf. *Novum organum* II, 5 (*Works*, vol. 1, p. 232).

54. Nicholas of Cusa, *Compendium* VIII. On this see H. Blumenberg, *The Legitimacy of the Modern Age* (Cambridge, MA: MIT Press, 1983), p. 536; or: *Die Legitimität der Neuzeit* (Frankfurt: Suhrkamp, 1966), p. 510.

55. Hobbes, *De corpore* VII, 1; *Works*, ed. W. Molesworth (London: 1839–1845), vol. 1, p. 81: "Doctrinae naturalis exordium, optime ... a privatione, id est, a ficta universi sublatione, capiemus. Supposita autem tali rerum annihilatione, quaeret fortasse aliquis, quid reliquum esset, de quo homo aliquis (quem ab hoc universo rerum interitu unicum excipimus) philosophari, vel omnino ratiocinari, vel cui rei nomen aliquod ratiocinandi causa imponere posset.— Dico igitur, remansuras illi homini, mundi et corporum omnium ... ideas, id est memoriam imaginationemque .... "

56. Henry More, *Psychathanasia and Democritus Platonissans*, critical edition by L. Haring, dissertation Columbia University, 1961. See C. A. Staudenbaur, "Galileo, Ficino, and Henry More's Psychathanasia," *Journal of the History of Ideas* 19 (1968): 565–578.

57. Diderot, article "Encyclopédie," in *Encyclopédie ou Dictionnaire raisonné des sciences, des arts at des métiers*, vol. 5 (Paris, 1755), pp. 640D–641A: "Et celui (sc. système) où l'on descendroit de ce premier être éternel, à tous les êtres qui dans le tems émanerent de son sein, ressembleroit à l'hypothese astronomique dans laquelle le philosophe se transporte en idée au centre du soleil, pour y calculer les phénomenes des corps célestes qui l'environnent .... " Heidegger described Hegel's philosophy with the metaphor of this Copernican 'transference' into the Sun: Kant, he wrote, was not able to construct a metaphysics, because such a task "demands nothing less than to jump over one's own shadow," but Hegel "apparently succeeded in jumping over this shadow, but only in such a way that he eliminated the shadow, i.e., the finiteness of man, and jumped into the sun itself." *What Is a Thing?*, trans. W. B. Barton, Jr., and V. Deutsch (Chicago: Regnery, 1967), p. 150; *Die Frage nach dem Ding* (Tübingen: Niemeyer, 1962), pp. 117–118.

58. Diderot, article cited in note 57, p. 641A: "... car quelle différence y auroit-il entre la

lecture d'un ouvrage où tous les ressorts de l'univers seroient développés, et l'étude même de l'univers? presqu'aucune . . . . ''

59. Diderot, article cited in note 57, p. 647D.

Chapter 4

60. Kant to Johann Friedrich Reichardt, 15 October 1790; *Werke*, ed. E. Cassirer (Berlin: B. Cassirer, 1912–1918), vol. 10, p. 55.

61. Kant, *Critique of Judgement*, section 91: "General Remark on Teleology"; trans. J. C. Meredith (Oxford: Oxford University Press, 1952), "Critique of Teleological Judgement," pp. 152–153.

62. Kant, *Critique of Practical Reason*, "Conclusion"; trans. L. W. Beck (Indianapolis: Bobbs-Merrill, 1956), p. 166; *Werke*, ed. E. Cassirer (Berlin: B. Cassirer, 1912–1918), vol. 5, p. 174.

63. Salomon Maimon, *Philosophisches Wörterbuch* (Berlin, 1791), pp. 30–31.

64. Kant, *Critique of Judgement*, section 23; trans. J. C. Meredith, "Critique of Aesthetic Judgement," p. 91.

65. Kant, *Critique of Judgement*, section 29: "General Remark upon the Exposition of Aesthetic Judgements"; trans. J. C. Meredith, "Critique of Aesthetic Judgement," p. 118 [subsequent quotations are from pp. 120–122].

Chapter 5

66. Friedrich Schlegel, *Transcendentalphilosophie, Erster Teil: Theorie der Welt*; in *Neue philosophische Schriften*, ed. J. Körner (Frankfurt: Schulte-Bulmke, 1935), p. 156.

67. F. Schlegel, *Sämtliche Werke* (Vienna: 1846), vol. 12, p. 149.

68. F. Schlegel, *Aus dem Nachlass*, ed C. J. H. Windischmann (Bonn: 1837), vol. 2, p. 201.

69. F. Schlegel, *Transcendentalphilosophie, Zweiter Teil: Theorie des Menschen; Neue philosophische Schriften* (cited in note 66), p. 187.

70. F. Schlegel, *Philosophische Vorlesungen aus den Jahren 1804/06, Sämtliche Werke* (Vienna: 1846), vol. 2, p. 235.

71. F. Schlegel, *Kritische Ausgabe*, ed. E. Behler (Munich: Schöningh, 1958–), vol. 12, pp. 456–457.

72. F. Schlegel, *Kritische Ausgabe*, vol. 12, p. 460. "Perfectibility is not the character of man alone, but also of the Earth, and it is also man's character only insofar as he himself belongs to the Earth" (same vol., p. 470).

73. F. Schlegel, *Kritische Ausgabe*, vol. 12, pp. 462, 466.

74. F. Schlegel, *Kritische Ausgabe*, vol. 12, p. 463.

75. H. Heine, *From the Memoirs of Herr von Schnabelewopski*, in *The Sword and the Flame*, trans. C. G.

Leland (New York and London: T. Yoseloff, 1960), p. 162. Original: Heine, *Sämtliche Schriften*, ed. K. Briegleb (Darmstadt: Wissenschaftliche Buchgesellschaft, 1968–), vol. 1, p. 515.

76. Heine, *The Sword and the Flame*, pp. 164–165; original: *Sämtliche Schriften*, vol. 1, pp. 517–518.

77. *Gespräche mit Heine*, ed. H. H. Houben (Frankfurt: Rütten & Loening, 1926), pp. 484–485. On this see W. Dilthey, *Heinrich Heine* (1876), in *Gesammelte Schriften* (Leipzig: Teubner, 1914–1982), vol. 15, pp. 209–210. Dilthey adds to the account, which he gives word for word, this observation: "Whatever basis this story may have, it at least represents vividly how he felt vis-à-vis the dominant philosophical school." As an authentic substratum for the Berlin scene we find in the fourth book of Heine's *Ludwig Börne*, built into a context dealing with Börne's Jewish descent, a remark of Hegel's about the ambiguity of nature that entirely bypasses the paradigm of the starry heavens, perhaps in favor of a more drastic effect that is meant to hit others besides the admirers of 'sublime' nature: "Nature, Hegel once said to me, is very peculiar; the same instruments that it uses for the most sublime purposes, it also employs for the lowliest functions, for example the member to which the highest mission, the reproduction of mankind, is entrusted, also serves for — — —. Those who complain about Hegel's obscurity will understand him here, and if he did not in fact utter the above words in reference to Israel, still they can be applied to it." *Sämtliche Schriften* (cited in note 75), vol. 4, p. 119.

78. Heine, *Religion and Philosophy in Germany*, trans. J. Snodgrass (Boston: Beacon Press, 1959), p. 117. Original: *Sämtliche Schriften*, vol. 4, p. 603.

79. Heine, *Sämtliche Schriften*, vol. 4, p. 55.

80. K. Löwith, "Hegels Aufhebung der christlichen Religion," *Hegel-Studien*, Beiheft 1 (Bonn, 1964), p. 233. In contrast to the other citations in this essay, Löwith gives no indication of the source either of the Hegel quotation about the "rash of light" or of the use that Marx made of Hegel's "shocking saying that all the wonders of the starry heavens were nothing in comparison to the most criminal thought of a human being." In a youthful, romantically sympathetic "Lied an die Sterne" ["Song to the Stars"], Marx had begun as follows:

Und euer Schein ist Höhnen,
Für Tat und Schmerz und Drang . . .

[And your light is mockery for action, pain and stress . . .].

On Engels, his biographer Gustav Meyer writes (in *F. Engels. Eine Biographie* [Berlin: Springer, 1920], vol. 2, p. 299), "It was not in keeping with his nature to look up to the starry heavens as something fixed and unattainable; his way of viewing things was more accurately reflected by the idea that the heavens, too, are in the process of coming into being. Thus in Kant he admired the *Universal Natural History and Theory of the Heavens* more than the *Critique of Reason* . . . . "

81. Goethe, "Prometheus-Fragment" I, *Gedenkausgabe der Werke*, ed. E. Beutler (Munich: Artemis, 1949–), vol. 4, p. 188.

82. J. G. Fichte, "Gerichtliche Verantwortung gegen die Anklage des Atheismus," *Sämmtliche Werke* (Berlin, 1845–1846), vol. 5, pp. 268–269. Fichte's philosophy, J. B. Erhard writes to Niethammer on 16 June 1796, is "the greatest aberration of reason that fails to recognize its limits, but it is an aberration that is correct, because it brought about a system that is directly opposed to that of Spinoza . . . . God grant that Fichte is not persecuted, otherwise a real Fichteism could arise, which would be ten times worse than previous 'isms' of this kind" (*Denkwürdigkeiten des Philosophen und Arztes J. B. Erhard*, ed. K. A. Varnhagen von Ense [Stuttgart, 1830], pp. 423–424). The principle of this system, which "results from the ideal of philosophy as art," can be comprehended, Erhard says, in the sentence, "The purpose of the universe is to be material for the knowledge and active power of man . . . " (to Niethammer, 31 January 1797; *Denkwürdigkeiten*, p. 433).

83. Nietzsche, *Gesammelte Werke* (Munich: Musarion, 1920–), vol. 20, pp. 222–223.

84. F. W. Schelling, *Einleitung in die Philosophie der Mythologie*, 21st lecture, manuscript outline; in *Philosophie der Mythologie* (Stuttgart: 1858; rpt. Darmstadt: Wissenschaftliche Buchgesellschaft, 1957), vol. 1, p. 490.

85. Schelling, op. cit., vol. 1, p. 491.

86. Schelling, op. cit., 18th lecture; vol. 1, p. 427.

87. Schelling, op. cit., vol. 2, p. 430.

88. Schelling, op. cit., vol. 1, p. 493.

89. Schelling, op. cit., vol. 1, p. 502.

90. Schelling, op. cit., vol. 1, p. 425.

Chapter 6

91. L. Feuerbach, *Sämtliche Werke*, ed. W. Bolin and F. Jodl (Stuttgart: Frommann, 1903–1911; rpt. 1960–64), vol. 2, pp. 135–137.

92. Galileo, *Dialogue Concerning the Two Chief World Systems—Ptolemaic and Copernican*, trans. Stillman Drake (Berkeley and Los Angeles: University of California Press, 1967), p. 339. On Feuerbach's use of this Gailieo quotation, see H. Blumenberg, *The Legitimacy of the Modern Age* (Cambridge, MA: MIT Press, 1983), p. 445; or *Der Prozess der theoretischen Neugierde*, suhrkamp taschenbuch wissenschaft #24 (Frankfurt: 1973), pp. 264–265.

93. Feuerbach, *Nachgelassene Aphorismen, Sämtliche Werke*, vol. 10, p. 298.

94. Feuerbach, *Sämtliche Werke*, vol. 2, p. 137.

95. Feuerbach, *Die Naturwissenschaft und die Revolution* (1850), *Sämtliche Werke*, vol. 10, pp. 9–10.

96. K. Marx and F. Engels, *The German Ideology* (part One), ed. C. H. Arthur (New York: International Publishers, 1970), pp. 61 and 51; original: *Frühe Schriften*, ed. H.-J. Leiber and P. Furth (Stuttgart: Cotta, 1971), vol. 2, pp. 53 and 32. Compare Voltaire, *Pensées sur le bonheur* (*Voltaire's Notebooks*, ed. Th. Besterman [Geneva: Institut et Musée Voltaire, 1952], vol. 2, p. 462): "Des astronomes observent des étoiles, un paysan dit, ils ont beau faire, ils n'en seront jamais plus près que nous. Ainsi des raisonneurs sur le bonheur."

97. Feuerbach, *Wider den Dualismus von Leib und Seele, Fleisch und Geist, Sämtliche Werke*, vol. 2, pp. 349–350.

98. Feuerbach, *Todesgedanken, Sämtliche Werke*, vol. 1, pp. 36–38.

99. Feuerbach, *Zur Kritik der Hegelschen Philosophie* (1839), final paragraph; translated by Z. Hanfi in *The Fiery Brook: Selected Writings of Ludwig Feuerbach* (Garden City, NY: Doubleday, 1972), p. 93.

100. Feuerbach, *Nachgelassene Aphorismen, Sämtliche Werke*, vol. 10, pp. 332–333.

101. Feuerbach, *Grundsätze der Philosophie der Zukunft* (1843), section 7; trans. Z. Hanfi in *The Fiery Brook* (cited in note 99), p. 182.

Chapter 7

102. Alexander von Humboldt to K. A. Varnhagen von Ense, 24 October 1834, translated (translator unnamed) in *Letters of Alexander von Humboldt, Written between the Years 1827 and 1858, to Varnhagen von Ense* (London: 1860) (cited henceforth as *Letters*), pp. 17–18. [I have revised and modernized this translation in this and all the following quotations.] Original: *Briefe von Alexander von Humboldt an Varnhagen von Ense* (3rd ed., Leipzig: 1860) (henceforth: *Briefe*), p. 22.

103. "And do also ease my mind about the title." *Letters*, p. 19; *Briefe*, p. 23.

104. H. Beck, *Alexander von Humboldt* (Wiesbaden: F. Steiner, 1959–1961), vol. 2, p. 18.

105. Humboldt to Varnhagen, 28 April 1841, *Letters*, pp. 69–70; *Briefe*, pp. 91–92. On Humboldt's concept of "physical cosmography" see F. Kaulbach, *Philosophie der Beschreibung* (Cologne: Böhlau, 1968), pp. 378–391.

106. Humboldt to Varnhagen, 24 October 1834, *Letters*, p. 17; *Briefe*, pp. 21–22.

107. Humboldt to Varnhagen, 28 April 1841, *Letters*, p. 69; *Briefe*, p. 92. Here the Romantic idea of the fragment's indicating the totality still appears as a possible principle by means of which to fulfill the claim to convey a "total impression."

108. Humboldt to Varnhagen, 9 September 1858, *Letters*, p. 314; *Briefe*, p. 398.

109. Alexander von Humboldt, *Gespräche*, ed. H. Beck (Berlin: Akademie Verlag, 1959), pp. 58, 286.

110. Humboldt to Varnhagen, 30 November 1845, *Letters*, p. 144; *Briefe*, p. 186. To the same addressee, 22 April 1846 (*Briefe*, p. 215): "The main thing I strive for is composition, the mastery of great quantities of information that are brought together painstakingly and with precise expert knowledge. The wielding of our noble, pliant, harmonious, descriptive language is only a secondary consideration."

111. Humboldt to F. Althaus, 1852, in *Gespräche* (cited in note 109), p. 334. Humboldt observing stars through the giant refractor: H. Beck, *Alexander von Humboldt* (Wiesbaden: Steiner, 1959–1961), vol. 2, p. 321.

112. A. Schopenhauer, *Parerga and Paralipomena* II, section 80; trans. E. J. F. Payne (Oxford: Oxford University Press, 1974), vol. 2, pp. 126–127.

113. Schopenhauer, *The World as Will and Representation* I, section 14, trans. E. J. F. Payne (Indian Hills, CO: Falcon's Wing Press, 1958; rpt. New York: Dover, 1966), vol. 1, pp. 64–67. Original: *Die Welt als Wille und Vorstellung, Sämtliche Werke*, ed. W. von Lohneysen (Stuttgart: Cotta, 1960–1965), vol. 1, pp. 113–115. See also Schopenhauer, *Handschriftlicher Nachlass*, ed. A. Hübscher (Frankfurt: W. Kramer, 1966–), vol. 1, pp. 455–456.

114. P.-S. Laplace, *Essai philosophique sur les probabilités* (Paris: 1814): "L'esprit humain offre dans la perfection qu'il a su donner à l'astronomie, une faible esquisse de cette intelligence .... Tous ses efforts dans la recherche de la vérité, tendent à le rapprocher sans cesse de l'intelligence que nous venons de concevoir, mais dont il restera toujours infiniment éloigné."

115. Laplace, *Exposition du systéme du monde* (Paris: 1796), vol. 2, pp. 288–291.

116. Nietzsche, from the literary remains, 1882–1888; *Gesammelte Werke* (Munich: Musarion, 1920–), vol. 16, p. 257.

117. Nietzsche, preliminary studies for a work on "The Philosopher" (summer of 1873), trans. D. Breazeale in *Philosophy and Truth. Selections from Nietzsche's Notebooks of the early 1870s* (Atlantic Highlands, NJ: Humanities Press, 1979) (henceforth: *Philosophy and Truth*), under the title "On Truth and Lies in a Nonmoral Sense," p. 88. Original: *Gesammelte Werke*, vol. 6, p. 87.

118. Nietzsche, "The Philosopher," *Philosophy and Truth*, p. 30 (translation revised); original: *Gesammelte Werke*, vol. 6, p. 32.

119. Nietzsche, "The Philosopher," *Philosophy and Truth*, p. 38; original: *Gesammelte Werke*, vol. 6, p. 42.

120. Nietzsche, *Beyond Good and Evil*, part 9, section 285; trans. W. Kaufmann in *Basic Writings of Nietzsche* (New York: Random House, 1966), p. 417 (translation slightly revised). Original: *Gesammelte Werke*, vol. 15, p. 252.

121. By making it convertible into "centuries," Nietzsche has still not departed very far from what the Göttingen professor Abraham Gotthelf Kästner had depicted a century earlier: "Since the light by means of which I saw a star last night may have left it more than half a century ago, and because during this time no inhabitant of the Earth can have had any information about the star except by means of the light that arrives from it, so yesterday I may have very painstakingly and accurately observed a star that was extinguished, burst asunder, perished, in some way unknown to me, before I ever heard the word *astronomy* pronounced." *Über sinnliche Wahrheit und Erscheinung* (1771), *Gesammelte Werke* (Berlin, 1841), vol. 3, p. 30.

122. Nietzsche, *Beyond Good and Evil*, part 5, section 196, trans. W. Kaufmann in *Basic Writings of Nietzsche*, p. 298. Original: *Gesammelte Werke*, vol. 15, p. 122.

123. I. Kant, *Critique of Practical Reason*, conclusion; trans. L. W. Beck (Indianapolis: Bobbs-Merrill, 1956), p. 166. Original: *Gesammelte Schriften*, Akademie ed. (Berlin, 1900–1942), vol. 5, p. 162.

124. Nietzsche, *Beyond Good and Evil*, part 4, section 71; trans. W. Kaufmann, *Basic Writings of Nietzsche*, p. 270. Original: *Gesammelte Werke*, vol. 15, p. 90.

125. Nietzsche, *Daybreak* 5, section 547; trans. R. J. Hollingdale (Cambridge: Cambridge Unversity Press, 1982), pp. 219–220. Original: *Gesammelte Werke*, vol. 10, pp. 339–340.

126. Nietzsche, "The Philosopher," *Philosophy and Truth*, p. 42 (trans. revised); original: *Gesammelte Werke*, vol. 6, p. 46.

127. Nietzsche, "On the Pathos of Truth" (1872), *Philosophy and Truth*, p. 65; original: *Gesammelte Werke*, vol. 4, pp. 145–146. The myth returns—in an abbreviated version, in which merely "someone" could invent the fable—in the fragment of the "Philosopher" book, from the summer of 1873, entitled "On Truth and Lies in a Nonmoral Sense," *Philosophy and Truth*, p. 79; original: *Gesammelte Werke*, vol. 6, p. 75.

128. Nietzsche, "On Truth and Lies in a Nonmoral Sense," *Philosophy and Truth*, p. 79; original: *Gesammelte Werke*, vol. 6, p. 76.

129. Nietzsche, from the literary remains, 1882–1888, *Gesammelte Werke*, vol. 16, p. 260.

130. Nietzsche, ibid., p. 64.

131. H. Blumenberg, *Schiffbruch mit Zuschauer*, suhrkamp taschenbuch wissenschaft 289 (Frankfurt: Suhrkamp, 1979), pp. 28–30.

132. Nietzsche to C. von Gersdorff, 21 June 1871; trans. C. Middleton in *Selected Letters of*

Friedrich Nietzsche (Chicago and London: University of Chicago Press, 1969), p. 81. Original: *Nietzsche's Briefe*, ed. R. Oehler (Leipzig: Insel, 1917), pp. 119–120.

Chapter 8

133. K. A. Varnhagen von Ense, *Tagebücher*, ed. L. Assing (Hamburg, 1869), vol. 11, p. 43.

134. P. Natorp, *Hermann Cohens philosophische Leistung unter dem Gesichtspunkt des Systems* (Berlin: Reuther und Reichard, 1918), p. 16.

135. A. Stifter, *Winterbriefe aus Kirchschlag*, no. 6; *Sämtliche Werke*, vol. 15, ed. G. Wilhelm (Reichenberg: Kraus, 1935), pp. 279–284.

136. J.-B. de Boyer, marquis d'Argens, *Lettres juives* (1738), no. 31; trans. (anon.) as *The Jewish Spy*, vol. 1 (London, 1776), pp. 229–230: Ninus "boasted that he had never seen the Stars, nor desired to see them, and that he despised the Sun, the Moon, and all the other Deities." Jean Paul, *Huldigungs-predigt vor und unter dem Regierungsantritt der Sonne, gehalten am Neujahrstag 1800*: "I know as well as anyone that King Ninus said that he had never seen the stars...." (*Werke*, ed. N. Miller [Munich: Hanser, 1960–], vol. 6, p. 140). Giovanni Pico della Mirandola takes the ancient report that Ninus, the opponent of observing the stars, was able to defeat in war the Bactrian king, Zoroaster, the inventor of all the magic and astrological arts, as an archaic argument against astrology (*Disputationes adversus astrologiam divinatricem* II 2, ed. E. Garin [Florence: Vallechi, 1946–1952], vol. 2, p. 108): "Quis Zoroastre inter antiquos in astrologia, in magicis et in omni superstitione potentior? Et tamen eum in bello et vicit et occidit Ninus, nec astrologus utique nec magus." Pico's positive interpretation of Ninus at the expense of Zoroaster, in view of the outcome of the war, may, however, be an inversion of the ancient typification, for Zoroaster was the great sage on the throne, to make war against whom and to kill him could only have occurred to the demonic despiser of higher realities, the crowned *banausos* [mechanic] Ninus. The ancient source is the quotation from a poem of Phoenix of Colophon given by Athenaeus, *Deipnosophistai* XII 40, ed. S. P. Peppinki (Leiden: E. J. Brill, 1936–1939), 530 c.

137. F. Hebbel, diary, no. 6279 (11 October 1862).

138. P. E. Schramm, introduction to *Hitlers Tischgespräche im Führer-Hauptquartier 1941–1942* (Stuttgart: Seewald, 1963), p. 79.

139. *Søren Kierkegaard's Journals and Papers*, ed. and trans. H. V. Hong and E. H. Hong (Bloomington and London: Indiana University Press, 1967–1975), vol. 2, p. 534 (translation slightly revised). Original: *Papirer* (Copenhagen: Gyldendal, 1968–1978) VII¹ A234. [The later quotation is from *Journals and Papers*, vol. 2, p. 533; *Papirer* VII¹ A 230.]

140. Mark Twain, *Letters from the Earth*, ed. B. DeVoto (New York and Evanston: Harper & Row, 1962), p. 15.

141. Max Planck, *Wege zur physikalischen Erkenntnis* (Leipzig: Hirzel, 1943), vol. 2, p. 107.

142. Op. cit., vol. 1, p. 169.

143. The [author's original] Haeckel quotations are taken from the presentation, indispensable here as elsewhere, of H. Lübbe, *Politische Philosophie in Deutschland* (Basel: Schwabe, 1963), p. 144. [The translation—E. Haeckel, *The Riddle of the Universe* (New York and London: Harper, 1950), p. 344—is by J. McCabe.]

144. *Hitler's Secret Conversations 1941–1944*, trans. N. Cameron and R. H. Stevens (New York:

Farrar, Straus and Young, 1953), p. 417 (translation revised). Original: *Hitlers Tischgespräche* (cited in note 138), p. 388.

145. Joseph Roth, *Der Antichrist*, in *Romane, Erzählungen, Aufsätze* (Cologne: Kiepenhauer & Witsch, 1970), pp. 616–706. See also the latter to René Schickele of 31 January 1934 (*Briefe 1911–1939*, ed. H. Kesten [Cologne: Kiepenhauer & Witsch, 1970], p. 312): "The book that I am writing now is called *The Antichrist*. And the individual sections contain, in fact, all the forms in which he appears. And just this is the purport of my book: The Antichrist is friend and enemy. And in the end a little bit of him is in me as well."

Part II

Chapter 1

1. On this note, which L. A. Birkenmajer first alluded to in 1900, we now have Z. Wardeska, "Copernicus und die deutschen Theologen des 16. Jahrhunderts," in *Nicolaus Copernicus zum 500. Geburtstag*, ed. F. Kaulbach, V. W. Bargenda, and J. Blühdorn (Cologne: Böhlau, 1973), pp. 155–156.

2. G. Gamauf, *Erinnerungen aus Lichtenbergs Vorlesungen uber Astronomie* (Vienna: 1814), p. 165: "Still more fortunate was Bradley's observation that light required just this time of 16 minutes and 15 seconds to traverse the diameter of the Earth's orbit. Finally he hit upon the happiest idea of all, of combining the motion of the light with the motion of the Earth according to the laws of the combination of motions, and the great discovery, according to which every star in the heavens proves that the earth revolves around the Sun and proclaims Copernicus's glory, was made."

3. Goethe, *Werke*, ed. E. Beutler (Munich: Artemis, 1948–), vol. 17, p. 755.

4. I. Bernard Cohen, *The Birth of a New Physics* (Garden City, NY: Doubleday, 1960), p. 37. This account is strictly constructed according to the basic idea of seeing Newton as the logical consequence of Copernicanism.

5. H. Blumenberg, "Self-Preservation and Inertia: On the Constitution of Modern Rationality," *Contemporary German Philosophy* 3 (1983): 209–256. Original: "Selbsterhaltung und Beharrung. Zur Konstitution der Neuzeitlichen Rationalität," *Abhandlungen der Akademie zu Mainz, Geistes- und sozialwiss.-kl.* (1969), no. 11.

Chapter 2

6. Jean Buridan, *Quaestiones de caelo et mundo*, ed. E. A. Moody (Cambridge, MA: Medieval Academy of America, 1942), pp. 226–233.

7. J. A. Weisheipl, "The Principle *Omne quod movetur ab alio movetur* in Medieval Physics," *Isis* 56 (1965): 26–45.

8. Franciscus de Marchia's Question on this subject was edited by A. Maier in her *Zwei Grundprobleme der scholastischen Naturphilosophie*, 2nd ed. (Rome: Edizioni di storia e letteratura, 1951), pp. 166ff. Cf. M. Clagett, *The Science of Mechanics in the Middle Ages* (Madison: University of Wisconsin Press, 1961), pp. 526–531 [a partial translation with commentary].

9. "Sed contra hoc arguo et ostendo, quod huiusmodi virtus primus sit in lapide vel in quocumque alio gravi moto quam in medio" [Maier, *Zwei Grundprobleme*, p. 170].

Notes to Pages 149–158

10. "Unde sit sciendum, quod est duplex virtus movens aliquod grave sursum, quaedam motum inchoans sive grave ad motum aliquem determinans et ista virtus est virtus manus; alia virtus est motum exequens inchoatum et ipsum continuans et ista est causata sive derelicta per motum a prima" [Maier, *Zwei Grundprobleme*, p. 172]. To understand the text we need to relate *causata* to *a prima* (namely, *virtute*), because either the second power is directly conditioned by the first or it is, as it were, incorporated as a result of the motion that the first produces.

11. "... iste motus violentus simpliciter per comparationem ad formam naturalem gravis, sed naturalis secundum quid comparando ipsum ad illam virtutem accidentalem derelictam in gravi a prima virtute motiva" [Maier, *Zwei Grundprobleme*, p. 176].

12. "... videtur posse concludi quod caelum motum ab animo recipiat aliquam virtutem sive formam ab ipso neutram, accidentalem, aliam a motu locali, caelo formaliter inhaerentem" [Maier, *Zwei Grundprobleme*, p. 177].

13. E. Gilson, *La Philosophie au Moyen Age*, 2nd ed. (Paris Payot, 1947), p. 271. Cf. M. Clagett, *The Science of Mechanics in the Middle Ages* (cited in note 8), p. 515.

14. S. Pinès, "Saint Augustin et la theorie de l'Impetus," *Archives d'histoire doctrinale et littéraire du Moyen Age* 44 (1969): 7–21. Only on account of the slightly Aristotelian character of its context does *impetus* in Augustine seem to be specific, that is, connected to a 'theory.' But, after all, Lucretius and Cicero had already used the term for the 'swinging around' of the heavenly bodies.

15. A. Maier has shown, in *An der Grenze von Scholastik und Naturwissenschaft* (Essen: Essener Verlagsanstalt, 1943), pp. 181ff., that Thomas Aquinas already prepares the way for considering the acceleration of a falling body from the point of view of its increasing distance from the point of departure of the fall, and that such a treatment is explicitly present in Aegidius Romanus.

16. Jean Buridan, *Supra libros Metaphysices* II q. 1: "Utrum de rebus sit nobis possibilis comprehensio veritatis."

17. Buridan, *Questions on the Heavens and the World* II q. 12, trans. M. Clagett, *The Science of Mechanics in the Middle Ages* (cited in note 8), p. 561 (translation slightly revised); original: *Quaestiones de caelo et mundo*, ed. Moody (cited in note 6), pp. 180–181: "Et sic aliquis posset imaginari, quod non oporteat ponere intelligentias moventes corpora caelestia, quia nec habemus ex scriptura sacra quod debeant poni. Posset enim dici quod quando deus creavit sphaeras caelestes, ipse incepit movere unamquamque earum sicut voluit; et tunc ab impetu quem dedit eis, moventur adhuc, quia ille impetus non corrumpitur nec diminuitur, cum non habeant resistentiam."

18. Loc. cit.: "... et forte si mola semper duraret sine aliqua eius diminutione vel alteratione, et non esset aliqua resistentia corrumpens impetum, mola ab illo impetu perpetue moveretur."

19. A. Maier, *Zwischen Philosophie und Mechanik* (Rome: Edizioni di storia e letteratura, 1958), pp. 223–224.

20. Buridan, *Commentary on the Physics* VIII q. 12, cited by A. Maier, op. cit., p. 365: "Sed hoc non dico assertive, sed ut a dominis theologis petam, quod in illis doceant me quomodo possunt haec fieri."

21. In his *Commentary on the Metaphysics* XII q. 9, Buridan expressed the way God distanced Himself from the world as follows: "Et in motibus caelestibus nullum est resistens contrarium, ideo cum in creatione mundi deus quamlibet sphaeram movit qua velocitate voluit, ipse cessavit a movendo, et per impetum illis sphaeris impressum semper postea duraverunt illi motus." (Quoted by A. Maier, *Die Vorläufer Galileis im 14. Jahrhundert* [Rome: Edizioni di storia e

letteratura, 1949], p. 147. There she also gives an important reference to Robert Kilwardby's very much earlier attempt to establish an analogy between the natural motions of falling objects and of the heavens: "Unicuique enim stellae vel orbi indidit deus inclinationem quasi proprii ponderis ad motum quem peragit . . . .")

22. Copernicus, *De revolutionibus* I, 8; *On the Revolutions of the Heavenly Spheres*, trans. A. M. Duncan (Newton Abbot, England: David & Charles; New York: Barnes and Noble, 1976), p. 43.

23. Nicole Oresme, *Le Livre du ciel et du monde* (completed 1377), ed. A. Menut and A. Denomy in *Medieval Studies* 3 (1941) and 5 (1943); rptd. with trans. by A. Menut (Madison: University of Wisconsin Press, 1968). See also F. Fellmann, *Scholastik und kosmologische Reform* (Münster: Aschendorff, 1971), pp. 20–38: "Das Problem der natürlichen Bewegung." Francis Bacon counts the Aristotelian axiom, "Motus singulis corporibus unicus et proprius," as one of the *idola theatri*, because Aristotle imposed it *arbitrio suo naturae rerum* (*Novum organum* I, 63).

24. F. Fellmann, op. cit., p. 42. [*Livre du ciel et du monde* (cited in note 23) 68 b, c.]

25. "La vertu qui ainsi meust en circuite ceste basse partie du monde, ce est sa nature, sa forme et est ce meisme qui meut la terre a son lieu quant elle en est hors" (Fellmann, op. cit., p. 33). [*Livre du ciel et du monde* 141 b, c.]

26. Buridan, *Quaestiones de caelo et mundo* II q. 10, ed. Moody (cited in note 6), p. 172: "Nos teneremus quod nulla est simpliciter necessaria subordinatio agentium nisi ad ipsum deum."

27. John Wilkins, *A Discourse Concerning a New Planet* (London: 1684 [first ed. 1640]), p. 158. [Later quotations from p. 159.] The German translation by J. G. Dopplemayr, *Vertheidigter Copernicus, oder curioser und gründlicher Beweiss der Copernicanischen Grundsätze* (Nuremberg: 1713) [which the author quoted in the original, German edition of this book], combines two works by Wilkins, *The Discovery of a World in the Moon* (London: 1638) and *Discourse Concerning a New Planet* (London: 1640). Wilkins's collected *Mathematical and Philosophical Works* were still found worthy of publication in London in 1708.

Chapter 3

28. *Voltaire's Notebooks*, ed. Th. Besterman (Geneva: Institut et Musée Voltaire, 1952), vol. 2, p. 512.

29. Copernicus, *De revolutionibus orbium caelestium*, praefatio authoris: " . . . coepit me taedere, quod nulla certior ratio motuum machinae mundi, qui propter nos, ab optimo et regularissimo omnium opifice conditus esset, philosophis constaret . . . ." Op. cit. I, 4: " . . . in optima sunt ordinatione constituta . . . ." I, 10: "Invenimus igitur sub hac ordinatione admirandam mundi symmetriam . . . ."

30. Op. cit., praefatio authoris: " . . . statim me explodendum cum tali opinione clamitent . . . tamen alienas prorsus a rectitudine opiniones figiendas censeo."

31. Ambrose, *De officiis ministrorum* I, 28; ed. J. Krabinger (Tübingen: 1857), pp. 132ff. (on Cicero, *De officiis* I 7, 22).

32. Ambrose, *De officiis ministrorum* I, 28: "Ergo omnia subiecta esse homini de nostris didicerunt et ideo censent propter hominem esse generata."

33. Cicero, *De natura deorum* I 2, 3.

34. Cicero, *De natura deorum* I 41, 116.

35. Cicero, *De natura deorum* II 13, 37. Cicero speaks of the *tributa* that men and gods perform for each other, the latter (to the former) that of serviceability, through nature, and the former (to the latter) those of *religio* (*cultus, honores, preces*). The basic idea of the 'same standards' for gods and men is already found in Plato, *Gorgias* 507E–508A. Accordingly, the natural-law *koinonia* [intercourse] between gods and men does not know the institution of punishment according to conventions, or the absolutist decree: The Stoics' cosmos perishes in the *ekpyrosis* [burning out] as a result of the exhaustion of its productive power. This eschatology is still teleological: The burning out has the same cause as the dissolution of order, namely, the degeneration of fire from a productive to a destructive power.

36. Cicero, *De natura deorum* II 14, 37: "Ipse autem homo ortus est ad mundum contemplandum et imitandum, nullo modo perfectus, sed est quaedam particula perfecti."

37. *Stoicorum veterum fragmenta*, ed. H. von Arnim (Leipzig: Teubner, 1921–1924), no. 245 (Zeno), no. 1153 (Chrysippus).

38. Lactantius, *Divinae institutiones* II 5, 6: "(Sidera) quae cum videant (sc. Stoici) divinis legibus obsequentia commodis atque usibus hominum perpetua necessitate famulari, tamen illa deos existimant esse, ingrati. . . . " On the serviceability of the stars: *De ira dei* 10, 35 and 13, 3; *Divinae institutiones* II 5, 24 and VI 2, 3–4. But at the same time an important criticism of the Stoics' manner of expression: "Quanquam in hoc ipso non mediocriter peccent, quod non hominis causa dicunt, sed hominum. Unius enim singularis appellatio totum comprehendit humanum genus. Sed hoc ideo, quia ignorant, unum hominem a deo esse formatum, putantque, homines in omnibus terris et agris tanquam fungos esse generatos."

39. Augustine, *De libero arbitrio* III, 13; "Quidquid enim tibi vera ratione melius occurrerit, scias fecisse Deum tanquam bonorum omnium conditorem . . . . Humana quippe anima naturaliter divinis ex quibus pendet connexa rationibus, cum dicit, Melius hoc fieret quam illud; si verum dicit, et videt quod dicit, in illis quibus connexa est rationibus videt. Credat ergo Deum fecisse quod vera ratione ab eo faciendum fuisse cognovit, etiamsi hoc in rebus factis non videt."

40. Augustine, *De diversis quaestionibus* LXXXIII, q. 30.

41. Augustine, *De diversis quaestionibus* q. 28.

42. Augustine, *De Genesi contra Manichaeos* I 2, 4.

43. Aristotle, *Politics* I 8; 1256 b 15–22.

44. Aristotle, *Nichomachean Ethics* VI 7; 1141 a 20–b 3; and *Metaphysics* XI 7; 1064 a 28–b 14. For the contrast to Socrates: Xenophon, *Memorabilia* IV 3, 8. Critique of Anaxagoras because he called man the most rational being: *De partibus animalium* 687 a 10. On the problem of teleology: W. Theiler, *Zur Geschichte der teleologischen Naturbetrachtung bis auf Aristoteles* (Zurich: K. Hoenn, 1924). Also see F. Dirlmeier's commentary in his translation of Aristotle, *Nikomachische Ethik* (Berlin: Akademie Verlag, 1956), pp. 327, 453, 520–521 (on *Rhetoric* II 17; 1391 b 2).

45. Aristotle, *De caelo* II 13; 293 a 21–b 14.

46. Blasius of Parma, *Quaestiones de anima* (1385): "utrum anima intellectiva possit a corpore separari" (cited by A. Maier, *Die Vorläufer Galileis im 14. Jahrhundert* [Rome: Edizioni di storia e letteratura, 1958], p. 287). Only in 1516 did Pietro Pomponazzi, in his *Tractatus de immortalitate animae*, again pronounce the Averroistic thesis of the mortality of the individual soul in a comparably blunt manner. Blasius, "qui inter philosophos est tamquam sol inter planetas" (according to a contemporary source quoted by Maier, p. 280), taught philosophy, mathematics, and astronomy in Pavia, Padua, and Bologna between 1374 and 1411, so he is hardly a marginal figure in Scholasticism.

47. Cl. Baeumker, "Petrus de Hibernia, der Jugendlehrer des Thomas von Aquino und seine Disputation vor König Manfred," *Sitzungsberichte der Bayerischen Akademie der Wissenschaften, Philosoph.-philolog. und historische Klasse*, 1920 series, no. 8. Baeumker characterizes Peter of Ireland as belonging "to the more advanced stage of the Aristotelian movement" (p. 33), and describes his approach to science as "exclusively philosophical" (p. 34). Cf. M. Grabmann, *Mittelalterliches Geistesleben* (Munich: Hueber, 1926–), vol. 1, pp. 249–265.

48. Peter of Ireland, in Baeumker, op. cit.: "Et ita ordinavit nautra universalis omnia propter aliquod iuvamentum et maxime propter iuvamentum et sustentamenta hominum."

49. Aristotle, *Physics* II 2; 194 a 32–36. (The passage is a quotation from his own, lost dialogue *On Philosophy*.)

50. John Argyropylos translates the *hōs hēmōn heneka*, which decides the meaning of the whole passage, with *hōs* as a causal conjunction: "et utimur, quia omnia sunt gratia nostri." Baeumker, who quotes both Aristotle and the *translatio arabico-latina prima* (according to A. Jourdain) on this passage, did not notice the differences of meaning (or did not find them worth remarking on). Averroes's commentary on the passage, which Peter knew, already formalizes the teleological idea in the juristic sense: "... sicut dicimus quod illud cuius est res est finis rei; et secundum hoc dicimus quod homo est finis rerum creatarum propter ipsum" (quoted by Baeumker in the work cited in note 47).

51. Thomas Aquinas, *In libros Physicorum Aristotelis expositio* II, 4; ed. P. M. Maggiòlo (Turin: Marietti, 1965), pp. 86ff.

52. Loc. cit.: "... nos utimur omnibus quae sunt secundum artem facta, sicut propter nos existentibus. Nos enim sumus quodammodo finis omnium artificialium .... Sicut finis domus ut 'cuius' est habitator, ut 'quo' est habitatio."

53. Thomas Aquinas, *Summa theologica* I q. 70 a.2: "Utrum convenienter causa productionis luminarium describatur." The passage in Lactantius is *Divinae institutiones* II 5, 12–14.

54. Thomas Aquinas, *Summa theologica* I q. 65 a.2: "Nihil tamen prohibet dici, quod dignior creatura facta est propter inferiorem, non secundum quod in se consideratur, sed secundum quod ordinatur ad integritatem universi."

55. Thomas Aquinas, *De veritate* q. 14 a.9 ad 8.

56. Theophrastus, *Metaphysics* 5 b 10ff. (Pseudo-) Aristotle, *De mundo* 6; 397 b–398 a 6, is a defense of the Aristotelian god, with His incapacity for an 'immediate' relation to the Earth (for *autourgein*), against the charge of ineffectiveness, but also an assertion that His *dynamis* reaches all the way down. Cf. Plotinus, *Enneads* V 3, 12, and VI 3, 6; Tatian, *Logos* II 8–9. The Dante reference is *Paradiso* 17, 37.

57. Thomas Aquinas, *Summa theologica* I q. 32 a.1 ad 2: "... sicut in astrologia ponitur ratio excentricorum et epicyclorum, ex hoc quod hac positione facta possunt salvari apparentia sensibilia circa motus coelestes; non tamen ratio haec est sufficienter probans, quia etiam forte alia positione facta salvari possent." See also *Commentaria in libros Aristotelis de caelo et mundo* XII, 17: "Illorum tamen suppositiones, quas adinvenerunt, non est necessarium esse veras: licet enim talibus suppositionibus factis, apparentia salvarentur, non tamen oportet dicere has supposi- tiones esse veras; quia fortasse secundum aliquem alium modum, nondum ab hominibus comprehensum, apparentia circa stellas salvantur. Aristoteles tamen utitur huiusmodi supposi- tionibus quantum ad qualitatem motuum tanquam veris." (The "quality" of the motions is, here too, that of the so-called Platonic axiom, that is, circularity and uniformity.) Even if Thomas could have taken the formula from Simplicius's commentary on the *De caelo*, as P. Duhem demonstrated, coordinating it with the specific epistemological concerns of Scholas- ticism was nevertheless original and of fundamental importance for the system.

58. Aristotle, *De caelo* II 2, 285 b 26; II 9, 291 a 14.

59. Thomas Aquinas, *Expositio de caelo* II lect. 4 n. 3: "Hanc autem elongationem dicit multo maiorem esse quam localem ... accidentia caelestium corporum sunt alterius rationis, et omnino improportionata accidentibus inferiorum corporum."

60. Thomas Aquinas, *Expositio de caelo* I lect. 8 n. 91 (on Aristotle, *De caelo* I 4; 271 a 33): "Est autem attendendum quod Aristoteles hic ponit Deum esse factorem caelestium corporum, et non solum causam per modum finis, ut quidam dixerunt."

61. Thomas Aquinas, *Summa theologica* I q. 70 a.2: "Utrum convenienter causa productionis luminarium describatur."

62. Loc. cit.: "Sed Moyses, ut populum ab idolatria revocaret, illam solam causam tetigit, secundum quod sunt facta ad utilitatem hominum." There follows a quotation from Deuteronomy 4 : 19. The fundamental idea of the Old Testament text, that using something and idolizing it are mutually exclusive, was already ineffective, as an argument, in the early Christian literature. That it made an impression on people who were outside the original Greek cultural formation is shown by the Syrian, Tatian, in his *Address to the Greeks*: "I refuse to adore that workmanship which He made for our sakes. The sun and moon were made for us: how, then, can I adore my own servants?" *Logos* IV, 4, translated in *Tatian, Theophilus and the Clementine Recognitions*, trans. B. P. Pratten, M. Dods, and T. Smith (Edinburgh: 1867), p. 9.

63. Robert Boyle (1626–1691), *Natural Philosophy, Works*, ed. Th. Birch (London, 1744), vol. 1, p. 431: "God was pleased to consider man, so much more than the creatures made for him, that he made the sun itself at one time stand still, and at another time to go back, and diverse times made the parts of the universe forget their nature, or act contrary to it . . . . "

64. Augustine, *Enchiridion* XVI, 62: "Instaurantur quippe, quae in caelis sunt, cum id quod inde in angelis lapsum est, ex hominibus redditur; instaurantur autem, quae in terris sunt, cum ipsi homines, qui praedestinati sunt ad aeternam vitam, a corruptionis vetustate renovantur."

65. John Duns Scotus, *Opus oxoniense* III dist. 32 q. un. n. 6; *Opera*, ed. Vivès (Paris, 1891–1895), vol. 15, p. 433: "Deus . . . vult propter illos (sc. praedestinatos) alia, quae sunt remotiora, puta hunc mundum sensibilem, ut serviat eis . . . Igitur quia deus vult mundum sensibilem in ordine ad hominem praedestinatum . . . homo erit finis mundi sensibilis." A modern monograph tells us that it would be "a degradation for the infinite will if something finite, even the highest, richest creation of thought, could serve as its motive" (J. Klein, *Der Gottesbegriff des Johannes Duns Scotus* [Paderborn: Schöningh, 1913], p. 68). Cf. *Opus oxoniense* IV d. 48 q. 2 n. 7.

66. Jean Buridan, *Quaestiones super libros Physicorum Aristoteles* II q. 7; *Kommentar zur Aristotelischen Physik* (rpt. of the Paris, 1509 edition, Frankfurt: Minerva, 1964), fol. 35. On the distinction between *finis quo* and *finis cuius*: A. Maier, *Metaphysische Hintergründe der spätscholastischen Naturphilosophie* (Rome: Edizioni di storia e letteratura, 1955), pp. 300–355.

Chapter 4

67. Thomas Aquinas, *Summa theologica* I q. 96 a.2: "Et sic etiam homo in statu innocentiae dominabitur plantis et rebus inanimatis, non per imperium, vel immutationem, sed absque impedimento utendo eorum auxilio."

68. "Quod non potest evidenter ostendi nobilitas unius rei super aliam." Cl. Baeumker, *Witelo* (Münster: Aschendorff, 1908), p. 427; J. Lappe, *Nicolaus von Autrecourt* (Münster: Aschendorff, 1908), p. 33.

Notes to Pages 204–211

69. Cicero, *De officiis* I 43, 153.

70. Cicero, *De officiis* I 44, 156: "... eloqui copiose, modo prudenter, melius est quam vel acutissime sine eloquentia cogitare, quod cogitatio in se ipsa vertitur, eloquentia complectitur eos, quibuscum communitate iuncti sumus."

71. Carolus Bovillus, *Liber de sapiente*, chapter 26; ed. R. Klibansky in E. Cassirer, *Individuum und Kosmos in der Philosophie der Renaissance* (Leipzig: Teubner, 1927), p. 353. See also Cassirer's commentary, in the same volume, pp. 93–97 (in English: *The Individual and the Cosmos in Renaissance Philosophy*, trans. M. Domandi [Philadelphia: University of Pennsylvania Press, 1963], pp. 88–92), and B, Groethuysen, "Die kosmische Anthropologie des Bovillus," *Archiv für Geschichte der Philosophie* 40 (1931): 66–89.

72. Bovillus, loc, cit.: "Est igitur et homo factus extra omnia, et omnia rursum extra hominem facta, homini ex adverso diametri praesentata .... Speculi autem natura est, ut extra omnia locatum sit, cunctis adversum et oppositum .... Verus igitur et speculi et hominis locus est in oppositione, extremitate, distantia et negatione universorum ...."

73. Bovillus, *De sensibus*, chapter 4: "Orbis solis est verum mundi medium. Firmamentum et terra extrema sunt mundi. Animae ... sedes est in medio ...." Quoted by D. Mahnke, *Unendliche Sphäre und Allmittelpunkt. Beiträge zur Genealogie der mathematischen Mystik* (Halle: Niemeyer, 1937), p. 113.

74. Bovillus, *Liber de sapiente*, Chapter 51: "Exhortatorius sermo." Here there appears the earliest formula for the possibility of human failure that has been characterized by the terms "alienation" and "inauthenticity": *divisus abs teipso* (*Liber de sapiente*, ed. Klibansky, p. 407).

75. G. C. Lichtenberg, *Nicolaus Copernicus*, in his *Schriften und Briefe*, ed. W. Promies, vol. 3 (Munich: Hanser, 1972), p. 159.

Chapter 5

76. Joachim Rheticus, *Narratio prima*, ed. M. Caspar, in J. Kepler, *Gesammelte Werke* (Munich: Beck, 1937–1982), vol. 1, pp. 114 and 126. Commentary in E. Rosen, *Three Copernican Treatises*, 2nd ed. (New York: Dover, 1959). [In translating Blumenberg's quotations from the *Narratio prima* I have drawn on Rosen's translation in *Three Copernican Treatises*, but I have followed the German translation that Blumenberg uses—Karl Zeller's *Des Georg Joachim Rhetikus Erster Bericht* (Munich and Berlin: Oldenbourg, 1943)—where there are significant differences between them.]

77. H. Blumenberg, "On a Lineage of the Idea of Progress," *Social Research* 41 (1974): 5–27, specifically pp. 18–23.

78. Ptolemy, *Syntaxis mathematica*, ed. J. L. Heiberg (Leipzig: Teubner, 1898–1903), XIII, 2. [Here Blumenberg cites the German translation, by K. Manitius, which he uses throughout this chapter: Ptolemy, *Handbuch der Astronomie* (Leipzig: Teubner, 1912/13; 2nd ed., 1963), vol. 2, pp. 333ff. The corresponding passage in G. J. Toomer's English translation, *Ptolemy's Almagest* (New York: Springer, 1984), is on pp. 600–601.] Kepler gives a translation of this epistemological excursus in his *Epitome astronomiae copernicanae* IV pars 2 cap. 2 (*Opera omnia*, ed. Ch. Frisch [Frankfurt: 1858–1871], vol. 6, p. 337): "Non enim aequum est, humana nostra Diis immortalibus aequiparare rerumque sublimium fidem ab exemplis petere rerum dissimillimarum. Nam quid cui magis dissimile, quam ea, quae semper eodem modo habent, iis, quae nunquam sibi constant, et ea, quae undiquaque ab omnibus, iis, quae ne a se ipsis quidem impediri possunt?"

79. The equivalence of transparency and penetrability in the spheres supports the general admissibility of the heavenly bodies' *implicationes* and *insertiones* (in Kepler's language): "... adeo ut omnia simpliciter ab omnibus penetrari non difficilius quam perspici possunt ... in coelo tamen videmus nequaquam obstare tam multiplicem motuum concursum, quo minus eveniant singuli .... Non igitur ex rebus nostratibus, sed ex ipsis naturis eorum, quae in coelo sunt, et ex motuum ipsorum immutabili tenore judicium est informandum. Ita fiet, ut hoc pacto motus omnes videantur simplices multoque simpliciores, quam ea, quae penes nos talia videntur esse: quippe nullum laborem, nullam difficultatem in circuitionibus eorum suspicari possumus" (*Epitome astronomiae copernicanae* [cited in note 78], p. 338).

80. Ptolemy, *Opera astronomica minora*, ed. J. L. Heiberg (Leipzig: Teubner, 1907), pp. 114–115.

81. Ptolemy, *Syntaxis mathematica* III, 4; translated by G. J. Toomer as *Ptolemy's Almagest* (New York: Springer, 1984), p. 153. The undecidability of the choice between epicyclic and eccentric constructions is still brought to bear by Andreas Osiander when he writes to Copernicus and Rheticus in 1540 and 1541 in order to press for a cautious and hypothetical presentation of the new world-system, such as he will finally suggest in his anonymous preface: "... quis enim nos certiores reddet, at Solis inaequalis motus nomine epicycli an nomine eccentricitatis contingat, si Ptolemaei hypotheses sequamur, cum id possit utrumque." Kepler, *Apologia Tychonis contra Ursum*, in *Opera omnia*, ed. Ch. Frisch (Frankfurt, 1858–1871), vol. l, p. 246.

82. Ptolemy, *Syntaxis mathematica* I, 7; trans. Toomer, p. 45.

83. Ptolemy, *Syntaxis mathematica* I, prooemium; trans. Toomer, p. 36; ed. Heiberg (cited in note 78), pp. 6–7.

84. Kepler, *Epitome astronomiae copernicanae* IV pars 2 cap. 2; loc. cit. (note 78), p. 338: "Etsi verum est non esse censendam facilitatem motuum coelestium ex difficultate motuum elementarium, propter causas bene multas, nondum tamen sequitur, motuum coelestium nulla in Terris exempla propinqua esse ...."

85. F. Krafft, "Sphaera activitatis—orbis virtutis. Das Entstehen der Vorstellung von Zentralkräften," *Sudhoffs Archiv* 54 (1970): 113–140 (relevant here: 132–134).

86. Kepler, *Epitome* (cited in note 78), p. 338: "... omnes omnino hypotheses in suspicionem falsitatis adducit, dum tantopere urget discrimen coelestium et terrestrium rerum, adeo ut etiam ratio ipsa errare ponatur in dijudicatione ejus, quod geometrice simplex est."

87. Kepler, *Epitome*, p. 339: "... illisque [namely, astris] plus tribuat, quam Deus ipse conditor habet, ut scilicet rationes geometricae simplices sint illis, quae sunt re vera compositae, quarum intellectum Deus homini, imagini suae, communem secum esse voluit."

88. Cicero, *Tusculanae disputationes* I, 62–63.

89. Copernicus, *Commentariolus*, ed. F. Rossmann (Munich: H. Rinn, 1948), p. 10: "Quapropter non satis absoluta videbatur huiusmodi speculatio, neque rationi satis concinna."

90. Copernicus, *De revolutionibus* I, 10.

91. Copernicus, *De revolutionibus* I, 4.

92. Most clearly and most beautifully where the confusion of the appearances is made perspicuous by the one complex motion of the Earth: "Huius igitur solius [namely, telluris] motus tot apparentibus in caelo diversitatibus sufficit" (*Commentariolus*—cited in note 89—p. 11). Or: "Posse autem haec omnia fieri mutabilitate telluris minus mirum est" (p. 14). The latter sentence is at the same time one of his cunning attempts, by equating *mobilitas* and *mutabilitas*, to make the novel aspect of the Earth seem long familiar.

Notes to Pages 222–229

93. Copernicus, *Commentariolus*, p. 26: "... attamen comprehendetur et ipse, modo altiori ingenio quispiam incumbat."

94. *Philippi Melanchthonis opera*, Corpus Reformatorum vols. 1–28 (Halle, 1834–1860), vol. 4, col. 847: "Indulsi aetati nostri Rhetici, ut ingenium quasi quodam Enthusiasmo incitatum ad hanc philosophiae partem, in qua versatur, proveheretur. Sed aliquoties ipsi dixi, me in eo plusculum socraticae philosophiae desiderare, quam fortassis adiunget, cum erit paterfamilias. Nam ea de re cogitare eum intellexi."

95. Rheticus, *Narratio prima* (cited in note 76), p. 97: "astronomicae veritatis restauratio." Compare p. 101: "renascens Astronomia"; p. 102: "ad certam rerum coelestium doctrinam exaedificandam"; p. 101: "atque exinde posito telluris motu in eccentrico, in promptu esse certam rerum coelestium doctrinam, in qua nihil mutandum, quin simul totum systema, ut consentaneum erat, de novo in debitas rationes restitueretur."

96. *Narratio prima*, p. 99: "... terrae mobilitate apparentias in coelo plerasque fieri posse, aut certe commodissime salvari ...."

97. *Narratio prima*, p. 100: "... quare cum hoc unico terrae motu, infinitis quasi apparentiis satisfieri videremus, Deo naturae conditori eam industriam non tribueremus, quam communes horologiorum Artifices habere cernimus?"

98. *Narratio prima*, p. 101: "... per suas hypotheses causam efficientem aequalis motus Solis geometrice deduci posse sentiebat ...."

99. *Narratio prima*, p. 103: "... ideo Deum tot eum orbem, nostra quippe causa, insignivisse globulis stellantibus, ut penes eos, loco nimirum fixos, aliorum orbium, et planetarum contentorum animadverteremus positus ac motus."

100. *Narratio prima*, p. 104: "Quare totam reliquam hanc naturam ... admirabimur, et contemplabimur, ad quam perscrutandam, et cognoscendam multis modis, infinitis instrumentis, et donis nos locupletavit, et idoneos nos effecit ...."

101. *Narratio prima*, p. 104: "... nihil extra concavum orbis stellati, quod inquiramus, erit, nisi quantum nos Sacrae literae de his scire voluerunt ...."

102. *Narratio prima*, p.115: "Et quidem si usquam alibi est videre, quomodo deus mundum nostris disputationibus reliquerit, hoc certe loco, ut quod maxime, est conspicuum."

103. *Narratio prima*, p. 101: "Illud autem hominum genus, quod omnes simul stellas pro suo arbitratu, haud secus ac iniectis vinculis, in aethere circumducere conatur ...."

104. *Narratio prima*, p. 104: "... animo citius concipi (propter affinitatem, quam cum coelo habet) quam ulla voce humana eloqui posse ...."

105. *Narratio prima*, pp. 114–115.

106. As K. H. Burmeister asserts in his *Georg Joachim Rhetikus. Eine Bio-Bibliographie* (Wiesbaden: Pressler, 1967–1968). vol. 3 (*Briefwechsel*), p. 45.

107. Rheticus, *Briefwechsel*, ed. Burmeister (cited in note 106), p. 43: "Nam mens humana, orta e coelo, cognitione rerum coelestium, velut conspectu patriae delectatur."

108. Rheticus, loc cit.: "Quam est enim absurdum, putare hanc miram et certissimam motuum varietatem frustra a Deo architecto institutam esse."

Chapter 6

109. Copernicus, *De revolutionibus* I, 10.

110. Ptolemy, *Syntaxis mathematica* IX, 1; in G. J. Toomer's translation, *Ptolemy's Almagest* (New York: Springer, 1984), pp. 419–420.

111. E. Goldbeck, *Der Mensch und sein Weltbild im Wandel vom Altertum zur Neuzeit. Gesammelte kosmologische Abhandlungen* (Leipzig: Quelle & Meyer, 1925), p. 22.

112. Plato, *Timaeus* 39B.

113. Plutarch, *Quaestiones platonicae* VIII, 1. Newton, who in any case regards heliocentrism as the *antiquissima sententia* of the philosophers, ascribes it to Plato *aetate maturiore* (*Opera*, ed. S. Horsley [London, 1779–1785], vol. 3, p. 179). Newton finds the explanation of the fact that the Greeks were able, at all, to lose or to abandon this insight, in their lack of a "philosophical" attitude—"Ab Aegyptiis ... ad Graecos, gentem magis Philologicam quam Philosophicam ..."— a remark that no doubt relates to the lack of a claim to truth in their astronomy.

114. *Procli Diadochi in Platonis Timaeum commentaria*, ed. E. Diehl (Leipzig: Teubner, 1903–1906), vol. 3, p. 66.

115. Diogenes Laertius V, 86.

116. Copernicus, *De revolutionibus* I, 8: "At si caelum fuerit infinitum et interiori tantummodo finitum concavitate ... permanebit caelum immobile."

117. Copernicus, *Commentariolus*, ed. F. Rossmann (Munich: H. Rinn, 1948), p. 12: "Eodem quoque ordine alius alium revolutionis velocitate superat .... "

118. A. Birkenmaier, commentary in Copernicus, *Über die Kreisbewegungen der Weltkörper, Erstes Buch*, ed. G. Klaus (Berlin: Akademie Verlag, 1959), p. 149.

119. Copernicus, *De revolutionibus* I, 10: "Quam vero causam allegabunt ii qui sub Sole Venerem, deinde Mercurium ponunt vel alio ordine separant, quod non itidem separatos faciunt circuitus et a Sole diversos, ut ceteri errantium, si modo velocitatis tarditatisque ratio non fallit ordinem?"

120. Copernicus, *De revolutionibus* I, 10: "Quapropter prima ratione salva manente, nemo enim convenientiorem allegabit, quam ut magnitudinem orbium multitudo temporis metiatur." In the manuscript of the work this sentence is on the same page as the diagram of the heliocentric system; the parentheses that surround the text from *nemo* to *metiatur* were evidently inserted later, and accordingly are also absent from the first edition (Nuremberg: 1543), which separates the diagram from the explanation.

121. My hypothesis, which uses this 'transition' in the text as the basis for an approach to the way Copernicus arrived at his theory, was first presented in "Kopernikus im Selbstverständnis der Neuzeit," *Abhandlungen der Akademie zu Mainz*, Geistes- und sozialwiss. Klasse, 1964 series, no. 5, pp. 355–356; then in *Die Legitimität der Neuzeit* (Frankfurt: Suhrkamp, 1966), pp. 367–368 (= *The Legitimacy of the Modern Age* [Cambridge, MA: MIT Press, 1983], pp. 367–368); and once again in *Science and History, Studies in Honor of Edward Rosen*, Studia Copernicana 16 (Wroclaw: Ossolineum, 1976), pp. 473–486 (submitted in March 1974).

122. Copernicus, *De revolutionibus* IV, prooemium: "Nos quidem in explicatione cursus lunaris non differimus a priscorum opinionibus in eo, quod circa terram fit."

123. Copernicus, *Commentariolus*, tertia petitio (ed. F. Rossmann [Munich: H. Rinn, 1948],

p. 10): "Omnes orbes ambire Solem, tanquam in medio omnium existentem, ideoque circa Solem esse centrum mundi."

124. *Commentariolus*, prima petitio: "Omnium orbium coelestium sive sphaerarum unum centrum non esse." Secunda petitio: "Centrum terrae non esse centrum mundi, sed tantum gravitatis et orbis Lunaris."

125. Nicholas of Cusa, *De docta ignorantia* II, 12. On this see H. Blumenberg, *The Legitimacy of the Modern Age* (Cambridge, MA: MIT Press, 1983), pp. 507–513; original: *Die Legitimität der Neuzeit* (Frankfurt: Suhrkamp, 1966), pp. 473–480.

126. Copernicus, *De revolutionibus* I, 9.

127. *De revolutionibus* I, 7: "Si igitur, inquit Ptolemaeus Alexandrinus, terra volveretur, saltem revolutione quotidiana . . ." (cf. Ptolemy, *Almagest* I, 7). On this passage see A. Birkenmaier's commentary (cited in note 118), p. 115. In Ptolemy the thesis of the Earth's diurnal rotation is part of a series including the discussion of a falling motion of the Earth, with which it would necessarily break through the spheres and leave the cosmos—so the context does not do much to stimulate continued reflection.

128. Copernicus chose a very similar connecting formula in *De revolutionibus* V, 2 also. He criticizes the representation of the apparent inequalities in the motions of the planets by the epicycle system: "Concedunt igitur (sc. prisci mathematici) et hic motus singularis fieri posse circa centrum alienum et non proprium . . .," and then continues, "Haec et similia nobis occasionem praestiterunt de mobilitate terrae aliisque modis cogitandi, quibus aequalitas et principia artis permanerent, et ratio inaequalitatis apparentis reddatur constantior."

129. *De revolutionibus* I, 9: "Ipse denique Sol medium mundi putabitur possidere. Quae omnia ratio ordinis, quo illa sibi invicem succedunt, et mundi totius harmonia nos docet . . . ." On the dating of the parts of book I: A. Birkenmaier, commentary (cited in note 118), p. 156.

130. F. Fellmann, *Scholastik und kosmologische Reform* (Münster: Aschendorff, 1971), p. 69.

131. Copernicus, *Commentariolus*, ed. F. Rossmann (Munich: H. Rinn, 1948), pp. 13–14: "Alius telluris motus est quotidianae revolutionis et hic sibi maxime proprius . . . ."

132. *Commentariolus*, p. 9. When Edward Rosen (in *Three Copernican Treatises* [New York: Dover, 1959], p. 57) suggests that the passage be amplified by the phrase "with uniform velocity in a perfect circle," he overlooks the fact that Copernicus reduces the requirement for astronomical construction to *one* element and regards the others as deducible from it, and also that he plans to use it to establish, for the first time, a motion that was hitherto unknown to astronomy: the rotation of a solid body around its axis.

133. Copernicus, *De revolutionibus* I, 10: "Quod facilius concedendum puto quam in infinitam pene orbium multitudinem distrahi intellectum, quod coacti sunt facere, qui terram in medio mundi detinuerunt."

Part III

Introduction

1. P. Bayle, *Dictionnaire historique et critique*, 3rd ed. (Rotterdam: 1720), vol. 1, pp. 804–806.

2. F. Hipler, *Die Vorläufer des Nikolaus Coppernicus*, Mitteilungen des Coppernicus-Vereins 4

(Thorn [Torun]: 1882), p. 55. The Göttingen mathematician Abraham Gotthelf Kästner (of whom Kant said that "in his hands everything becomes exact, intelligible, and pleasant") acquired a curious example of the reversal of accreditations: an edition, printed in Madrid in 1641, of the *Tabulae Alphonsinae perpetuae motuum coelestium denuo restitutae et illustratae a Francisco Garcia Ventanas*. The editor justifies his publication with the praise of the astronomers Peurbach, Regiomontanus, Copernicus, Tycho, and Kepler for the Castilian king: "Que abstrayando de la verdad, las Tablas Alfonsinas son las que mas concuerdan con la perpetuidad de los Tiempos." In a significant anachronism, Copernicus appears as an authority for Scholastic piety. Kästner mentions this in passing in his essay, "Worin mag König Alphons des Weisen Gotteslästerung bestanden haben?" *Gesammelte Werke* (Berlin: 1841), vol. 2, pp. 133–134.

3. Bayle, article "Castille" (loc. cit., note 1): "Encore que la silence d'un si sage Historien par raport au Système de Ptolomée doive être de quelque poids, je ne laisse pas de croire que si Alfonse porta sa critique audacieuse sur quelque partie de l'Univers, ce fut sur les Spheres célestes. Car, outre qu'il n'étudia rien tant que cela, il est sûr que les Astronomes expliquoient alors le mouvement des cieux par des Hypotheses si embarrassées et si confuses, qu'elles ne faisoient point d'honneur à Dieu, et ne répondoient nullement a l'idée d'un habile ouvrier. Il y a donc aparence que ce fut en considérant cette multitude de Spheres dont le Système de Ptolomée est composé, tant de cercles eccentriques, tant d'épicycles, tant de librations, tant de déférans, qu'il lui échapa de dire, que si Dieu l'eût appellé à son conseil, quand il fit le monde, il lui eût donné de bons avis."

4. Fontenelle, *Entretiens sur la pluralité des mondes* I; ed. A. Calame (Paris: Didier, 1966), pp. 25–26.

5. Leibniz, "Lettre sur ce qui passe les sense et la matière," *Philosophische Schriften*, ed. C. J. Gerhardt (Berlin: 1875–1890), vol. 6, p. 498: "... toutes nos plaintes viennent de nostre peu de connoissance, à peu prés comme le Roy Alphonse à qui nous devons des tables Astronomiques, trouvoit à redire au systeme du monde, faut de connoistre celuy de Copernic seul capable de faire juger sainement de la grandeur et de la beauté de l'ouvrage de Dieu." Gottsched's memorial address on the two hundredth anniversary of the astronomer's death falls right into the theodicy pattern of connecting Copernicus with the Alfonso anecdote: "Even King Alphonsus in Castile, who was immortalized more by his love of astronomy than by his kingdom, was heard to say, when he considered the strange circles that were ascribed to the planets, that if he had been consulted by the Author of the heavens and the Earth when He was creating the world structure he could have given Him much better suggestions. Admittedly, such a verdict on God's works does not seem to be in sufficient accord with the veneration that one should have for the Highest Being. But what could Alphonsus do about it when the doctrines and the strange opinions of the astronomers of his time presented the world to him in such absurd figures? ... But Copernicus, possessing common sense, and having these same muddled pictures before his eyes, realized that the responsibility for these disorders should much rather fall on men than on the wise Creator of the world, and indeed that the reason for these confusions must be simply in the imperfection of the human understanding. So now he began to open all the writings of ancient and recent astronomers with an eager hand. His impatient eye ran through everything that the admirers of the heavens had ever written ...." ("Gedächtnisrede auf Nicolaus Coppernicus," 1743; *Gesammelte Schriften*, ed. E. Reichel [Berlin: Gottsched-Verlag: 1903–], vol. 4, pp. 137–139).

6. Goethe, *Maximen und Reflexionen*, No. 835. The criticism of the Creation has lost all its blasphemous quality when it becomes the premise of man's aesthetic authenticity. On 6 December 1895, Paul Valéry thanks André Gide for the gift of a box of compasses: "Tu me donne là de quoi construire le monde. Ah! si Dieu avait été outillé comme ça! quelles orbites, seigneur! quelles électricités! Les hommes, peut-être, auraient la tête un peu plus précise" (*André Gide-Paul Valéry. Correspondance*, ed. R. Mallet [Paris: Gallimard, 1955], p. 252).

Notes to Pages 264–271

Chapter 1

7. Voltaire, *Correspondance*, ed. Th. Besterman (Geneva: Institut et musée Voltaire, 1953–1965), vol. 85, p. 211 (no. 17417): "Thorn ne se trouve point dans la partie qui m'est échue de la Pologne . . . mais j'érigerai dans une petite ville de la Warmie un monument sur le tombeau du fameux Copernic, qui s'y trouve enterré."

8. K. A. Varnhagen von Ense, *Tagebücher*, vol. 9 (Hamburg: 1868), p. 5: 3 January 1852.

9. *Vindiciae sedis apostolicae adversus Neo-Pythagoreos terrae motores et solis statores* (1635), in D. Berti, *Il processo originali di Galileo Galilei pubblicato per la prima volta* (Rome: 1876), par. CXXXV. P. Duhem (*Le Système du monde* [Paris: Hermann, 1913–1954], vol. 1, p. 21) insists that Copernicus cited the Pythagorean Philolaus only for the proposition that the Earth is a star, while Gassendi, in a *Vita Copernici* of 1654, was the first to ascribe heliocentrism, also, to the Pythagoreans. After all, it was sufficient for the stellar quality that the Earth should have one motion, namely, its rotation around its axis, in its old place. G. Schiaparelli, in his *Precursori di Copernico nell'Antichità* (Milan: 1873; German trans. M. Kurtze, Leipzig: 1876), was the first to draw the ultimate conclusion that without the Pythagoreans, Copernicus and his successors would not have been possible at all. In his preliminary work (done before 1870) for an unfinished treatise on the question of the authenticity of the catalogue of titles of works of Democritus, Nietzsche, for whom Copernicus was later to take on an entirely different significance in relation to the specificity of the post-Christian idea of science, noted, "Copernicus relies on Pythagorean tradition; the Congregation of the Index calls his doctrine a *doctrina Pythagorica*." (*Gesammelte Werke* [Munich: Musarion, 1920–], vol. 2, p. 139.)

10. John Donne, *Complete Poetry and Selected Prose*, ed. J. Hayward (London: Nonesuch Press, 1955), p. 363: ". . . beat the dores, and cried: 'Are these shut against me, to whom all the Heavens were ever open, who was a soul to the Earth, and gave it motion?'" Copernicus has to refer to the most extreme misdeed that the author thinks he can formulate: He withdrew Lucifer from God's punishment by raising him, together with his prison, the Earth, from the center of the world into the heavens, and by thrusting the Sun, God's eye, into the lowest place instead of them. As in Dante, geocentrism here is not an indicator of anthropocentrism but of infernocentrism, and Copernicus has violated this. "Shall these gates be open to such as have innovated in small matters? and shall they be shut againt me, who have turned the whole frame of the world, and am thereby almost a new Creator?"

11. Galileo, *Opere*, ed. S. Timpanaro (Milan and Rome: Rizzoli, 1936–), vol. 1, p. 258.

12. Machiavelli, *Opere complete* (Milan, 1850), vol. 1, p. 327: ". . . come se il cielo, il sole, gli elementi, gli uomini fossero variati di moto, di ordine e di potenza da quello ch'egli erano anticamente."

13. Christopher Heydon, *A Defence of Judiciall Astrologie* (1603): Copernicus "altered the whole of nature in order to rectifie his Hypothesis" (quoted in P. Meissner, *Die geistesgeschichtlichen Grundlagen des englischen Literaturbarocks* [Munich: Hueber, 1934], pp. 62–63: "What stands behind all the criticism of Copernicus is fear of the cultural revolution . . . ").

14. E. Rosen, "Copernicus' Attitude Toward the Common People," *Journal of the History of Ideas* 32 (1971): 281–289.

15. Tertullian, *Apologeticum* XL, 2 (with the dreadfully magnificent apostrophe to the "Ad leonem!": "Tantos ad unum?")

16. Kepler, Introduction to *Commentaria de motibus stellae Martis* (*Werke*, ed. Ch. Frisch [Frankfurt, 1858–1871], vol. 3, p. 154): "Facile autem Deus ex Josuae verbis quid is vellet intellexit

Notes to Pages 272–278

praestititque inhibito motu Terrae; ut illi stare videretur Sol. Petitionis enim Josuae summa huc redibat, ut sic sibi videri posset, quicquid interim esset."

17. Gunzo of Novara (*Patrologia latina*, ed. J. P. Migne [Paris, 1844–1864], vol. 136, col. 1299), cited after H. M. Klinkenberg, "Der Verfall des Quadriviums im frühen Mittelalter," in *Artes Liberales*, ed. J. Koch (Leiden: Brill, 1955), pp. 25–26.

18. Nicholas of Cusa, *Excitationes* VII [*Opera*, ed. J. Lefèvre d'Étaples (Paris: 1514)].

19. Nicole Oresme, *Le Livre du ciel et du monde*, fols. 91d–92b (trans. A. D. Menut [Madison: University of Wisconsin Press, 1968], p. 365): "Dieu est fin principal de Lui meisme .... Il ne s'ensuit pas se Dieu est que le ciel soit et, par consequent, il ne s'ensuit pas que le mouvement du ciel soit, car selon verité, tout ce depent de la volenté de Dieu franchement sanz ce que il soit aucune neccessité .... Et selon verité, ce monstra Il ou temps de Josué, quant le soleil se arresta." On the regularization of miracles: "Also, when God performs a miracle, we must assume and maintain that He does so without altering the common course of nature, inso far as possible" (Oresme, op. cit., fol. 144a; trans. A. D. Menut, p. 537.)

20. H. S. Reimarus, *Apologie*, ed. G. Alexander (Frankfurt: Insel, 1972), vol. 1, pp. 494–495.

21. Herder, *Adrastea* III, 6; *Sämtliche Werke*, ed. B. Suphan (Berlin: Weidmann, 1877–1913), vol. 23, p. 551.

22. Voltaire, *Oeuvres complètes* (Basel, 1792), vol. 40, p. 190.

23. Luther, *Tischreden*, vol. 4 (Weimar: Bohlau, 1916), No. 4638. Melanchthon is the first to apply the 'perpetrator' language, with a negative evaluation, to the "Sarmatian astronomer," who thinks he has to distinguish himself with his *res absurda*, and *qui movet terram et figit Solem* (to Burkard Mithobius, 16 October 1541; *Opera*, Corpus Reformatorum vols. 1–28 [Halle, 1834–1860], vol. 4, p. 679). Recently Z. Wardeska ("Copernicus und die deutschen Theologen des 16. Jahrhunderts," in *Nicolaus Copernicus zum 500. Geburtstag*, ed. F. Kaulbach, U. W. Bargenda, and J. Blühdorn [Cologne: Böhlau, 1973], pp. 165–169) has thought that the two versions of Luther's remark in the *Tischreden* exhibit contradictory evaluations—indeed that they differ in principle—and that "the fact that there exists a positive remark by Luther about Copernicus" is a "discovery that has not been taken advantage of in previous investigations of the reception of Copernicanism in Germany, and has been undervalued." This thesis must be based on a misunderstanding of the texts—on a failure to appreciate both the irony directed at the passion for innovation, and the connection of the formula "qui totam astrologiam invertere vult" with the following concessive "Etiam illa confusa tamen ego credo...," which by no means refers to a knowledge of the sad state of the old astronomy, but rather to the inversion carried out by the new one.

24. Petrus Ramus, *Scholae mathematicorum* (Paris: 1565), quoted by Kepler, *Astronomia nova* (*Gesammelte Werke*, ed. W. von Dyck and M. Caspar [Munich: Beck, 1937–1982], vol. 3, p. 6): "Longe enim facilius ei fuisset, Astrologiam, astrorum suorum veritati respondentem describere, quam gigantei cuiusdam laboris instar, Terram movere, ut ad Terrae motum quietas stellas specularemur."

25. Nikodemus Frischlin (1547–1590), *Carmen de astronomico horologio Argentoratensi* (Strasbourg: 1575). On the portrait: E. Schwenk zu Schweinsberg, "Kopernikus-Bildnisse," in *Nikolaus Kopernikus*, ed. F. Kubach (Munich: 1943), pp. 278–279.

26. N. Frischlin, op. cit., praefatio (1574): "Videmus enim hac extrema mundi senecta politiorem literaturam, et sublimiora artium ac disciplinarum studia paulatim vilescere; et hominum nostrorum animos aut ad luxum et voluptates rapi, aut ad quaestuosas artes et splendida pecuniarum aucupia converti .... Artes autem humanitatis, et Mathematica studia

minus ubique florere, nec nisi a paucissimis excoli videmus. Solet enim pueritia ea, quae a superiorum ordinum hominibus negligi et contemptim habere videt, ipsa etiam negligere atque contemnere . . . ."

27. Loc. cit.: "Illum scrutanti similem, similemque docenti/ Aspiceres: qualis fuerat, cum sidera iussit/ Et coelum constare loco: terramque rotari/ Finxit, et in medio mundi Titana locavit."

28. Simon Starowolski, *Vita Copernici*, ed. F. Hipler in *Zeitschrift für die Geschichte und Alter-tumskunde Ermlands* 4 (1869): 539: "Juppiter ut vidit quod mente Copernicus orbem/ Contra naturae iura creasset homo,/ Ut vidit coelum firma statione teneri/ Currente et terra sydera stare bene . . . ."

29. Peter Damian, *De divina omnipotentia*, chapters 10–12; *Patrologia latina*, ed. J. P. Migne, vol. 145, cols. 610–615.

30. F. Hipler, "Die Vorläufer des Nikolaus Coppernicus insbesondere Celio Calcagnini," *Mitteilungen des Coppernicus-Vereins* 4 (Torun: 1882): 61.

31. Calcagnini, *De perenni motu terrae*, ed. F. Hipler (in the paper cited in note 30), p. 70. On this see H. Blumenberg, "Der archimedische Punkt des Celio Calcagnini," in *Studia Humanitatis. Festschrift E. Grassi*, ed. E. Hora and E. Kessler (Munich: Fink, 1973), pp. 103–112.

32. *De perenni motu terrae*, p. 72: "Nec enim tantisper rotari desinit, donec obiicem aut lacunam invenerit ubi sistat."

33. Loc. cit.: "Hic ea sunt quae maximus ille philosophorum Plato nunquam fieri, semper esse praedicavit."

34. *De perenni motu terrae*, p. 76: " . . . ita terra in medio collocata est mundo, que suo motu reliqua excitet ad generationem elementa, eaque animantia sustineat, quae ad rerum divinarum contemplationem deus immortalis procreavit."

35. *De perenni motu terrae*, p. 74: "ausim dicere, uix aliud quicquam absurdius a philosophis commentatum esse, quam caelum moueri."

36. Calcagnini's text contains, at this point, a misunderstanding of the ancient anecdote, when he speaks of a solid base onto which the Earth could be moved ("si modo basem in quam transferretur invenisset"), whereas the intended reference of Archimedes's remark is after all to the fixed point on which he would prop his lever. Even if Calcagnini did not understand the original sense of the anecdote, he nevertheless brings it under the premises of the cosmology that he accepts: Only in the center of of the universe, as its "natural place," does the Earth rest on its own and unsupported in space, while if it were moved (in accordance with the thought experiment) to another location, it would have to have a foundation underneath it if it were not to fall back to its original central position. The idea that the center of the world could also be a 'contingent' location for the Earth presupposes the abandonment of the Aristotelian premise that the center of the sphere is at the same time the 'lowest place of all' for weight. Calvin's perspective on geocentrism was that the Earth's resting in the center, "floating in the air," could only result from a special exertion of power on God's part. In a sermon on Job 26:7 (" . . . et appendit terram super nihilum") he says, "Or il est vrai que les Philosophes disputent bien pourquoi c'est que la terre est ainsi demeuree, veu qu'elle est au plus profond du monde: et qu c'est merveille comme ellse n'est abysmee, veu qu'il n'y a rien qui la soustienne . . . " (*Corpus Reformatorum*, vol. 62, p. 430). And on psalm 104:5 ("Fundavit terram super bases suas . . . ") Calvin explains, "Hic in terrae stabilitate praedicat Dei gloriam, quomodo enim locum suum tenet immobilis, quum in medio aere pendeat, et solis aquis fulciatur? Non caret hoc ratione fateor, quia terra infimum locum occupans, ut est centrum mundi, naturaliter illic subsidit, sed

in hoc quoque artificio relucet admirabilis Dei potentia" (*Corpus Reformatorum*, vol. 32, p. 86). That what happens *naturaliter* can nevertheless be *artificium* is, for its part, highly artificial.

37. John Wilkins, *A Discovery of a New World*, book 2, proposition 9; 5th ed. (London: 1684), p. 156. [The title given in the text is that of the first edition, published in 1638.]

38. Gottsched, *Gesammelte Schriften*, ed. E. Reichel (Berlin: Gottsched-Verlag, 1903–), vol. 4, pp. 141–142.

39. Lichtenberg, *Vermischte Schriften* (Göttingen: 1800–1806), vol. 7, pp. 155ff.

40. Synesius of Cyrene, *De somniis* 133 A [following the German translation of W. Lang, *Das Traumbuch des Synesios von Kyrene. Übersetzung und Analyse der philosophischen Grundlagen* (Tübingen: Mohr, 1926), p. 6, which the author uses in the original].

41. Herder, *Adrastea* III, 6; *Sämtliche Werke*, ed. B. Suphan (Berlin: Weidmann, 1877–1913), vol. 23, pp. 511–512.

42. Feuerbach, *Die Naturwissenschaft und die Revolution* (1850); *Sämtliche Werke*, ed. W. Bolin and F. Jodl (Stuttgart: Frommann-Holzboog, 1960–1964), vol. 10, pp. 9–10.

Chapter 2

43. *Nikolaus Kopernikus Gesamtausgabe*, vol. 2, ed. F. Zeller and C. Zeller (Munich: Oldenbourg, 1949), p. 453: "De hypothesibus ego sic sensi semper, non esse articulos fidei, sed fundamenta calculi, ita ut, etiamsi falsae sint, modo motuum 'phaenomena' exacte exhibeant, nihil referat; quis enim nos certiores reddet . . . . "

44. Loc. cit.: "Quare plausibile fore videretur, si hac de re in praefatione nonnihil attingerem. Sic enim placidiores redderes peripateticos et theologos, quos contradicturos metuis."

45. Loc. cit.: "Peripatetici et theologi facile placabuntur, si audierint, eiusdem apparentis motus varias esse posse hypotheses nec eas afferri, quod certo ita sint, sed quod calculum apparentis et compositi motus quam commodissime gubernent, et fieri posse, ut alius quis alias hypotheses excogitet, et imagines hic aptas, ille aptiores, eandem tamen motus apparentiam causentes . . . ita a vindicandi severitate ad exquirendi illecebras avocati ac provocati primum aequiores, tum frustra quaerentes pedibus in actoris sententiam ibunt."

46. E. Zinner, *Entstehung und Ausbreitung der coppernicanischen Lehre*, Sitzungsberichte der Physikalisch-medizinischen Sozietät 74 (Erlangen: Mencke, 1943), p. 256. Zinner nevertheless considers it possible that the words *orbium mundi* in the manuscript could be seen to have been added later, since in his own references Copernicus speaks (Zinner says) only of the *Liber revolutionum*. Here, however, he has overlooked the instance that is the most important, because it represents the latest phase in the history of the genesis of the work—the instance that is found right in the first sentence of the dedicatory preface to Paul III: " . . . hisce meis libris, quos de Revolutionibus sphaerarum mundi scripsi . . . . " Also, for Osiander an addition to the copy of the manuscript that lay before him could only come from Copernicus or Rheticus, so it would be no less binding for him than everything else.

47. E. Zinner, *Entstehung und Ausbreitung* (cited in note 46), pp. 448ff.

48. A. Heller, *Geschichte der Physik* (Stuttgart: 1882; rpt. Wiesbaden: Sändig, 1965), vol. 1, p. 267.

49. Copernicus, *De revolutionibus orbium coelestium*, facsimile reprint of the first edition of 1543 with an introduction by J. Müller (New York and London: Johnson Reprint Corporation, 1965).

50. M. Curtze, in F. Hipler, *Spicilegium Copernicanum* (Braunsberg: 1873), pp. 361–362: "... an earlier revision, indeed even remnants of an even earlier one...." The deviations of the version in the first printed edition from the autograph are so great "that only one page (sheet 103a of the published editions), and in fact one table, agree completely in the manuscript and the printed text" (M. Curtze, loc. cit., pp. 363–364). The explicit of the fifth book of the manuscript would be that of the fourth book in the printed *editio princeps*.

51. J. Voigt, *Briefwechsel der berühmtesten Gelehrten des Zeitalters der Reformation mit Herzog Albrecht von Preussen* (Königsberg: 1841), p. 519.

52. J. Voigt, *Briefwechsel*, pp. 541–543.

53. Giese to Rheticus, in F. Hipler, ed., *Spicilegium Copernicanum* (Braunsberg: 1873), pp. 354ff. Now in K. H. Burmeister, *G. J. Rhetikus. Eine Bio-Bibliographie* (Wiesbaden: Pressler, 1967–1968), vol. 3, pp. 54–59.

54. E. Zinner, *Entstehung und Ausbreitung* (cited in note 46, above), pp. 216–217.

55. Kepler to Heydonus, summer 1605 (*Gesammelte Werke*, ed. M. Caspar [Munich: Beck, 1937–], vol. 15, p. 232; German trans. M. Caspar and W. von Dyck, *Kepler in seinen Briefen* [Munich: Oldenbourg, 1930], vol. 1, p. 246): "Thus in this regard [that is, in regard to the investigation of the orbit of Mars] I can boast that I have established an astronomy without hypotheses." Kepler to D. Fabricius, 10 November 1608 (*Ges. Werke*, vol. 16, p. 205; *Kepler in seinen Briefen*, vol. 1, p. 316): "Naked nature has led me, free of any garment of hypotheses."

56. Newton to Bentley, February 25 1692/3; *Correspondance*, ed. H. W. Turnbull and J. S. Scott (Cambridge: Cambridge University Press, 1959–1977), vol. 3, pp. 253–254: "That gravity should be innate inherent and essential to matter so yt [that] one body may act upon another at a distance through a vacuum wthout the mediation of anything else by and through wch their action or force may be conveyed from one to another is to me so great an absurdity that I believe no man who has in philosophical matters any competent faculty of thinking can ever fall into it. Gravity must be caused by an agent acting constantly according to certain laws, but wether this agent be material or immaterial is a question I have left to ye consideration of my readers."

57. Newton, *Principia mathematica* III: "Scholium generale"; *Opera omnia*, ed. S. Horsley (London, 1779–1785), vol. 3, p. 173: "... sed causam gravitatis nondum assignavi." Op. cit., p. 174: "Rationem vero harum Gravitatis proprietatum ex Phaenomenis nondum potui deducere, et hypotheses non fingo." (On this see I. B. Cohen, "The First English Version of Newton's *Hypotheses non fingo*," *Isis* 53 [1962]: 379–388.) The connection that Osiander set up between the basic ideas of hypothesis and fiction can also be found in the letter Petrus Ramus wrote to Joachim Rheticus in 1563, though here it is combined with the challenge to dissolve this connection—following the example of the Gordian knot—by means of an "astronomy without hypotheses": *astrologia figmentis hypothesium per te liberata* ... (ed. by M. Delcourt in *Bulletin de l'Association Budé* 44 [1934]: 15).

58. I. B. Cohen, *Introduction to Newton's "Principia"* (Cambridge: Cambridge University Press, 1971), pp. 152–156.

59. Leibniz. *Promemoria*, printed by the editor as a note to the *Tentamen de motuum coelestium causis; Mathematische Schriften*, ed. C. J. Gerhardt (Berlin and Halle: 1850–1863), vol. 6, p. 146n.

60. Loc. cit.: "Ut vero res intelligatur exactius, sciendum est motum ita sumi, ut involvat aliquid respectivum et non posse dari phaenomena ex quibus absolute determinetur motus aut quies;

consistit enim motus in mutatione situs seu loci. Et ipse locus rursus aliquid relativum involvit . . . . Hinc in rigore omne systema defendi potest . . . . "

61. "Copernici opera permiserunt, iis tamen correctis iuxta subiectam emendationem locis, in quibus non ex hypothesi sed asserendo de situ et motu terrae disputat." *Librorum prohibitorum decreta* (Rome: 1624), p. 144.

62. This is made plausible by E. J. Aiton's analysis, "The Celestial Mechanics of Leibniz: A New Interpretation," *Annals of Science* 18 (1962): 31–41, which supposes, instead, a dependence on Borelli's *Theoricae mediceorum planetarum ex causis physicis deductae* (Florence: 1666). I. B. Cohen (op. cit. [note 58], p. 154) pointed out that for the *Tentamen* (". . . certainly an ingenious and original piece of research") Leibniz would not have needed anything more than an acquaintance with Kepler's second law. Leibniz first heard of Newton's *Principia* through a review in the *Acta eruditorum* of June 1688. The copy of the *Principia* that he brought with him from Rome to Hanover, and in which he had made numerous marginal comments, found its way from Hanover to Göttingen in 1749 and there was culled out, in 1926, as a duplicate, by someone who failed to recognize the author of the disfiguring handwritten additions. Now in the possession of a Swiss collector, the marginalia have been edited by E. A. Fellmann: *Marginalia in Newtoni Principia mathematica* (Paris: J. Vrin, 1973).

63. Nevertheless, the appeal to the systems' kinematic equivalence is not a mere trick. In Leibniz's *Dynamics*, no particular diagram of the distribution of the bodies in the universe is true (*Dynamica* II 2, 16; *Mathematische Schriften*, ed. C. J. Gerhardt [Berlin and Halle: 1850–1863], vol. 6, p. 484). Kinematics is the "logic of physics."

64. Leibniz, *Elementa philosophiae arcanae de summa rerum* (1675), ed. I. Jagodinski (Kazan, Russia: 1913), p. 126.

65. M. Born, *Einstein's Theory of Relativity*, trans. H. L. Brose (London: Methuen, 1924), p. 277; original: *Die Relativitäts-theorie Einsteins*, 3rd ed. (Berlin: Springer, 1922), p. 251. Compare also M. Planck, *Wege zur physikalischen Erkenntnis*, 3rd ed. (Leipzig: Hirzel, 1943), vol. 1, p. 169: "Accordingly, Copernicus is to be valued not as a pathbreaking discoverer but as a highly gifted inventor." In the same way, Osiander had called Copernicus *hic artifex*.

66. D. I. Blokhintsev, "Kritik der philosophischen Ansichten der sogenannten 'Kopenhager Schule,'" in *Philosophische Fragen der Gegenwartsphysik* (Moscow: 1952), pp. 358–395.

67. H. Jonas, "Is God a Mathematician?" in *The Phenomenon of Life* (New York: Harper and Row, 1966), pp. 66–92.

68. Lichtenberg, "Nicolaus Copernicus," in *Pantheon der Deutschen*, vol. 3 (Leipzig, 1800); *Schriften und Briefe*, ed. W. Promies, vol. 3 (Munich: Hanser, 1972), pp. 138–188.

69. Descartes, *Principia philosophiae* I, 28: "Ita denique nullas unquam rationes circa res naturales, a fine, quem Deus aut natura in iis faciendis sibi proposuit, desumemus; quia non tantum nobis debemus arrogare, ut eius consiliorum participes esse putemus . . . "

70. Descartes, *Principia* III, 1–2.

71. Descartes, *Principia* III, 15: "In quem finem inventae sunt ab Astronomis tres diversae hypotheses, hoc est, positiones quae non ut verae, sed tantum ut phaenomenis explicandis idoneae considerantur."

72. Descartes, *Principia* III, 17: " . . . adeo ut Tycho non habuerit occasionem illam mutandi, nisi quia non hypothesin dumtaxat, sed ipsam rei veritatem explicare conabatur."

Notes to Pages 314–321

73. Descartes, *Principia* III, 18: "... verbo tantum asseruit terram quiescere."

74. Descartes, *Principia* III, 19: "... ipsamque tamtum pro hypothesi, non pro rei veritate haberi velim."

75. Descartes, *Discours de la méthode* V; ed. E. Gilson (Paris: J. Vrin, 1947), p. 42. Compare *Le Monde (Oeuvres*, ed C. Adam and P. Tannery [Paris: Cerf, 1897–1913], vol. 11, pp. 31–32): "Permettez donc pour un peu de temps à votre pensée de sortir hors de ce Monde, pour en venir voir un autre, tout nouveau, que je ferai naître en sa présence dans les espaces imaginaires."

Chapter 3

76. H. Bornkamm, "Kopernikus im Urteil der Reformatoren," *Archiv für Reformationsgeschichte* 40 (1943): 175.

77. As H. Bornkamm has it: op cit., p. 176.

78. K. H. Burmeister, *Georg Joachim Rheticus* (Wiesbaden: Pressler, 1967–1968), vol. 1, p. 47. Even if we presume that in Wittenberg it was, besides Erasmus Reinhold, above all Caspar Cruciger—preacher in the palace church, expert in Euclid, founder of botanical gardens, and practicing astronomer—who was the moving force in Rheticus's journeying to Frauenburg, nevertheless Rheticus, as a Master, could not have been given his leave of absence without Melanchthon's concurrence.

79. In this form, in which it is usually cited, the dictum comes from Aurifaber's adaptation of the table talk in the first printed edition of 1566 (Martin Luther, *Werke, Tischreden* [Weimar: Böhlau, 1912–1921], vol. 1, no. 855). The original transcript, from 4 June 1539 (same ed., vol. 4, no. 4638), gives the following text: "De novo quodam astrologo fiebat mentio, qui probaret terram moveri et non caelum, solem et lunam, ac si quis in curru aut navi moveretur, putaret se quiescere et terram, arbores moveri. [Luther:] But nowadays this is what happens: Anyone who wants to be clever must not be content with anything that other people respect. He must do something of his own, sicut ille facit, qui totam astrologiam invertere vult. Etiam illa confusa tamen ego credo sacrae scripturae, nam Josua iussit solem stare, non terram."

80. A. Deissmann, *Johann Kepler und die Bibel. Ein Beitrag zur Geschichte der Schriftautorität* (Marburg: 1894), pp. 10–11. The religious legitimacy of opposition to Copernicus consisted, we are told, in the fact that for his opponents "heaven and Earth [were], precisely in the old biblical sense, religiously valuable. Every morning and evening the worshipper's eyes saw the rise and the setting of the Sun, its continual revolution around the Earth, which was at rest. This sense perception was confirmed for him by faith, which assigned to the Earth the fixed central position in the Almighty's loving plan." In view of this sheltered situation in the world, the Copernican event has to take on the character of theological usurpation: "A pious member of the Old Covenant shrank from the sight of God, for fear of perishing; a similar feeling of infinite smallness grasps us in view of the structure of the world, throws doubt into our prayers, and gnaws at the roots of our moral conviction." In that case the Copernican universe would be counterbiblical even without Joshua 10:13 and apart from the principle of the sole authority of Scripture.

81. Luther, *Tischreden*, no. 290; *Sämmtliche Werke* (Erlangen: 1826–1857), vol. 57, p. 244. Here we should also mention how Luther, in the *Table Talk* (no. 17), once (in 1531) characterized Melanchthon's astronomical/astrological interests: "Ego puto, quod Philippus astrologica tractat, sicut ego bibo [German:] a strong drink of beer, quando habeo graves cogitationes."

82. *Luthers Randbemerkungen zu Gabriel Biels Collectorium*, ed. H. Degering (Weimar: Böhlau, 1933), p. 14.

83. "Domine ne in furore tuo arguas me. Hoc dicit trepida conscientia, quae semper timet, ne coelum super se ruat et in infernum descendat" *Werke* (Weimar: Böhlau, 1883–), vol. 3, p. 168.

84. Melanchthon, preface to Georg Peurbach, *Theoricae novae planetarum* (Wittenberg: 1535), p. 3: "Recte enim iudicant, reliquam philosophiam mancam atque mutilam esse, nisi rerum coelestium cognitio accedat."

85. "... bellum gerunt cum humana natura, quae praecipue ad has divinas res aspiciendas condita est ... " (op. cit., p. 4).

86. "... gubernare haec inferiora (sc. deus), ita nos vicissim huius summi artificis lineas considerantes" (op. cit., p. 5).

87. "Si ob hanc causam praecipue condita est coelestis natura, ut certe est, ut Deum nobis monstret, satis constat voluntati Dei non parere istos, qui haec divinitatis vestigia non aspiciunt neque inquirunt" (op. cit., p. 5).

88. Melanchthon, preface to the *Libellus de sphaera* of John of Sacrobosco (Wittenberg: 1540), p. 2.

89. This first edition of Melanchthon's *Initia doctrinae physicae* is the basis of the edition of C. G. Bretschneider in the *Corpus Reformatorum*, vol. 13 (Halle: 1846), from which (where not otherwise indicated) I quote here.

90. *Initia doctrinae physicae* I; *Corpus Reformatorum*, vol. 13, p. 216: "Bonae mentis est veritatem a Deo monstratam reverenter amplecti, et in ea acquiescere .... "

91. *Initia*, p. 216: "Quanquam autem rident aliqui physicum testimonia divina citantem, tamen nos honestum esse censemus philosophiam conferre ad coelestia dicta, et in tanta caligine humanae mentis autoritatem divinam consulere, ubicumque possumus."

92. *Initia*, p. 218: "Postremo generaliter, ubicumque ponatur terra extra medium, confunde-retur ratio aequalis incrementi et decrementi dierum et noctium."

93. Melanchthon to Duke Albert of Prussia, 18 October 1544.

94. Emil Wohlwill, "Melanchthon und Copernicus," *Mitteilungen zur Geschichte der Medizin und der Naturwissenschaften* 3 (1904): 260–267.

95. Melanchthon to Burkard Mithobius, 16 October 1541; *Opera* (*Corpus Reformatorum* vols. 1–28 [Halle: 1834–1860]), vol. 4, p. 679: "... sed quidam putant esse egregium KATORTHOMA rem tam absurdam ornare, sicut ille Sarmaticus Astronomus, qui movet terram et figit Solem. Profecto sapientes gubernatores deberent ingeniorum petulantiam cohercere." ("KATORTHOMA" is transliterated from the Greek.)

96. *Corpus Reformatorum* (cited in note 95), vol. 4, p. 847.

97. *Corpus Reformatorum*, vol. 4, pp. 810, 839. On the question of the relation between Melanch-thon and Rheticus see H. Bornkamm, "Kopernikus im Urteil der Reformatoren," *Archiv für Reformationsgeschichte* 40 (1943):171–183. What 'forms' Melanchthon was still able to maintain, even in this polemic, can be seen from Bornkamm's significant observation that Copernicus's name is used, in Melanchthon's remarks, only in those that contain positive, approving judgments (*Corpus Reformatorum*, vol. 11, p. 839; vol. 13, p. 241; vol. 20, p. 808), and is left out in those that contain negative evaluations (*Corpus Reformatorum*, vol. 4, p. 679; vol. 13, p. 216).

98. *Corpus Reformatorum*, vol. 11, p. 839: "His et similibus observationibus moti Copernicum magis admirari et amare coepimus. Eumque secuti nostras observationes cum omnibus praece-

dentibus in hoc genere contulimus. In qua consideratione animadvertimus anni veram seu adparentem magnitudinem ... nunc iterum maiorem esse, quam vel Albategnii vel Alphonsi temporibus: ac pene rursus aequalem factam ei, quam Ptolemaeus prodidit .... " So in the question of the periodicity of the variation in the length of the year as well, people in Wittenberg were already 'Copernican': The preferred, intermediate value is ascribed to Ptolemy.

99. The omitted or altered passages are printed in contrasting type:

First edition	Second edition
"Sed hic aliqui *vel amore vel novitatis, vel ut ostentarunt ingenia*, disputarunt moveri terram, et *contendunt* nec octavam sphaeram, nec Solem moveri, cum quidem caeteris coelestibus orbibus motum tribuant, Terram etiam inter sidera collocant. *Nec recens hi ludi conficti sunt.* Extat adhuc liber Archimedis	"Sed hic aliqui disputarunt moveri terram, et *dicunt* nec octavam sphaeram ....
Etsi autem artifices acuti multa exercendorum ingeniorum causa quaerunt, tamen *adseverare palam absurdas sententias, non est honestum et nocet exemplo. Bonae mentis est,* veritatem a Deo monstratam reverenter amplecti, et in ea acquiescere	Etsi ... quaerunt, tamen *sciant iuniores, non velle eos talia adseverare. Ament autem in prima institutione sententias receptas communi artificum consensu, quae minime sunt absurdae, et ubi intelligunt* veritatem a Deo monstratam *esse* reverenter ....
His divinis testimoniis confirmati, veritatem amplectamur, *nec prestigiis eorum qui decus ingenii esse putant, conturbare artes, abduci nos ab ea sinamus.*"	His divinis testimoniis confirmati, veritatem amplectamur."

100. W. Dilthey, *Weltanschauung und Analyse des Menschen seit Renaissance und Reformation, Gesammelte Schriften* (Leipzig: Teubner, 1914–1982), vol. 2, p. 179.

101. Melanchthon, *Loci communes*, ed. H. Engelland, in *Melanchthons Werke. In Auswahl*, ed. R. Stupperich (Gütersloh: Bertelsmann, 1951–), vol. 2, pt. 1, p. 7.

102. *Philippi Melanchthonis Commentarii in epistolam Pauli ad Romanos* (1540), ed. Th. Nickel (Leipzig: 1861), p. 70: "Nihil dubium est homines praecipue conditos esse, ut in eis luceret noticia Dei .... Nec dubium est hunc totum mundum conditum esse, ut dicat testimonium de Deo."

103. Op. cit., pp. 70ff.: "Sed tamen, quia non prorsus deleta est in animis noticia naturalis, excitemus eam et aspiciamus vestigia Dei in natura rerum .... Nec frustra impressa sunt naturae tot vestigia Dei .... Et eruditus physicus multa apte colligere potest."

104. Melanchthon, *Commentarius in Genesin*, praefatio (*Corpus Reformatorum*, vol. 13, p. 762): "Et video non alium locum esse, in quo se mirabilius fides exerceat, quam hunc." On the importance of the idea of the Creation for Melanchthon, Otto Riemann asserts, in his dissertation (which is even more instructive in connection with the nineteenth century than it is in connection with the sixteenth), *Philippi Melanchthonis studia philosophica, quam rationem et quid momenti ad eius theologiam habuerint, quaeritur* (Halle: 1885), p. 25: "hoc est fundamentum, in quo universa huius viri philosophia posita videtur."

105. Melanchthon, *Commentarius in Genesin*, I (*Corpus Reformatorum*, vol. 13, p. 765): "Philosophi docuerunt eam esse terrae naturam, ut si extra suum naturalem locum dimoveatur, sua sponte relabatur ad locum suum, ut in saxo apparet. Hic ostendit Moses non esse naturam terrae hoc vel illo loco subsidere: sed voluntatem Dei, qui ei definivit locum .... "

106. Op. cit., p. 774; "Sic scriptura per omnia nobis inculcat non naturam aliquam esse, quae per sese gignat, sed Dei beneficia esse omnia quae gignuntur."

107. Op. cit., p. 767: "Imo verbo Dei consistunt coeli, non natura sua."

108. *Corpus Reformatorum*, vol. 12, p. 690: "Sicut astronomia est cognitio motuum coelestium, qui divinitus ordinati sunt, ita philosophia moralis est cognitio operum, videlicet causarum et effectuum, quos Deus ordinavit in mente hominis."

109. Melanchthon, *Commentarius in Ecclesiastem* (*Corpus Reformatorum*, vol. 14, p. 103): "Obiter autem observent hic studiosi dicta, quae affirmant, Terram stare, et Solem moveri. Caeteras physicas disputationes omitto."

110. Melanchthon, *Commentarius in Genesin*, II (*Corpus Reformatorum*, vol. 13, p. 774): "Nam omnia, quae de hominis dignitate dicuntur, (sc. dicuntur de Christo) in quo recuperavimus dignitatem amissam in Adam. Magnum est autem credere Christum dominari, credere nobis subiecta esse omnia."

111. Melanchthon, *Initia doctrinae physicae* I (*Corpus Reformatorum*, vol. 13, pp. 213–214): "Mundus est compages coelestium et inferiorum corporum arte distributorum .... Haec est vetus definitio, cui vere addi potest, conditum esse hoc tantum et tam mirandum opus a Deo, ut sit domicilium humanae naturae, in qua Deus innotescere et conspici voluit .... Et quanquam interdum necesse est Stoicorum dicta reprehendi, tamen hoc dictum probandum est, quod ait, omnia in natura rerum propter hominem nasci, homines autem natos esse propter Deum, videlicet, ut innotescat et celebretur Deus."

112. In these voluntaristic formulations of Melanchthon's, the Cartesian problem of certainty seems still to slumber in touching innocence—for example, in the following question at the beginning of the *Initia* (*Corpus Reformatorum*, vol. 13, p. 185): "Estne certitudo aliqua doctrinae physicae?—Vult Deus artes aliquas vitae restrices, imo ipsum quoque aliquo modo monstrantes, certas et firmas esse .... " Of course the problem of certainty also still sleeps right in the bosom of the principle of teleology, which provides everything for Melanchthon: If an implication of God's will can be perceived in a cognition, then the element of certainty is guaranteed along with it. Thus, very concisely, in the same work (p. 189): "Haec experientia tanquam ordinatio Dei indicetur esse certa, nam Deus non frustra res condidit."

113. Melanchthon, preface to Johannes Voegelin, *Elementa geometriae* (Wittenberg: 1536): " ... haec ars aditum patefacit ad illam praestantissimam philosophiam de rebus celestibus, quae quantum habeat dignitatis, qua multipliciter prosit hominum vitae, minime obscurum est .... "

114. Erasmus Reinhold, *Themata quae continent methodicam tractationem de horizonte rationali* (Wittenberg: 1541), thema 17: "Si enim intelligamus terram esse perfecte rotundam ac aequabili superficie sine hisce montibus ac vallibus, quae Deus, ut caetera, condidit ad usus hominum, geometrice ostendi potest, quod .... "

Chapter 4

115. Hegel, "Über das Wesen der philosophischen Kritik," *Gesammelte Werke* (Hamburg: Meiner, 1968–), vol. 4, p. 124: " ... the Enlightenment expresses already in its origin, and in general, the commonness of the understanding, and its vain assumption of superiority over reason .... " Volume 4, p. 126: " ... the tedium of the sciences—of this edifice of an understanding abandoned by reason .... "

116. J. Voigt, *Briefwechsel der berühmtesten Gelehrten des Zeitalters der Reformation mit Herzog Albrecht von Preussen* (Königsberg: 1841), p. 112.

117. Rheticus to Joachim Camerarius, Cracow, 29 May 1569; in K. H. Burmeister, *Georg Joachim Rhetikus. Eine Bio-Bibliographie* (Wiesbaden: Pressler, 1967–1968) (henceforth: *Rheticus*), vol. 3, pp. 190–191. Also, to Petrus Ramus, Cracow, 1568 (*Rheticus*, vol. 3, p. 188): "Tot et tanta sunt, quae tracto, et ad quae mihi hactenus ars medica, meus Maecenas, sumptus suppeditavit . . . . "

118. Rheticus to Ferdinand I (*Rheticus*, vol. 3 p. 134): "Horum siderum motus, ortui et occasui, respondet transitus imperiorum, ab ortu ad occasum et inde in septentrionem."

119. Rheticus to Ferdinand I (*Rheticus*, vol. 3, p. 138): "Diximus astra naturae ordine haec inferiora gubernare. Sed coeli conditor, qui astra nomine vocat, eis modum et terminum praescribit, ubi vult, cursum sistit, effectus ut vult, moderatur. Sicut Josuae solum in coelo sistebat . . . . "

120. Rheticus to Paul Eber in Wittenberg, Cracow, 1 March 1562 (*Rheticus*, vol. 3, pp. 162–163): "Pono autem meum principium 60 annis ante Christi nativitatem, sub initium anomaliae praecessionis. Hoc posito progredior per anomaliae revolutiones ad mundi creationem et ad mundi finem. Videbis multa admiratione digna." ("Sub initium anomaliae praecessionis" is incorrectly translated in the cited edition.)

121. Rheticus to Tiedemann Giese, Leipzig, 14 October 1549 (*Rheticus*, vol. 3, p. 86).

122. (Rheticus), *Iudicium de successoribus in regno Poloniae; Rheticus*, vol. 3, pp. 193–195. The conclusion of the horoscope reads, "Ultra autem me prognosticare sidera non permittunt."

123. Rheticus, dedicatory preface to Euclid's *Elementa*, Leipzig, 11 November 1549 (*Rheticus*, vol. 3, p. 90): "Etsi autem hoc . . . liberale et usu compertum et veterum cura excultum studium est, tamen in praesentia mentionem istam omittam, propterea quod aliter aliis sentientibus et iudicantibus de hac parte, ad explicationem rei nova esset disputatione opus, quam alio magis idoneo tempore ac loco fortasse exequemur."

124. Rheticus to Petrus Ramus, Cracow, 1568 (*Rheticus*, vol. 3, p. 188): "In ea vero parte, quae est de effectibus siderum, pandectas astrologiae in ordinem redigo. Sed et eius propriam condidi artem, antiquissimis artis fundamentis exquisitis."

125. Girolamo Cardano, *De propria vita*, chapter 15: " . . . velut et Georgium Porrum Rheticum . . . . " See Burmeister, *Rheticus*, vol. 1, pp. 90 and 92.

126. Kaspar Brusch to Joachim Camerarius, Lindau, November 1547 (*Rheticus*, vol. 3, p. 73): "Horum ipse spectator fui et auditor quotidianus . . . . " A note by Luca Gaurico on Rheticus's horoscope reads, "After his return from Italy he was attacked by madness. In April 1547 he died" (*Rheticus*, vol. 1, pp. 95–96).

127. Dedication to Heinrich Widnauer in Feldkirch, 13 August 1542 (*Rheticus*, vol. 3, p. 49).

128. However, one should not interpret the reduced level of truth-fervor, as one proceeds from the first book of the *De revolutionibus* to the later books, as a resignation on Copernicus's part, which would then have been continued in Rheticus. In that case not only would the dedicatory preface and the resistance to Osiander be unintelligible, but Rheticus's *Narratio prima* (which, after all, is already based on the completed manuscript of the whole book) would be unthinkable. In fact that 'reduction' is due to the specific function of the first book, which is to shield the work from criticisms based on natural philosophy. There is no special "optimism of the first book" (J. Mittelstrass, *Die Rettung der Phänomene* [Berlin: DeGruyter, 1962], p. 201), but only its heterogeneous argumentative intention.

129. Rheticus, dedicatory preface to Ferdinand I, Cracow, 1557 (*Rheticus*, vol. 3, p. 139): "Nam

Notes to Pages 344–361

meo iudicio nullum aliud instrumentum praestantius obelisco fuerit."

130. Dedicatory preface to Ferdinand I (*Rheticus*, vol. 3, p. 149).

131. Op. cit., p. 139: "Non igitur obeliscus humanum est inventum, sed Deo auctore institutus, non ut satisfaceret curiositati humanae, sed ut Dei in coelo et terra geometriam doceret; armillae, regulae, astrolabia, quadrantes sunt humana inventa. Ideo et laboriosa et maxime erroribus oportuna. Obeliscus Dei monitu aedificatus facile omnia haec praestat et exacte."

132. Rheticus to Hans Crato (*Rheticus*, vol. 3, p. 123): "Per hunc, Deo dante, totum orbem stellarum fixarum denuo conscribam."

133. Dedicatory preface to Ferdinand I (*Rheticus*, vol. 3, p. 138): "Cum vero in Prussia plus minus triennio egissem, discedenti mihi optimus senex iniunxit, ut eniterer ea perficere, quae ipse senio et suo quodam impeditus fato minus potuisset absolvere."

134. Op. cit., p. 139: "... et hanc provinciam dominus Copernicus nobis iniunxerit, quem non solum tanquam praeceptorem, sed ut patrem colui .... "

135. Petrus Ramus to Rheticus; ed. M. Delcourt in *Bulletin de l'Association Guilleaume Budé* 44 (1934).

136. Ramus to Rheticus (cited in previous note), p. 9: "Id vero te assecuturum arbitrarer si, sublatis hypothesibus omnibus, tam simplicem astrologiam faceres, quam simplicem astrorum essentiam natura ipsa fecerit."

137. Rheticus to Ramus, 1568; *Rheticus*, vol. 3, p. 188.

138. Rheticus, dedicatory preface to the ephemeris for the year 1550 (*Rheticus*, vol. 3, p. 107): "Quid dicam quantum operae huic cognitioni ac scientiae impenderim et ... tamquam Hercules huc illuc oberrans et non quidem cingulum aliquod aut poma quaerens, quae referrem ad meos, sed de maximis et necessariis rebus veritatem, quae iam tamquam monstris errorum profligatis conspici et apprehendi posset." Almost simultaneously, Rheticus publishes in Leipzig, as an introduction to his trigonometry, the *Dialogus de canone doctrinae triangulorum*, in which he depicts himself as Philomathes, who conveys the fruits from Copernicus's marvelously lovely gardens (*Rheticus*, vol. 1, pp. 101–102, and vol. 2, p. 75).

Chapter 5

139. Here I append to my presentation of the Nolan in *The Legitimacy of the Modern Age* (Cambridge, MA: MIT Press, 1983), pp. 549–596 (original: *Die Legitimität der Neuzeit* [Frankfurt: Suhrkamp, 1966], pp. 524–585; rev. ed., *Aspekte der Epochenschwelle* [Frankfurt: Suhrkamp, 1976], pp. 109–163) the aspect that deals with the history of his influence. [Quotes from *La Cena de le Ceneri—The Ash Wednesday Supper*—have been translated from the original in Bruno, *Dialoghi italiani*, ed. G. Gentile and G. Aquilecchia (Florence: Sansoni, 1958), pp. 5–171. In the original German edition of the present book the translation by F. Fellmann: G. Bruno, *Das Aschermittwochsmahl*, sammlung insel 43 (Frankfurt: 1969) and insel taschenbuch 548 (Frankfurt: 1981), with an introduction by Hans Blumenberg, was used.]

140. Bruno, *La Cena de le Ceneri* I (*Dialoghi italiani*, pp. 26–27): "... per intender il suo Copernico, et altri paradossi di sua nuoua philosophia." In fact Goethe, with his faculty for finding *le mot juste*, in 1770 called Bruno *cet homme paradoxe*, or else characterized him by an untraceable excerpt (*Werke*, ed. E. Beutler [Munich: Artemis, 1949–], vol. 4, p. 960). But the French *Encyclopedia* (vol. 11, col. 895; author: d'Alembert) still refers to the Copernican system as a paradox *au sentiment du peuple*, even though all scholars agreed that it was true.

141. H. Blumenberg, "On a Lineage of the Idea of Progress," *Social Research* 41 (1974): 23–25.

142. Bruno, *Cabala del cavallo Pegaseo* II, i; *Opere italiane*, ed. P. de Lagarde (Göttingen: 1888), p. 589.

143. *Il Sommario del processo di Giordano Bruno*, ed. A. Mercati, Studi e Testi 101 (Rome: Bibliotheca apostolica vaticana, 1942): "Natura dei est finita, si non producit de facto infinitum, aut infinita."

144. Not accidentally, and not without a relation to Bruno's conflict, much later, with the dogma of the Incarnation, the idea first appears in the struggle over Arianism, as an argument against the "begetting" (*generatio*) of the Son, and in favor of His having been produced by volition (*creatio*): Candidus Arianus, *De generatione divina ad Marium Victorinum*, chapter 7; in Marius Victorinus, *Traités Theologiques sur la Trinité*, Sources Chrétiennes vol. 68, ed. P. Henry and P. Hadot (Paris: Cerf, 1960), p. 118 (against the Neoplatonic metaphors of the 'source,' for God's activity): "Sed effundit et semper effundit (sc. deus), semper enim superfluit. Ergo et novi angeli et novi mundi." Other formulations of the Arian argument can be found in (pseudo-)Rufinus, *De fide* (*Patrologia latina*, ed. J. P. Migne [Paris: 1844–1864], vol. 21, col. 1131 a): "Si natura deus crearet et non voluntate, infiniti mundi creati reperirentur, innumerabiles vero qui per dies singulos crearentur" (cited by P. Hadot in his *Commentaire* to Marius Victorinus, Sources Chrétiennes, vol. 69 [Paris: Cerf, 1960], p. 680).

145. Lessing, *Das Christentum der Vernunft* (1753), sections 2, 3, 13, 15, 18.

146. V. Spampanato, *Documenti della vita di Giordano Bruno* (Florence: L. S. Olschki, 1933), p. 169: "Deinde fuit admonitus ad reliquendum huiusmodi eius vanitates diversorum mundorum, atque ordinatum quod interrogetur stricte. Postea detur ei censura." It may not be superfluous to point out that while Bruno's conflict was in fact a conflict with Christianity, it did not essentially have to be this. The post-Copernican hyperbole of "infinity" and of "worlds" is intolerable for any implicitly or explicitly anthropocentric religion. In a letter written in 1917, the Jewish philosopher of religion Franz Rosenzweig describes Copernicanism as radically incompatible with any idea of revelation: " . . . the difference is that for pagan thought there are many worlds and possibilities, reasons and accidents, goals and goods, and for revealed thought everything exists only 'in *one* instance.' For revelation establishes an above and a below, a Europe and an Asia, etc., on the one hand, and an earlier and a later, a past and a future, on the other. The boundless . . . descends to the earth and from here, from the place of its descent, draws boundaries in the ocean of space and the stream of time. That is the thing, an above and a below, in spite of Copernicus . . . " (Franz Rosenzweig, *Briefe* [Berlin: Schocken, 1935], p. 211).

147. V. Spampanato, *Vita di Giordano Bruno* (Messina: Principato, 1921), p. 784 (documenti romani VIII).

148. Aristotle, *De caelo* I, 5; 272 a 4–5. Also important for the transformations that Copernicanism entails for Aristotelianism is the argument in *De caelo* I, 7; 275 b 12–15: Something infinite cannot rotate because it has no center, which is necessary in order for a rotation around it to be definable.

149. Bruno, *Oratio valedictoria . . . in academia Witebergensi* (1588), in *Opera latina*, ed. F. Fiorentino et al. (Naples: 1879–1891), I i, 19: "Tellurem matrem nostram unum ex astris . . . . "

150. *Die Hauptschriften zum Pantheismusstreit zwischen Jacobi und Mendelssohn*, ed. Heinrich Scholz (Berlin: Reuther & Reichard, 1916).

151. Bayle, *Dictionnaire historique et critique*, 3rd ed. (1720), vol. 1, p. 671: " . . . il y fut brûlé, dit-on, comme un impie . . . . On ne sait point au bout de quatre vints ans, si un Jacobin a été brûlé à Rome, en place publique, pour ses blasphêmes. I n'y a pas loin de l'incertitude à la fausseté dans des faits de cette nature."

152. Marcellus Stellatus Palingenius (Pier Angelo Manzoli), *Zodiacus vitae* (Venice: 1534) (placed on the Index in 1558). English translation by Barnaby Goodge, 1560 and 1565. Bruno's acquaintance with this didactic poem is demonstrated by his Wittenberg *Oratio valedictoria*, in which he says that Palingenius was more of a philosopher than all the Greeks and Romans. cf. F. W. Watson, *The Zodiacus Vitae of Marcellus Palingenius: An Old School Book* (London: 1907).

153. Kant, *Allgemeine Naturgeschichte und Theorie des Himmels*, 3rd Anhang; *Gesammelte Schriften*, Akademie ed. (Berlin, 1900–1942), vol. 1, p. 354.

154. Heine, *Zur Geschichte der Religion und Philosophie in Deutschland* II; *Sämtliche Schriften*, ed. K. Briegleb (Darmstadt: Wissenschaftliche Buchgesellschaft, 1968–), vol. 3, pp. 570–571. Translation is based in part on that of J. Snodgrass: Heine, *Religion and Philosophy in Germany* (Boston: Beacon Press, 1959), pp. 78–80.

155. Feuerbach, *Towards a Critique of Hegel's Philosophy* (1839), trans. Z. Hanfi, in *The Fiery Brook: Selected Writings of Ludwig Feuerbach* (Garden City, NY: Doubleday, 1972), p. 57. Original: *Zur Kritik der Hegelschen Philosophie*, in *Kleine Schriften*, ed. K. Löwith (Frankfurt: Suhrkamp, 1966), p. 82. In 1843, in *Principles of the Philosophy of the Future* (section 14; *The Fiery Brook*, p. 192; *Kleine Schriften*, p. 161), Feuerbach writes that "all the conceptions of theism, if taken seriously, carried out, and realized, must necessarily lead to pantheism." Here Feuerbach describes the process as a historical one, to which Bruno had subjected himself as a process of personal rigor, without entirely taking the final step, which Feuerbach defines as follows: "Material objects can be derived from God only if God Himself is determined as a material being." That the realization of Christianity is its negation is Feuerbach's still more concise formula for the state of affairs in relation to which Bruno's name and fate can claim priority over the founder-figures, Descartes and Bacon.

Chapter 6

156. E. Husserl, *The Crisis of European Sciences and Transcendental Phenomenology*, trans. D. Carr (Evanston: Northwestern University Press, 1970), p. 227; *Die Krisis der europäischen Wissenschaften und die transzendentale Phänomenologie*, Husserliana, vol. 6 (The Hague: Nijhoff, 1954) p. 230.

157. This and the quotations in the next two paragraphs are from Husserl, *The Crisis*, pp. 48–52; *Die Krise*, pp. 49–53.

158. Husserl, "Die Reaktion des Empirismus gegen den Rationalismus" (a sketch from the period of the *Crisis* treatise), Husserliana, vol. 6 (The Hague: Nijhoff, 1954), p. 448.

159. Vincenzo Viviani, *Racconto istorico della vita di Galileo* (1654), in Galileo, *Opere*, ed. A. Favaro (Florence, 1890–), vol. 19, p. 620.

160. Goethe, *Werke*, ed. E. Beutler (Zurich: Artemis, 1948–), vol. 17, p. 758.

161. E. Cassirer, "Wahrheitsbegriff und Wahrheitsproblem bei Galilei," *Scientia* 62 (Milan, 1937): 129. See also E. Cassirer, "Galileo's Platonism," in *Studies in the History of Science, Offered to G. Sarton* (New York: Schumann, 1946), pp. 277–297, and H. Blumenberg, "Neoplatonismen und Pseudoplatonismen in der Kosmologie und Mechanik der frühen Neuzeit," in *Le Neo-Platonisme* (Paris: Editions de C.N.R.S., 1971), pp. 447–474.

162. Galileo, *Opere*, ed. S. Timpanaro (Milan-Rome: Rizzoli, 1936–), vol. 2, p. 85.

163. Galileo, *Dialogue Concerning the Two Chief World Systems*, trans. S. Drake (Berkeley and Los Angeles: University of California Press, 1970), p. 207. Original: *Opere*, ed. Timpanaro, vol. 1,

pp. 282–283. Additional material on this subject can be found in F. Klemm, *Technik. Eine Geschichte ihrer Probleme* (Freiburg: Alber, 1954), pp. 170–176.

164. Aristotle, *Physics* VI 10; 240 b 15–17.

165. (Pseudo-) Aristotle, *Quaestiones mechanicae* 847a 11–19. On this see F. Krafft, "Die Anfänge der theoretischen Mechanik und die Wandlung ihrer Stellung zur Wissenschaft von der Natur," in *Beiträge zur Methodik der Wissenschaftsgeschichte*, ed. W. Baron (Wiesbaden: Steiner, 1967), pp. 12–33.

166. Aristotle, *Quaestiones mechanicae* 847 a 21–28 and 851 b 15–24.

167. Galileo, *Opere*, ed. Timpanaro, vol. 1, pp. 64–65.

168. E. Cassirer, "Galileo's Platonism" (cited in note 161). Cassirer points to the possibility (or impossibility) of using the metaphor of the "book of nature" as an illustration of the difference between Galileo and Plato: "All this is implied in the saying of Galileo's that Philosophy is written in the great book of nature. For Plato philosophy was not written in nature. It was written in the minds of men" (op. cit., p. 284).

169. Galileo, *Opere*, ed. Timpanaro, vol. 1, pp. 147–152. On this see H. Blumenberg, *The Legitimacy of the Modern Age* (Cambridge, MA: MIT Press, 1983), pp. 392–393; original: *Die Legitimität der Neuzeit* (Frankfurt: Suhrkamp, 1966), pp. 395–397.

170. M. Nicolson, "The Telescope and Imagination," *Modern Philology* 32 (1934/5): 233–260; here, pp. 254–255.

171. A. Oregio, *Praeludium philosophiae ad tractatus theologiae* (Rome: 1637), p. 119 (quoted in H. Grisar, *Galileistudien* [Regensburg, 1882], p. 304).

172. Galileo, *Opere*, ed. Timpanaro, vol. 1, p. 551. The liberation by the principle of omnipotence, a liberation from which the modern idea of science received propelling stimuli, makes it to be expected that its losses, too—especially its losses of intuitiveness—would occasionally be formulated as negations of the principle of omnipotence. Goethe expressed this, quite naturally, in his poem of 1806, "Metamorphose der Tiere": "Diese Grenzen erweitert kein Gott, es ehrt die Natur sie:/ ... Und daher ist den Löwen gehörnt der ewigen Mutter/ Ganz unmöglich zu bilden, und böte sie alle Gewalt auf ... " ["These are bounds which no god can extend, and Nature honors them:... and it would therefore be quite impossible for the eternal Mother to fashion a lion with horns, whatever efforts she might make ... " (trans. D. Luke in his *Goethe. Selected Verse* [Baltimore: Penguin, 1964], pp. 153–154)]. Goethe repeats the idea as late as 28 March, 1830, to Chancellor von Müller: "God himself could not create a lion with horns .... "

173. Galileo, *Opere*, ed. Timpanaro, vol. 1, pp. 484–485.

Part IV

Chapter 1

1. H. Diels, ed., *Fragmente der Vorsokratiker* (Berlin: Weidmann, 1934, 1951–1952, 1954), 58 B 37, which is Simplicius, *In Aristotelis de caelo* II, 13, in J. L. Heiberg, ed., *Commentaria in Aristotelem graeca*, vol. 7 (Berlin: 1894), pp. 511ff.

2. Jakob Brucker, *Kurtze Fragen aus der Philosophischen Historie*, vol. 2 (Ulm: 1731), pp. 136–137: "These two doctrines, which in recent times Copernicus has hunted out again, and almost the whole mathematical world has accepted .... "

3. J. Brucker, op. cit., vol. 1 (Ulm: 1731), p. 24: "Are there also examples of paradoxes, of truths that seemed absurd, and therefore were spurned, but were nevertheless finally found to be well-founded and true? There are plenty of them, but one will make the matter clear . . . ."

4. Aristotle, *De caelo* II 13; 293 b 25–30. Regarding the dynamic function of the central fire in the Pythagorean cosmology we have the testimony of Chalcidius, who reports in his (4th century A.D.) commentary on the *Timaeus* (a commentary that was much used throughout the entire Middle Ages), "The Pythagoreans assert that as the primary ground of all matter, fire occupies the center of the world, and they call it Zeus's guardian. By its power the Earth and the counter-Earth are moved in a circle like the stars" (*Fragmenta philosophorum graecorum*, ed. F. W. A. Mullach [Paris: 1860–1881], vol. 2, paragraph 209).

5. Kepler's text is given in an appendix to *Nikolaus Kopernikus, Erster Entwurf seines Weltsystems*, ed. F. Rossman (Munich: H. Rinn, 1948), pp. 57–90. The dating of Kepler's marginal and interpolated remarks is limited by the *terminus a quo* of his mention of the telescope.

6. F. M. Cornford (*Plato's Cosmology* [London: Routledge and Kegan Paul, 1937; rpt. Indianapolis and New York: Bobbs-Merrill, n.d.], p. 115, n. 4), who decides in favor of Archer-Hind's reading, gets into conflict, as a result, with his own general thesis, which is that Plato outlines a cosmos without an anthropocentric teleology—in which case it would be consistent that the information provided by the cosmic time indicator is only contingently and partially accessible to men, and not accessible at all in the case of the greatest and most perfect unit, the "Great Year." If Cornford's conjecture is correct, that Plato could have written the myth of the demiurge with an armillary sphere before him, so clearly does his cosmology resemble the setting up of a model of the orbits of the Sun and the planets (op. cit., p. 74)—if this conjecture is correct, then one can understand from what point of view he contemplated this structure, which at any rate was not the point of view of man, but more nearly that of an external onlooker who makes the whole an object for himself.

7. Jean Paul, *Hesperus*, Second Notebook, Sixth Intercalary Day: "Uber die Wüste und das gelobte Land des Menschengeschlechts."

Chapter 2

8. Thus, for example, Tatian, *Ad Graecos* VI, 1; XXV, 6.

9. Augustine, *De Genesi ad litteram* IV, 35: ". . . ubi et ipsa requiescere posset invenit, tanto stabilius atque firmius, quanto ipsa illius, non ille huius eguit ad quietem suam."

10. *De Genesi* II, 28–29. See also *Epist.* XI, 3: *causa, species, manentia* as a trinity of determining elements for every substance.

11. *De Genesi* IV, 37.

12. *De Genesi* V, 41.

13. A. Maier, "Scholastische Diskussionen uber die Wesensbestimmung der Zeit," *Scholastik* 26 (1951): 536–537: "It is an original and daring theory, and Olivi knew very well that it was not likely to meet with approval, and was more likely to produce new difficulties. So he preferred to retract everything in a final note and ceremonially to embrace the usual view: "Ista autem via longe est ab aula, quia a nullo magno communiter hodie, quod sciam, tenetur . . . ."

14. My presentation does not follow a chronological order, but rather obeys the requirements of clarifying the state of the problem, which resulted from the various approaches to the topic of time, and which finally constitutes the 'pre-Copernican situation.' In connection with this late

convergence point and the resources that were available then, the original sequences and relations of influence are not important.

15. Plotinus, *Enneads* III, 7, 12. See also H. Jonas, "Plotin über Zeit und Ewigkeit," in *Politische Ordnung und menschliche Existenz. Festgabe fur Eric Voegelin* (Munich: Beck, 1962), pp. 295–319, and W. Beierwaltes's commentary on the passage in Plotinus, *Über Ewigkeit und Zeit*, Quellen der Philosophie 3 (Frankfurt: Klostermann, 1967), pp. 62–71.

16. W. Gundel, "Planeten," in *Paulys Realenzyklopädie der Classischen Altertumswissenschaften*, ed. G. Wissowa (Stuttgart: Metzler, 1894–1963), vol. 20, p. 2082. Giordano Bruno extended the comparison into a total myth of his universe: *Cabala del cavallo Pegaseo* II, 1 (*Opere italiane*, ed. P. de Lagarde [Göttingen: 1888], p. 589): "It seems to me that this opinion is not very distant from, nor does it contradict, that prophetic dogma according to which the whole is in the hand of the universal maker like a single lump of clay in the hand of a single potter, to be made and unmade—on the wheel of this spinning of the heavenly bodies—in accordance with the vicissitudes of the generation and the corruption of things, now as a good vessel and now as a bad one, from the same material." Augustine introduced the comparison of the motion of the heavens to the *rota figuli* into his argument against astrology when he turned Nigidius Figulus's argument from the potter's wheel—that on account of the speed of the rotation of the heavens, twins would have different astrological aspects—against the possibility of any horoscope at all, because it could never hit the moment of birth with sufficient accuracy (*De civitate dei* V, 3).

17. Augustine, *De Genesi ad litteram* II, 23.

18. A. Schneider, *Die abendländische Spekulation des zwölften Jahrhunderts in ihrem Verhältnis zur aristotelischen und jüdisch-arabischen Philosophie* (Münster: Aschendroff, 1915), p. 36.

19. "... idcirco ab ipso primo momento creationis suae coepit circulariter converti, ita ut illa prima conversio integre perfecta esset in spatio quod prima dies appelatum est" (loc. cit.).

20. "Sed omnem circularem motum necesse est habere aliquid immobile circa quod innitatur; motus igitur ignis et aeris non potest esse sine medio centro cui innitatur. At illud medium solidum est et a motu circumstrictum; non potest igitur eorum motus esse nisi solido innitatur" (A. Schneider, op. cit., p. 37).

21. William of Conches, *Philosophia* II, 5 (*Patrologia latina*, ed. J. P. Migne [Paris: 1844–1864], vol. 172, pp. 59D–60A): "Omnis motus discernitur vel per immobile vel minus mobile .... Motus ergo stellarum vel per immobile vel minus mobile superpositum sentitur, nunquam vero per suppositum."

22. Augustine, *De Genesi ad litteram* V, 42.

23. Thomas Aquinas, *Summa contra gentiles* II, c. 16, n. 8: "Non est igitur una materia quae sit in potentia ad esse universale. Ipse autem Deus est totius esse causa universaliter. Ipsi igitur nulla materia proportionaliter respondet; non igitur materiam ex necessitate requirit."

24. Giacomo Zabarella, *De natura caeli*, *Opera* (Cologne: 1597), pp. 270ff.

25. Zabarella, *De natura caeli*, cap. 8, p. 283.

26. Zabarella, *De natura caeli*, cap. 9, p. 284: "Corpus coeleste est materia existens per se, quoniam ex sua natura proprium habet complementum, quo existit actu sine ope alicuius externi."

27. Zabarella, *De natura caeli*, cap. 11, p. 290.

28. A. Mansion, "La théorie aristotélicienne du temps chez les péripatéticiens médiévaux," in

Hommage à Maurice de Wulf (Louvain: Institut superieur de philosophie, 1934), pp. 275–307; A. Maier, *Metaphysische Hintergründe der spätscholastischen Naturphilosophie* (Rome: Edizioni di storia e letteratura, 1955), pp. 47–137; P. Ariotti, "Celestial Reductionism of Time," *Studi Internazionali di Filosofia* 4 (Turin: 1972): 91–120.

29. Averroes, *Commentary on the Physics* book IV, comm. 98 (in Mansion, op. cit., note 28, p. 282): "Et manifestum est quod nos non sentimus nos esse in esse transmutabili, nisi ex transmutatione caeli."

30. A. Maier, "Scholastische Diskussionen über die Wesensbestimmung der Zeit," *Scholastik* 26 (1951): 523.

31. Thomas Aquinas, I *Sent.* dist. 19, q. 2, a.1 ad quartum: "Tempus per se est mensura motus primi: unde esse rerum temporalium non mensuratur tempore nisi prout subiacet variationi ex motu caeli. Unde dicit Commentator .... "

32. Averroes, loc. cit., note 29 (Mansion, p. 282): "Et, si esset possibile ipsum (sc. caelum) quiescere, esset possibile nos esse in esse non transmutabili. Sed hoc est impossibile. Ergo necesse est ut sentiat hunc motum qui non sentit motum corporis caelestis, scilicet per visum."

33. Richard of Mediavilla, *Quaestiones disputatae*, q. 2 (in A. Maier, "Scholastische Diskussionen .... " *Scholastik* 26 [1951]: 541): "Ratio mensurae respectu mensurati habet rationem causalitatis, maioris simplicitatis et maioris uniformitatis, cognoscibilitatis universalioris et cognoscibilitatis certioris."

34. Villard de Honnecourt, *Bauhüttenbuch*, ed. H. R. Hahnloser (Vienna: A. Schroll, 1935), plate 9. Compare F. Klemm, *Technik. Eine Geschichte ihrer Probleme* (Freiburg: Alber, 1954), p. 79: "Perhaps people's efforts, since the thirteenth century, to produce such a perpetual motion here on this Earth can be interpreted as a 'profanation' of the Aristotelian idea of the eternal circular motion that was reserved for the heavens alone. And just as the whole of Aristotle, especially the Aristotelian cosmology and physics, are becoming known in the West thanks to their transmission by the Arabs—beginning, that is, just with the thirteenth century—we also already encounter the profanation of this idea of celestial circular motion."

35. Petrus Peregrinus, *Epistola de magnete*, ed. G. Hellmann (Berlin, 1898). Translated into German by H. Balmer, *Beiträge zur Geschichte der Erkenntnis des Erdmagnetismus* (Aarau: Sauerländer, 1956), pp. 261–277.

36. Thomas Aquinas, *Summa contra gentiles* III, 23: "Impossibile est igitur quod natura intendat motum propter seipsum." Compare his *De potentia*, q. 5, a.5.

37. Thomas Aquinas, I. *Sent.* dist. 19, q. 5, a.1: " ... de tempore, quod habet fundamentum in motu, scilicet prius et posterius ipsius motus, sed quantum ad id quod est formale in tempore, scilicet numeratio, completur per operationem intellectus numerantis." Compare Cl. Baeumker, *Die Impossibilia des Siger von Brabant* (Münster: 1898), pp. 148–165.

38. The 219 propositions of Siger of Brabant, Boetius of Dacia, and others are condemned with the full weight of the authority of the Bishop of Paris: " ... ea totaliter condemnamus, excommunicantes omnes illos, qui dictos errores vel aliquem ex illis dogmatizaverint, aut defendere seu sustinere praesumpserint quoquomodo .... " Text of the decree: *Chartularium universitatis Parisiensis*, eds. H. Denifle and A. Chatelain (Paris: 1889–1897), vol. 1, pp. 543–555. Propositio 190: "Quod prima causa est causa omnium remotissima. Error, si intelligatur ita scilicet, quod non propinquissima."

39. Op. cit., p. 546 (Prop. 49): "Quod Deus non possit movere celum motu recto."

40. Op. cit., p. 552 (Prop. 156): "Quod si celum staret, ignis in stupam non ageret, quia Deus

Notes to Pages 481–489

non esset." Prop. 186 (p. 553): "Quod celum nunquam quiescit quia generatio inferiorum, que est finis motus celi, cessare non debet; alia ratio, quia celum suum esse et suam virtutem habet a motore suo; et hec conservat celum per suum motum. Unde si cessaret a motu, cessaret ab esse."

41. William of Ware, II. *Sent.* dist. 2, q. 1 (A. Maier, *Metaphysische Hintergründe* [cited in note 28], p. 119): "Non omnis motus inferior dependet a motu primo effective, quia potest fieri motus inferior stante sole vel caelo."

42. William of Ockham, *Quaestiones in librum Physicorum*, q. 40 (A. Maier, "Scholastische Diskussionen ...," *Scholastik* 26 [1951]: 551). Compare L. Baudry, *Lexique philosophique de Guilleaume d'Ockham* (Paris: P. Lethiellux, 1958), pp. 265–268.

43. William of Ockham, op. cit., q. 47; L. Baudry, *Lexique*, p. 266: "Et sic quilibet motus posset vocari tempus quo possunt alii motus mensurari et certificari."

44. William of Ockham, op. cit., q. 43 (A. Maier, "Scholastische Diskussionen ...," p. 551): "Dico quod motus inferior, per cuius notitiam possumus devenire in cognitionem alicuius motus caelestis nobis ignoti, potest dici tempus."

45. William of Ockham, I. *Sent.* 27, 3 K. It appears that the suspension of the reality of motion as a change in relations has a theological background insofar as only the intratrinitarian *respectus* now continue to have the dignity of real predicates (I. *Sent.* 30, 4 EF), whereas the Aristotelian category of *ubi* loses it (I. *Sent.* 30, 2 C). Therefore, to reach or to occupy a certain 'place' cannot be a positive characteristic of an object and cannot be interpreted teleologically; the only positive aspect is the *nihil medium inter ipsum et locum* (I. *Sent.* 30, 2 G). Seen from this point of view, the circular motion of the outermost sphere can no longer be understood as a real distinction either: "Ultima sphaera per hoc quod movetur nihil acquirit vel saltem hoc est possibile."

46. Denifle and Chatelain, *Chartularium* (cited in note 38), p. 545 (Prop. 34): "Quod prima causa non posset plures mundos facere."

47. A. Maier, *Metaphysische Hintergründe*, pp. 134–137. William of Ockham admits at least a time *after* the world—a time that he does not need *before* the world—when he makes the blessed, after the Last Judgment, measure the duration of the punishment of the damned by (in the absence of a motion of the heavens) a merely imagined motion: " ... et hoc modo beati mensurabunt post iudicium poenas damnatorum secundum magis et minus per motum ymaginatum" (II. *Sent.* q. 12 SS).

48. William of Ockham, II. *Sent.* q. 12 SS.: " ... et sic fuisset tempus Josuae si omnia corpore coelestia stetissent sicut stetit sol, adhuc per motum ymaginatum posset quis mensurare motus inferiores."

49. William of Ockham, II. *Sent.* q. 12 OO.

50. Nicholas of Cusa, *De ludo globi* II: "Annus, mensis, horae sunt instrumenta mensurae temporis per hominem creatae. Sic tempus, cum sit mensura motus, mensurantis animae est instrumentum."

Chapter 3

51. Galileo, *Dialogo* II; *Opere*, ed. S. Timpanaro (Milan-Rome: Rizzoli, 1936–), vol. 1, p. 173.

52. Copernicus, *De revolutionibus* I, 4: "Apertissima omnium est cotidiana revolutio ... cum etiam tempus ipsum numero potissime dierum metimur." In the Prooemium to the first book Copernicus had quoted from Plato's *Nomoi*: "Dierum ordine in menses et annos digesta

Notes to Pages 490–495

tempora." The choice between a tropical and a sidereal basis for time had already been decided in the *Commentariolus*: "Rectus igitur agit, quicumque annuam aequalitatem ad stellas fixas referet."

53. Copernicus, *De revolutionibus* II, Prooem.: "Nihilque refert, si quod illi per quietam terram, et mundi vertiginem demonstrant, hoc nos ex opposito suscipientes ad eandem concurramus metam: quoniam in his, quae ad invicem sunt, ita contingit, ut vicissim sibi ipsis consentiant."

54. Aristotle, *De caelo* II, 14; 297 a 30–b 14.

55. Richard of Mediavilla, II. *Sent.* d 14, q. 6 (A. Maier, *Zwei Grundprobleme der scholastischen Naturphilosophie*, 2nd ed. [Rome: Edizioni di storia e letteratura, 1951], pp. 190–191): "Dico ergo quod caelum movetur ad intelligentia per hoc quod aliqua potentia ipsius intelligentiae ... aliquam virtutem educit de potentia caeli ad actum ... et per illam virtutem ipsum caelum movet se .... Forma autem naturalis, tam accidentalis quam formalis ipsius caeli est illud quod est in caelo naturalis aptitudo, ut sic moveatur." The source of the formula, "aptitudo ut sic moveatur," seems to me to be Thomas Aquinas's commentary on Aristotle's *On the Heavens: Expositio de caelo* I, 3, n. 21 (ed Spiazzi [Turin: Marietti, 1952], p. 14). There we also find this distinction—"... principium activum motus caelestium corporum est intellectualis substantia: principium autem passivum est natura illius corporis, secundum quam natum est tali motu moveri"—to which is added a comparison between the celestial and the human body: "Et esset simile in nobis si anima non movet corpus nostrum nisi secundum naturalem inclinationem eius, scilicet deorsum" (loc. cit., n. 22).

56. Ptolemy, *Syntaxis mathematica* I, 7 (ed. J. L. Heiberg [Leipzig: Teubner, 1898–1903], p. 24).

57. Copernicus, *De revolutionibus* I, 2: "Terra quoque globosam esse patet, quoniam ab omni parte centro suo innititur. Tametsi absolutus orbis non statim videatur ...." The terminology of the *orbis absolutus* points, for the whole complex of the characteristics of the perfect shape of the Earth, to Pliny, *Historia naturalis* II, 3, where, however, these are attributes of the universe as a whole (*mundus ipse*): "Forman eius in speciem orbis absoluti globatam esse nomen in primis et consensus in eo mortalium orbem appelantium, sed et argumenta rerum docent, non solum quia talis figura omnibus sui partibus vergit in sese ac sibi ipsi toleranda est seque includit et continet nullarum egens compagium nec finem aut initium ullis sui partibus sentiens, nec quia ad motum, quo subinde verti mox adparebit, talis aptissima est, sed oculorum quoque probatione ...." His possession of Pliny as early as his stay in Italy—as demonstrated by the marginal note on Hicetas—is after all the earliest and most certain source for Copernicus that we can document.

58. Copernicus, *De revolutionibus* I, 3: "... qualem (sc. figuram) umbra ipsius ostendit: absoluti enim circuli amfractibus Lunam deficientem efficit."

59. Copernicus, *Commentariolus*, ed. F. Rossmann (Munich: H. Rinn, 1948), p. 9: "Valde enim absurdum videbatur coeleste corpus in absolutissima rotunditate non semper aeque moveri."

60. Aristotle, *Physics* VI, 10; 240 b 15–17. On this see W. Breidert, *Das Aristotelische Kontinuum in der Scholastik* (Münster: Aschendorff, 1970), p. 47.

61. G. von Bredow, "Figura mundi. Die Symbolik des Globusspiels von Nikolaus von Kues," in *Urbild und Abglanz*, Festschrift Herbert Doms, ed. J. Tenzier (Regensburg: J. Habbel, 1972), pp. 195–196.

62. Nicholas of Cusa, *De ludo globi* I: "Nam rotunditas, quae rotundior esse non posset, nequaquam est visibilis ... non enim rotunditas ex punctis potest esse composita." A different interpretation of the passage, which W. Breidert has proposed, is based on the principle of the identity of indiscernibles (*identitas indiscernibilium*), which was introduced shortly before: The perfection of the globe would be reached when its radii are absolutely equal, but this would have

to lead to the identity of all the radii, and thus to the impossibility of the globe. Of course this interpretation would make not only the appearance, but even the idea of the globe impossible—or would require that there could only be ideas of inexact globes. Such an intensification of the Cusan's 'exploding' metaphors cannot be excluded, if one considers that what is supposed to be gained in the *coincidentia oppositorum* is not the idea *as* transcendence, but, beyond that, the transcendence *of* the ideas.

63. Nicholas of Cusa, loc. cit.: "Perfecte igitur rotundus ... postquam incepit moveri, quantum in se est nunquam cessabit, cum varie se habere nequeat. Non enim id, quod movetur aliquando, cessaret nisi varie se haberet uno tempore et alio."

64. Nicholas of Cusa, loc. cit.: "Forma igitur rotunditatis ad perpetuitatem motus est aptissima. Cui si motus advenit naturaliter numquam cessabit."

65. Copernicus, *De revolutionibus* I, 2.

66. *De revolutionibus* I, 3 ("Quomodo terra cum aqua unum globum perficiat"): "Ex his demum omnibus puto manifestum terram simul et aquam uno centro gravitatis inniti nec esse aliud magnitudinis terrae, quae cum sit gravior, dehiscentes eius partes aqua expleri; et idcirco modicam esse comparatione terrae aquam, etsi superficie tenus plus forsitan aquae appareat. Talem quippe figuram habere terram cum circumfluentibus aquis necesse est, qualem umbra ipsius ostendit ...."

67. *De revolutionibus* I, 4: "Mobilitas enim sphaerae est in circulum volvi, ipso actu formam suam exprimentis in simplicissimo corpore, ubi non est reperire principium et finem nec unum ab altero secernere, dum per eadem in seipsam movetur."

68. Ptolemy, *Syntaxis mathematica* I, 7; translated by G. J. Toomer as *Ptolemy's Almagest* (New York: Springer, 1984), pp. 44–45; original: ed. Heiberg (see note 56), pp. 23–25.

69. Copernicus, *De revolutionibus* I, 8. As was to become evident, this argument drives a wedge between nature and culture, between the Earth and the cultivator of its surface. John Wilkins was still embarrassed by the fact that man's artifacts do not enjoy the protection of this assertion, since "supposing ... that this motion were natural to the Earth, yet it is not natural to Towns and Buildings, for these are Artificial." What does one do when faced with such an embarrassment? Wilkins continues, "To which I answer: Ha, ha, ha." (*Discourse concerning a New Planet* [5th ed., London: 1684], p. 111.) But when Wilkins wrote his *Discourse* (it was first published in 1640), there was as yet nothing to laugh about in this.

70. *De revolutionibus* I, 4: "Fateri nihilominus oportet circulares esse motus vel ex pluribus circulis compositos, eo quod inaequalitates huiusmodi certa lege statisque observant restitutionibus: quod fieri non posset, si circulares non essent. Solus enim circulus est, qui potest peracta reducere."

71. *De revolutionibus* I, 4: "Cum vero ab utroque abhorreat intellectus sitque indignum tale quiddam in illis existimari, quae in optima sunt ordinatione constituta ...."

72. *De revolutionibus* I, 5: "Iam quidem demonstratum est terram quoque globi formam habere. Videndum arbitror an etiam formam eius sequatur motus et quem locum universitatis obtineat: sine quibus non est invenire certam apparentium in caelo rationem."

Chapter 4

73. Petrus Ramus, *Scholae Physicae* (Paris: 1565) IV 11–14 (reprint of the edition of 1606 [Frankfurt: Minerva, 1967], pp. 119–123): "Omnis motus fit in tempore, ait Physicus noster.

At vero quoties in hac thesi dixisti, tempus esse mensuram motus? quid igitur tautologia ista nobis opus est? . . . O philosophe praestantissime, ubi est illud alioqui summum acumen tuum? . . . Etenim Copernicus astrologus aetatis nostrae summus, caelo detraxit omnem motum; tempusque, solo terrae motu longe exactius metitur, quam Astrologus adhuc ullus dimensus est . . . . Locus et tempus, si, quid esset, quaereretur, paucis verbis tota veritas erat contenta: Locus est spatium locati; Tempus est duratio rei." On the preparation of the concept of absolute time see P. E. Ariotti, "Toward Absolute Time: Continental Antecedents of the Newtonian Concept of Absolute Time," *Studi Internazionali di Filosofia* 5 (1973): 141–168.

74. Newton, *Mathematical Principles of Natural Philsophy*, trans. A. Motte, Great Books of the Western World vol. 34 (Chicago: Encyclopedia Brittanica, 1952), p. 9; original: Newton, *Opera Omnia*, ed. S. Horsley (London: 1779–1785), vol. 2, p. 8: "Possibilis est, ut nullus sit motus aequabilis, quo tempus accurate mensuretur . . . . "

75. Plotinus, *Enneads* II, 2: "On the Movement of Heaven."

76. Newton to Bentley, 17 January 1692/93; *Correspondence*, ed. H. W. Turnbull and J. S. Scott (Cambridge: Cambridge University Press, 1959–1977), vol. 3, p. 240.

77. Loc. cit.: " . . . but then ye divine power is here required in a double respect; namely to turn ye descending motion of ye falling planets into a side motion, and at ye same time to double ye attractive power of ye sun."

78. Loc. cit.: "Blondel tells us somewhere in his book of Bombs that Plato affirms that ye motion of ye planets is such as if they had all of them been created by God in some region very remote from our Systeme and let fall from thence towards ye Sun . . . . "

79. Galileo, *Discorsi* IV; *Opere*, ed. S. Timpanaro (Milan-Rome: Rizzoli, 1936–), vol. 2, pp. 357–358: "Il concetto è veramente degno di Platone . . . . "

80. Galileo, *Dialogo* I; *Opere*, ed. Timpanaro, vol. 1, p. 42: "In oltre, essendo il moto retto di sua natura infinito . . . , è impossibile che mobile alcuno abbia da natura principio di muoversi per linea retta, cioè verso dove è impossibile di arrivare, non vi essendo termine prefinito . . . . "

81. Loc. cit.: " . . . se però noi non volessimo dir con Platone, che anco i corpi mondani, dopo l'essere stati fabbricati e del tutto stabiliti, furon per alcun tempo dal suo Fattore mossi di moto retto, ma che dopo l'esser parvenuti in certi e determinati luoghi, furon rivolti a uno a uno in giro, passando dal moto retto al circolare, dove poi si son mantenuti e tuttavia si conservano: pensiero altissimo e degno ben di Platone . . . . " The customary reference to *Timaeus* 30 A is not specific [to this idea], since the talk there is only of the transformation of an originally confused and disordered motion of everything visible into an order.

82. Ptolemy, *Syntaxis mathematica* IX, 2. In his biography of Copernicus, Lichtenberg described this as follows: "According to them [that is, the ancients], the perfection of nature required perfect circular motions everywhere, and uniformity in these motions. To them the circle was the most perfect line, indeed it was a symbol of perfection itself, and in these hypotheses it was inviolable for them, as though it were sacred." He added in a note that the foundation of this idea "evidently lies in human nature itself, which of course is very old. How natural this idea must be one sees also from the fact that our great Copernicus, who was entirely natural, was unable to free himself from it, and consequently blundered." *Schriften und Briefe*, ed. W. Promies, vol. 3 (Munich: Hanser, 1972), p. 155.

83. Kepler to Crüger, 9 September 1624; Kepler, *Gesammelte Werke*, ed. M. Caspar (Munich: Beck, 1937–), vol. 18, p. 199.

84. E. Cassirer, *Substanzbegriff und Funktionsbegriff* (Berlin: B. Cassirer, 1910), pp. 100ff.

Notes to Pages 514–526

85. Newton, *Principia*, book III, prop. 18, theorem 16, and prop. 19, problem 3: *Opera*, ed. S. Horsley (London, 1779–1785), vol. 3, pp. 34–35.

86. Newton, *Mathematical Principles* (cited in note 74), Definitions, Scholium, p. 9; original: *Opera*, vol. 2, p. 8.

87. *Mathematical Principles*, p. 11; *Opera*, vol. 2, p. 9: "Loca autem immota non sunt, nisi quae omnia ab infinito in infinitum datas servant positiones ad invicem; atque adeo semper manent immota, spatiumque constituunt quod immobilem appello."

88. *Mathematical Principles*, p. 9; *Opera*, vol. 2, pp. 7–8: "Inaequales enim sunt dies naturales, qui vulgo tanquam aequales pro mensura temporis habentur. Hanc inaequalitatem corrigunt Astronomi, ut ex veriore tempore mensurent motus coelestes. Possibile est, ut nullus sit motus aequabilis, quo tempus accurate mensuretur."

89. A. Harnack, *Geschichte der Königlich Preussischen Akademie der Wissenschaften zu Berlin* (Berlin: 1900), vol. 2, p. 306: "Si le mouvement diurne de la Terre a été de tout tems de la même rapidité ou non? Par quels moyens on peut s'en assurer? Et en cas qu'il y ait quelque inégalité, quelle en est la cause?"

90. Maupertuis, *Discours sur les différentes figures des astres* (Paris: 1732); *Essai de cosmologie* (Berlin: 1750), pp. 117–118: "In these recent times, people conjectured that the Earth was not perfectly spherical . . . . This is how uncertain the matter was when the greatest king that France has had ordered the most magnificent undertaking that was ever fashioned on behalf of the sciences." On Maupertuis's faculty for thinking up prolems see H. Blumenberg, *The Legitimacy of the Modern Age*, pp. 407–416; *Der Prozess der theoretischen Neugierde*, suhrkamp taschenbuch wissenschaft 24 (Frankfurt: Suhrkamp, 1973), pp. 219–230.

91. Giordano Bruno, *Articuli de natura et mundo*, no. 38; *Opera latina*, ed. F. Fiorentino et al. (Naples: 1879–1891), vol. 1, part 1, pp. 143–146: "Tempus, quod est mensura motus, non est in coelo, sed in astris, et primus ille motus, quem concipimus, non est alibi, quam in terra subjective." From the *ratio articuli* (p. 145): " . . . primus motus Aristoteli habebatur omnium regulatissimus . . . at quid nunc diceret, si alios motus comperiret diurnique motus mensuram millegeminis irregulatisque commotionibus turbari videret?"

92. Robert Hooke, *Posthumous Works* (London: 1705), p. 322. See F. N. Egerton, "The Longevity of the Patriarchs: A Topic in the History of Demography," *Journal of the History of Ideas* 27 (1966): 578.

93. R. Vieweg, "Ephemeridenzeit und Atomzeit," *Abhandlungen der Akademie zu Mainz*, Math.-naturwiss. Klasse, 1965, no. 11.

Part V

Introduction

1. J. C. P. Erxleben, *Anfangsgründe der Naturlehre*, XII, section 638; 6th ed. (Göttingen: 1794), pp. 617–618. The supplement signed "L." is already contained in the fourth edition, of 1787. On the importance of the transits of Venus for astronomy and (beyond that) for the eighteenth century's form of science, see H. Woolf, *The Transits of Venus* (Princeton: Princeton University Press, 1959). In its next volume as well, the *Astronomisches Jahrbuch* for the year 1778 (Berlin: 1776; pp. 186–191) contains another note by Lambert about Venus's satellite, in which he passes on communications that had reached him after the first publication. What is more

important is that he provides an account of his own observation of Venus's transit of the Sun in 1761, in Augsburg, and in doing so admits that he could have thoughtlessly let the discovery of Venus's satellite slip by him. "A little before 9 o'clock thin clouds moved in front of the Sun, so that one could look at the Sun with one's bare eyes. Some onlookers [whom Lambert had gathered] not only saw Venus, but also said that they saw a smaller one as well. At that time I knew nothing about the satellite. And the result was that I only answered that the smaller Venus would be a Sun spot, the likes of which were often present. Nor did I continue to look, because I had already seen enough Sun spots, and I did not know that there was something else as well to look for and to observe. Now, I admit, I wish I had paid more attention to this . . . . " One can sense his feeling of having let a unique moment escape him. But the concluding sentence of the second note about the satellite is still distanced, almost ironical: "And if it should really be seen, this will make a very good contribution to deciding the chief question, whether Venus is circled by a moon." Lambert's "chief question" was not at all that of the astronomy of his time, which— at great expense—observed the transits of Venus in order to determine the mean distance between the Sun and the Earth more precisely. This divergence of interest is instructive enough regarding Lambert's style of theory.

2. Max Steck indicates this in his edition of Lambert's *Schriften zur Perspektive* (Berlin: G. Lüttke, 1943), pp. 7–8, following the testimony of C. H. Müller in his edition of Lambert's *Logische und philosophische Abhandlungen* (Berlin: 1787), p. 348. The biography appeared again in the widely read volume *Leben der berühmtesten vier Gelehrten unseres philosophieschen Jahrhunderts: Rousseaus, Lamberts, Hallers und Voltaires* (Frankfurt and Leipzig: 1779). However, doubts about Lichtenberg's authorship arise from the observation that the day of Lambert's death is not given correctly in the note to "Erxleben" that I quoted, whereas the biography designates it correctly as 25 September 1777. As late as 8 September Lambert had attended a session of the Academy. The assumption that Lichtenberg was the author is also weakened by the way in which Lambert's affair of the moon of Venus is mentioned and almost evaluated positively—at least mentioned in one breath with the *Cosmological Letters*: "How easy it was for him to abstract a theory from one case or from a few data and to refine it to a high degree of probability and completeness is shown eloquently by his *Cosmological Letters* and his calculations about the conjectured satellite of Venus." Would not a Lichtenberg have adduced with much more justification, as an example of this outstanding gift, Lambert's methods for determining the paths of comets? Finally, one will look for examples of Lichtenberg's skill in expression. In vain, in my opinion. Or is the following sharp remark sufficient, which concludes the discussion of Lambert's *Architectonics*: a work of such qualities "that it would be immortal, if immortality were possible in a work of metaphysics"?

3. Lambert to M. Holland, 6 November 1768; *Johann Heinrich Lamberts deutscher gelehrter Briefwechsel*, ed. J. Bernoulli (Berlin: 1782–1784), vol. 1, p. 299.

4. M. Holland to Lambert, 29 November 1768; op. cit., pp. 306–308.

5. Lambert to M. Holland, 11 December 1768; op. cit., pp. 314–315. The fascinating history of imaginary libraries begins with T. J. van Almeloveen's *Bibliotheca promissa et latens* (Gouda: 1692), and proceeds by way of the one, copied from a fair catalogue, that belonged to the little schoolmaster Wuz, to the library of Benjamin's and Scholem's "University of Muri."

Chapter 1

6. J. H. Lambert, *Schriften zur Perspektive*, ed. M. Steck (Berlin: G. Lüttke, 1943), p. 161. The expression "free" perspective refers to the method, which Lambert indicates, of drawing perspectivally without a pregiven ground plan. In the edition printed in 1759, section 1 begins, "Visible things present themselves to the eye entirely differently from the way they are in reality . . . . The most accurate plan of them is often so different from their appearance that one can

hardly compare them" (p. 195). The initial statement, that the way things really are is not also the way they appear, originates as early as ancient "scenography," which was founded by Agatharchus for Aeschylus's plays and was carried over by Democritus and Anaxagoras to cosmological questions such as the shape of the Earth, estimating the size of the Sun and the Moon, constructing the Earth's shadow, and the relative distances of the heavenly bodies (H. Diels and W. Kranz, eds., *Fragmente der Vorsokratiker* [Berlin: Weidmann, 1951–1952], 59 A 39). E. Frank says that Anaxagoras introduces an "optics of the universe," and characterizes Democritus's natural philosophy, as a whole, as a "philosophy of perspective" (*Plato und die sogenannten Pythagoreer* [Halle: Niemeyer, 1923], pp. 22–23).

7. The passion for perspective, in which the person who is seen becomes a seer, and thus also sees his own point of view, has the appearance of "a sort of technical frenzy," of which Goethe speaks in connection with the influence of Brunelleschi on his Cellini, and which consists in "working through all the conditions of the one thing one has discovered" (Supplement to *Benvenuto Cellini*, *Werke*, ed. E. Beutler [Munich: Artemis, 1949–], vol. 15, p. 863). On the connection between the aesthetic and the cosmological concepts of space, see Dagobert Frey, "Kunst und Weltbild der Renaissance," *Studium Generale* 6 (1953): 420, and G. Boehm, *Studien zur Perspektivität* (Heidelberg: C. Winter, 1969).

8. E. F. Apelt, *Die Reformation der Sternkunde* (Jena: 1852), p. 116; E. Zinner, *Entstehung und Ausbreitung der coppernicanischen Lehre*, Sitzungsberichte der Physikalisch-medizinischen Sozietät 74 (Erlangen: Mencke, 1943), pp. 146, 149.

9. The remains of Copernicus's library, in Uppsala, are listed in E. Zinner, op. cit. (previous note), pp. 404–408. On Rheticus's gift of books: K. H. Burmeister, *Georg Joachim Rheticus* (Wiesbaden: Pressler, 1967–1968), vol. 1, pp. 56–57.

10. Lambert to M. Holland, 14 December 1766 (*Johann Heinrich Lamberts deutscher gelehrter Briefwechsel*, ed. J. Bernoulli [Berlin, 1782–1784], vol. 1, pp. 171–172): "So we cannot very well predict in what direction the current of the times will turn. Wherever belles lettres become prominent, they always make the biggest stir. But it is easy for their intoxication to be over quickly, because their material is far from inexhaustible. Whether this will be followed by a barbarism, or whether people will turn to the sound sciences again, remains to be seen . . . ."

11. Lambert to M. Holland, 11 December 1768; *Briefwechsel* (cited in previous note), vol. 1, p. 316.

12. Lambert reports his submission to the Academy of an essay *Sur les secours mutuels que peuvent se prêter les sciences solides et les belles lettres* in a letter to M. Holland on 1 September 1767. This essay is evidently not identical with Lambert's inaugural address to the Academy, *Sur la liaison des connaissances qui sont l'objet des quatres classes de l'académie*.

13. Lambert, *Briefwechsel*, vol. 1, p. 400.

14. Abraham Gotthelf Kästner, *Gesammelte poetische und prosaische schönwissenschaftliche Werke* (Berlin, 1841), vol. 1, p. 80: "Der Venustrabant."

. . .

Sie haben freilich stets die Weiber im Verdacht;
Manch Sternrohr hat umsonst den Cicisbee bewacht.
Zu zeigen hat sich ihn einst Lambert unterstanden,
Und die Verläumdung ward zu Schanden.
So ist's am Himmel nur; man sieht Trabantenheere
Auf Erden leicht um jede Cythere.

[Of course they always hold the women in suspicion./ Many a telescope has watched over a cicisbeo in vain./ Lambert once ventured to show his [the satellite's] presence,/ and the calumny

Notes to Pages 538–544

became a disgrace./ Only in the heavens is it like that; armies of satellites are easily seen around every Cytherea on Earth.]

Kästner's late criticism of Lambert coincides in one essential point with that of Lambert's anonymous biographer, according to whom he was "an example of how, despite having a great intellect, a person can preserve prejudices that are difficult to reconcile with his insights. Examples of this sort of thing were his preference for small, poor implements over larger ones, and for determining by means of drawing things that he could more easily have calculated. And in opposition to every type of calculation, he made his own tables, where one can get by with universal ones, so that even in matters of his own science he operated in such unexpected ways" (letter to Friedrich Nicolai, 17 June 1784, in A. G. Kästner, *Briefe aus Sechs Jahrzehnten* [Berlin: B. Behr, 1912], pp. 150–151). Kästner had also already spoken his mind to Lambert himself, opposing his investing his powers in tables of the ecliptic (which Lambert entitled his *Machina eccliptica*), though the reproach contains high praise: "Il y a assez de calculateurs mais peu de mathematiciens philosophes" (letter of 26 July 1770, in *J. H. Lamberts und A. G. Kästners Briefe*, ed. K. Bopp, *Sitzungsberichte der Heidelberger Akademie*, Abhandlungen der Math.-Nat. Klasse, Jahrgang 1928, no. 18 [Berlin: 1928], p. 22).

15. *Lamberts und Kästners Briefe* (see previous note), p. 15: "Aparemment serez Vous surpris de la hardiesse avec laquelle j'y depeins la structure de l'univers et peut être aussi de la fécondité de toute espece d'argumens, que j'y entasse. Je suis extremment curieux de voir quelle fortune fera cet ouvrage chez les Lecteurs et de lui dresser l'horoscope. A ce qu'il me semble il est impossible de le refuter solidement, et peut être ce systeme deviendra dominant dans la suite, et après plusieurs débats, mais je ne crois pas quelqu'un s'avisera tout de bon à en faire l'examen, que je propose dans quelques unes de ces lettres, puisqu'il demande de longs Calculs et une suite d'observations."

16. *Jean Paul und Herder; Der Briefwechsel Jean Pauls und Karoline Richters mit Herder und der Herderschen Familie in den Jahren 1785 bis 1804*, ed. P. Stapf (Bern: Francke, 1959), p. 47.

17. Jean Paul, *Vorschule der Ästhetik* I, 4 s. 19; I, 5 s. 24; translated by Margaret R. Hale, *Horn of Oberon. Jean Paul Richter's School for Aesthetics* (Detroit: Wayne State University Press, 1973), pp. 52, 66–67.

18. Herder, *Adrastea* III, 6; *Sämmtliche Werke*, ed. B. Suphan (Berlin: Weidmann, 1877–1913), vol. 23, pp. 523–530.

Chapter 2

19. Francis Bacon, *Novum organum* II, 46; *Works*, ed. J. Spedding, R. Ellis, and D. Heath (London: 1857–1874; rpt. Stuttgart-Bad Cannstatt: Frommann-Holzboog, 1963), vol. 1, p. 326. How important the reassuring outcome of this train of thought was for Bacon can be seen from the fact that in this passage he gives what is for him (in contrast to Descartes) one of the very uncommon glimpses into the development of his own thinking. The thesis of the chapter— that all natural processes are temporal ("omnis motus sive actio naturalis transigitur in tempore")—is maintained by assuming the speed of light to be so great that it escapes measurability (as opposed to *in tempore aliquo notabili*). As it stands there, the reason for the change in the initial idea is incomprehensible: A finite speed of light would have to mean an immense loss of mass by the stars. Bacon does not explicitly supply his premise, that for him the assumption of a finite speed of light only follows from the corpuscular nature of light.

20. Lambert to Johann Lorenz Böckmann, 7 March 1773; *Johann Heinrich Lambert's deutscher gelehrter Briefwechsel*, ed. J. Bernoulli (Berlin: 1782–1784), vol. 1, pp. 418–421. In this letter Lambert also describes the process by which the *Cosmological Letters* came into being: "First of all I have to note that at the time when I was writing my *Cosmological Letters* (in June, July, and

Notes to Pages 545–565

October, 1760) I had neither my books with me nor more than a quarto page of prepared material. The result was that I had to leave various things vague that I could not remember precisely. The *Orbitae cometarum*, which I composed at the same time, as well as my stock of tables did help considerably, and I was even able to apply various things from the *Photometria*, which had been sent to the printer shortly before. So the way these letters came into being was that each one gave rise to the next, and I did not foresee how far I would get in that way .... " Lambert's *Monatsbuch* [a sort of diary, edited by K. Bopp in *Abhandlungen der Königlichen Bayerischen Akademie der Wissenschaften*, Math.-Phys. Klasse, vol. 27, 6th contribution (Munich: 1915)] has as its first entry in June 1760: "Commercium epistolicum de systemate mundi inchoavi."

21. H. Blumenberg, *Selbsterhaltung und Beharrung. Zur Konstitution der neuzeitlichen Rationalität*, Abhandlungen der Akademie zu Mainz, Geistes- und sozialwiss. Kl., 1969 series, no. 11 (Mainz: 1970).

22. Perhaps one may also relate to these metaphors of Lambert's a remark he made in a letter to Aloysius Havichorst on 31 May 1777 (*Briefwechsel*, vol. 1, p. 428): "I gave some notions of the greatness of God in my *Cosmological Letters*. The *civitas dei* is also clarified by them in some respects."

23. Lambert, *Briefwechsel*, vol. 1, pp. 377–379.

24. M. Holland to Lambert, 19 March 1769 (*Briefwechsel*, vol. 1, pp. 321–322): "Some natural philosophers are comparable in this respect to certain Wolffian theologians, who demonstrate all the doctrines of revealed religion from the essence of God, but who could also have derived the opposite from it in just the same manner and with just as much conclusiveness .... The inference, 'The thing is, therefore it is good,' is always more certain than the reverse inference, 'The thing is good, therefore it is.' "

25. H. S. Reimarus, *Abhandlungen von den vornehmsten Wahrheiten der natürlichen Religion* (1st ed., 1754) (6th ed., Hamburg: 1791), p. 277, Abhandlung 5, section 3.

26. T. W. Adorno, *Negative Dialectics*, trans. E. B. Ashton (New York: Seabury, 1973), p. 400; original: *Negative Dialektik* (Frankfurt: Suhrkamp, 1966), p. 390.

27. Lambert, *Briefwechsel*, vol. 2, pp. 46–47.

28. Lambert to G. F. Brander, 11 November 1769 (*Briefwechsel*, vol. 3, pp. 189–190): "I observed the comet in my room; which was more comfortable .... "

29. Lambert to the Abbot of Felbinger, 22 March 1771; *Briefwechsel*, vol. 4, pp. 120–121.

30. Lambert's response to a letter of J. Wegelin to [Johann Jakob] Bodmer (*Briefwechsel*, vol. 1, pp. 375–376) (undated; ca. 1762): "What I think about the type of teleological proof that I used .... We are not yet in a position to measure their sum total, as to whether it makes up a whole. Consequently one can always overturn them again, if one invalidates each one individually, by itself. This is possible *ex hypothesi*, just because each individual proof, viewed by itself, is too weak. But I do not think that it would be correct to proceed in that way, because the argument that because every proof is too weak therefore all taken together are also too weak is not valid. One has to see whether one does not fill the gap in the other—which, however, is difficult."

Chapter 3

31. Karl Schwarzschild, in *Nachrichten der Königlichen Gesellschaft der Wissenschaften zu Göttingen* (Berlin: 1907), pp. 88–102.

32. Lambert to Karsten, 26 September 1771; *Briefwechsel*, vol. 4, p. 306.

Chapter 4

33. Lambert to M. Holland, 7 April 1766; *Briefwechsel* (cited in note 3), vol. 1, p. 136. In this instance the remark relates to Kant's just published "little treatise," *Träume eines Geistersehers, erläutert durch Träume der Metaphysik* [*Dreams of a Visionary, Explained by Means of Metaphysical Dreams*]. Lambert had already used the formula of "similarity in our type of thinking" toward Kant himself, when he wrote to him for the first time, 13 November 1765.

34. Schelling, *Immanuel Kant* (1804); *Schriften* (1801–1804) (Darmstadt: Wissenschaftliche Buchgesellschaft, 1968), p. 593.

35. Schopenhauer, *The World as Will and Representation*, trans. E. F. J. Payne (New York: Dover, 1966), vol. 2, p. 53. [The next quotation is from the same page.] Original: *Sämtliche Werke*, ed. W. von Lohneysen (Stuttgart: Cotta, 1960–1965), vol. 2, pp. 73–74.

36. Kant to Johann Erich Biester, 8 June 1781; *Briefe von und an Kant*, ed. E. Cassirer (Berlin: B. Cassirer, 1922–1923), vol. 1, pp. 200–203. [Or in Kant's *Werke*, ed. E. Cassirer (Berlin: B. Cassirer, 1912–1918), vol. 9, pp. 200–203.]

37. Kant to Lambert, 31 December 1765; *Briefe*, ed. E. Cassirer, vol. 1, p. 48.

38. Lambert to Kant, 13 November 1765; *Briefe*, ed. E. Cassirer, vol. 1, pp. 42–45.

39. According to his own statement, Kant was acquainted with Wright's *An Original Theory and New Hypothesis of the Universe*, which had appeared in 1750, through an account of it that had been published in 1751 in the *Freye Urtheile und Nachrichten zum Aufnehmen der Wissenschaften und Historie überhaupt* [*Free Opinions and News to Aid in the Assimilation of the Sciences and History in General*]—that is, in the same and the only organ that was also to take notice of the appearance of his *Theory of the Heavens*. As for Buffon, he had expressed his basic idea in the widely disseminated *Histoire naturelle*, vol. 1 (Paris: 1749), p. 133. Laplace still took Buffon as his point of departure, who had, of course, in the meantime received powerful support from William Herschel, who had advanced a succinct conjecture about the evolution of systems such as nebulas and star clusters in his *Catalogue of a Second Thousand of New Nebulae and Clusters of Stars, with a Few Introductory Remarks on the Construction of the Heavens* (1789).

40. Kant, *Gesammelte Schriften*, Akademie ed. (Berlin, 1902–), vol. 1, p. 548: note inserted by Gensichen "at Kant's request."

41. Gottlieb Gamauf, *Erinnerungen aus Lichtenbergs Vorlesungen über die physikalische Geographie* (Vienna and Trieste: 1818), pp. 360–361: "There are already about 50 theories of the Earth, of which certainly 9/10 are more important for the history of the human spirit than for the history of the Earth. It is unbelievable what consequences the revolutions on Earth have had for the revolutions in people's heads . . . . "

42. G. Gamauf, *Erinnerungen aus Lichtenbergs Vorlesungen über Astronomie* (Vienna and Trieste: 1814), pp. 378–380, 384–387, and 390–391.

43. Lichtenberg, "Etwas von Hrn. Herschels neuesten Entdeckungen," *Vermischte Schriften* (Göttingen: 1800–1806), vol. 6, p. 337.

44. As A. Buchenau says in Ernst Cassirer's edition of Kant's *Werke* (Berlin: B. Cassirer, 1912–1918), vol. 1, p. 526.

45. Lichtenberg, *Sudelbücher* E 365 (ca. 1775), *Schriften und Briefe*, ed. W. Promies (Munich: Hanser, 1967–), vol. 1, pp. 424–425.

Chapter 5

46. N. R. Hanson, "Copernicus's Role in Kant's Revolution," *Journal of the History of Ideas* 20 (1959): 274–281: "This is not only a matter of Kantian philology. Reference to the 'Copernican revolution' has carried the burden of the most important expositions of Kant's philosophy . . . ."

47. H. Blumenberg, "Kant und die Frage nach dem 'gnädigen Gott,'" *Studium Generale* 7 (1954): 554–570; here, pp. 568–569.

48. In G. B. Jäsche's authorized compendium of Kant's lectures on logic, published in 1800, Tycho Brahe's system is cited as an example of a hypothesis that violates the requirement of logical unity, i.e., not needing auxiliary hypotheses, and consequently cannot compete with the Copernican hypothesis: "In contrast, the Copernican system is a hypothesis from which everything that is supposed to be explainable by it (that we have encountered so far) can be explained" (*Gesammelte Schriften*, Akademie ed. [Berlin: 1902–], vol. 9, pp. 85–86). That "hypotheses always remain hypotheses," and can be increased in their probability only to an "analogue of certainty," is definitely implied in the second preface as a more precise formulation of the relation between Copernicus and Newton.

49. Kant, *Gesammelte Schriften*, Akademie ed., vol. 22, pp. 68, 353, 479, 483, 519–520, 538. On Kant's 'horizon' in relation to the history of science, see H. Heimsoeth, *Studien zur Philosophie Kants*, vol. 2 (Bonn: Bouvier, 1970), pp. 1–85: "Kants Erfahrung mit den Erfahrungswissenschaften." Heimsoeth also refers to the sequence Archimedes, Newton, Lavoisier—"inventers of new methods" (*Transzendentale Dialektik. Ein Kommentar zu Kants Kritik der reinen Vernunft* [Berlin: deGruyter, 1966–1971], vol. 3, p. 562).

50. Newton, *Opera omnia*, ed. S. Horsley (London: 1779–1785), vol. 2, p. XIV. In his *Opus postumum* Kant returned to the basic idea of an analogy between the physics of nonappearing forces and practical knowledge based on the principle of nonappearing freedom: "The Newtonian attraction, which operates across empty space, and man's freedom are analogous concepts; they are categorically imperative ideas" (*Gesammelte Schriften*, Akademie ed., vol. 21, p. 35).

51. W. Bröcker, *Kant über Metaphysik und Erfahrung* (Frankfurt: Klostermann, 1970), p. 29.

52. The Kant commentators have made it into a commonplace that Kant himself described his transcendental turning as "Copernican": N. Kemp Smith, *Immanuel Kant's Critique of Pure Reason* (London: Macmillan, 1918), p. 22; H. J. Paton, *Kant's Metaphysic of Experience* (New York: Macmillan, 1936), vol. 1, p. 75. Paton took into account (in his *Defense of Reason* [London: Hutchinson's, 1951], pp. 91–98) the correction that F. L. Cross brought forward in *Mind* (1937), but he caused new confusion by contaminating the sentence in the text of the second preface with this footnote: "Personally I believe that Kant makes in the text the same point which he elaborates in the footnote . . . ."

53. H. Cohen, *Kommentar zu Immanuel Kants Kritik der reinen Vernunft* (Leipzig: Dürr, 1907), p. 2: ". . . the author has to have developed, if he can become a reader of his own work. This preface is the ideal preface."

54. F. Kaulbach, "Die Copernicanische Denkfigur bei Kant," *Kant-Studien* 64 (1973), p. 34n: ". . . Copernicus's accomplishment is itself a paradigmatic, practical/theoretical step, which

includes a method—that is, a mental diagram of a path—that Kant had in mind in drawing an analogy between himself and Copernicus."

55. H. Heimsoeth, *Transzendentale Dialektik* (cited in note 49), vol. 3, p. 549. This interpretation is first expressed, obscurely, by Schelling, "Immanuel Kant" (1804) (*Schriften 1801–1804* [Darmstadt: Wissenschaftliche Buchgesellschaft, 1968], p. 591): "Like his fellow countryman Copernicus, who transferred the motion from the center to the periphery, Kant completely reversed the idea according to which the subject is inactive and calmly receptive, and the object is active: a reversal that was transmitted to all the branches of knowledge as though by an electric current." Neokantianism took up the interpretive notion of a "change of standpoint": "Just as the former [namely, Copernicus] brought unity and order into the theory of the motions of the planets by means of a complete change of the standpoint from which he regarded them, so Kant wanted, by means of an analogous change in his standpoint, to produce unity and order in the theory of human knowledge" (P. Natorp, "Die kosmologische Reform des Kopernikus in ihrer Bedeutung für die Philosophie," *Preussische Jahrbücher* 49 [1882]: 356.)

56. F. Kaulbach, "Der Begriff des Standpunktes im Zusammenhang des Kantischen Denkens," *Archiv für Philosophie* 12 (1963): 22. One will want to concur even less when Kaulbach sharpens his thesis into the proposition that "Kant radicalized the Copernican turning by displaying in pure form the principle of freedom that is contained in it" (*Philosophie der Beschreibung* [Cologne: Böhlau, 1968], p. 276n).

57. Kuno Fischer, *Geschichte der neueren Philosophie*, vol. 1 (5th ed., Heidelberg: C. Winter, 1912 [1st ed., 1854]), p. 123.

58. B. Russell, *Human Knowledge* (New York: Simon and Schuster, 1948), p. 9: "Kant spoke of himself as having effected a 'Copernican revolution,' but he would have been more accurate if he had spoken of a 'Ptolemaic counter-revolution,' since he put Man back at the centre from which Copernicus had dethroned him."

59. R. Carnap, "Psychologie in physikalischer Sprache," *Erkenntnis* 3 (1932/1933): 109–110.

60. *Kant's Critique of Aesthetic Judgement* (section 28), trans. J. C. Meredith (Oxford: Clarendon Press, 1911), p. 111; *Gesammelte Schriften*, Akademie ed., vol. 5, p. 261.

61. S. Morris Engel, "Kant's Copernican Analogy: A Re-examination," *Kant-Studien* 54 (1963): 243–251. Unfortunately the author thought that he was finally saying what Kant had meant. Hard on the heels of his paper came the (in its turn, limping) judgment: J. W. Olivier, "Kant's Copernican Analogy: An Examination of a Re-examination," *Kant-Studien* 55 (1964): 505–511.

Part VI

Introduction

1. Schopenhauer, *The World as Will and Representation*, book 1, section 14; trans. E. F. J. Payne (Indian Hills, CO: Falcon's Wing Press, 1958), vol. 1, p. 67.

2. E. F. Apelt, *Die Reformation der Sternkunde* (Jena: 1852), p. 154.

3. G. C. Lichtenberg, *Vermischte Schriften* (Göttingen: 1800–1806), vol. 9, p. 305 (from his unpublished papers).

4. Kepler, *De macula in Sole observata*: "O multiscium et quovis Sceptro pretiosius perspicillum! an qui te dextra tenet, ille non dominus constituatur operum Dei?"

5. Lichtenberg, "Über das Weltgebäude," *Vermischte Schriften* (Göttingen: 1810–1806), vol. 6, pp. 193–194. Lichtenberg carried out the most radical relativization of technical optics with his idea that more discoveries had been made by diminishing objects than by magnifying them: "The most beautiful starry heavens look empty to us through a reversed telescope" (*Vermischte Schriften*, vol. 1, pp. 164–165). We still see too much in order to be able to see what matters. That is the announcement of the theory of "stimulus overload," and of its correlate, that a reduction of the intake of information is a condition of the possibility of man's existence.

6. E. Du Bois-Reymond, *Die Humboldt-Denkmäler vor der Berliner Universität* (address delivered in 1883) (Leipzig: 1884), p. 70. As Herder put it, "The heavens presented themselves to him like an immense garden ... " (*Adrastea* III, p. 6; *Sämtliche Werke*, ed. B. Suphan [Berlin: Weidmann, 1877–1913], vol. 23, p. 524).

7. Louis-Sébastien Mercier, *Memoirs of the Year Two Thousand Five Hundred* (London: 1772), p. 153, note (a); translated (by W. Hooper) from *L'an deux mille quatre cent quarante. Rêve s'il en fût jamais* (London: 1772) (written in 1768). [Blumenberg quotes from the German translation, *Das Jahr Zwey tausend vier hundert und vierzig. Ein Traum aller Träume* (London: 1772), p. 162, n. 1.]

8. Leibniz, *Die philosophischen Schriften*, ed. C. J. Gerhardt (Berlin: 1875–1890), vol. 7, p. 187.

9. Johann Wilhelm Ritter, *Fragmente aus dem Nachlass eines jungen Physikers* (Heidelberg: 1810), vol. 1, pp. 161–162 (fragment 249).

10. Montesquieu, *Lettres Persanes*, #59: "Il me semble, Usbek, que nous ne jugeons jamais des choses que par un retour secret que nous faisons sur nous-mêmes."

Chapter 1

11. Fontenelle, *Histoire de l'Académie des Sciences*, for the year 1708 (pub. 1730), pp. 112ff. (quoted in J. F. Carré, *La Philosophie de Fontenelle ou le Sourire de la raison* [Paris: Alcan, 1932], pp. 367–368).

12. Fontenelle, *Dialogues des morts anciens et modernes* (1683), *Oeuvres completes* (Paris: 1818), vol. 2, pp. 232–233. [Blumenberg quotes from the German translation of Johann Christoph Gottsched (Leipzig: 1760), pp. 337–340.]

13. Montaigne, *Essais* II, 12 (Paris: Didot, 1870), pp. 270B, 273AB [ed. P. Villey (Paris: Presses Universitaires de France, 1965), p. 531]. German translation by J. J. C. Bode, *Montaigne's Gedanken und Meinungen, ins Teutsche übersetzt* (Vienna and Prague: 1797), vol. 3, pp. 406–416.

14. G. Gamauf, *Erinnerungen aus Lichtenbergs Vorlesungen über Astronomie* (Vienna and Trieste: 1814), pp. 484–485.

15. Feuerbach, *Der rationalistische und ungläubige Unsterblichkeitsglaube*, *Sämtliche Werke*, ed. W. Bolin and F. Jodl (2nd. ed. Stuttgart: Frommann-Holzboog, 1960–1964), vol. 1, pp. 173–174. Cf. H. Blumenberg, "Neugierde und Wissenstrieb," *Archiv für Begriffsgeschichte* 14 (1970): 23–31. [And *The Legitimacy of the Modern Age*, pp. 440–447; = *Der Prozess der theoretischen Neugierde*, pp. 258–267.]

16. "Lorsque Copernic proposa son système, dans un tems où les lunettes d'approche n'étoient pas inventées, on lui objectoit la non existence de ces phases. Il prédit qu'on les découvriroit un jour, et les télescopes ont vérifié sa prédiction": *Encyclopédie*, vol. 4 (1754), p. 174, col. B. See Edward Rosen, "Copernicus on the Phases and the Light of the Planets," *Organon* 2 (1965): 61–78. There is a remarkable parallel from antiquity of an unverifiable astronomical predic- tion, in the report of the historian Cassius Dio (60, 26, 1) that the Roman emperor Claudius

made public the prediction that there would be an eclipse of the Sun on his birthday, 1 August 45. The motive for publicizing this was to demonstrate the natural character of the eclipse and to avert fear and supernatural interpretations. As Alexander Demandt has shown ("Verformungstendenzen in der Überlieferung antiker Sonnen- und Mondfinsternisse," *Abhandlungen der Akademie zu Mainz* [Geistes- und sozialwiss. Kl. 1970], no. 7, p. 31), this is the only example of a correct ancient eclipse prediction of which it is certain that it was not invented; "the proof that the prediction was not invented retrospectively lies in the fact that the eclipse was so small that it remaind invisible." So as far as the appearances were concerned, the emperor's astronomy was refuted, and did not achieve its goal of enlightenment.

17. Herder, *Adrastea* III, 6; *Sämtliche Werke*, ed. B. Suphan (Berlin: Weidmann, 1877–1913), vol. 3, pp. 508–509.

18. Bacon, *Novum organum* II, 39; *Works*, ed. J. Spedding, R. Ellis, and D. Heath (London: 1857–1874; rpt. Stuttgart, Bad Canstatt: Frommann-Holzborg, 1963), vol. 1, p. 308: "... quorum ope, tanquam per scaphas aut naviculas, aperiri et exerceri possint propiora cum coelestibus commercia."

19. This was written in 1620. About eight years earlier, in the *Descriptio globi intellectualis*, Bacon still had undiminished expectations: "Superest tantum constantia, cum magna judicii severitate, ut et instrumenta mutent, et testium numerum augeant, et singula et saepe experiantur et varie ...."

20. J. Glanvill, *Plus ultra or the Progress and Advancement of Knowledge since the Days of Aristotle* (London; 1668; rpt. Gainesville: Scholars' Facsimiles and Reprints, 1958), chapter 15, pp. 112ff. As an example of the kind of writing that appeared in opposition to Glanvill: Henry Stubbe, *Plus ultra Reduced to a non-plus* (London: 1670).

21. J. W. Ritter, *Fragmente aus dem Nachlasse eines jungen Physikers* (Heidelberg: 1810), vol. 1, pp. 49–50 (fragment #83).

22. Glanvill, *Plus ultra*, chapter 11, p. 77.

23. Glanvill, *Plus ultra*, chapter 7, p. 55: "I dare not therefore mention our greatest hopes: but this I adventure, that 'tis not unlikely but Posterity may by those Tubes, when they are brought to higher degrees of perfection, find a sure way to determine those mighty Questions, whether the Earth move? or, the Planets are inhabited?"

24. Pascal, *Pensées*, ed. L. Brunschvicg (Paris: Hachette, n.d.), fr. [fragment] 266.

25. Voltaire, *Remarques sur les Pensées de Pascal*, XXXII.

26. Pascal, *Fragment d'un traité du vide*, ed. E. Havet, p. 588: "Mais comme les sujets de cette sorte sont proportionnés á la portée de l'esprit, il trouve une liberté tout entière de s'y étendre: sa fécondité inépuisable produit continuellement, et ses inventions peuvent être tout ensemble sans fin et sans interruption ...."

27. Pascal, *Pensées*, ed. Brunschvicg, fr. 72: "Manque d'avoir contemplé ces infinis, les hommes se sont portés témérairement à la recherche de la nature, comme s'ils avaient quelque proportion avec elle ... car il est sans doute qu'on ne peut former ce dessein sans une présomption ou sans une capacité infinie, comme la nature."

28. I. O. Wade, *Voltaire's Micromégas. A Study in the Fusion of Science, Myth and Art* (Princeton: Princeton University Press, 1950), p. 65: "... et quelques-uns font des lunettes d'approche qui ne laissent pas de servir parmi ces petits hommes, parce que leurs yeux et tous leurs sens sont proportionnés à ce petit monde ...."

29. G. Simmel, *The Philosophy of Money*, trans. T. Bottomore and D. Frisby (London, Henley, and Boston: Routledge & Kegan Paul, 1978), p. 475. Original: *Die Philosophie des Geldes* (Munich: Duncker & Humblot, 1900), p. 540.

30. W. Breidert, "Die nichteuklidische Geometrie bei Thomas Reid," *Sudhoffs Archiv* 58 (1974): 235–253.

Chapter 2

31. Galileo, *Dialogo* II; *Opere*, ed. S. Timpanaro (Milan-Rome: Rizzoli, 1936–), vol. 1, p. 157. Aristotle, *De generatione animalium* V 1; 780 b 21.

32. Petrus Borellus, *De vero telescopii inventore cum brevi omnium conspiciliorum historia* (The Hague: 1655), vol. 1, pp. 2–3: "At insurgent aliqui, dicentes, stupendam fuisse hominum arrogantiam, quod limites sibi a Deo praescriptos transilire, seiusque opera aliquo modo reformare ausi fuerint.... Hi apud acriores censores videntur Dei Optimi Max. opera defectus accusasse, et eo meliora facere voluisse, sed procul haec atheorum opinio: nonne praestat dicere, Deum ob peccata primi nostri Parentis Adami nos hisce artibus orbari per multa saecula voluisse, nunc vero speciali ac paterna clementia, nobis mentes obtenebratas ob peccatum aliquo modo aperire dignatum esse?"

33. V. Viviani, *Racconto istorico della vita di Galileo* (1654), in Galileo, *Opere*, ed. A. Favaro (Florence: 1890–), vol. 19, p. 609.

34. Petrus Borellus, *De vero telescopii inventore*, vol. 1, p. 3.

35. Petrus Borellus, op. cit., vol. 1, p. 21.

36. Borelli is the first to quantify the leap that Galileo had accomplished in comparison to natural optics: Galileo had immediately increased the number of stars (which had been regarded as constant since Hipparchus's catalogue) from 1,022 to 2,000—and since then it had increased into the infinite (*nunc in infinitum*) (Petrus Borellus, op. cit., vol. 1, p. 17).

37. *The Starry Messenger*, trans. Stillman Drake in *Discoveries and Opinions of Galileo* (Garden City, New York: Doubleday Anchor, 1957), p. 29.

38. E. Rosen, *Kepler's Conversation with Galileo's Sidereal Messenger* (New York: Johnson Reprint Corp., 1965), pp. 71–72.

39. In the second edition, a quarter of a century later, Kepler gave his early work a self-interpretation that bordered on the idea of inspiration: It would be an error, he says, to regard it as an invention of his mind. "Instead, as though the voice of the heavens had dictated it to me, the little book was immediately recognized as excellent and entirely true in all of its parts (as is the rule with unmistakeable works of God)."

40. Kepler, *Dissertatio cum nuncio sidereo nuper ad mortales misso a Galilaeo Galilaeo* (Prague: 1610); *Gesammelte Werke*, ed. W. von Dyck and M. Caspar (Munich: Beck, 1938–), vol. 4, p. 289.

Chapter 3

41. Francesco Sizzi, *Dianoia astronomica, optica, physica, qua siderei nuncii rumor de quatuor planetis a Galilaeo Galilaeo mathematico celeberrimo recens perspicilli cuiusdam ope conspectis vanus redditur.*

Notes to Pages 661–670

42. Kepler's *Dissertatio*, which appeared in Prague in May 1610, and was immediately reprinted in Florence, was already known to Sizzi, who, however, does not allow himself to be disturbed in the least, in his optical objections to Galileo's observations, by their acceptance by the author of an *Optics* that was authoritative for him: "... suas Opticorum liber me ad hanc scientiam excitavit; in sua cum sydereo nuncio dissertatione probat per aspectus astra et planetas in haec inferiora agere ..." (Sizzi, *Dianoia*, p. 53).

43. "Primum cum ad me pervenisset, alios novos planetas repertos pluresque in caelo praeter septem extare erraticas stellas ... maxima mentis perturbatione angebar" (Sizzi, *Dianoia*, "Prothesis ad lectorem").

44. *Dianoia*, p. 40: "... nunc ad huius hallucinationis causas transeundum, et planetas fictos imaginarios esse demonstrandum et oculos inspicientium eludi confirmandum est, ut penitus a caelo haec in rerum natura portenta et ludibria exterminentur ...."

45. Goethe to F. von Müller, 24 April 1819 (*Werke*, ed. E. Beutler [Munich: Artemis, 1949–], vol. 23, p. 52): "Then he talked about the art of seeing ...."

46. Sizzi, *Dianoia*, p. 21: "... tamen haec non tollit veterem domiciliorum rationem ...."

47. Dianoia, p. 52: "Deum et naturam nil frustra efficere assero, at frustraneus est actus, qui potentiam et virtutem in se includit nullam ...."

48. *Dianoia*, p. 55: "Si hi planetae reales existerent, saltem aliquis rumor, ut de maculis in facie Lunae, de nebulosis, Lacteique circuli stellis ad nos pervenisset; sed in omnibus Astronomorum et Philosophorum Historiographorumque libris nulla mentio eorum facta est."

49. *Dianoia*, p. 23: "Adde quod sicut planetae syderei nuncii indigent perspicillo, ut cognoscantur et videantur, necessario et hoc perspicillo opus esset, cum suas influentias et effectus in haec inferiora eiaculari vellent: etenim sydera motu et lumine operari et influere certum est, an hoc absurdum? Quod probari potest."

50. *Dianoia*, p. 48: "... mihi certissime asseverare liceat omnes observationes a sydereo nuncio allatas et notatas nec ordinem nec motus regulam servare ...." Op. cit., p. 66: "... ita ut mirum esse non debeat si in 65 observationibus, quas sydereus nuncius Mathematicis et Philosophis exhibet, duas similes et aequales situ forma et figura videre non contingat. Perdifficile enim est, imo impossibile oculum in eadem constitutione et positione ... collocari, uti et in eodem flumine bis nos abluere."

51. *Dianoia*, p. 67: "Recordare Galilaee, quod illa nocte, in qua cum plurimis aliis doctissimis et nobilibus viris tecum Iovem contemplatus sum; Tu ipsemet priori observatione unam Iovis imaginem conspexisti, cum vere aliquis illustrium virorum adstantium, qui post tuam primam observationem Iovem inspexit, duas videre fassus est; tu iteratis observationibus duas etiam Iovis imagines conspexisti, quod pluribus adstantibus viris eadem nocte contigit. Ego vero numquam Iovis imaginem nisi simplicem intuitus sum ...." Galileo's *Sidereus nuncius* contains two observations in which Jupiter showed only one moon and which would come into question in an attempt to identify the night that Sizzi is remembering: On 23 January it is the second observation, which does not fit Sizzi's account, and on 27 January it is the only observation, which also does not fit well.

52. *Dianoia*, pp. 55ff.: "Qualis autem syderum indagator extiterit ipse Ptolemaeus, cui Astronomiam acceptam referre debemus, a nemine ignoratur; asseverare an debemus hos planetas, si reales existerent eum latuisse, ex opinione talis viri, qualis Galilaeus? abeat quaeso haec cogitatio."

Notes to Pages 670–673

53. *Dianoia*, pp. 57ff.: "... ne tantum inventum vulgo innotesceret, neve per omnium mortalium manus diffunderetur ... ne vulgus imperitum ... ludibrio habuisset."

54. *Dianoia*, p. 35: "... superius allatum Philosophorum axioma in caelestibus minime locum habere demonstro, nam quidquid est in caelo non sensus beneficio, sed ipsa rationis potentia cognoscimus et ratiocinationis intervenientis beneficio haurimus ...."

55. *Dianoia*, p. 35: "At visus est causa erronea et fallax, qui res, quas videt, aliter se habere non posse, comprehendit, etsi decipiatur."

56. John Wilkins, *Discovery of a New World* (London, 1684 [first published, 1638]), second book, "A Discourse concerning a New Planet," p. 16.

57. Galileo, *Opere*, ed. E. Albèri (Florence: 1842–1856), vol. 8, p. 110.

58. Christopher Scheiner, *Rosa ursina sive sol ex admirando facularum et macularum suarum phoenomeno varius nec non circa centrum suum et axem fixum ... mobilis ostensus*, Lib. II (Bracciano, 1630), par. 68.

59. Loc. cit.: "Nunc autem post nobis patefactum divino beneficio ex Germania primitus oculum illum perspicacissimum, tubum inquam opticum, fas est aetherias intrare domos et rutilantia illa corpora viciniore obtutu obire, quin et ipsa penetralia et adyta caelestia secretius aliquanto lustrare intimiusque pervadere, ut quae priora saecula neque sensu aliquo stabili neque ratione certa aut evidenti attigerunt, velut praesente aspectu palpabilique contactu liceat nobis attrectare."

Name Index

Name Index

*Pages on which Copernicus has been mentioned merely in passing have not been indexed in order to keep this entry to a reasonable length.

Name Index

Name Index